The Book of Michael of Rhodes

The Book of Michael of Rhodes

A Fifteenth-Century Maritime Manuscript

edited by Pamela O. Long, David McGee, and Alan M. Stahl

Volume 2: Transcription and Translation
edited by Alan M. Stahl

transcription by Franco Rossi
translation by Alan M. Stahl

The MIT Press Cambridge, Massachusetts London, England

MIT Press books may be purchased at special quantity discounts for business or sales promotional use. For information, please email special_sales@mitpress.mit.edu or write to Special Sales Department, The MIT Press, 55 Hayward Street, Cambridge, MA 02142.

This book was set in Garamond Pro on 3B2 by Asco Typesetters, Hong Kong and was printed and bound at Grafos, Barcelona, Spain.

Library of Congress Cataloging-in-Publication Data

Michael, of Rhodes, d. 1445.
The book of Michael of Rhodes : a fifteenth-century maritime manuscript / edited by Pamela O. Long, David McGee, and Alan M. Stahl.
v. cm.
Contents: v. 1. Facsimile / edited by David McGee — v. 2. Transcription and translation / edited by Alan M. Stahl ; transcription by Franco Rossi and translated by Alan M. Stahl — v. 3. Studies / edited by Pamela O. Long.
Text in English and Venetian Italian.
ISBN 978-0-262-13503-0 (v. 1 : hbk. : alk. paper) — ISBN 978-0-262-19590-4 (v. 2 : hbk. : alk. paper) —
ISBN 978-0-262-12308-2 (v. 3 : hbk. : alk. paper)
1. Michael, of Rhodes, d. 1445. 2. Naval art and science—Early works to 1800. 3. Navigation—Early works to 1800. 4. Mathematics—Early works to 1800. 5. Astrology—Early works to 1800. 6. Calendars—Italy—Early works to 1800. 7. Shipbuilding—Early works to 1800. I. Long, Pamela O. II. McGee, David, 1955– III. Stahl, Alan M., 1947–
IV. Rossi, Franco. V. Title.
V46.M56 2009
623.80945′309023—dc22 2008008611

10 9 8 7 6 5 4 3 2 1

Contents

Volume 1: Facsimile

Volume 2: Transcription and Translation

Preface to Volume 2

The manuscript published here is in private hands and has never before been available for consultation by scholars or the general public. For that reason, we have sought in this edition to convey the exact contents of the manuscript while making the material in it accessible to as wide a variety of specialists and interested readers as possible. In the cause of preserving the literal contents of the manuscript, we are greatly aided by volume 1 of this publication, which contains a high-quality, color facsimile of the entire manuscript. The studies in volume 3 present an introduction to the subjects treated in the manuscript that should aid its interpretation.

We have structured volume 2 with a transcription of the manuscript on the left-hand pages facing the corresponding translation into English on the right, to facilitate comparison even for readers without prior experience with the medieval Venetian language. Notes to the transcription, chiefly of a technical nature, are found at the foot of the left-hand page referenced by letter, while those to the translation, sometimes more general in nature, are referenced by number and begin immediately after them to keep the pages in tandem without wasted space. Folio numbers in the outside margins indicate the manuscript folio that begins at that point; where the folio break occurs within a line of text, the end of each manuscript folio is indicated by a vertical line [|] inserted into the transcription at the exact point of break. (The corresponding point of the translation, likewise indicated, is necessarily more approximate.) We have followed the original numeration of the manuscript pages, which assign the same folio number to left and right pages of each opening; we have assigned the letters a and b to the left and right pages respectively. We have omitted the word "Jesus" that appears at the top of most pages from both the transcription and the translation.

In his introduction to the manuscript below, Franco Rossi sets forth the principles for the transcription in detail. Of note is the fact that all words are spelled exactly as in the manuscript, with inconsistencies and even errors maintained (though abbreviations are spelled out). On the other hand, in the interest of comprehensibility (and with the expectation that the reader can always consult the facsimile), capitalization and punctuation have been standardized to modern Italian usage. In a similar way, the translation has adopted the capitalization and punctuation norms of modern English. In a very few cases, differences in punctuation of corresponding passages in the transcription and translation go beyond those between modern Italian and English and reflect differences of interpretation between Franco Rossi and me.

The table of contents at the beginning of the manuscript gives titles for most of its sections; the actual text lacks such titles. We have taken these titles from the table of contents and inserted them at the appropriate places in the text, using square brackets in the transcription to indicate that they are interpolated.

In the translation, Venetian personal names have been rendered following the convention of using the equivalent modern Italian forename and a standardized, usually early modern form of the Venetian family name. Names of saints and historical individuals appear in the translation in their common English forms. Common geographical names are given in their English form (e.g., Venice, Constantinople), and others are given in the language of the country in which they now are contained.

Some parts of this manuscript have never been published before and many have never been translated into a modern language. Almost none have ever been translated into a non-Romance language. This translation is then, of necessity, a rough first attempt to make at least the basic meaning of the text intelligible to the reader of English. In cases where there was a choice between presenting what the text actually says and what I interpret it to mean, I have chosen to use the former in the translation, saving interpretations for the notes. I have maintained the tenses used in the original text, which are especially inconsistent in the mathematical sections at the beginning of the manuscript.

This translation is, to a very great extent, the result of a collaborative process. David McGee, coeditor of this publication, set up a private website for use by the Michael of Rhodes team, which comprised the contributors of the studies published in volume 3. Each page of the manuscript was represented by a file that included a scan of the page, Franco Rossi's preliminary transcription, and my preliminary translation. Team members posted comments to these pages, offering suggestions and raising questions for group discussion. The website was eventually opened up to the additional scholars invited to participate in the December 2005 public conference at the Dibner Institute, who also added comments to the pages.

The manuscript's technical vocabulary and concepts have presented the greatest challenges for translation, especially those related to shipbuilding. Some terms, such as *paraschuxula*, have been the subject of debate for over a century and a half; others, like *poselexe del choltro*, appear never to have been explicated; while words such as *sesto* had many discrete meanings in the Venetian maritime vocabulary, with corresponding possibilities of translation. At a certain point arbitrary decisions have been made, such as whether to translate the Venetian name for a specific rope with an approximately equivalent one from English sailing vocabulary (in itself archaic and obscure) or to leave it untranslated; in most such cases the Venetian name appears in the translation with an explanation of its significance in a note. Following specialized sessions at the 2005 Dibner conference, an email discussion group was constituted specifically to discuss the translation of this technical maritime terminology, comprising Mauro Bondioli, Claire Calcagno, Filipe Vieira de Castro, John Dotson, Jeff Gedney, Matthew Harpster, Alan H. Hartley, Brad Loewen, Alex Medico, and John Pryor as well as the coeditors. I am grateful to all of them for their contributions to the discussions and hence to this translation.

When the time came for producing continuous transcriptions and translations of the manuscript pages, Pamela O. Long, coeditor of this publication, compared each page of the transcription on the website with the facsimile of that page and each page of the translation with the corresponding transcription and facsimile, thereby aiding Franco and me immeasurably in assembling our revised drafts. Members of the study team read and critiqued sections of the manuscript relating to their own specialty; Piero Falchetta's contribution of a full listing of the modern names of all places referred to in the portolans was especially valuable. For assistance in interpreting the Greek prayers that appear in the manuscript in transliterated form and reconstructing their original text, I am very

grateful to the following scholars: Eleni Kalkani-Passali, a specialist in the medieval Rhodian dialect; Anna-Maria Kasdagli of the fourth Ephorate of Byzantine Antiquities in Rhodes; and Diana Wright, a specialist in Byzantine Greek. Alan H. Hartley, the specialist in maritime terminology for the *Oxford English Dictionary*, has provided invaluable help by constructing working glossaries for the shipbuilding and rigging terminology and then reading and emending drafts of each of the relevant sections; it is no exaggeration to say that a meaningful translation of these passages would not have been possible without his unstinting assistance. Also of tremendous importance for the translation has been the assistance of Linda L. Carroll, a scholar of the Venetian language, who read the entire draft of the translation and offered countless suggestions for improving its accuracy and tone. Perhaps no individual associated with this publication has gotten more deeply into the thinking and writing of Michael of Rhodes than our intrepid copyeditor, Matthew Abbate of the MIT Press, who reworked all of the mathematical calculations and established the current names of all the locations while keeping a keen lookout for inconsistencies on the part of the medieval and modern participants in this collaboration. Chryseis Fox mastered many layers of complexity to make a beautifully legible design.

Just as the text of Michael's manuscript represents the compilation of a large number of individual contributions, so this edition is the result of the work of many scholars. It is our hope that we do justice in conveying Michael's writings to a modern audience.

Alan M. Stahl
Princeton, New Jersey
June 9, 2008

Introduction to the Manuscript

Franco Rossi

This essay provides a detailed description of the manuscript of Michael of Rhodes that is the subject of the present edition. It includes a description of the subject contents, a discussion of the manuscript's position vis-à-vis related manuscripts, and an analysis of the manuscript's material properties, including the papers and inks used. It contains as well an analysis of the hands and morphology of scripts, a discussion of the dating of the manuscript, and an account of its composition and fasciculation.

A second essay by the present author in volume 3 of this edition contains an assessment of Michael of Rhodes as a writer and also describes the illustrations of the manuscript in detail. Finally, it discusses the discovery, made during the course of this research, that Michael of Rhodes wrote a second manuscript book, currently in the Biblioteca Nazionale Marciana and formerly attributed to Pietro di Versi.[1]

Contents of the Manuscript

The manuscript is miscellaneous and composite by nature even with the table of contents provided by Michael of Rhodes. However, it can be usefully subdivided into sections corresponding to the subjects treated. Additional subdivisions can then be identified within these sections without affecting the continuity of the text as a whole. A suitably detailed description follows:

• Summary	fols. TOC 1b–TOC 4a
• Arithmetic and algebra[2]	fols. 1b–90-2a, 194a–203a
Problems related to the commerce of pepper	fols. 1b–4a, 64b–65a, 68a–69a
Calculation with fractions	fols. 4b–11b
The rule of three	fols. 9a–10b
Rules of algebra	fols. 12a–19b
Problems of alligation	fols. 19b–20b, 197b–198b

1. This manuscript, Venice, Biblioteca Nazionale Marciana, Ms. It. IV, 170 (= 5379), is available in a recent edition. See Pietro di Versi, *Raxion de' marineri: Taccuino nautico del XV secolo*, ed. Annalisa Conterio (Venice: Comitato per la Pubblicazione delle Fonti relative alla Storia di Venezia, 1991).

2. For the subdivisions in this section I am very much indebted to the valuable contribution by Raffaella Franci, "Mathematics in the Manuscript of Michael of Rhodes," vol. 3, pp. 115–146.

Problems of barter	fols. 20b–27b, 63b–64a, 71b–72a
Problems of partnership	fols. 28a–30b, 43a–44a, 49a, 194a–196b
Problems of freight	fols. 30b–32b
Playing dice	fols. 33a–35b, 90-1b
Buying jewels in a partnership	fols. 35b–37a, 65b–67b
Recreational problems	fols. 37b–39b, 44b–45b, 50a–54a, 57b–61a, 90-1b–91-1a, 199b–201b
Finding numbers in a given proportion; finding a number such that . . . ; dividing a number into two parts	fols. 40a–42b, 46a–46b, 54b–56a, 61b–63b
Problems of the *marteloio*	fols. 47a–48b
Squared numbers	fols. 56b–57a
Various commercial problems	fols. 69b–71a
Various algebraic problems	fols. 72b, 74b–76b, 89b–90-1b, 91-1b–90-2a, 199a, 203a
Problems involving travel	fols. 73a–74a
Calculating square roots	fols. 77a–79b
Calculating cubed roots	fols. 79b–82a
Calculating with radicals	fols. 82b–90-1a
Problems of geometry	fols. 196b–197a
• Michael's professional *curriculum vitae*	fols. 90-2b–93b, 204a
• Astrology, astronomy, and chronological computations	fols. 95a–111b, 129b–135a, 185a–190a
Solar calendar for the twelve months of the year	fols. 95a–102b
Instructions for drawing blood in all the months of the year	fols. 102b–103a
Description of the signs of the zodiac	fols. 103a–110a
Properties of the signs of the zodiac dominating hours of the day and days of the week	fol. 110a
List of stars and information about the day they rise	fols. 110a–111a
Odious and perilous days	fols. 111a–111b
The four times to avoid	fol. 111b
Table of the Christian and Jewish Easter from 1401 to 1500	fol. 129b
Table of the signs of the zodiac	fols. 130a–130b
Rules of the tables of Solomon for the Jewish moon	fols. 131a–135a
Instructions on how to know when the moon turns, by means of mariners' rules	fols. 185a–186b

Position of the moon in relation to the sun	fol. 187a
Calculation of the epact	fols. 187b–188a
Instructions on knowing when the month begins and numerical names of the months for hand calculations	fols. 188a–189b
Calculating on fingers to find the Jewish Passover and from this the Christian Passover (Easter)	fols. 189b–190a
• Orders given by the captain general of the sea, Andrea Mocenigo, to the Venetian galleys in 1428	fols. 111b–118b
• Instructions for navigation	fols. 118b–127a, 190b–193b
To enter the port of Venice	fols. 118b–119b
Portolan made by Zuan Pires, pilot of the Flanders sea	fols. 120a–121a
Crossings of Spain	fols. 121a–121b
Crossings from Ouessant to Calais in the Flanders channel	fols. 121b–122b
Waters and tides of Flanders	fols. 122b–123b
Tides and waters of Ireland and Wales and of the island of England	fols. 123b–125a
Names of the winds in Spanish	fol. 125a
To know how to enter into Sandwich	fol. 125b
To enter the port of Sluys	fol. 125b
To enter Santander	fols. 125b–126a
Soundings of the channels of Flanders	fols. 126a–127a
Portolan for the coast of Apulia	fols. 190b–192b
Portolan for the Gulf of Salonika	fols. 192b–193b
• Instructions for sail making	fols. 127a–129a
• Shipbuilding	fols. 135b–182b
Galley of the Flanders design	fols. 135b–147b, 202b
Galley of the Romania design	fols. 148a–156a
Light galley	fols. 156b–164a
Lateen-sailed ship	fols. 164b–168a
Square-sailed ship	fols. 168b–180a
Instructions on making masts and yards	fols. 180a–181b
How to make rigging	fols. 181b–182a
Ship under sail (illustration)	fol. 182b
• Pseudo-heraldic coat-of-arms (illustration)	fol. 147b

• Prayers, invocations, ritual and magical formulas	fols. 183a–185a, 193b
• St. Christopher (illustration)	fol. 202a
• Portolans added later by different hands	fols. 205a–210b
Portolan from Venice to Constantinople along the coast as the galleys go	fols. 205a–206a
Portolan for the crossings of the Gulf of Venice	fols. 206b–207a
Portolan from Cape Maléas to the island of Famagusta	fols. 207a–208a
Portolan from Venice to Tana, on the route of the galleys by the coast	fols. 208a–210b
• Last wishes of Giovanni da Drivasto, *paron zurado* of Marino Dandolo, August 29, 1473	fols. 238a–238b

Comparison with Other Manuscripts

Not very many late medieval Venetian manuscripts of "nautical" interest have come down to us.[3] In fact, there are fewer than ten. Several of these manuscripts seem to be unique and unrelated to the others; others instead reveal relationships or affinities that are altogether astonishing.

Based on a rigorous analysis of the intrinsic and extrinsic elements of these manuscripts, we can identify at least six generational lines that are sufficiently independent from each other, each of which, as is customary, has been assigned a letter of the alphabet:

A) *Zibaldone da Canal.* Mercantile manuscript of the fourteenth century.[4]

B) *Michael of Rhodes.*

- B^{1}) *Raxion de' marineri.*[5]
- B^{2}) *Libro di marineria.*[6]
- B^{2a}) *Arte de far vasselli.*[7]
- B^{2b}) *Trattato de re navali cavato dall'esemplar di G. B. R.*[8]

3. To be sure, there may be others still hidden in private collections, or lost and unrecognized in libraries, perhaps even public ones.

4. New Haven, Yale University, Beinecke Rare Book and Manuscript Library, Ms. 327. This text was edited over forty years ago in an edition containing contributions by Frederic C. Lane, Thomas E. Marston, and Oysten Ore that are fundamental for the period under study. See Alfredo Stussi, ed., *Zibaldone da Canal: Manoscritto mercantile del sec. XIV* (Venice: Comitato per la Pubblicazione delle Fonti relative alla Storia di Venezia, 1967). See also John E. Dotson, trans., *Merchant Culture in Fourteenth Century Venice: The Zibaldone da Canal* (Binghamton, N.Y.: Medieval and Renaissance Texts and Studies, 1994).

5. Venice, Biblioteca Nazionale Marciana, Ms. It. IV, 170 (= 5379); published as Pietro di Versi, *Raxion de' marineri.*

6. Also *Fabrica di galere*, Florence, Biblioteca Nazionale Centrale, Ms. Magliabechiano, cl. XIX, cod. 7.

7. Vienna, Österreichische Nationalbibliothek, Collezione Marco Foscarini, cod. CCCXVIII, n. 6391. In the manuscript catalog of the Foscarini collection is the following description: "È del sec. XVI, di 116 carte in 8° ben conservate" ("It is from the sixteenth century, of 116 well-preserved leaves in octavo"). See Tommaso Gar, "I codici storici della collezione Foscarini, che si conservano nell'Imperiale Biblioteca di Vienna," *Archivio Storico Italiano*, ser. 1, 5 (1843): 281–505, at 426.

8. Gian Battista Ramusio, also known as *Trattato dell'arte di fabbricar navi*, Milan, Biblioteca Ambrosiana, Ms. H. 149 inf. I am grateful to Mauro Bondioli for bringing this reference to my attention.

C) *Algune raxion per marineri li quali serano utile a saver.*[9]
D) Manuscript of Giorgio "Trombetta" da Modone.[10]
E) *Ragioni antique spettanti all'arte del mare et fabriche de vasselli.*[11]
Y) *Libro da navegar.*[12]

Bearing in mind that the *Zibaldone da Canal* and the *Ragioni antique* have their own story to tell with respect to Michael, the relationship between his manuscript and the remaining texts could be visualized in the following manner:

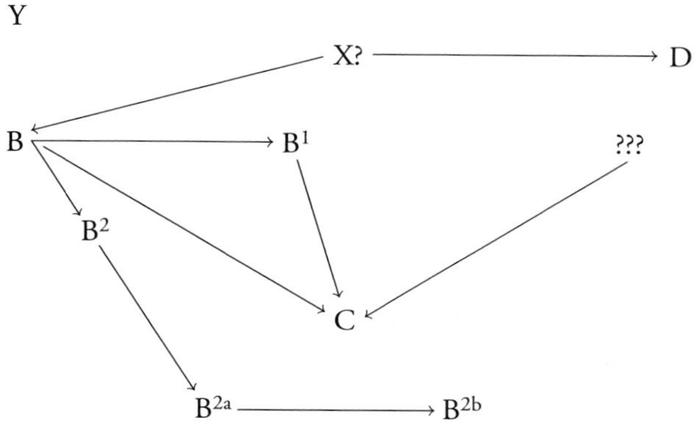

The comparative textual analysis of the individual manuscripts[13] can assist us in illuminating what was transferred by Michael of Rhodes from B to B^1, as well as the debts owed by C with respect to B and B^1, and the debts of B^2 with regard to B. Similarly it allows the emergence of the "genetic"

9. Padua, Biblioteca del Museo Civico, Ms. C.M. 17. Also known as *Arte veneziana del navigare*, or perhaps better as *Algune raxion per marineri li quali serano utelle a saver*, as suggested by Ornella Pittarello, who is studying it in view of an upcoming new edition. I would like to thank Pittarello, who very kindly made available to me a copy of her thesis in which she transcribes and introduces the manuscript. See Ornella Pittarello, "Testimonianza di una civiltà mercantile: Il 'Libro veneziano del navigare', ovvero 'Algune raxion per marineri li quali serano utelle a saver' (Padova, Biblioteca Civica, MS C.M. 17)," Laurea thesis, Venice, Università Ca' Foscari, 2002–2003. Now "outdated" but still useful in several ways is Mirella Blason, "Il C.M. 17 della Biblioteca Civica di Padova e la rotta veneziana delle galee di Fiandra (1428)," *Bollettino del Museo Civico di Padova* 73 (1984): 163–178.
10. London, British Library, Cotton Ms. Titus A. XXVI.
11. Greenwich, National Maritime Museum, Ms. NVT 19, which is available in a published edition: Giorgetta Bonfiglio Dosio, ed., *Ragioni antique spettanti all'arte del mare et fabriche de vasselli: Manoscritto nautico del sec. XV* (Venice: Comitato per la Pubblicazione delle Fonti relative alla Storia di Venezia, 1987).
12. This is a recently identified manuscript of "nautical" character from Venice, now held in Bergamo, Civica Biblioteca Angelo Mai, MA 334. This manuscript has not yet been adequately studied and, in order to avoid superficial conclusions, will not be discussed here in relation to the others just mentioned. It is my intention to study the manuscript in the near future. I thank Raffaella Franci for kindly drawing my attention to it.
13. For an analytical comparison of the texts of the Marciana codex once attributed to Pietro di Versi (B^1) and the C.M. 17 of Padua (C), see Annalisa Conterio, preface to *Raxion de' marineri*, xxxviii–xli, with the proviso that the conclusions she reaches are not always shared, and occasionally should be rejected, in particular because Conterio did not have the opportunity to consult the manuscript of Michael of Rhodes.

link between B and B^{2b} through the intermediary variants of B^2 and B^{2a}. Once this analysis is extended to A, D, E, and Y, we can recognize possible affinities or hierarchical relationships within the entire group.

In any case, the comparison highlights the complex relationships among the manuscripts being studied here, through which we can perceive the intensive circulation of the knowledge they contained. This circulation of knowledge developed through a dense weave of borrowings and contaminations that are not always easily distinguishable or quantifiable, but are nevertheless deserving of study.

The manuscripts of Michael of Rhodes and of Giorgio "Trombetta" da Modone may have derived their respective shipbuilding sections—which are complementary and not repetitive—from a common source (X) that at this moment might perhaps be intuited though not definitively identified, one that ties these texts together with a strong bond of affinity.

Having already remarked that the *Raxion de' marineri* derives directly from Michael's first effort, it remains for us to clarify the relationship between both manuscripts coming from the hand of the Rhodian (respectively B and B^1) and the work of the Paduan manuscript C.M. 17 (C). It is incontestable that C, composed between 1445 and 1446, was transcribed directly from B^1, to the point that in many parts it appears to be an actual copy (albeit not always respectful of the systematic order of the original). Nevertheless several parts of it—such as the "Portolan fatto per Zuan Pires"—are better linked to B rather than to B^1. Thus we can suppose that the anonymous compiler of C had had at least the opportunity to gain access also to Michael's major work. Other parts, instead, reveal no particular connection (not even indirect ones) with that manuscript.[14]

On the other hand, B^2, B^{2a}, and B^{2b} are partial copies of Michael's manuscript. To put it more precisely, in addition to other texts on shipbuilding that cannot be directly linked to him,[15] these manuscripts contain materials he prepared on naval architecture including illustrations, drawings, diagrams, and renderings, as can be easily grasped in the summary below. Apart from the sail-making instructions (which actually appear here within a more systematically logical context), these items are also presented in exactly the same order that Michael followed. The only discrepancies seem to be lexical in kind: in transcribing Michael's texts the copyist of B^2 felt it opportune to partly update the vocabulary, pruning it of any overly archaic-sounding elements and bringing it closer to the contemporary usage of the first half of the sixteenth century. In so doing, the copyist in some ways rewrote Michael's texts, modernizing only what was necessary to make them easier to read.

The manuscript of B^2 is almost certainly written in the hand of Giovanni Battista Ramusio, a singular figure of great erudition, an enthusiast of voyages and geographical explorations as well as of anything that could be remotely connected to navigation, and the author of the collection *Delle navigationi et viaggi* ("Of navigations and voyages") published in three volumes in Venice between

14. This is certainly not the venue in which to identify all the sources of C.M. 17, not all of which can be attributed to B and B^1. In any case, see Conterio, preface to *Raxion*, and especially Pittarello, "Testimonianza di una civiltà mercantile."
15. For the identification of these sources, see Mauro Bondioli, "Early Shipbuilding Records and the Book of Michael of Rhodes," vol. 3, pp. 243–279.

1550 and 1559.[16] For one, B^{2b} explicitly provides the name of Ramusio, the manuscript's owner, as an integral part of the title itself: *Trattato de re navali cavato dall'esemplar di G.B.R.* (Treatise of things nautical derived from the exemplar of G.B.R.). In addition, a comparison with Ramusio's will, written in his hand,[17] seems to fully confirm the hypothesized attribution. Thus B^2 would have been written by Giovanni Battista Ramusio, and, given the "freshness" of the writing, very likely went back to the 1520s to 1530s. B^{2a} was derived at a later stage from B^2 and not directly from B. Some time later, B^{2b} was derived from B^{2a}, as is suggested by certain minor omissions in the iconographic commentary that differentiates B^{2a} and B^{2b} with respect to B^2. In effect, the two copies B^{2a} and B^{2b} are absolutely identical in terms of their written texts, illustrations, drawings, renderings, diagrams, and even in the number of pages left deliberately blank.[18] Furthermore, based on the results of paleographic analysis we can hypothesize that B^{2a} was copied no later than the middle of the sixteenth century, and that B^{2b} was written at the latest during the decade between 1570 and 1580.

From all of this it seems manifestly evident that Giovanni Battista Ramusio, at least for a certain period of time, had the manuscript of Michael of Rhodes at his disposal. From whom he received it, and under what terms, is information that eludes even the most nuanced hypothesis.

Several subjects of the Michael of Rhodes manuscript—the instructions for knowing when the month begins and numerical names of the months for calculation by fingers; the instructions for sail making; and the *raxion del martoloio*, to name a few—can also be found in D. However, the level of overlap among these shared topics is really quite minimal, as if the substantial shared interest in particular topics had been derived from sources that were entirely distinct and unrelated to one another. And yet, at least with regard to shipbuilding, both Michael of Rhodes and Giorgio "Trombetta" da Modone seem to have drawn from the same source, as amply demonstrated by linguistic analysis of the two texts.

Manuscript E, on the other hand, is a different case. In fact, several sections of this text correspond perfectly with the texts that Michael presents, so much so that we might hypothesize that the anonymous compiler of *Ragioni antique* copied directly from Michael's manuscript. Or else we might suppose that they transcribed their respective texts, unknowingly, from the same common source, although for obvious reasons this seems less likely. This is the case, for example, with the solar calendar for twelve months of the year, or the properties of the signs of the zodiac that govern the hours of the day and the days of the week. Elsewhere the correspondences are much more nuanced, being limited to the simple presence of the same topics; thus it seems very difficult to support the idea that E was derived from B. Examples include instructions for calculating the epact (by calculating on one's fingers) useful for finding the dates of the Jewish Passover, which were

16. His family originally came from Rimini; he was born in Treviso in 1485 and died in Padua in 1557. At Venice he was secretary of the Senate and the Council of Ten, and an envoy of the Republic to the court of Louis XII. See Giovanni Battista Ramusio, *Navigazioni e viaggi*, ed. Marica Milanesi, 6 vols. (Turin: Einaudi, 1978–1988); in particular see vol. 1, xi–xxxvi. Still relevant is Antonio Del Piero, *Della vita e degli studi di Gio. Battista Ramusio* (Venice: Visentini, 1902). Additional biographical details can be found in Massimo Donattini, "Giovanni Battista Ramusio e le sue 'Navigazioni': Appunti per una biografia," *Critica Storica*, n.s., 17 (1980): 55–100.

17. Venice, Archivio di Stato, Notai di Venezia, Testamenti, B. 211, n. 420 (Angelo Canal).

18. I do not believe we need to give too much weight to the fact that B^{2a}, unlike B^{2b}, lacks an annotation about the copy's source.

undoubtedly derived from a source other than B. Furthermore, every portolan in E can be linked to sources other than those used by B.

Subject	B	B¹	C	B²	B²ᵃ	B²ᵇ	D	E
Arithmetic and algebra	■							
Professional vita	■							
Solar calendar for the twelve months of the year	■[19]	■	■					■[20]
"Amaistramento a tuor sangue per tuti li mexi del'ano" (instructions for drawing blood in all the months of the year)	■[21]	■						
Description of the signs of the zodiac	■							■[22]
Properties of the signs of the zodiac dominating hours of the day and days of the week	■[23]	■	■					■[24]
List of the stars and information about the day they rise	■	■	■					■[25]
"Dì uziagi e zorny pericholoxi" (odious and perilous days)	■[26]	■	■					■[27]
"4 tenpore che se die vardar" (the four times to avoid)	■	■	■					■[28]
Table of the Christian and Jewish Passover from 1401 to 1500	■	■[29]	■[30]					■[31]
Table of the signs of the zodiac	■	■	■					■[32]
"Raxion dela taula de Salamon ala luna zudescha" (method of the table of Solomon for the Jewish moon)	■	■[33]	■[34]					
"Amaistramento di saver quando fa la luna a raxion uxa i marineri" (instructions on knowing when the moon is new, by means of mariners' usage)	■	■[35]	■[36]					

19. Also partially in A. Days of the month are represented by symbols.
20. Text largely coincides with B, with slight lexical variations.
21. Also partially in A, with slight lexical variations. In A only the days from the 26th to the 30th.
22. Text largely coincides with B, albeit with slight lexical variations. Lacks the graphical representations of the signs of the zodiac.
23. Also partially in A.
24. Texts largely coincide with B, with slight lexical variations.
25. List repeated twice within a few folios, the first in the same sequence with respect to other chapters as in B. Texts largely coincide with B, with slight lexical variations.
26. Partially also in A. The days of the month are represented by symbols.
27. List repeated twice within only a few folios, the first in the same sequence with respect to other chapters as in B. Texts largely coincide with B, with slight lexical variations and apparent omissions due to distraction.
28. Text largely coincides with B, with slight lexical variations.
29. From 1444 to 1533.
30. From 1444 to 1543.
31. Table for the years 1411–1498. Data for the epact and the Jewish Passover are absent.
32. Text largely coincides with B, with slight lexical variations.
33. Coincides only partially with B. The table covers from 1444 to 1455; the table in B covers the years from 1435 to 1530.
34. Coincides only partially with B; the table covers from 1444 to 1455.
35. Coincides only partially with B. Exemplified for the years 1444–1445; in B exemplified for the years 1435–1436.
36. Coincides only partially with B; exemplified for the years 1445–1446.

Subject	B	B^1	C	B^2	B^{2a}	B^{2b}	D	E
Position of the moon in relation to the sun	■	■[37]	■[38]					■[39]
Calculation of the epact	■	■[40]	■[41]					■[42]
"Amaistramento a saver quando intra el mese" (instructions for knowing when the month begins) and numerical names of the months for calculations on the hand	■	■[43]	■[44]				■[45]	■[46]
Calculation on fingers to determine the Jewish Passover and from this the Christian Passover (Easter)	■	■[47]	■[48]					■[49]
Orders given by the *chapetagno generalle da mar* (captain general of the sea), Andrea Mocenigo, to the Venetian galleys in 1428	■	■	■					
Entry to the port of Venice	■	■	■					■[50]
"Portolan fatto per Zuan Pires, pedotta del mar di Fiandria" (portolan made by Zuan Pires, pilot of the Flanders sea)	■	■	■[51]					■[52]
"Traversse de Spagna" (crossings of Spain)	■	■	■					■[53]
"Traversse da Ossente a Chales in chanal di Fiandres" (crossings from Ouessant to Calais in the Flanders channel)	■	■	■					■[54]
"Aque e marie de Fiandria" (waters and tides of Flanders)	■	■	■					■[55]
"Marie e aque d'Erlanda e de Gaules e del'ixola de Ingletera" (tides and waters of Ireland and Wales and of the island of England)	■	■	■					■[56]
Names of the winds in Spanish	■	■						
"Per saver entrar in Sentuzi" (to know how to enter into Sandwich)	■							

37. Text differs from B.
38. Text differs from B. Almost identical to B^1. In B^1 exemplified for the year 1444, in C for the year 1445.
39. Text is essentially the same with regard to its methodological formulation, with slight lexical variations.
40. Text differs from B. Exemplified for the years 1444–1445.
41. Text differs from B. Closer to B^1. Exemplified for the years 1445–1447.
42. Text differs from B, although the same with regard to its methodological formulation.
43. Text differs from B. Exemplified for the year 1444; in B exemplified for the year 1436.
44. Text differs from B. Closer to B^1. Exemplified for the year 1444.
45. The exemplification is extremely reduced with respect to B.
46. Text is largely the same as B, though with significant lexical variations.
47. Coincides only partly with B. Exemplified for the year 1444; in B exemplified for the year 1436.
48. Coincides only partly with B. Closer to B^1. Exemplified for the year 1444.
49. The subject is the same, but the texts do not present significant correspondences other than for the numbers of the joints of the right hand.
50. Text largely the same as B, though with significant lexical variations.
51. Closer to B.
52. Text differs from B.
53. Text differs from B.
54. Text differs from B.
55. Text differs from B.
56. Text differs from B.

Subject	B	B^1	C	B^2	B^{2a}	B^{2b}	D	E
"Per intrar al porto del'Eschioza" (to enter the port of Sluys)	■							
"Per voler intrar in Sancto Ander" (to enter Santander)	■							
"Sonde d'i chanalli de Fiandria" (soundings of the channels of Flanders)	■							
"Portolan per la riviera de Poya" (portolan for the coast of Apulia)	■	■	■					■[57]
"Portolan per lo golfo de Salonychi" (portolan for the Gulf of Salonika)	■	■	■					■[58]
Instructions for sail making	■	■	■[59]	■[60]	■[61]	■[62]	■[63]	■[64]
"Raxion del martoloyo" (problems of the *marteloio*)	■	■[65]	■[66]				■[67]	■[68]
Galley of Flanders	■			■[69]	■[70]	■[71]		■[72]
Galley of Romania	■			■[73]	■[74]	■[75]		■[76]
Light galley	■			■[77]	■[78]	■[79]	■[80]	■[81]
Lateen-sailed ship	■			■[82]	■[83]	■[84]		

57. Text largely the same as B, but with slight lexical variations.
58. Text largely the same as B, but with slight lexical variations.
59. In inverse order.
60. Inserted within the chapter on the galley of Romania.
61. Coincides with B^2.
62. Coincides with B^2.
63. Text differs from B. Measurements do not match. Inverse order.
64. Text differs from B. Measurements do not match. Inverse order.
65. Text differs from B.
66. Text differs from B. Closer to B^1.
67. Text differs from B, but similar with respect to structural elements.
68. Text differs from B, but similar with respect to structural elements.
69. Coincides with B, but with slight lexical variations. Fol. 8r of B^2 which begins "Questa galia del sexto de Fiandra" ("This galley of the Flanders design") corresponds, with slight lexical variations, to fol. 202b of B that dates from the second phase of the manuscript's composition, i.e., 1444–1445. Because of this, the *incipit* of fol. 8v of B^2, "La galia antedita," necessarily differs from the *incipit* of fol. 142b of B, "Quista galia inchontro del sesto de Fiandria." In fact the preceding folio of B^2 contains a written text, while the folio of B contains the drawing of the hull of the galley of Flanders. Lacunae as in B.
70. Coincides with B^2.
71. Coincides with B^2.
72. Text differs from B, but can be partially superimposed in occasional points.
73. Coincides with B, but with slight lexical variations. Lacunae as in B.
74. Coincides with B^2.
75. Coincides with B^2.
76. Text differs from B, but can be partially superimposed in occasional points.
77. Coincides with B, but with slight lexical variations. Initial part mutilated as in B (fols. 156b and 157a torn out). Lacunae as in B.
78. Coincides with B^2.
79. Coincides with B^2.
80. Text differs from B.
81. Text differs from B, but can be partially superimposed in occasional points.
82. Coincides with B, but with slight lexical variations.
83. Coincides with B^2, but with slight lexical variations.
84. Coincides with B^{2a}.

Subject	B	B^1	C	B^2	B^{2a}	B^{2b}	D	E
Square-rigged ship	■			■[85]	■[86]	■[87]		
"Amaistramento de far albori e antene" (instructions on making masts and yards)	■			■[88]	■[89]	■[90]	■[91]	
"Raxion de far sartia" (how to make rigging)	■			■[92]	■[93]	■[94]	■[95]	
"Nave a velo" (ship under sail; illustration)	■			■[96]	■[97]	■[98]		
Prayers, ritual, and magic formulas	■[99]							
Pseudo-heraldic ensign (illustration)	■							
St. Christopher (illustration)	■							

HISTORY OF THE MANUSCRIPT

The proven, undisputed information we possess regarding the vicissitudes of the Michael of Rhodes manuscript is extremely limited. We can only say with absolute certainty that its history must not have been particularly serene.

Already in 1473, just a few decades after it was composed, the manuscript could be found outside of Venice in a mariner's sack embarked on a Venetian galley. At least, this is what we can deduce from the presence on fols. 238a and 238b of the presentation of the last wishes of Giovanni da Drivasto, "*paron zurado* of the magnificent sir Marino Dandolo."

The three hands represented by brief notes on the pasted-down front endpaper (numbered fol. Ala in this edition) probably represent three owners of the manuscript, in the second half of the fifteenth century, the beginning of the sixteenth century, and the end of the sixteenth century, respectively. The manuscript's location in Venice in the early sixteenth century seems well attested, in view of the fact that Giovanni Battista Ramusio was able to extract a partial copy from it. The Magliabechiano XIX.7 codex, better known under the somewhat infelicitous title *Fabrica di galere*, which as noted earlier was almost certainly by Ramusio, contains among its texts a copy of the shipbuilding section of Michael's manuscript.[100]

85. Coincides with B, but with slight lexical variations.
86. Coincides with B^2, but with slight lexical variations.
87. Coincides with B^{2a}.
88. Coincides with B, but with slight lexical variations.
89. Coincides with B^2, but with slight lexical variations.
90. Coincides with B^{2a}.
91. Text differs from B.
92. Coincides with B, but with slight lexical variations.
93. Coincides with B^2, but with slight lexical variations.
94. Coincides with B^{2a}.
95. Text differs from B.
96. Coincides with B.
97. Coincides with B and with B^2.
98. Coincides with B, B^2, and B^{2a}.
99. Texts of similar interests, but different actual content, are also found in A.
100. Cf. Augustin Jal, "Mémoire no. 5," in Jal, *Archéologie navale* (Paris: Arthus Betrand, 1840), 2: 1–133; Roger Charles Anderson, "Jal's 'Memoire No. 5' and the Manuscript 'Fabrica di Galere,'" *Mariner's Mirror* 31 (1945): 160–167; Frederic C. Lane, *Venetian Ships and Shipbuilders of the Renaissance* (Baltimore: Johns Hopkins University Press, 1934), 56–57.

For several decades during the first half of the twentieth century the Michael of Rhodes manuscript was in the private collection of Federico Patetta, docent of history of Italian law at various Italian universities; *Accademico d'Italia* from 1933; member of the Royal Academy of Sciences of Turin; member of the Accademia dei Lincei and of numerous national historical delegations and academies; and indefatigable collector of manuscripts and autographs, largely of historical-juridical interest. Patetta cataloged the manuscript as no. 32 in his collection, as attested by the note of ownership in his own hand on the first folio.

At his death (October 28, 1945), all of Patetta's manuscripts and autographs were acquired by the Vatican Library (Biblioteca Apostolica Vaticana), as per his testamentary disposition: "Lego alla Biblioteca Vaticana in Roma tutti i codici manoscritti, autografi, pergamene, documenti di mia proprietà . . ." (I bequeath to the Vatican Library in Rome all the manuscript codices, autographs, parchments, documents belonging to me).[101]

The manuscript of Michael of Rhodes, however, appears not to have ever arrived at the Vatican Library, perhaps because in the meantime Patetta himself may already have parted with it, or perhaps for other reasons that cannot be determined at the moment. In any case Paul Oskar Kristeller made no mention of it in his *Iter Italicum* in which he treats the Patetta Estate of the Vatican Library.[102]

However, the codex appears inexplicably in the catalog of Nicolas Rauch's Swiss bookstore Beaux Livres (p. 123), printed in 1,350 copies in 1949, in which it features as being for sale at the price of 70,000 Swiss francs. In this catalog the manuscript is described as "Manuscrit italien d'une importance capitale pour l'histoire de la marine et de la construction des bateaux au XVe siècle" (Italian manuscript of capital importance for naval history and shipbuilding in the fifteenth century) and further below as "le seul document authentique existant sur l'archéologie navale de cette époque" (the only extant authentic document on shipbuilding of this period). Its description, together with the reproduction of several folios, highlights with particular effectiveness its rarity and significance.[103]

In 1966 it made a fine show in the Sotheby's catalog of July 11, 1966, as lot no. 254, which was purchased for 5,500 pounds by a certain Berthier.[104]

Despite its extreme importance, the manuscript remained inaccessible to scholars, who during this time lamented its absence to no avail; it reappeared in a later auction at Sotheby's on December 5, 2000, as lot no. 54.[105] Thanks to the disinterested generosity of its purchaser, the current owner, the manuscript has been edited in this venue and thus put at the disposal of the international scholarly community.

101. Federico Patetta, holographic will, May 6, 1935, published by the notary Mario Bordon, of the Notary College of Savona, December 20, 1945.

102. Paul Oskar Kristeller, *Iter Italicum: A Finding List of Uncatalogued or Incompletely Catalogued Humanistic Manuscripts of the Renaissance in Italian and Other Libraries* (London: Warburg Institute, 1963–1966), 6: 400–406. Nor is it mentioned in the Vatican Library's typescript catalog of the collection.

103. Nicolas Rauch, *Livres précieux et autographes des XVe et XVIe siècles*, catalog no. 2, lot no. 123, auction cat. (Basel: Benno Schwabe, 1949), 123–126. I thank Mauro Bondioli for his valuable recommendation of this catalog and Gilberto Penzo who kindly lent me a copy.

104. Sotheby's London, *Catalogue of Important Western and Oriental Manuscripts and Miniatures*, July 11, 1966, lot 254, auction cat., entry by Andreas Meyer (London: Sotheby and Co., 1966), 89–93.

105. Sotheby's London, *Western Manuscripts and Miniatures*, December 5, 2000, lot 54, auction cat. (London: Sotheby and Co., 2000), 60–72.

Description of the Manuscript

The manuscript, which is made entirely of paper, is currently in a rather mediocre state of preservation. A questionable rebinding procedure likely dating to the first decades of the twentieth century has to a large degree compromised the legibility of the volume's exterior appearance, although in no way modifying what is assumed to have been the codex's original composition. The procedure involved covering (or rather, clumsily wrapping) the spine and slightly less than half of the front and back boards in low-quality red leather. The remaining half of the boards not covered in red leather—actually more on the front than the back—preserves a considerable portion of what must have been the wooden boards' original external cover: a musky green leather, quite worn over time and especially by evident handling. Nevertheless, a significant part of the back board, clearly the portion not treated by the binding procedure described above, remains uncovered. The codex is kept closed by a leather strap ending in a metal clasp, which starts from the back board and is attached to the front board by an eyelet also of metal.

The dimensions of the individual folios vary in height between 19.5 and 19.7 cm, and in width between 13.5 and 14.3 cm. The thickness of the whole volume varies between 5.5 and 6.1 cm, based on the amount of pressure applied while measuring. The boards measure respectively 20.5 cm in height, 14.3 cm in width, and 0.5 cm in thickness.

The writing area extends in height from 15.5 to 16.5 cm, and between 10.5 and 11.0 cm in width. Each folio averages 25–27 lines of writing. The number of lines[106] can nevertheless vary, even significantly, where there are arithmetic and algebraic operations, drawings, full-page illustrations and geometric diagrams, decorative dividers, and line spacing more or less widened from one paragraph to another. Clearly all this creates a specific individuality to each folio of the volume. Thus the number of lines and the broadening of the writing area cannot be seen in themselves as particularly significant elements.

The writing area is delineated on the left and right sides of each folio by two vertical lines marked by lead point. The upper margin is marked by a hole made with a pointed instrument that was pressed with enough strength to puncture several folios, at about 1.5 cm from the edge, along the line that defines the external limit of the writing area. The lower margin, which is not always rigorously respected, tends to be located at about 2.5–3.5 cm from the edge of the folio, and is clearly affected by the presence of the arithmetic operations, drawings, and diagrams mentioned earlier.

There are no noticeable traces of other marks either for line spacing or for ruling. Nevertheless, the lines of writing maintain a singularly regular and uniform arrangement on the folio and run parallel one after the other, albeit with a slight tendency to stray up or down with respect to the median horizontal writing axis and toward the right as they get closer to the lower margin. The paper's thickness and its evident opacity, however, lead us to strongly doubt that the *scriptor* would have used a guide sheet of ruled paper underneath.

The *incipit* of each paragraph is indicated by a capital letter that sticks out significantly past the left margin, and by a notable widening of the line spacing.

---

106. And consequently also the writing area.

The most significant separations between individual paragraphs (i.e., those marking the chapters and sections that constitute the manuscript's supporting structure) tend to be suggested graphically by dividers that have both a decorative and separating function. These dividers can be found quite frequently in certain Venetian writing centers that can generally be linked to the cultivated spheres of Greeks who had chosen Venice, *alterum Bisantium*, as a new and more comfortable homeland of choice.[107] While each design has its own graphical individuality, these dividers can be grouped into four distinct typological variants:

A) braided with thorns;
B) braided without thorns;
C) chained with thorns;
D) chained without thorns.

In its current configuration the manuscript is composed of 253 folios, in turn divided into 17 fascicules or gatherings. Of these 253 folios, the first 10 were undoubtedly added at a later date after the codex was originally bound, perhaps during the course of the more recent reconditioning procedure.[108]

The paper of these first 10 folios clearly seems to have been made considerably later (well into the sixteenth century), and is of undoubtedly higher quality than the remaining original folios, to which it is certainly unrelated. The folios are of limited thickness, made of regular and uniform pulp tending in color toward a light ochre, and entirely free of the spots of grease and dirt that mark many other parts of the manuscript. These stains amply attest to the repeated handling of the manuscript by hands not accustomed to cleanliness—and thus, in their own way, to the extraordinary fortune of the manuscript. Nevertheless, these additional folios do not fit well within the context of the structure of the volume.

A faint, partial trace of a watermark, in which it might easily be possible to discern some sort of representation, does not correspond at all to what was described in Sotheby's "Description of the Manuscript"—"watermarks of three hills in a circle (of the type of Briquet 11851–11888, extensively used throughout the late 14th and 15th centuries) and a sun (of the same type as Briquet

107. In this regard, see Venice, Biblioteca Nazionale Marciana, Cod. Gr.Z.301 (coll. 635), a collection of mathematical and astronomical treatises in Greek; and also Cod. Gr.Z.263 (coll. 1025), the *Pneumatics* of Hieron in Greek. The codices are also described in *La scienza a Venezia tra Quattrocento e Cinquecento: Opere manoscritte e a stampa*, exh. cat., Biblioteca Nazionale Marciana, October 3–15, 1985 (Venice: Stamperia di Venezia, 1985), 14, 21, 35, 38. I would like to thank Elisabetta Barile for her generous and valuable recommendation. See also Alexander Turyn, *Dated Greek Manuscripts of the Thirteenth and Fourteenth Centuries in the Libraries of Italy*, 2 vols. (Urbana: University of Illinois Press, 1972), plates 2: 49, 237, and 246; and Elpidio Mioni, *Introduzione alla paleografia greca* (Padua: Liviana, 1973), plate XVI. The *scriptor*'s connection to the cultivated sphere of Greeks in Venice was also suggested by Conterio, preface to *Raxion de' marineri*, xiv.

108. Oddly, there is no mention of these added folios in the codex's description in the printed catalog of the Swiss antiquarian bookseller Nicolas Rauch: "Michalli Daruodo.—Traité de la construction des galères et des nefs latines. Manuscrit autographe en italien, sur papier, daté de la première moitié du XVe siècle (1444). In -4 de 3 ff. n. ch. (ch. par erreur 204, les ff. 90 et 91 en double, les 2 feuillets manquent), 35 ff. n. ch., la plupart blancs, peau verte sur ais de bois, le dos nouvellement recouvert d'une peau rouge." (Michael of Rhodes: Treatise on the construction of galleys and lateen ships. Autograph manuscript in Italian, on paper, dated to the first half of the fifteenth century (1444). In -4 of 3 ff unnumbered (numbered by mistake 204, folios 90 and 91 in duplicate, two pages missing), 35 leaves unnumbered, the majority blank; [bound in] green leather over wooden boards, the spine recently rebacked with red leather.) Rauch, *Livres précieux*, 123.

13903–13982 but not closely resembling any particular example)."[109] Unfortunately it does not allow us any greater precision with regard to chronological attribution. We can only offer the hypothesis that unused waste sheets, likely recovered from another manuscript in which they had not been used, were reutilized as guard leaves in order to strengthen the volume. The thread used for sewing and attaching this quire, which is quite extraneous to the original architecture of the codex, is thin and of perfect caliber, very different from the thicker and mediocre quality used for the original binding. It further confirms the hypothesis that the invasive intervention is modern. This last procedure is so botched and clumsy that we can attribute it to someone utterly unskilled in binding and book restoration.

In contrast, the paper of the manuscript in its original composition demonstrates an entirely different workmanship. It is fairly mediocre, somewhat thick but at the same time extremely fragile and easily fragmented, of irregular grain and quite unhomogeneous texture. In color the paper tends toward a light ochre (although it is considerably darkened from repeated handling) and is of the type commonly used between the second half of the fourteenth century and the first half of the following century. The absence of a watermark—of which not even a minimal trace could be discerned—would lead us to assume that this paper might come from an eastern source rather than from a northern Italian paper factory.

As stated earlier, the manuscript is in mediocre condition. Undoubtedly its material condition has been partly compromised by repeated handling, the physical damage sustained, and especially the modest quality of the writing support. Nevertheless, some responsibility for its condition can be attributed to the nature and actual dimensions of the manuscript; while it cannot be called a "knapsack book" (*libro da bisaccia*) in the fullest sense of the term, it could always find its way into the baggage of traveling merchants and mariners. Several rudimentary efforts at restoration, such as those still noticeable today at fols. 103b/104a, 104b/105a, 140b/141a, 146b/147a, 183b/184a, and 193b/194a, reflect the damage that occurred early on within the more frequently consulted sections, subjected to inevitable wear and inexorable degradation. From the materials used, these "restoration" efforts seem to date to a period not too distant from the time when the codex was first written. Tears and lacerations occur more frequently especially in the central part of the manuscript—i.e., the part of the codex that is most easily opened quickly and carelessly to full opening, but also the part that was most sought after and consulted by mariners and seamen in general. On the other hand, this is not the case for the first half of the work, which is a veritable *liber abbaci* (abbacus book), by its very nature reserved for less frequent and less casual consultation.

Examination of the manuscript's foliation, and in particular several significant peculiarities, allows us to hypothesize with a fair degree of certainty about the procedures followed by the *scriptor* during the various phases of his work, from the preparation of the writing surface to the arrangement of gatherings for final binding.

First of all it should be observed that the foliation, in the same hand to which a good part of the manuscript can be attributed, almost certainly preceded the writing of the text and constituted an integral part of the preliminary procedures for the preparation of the writing area. In other words, we can surmise that before the text was written, Michael of Rhodes saw to it that the folios were

109. Sotheby's London, *Catalogue*, July 11, 1966, 89.

numbered in a systematic progressive order, after having folded and recut the available paper in quartos and packaged everything into gatherings composed on average of 8–9 bifolia each. The *scriptor* followed a page numeration method that was fairly typical for manuscript books: he gave the same number successively to the verso of each folio (in this edition identified with the letter "a") and to the recto of the folio immediately following in the gathering sequence (identified with the letter "b"), writing the number at the upper extremity of the left and right margin of each folio.

This statement is further supported by a rather interesting detail. The verso of several folios (i.e., in this edition indicated with "a"), from fol. 147a up to 151a and from 156a up to 161a, presents traces of another foliation on its lower external margin. In this case it is in an upside-down position with respect to the normal orientation of the writing, and numerated in descending order with respect to the progression of the folios. This numeration begins with the number 160 and continues with the numbers 159, 158, 157, 156, 151, 149, 148, 147, and 146. Originally there may also have been 150, but the absence of the folio numbered 156b/157a, unfortunately missing today, allows us only to hypothesize its earlier existence.[110] Most likely Michael of Rhodes made an error in the preliminary numeration of his folios, perhaps out of distraction, repeating numbers that had already been used earlier. Having realized his mistake, he did not throw the folios away but simply inverted their arrangement in the gathering. Not too concerned to cancel the first numeration, he reutilized the folios after having renumbered them, this time according to the correct succession of folios within the gathering.

Not every folio appears to have been numbered. However, it is not possible to distinguish with absolute certainty the cases in which foliation was actually deliberately omitted by the *scriptor* from those in which it has simply become materially impossible to read. Several folios may have been excessively trimmed at the time of binding—whether for the original or the later binding procedure it is impossible to know—and consequently lost the foliation when it was too close to the folio's upper margins.

The 10 added folios do not present any obvious traces of numeration. On the other hand, the lack of numeration on the first three folios of the manuscript in its original form, which contained the index of materials or "table of contents," could not have been anything other than intentional. In this edition, in order to avoid confusion with the remaining folios, these initial three folios have been renumbered from TOC [table of contents] 1b to TOC 4a. The original foliation is interrupted at folio 204b. The verso of this and another 35 folios following have been numbered for this edition from 205a to 241a; some of these contain various texts added at a later date in various hands that are distinct from the prevailing one, while some are blank. The folios assigned the numbers 240a and 240b had already been numbered 141a and 141b by the same hand to which the original foliation of the manuscript has been attributed.

Nevertheless, the *scriptor* made several accidental errors, such as repeating numbers or superimposing corrections. While they may be fairly negligible in quantity and practical consequences, they are not insignificant for the aims of an effective and comprehensive analysis of the codex, because they constitute very clear traces of the *scriptor*'s method of operation. Foliation errors encountered are as follows:

110. The numbered folio 156b/157a was almost certainly torn out subsequent to the binding of the manuscript, as the current folio 157b clearly contains an acephalous text.

- 40a corrected ("3" crossed out);
- 47a/b corrected from "147" by crossing out the first numeral;
- 79b superimposed on "78," with "9" constructed on the lower eyelet of "8";
- 80b, multiple foliation with "79," "80," "90";
- 90a/b and 91a/b repeated;
- 113a/b repeated;
- 132a/b both numbered "32";
- 144b corrected from "145," with final "4" written over "5";
- 151b corrected from "152," with final "1" written over "2."

The ink used for the foliation is a rather dark brown, tending almost toward black, and never varies in color or intensity throughout the folios. In some cases it is decidedly different from the ink used in writing the text of these folios.

The generalized repetition of the semantic and verbal invocation "Ihesus" on each folio, recto and verso, can also be considered as an integral part of the preparatory phase of the writing area.

Despite its apparent vicissitudes, amply reflected in its mediocre state of preservation, the manuscript has reached us almost essentially intact with regard to composition. Although today several folios are clearly missing, having been carelessly torn out (in particular fols. 9b/10a, 156b/157a, and 215b/216a, the last of which was almost certainly blank), none of the gaps goes so far as to hinder the understanding and significance of the work. Folios 9b/10a and 156b/157a were undoubtedly lost subsequent to the binding of the manuscript. In fact there is a lack of continuity in contents between the closing words (the explicit) of fol. 9a and the opening words (incipit) of fol. 10b; the same holds for fols. 156a and 157b.

Moreover, there are occasional breaks between the explicit of one folio and the incipit of the following one (for example between fol. 65a and fol. 65b, and between fol. 70a and fol. 70b). These can be attributed not to the loss of intermediary folios but rather to errors Michael made while copying texts from which he directly drew his material, or to gaps that were already present in these texts.

In the manuscript we can essentially discern five principal hands, labeled according to their order of appearance A, B, C, D, and E.[111]

Hand A clearly comes from the mercantile sphere,[112] although corrupted and practically softened by the writer's own particular word usage and especially by his being essentially outside this sphere and its associated genre of writing;[113] this is the hand of Michael of Rhodes. The manuscript foliation, the index of various subjects into which this is articulated, and, clearly, the manuscript itself from fol. 1b to fol. 204a are all to be attributed to this hand.

The writing is somewhat rounded and fluid and generally moderately cursive (more so than the design of each letter); it is also particularly small and rather closed within itself, at least up to fol.

111. There are an additional three hands on the pasted-down front endpaper, numbered A1a in this edition.

112. Gianfranco Orlandelli, "Osservazioni sulla scrittura mercantesca nei secoli XIV e XV," in *Studi in onore di Riccardo Filangieri* (Naples: L'Arte Tipografica, 1959), 1: 445–460; reprinted in his *Scritti di paleografia e diplomatica,* ed. Roberto Ferrara and Giovanni Feo (Bologna: Istituto per la Storia dell'Università di Bologna, 1994), 147–178.

113. A fundamental work on this topic is Federigo Melis, *Documenti per la storia economica dei secoli XIII–XVI (con una nota di paleografia commerciale a cura di Elena Cecchi)* (Florence: Leo S. Olschki, 1972).

199a. From 199b onward, oddly coinciding with the change in ink, the module clearly becomes larger and the letter design less careful. Individual letters tend to be placed next to one another rather than actually being linked, and thus do not give rise to the usual morphological deformations produced by linking strokes. Exceptions, of course, occur in the classic and almost obligatory cases of ligatures such as "ch," "sc," and "st," and the links "de" and "di." While it is precise, adequately uniform, and regular, Michael's hand occasionally reveals a particular precious affectation, more apparent in certain individual letters rather than in the overall layout. This affectation is easier to recognize in capital letters, especially when they occur at paragraph headings as a paragraph marker.

The *ductus*,[114] which is intentionally poised at least at the beginning of a page, by the end of it indulges inevitably in a certain cursiveness perhaps due to tiredness of the hand. Similarly words are, as a rule, well separated from one another, although occasionally articles and prepositions become joined to the adjectives and nouns that they introduce.

Capital letters are usually reserved for paragraph headings, since each sentence essentially comprises the entire paragraph and contains no obvious internal breaks. The punctuation, which is very limited, does not correspond to any syntactic rule; thus there are no traces at all of a consistent and orderly graphical decoration of logical separators. If anything, it constitutes in the intentions of the *scriptor* an additional instrument utilized to separate individual words rather than different parts of the discussion. Occasionally (but according to no clear or rigorously observed rule) the most significant separation with regard to logical-syntactic value is represented by a transverse bar. Overall, the punctuation certainly does not facilitate our understanding of the concepts, which are often obscure and virtually reserved to a restricted group of experts, subject by subject.

With regard to the salient characteristics of the most significant capital letters, we can easily observe how the "A," which always lacks the horizontal bar, is simply constructed by crossing two inclined strokes, and thus can be easily confused with a fairly enlarged "X." At any rate, it does not distinguish itself from the latter other than by the greater size of the module—in addition, of course, to its different phonetic restitution. The "C," "F," "N," "O," and "P" are capitalized simply by enlarging their respective lower-case letters. The "D," although distorted by the evident cursiveness and by the similarly apparent expansion of the module, tends to evoke models of the capital letter more specifically rustic than epigraphic. The "E," "M," "Q," and "S" instead betray a clearly uncial origin. The "I" and "L" are quite characteristic, singularly similar in aspect to the capital forms of corresponding cursive models in use today. The "S" very closely resembles the number "6," similarly to what can be found in the best examples of "modern" chancery. It is sometimes easily confused with "E," especially when the latter, in its peculiar uncial typology, closes the intermediary horizontal line with the lower line almost like an eyelet. The latter is strongly curved toward the top and constructed without a break such as the rounded, right-leaning prolongation of the vertical line. The doubling of the internal segment of the "Q," which occasionally also appears in the letters "C," "E," and "T," is typical and constant, with clearly decorative rather than structural aims. The "T" is also quite particular, very similar to an overlapping "Z" and "T." There is no substantial difference, on the other hand, between the initial "U" and "V."

---

114. I.e., the distinctive manner in which the strokes are traced on the writing surface.

Turning to the lower-case letters, the dimorphism of "d" (which is sometimes uncial and sometimes straight) is not particularly noteworthy. On the other hand the prepositions "de" and "di" take on significant importance, as their correct interpretation is essential to an accurate reading of the text, especially in passages where the meaning is more ambiguous. Even the "s" displays the usual dimorphism, although without following a regular rule: sometimes it is round, when it occurs as an initial letter, almost like a slightly miniaturized capital letter, whereas it is always straight when it occurs within the body of a word. The "e," which lacks an eyelet, is constructed by bringing together a short vertical stroke and a barely visible horizontal stroke. The "i," usually elongated at the bottom when at the end of a word, occasionally appears with the diacritical mark of a dot, and other times not; it seems to follow no rules, almost according to the whim of the *scriptor*. The ligature "ch," as with the upper ligature "st," is regular and typified. The latter is not always easily distinguishable from the similar "sc." Only the context can assist us in more problematic cases. The "u" and "v" occur in the two morphological variants with equivalent phonetic value. The "z" always takes the form of a "3" slightly raised above the line. It is often confused with the number, and so it becomes very easy to read "*zo*" (in the sense of "ciò" [this] or "giù" [down]) in place of "30." In this case as well, only the context of the discussion can help resolve the ambiguity.

Abbreviations are used in moderation and conform to the typical records of mercantile writings, and almost never pose troublesome difficulties of interpretation:[115] abbreviations can be made by truncation, by contraction, by letter in superscript, by specific letter, by tachigraphic note, by proper and conventional symbol, and with general or particular reference. One particularity concerns the consonant "q" abbreviated by a horizontal line across the descender, used both for "qui" and for "que," whether in the whole word (usually "qui" in the meaning of "who") or in the case of a prefix ("que-sto," "que-sta" [this], "que-sti," "que-ste" [these]).

The "technical" abbreviations in the first section of the manuscript, which is entirely dedicated to mathematics, are specific to that particular discipline. More will be said of these below, in our discussion of the transcription criteria that have been adopted.

Thus the hand of Michael of Rhodes, while adhering to the customs and writing canons of its time and place, reveals more than a few personal contributions, and in particular a significant degree of personalization—elements demonstrating that it cannot be attributed to a professional scribe. Moreover, the earlier observations regarding the preliminary phases of preparation of the writing area, the pagination methods, and above all the general organization of the codex lead us to believe without a doubt that the primary occupation of Michael of Rhodes (whom Armando Petrucci would certainly define as "alfabeta dell'uso"[116] [literate through practice]) is certainly not that of a copyist.

Hand B appears for the first time at the head of fol. TOC 1b ("Scritti e ricordi di Michele Daruodo" [Writings and remembrances of Michael of Rhodes]), and a second time at fol. 204a: "Questi sono i Ricordi e Scritti d'un tal Michele Daruodo Veneziano il di cui Nome si vede di sopra e f. 90" (These are the remembrances and writings of a certain Michael of Rhodes, a Venetian, whose name appears above on fol. 90), immediately beneath Michael's last handwritten autobiographical annotation. Thus it repeats what had already been stated as a title at the beginning of

115. See Armando Petrucci, *Breve storia della scrittura latina* (Rome: Bagatto Libri, 1989), 161.
116. Armando Petrucci, *Prima lezione di paleografia* (Rome: Laterza, 2002), 20.

the manuscript, almost as if to mark the end of Michael's autographical text at that particular spot. The design of the individual letters, posed with affectation and intentionally calligraphic,[117] does not facilitate the dating of these few lines, which likely go back to a period around the end of the nineteenth century and the beginning of the twentieth. This hypothesis is actually supported by the language itself, which vaguely recalls the flavor of late nineteenth-century Italian even in its brevity and extreme concision. If anything, this might perhaps narrow it down more precisely to the later second half of the nineteenth century, i.e., before the manuscript underwent the invasive and clumsy intervention of rebinding and the addition of those initial 10 folios concerning which we commented earlier. One notable orthographic peculiarity is the symbol used to divide syllables at the end of a line: "Vene-ziano." Another linguistic curiosity is the transformation of an indication of provenience "da Ruodo" (from Rhodes) into a veritable surname, with a capital letter "Daruodo," in line with what the anonymous extensor of the note in hand B (being perhaps as unfamiliar with geography as he was with paleography) could have read correctly at fol. 90-2b: "Michalli da Ruodo." Thus it is likely that one of the more recent owners of the manuscript, precisely because it was anonymous and lacked a heading, felt practically obliged to confer title, content, and authorship at the beginning and end of the manuscript, demonstrating a complete lack of conservational sensitivity for it.[118]

The preparation of the four portolans located from fol. 205a to fol. 210b can be attributed to a third hand, C: "Portulan da Venesia fina a Constantinopoli pe[r] Rivera como le galie vano"; "Portolan per i traversi del Colpho de Venexia"; "Portolan da Cavo Malio fina al'isola de Famagosta"; "Portolan da Venesia infina ala Tana, ala via dele galie per staria" (Portolan from Venice to Constantinople along the coast as the galleys go; Portolan for the crossings of the Gulf of Venice; Portolan from Cape Maléas to the island of Famagusta; Portolan from Venice to Tana, on the route of the galleys by the coast). The writing, which most likely goes back to the later second half of the fifteenth century, is undoubtedly more typical of a book hand than a chancery (i.e., documentary) hand: calligraphic in its own way, orderly, without smudges or corrections of any kind, lacking ligatures between individual letters, sufficiently composed, airy, if anything marked by a precious and persistent affectation. In any case the ink used, which lacks even minimal chromatic variations, would attest to the speed and continuity with which the entire work was carried out.

The writing area of the folios in this section of the manuscript, which is definitely extraneous to Michael's text, was prepared in a manner somewhat different from the preceding folios. Each folio, lacking the *invocatio* "Ihesus," has been preliminarily given margins by lead point, and each one was ruled, again by lead point. It was lined horizontally, so as to allow a constant number of 29 lines of writing perfectly parallel with each other (at least when there are no skips in the units of ruling due to the *incipit* of a new portolan or to other intermediary breaks). Each folio was also lined vertically, just like the grids in a modern square-ruled notebook, to allow the perfect columnar arrangement of abbreviations for wind directions and distances in miles between points along the coast. Furthermore, portolan titles are always underlined in red ink to facilitate their immediate identification

117. The individual letters are written with extreme care and are rigorously separated from one another.

118. The same criticism can also be leveled at Federico Patetta. In fact he had no compunction about adding, in his own hand at the bottom of the same fol. TOC 1b, the note of ownership "Federico Patetta / Ms. n° 32"—thus further expanding (albeit by only a little) the set of hands found in the manuscript itself.

within the writing area, which itself was already aesthetically well arranged, almost as if to confirm the great care given to the formal aspect of the texts by the *scriptor*, certainly more practiced in book script than Michael of Rhodes.

The writing attributed to hand D appears in fol. 225b: "Chi vol far una tavola de Salamon die saver quanto / son una ora, che son ponti 1000 e 80, son una ora, / e una luna son dì 29, ore 12, ponti 793" (Whoever wishes to make a table of Solomon should know how much an hour is, which is 1,080 points, which is an hour, and a moon is 29 days, 12 hours, 793 points). Three lines in all, written almost as a note, perhaps as a written reflection, a mnemonic recollection, a personal musing, an extrapolation in the form of a general rule about what Michael of Rhodes states in fol. 131a regarding the table of Solomon, and later in fol. 186a regarding the duration of the lunar cycle: "dichotte che una luna sie dì 29, ore 12, ponti 793, arigordandotti che 1080 punti serà ora 1, e ore 24 serà dì 1" (I tell you that a moon is 29 days, 12 hours, 793 points, remembering that 1,080 points will be an hour, and 24 hours will be 1 day).

The last two written folios of the codex belong to hand E, at fols. 238a and 238b. This text, which has nothing remotely to do with the contents of the rest of the manuscript, is a sort of presentation of a last will by a certain Giovanni da Drivasto, *paron zurado* of Marino Dandolo, written on August 29, 1473, almost certainly on board a Venetian galley docked "al Chiarcho" (Charchi, now Khálkhi, a small island near Rhodes). The testator (about whom it has not been possible to find additional information) likely found himself out of writing paper and resorted to using a couple of folios that had been left unutilized toward the end of the manuscript in order to write down his last wishes. If nothing else this event, quite significant in itself, documents the fact that several decades after Michael's manuscript was composed, it was no longer located in Venice but was perhaps traveling with its current owner.

The writing, in a strongly personalized lower-case cursive, appears initially difficult to read especially because the words, more than the individual letters, are rarely separated from each other but run together almost without a break. There are no traces of capital letters or of punctuation marks other than an occasional transverse bar that certainly does not help to separate phrases into intelligible sentences. Apart from these the writing seems to flow uninterrupted, in places is confused and uncertain, does not lack errors and changes of mind, and is seriously inaccurate from morphological, grammatical, and syntactical points of view. To summarize (quoting another definition of Armando Petrucci), the author of these two final folios cannot be other than a "functional semiliterate" (*semialfabeta funzionale*), someone who certainly had not had many opportunities to develop adequate confidence in writing.[119]

With regard to the inks used in the manuscript, we must highlight how from its beginning (i.e., from fol. 1b) at least right through fol. 199a Michael of Rhodes used a single ink, which was a dark brown color that tended toward black, with the significant exceptions of annotations at fol. 93b for which he utilized inks that are substantially different from one another. It appears that Michael also used the same ink for the initial table of contents and for the foliation, up to and including fol. 204a. Starting with fol. 199b, almost to underscore a sort of temporal break occurring between two different times of writing—also indicated by the widening of the writing spacing that occurs precisely from fol. 199 onward—the ink used is quite different, of a rather light and faded sepia

119. Petrucci, *Prima lezione*, 20.

color. Further confirming the existence of this break, it should also be noted how the last three items in the table of contents—covering the contents of folios 199b through 201b, to which we should also add the contents of folios 102b and 103a, which are completely ignored in the first version of the table of contents—are written in the same light sepia-colored ink. Michael used this same ink to write the additions to his autobiographical digression regarding the death of his first wife Dorotea at fol. 91-1b, the death of his son Teodorino at fol. 92a, the death of his second wife Cataruccia at fol. 93a, and lastly the privilege of the steelyard, a first time at fol. 93b and a second time at fol. 204a.

As the inks used by other hands in the manuscript are not relevant to the aims of its analysis, we see no reason to discuss them here.

The Date of the Manuscript

Thanks to several explicit comments by the author, it is possible to estimate fairly closely the stages in which the manuscript was written. Indeed, in the first folio of the table of contents (TOC 1b) Michael makes certain to add the date 1434 with the invocation. Despite its being placed in the table of contents (which is usually among the last sections to be undertaken when writing a book), this date can in no way be interpreted as the date when the manuscript's composition was completed (i.e., the *terminus ad quem* of Michael's efforts). For a variety of reasons that emerge from several intrinsic peculiarities of the manuscript itself, it should instead be understood as the date of the beginning of its composition—i.e., its *terminus a quo*.

To begin with, a careful examination of the handwriting of the autobiographical annotations, and especially the arrangement on the page of paragraphs and the spaces between them, clearly allow us to grasp how these were surely written without a break to cover the more salient episodes of Michael's *cursus honorum*, starting with his first deployment as an oarsman on a Venetian galley—"In the name of God. I, Michael of Rhodes, shall write below about the time I came to Venice. It was on June 5, 1401. / And first, I signed on in Manfredonia as an oarsman with the nobleman Pietro Loredan, son of the late Alvise Loredan"[120]—up to the annotation about his enlistment as *armiraio* of the galleys of Flanders in 1436 at fol. 93a: "I signed on as *armiraio* with the noteworthy Francesco Capello, my *comito* Lazaro Parizotto, *paron* Corzulla, in 1436 on the voyage to Flanders."[121]

Following this entry, Michael's handwriting changes radically, at least with regard to its general arrangement (the module, *ductus*, and especially the inks change significantly). What is more important, however, is that he no longer seems capable of organizing the salient stages of his career with the same reliability, the same attention to extrinsic forms, the unvarying attention to symmetry, that he had demonstrated earlier. This was precisely because they were no longer being entered in a single writing sequence (as had occurred at least up to the note regarding 1436), but only once

120. MOR, fol. 90-2b: "Qui de sotto scriverò mi Michalli da Ruodo el tenpo veny in Veniexia. Zò fu 1401 adì 5 zugno. / E primo m'achordiè in Manfredonya per homo da remo chon el nobille homo miser Piero Loredan fu de miser Alvixe Loredan."

121. MOR, fol. 93a: "M'achordiè per armiragio chon el spectabille homo miser Franzescho Chapello, el mio chomitto Lazaro Parixotto, paron Chorzulla, del 1436 al viazio de Fiandria." And see Venice, Archivio di Stato, Notatorio di Collegio, R. 6, fol. 157r.

a specific appointment had been attained. Michael notes these appointments in the space he intentionally left blank for this specific purpose (just under three and a half folios), with a truly unusual carelessness that is certainly out of place and unjustifiable in a written project with clearly appreciable aesthetic characteristics.

It is possible to conjecture a first phase of intensive writing that would have extended from 1434 to 1436, i.e., up to the moment of departure of the convoy of Flanders.[122] This phase would have comprised the first section of the arithmetic-algebraic instructions, the autobiographical annotations up to the mention of his election as *armiraio* of the galleys of Flanders in 1436, and all the subsequent sections up to and including fol. 199a.

An even more precise definition of the writing periods might be suggested if we wish to give particular weight to the fact that the chronological computations that Michael proposes always refer to the years 1435 and 1436: for instance the table of Solomon that begins at fol. 131b, or the calculation of the epact and the duration of the lunar cycle at fols. 185a and 185b, as well as the entire remaining chronological-astronomical examination at fols. 188a–190a. In other words, we could think of the biennium 1434–1435 as the period of composition of what, by weight, appears to be the first half of the manuscript (the portion dedicated to arithmetic and algebra).[123] The first months of 1436 would have been dedicated to the remaining sections up to fol. 199a. Since Michael's election as *armiraio* of the galleys of Flanders is dated February 14, 1436, we can justifiably believe that this—give or take a day—was the *terminus a quo* for the composition of the portion of the manuscript from fol. 90-2b to fol. 199a. Thus during the months of forced inactivity while waiting to embark on his next voyage, Michael of Rhodes might have managed to organize a systematic and up-to-date *curriculum vitae*, perhaps based on the various contracts granted him by the Camera dell'Armamento. He would have reworked his written and oral sources on shipbuilding into a sufficiently methodical text, obviously postponing laying out the iconographic decoration to a later date precisely in order not to slow down his writing rhythm. He would have focused on writing the portolans, and presented his—or someone else's?—astrological knowledge, leaving the necessary spaces blank for the graphical representation of the signs of the zodiac. He used the current year (1436) and the year just passed (1435) as preferred references for his complex chronological-astronomical explanations, as indeed was common practice in contemporary

122. In that year the convoy of four galleys, two of which were heading directly for London and two to Bruges, left Venice on Sunday, April 22, after much resistance by the *patroni* and the partners, who were rightly concerned by the news that was arriving about the war between the king of England and the duke of Burgundy. The ships returned no earlier than the end of April of the following year. On this occasion the Senate was obliged to authorize all who wished to do so to unload merchandise that belonged to them from the galleys on which it had already been stowed. Venice, Archivio di Stato, Senato, Deliberazioni Miste, R. 59, fols. 142r–142v, 149v, and 154v.

123. The hypothesis that the time invested in writing the mathematics section was undoubtedly more substantial than the time required for the remaining sections is by no means devoid of merit. Indeed, as Raffaella Franci rightly points out in her essay in this edition, Michael of Rhodes did not limit himself to transcribing the sources at his disposal in a slavish and mechanical manner. Instead he carefully analyzed the entire system of calculations, which he clearly verified one at a time, and repeatedly presented solutions in multiple and not always justifiable methodological variations. In so doing, Michael demonstrated an unusual familiarity with the subject, which he had undoubtedly mastered and was not merely reproposing superficially as an easily grasped and well-received literary genre. This section, therefore, could be considered the most original section of the entire manuscript. See Raffaella Franci and Laura Toti Rigatelli, *Introduzione all'aritmetica mercantile del Medioevo e del Rinascimento: Realizzata attraverso un'antologia degli scritti di Dionigi Gori (sec. XVI)* (Urbino: Quattro Venti, 1982).

treatments of the subject. In other words, Michael would have taken the time when not at sea to bring a large portion of his entire work to completion.

At any rate, starting at fol. 199b, the variation in color and quality of the ink clearly reflects a gap in the writing continuum. The later phase of writing, which can be dated to the last months of 1444, could have ended naturally in the very first months of 1445, as suggested not only by the entry related to the achievement of the *grazia della stadera* (privilege of the steelyard), but also by the perfect correspondence between the ink used for this entry, the ink used in all the texts included between fol. 199b and fol. 204a, and, in a singular coincidence, the ink used for the other work of Michael of Rhodes once attributed to Pietro di Versi, which can be dated to 1444–1445. During the course of the intermediate phase, i.e., between 1437 and 1444, Michael may have been able to make entries regarding his *cursus honorum* after 1436, as well as perhaps the apparent integration of the magic-ritual formula to soothe toothaches at fol. 184b, for which he had only been able to write the title in a first round, perhaps because he didn't have the source immediately available to him. He could also have prepared the illustrations, tables, drawings, and copies of the diagrams—which certainly required longer periods of time to be made, especially for a hand that was objectively less skilled in drawing. For these, Michael had left the needed space near the respective captions.

The texts from fol. 199b onward, however, constitute a sort of meaningless resumption of previously treated subjects, sometimes even useless and overabundantly repeated. For example, the mathematical problems that Michael of Rhodes presents in these last folios (fols. 199b–201b, 203b) add nothing new to what he had presented earlier. The same riddle with the preestablished result of the 15 Christians versus 15 pagans at fol. 201b is merely a different formulation on the basis of 10 (though much reduced and almost banalized with regard to the explanation) of the same question presented at fols. 91-1a–91-1b with 15 Christians and 15 Jews, that time calculated on the basis of 9. Similarly, the algebraic rule presented at fol. 203a: "The topic of the fifth chapter is squared unknowns and numbers equal to an unknown"[124] had already been presented—and certainly more clearly and with plenty of examples—starting at fol. 13a.

And finally, the material added to fol. 202b—regarding the quantity and quality of timber, of artisans, of ironware, pitch, and oakum needed to construct a galley of Flanders—aside from being absolutely isolated from the context in which it was inserted, appears really to be a later integration of information already presented in the first section. Nevertheless, an interesting feature is the similar disposition of writing lines, the design of the letters, the *ductus* itself: quite surprisingly, these can be linked to the more generally extrinsic characteristics of the folios dedicated earlier to shipbuilding, more than to the folios immediately preceding and following it, with respect to which this folio seems to remain essentially extraneous.

Several autobiographical additions that occur after fol. 90-2b can also be attributed to this last phase, enriching the cold and bureaucratic list of appointments with some human depth. These additions related to the death of Michael's first wife Dorotea at fol. 91-2b: "And at this point I found that my wife Dorotea had died"; the death of his son Teodorino at fol. 92a: "And my son Teodorino died on this voyage"; and the death of his second wife Cataruccia at fol. 93a: "And on this voyage I found my wife Cataruccia dead."[125]

---

124. MOR, fol. 203a: "La natura del quinto chapitulo ssie zensso e numero ingual a chossa."

125. MOR, fol. 91-2b: "E in questo troviè mia moier Dorattia morta"; fol. 92a,"Et in questo viazio morì mio fio Thodorin"; fol. 93a: "Et in questo viazio truovyè mia moier Chataruza morta."

COMPOSITION OF THE MANUSCRIPT

To better understand the actual composition of the manuscript and its fasciculation, its structure is analyzed below.[126] As noted above, the manuscript is made up of 17 gatherings, not all of which contain the same number of folios. In fact this number can vary from a minimum of 1 + 1 folios to a maximum of 10 + 10 folios, although on average each gathering, or fascicule, is composed of 8 + 8 folios.

		[fol. A1a]	*[pasted-down endpaper]*
fasc. 0	fols. 5 + 5	[fols. A1b–A11a]	*[blank]*
fasc. 1	fols. 1 + 1	[fol. A11b]	*[blank]*
		[fol. A12a]	*[blank]*
		[fol. TOC 1b]	
		[fol. TOC 2a]	
fasc. 2	fols. 4 + 7	[fol. TOC 2b]	
		[fol. TOC 3a]	
		[fol. TOC 3b]	
		[fol. TOC 4a]	
		fol. 1b	
		[fol. 2a]	*[not numbered]*
		[fol. 2b]	*[not numbered]*
		fol. 3a	
		fol. 3b	
		fol. 4a	
		fol. 4b	
		fol. 5a	
		fol. 5b	
		fol. 6a	
		fol. 6b	
		fol. 7a	
		fol. 7b	
		fol. 8a	
		[fol. 8b]	*[not numbered]*
		fol. 9a	
		[fol. 9b]	*[missing folio, torn out subsequent to binding]*
		[fol. 10a]	
		fol. 10b	
		fol. 11a	

126. Square brackets and italics indicate editorial interventions in the foliation.

fasc. 3	fols. 8 + 8	fol. 11b fol. 12a
		fol. 12b fol. 13a
		fol. 13b fol. 14a
		fol. 14b fol. 15a
		fol. 15b fol. 16a
		fol. 16b fol. 17a
		fol. 17b fol. 18a
		fol. 18b fol. 19a
		fol. 19b fol. 20a
		fol. 20b fol. 21a
		fol. 21b fol. 22a
		fol. 22b fol. 23a
		fol. 23b fol. 24a
		fol. 24b fol. 25a
		fol. 25b fol. 26a
		fol. 26b fol. 27a
fasc. 4	fols. 8 + 8	fol. 27b fol. 28a
		fol. 28b fol. 29a
		fol. 29b fol. 30a
		fol. 30b fol. 31a
		fol. 31b fol. 32a
		fol. 32b fol. 33a

		fol. 33b	
		fol. 34a	
		fol. 34b	
		fol. 35a	
		fol. 35b	
		fol. 36a	
		fol. 36b	
		fol. 37a	
		fol. 37b	
		fol. 38a	
		fol. 38b	
		fol. 39a	
		fol. 39b	
		fol. 40a	*["3" crossed out]*
		fol. 40b	
		fol. 41a	
		fol. 41b	
		fol. 42a	
		fol. 42b	
		fol. 43a	
fasc. 5	fols. 9 + 9	fol. 43b	
		fol. 44a	
		fol. 44b	
		fol. 45a	
		fol. 45b	
		fol. 46a	
		fol. 46b	
		fol. 47a	*["1" crossed out]*
		fol. 47b	*["1" crossed out]*
		fol. 48a	
		fol. 48b	
		fol. 49a	
		fol. 49b	*[blank, only invocation and border framing at the end of the page]*
		fol. 50a	
		fol. 50b	
		fol. 51a	
		fol. 51b	
		fol. 52a	
		fol. 52b	
		fol. 53a	
		fol. 53b	
		fol. 54a	
		fol. 54b	
		fol. 55a	

		fol. 55b	
		fol. 56a	
		fol. 56b	
		fol. 57a	
		fol. 57b	
		fol. 58a	
		fol. 58b	
		fol. 59a	
		fol. 59b	
		fol. 60a	
		fol. 60b	
		fol. 61a	
fasc. 6	fols. 8 + 8	fol. 61b	
		fol. 62a	
		fol. 62b	
		fol. 63a	
		fol. 63b	
		fol. 64a	
		fol. 64b	
		fol. 65a	
		fol. 65b	
		fol. 66a	
		fol. 66b	
		fol. 67a	
		fol. 67b	
		fol. 68a	
		fol. 68b	
		fol. 69a	
		fol. 69b	
		fol. 70a	
		fol. 70b	
		fol. 71a	
		fol. 71b	
		fol. 72a	
		fol. 72b	
		fol. 73a	
		fol. 73b	
		fol. 74a	*[no "Ihs" (fols. 74a–77a)]*
		[fol. 74b]	*[not numbered]*
		fol. 75a	
		[fol. 75b]	*[not numbered]*
		fol. 76a	
		fol. 76b	
		fol. 77a	

fasc. 7	fols. 10 + 10	fol. 77b	*["Ihs" resumes]*
		fol. 78a	
		fol. 78b	
		fol. 79a	*[no "Ihs" (fols. 79a–91-1b)]*
		fol. 79b	*[corrected from "78"]*
		[fol. 80a]	*[not numbered]*
		fol. 80b	*[also numbered "79" and "90"]*
		fol. 81a	
		fol. 81b	
		fol. 82a	
		fol. 82b	
		fol. 83a	
		fol. 83b	
		fol. 84a	
		fol. 84b	
		fol. 85a	
		fol. 85b	
		fol. 86a	
		fol. 86b	
		fol. 87a	
		fol. 87b	
		fol. 88a	
		fol. 88b	
		fol. 89a	
		fol. 89b	
		fol. 90[-1]a	
		fol. 90[-1]b	
		fol. 91[-1]a	
		fol. 91[-1]b	
		fol. 90[-2]a	*[number repeated by scriptor; "Ihs" resumes]*
		fol. 90[-2]b	*[number repeated by scriptor]*
		fol. 91[-2]a	*[number repeated by scriptor]*
		fol. 91[-2]b	*[number repeated by scriptor]*
		fol. 92a	
		fol. 92b	
		fol. 93a	
		fol. 93b	
		fol. 94a	*[blank, only invocation]*
		[fol. 94b]	*[blank; not numbered; no "Ihs"]*
		fol. 95a	
fasc. 8	fols. 10 + 10	fol. 95b	
		fol. 96a	
		fol. 96b	
		fol. 97a	

		fol. 97b fol. 98a	
		fol. 98b fol. 99a	
		fol. 99b fol. 100a	
		fol. 100b fol. 101a	
		fol. 101b fol. 102a	
		fol. 102b fol. 103a	
		fol. 103b fol. 104a	
		fol. 104b fol. 105a	
		fol. 105b fol. 106a	
		fol. 106b fol. 107a	
		fol. 107b fol. 108a	
		fol. 108b fol. 109a	
		fol. 109b fol. 110a	
		fol. 110b fol. 111a	
		fol. 111b fol. 112a	
		fol. 112b fol. 113[-1]a	
		fol. 113[-1]b fol. 113[-2]a	 *[number repeated by scriptor]*
		fol. 113[-2]b fol. 114a	*[number repeated by scriptor]*
fasc. 9	fols. 8 + 8	fol. 114b fol. 115a	
		fol. 115b fol. 116a	
		fol. 116b fol. 117a	
		fol. 117b fol. 118a	

		fol. 118b	
		fol. 119a	
		fol. 119b	
		fol. 120a	
		fol. 120b	
		fol. 121a	
		fol. 121b	
		fol. 122a	
		fol. 122b	
		fol. 123a	
		fol. 123b	
		fol. 124a	
		fol. 124b	
		fol. 125a	
		fol. 125b	
		fol. 126a	
		fol. 126b	
		fol. 127a	
		fol. 127b	
		fol. 128a	
		fol. 128b	
		fol. 129a	
		fol. 129b	
		fol. 130a	
fasc. 10	fols. 8 + 7	fol. 130b	
		fol. 131a	
		fol. 131b	
		fol. 132a	*[numbered "32" by scriptor]*
		fol. 132b	*[numbered "32" by scriptor]*
		fol. 133a	
		fol. 133b	
		fol. 134a	
		fol. 134b	
		fol. 135a	
		fol. 135b	
		fol. 136a	
		fol. 136b	
		fol. 137a	
		fol. 137b	
		fol. 138a	
		fol. 138b	
		fol. 139a	
		fol. 139b	
		fol. 140a	

		fol. 140b fol. 141a	
		fol. 141b fol. 142a	
		fol. 142b fol. 143a	
		fol. 143b fol. 144a	
			[stub between with some writing and drawing]
		fol. 144b fol. 145a	*[superimposed over "145"]*
fasc. 11	fols. 8 + 7	fol. 145b fol. 146a	
		fol. 146b fol. 147a	*[numbered "160" upside down at bottom right]*
		fol. 147b fol. 148a	*[numbered "159" upside down at bottom right]*
		fol. 148b fol. 149a	*[numbered "158" upside down at bottom right]*
		fol. 149b fol. 150a	*[numbered "157" upside down at bottom right]*
		fol. 150b fol. 151a	*[numbered "156" upside down at bottom right]*
		fol. 151b fol. 152a	*[superimposed over "152"]*
		fol. 152b fol. 153a	
		fol. 153b fol. 154a	
		fol. 154b fol. 155a	
		fol. 155b fol. 156a	*[numbered "151" upside down at bottom right]*
		[fol. 156b] *[fol. 157a]*	*[missing folio, torn out subsequent to binding]*
		fol. 157b fol. 158a	*[numbered "149" upside down at bottom right]*
		fol. 158b fol. 159a	*[numbered "148" upside down at bottom right]*
		fol. 159b fol. 160a	*[numbered "147" upside down at bottom right]*
		fol. 160b fol. 161a	*[numbered "146" upside down at bottom right]*

fasc. 12	fols. 8 + 8	fol. 161b fol. 162a
		fol. 162b fol. 163a
		fol. 163b fol. 164a
		fol. 164b fol. 165a
		fol. 165b fol. 166a
		fol. 166b fol. 167a
		fol. 167b fol. 168a
		fol. 168b fol. 169a
		fol. 169b fol. 170a
		fol. 170b fol. 171a
		fol. 171b fol. 172a
		fol. 172b fol. 173a
		fol. 173b fol. 174a
		fol. 174b fol. 175a
		fol. 175b fol. 176a
		fol. 176b fol. 177a
fasc. 13	fols. 8 + 8	fol. 177b fol. 178a
		fol. 178b fol. 179a
		fol. 179b fol. 180a
		fol. 180b fol. 181a
		fol. 181b fol. 182a
		fol. 182b fol. 183a

		fol. 183b fol. 184a	
		fol. 184b fol. 185a	
		fol. 185b fol. 186a	
		fol. 186b fol. 187a	
		fol. 187b fol. 188a	
		fol. 188b fol. 189a	
		fol. 189b fol. 190a	
		fol. 190b fol. 191a	
		fol. 191b fol. 192a	
		fol. 192b fol. 193a	
fasc. 14	fols. 8 + 8	fol. 193b fol. 194a	
		fol. 194b fol. 195a	
		fol. 195b fol. 196a	
		fol. 196b fol. 197a	
		fol. 197b fol. 198a	
		fol. 198b fol. 199a	
		fol. 199b fol. 200a	
		fol. 200b fol. 201a	
		fol. 201b fol. 202a	
		fol. 202b fol. 203a	
		fol. 203b fol. 204a	*[blank; no "Ihs"]*
		fol. 204b [fol. 205a]	*[blank]* *[no "Ihs"; and henceforth]*

		[fol. 205b]	
		[fol. 206a]	
		[fol. 206b]	
		[fol. 207a]	
		[fol. 207b]	
		[fol. 208a]	
		[fol. 208b]	
		[fol. 209a]	
fasc. 15	fols. 7 + 8	[fol. 209b]	
		[fol. 210a]	
		[fol. 210b]	
		[fol. 211a]	*[blank]*
		[fol. 211b]	*[blank]*
		[fol. 212a]	*[blank]*
		[fol. 212b]	*[blank]*
		[fol. 213a]	*[blank]*
		[fol. 213b]	*[blank]*
		[fol. 214a]	*[blank]*
		[fol. 214b]	*[blank]*
		[fol. 215a]	*[blank]*
		[fol. 215b]	*[missing folio, torn out subsequent to binding]*
		[fol. 216a]	
		[fol. 216b]	*[blank]*
		[fol. 217a]	*[blank]*
		[fol. 217b]	*[blank]*
		[fol. 218a]	*[blank]*
		[fol. 218b]	*[blank]*
		[fol. 219a]	*[blank]*
		[fol. 219b]	*[blank]*
		[fol. 220a]	*[blank]*
		[fol. 220b]	*[blank]*
		[fol. 221a]	*[blank]*
		[fol. 221b]	*[blank]*
		[fol. 222a]	*[blank]*
		[fol. 222b]	*[blank]*
		[fol. 223a]	*[blank]*
		[fol. 223b]	*[blank]*
		[fol. 224a]	*[blank]*
		[fol. 224b]	*[blank]*
		[fol. 225a]	*[blank]*
fasc. 16	fols. 8 + 7	[fol. 225b]	
		[fol. 226a]	*[blank]*
		[fol. 226b]	*[blank]*
		[fol. 227a]	*[blank]*

[fol. 227b]	*[blank]*
[fol. 228a]	*[blank]*
[fol. 228b]	*[blank]*
[fol. 229a]	*[blank]*
[fol. 229b]	*[blank]*
[fol. 230a]	*[blank]*
[fol. 230b]	*[blank]*
[fol. 231a]	*[blank]*
[fol. 231b]	*[blank]*
[fol. 232a]	*[blank]*
[fol. 232b]	*[blank]*
[fol. 233a]	*[blank]*
[fol. 233b]	*[cut folio, only residual margin]*
[fol. 234a]	
[fol. 234b]	*[blank]*
[fol. 235a]	*[blank]*
[fol. 235b]	*[blank]*
[fol. 236a]	*[blank]*
[fol. 236b]	*[blank]*
[fol. 237a]	*[blank]*
[fol. 237b]	*[blank]*
[fol. 238a]	
[fol. 238b]	
[fol. 239a]	*[blank]*
[fol. 239b]	*[blank]*
[fol. 240a]	*[original numeration "141"; blank, only invocation]*
[fol. 240b]	*[original numeration "141"; blank, only invocation]*
[fol. 241a]	*[pen trial]*
[fol. 241b]	*[pasted-down endpaper]*

Transcription Criteria

As is customary, several fundamental standards[127] were observed in preparing the transcription, with the aim of facilitating textual reading and comprehension while making every effort to respect the original intentions of the *scriptor* as closely as possible:

127. See "Norme per la pubblicazione dell'Istituto Storico Italiano," *Bullettino dell'Istituto Storico Italiano* 28 (1906), vii–xxiv; Alessandro Pratesi, "Una questione di metodo: L'edizione delle fonti documentarie," *Rassegna degli Archivi di Stato* 17 (1957): 312–333, which he subsequently merged into the chapter "L'edizione delle fonti documentarie" in his *Genesi e forme del documento medievale* (Rome: Jouvence, 1987), 99–109; Giampaolo Tognetti, *Criteri per la trascrizione di testi medievali latini e italiani*, Quaderni della "Rassegna degli Archivi di Stato," 51 (Rome: Ministero per i Beni Culturali e Ambientali, 1982); Armando Petrucci, *La descrizione del manoscritto: Storia, problemi, modelli* (Rome: La Nuova Italia Scientifica, 1984); and Philippe Contamine, "La noblesse et les villes en France, XIVe–XVe siècle: Progetto di norme per l'edizione delle fonti documentarie," *Bullettino dell'Istituto Storico Italiano per il Medioevo e Archivio Muratoriano* 91 (1984): 491–503.

1) Even in the case of an autographic text that can be attributed to the hand of a single *scriptor*, we cannot find the kind of morphological and linguistic consistency that might allow us to identify the most prevalent usages, in order to establish well-organized and regular standards with which to expand abbreviations. Thus it is clear that no single solution allows for a disciplined respect of the original reading. Consequently, no matter what criterion is preselected, there will always be criticisms and reproofs, even if the effort is configured as an attempt, not always guaranteed success, to mediate between two opposed subjectivities: that of the author of the texts and that of the reader.

   Every abbreviation has been written out in full. The only exceptions regard specific abbreviations in the mathematical section that represent algebraic operators ("chu" for cubed,[128] "co" or "C" for "chosa,"[129] "n" or "N" for number, "Rx" for root, "□" for square, "Ç" for "zensso," "zenso," or "censo"[130]), as these clearly function within the forms of the abbacus tradition and in the *scriptor*'s mathematical language.[131]
2) Articulated prepositions have been made uniform according to modern Italian usage, linking preposition and article.
3) Paragraph headings have been maintained; they are usually indicated by the *scriptor* by a slightly enlarged capital letter that protrudes significantly into the left margin, and follow a wider line spacing than usual.
4) Punctuation has been converted to modern usage, especially with the intention of making the logico-syntactic structure of the sentences easier to understand, making sure to avoid changes that might significantly distort the *scriptor*'s own style.
5) Capital letters are used only for sacred persons, proper names, family names, and always after a period, following modern usage. The terms "santo," "santa," "santi" (saint/saints) are transcribed with the initial capitalized only when they form part of a toponym, and not when used as personal predicates. A capital letter is used to distinguish "Luna" (Moon) as an astrological name from "luna" (moon) in other contexts; and likewise for "Sol"/"sol" and "Sul"/"sul."
6) The letter "j" is not distinguished from the letter "i." Instead we have respected the *scriptor*'s use of "y," clearly only in cases where it does not correspond to "ij." Similarly, the forms "ch" and the extremely precise "tt" and "ss" at the beginning of a word are maintained, in accordance with the *usus scribendi* (i.e., the writer's personal word usage). The vowel "u" is always distinguished from the consonant "v." In addition the morphology of the letter "ç" is maintained even when it is not an algebraic operator.
7) The vowels "a" and "o" have been accented when they have a value as a word ("à" corresponding to the modern Italian "ha" ([he/she/it] has) and "ò" corresponding to "ho" ([I] have).
8) Numerals are written just as they are found in the manuscript and consistent with the *scriptor*'s intentions, preserving their specific morphology: "1," "1°," "1ª," "un," "uno," "una" (1, 1st, a/an).

128. Third power.
129. The unknown in a problem.
130. The square of the "chosa" (unknown).
131. See MOR, fol. 18b. In this regard see also Gilio da Siena, *Questioni d'algebra: Dal Codice L.IX.28 della Biblioteca Comunale di Siena*, ed. Raffaella Franci (Siena: Servizio Editoriale dell'Università di Siena, 1983), xii.

9) Intentional lacunae in the text are represented by three periods within square brackets [. . .]. Areas that are impossible to read due to tears, lacerations, erasures, etc., are represented instead by three asterisks ***. Every such case is accounted for in a note.
10) Addenda to the reading, to amend possible omissions by the *scriptor* due to an oversight or apparent slip of the pen (*lapsus calami*), are suggested in italics within square brackets. In the more significant cases, justification for the addendum is provided in a note.
11) An apostrophe is used at the end of a word to indicate the dropping of a final vowel or syllable. In addition, the apostrophe is used at the beginning of a word to indicate the elision of the initial vowel.
12) The criterion for folio numbering is respected according to the book hand adopted by the *scriptor* [e.g., fols. 2a–2b instead of fols. 1(verso)–2(recto)].
13) When it falls in the middle of a line of text, the end of the folio is represented by a single vertical bar [|].

Translated from the Italian by Claire Calcagno

Note on the Recent Restoration of the Manuscript

The description of the manuscript presented by Franco Rossi in the preceding pages was made in the course of several examinations of the manuscript, the most recent in 2006. When the editors of this publication compared proofs against the original manuscript in the fall of 2008, we discovered that it had been restored since our last view of it.[1] All of the restorations have been professionally done, and most were to repair tears and other problem areas.

One aspect of the restoration is worth noting, however, as it changes the manuscript substantially from the way it was when Rossi and the previous catalogers described it: the removal of eleven blank sheets at the beginning of the volume. These are the watermarked papers that appear to have been later than those used in the rest of the manuscript and were sewn in with thin threads, rather than the thick ones that secure the rest of the volume. We have chosen to leave the images of these pages in the facsimile in volume 1 of this publication and their description in Franco Rossi's essay. The removed sheets, along with the threads that bound them and scraps of parchment, remain housed in the modern box that holds the manuscript.

The first ten of the removed sheets comprised a gathering of five folios, which we have numbered as folios A1b through A11a. The last removed sheet was a single sheet, which we have numbered A11b and A12a. The first page of the manuscript with writing, the beginning of the table of contents in Michael's hand (as well as later annotations), which we have numbered TOC1b and its verso TOC2a, is now a single leaf attached with a guard tissue; it now appears at the start of the manuscript, facing the pasted-down front endpaper.

Alan M. Stahl
October 25, 2008

---

1. We are most grateful to Consuelo Dutschke, Curator of Medieval and Renaissance Manuscripts, and the staff of the Rare Book and Manuscript Library, Butler Library, Columbia University for making their facilities available to us for a week for this purpose, and to Alexis Hagadorn, Head of Conservation, Columbia University Libraries, for examining the manuscript with us and discussing issues of its collation and binding.

Principal Sections of the Manuscript

Transcription and Translation

Transcription of the Text

f. A1a +acb[us]aim+lanim+[a]
fusta, galia, falconi[b]
Aritmethica[c]
Cattalogo di alcuni sopracomiti Venetiani di galera. 90
De' segni celesti
Dissegni de gallie

f. TOC 1b + In Christi nomine, 1434 + Ihesus +

Scritti e ricordi di Michele Daruodo[d]

Qui vuol veder e lezer zò che die 'ser scritto. In questo libro pono qui per singollo de ttutte raxion:

a[e]	chargo de piper	charta 1, 3
a	sporta de piper	charta 2, 4
a	moltipichar rutti	charta 4, 0
a	zonzer rutti chon rutti	charta 5, 6
a	partir rutti chon rutti	charta 6, 7
a	trazer rutti da rutti	charta 7, 8
a	dar e tuor rutti	charta 8, 0
a	l'alt[*r*]a guixa dela riegola del 3	charta 9, 10
a	trar rutti e poner rutti	charta 9, 10, 11
a	chapitulli d'alzebra	charta 12, 13, 14, 15, 16
a	chose, zensi, chubi, radixe	charta 17, 18

a. *Hand of the beginning of the second half of the fifteenth century.*
b. *Hand of the late fifteenth or early sixteenth century.*
c. *This and remaining lines in a hand of the late second half, perhaps the last decade, of the sixteenth century.*
d. *In another, somewhat later hand.*
e. *The entries in this list are all preceded by a lower-case* a. *While this can be read as a word when the following word is an infinitive form of a verb, in other cases it makes no sense syntactically, so it is probably best thought of as the medieval equivalent of a list bullet.*

Translation of the Text

Acbusaim lanim[1] f. A1a
Small ship, galley, falcons[2]
Arithmetic
Catalog of some Venetian galley *sopracomiti*, 90
On the heavenly signs
Drawings of galleys

[Table of contents]

In the name of Christ, 1434 + Jesus + f. TOC 1b

Writings and remembrances of Michael of Rhodes.

Here one can see and read what should be written. In this book, I set forth here one by one all kinds of instructions:

1. Uncertain meaning, perhaps some sort of incantation.
2. Small ships constructed in the Venetian arsenal, described in Giorgetta Bonfiglio Dosio, ed., *Ragioni antique spettanti all'arte del mare et fabriche de vasselli* (Venice: Comitato per la Pubblicazione delle Fonti relative alla Storia di Venezia, 1987), pp. 20–23, in a hand dated to c. 1470.

a. *At the bottom of the page, in a modern hand:* Federico Patetta / Ms. n°. 32.

f. TOC 2a

1. At the bottom of page, in a modern hand: *Federico Patetta, Manuscript no. 32.*

	a	piper in Alesandria vadagnar e perder, chondur in Venyexia	charta 68, 69
f. TOC 2b	a	chonprar braza 5 de fustagno per duchato e dar la mitade braza 4 per duchato e l'oltra mittade braza 7 per 2 duchati e vadagnar 1 duchato, quanti fu le braza e quanto el chavedal	charta 69, 70
	a	braza de pano e tela e soldi e grosi mesidadi	charta 65
	a	partir 4, 3, roman 1, qual fu el partior	charta 66
	a	moltipichar per radixe	charta 67
	a	veder perché se sedemeza le chosse e partir per zenssi	charta 72
	a	veder 1 marchadante a far 2 viazi e non saver chon quanti denari, quanto fu el chavedal	charta 73, 74
	a	truovar numeri	charta 74
	a	partir ducati 54 alguny e domandar uno la radixe de ttuti	charta 74, 75
	a	far de 14 3 parte in proporzion che moltipichada l'una per 2, l'oltra per 3, l'oltra per 4, e zonto le moltipichazion faza 48	charta 75, 76
	a	truovar 2 homeny a denari; dixe el primo: "Se tu me dà $\frac{1}{3}$ d'i tto averò 24"; el segondo: "Se tu me dà $\frac{1}{5}$ d'i tto arò 56"	charta 76
f. TOC 3a	a	truovar numeri quadrati de radixe d'ogni numero chubicho	charta 77
	a	moltipicar radixe chon numero o radixe de zò che tu vuol; a partir radixe chon numero; a zonzer, a sutrar numeri per radixe	charta 77 e 90
	a	trar el dado, a saver quanti ponti; a penssar, a tuor denari e saver quanti	charta 90, 91
	a	eser 15 christiany e 15 zudie in un ballo, a nonbrar e getar ogny 9 un e gettar fuora i zudiei	charta 91
	e	gettar d'ogni 10 1[a]	charta 201
	a	chavar denari d'una quantità e romagnir in man alguny, a saver che fu el chavedal	charta 91, 92
	a	saver che tenpo veny in Veniexia e chon chi son stado in viazio	charta 90, 94
	a	veder chalandario	charta 95, 102

a. *Added later than the preceding passage, probably in another hand, with another ink.*

1. This and the following two folio numbers, though out of order, are accurate.
2. This section actually begins on fol. 82b.

1. The page that would have carried this heading is lacking in the MS.
2. This section actually ends on fol. 180a.
3. This and the next listing do not refer to subjects in the manuscript. There do not seem to be any pages missing from this section.

f. 1b

[Chargo de piper][a]

El chargo del piper, el qual è libre 400, val ducati $49\frac{1}{2}$, per libre 315 quanti duchati averò? E per farla per la riegola del 3 diremo: "Se 400 libre me dà duchati $49\frac{1}{2}$, che me darà libre 315?" Primo moltipicha per lo muodo vedirì qui de sutto e puo' parti.

---

a. *The section headings given in brackets are taken from the table of contents at the beginning of the manuscript.*

THE LOAD OF PEPPER[3]

f. 1b

If a load[4] of pepper which weighs 400 pounds is worth $49\frac{1}{2}$ ducats, how many ducats will I have for 315 pounds? And to solve this by the rule of three, we will say: "If 400 pounds gives me $49\frac{1}{2}$ ducats, what will 315 pounds give me?" First multiply by the method you see below and then divide.

---

1. Actually, fol. 196a.
2. Actually, fols. 196b–197a.
3. The section headings given in boldface are taken from the table of contents at the beginning of the manuscript. For an introduction to the mathematical sections of the manuscript, see Raffaella Franci, "Mathematics in the Manuscript of Michael of Rhodes," vol. 3, pp. 115–146.
4. The *cargo* was a measure of weight used for spices equal to 400 of the Venetian light pound, or about 120 kilograms.

400 $49\frac{1}{2}$ 315 / $\frac{400}{1} \times \frac{99}{2}$ — $\frac{315}{1}$ / $\frac{315}{99}$ | $\frac{2835}{2835}$ / 31185

0
0~~2~~4
~~1~~~~8~~~~8~~40
~~8~~000] 23
~~8~~0

| 18840 | $\frac{3140}{1570}$ | $\frac{785}{24}$

$\frac{440}{32}$ | $\frac{880}{1320}$ | 14080

0
0~~7~~7
~~3~~~~1~~~~1~~85
~~8~~000] 38
~~8~~0

Monta duchati 38, denari 23, pizulli $17\frac{480}{800}$. 18840

0
0~~6~~4
~~1~~~~4~~~~0~~80
~~8~~000] $17\frac{480}{800}$
~~8~~0

E per voler far la ditta raxion per la chossa, questo intra in lo primo chapitolo dove dize: "Chosa ingual al numero, se die partir lo numero per le chosse." Esempio dela raxion sovra scritta. Pony che valese 1^{co}. Adoncha se die dir: "Se 1^{co} me dà libre 315, che me darà duchati $49\frac{1}{2}$, che son $\frac{99}{2}$?" Adoncha moltipicha $\frac{99}{2}$ chon $\frac{315}{1}$, fano 31185, e può moltipicha 400 fia 1^{co}, fa 400^{co}. Puo' parti 31185 per 2, serano $15592\frac{1}{2}$. Parti li numeri per le chose, zoè per 400^{co}, insirà duchati 38. Lo romagnente moltipichado per 24 e partido per 400^{co} serano denari 23. E lo ruto moltipichado per 32 e partido per 400^{co} insirà pizulli 17. E questa serà ingual a quela de sovra.

E per far la ditta raxion per impoxizion pony che avesemo posto ala prima impoxizion 10, e diremo: "Se $\frac{400}{1}$ me dà $\frac{99}{2}$, che me darà $\frac{315}{1}$?" Se moltipicho $\frac{315}{1}$**a** unità chon $\frac{99}{2}$ farano $15592\frac{1}{2}$. Mo' si moltipicho 400 ch'è lo partidor chon 10 che ò posto a l'inpoxizion, me die venir $15592\frac{1}{2}$, e mi so che non ò se non 4000. Adoncha sotrazi 4000 de quello, vien a romanir $11592\frac{1}{2}$, e meto che questo sia men.

E per la segonda inpoxizion mettemo che la fose 20, e per lo simile muodo moltipichemo $\frac{99}{2}$ chon
f. 2a $\frac{315}{1}$, fano $15592\frac{1}{2}$, e moltipichà 20 fia 400 | sì fa 8000. Se sotrazo 8000 de $15592\frac{1}{2}$, romagnerà $7592\frac{1}{2}$. Adoncha questo è men a l'inpoxizion. Arigordati a moltipicar in cruxe, men e men sutrazi, più e più sutrazi, men e più s'azonze, e più e men s'azonze, e chosì per lo simille el partidor, chomo vedirì qui de sutto per singollo.

---

a. 400 *crossed out,* 315 *written above.*

400 $49\frac{1}{2}$ 315 / $\frac{400}{1} \times \frac{99}{2}$ — $\frac{315}{1}$ / 315 99 | 2835 2835 / 31185

0 0~~2~~4 ~~1~~~~8~~840 ~~8~~000 ~~8~~0	23	18840	3140 1570	785 24
		440 32	880 1320	14080

0
0~~7~~7
~~3~~~~1~~~~1~~85
~~8~~000
~~8~~0] 38

It results in 38 ducats, 23 denari,[1] $17\frac{480}{800}$ piccoli.[2] 18840

0
0~~6~~4
~~1~~~~4~~~~0~~80
~~8~~000
~~8~~0] $17\frac{480}{800}$

And to do this problem by an unknown, this is covered in the first chapter where it says: "Unknown equal to a number, the number should be divided by the unknowns." The example of the problem given above. Put that it is worth $1x$. Then you should say: "If $1x$ gives me 315 pounds, what will $49\frac{1}{2}$ ducats give me, which is $\frac{99}{2}$?" Then multiply $\frac{99}{2}$ by $\frac{315}{1}$, which makes 31,185, and then multiply 400 by the unknown, which makes $400x$. Then divide 31,185 by 2, which will be $15{,}592\frac{1}{2}$. Divide the numbers by the unknowns, that is by $400x$, the result will be 38 ducats. The remainder multiplied by 24 and divided by $400x$ will be 23 denari. And the fraction multiplied by 32 and divided by $400x$ will result in 17 piccoli. And this will be equal to the solution above.

And to do this problem by false position, suppose that we had put 10 at the first false position, and we will say: "If $\frac{400}{1}$ gives me $\frac{99}{2}$, what will $\frac{315}{1}$ give me?" If I multiply $\frac{315}{1}$ together with $\frac{99}{2}$ it will make $15{,}592\frac{1}{2}$. Now if I multiply 400, which is the divisor, with 10 which I have placed at the first false position, I should get $15{,}592\frac{1}{2}$, and I know that I have only 4,000. So subtract 4,000 from that, which leaves $11{,}592\frac{1}{2}$, and I put that this is a minus.

And for the second false position let's say that it was 20, and by a similar method we multiply $\frac{99}{2}$ by $\frac{315}{1}$, which makes $15{,}592\frac{1}{2}$, and 20 multiplied by 400 | makes 8,000. If I subtract 8,000 from f. 2a
$15{,}592\frac{1}{2}$, there will remain $7{,}592\frac{1}{2}$. So this is a minus at the false position. Remember to multiply across, subtract minus and minus, subtract plus and plus, add minus and plus, and add plus and minus, and the same for the divisor, as you see set out below step by step.

1. By denaro is meant the Venetian grosso of account, which was defined as $\frac{1}{24}$ of a ducat.
2. The piccolo was the Venetian penny, 32 of which were worth a grosso of account.

10	20
men	men
$11592\frac{1}{2}$	$7592\frac{1}{2}$

$$\begin{array}{r} 7592\frac{1}{2} \\ 10 \\ \hline 75925 \end{array} \qquad \begin{array}{r} 11592\frac{1}{2} \\ 20 \\ \hline 231850 \\ 75925 \\ \hline 155925 \end{array}$$

Se die partir | 155925

$$\begin{array}{r} 11592\frac{1}{2} \\ 7592\frac{1}{2} \\ \hline \end{array}$$

partidor 4000

$$\begin{array}{r} 0 \\ 0\not{3}3 \\ \not{1}\not{5}\not{5}925 \\ \not{4}0000 \\ \not{4}00 \end{array} \Bigg] \; 38$$

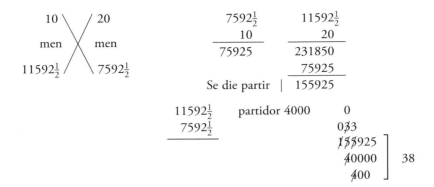

E lo romagnente moltipichado per 24 e lo romagnente per 32, serà ala simille qualitade chomo dixe la riegola del 3, chomo dixe qui de sutto duchati 38, denari 23, pizulli $17\frac{480}{800}$.

E per farla per inpoxizion per un'oltra via se die far per questo muo[*do*] seguirò qui de sutto. Mettemo che ala prima inpoxizion fusse 10, e noi diremo: "Se 315 tornase 10 de l'inpoxizion, che serave 400?" Moltipichado l'una per l'oltra, partida per 315, tu ssa che me die venir duchati $49\frac{1}{2}$, che son $\frac{99}{2}$. Adoncha varda che diferenzia è da[*da*] da quel che serà insio dal partior e sutrazi e mette sutto al'inpoxizion manchar tanto, moltipicha in croxe e parti. Serà duchati 38, denari 23, pizulli $17\frac{480}{800}$.

10	20
men	men
$36\frac{505}{630}$	$24\frac{65}{630}$

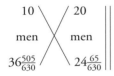

E si moltipichi in croxe e trazi e parti, inssirà chomo de sovra avemo ditto per le oltre raxion.

Insirà duchati 38, denari 23, pizulli $17\frac{480}{800}$.

f. 2b

[Sporta de piper]

El chargo de piper, che son libre 400, val in Veniexia duchati 50, e si me truovo aver duchati 264 quanto piper averò? E primo per farla per la riegola del 3 questo serà lo muodo. Tu die dir: "Se 50 duchati me dà libre 400 de piper, che me darà duchati 64?" Se moltipichi 400 fia 64 die venir 25600, e partir per 50 venirà libre 512 de piper.

10	20		
minus	minus		
$11592\frac{1}{2}$	$7592\frac{1}{2}$		

$$\begin{array}{r} 7592\frac{1}{2} \\ 10 \\ \hline 75925 \end{array} \qquad \begin{array}{r} 11592\frac{1}{2} \\ 20 \\ \hline 231850 \\ 75925 \\ \hline 155925 \end{array}$$

should be divided | 155925

$$\begin{array}{r} 11592\frac{1}{2} \\ 7592\frac{1}{2} \\ \hline \end{array}$$

divisor 4000

$$\begin{array}{r} 0 \\ 0\not{3}3 \\ \not{1}\not{5}\not{5}925 \\ \not{4}0000 \\ \not{4}00 \end{array} \Bigg] \; 38$$

And the remainder multiplied by 24 and the resulting remainder by 32, will be the same sum as derived by the rule of three, as is said below, 38 ducats, 23 denari, $17\frac{480}{800}$ piccoli.

And to do it by false position by another way, it should be done by the method that will follow below. Let's suppose that at the first false position it was 10, and we will say: "If 315 gives 10 of the false position, what will 400 be?" If the one is multiplied by the other and divided by 315, you know that I should get $49\frac{1}{2}$ ducats, which is $\frac{99}{2}$. So watch what difference is given from the result of the division and subtract and put below at the false position that so much is lacking, multiply across and divide. It will be 38 ducats, 23 denari, $17\frac{480}{800}$ piccoli.

10	20
minus	minus
$36\frac{505}{630}$	$24\frac{65}{630}$

And if you multiply across and subtract and divide, it will turn out as we have said above for the other method.

The result will be 38 ducats, 23 denari, $17\frac{480}{800}$ piccoli.

Concerning the basket of pepper f. 2b

A load of pepper weighing 400 pounds is worth 50 ducats in Venice, and if I find that I have 64[1] ducats, how much pepper will I have? And first to do it by the rule of three, this will be the way. You should say: "If 50 ducats gives me 400 pounds of pepper, what will 64 give me?" If you multiply 400 by 64 you should get 25,600, and divided by 50 it will come to 512 pounds of pepper.

1. The manuscript says 264 here, but the various solutions are for a problem of 64 ducats.

Se	50	400	64	400	25600	0	
				64		001̸0	
						2̸5̸6̸0̸0	
						5̸000	512 libre
						5̸5̸	

E per far ritorno, se	400	64	512	0	
			50	01̸0	
				2̸5̸6̸00	
				4̸000	64
				4̸0	

E per far la ditta raxion per la chosa pony che fusse vendudo 1^{co}. Adoncha diremo: "Se 1^{co} me dà libre 400 de piper, che me die venir per duchati 64?" Se moltipicho 400 fia 64 me die venir 25600^{n}, e questo è in lo primo chapitolo, dove dize chosa ingual a numero. A partir per 50 fia 1^{co} fa 50^{co}, a partir 25600^{n} die venir libre 512 de piper.

E per far la ditta raxion de sovra per inpoxizion pony che ala prima inpoxizion fusse 10, e diremo: "Se 64 me torna 10, che me tornerà 50?" Se moltipicho 10 fia 50 fa 500, se parto per 64 fa $7\frac{52}{64}$, e my voyo che ssia 400. Adoncha sutrazi $7\frac{52}{64}$ de 400, roman $392\frac{12}{64}$, e questi pony men ala prima inpoxizion. E per voler far la segonda inpoxizion pony che fusse 20. Adoncha diremo: "Se 64 me torna 20, che me tornerave 50?" Moltipicha 20 fia 50, fa 1000, e questi partidi per 64 serave $15\frac{40}{64}$, e my vorave che fuse 400. Adoncha sutrazi $15\frac{40}{64}$, romagnerà $384\frac{24}{64}$, e questi pony men ala segonda inpoxizion. E moltipichada in croxe, 10 fia $384\frac{24}{64}$ fa $3843\frac{48}{64}$, puo' moltipicha 20 fia $392\frac{12}{64}$, fa $7843\frac{48}{64}$. Sutratto l'un de l'oltro die romanir 4000, e sutratto de $392\frac{12}{64}$ $384\frac{24}{64}$ die romanir $7\frac{52}{64}$, e questo è lo so partidor. A partir 4000 unytta[a] e moltipichado serano 256000, partido per 500 insirà fuori libre 512, e tanto piper arai per duchati 64, chomo vedirì qui indriedo per figura.

---

a. e partido *crossed out.*

If	50	400	64	400	25600	0	
				64		$00\not{1}0$	
						$\not{2}\not{5}\not{6}\not{0}0$	
						$\not{5}000$	512 pounds
						$\not{5}\not{5}$	

And to do it backward, it's	400	64	512	0	
			50	$0\not{1}0$	
				$\not{2}\not{5}\not{6}00$	
				$\not{4}000$	64
				$\not{4}0$	

And to do the same problem by an unknown, put that $1x$ was sold. Then we will say: "If $1x$ gives me 400 pounds of pepper, what should come to me for 64 ducats?" If I multiply 400 by 64, it should come to 25,600, and this is in the first chapter, where it says an unknown is equal to a number. To divide by 50 times $1x$ makes $50x$, to divide 25,600 should come to 512 pounds of pepper.

And to do the above problem by false position, suppose that at the first false position it was 10, and we will say: "If 64 gives me 10, what will 50 give me?" If I multiply 10 by 50 it makes 500, if I divide by 64 it makes $7\frac{52}{64}$, and I want it to be 400. So subtract $7\frac{52}{64}$ from 400, the remainder is $392\frac{12}{64}$, and put this as a minus at the first false position. And to do the second false position suppose that it was 20. So we will say: "If 64 gives me 20, what will 50 give me?" Multiply 20 times 50, it makes 1,000, and this divided by 64 will be $15\frac{40}{64}$, and I wanted it to be 400. So subtract $15\frac{40}{64}$, there will remain $384\frac{24}{64}$, and put this as a minus at the second false position. And multiplied across, 10 times $384\frac{24}{64}$ makes $3{,}843\frac{48}{64}$, then multiply 20 times $392\frac{12}{64}$, which makes $7{,}843\frac{48}{64}$. One subtracted from the other should leave 4,000, and $384\frac{24}{64}$ subtracted from $392\frac{12}{64}$ should leave $7\frac{52}{64}$, and this is its divisor. To divide 4,000 units and multiplied will be 256,000, divided by 500 will result in 512 pounds, and that's the amount that you will have for 64 ducats, as you will see in the diagram following this.

f. 3a

5
0~~8~~2
|| 64 10 50 | ~~500~~]
~~64~~] $7\frac{52}{64}$

400 || $392\frac{12}{64}$ | 1 |
$7\frac{52}{64}$ ||

0
~~3~~4
0~~4~~~~6~~0
64 20 50 ~~1000~~]
~~644~~] $15\frac{40}{64}$
~~6~~]

400 || $384\frac{24}{64}$ | 2 |
$15\frac{40}{64}$[a] ||

10 \ / 20
men X men
$392\frac{12}{64}$ / \ $384\frac{24}{64}$
20 10

$7843\frac{48}{64}$ $3843\frac{48}{64}$
$3843\frac{48}{64}$ $7843\frac{48}{64}$
4000 questo die 'ser partido

$392\frac{12}{64}$ $7\frac{52}{64}$ questo è lo partidor
$384\frac{24}{64}$

$\frac{500}{64}$ $\frac{4000}{1}$ 4000 256000 questo die 'ser partido per 500
64

E moltipichado e partido per lo muodo ditto de sovra te darà libre 512 de piper.

E per farla ancuor per inpoxizion per un altro muodo vui vedirì qui de sutto. Pony ala prima inpoxizion che fusse 10, e noi diremo: "Se 50 fusse 400, che seria 64?" Adoncha moltipicha 400 fia 64, fa 25600, poi moltipicha 10 che tu à posto al'inpoxizion chon 50, ch'è lo partidor, e tuo' la diferenzia ch'è da 25600, zoè a sutrar 500 de quello die romanir 25100, e questo poni men ala prima inpoxizion.

E per far la segonda inpoxizion pony che fusse 20, e tu die moltipicar per lo muodo ditto de sovra: moltipicar 400 fia 64 fano 25600, e puo' moltipicha 20 fia 50, fano 1000, e questi sutratti da 25600 die romanir 24600, e questi poni men ala segonda inpoxizion. Puo' moltipicha in croxe e dira': "20 fia 25100 fano 502000," e puo' moltipicha 10 fia 24600, fano 246000, e questi sutrati roman 256000, e puo' sutra' de 25100 24600, roman 500, e questo è lo partidor. Parti 256000 per 500, venirà piper libre 512, chomo vedirì incontro.

---

a. *Corrected over* $45\frac{40}{64}$.

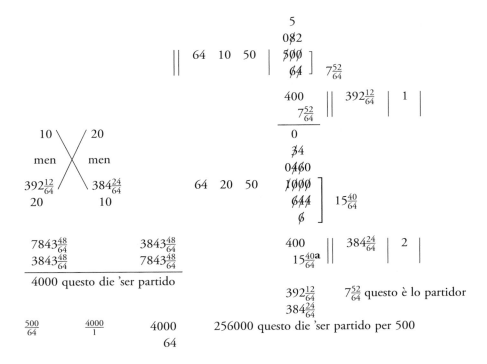

f. 3a

5
0~~8~~2
|| 64 10 50 | ~~500~~]
~~64~~] $7\frac{52}{64}$

400 || $392\frac{12}{64}$ | 1 |
$7\frac{52}{64}$ ||

0
~~3~~4
0~~4~~~~6~~0
~~1000~~]
~~644~~] $15\frac{40}{64}$
~~6~~]

10 \ / 20
minus X minus
$392\frac{12}{64}$ / \ $384\frac{24}{64}$
20 10

64 20 50

400 || $384\frac{24}{64}$ | 2 |
$15\frac{40}{64}$ ||

$7843\frac{48}{64}$ $3843\frac{48}{64}$
$3843\frac{48}{64}$ $7843\frac{48}{64}$
4000 this should be divided

$392\frac{12}{64}$ $7\frac{52}{64}$ this is the divisor
$384\frac{24}{64}$

$\frac{500}{64}$ $\frac{4000}{1}$ 4000
64

256000 this should be divided by 500

And multiplied and divided in the way said above will give you 512 pounds of pepper.

And you will see below how to do it again by false position by another method. Suppose at the first false position that it was 10, and we will say: "If 50 was 400, what would 64 be?" Then multiply 400 by 64, it makes 25,600, then multiply the 10 that you have put as the false position by 50, that is the divisor, and take the difference from 25,600, that is subtract 500 from that, there should remain 25,100, and put this as a minus at the first false position.

And to do the second false position, suppose that it was 20, and you should multiply as said above: multiply 400 by 64 makes 25,600, and then multiply 20 by 50, which makes 1,000, and this subtracted from 25,600 should leave 24,600, and put this as a minus at the second false position. Then multiply across and say: "20 times 25,100 makes 502,000," and then multiply 10 times 24,600, which makes 246,000, and when these are subtracted the remainder is 256,000, and then subtract 24,600 from 25,100, and 500 remains and this is the divisor. Divide 256,000 by 500, which will result in 512 pounds of pepper, as you will see opposite.

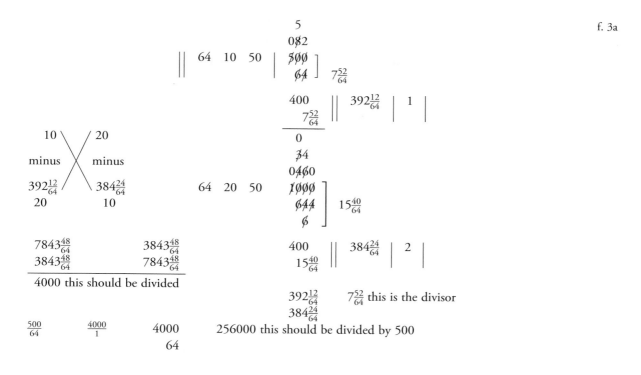

f. 3b

10 men 25100 20 — 20 men 24600 10

50	400	64	400	25600
10			64	500
500 questo die sotrar				25100
50	400	64	400	25600
20			64	1000
1000 questo se die sotrar				24600

502000	246000	25100
246000	sutrato	24600
256000		0500

0
0010
256000
50000 | 512
500
5

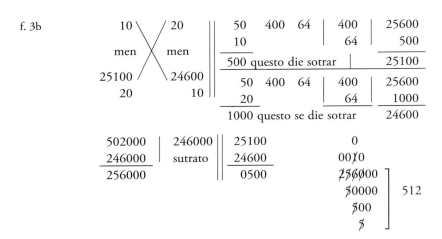

E questo è per la segonda riegola ditto de sovra, partido e moltipichado.

El piper in Alessandria. Per quanto comprerò la sporta, la qual è libre 700, che chondutto in Veniexia, là dove val el chargo duchati 50, zoè è el chargo libre 400, vadagna 15 per zentto? E per farlo per la riegola del 3 noi diremo: "Se libre 400 de piper val duchati 50, che die valer in Allesandria libre 700?" A moltipichar 50 fia 700 fano 35000, e questi partidi per 400 serano duchati $87\frac{1}{2}$. Poi diremo: "Se 115 eran 100, che tornerà $87\frac{1}{2}$?" Moltipicha $87\frac{1}{2}$ chon 100, serano 17500, e questo partido per 2 fiade, 115 val 230, insirà duchati $76\frac{20}{230}$.

E per far la ditta raxion per la chosa pony che vallesse 1^{co}. Se moltipicho 50 fia 700, val 35000, so che me die venir 400^{n}. Adoncha se moltipicho 1^{co} fia 400 fa 400^{co}, che son ingual a 35000^{n}, adoncha parti per 400^{co}, insirà fuor duchati $87\frac{1}{2}$. Puo' diremo: "Se 115 torna 100, che tornerà $87\frac{1}{2}$?" Pony che fuse 1^{co}, moltipicha 100 fia $87\frac{1}{2}$, serano 17500. Puo' moltipicha 1^{co} fia 115^{n}, farano 115^{co}, parti 17500^{n} per le chosse, insirà fuora del partidor duchati $76\frac{20}{230}$.

f. 4a E per far la raxion in driedo per inpoxizion metemo che la prima inposizion fusse 50, e diremo: "Se 400 me dà 50, che me daria 700?" Adoncha se moltipicho 50 fia 700 fano 35000, e questo salva, e moltipicha 50 che tu a' posto al'inpoxizion chon 400 che fu lo partidor, fano 20000. Adoncha varda che diferenzia serà dal'una parte al'oltra, sutra', e sutratto romanirà 15000, e questi pony a manchar ala prima inpoxizion. Puo' metemo la segonda inpoxizion che fusse 100, e per lo muodo

f. 3b

10	20	50	400	64	400		25600
minus	minus	10			64		500
		500 this you should subtract					25100
25100	24600	50	400	64	400		25600
20	10	20			64		1000
		1000 this should be subtracted					24600

502000	246000	25100	0	
246000	subtracted	24600	0010	
256000		0500	256000	
			50000	512
			500	
			5	

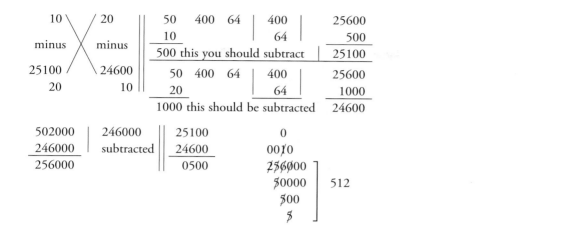

And this is by the second rule given above, divided and multiplied.

Pepper in Alexandria. How much will I pay for a bag full,[1] which is 700 pounds, which is brought to Venice, where it is worth 50 ducats per load, that is a load of 400 pounds, for a profit of 15 percent? And to do this by the rule of three, we will say: "If 400 pounds of pepper is worth 50 ducats, what should 700 pounds in Alexandria be worth?" Multiply 50 by 700 which makes 35,000, and this divided by 400 will be $87\frac{1}{2}$ ducats. Then we will say: "If 115 was 100, what would $87\frac{1}{2}$ return?" Multiply $87\frac{1}{2}$ by 100, it will be 17,500, and this divided by 2 times 115, that is 230, will result in $76\frac{20}{230}$ ducats.

And to do this problem by an unknown, put that it would be worth $1x$. If I multiply 50 times 700, it's worth 35,000, I know I should get 400. Then if I multiply $1x$ by 400 it makes $400x$, which is equal to 35,000, then divide by $400x$, it will result in $87\frac{1}{2}$ ducats. Then we will say: "If 115 returns 100, what would $87\frac{1}{2}$ return?" Put that it was $1x$, multiply 100 times $87\frac{1}{2}$, it will be 17,500.[2] Then multiply $1x$ times 115, it will make $115x$, divide 17,500 by the unknowns, the result of the division will be $76\frac{20}{230}$ ducats.

And to solve the problem on the previous page by false position, we suppose that the first false position would be 50, and we will say: "If 400 gives me 50, what would 700 give me?" Then if I multiply 50 times 700, which makes 35,000, and save this, and multiply the 50 that you put as the false position with 400 which was the divisor, it makes 20,000. Then look what the difference will be from one part to the other, subtract, and once you have subtracted there will remain 15,000, and put these as a minus at the first false position. Then we do the second false position, which would be 100, and in the same way as above, 50 multiplied by 700 makes 35,000, and save this. f. 4a

---

1. The *sporta* was a weight measure of 700 Venetian light pounds (about 210 kilograms) used for Egyptian commodities.
2. I.e., $\frac{17500}{2}$.

de sovra, moltipichado 50 fia 700 fano 35000, e questo salva. E puo' moltipicha 100 fia 400, fano 40000, e questo sutrato dal'altra parte serà più 5000, e questi pony più ala segonda inpoxizion. Può moltipicha in croxe, e diremo 50 fia 5000 fano 250000. Puo' moltipicha 100 fia 15000, fano 1500000. Puo' fara' chomo dixe la riegola che più e men s'azonze, serano 1750000. Puo' azonzi quel de sutto per far partidor, serano 20000, e questo è lo partidor. Parti 1750000 per 20000, insirà ducati $87\frac{1}{2}$.

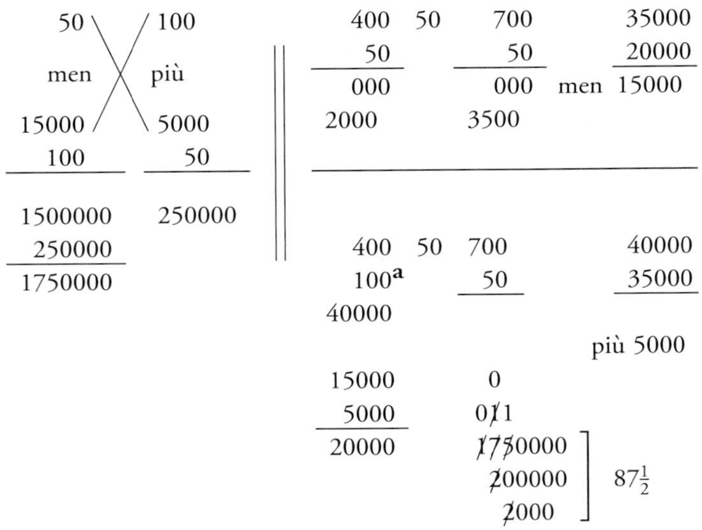
50 100
men più
15000 5000
100 50
1500000 250000
250000
1750000
400 50 700 35000
50 50 20000
000 000 men 15000
2000 3500
400 50 700 40000
100[a] 50 35000
40000
più 5000
15000 0
5000 0/1
20000 1750000
200000 87½
2000

E tanti duchati varà la sporta in Allesandria, duchati $87\frac{1}{2}$.

---

f. 4b

[Moltipichar rutti]

Al nomen de Dio. Questo è l'amaistramento de tute le raxion d'i rutti, zoè a moltipicar rutto chon rutto, e sutrar un rutto chon rutto, e partir rutto con lutto,[b] e azonzer ruto chon rutto, e far chadauna raxion d'i rutti. E prima, sichomo serave a dir che parte $\frac{1}{2}$ de $\frac{1}{2}$, e queste raxion sì facte se vuol moltipicar i numeri de sovra le verge l'un per l'altro, e queli partir per lo numero de sutto moltipichado l'un per l'oltro.

Or tu dirai: "Un via 1 fa 1," e quelo metti de sovra la verga in tal forma $\frac{1}{}$, poi tu dirai quel de sutto: "2 fia 2 fa 4," e queli metti de sutto la verga in tal forma $\frac{1}{4}$, e chosì serà chiamatto $\frac{1}{4}$, et a simile raxion fa tute le oltre. E per più chiareza farò algune qui de sutto.

---

a. 100 *corrected over* 110.
b. *Sic in MS, evidently for* rutto.

Then multiply 100 times 400, which makes 40,000, and this subtracted from the other part will be plus 5,000, and then put this as a plus at the second false position. Then multiply across, and we will say that 50 times 5,000 makes 250,000. Then multiply 100 times 15,000, which makes 1,500,000. Then do as the rule says, that minus and plus are added, which will be 1,750,000. Then add the lower one to be the divisor, which will be 20,000, and this is the divisor. Divide 1,750,000 by 20,000, and $87\frac{1}{2}$ ducats will result.

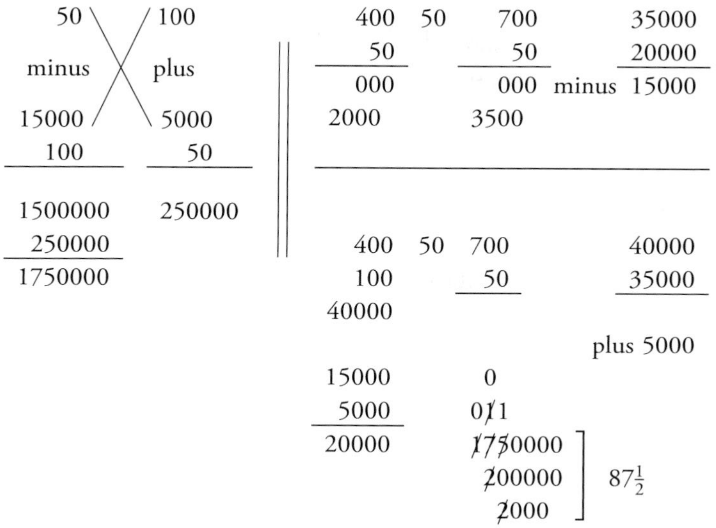

50 100
minus plus
15000 5000
100 50
1500000 250000
250000
1750000
400 50 700 35000
50 50 20000
000 000 minus 15000
2000 3500
400 50 700 40000
100 50 35000
40000
plus 5000
15000 0
5000 011
20000 1750000
200000 87½
2000

So the bag will be worth this much in Alexandria, $87\frac{1}{2}$ ducats.[1]

To multiply fractions

f. 4b

In the name of God. These are instructions for all of the problems concerning fractions, that is to multiply a fraction by a fraction, and subtract a fraction from a fraction, and divide a fraction by a fraction, and add a fraction to a fraction, and do any problem of fractions. And first would be to say divide $\frac{1}{2}$ by $\frac{1}{2}$, and this problem is done if you multiply the numbers above the line one by the other, and divide them by the number below multiplied one by the other.

So you will say: "One times 1 makes 1," and put this over the line in the form $\frac{1}{}$, then you will say for that below: "2 times 2 makes 4," and put this below the line in the form $\frac{1}{4}$, and thus it will be called $\frac{1}{4}$. And in the same way do all the others, and to make things clearer, I will do some below.

1. This is the value of a bag of pepper in Venice; Michael has not completed this solution by figuring in the 15 percent profit.

$\frac{1}{2}$ via $\frac{1}{2}$ fa $\frac{1}{4}$. || $\frac{1}{7}$ via $\frac{1}{7}$ fa $\frac{1}{49}$. || $\frac{4}{19}$ via $\frac{7}{15}$ fa $\frac{28}{285}$. ||
$\frac{1}{3}$ via $\frac{1}{3}$ fa $\frac{1}{9}$. || $\frac{1}{7}$ via $\frac{1}{9}$ fa $\frac{1}{63}$. || $\frac{19}{21}$ via $\frac{101}{99}$ fa $\frac{1919}{2079}$. ||
$\frac{1}{4}$ via $\frac{1}{4}$ fa $\frac{1}{16}$. || $\frac{1}{9}$ via $\frac{1}{12}$ fa $\frac{1}{108}$. || $\frac{1}{10}$ via $\frac{2}{13}$ fa $\frac{2}{130}$. ||

Anchora moltipicha $3\frac{1}{3}$ via $\frac{1}{2}$, tu die dir: "3 via 3 fa 9," e azonzi un ch'è de sovra la verga, fa $\frac{10}{3}$, e puo' metti $\frac{1}{2}$. Mo' tu die moltipicar 10 via 1, fa 10, i qual metti de sovra la verga, chosì $\frac{10}{}$, e puo' moltipica 2 via 3, fa 6, i qual son de sutto la verga, e mettti de sutto e starà chosì $\frac{10}{6}$, che vien eser $1\frac{2}{3}$.

Moltipicha $\frac{1}{2}$ via $3\frac{1}{3}$. Fa' chosì chomo sta de sopra e men $1\frac{2}{3}$. Moltipicha $3\frac{1}{3}$ via $3\frac{1}{3}$. Fa' chossi: 3 via 3 fa 9 e un fa $\frac{10}{3}$, puo' moltipicha 3 via 3 fa 9 e azonto 1 ch'è de sovra la verga, fa 10, che sono $\frac{10}{3}$. Puo' moltipicha 10 via 10, fa 100, e parti per 3 via 3, che fano 9, sì che te vien a partir 100 per 9 parte, che sono $11\frac{1}{9}$. E chosì fa tute le oltre raxion.

---

f. 5a Tanto vien a dir moltipicha $\frac{3}{5}$ via $23\frac{1}{2}$ quanto serave a dir che xe $\frac{3}{5}$ de $23\frac{1}{2}\ \frac{47}{2}$, poi moltipicha 3 via 47, fa 141, e metti de sovra la verga, e puo' moltipicha 2 via 5, fa 10, e metti de sutto la verga, che vien a partir 141 per 10, fa $14\frac{1}{10}$. ||

[Zonzer rutti chon rutti]

Questo sie l'amaistramento di saver azonzer un rutto chon rutto, chomo t'insegnerò qui de sutto. E primo:

azonzi $\frac{1}{2}$ chon $\frac{1}{2}$, fa 1
azonzi $\frac{1}{5}$ chon $\frac{1}{7}$, fa $\frac{12}{35}$
azonzi $\frac{1}{7}$ chon $\frac{1}{9}$, fa $\frac{16}{23}$
azonzi $\frac{2}{3}$ chon $\frac{3}{4}$, fa $1\frac{5}{12}$
azonzi $\frac{3}{4}$ chon $\frac{4}{5}$, fa $1\frac{11}{20}$
azonzi $1\frac{1}{2}$ chon $2\frac{1}{3}$, fa $3\frac{5}{6}$
azonzi $4\frac{1}{7}$ chon $9\frac{1}{9}$, fa $13\frac{16}{63}$
azonzi $9\frac{5}{6}$ chon $1\frac{6}{7}$, fa $11\frac{20}{42}$
azonzi 3[a]$\frac{6}{7}$ chon $4\frac{7}{9}$, fa $6\frac{40}{63}$
azonzi $5\frac{1}{2}\ \frac{1}{3}$ chon $9\frac{1}{4}\ \frac{1}{5}$, fa $15\frac{17}{60}$
azonzi $9\frac{6}{7}\ \frac{7}{8}$ chon $1\frac{8}{9}$, fa $13\frac{26}{50}\ \frac{26}{40}$

---

a. 3 *corrected over* 6.

$\frac{1}{2}$ times $\frac{1}{2}$ makes $\frac{1}{4}$. || $\frac{1}{7}$ times $\frac{1}{7}$ makes $\frac{1}{49}$. || $\frac{4}{19}$ times $\frac{7}{15}$ makes $\frac{28}{285}$. ||
$\frac{1}{3}$ times $\frac{1}{3}$ makes $\frac{1}{9}$. || $\frac{1}{7}$ times $\frac{1}{9}$ makes $\frac{1}{63}$. || $\frac{19}{21}$ times $\frac{101}{99}$ makes $\frac{1919}{2079}$. ||
$\frac{1}{4}$ times $\frac{1}{4}$ makes $\frac{1}{16}$. || $\frac{1}{9}$ times $\frac{1}{12}$ makes $\frac{1}{108}$. || $\frac{1}{10}$ times $\frac{2}{13}$ makes $\frac{2}{130}$. ||

Then multiply $3\frac{1}{3}$ by $\frac{1}{2}$, you should say: "3 times 3 makes 9," and add one, which is above the line, makes $\frac{10}{3}$, and then put $\frac{1}{2}$. Now you should multiply 10 by 1, which makes 10, which you put above the line, thus $\frac{10}{\ }$, and then multiply 2 times 3 makes 6, which go below the line, and put it below and it will be $\frac{10}{6}$, which becomes $1\frac{2}{3}$.

Multiply $\frac{1}{2}$ by $3\frac{1}{3}$. Do as above and $1\frac{2}{3}$ remains. Multiply $3\frac{1}{3}$ by $3\frac{1}{3}$. Do this: 3 times 3 makes 9 and one makes $\frac{10}{3}$, then multiply 3 times 3 makes 9 and added to 1, which is above the line, makes 10, which is $\frac{10}{3}$. Then multiply 10 by 10, which makes 100, and divide 3 by 3, which makes 9, so you're going to divide 100 by 9 parts, which is $11\frac{1}{9}$, and do all the other problems the same way.

To say multiply $\frac{3}{5}$ by $23\frac{1}{2}$ is as much as to say what is $\frac{3}{5}$ of $23\frac{1}{2}$, $\frac{47}{2}$, then multiply 3 times 47, which f. 5a
makes 141, and put that above the line, and then multiply 2 times 5, which makes 10, and put that below the line, which is to divide 141 by 10, which makes $14\frac{1}{10}$. ||

To add fractions to fractions

These are the instructions to know how to add a fraction to a fraction, as I will teach you below. And first:

add $\frac{1}{2}$ to $\frac{1}{2}$ makes 1
add $\frac{1}{5}$ to $\frac{1}{7}$ makes $\frac{12}{35}$
add $\frac{1}{7}$ to $\frac{1}{9}$ makes $\frac{16}{63}$
add $\frac{2}{3}$ to $\frac{3}{4}$ makes $1\frac{5}{12}$
add $\frac{3}{4}$ to $\frac{4}{5}$ makes $1\frac{11}{20}$
add $1\frac{1}{2}$ to $2\frac{1}{3}$ makes $3\frac{5}{6}$
add $4\frac{1}{7}$ to $9\frac{1}{9}$ makes $13\frac{16}{63}$
add $9\frac{5}{6}$ to $1\frac{6}{7}$ makes $11\frac{20}{42}$[1]
add $3\frac{6}{7}$ to $4\frac{7}{9}$ makes $6\frac{40}{63}$[2]
add $5\frac{1}{2}\frac{1}{3}$ to $9\frac{1}{4}\frac{1}{5}$ makes $15\frac{17}{60}$
add $9\frac{6}{7}\frac{7}{8}$ to $1\frac{8}{9}$ makes $13\frac{26}{50}\frac{26}{40}$[3]

---

1. This should be $11\frac{29}{42}$ (as Michael correctly solves it on the next page).
2. This should be $8\frac{40}{63}$.
3. This should be $12\frac{313}{504}$.

Notta che senpre l'azonzer se die moltipicar in croxe l'un per l'oltro e quello azonzer insenbre, poi portarlo per lo moltipicho de quelo ch'è sutto le verge.

Sì chomo chi volese zonzer $\frac{1}{2}$ con $\frac{1}{2}$ tu die moltipicar 1 via 2 fa 2, e puo' anchora in croxe 1 via 2 fa 2, e parti 4 per 4, farà 1. Adoncha azonzi $\frac{1}{2}$ chon $\frac{1}{2}$, farà 1, e 2 e 2 fa 4, puo' moltipicha 2 via 2, fa 4.

f. 5b Azonzi $\frac{2}{3}$ chon $\frac{3}{4}$, fa' 3 via 3 fa 9, e 2 via 4 fa 8, e questi insenbre | 8 e 9 fa 17 e puo' 3 via 4 fa 12, ch'è sutto la verga, e parti 17 in 12 che ne vien $1\frac{5}{12}$, e per simile muodo fara' le oltre chi fuse rutti senza sanni.

Azonzi $1\frac{1}{2}$ chon $2\frac{1}{3}$. Si di' chosì: "Un e 2 fa 3," e mete 3 sani, e puo' di' in croxe: "1 fia 3 fa 3, e 1 via 2 fa 2, e 3 e 2 fa 5," e metti 5. E puo' moltipicha quelo ch'è suto la verga, zoè 2 via 3 fa 6, e parti 5 per 6, sì che te ne vien $3\frac{5}{6}$. Adoncha se tu azonzi $1\frac{1}{2}$ chon $2\frac{1}{3}$ serà $3\frac{5}{6}$.

E per lo simille muodo $9\frac{5}{6}$ chon $1\frac{6}{7}$ si di' chosì: "9 e 1 fa 10," e metti 10 sany, puo' 5 via 7 fa 35 e 6 via 6 fa 36. Azonzi 35 e 36 fa fa[a] 71, e puo' 6 via 7 fa 42, e metti 42 che ne vien $1\frac{29}{42}$ e 10 che tu avevy in prima, che fa in tuto $11\frac{29}{42}$, et è fatta la raxion.

Azonzi $8\frac{1}{4}$ e $\frac{1}{7}$ chon $3\frac{1}{8}\,\frac{1}{9}$, e sun $11\frac{632}{1008}$.
Azonzi $4\frac{2}{3}\,\frac{3}{4}$ chon $8\frac{4}{5}\,\frac{5}{6}$, e sun $15\frac{1}{20}$.
Azonzi $8\frac{4}{5}\,\frac{5}{6}$ chon $3\frac{6}{7}\,\frac{7}{8}$, e sun $14\frac{614}{1680}$.

Se tu volesti azonzer $8\frac{1}{4}\,\frac{1}{7}$ chon $3\frac{1}{8}\,\frac{1}{9}$ si fa in questo muodo, chomo te mostrarò. Prima azonzi 8 e 3, fano 11, puo' ne sta $\frac{1}{4}\,\frac{1}{7}$ chon $\frac{1}{8}\,\frac{1}{9}$. Fa' primo $\frac{1}{4}\,\frac{1}{7}$. Di' chosì: "Un via 7 fa 7 e 1 via 4 fa 4." Azonzi insenbre, farà 11. Puo' moltipicha i nomeri de sutto le verge, zoè 4 via 7 fa 28, e metti chosì $\frac{11}{28}$. E puo' fa $\frac{1}{8}\,\frac{1}{9}$, moltipicha in croxe 1 via 9 fa 9, 1 via 8 fa 8, azonzi insenbre, fa 17. Puo' moltipicha 8 via 9 fa 72, che serà $\frac{17}{72}$, azonzi ancho insempre $\frac{11}{28}$ chon $\frac{17}{72}$, fa 11 via 72 fa 792. E puo' fa' 28 via 17 fa 476, i quali parti per 28 via 72. Azonzi primo insenbre che fa 2016,[b] e starà chosì $\frac{1268}{2016}$, schixa per 2 e serà $\frac{634}{1008}$, sì che a voler azonzer $8\frac{1}{4}\,\frac{1}{7}$ chon $3\frac{1}{8}\,\frac{1}{9}$ sono $11\frac{634}{1008}$. Anchora se porà scixar per 2 e farà $\frac{317}{504}$. E per lo simylle fara' le oltre.

f. 6a Anchora azonzi $4\frac{2}{3}\,\frac{3}{4}$ chon $8\frac{4}{5}\,\frac{5}{6}$. Fa' chomo ai fatto de supra. Primo 4 e 8 fa 12, e puo' 2 via 4 fa 8 e 3 via 3 fa 9. Azonzi chon 8, farano 17. Parti per 3 via 4 fa 12, serà $1\frac{5}{12}$. Azonzi chon 12 e puo' 4 via 6 fa 24, e 5 via 5 fa 25, azonti inse[*n*]bre fano 49. Parti per 5 via 6, che fa 30, serà $1\frac{19}{30}$. Azonzi insenbre tuto, serà $14\frac{5}{12}$ e $\frac{19}{30}$. Azonti insenbre sarà $15\frac{18}{360}$. Schixadi serano $15\frac{1}{20}$.

---

a. *Repeated thus in the MS.*
b. 2016 *corrected over* 2018.

Note that the terms to be added should always be multiplied crosswise one with the other and that to be added together, then divide that by the product of what is below the line.

Thus as whoever would like to add $\frac{1}{2}$ to $\frac{1}{2}$, you should multiply 1 times 2 makes 2, and then again crosswise 1 times 2 makes 2, and divide 4 by 4, will make 1. Then add $\frac{1}{2}$ to $\frac{1}{2}$, it will make 1, and 2 and 2 make 4, then multiply 2 times 2 makes 4.

Add $\frac{2}{3}$ to $\frac{3}{4}$, it makes 3 times 3 makes 9, and 2 times 4 makes 8, and these together | 8 and 9 make f. 5b
17 and then 3 times 4 makes 12, which is below the line, and divide 17 into 12 which results in $1\frac{5}{12}$, and the others that are fractions without an integer will go by a similar method.

Add $1\frac{1}{2}$ with $2\frac{1}{3}$. Say to yourself like this: "One and two make 3," and put the integer 3, and then say crosswise "1 times 3 makes 3, and 1 times 2 makes 2, and 3 and 2 make 5," and put 5. And then multiply what is below the line, that is 2 times 3 makes 6, and divide 5 by six, which gives you $3\frac{5}{6}$. So, if you add $1\frac{1}{2}$ with $2\frac{1}{3}$ it will be $3\frac{5}{6}$.

And in the same way for $9\frac{5}{6}$ with $1\frac{6}{7}$ you say this: "9 and 1 make 10," and put the integer 10, then 5 times 7 makes 35 and 6 times 6 makes 36. Add 35 and 36 makes 71, and then 6 times 7 makes 42, and put 42 which results in $1\frac{29}{42}$ and 10 that you had before, which makes in all $11\frac{29}{42}$, and the problem is done.

Add $8\frac{1}{4}$ and $\frac{1}{7}$ to $3\frac{1}{8}\frac{1}{9}$,[1] and it is $11\frac{632}{1008}$.
Add $4\frac{2}{3}\frac{3}{4}$ to $8\frac{4}{5}\frac{5}{6}$, and it is $15\frac{1}{20}$.
Add $8\frac{4}{5}\frac{5}{6}$ to $3\frac{6}{7}\frac{7}{8}$, and it is $14\frac{614}{1680}$.

If you would like to add $8\frac{1}{4}\frac{1}{7}$ with $3\frac{1}{8}\frac{1}{9}$, it is done in this way, as I shall show you. First add 8 and 3, which makes 11, then there remains the $\frac{1}{4}\frac{1}{7}$ and $\frac{1}{8}\frac{1}{9}$. Do first $\frac{1}{4}\frac{1}{7}$. Say this: "One times 7 makes 7 and 1 times 4 makes 4." Add together, it will make 11. Then multiply the numbers below the line, that is 4 times 7 makes 28, and put so: $\frac{11}{28}$. And then do $\frac{1}{8}\frac{1}{9}$, multiply across 1 times 9 makes 9, 1 times 8 makes 8, add together, makes 17. Then multiply 8 times 9 makes 72, which will be $\frac{17}{72}$, now add together $\frac{11}{28}$ with $\frac{17}{72}$, which is 11 times 72 makes 792. And then do 28 times 17 makes 476, which you divide by 28 times 72. Combine them first together which makes 2016 and it will be like this $\frac{1268}{2016}$, factor by 2 and it will be $\frac{634}{1008}$, and if you want to add $8\frac{1}{4}\frac{1}{7}$ with $3\frac{1}{8}\frac{1}{9}$ it is $11\frac{634}{1008}$. You can factor again by 2 and it will be $\frac{317}{504}$. And the others will be done the same way.

Again, add $4\frac{2}{3}\frac{3}{4}$ with $8\frac{4}{5}\frac{5}{6}$. Do as you have done above. First 4 and 8 make 12, and then 2 times 4 f. 6a
makes 8 and 3 times 3 makes 9. Add to 8, it will make 17. Divide by 3 times 4 makes 12, it will be $1\frac{5}{12}$. Add to 12 and then 4 times 6 makes 24, and 5 times 5 makes 25, added together make 49. Divide by 5 times 6, which makes 30, it will be $1\frac{19}{30}$. Add everything together, it will be $14\frac{5}{12}$ and $\frac{19}{30}$. Added together it will be $15\frac{18}{360}$. Factored it will be $15\frac{1}{20}$.

---

1. That is, $3\frac{1}{8}$ plus $\frac{1}{9}$.

Se tu volesti azonzer $\frac{1}{2}\,\frac{1}{3}\,\frac{1}{4}\,\frac{1}{5}\,\frac{1}{6}$ insenbre, questi se truova in 60, e di' lo mezo de 60. El mezo de sesanta sie 30, el terzo de 60 sie 20, el $\frac{1}{4}$ de 60 sie 15, el $\frac{1}{5}$ de 60 sie 12, el $\frac{1}{6}$ de 60 sie 10, e farà $\frac{87}{60}$, che sono $1\frac{9}{20}$, e tanto farà azonzer $\frac{1}{2}\,\frac{1}{3}\,\frac{1}{4}\,\frac{1}{5}\,\frac{1}{6}$. E per questo muodo se azonze tuti i rutti.

Azonzi $|\frac{2}{3}\,\frac{3}{4}\,\frac{4}{5}|$, i qualli se truova in 60. Di' chosi: "$\frac{2}{3}$ de 60 sie 40, $\frac{3}{4}$ de 60 sono 45, $\frac{4}{5}$ de 60 sono 48." Suma insenbre 40, 45, 48, fano 133, che sono $\frac{133}{60}$. Partidi, me vien $2\frac{13}{60}$. E chosì è fata.

E per un oltro muodo $\frac{2}{3}\,\frac{3}{4}$ azonti insenbre fa 2 via 4 fa 8, e 3 via 3 fa 9. Azonti, 9 e 8 fano 17, puo' 3 via 4 fa 12. Parti chosì $\frac{17}{12}$, e puo' azonzi $\frac{4}{5}$ chon $\frac{17}{12}$. Fa' 5 fia 17 fa 85, e po' 4 via 12 fa 48. Azonti insempre fa 133. Parti per 5 via 12 fa 60, che me vien $2\frac{13}{60}$. E questo muodo.

[Partir rutti chon rutti]

Inchina mo' de sopra abiamo dito del'azonzer d'i rutti l'un chon l'oltro. Ora diremo chomo se die partir 1 ruto per 1 oltro rutto, e sani e ruti per rotti, e sany e rotti per sany e rotti, chomo vedirì qui de sutto.

Partimo[a] $\frac{1}{4}$ per $\frac{1}{7}$. Sie $1\frac{3}{4}$.
Partime $\frac{1}{7}$ per $\frac{1}{9}$. Sie $1\frac{2}{7}$.
Partime $\frac{1}{9}$ per $\frac{1}{11}$. Sie $1\frac{2}{9}$.
Partime $\frac{2}{3}$ per $\frac{3}{4}$. Sie $1\frac{8}{9}$.
Partime $\frac{3}{5}$ per $\frac{5}{7}$. Sie $\frac{21}{25}$.
Partime $\frac{5}{7}$ per $\frac{7}{9}$. Sie $\frac{45}{49}$.
Partime 6 per $\frac{4}{5}$. Sie $7\frac{1}{2}$.
Partime $\frac{4}{5}$ per 9 Sie $11\frac{1}{4}$.

f. 6b Partime $8\frac{2}{3}$ per 13. Sie $1\frac{1}{2}$. ____________
Partime 15 per $9\frac{6}{7}$. Sie $1\frac{12}{23}$.
Partime $6\frac{5}{7}$ per $8\frac{7}{9}$. Sie $1\frac{130}{423}$. ____________

Questo sie lo muodo da partir $\frac{1}{5}$ per $\frac{1}{7}$ e tutti li so similli. Se tu vuol partir $\frac{1}{5}$ per $\frac{1}{7}$ moltipicha in croxe e di': "1 via 7 fa 7," el qual parti per 1 via 5, serà $1\frac{2}{5}$. E per lo simille $\frac{1}{7}$ per $\frac{1}{9}$ fa' 1 via 7, e sarà $1\frac{2}{7}$. E così fa' le oltre simille raxion.

Anchora se può far per un altro muodo a partir $\frac{1}{5}$ per $\frac{1}{7}$. Si di' chosì: "$\frac{1}{5}\,\frac{1}{7}$ se truova in 35 che è 7," e puo' tuo' $\frac{1}{7}$ per $\frac{1}{9}$ de 35 che è 5. Mo' parti 7 per 5, che ne vien $1\frac{2}{5}$. Et è fata.

E per lo simile $\frac{1}{7}$ per $\frac{1}{9}$. Di': "Setimo e nono se truova in 63." Mo' tuo' $\frac{1}{7}$ de 63, el qual è 9, e puo' tuo' $\frac{1}{9}$ de 63, el qual è 7. Mo' parti 9 per 7, te ne vien $1\frac{2}{7}$, et è fatta. E per questo muodo se può far ogni simile raxion, sì simile chomo grande pizole.

---

a. *Sic in MS.*
1. This should be $\frac{8}{9}$ (as correctly solved below).
2. This should be $\frac{4}{45}$ (as correctly solved below).
3. This should be $\frac{2}{3}$ (as correctly solved below).

If you would like to add together $\frac{1}{2}\,\frac{1}{3}\,\frac{1}{4}\,\frac{1}{5}\,\frac{1}{6}$, these are all found in 60 and say the half of 60. And half of sixty is 30, a third of 60 is 20, $\frac{1}{4}$ of 60 is 15, $\frac{1}{5}$ of 60 is 12, $\frac{1}{6}$ of 60 is 10, and it will make $\frac{87}{60}$, which is $1\frac{9}{20}$, and this is the sum of $\frac{1}{2}\,\frac{1}{3}\,\frac{1}{4}\,\frac{1}{5}\,\frac{1}{6}$. And you add all fractions this way.

Add $\frac{2}{3}\,\frac{3}{4}\,\frac{4}{5}$, which are found in 60. Say this: "$\frac{2}{3}$ of 60 is 40, $\frac{3}{4}$ of 60 is 45, $\frac{4}{5}$ of 60 is 48." Add together 40, 45, 48, it makes 133, which is $\frac{133}{60}$. Divided, I get $2\frac{13}{60}$. And so it is done.

And by another method $\frac{2}{3}\,\frac{3}{4}$ added together make 2 times 4 makes 8, and 3 times 3 makes 9. Added, 9 and 8 make 17, then 3 times 4 makes 12. Divide this way, $\frac{17}{12}$, and then add $\frac{4}{5}$ to $\frac{17}{12}$. It makes 5 times 17 is 85 and then 4 times 12 makes 48. Added together they make 133. Divide by 5 times 12 makes 60, which gives me $2\frac{13}{60}$. And this is the method.

To divide fractions by fractions

Up to now, we've said above how to add the fractions to each other. Now we will say how to divide one fraction by another fraction, and integers and fractions by fractions, and integers and fractions by integers and fractions, as you will see below.

Divide for me $\frac{1}{4}$ by $\frac{1}{7}$; it's $1\frac{3}{4}$.
Divide for me $\frac{1}{7}$ by $\frac{1}{9}$; it's $1\frac{2}{7}$.
Divide for me $\frac{1}{9}$ by $\frac{1}{11}$; it's $1\frac{2}{9}$.
Divide for me $\frac{2}{3}$ by $\frac{3}{4}$; it's $1\frac{8}{9}$.[1]

Divide for me $\frac{3}{5}$ by $\frac{5}{7}$; it's $\frac{21}{25}$.
Divide for me $\frac{5}{7}$ by $\frac{7}{9}$; it's $\frac{45}{49}$.
Divide for me 6 by $\frac{4}{5}$; it's $7\frac{1}{2}$.
Divide for me $\frac{4}{5}$ by 9; it's $11\frac{1}{4}$.[2]

Divide for me $8\frac{2}{3}$ by 13. It's $1\frac{1}{2}$.[3] f. 6b
Divide for me 15 by $9\frac{6}{7}$. It's $1\frac{12}{23}$.
Divide for me $6\frac{5}{7}$ by $8\frac{7}{9}$. It's $\frac{130}{423}$.[4]

This is the way to divide $\frac{1}{5}$ by $\frac{1}{7}$ and for all similar ones. If you want to divide $\frac{1}{5}$ by $\frac{1}{7}$, multiply across and say: "1 times 7 makes 7," which you divide by 1 times 5, it will be $1\frac{2}{5}$. And by the same way $\frac{1}{7}$ by $\frac{1}{9}$ makes 1 times 7,[5] and it will be $1\frac{2}{7}$. And in this way the other similar problems are done.

Moreover you can use another method to divide $\frac{1}{5}$ by $\frac{1}{7}$. You say this to yourself: "$\frac{1}{5}\,\frac{1}{7}$ are found in 35, which is 7," and then take $\frac{1}{7}$ by $\frac{1}{9}$ of 35 which is 5. Now divide 7 by 5 which comes to $1\frac{2}{5}$. And it's done.

And the same way for $\frac{1}{7}$ by $\frac{1}{9}$. Say: "A seventh and a ninth are found in 63." Now take $\frac{1}{7}$ of 63 which is 9, and then take $\frac{1}{9}$ of 63, which is 7. Now divide 9 by 7, which gives you $1\frac{2}{7}$, and it's done. And by this method you can do all similar problems, similar large ones as well as small.

---

4. This should be $\frac{423}{553}$ (as correctly solved below).
5. A step is omitted here.

Partime $\frac{2}{3}$ per $\frac{3}{4}$. Si di' chosì: "2 via 4 fa 8, e 3 via 3 fa 9." Parti 8 per 9, e starà chosì $\frac{8}{9}$.

Partime $\frac{3}{5}$ per $\frac{5}{7}$. Si di' chosì: "3 via 7 fa 21, e 5 via 5 fa 25," e parti 21 per 25, e starà chosì $\frac{21}{25}$. E per questa fari' le simile raxion.

Partime 6 per $\frac{4}{5}$. Se die meter chomo sta qui de sotto, e di' chosì: "5 via 6[a] fa 30,[b] e 1 via 4 fa 4." Adoncha parti 30 per 4. Serà $7\frac{1}{2}$.

$$\Bigg\| \quad \frac{6}{1} \times \frac{4}{5} \qquad \begin{array}{r} 6 \\ \underline{5} \\ 30 \end{array} \qquad \left.\begin{array}{r} 02 \\ \not{3}\not{0} \\ \not{4} \end{array}\right] \quad 7\tfrac{2}{4} \quad 7\tfrac{1}{2} \quad \Bigg\|$$

Partime $\frac{4}{5}$ per 9 e metti chomo sta qui de sotto, e di' chosì: "5 via 9 45, e 1 via 4 fa 4." Adoncha parti 4 per 45, e farano $\frac{4}{45}$.

$$\frac{4}{5} \times \frac{9}{1} \qquad \begin{array}{r} 9 \\ \underline{5} \end{array} \qquad \tfrac{4}{45} \quad \tfrac{4}{45}$$

f. 7a Partime $8\frac{2}{3}$ per 13. Si di' chosì: "3 via 8 fa 24 e 2 ch'è de sovra la verga fa 26," e metti $\frac{26}{3}$, e puo' moltipicha 1 via 26 fa 26, el qual parti per 3 via 13, sarà $\frac{26}{39}$. Et è fatta. $8\frac{2}{3}$ per $\frac{13}{1}$. | $\frac{26}{3} \times \frac{13}{1}$.

Partime 15 per $9\frac{6}{7}$. Metti chomo starà qui de sutto e di' chosì: "7 fia 9 fa 63, e 6 ch'è de sovra la verga fa $\frac{69}{7}$." E puo' metti $\frac{15}{1}$, e puo' 7 via 15 fa 105 e 1 via 69 fa 69. Parti per 105 per 69 e te ne intrarà $1\frac{36}{69}$, che son schixadi $\frac{12}{23}$. E chusì è fatta. || $\frac{15}{1}$ per $9\frac{6}{7}$ || $\frac{15}{1} \times \frac{69}{7}$ ||

Partime $6\frac{5}{7}$ per $8\frac{7}{9}$. Si di' chusì: "6 via 7 fa 42, e 5 fa 47, che son $\frac{47}{7}$." E puo' di': "8 via 9 fa 72, e 7 fa 79," e metti $\frac{79}{9}$. Puo' di': "7 via 79 fa 553," e puo' fa' 9 via 47 fa 423, e parti 423 per 553 e sarano e sarano[c] $\frac{423}{553}$. || $6\frac{5}{7}$ per $8\frac{7}{9}$ || $\frac{47}{7} \times \frac{79}{9}$ ||

Ponamo che noi voiamo savere $7\frac{2}{3}$ che parte è de $9\frac{1}{4}$. Questa è chotal raxion chomo serave a dir parti $7\frac{2}{3}$ per $9\frac{1}{4}$. Rechordate che noi dovemo dir: "3 via 7 fa 21, e 2 ch'è sovra la verga fa 23," e metti $\frac{23}{3}$ e puo' di': "4 via 9 fa 36, e 1 ch'è de sovra la verga fa 37," e metti $\frac{37}{4}$. E puo' di: "4 via 23 fa 92," i quali parti per $\frac{4}{3}$ via 37 che fa 111, che ne vien $\frac{92}{111}$. Et è fatta la raxion. || $7\frac{2}{3}$ per $9\frac{1}{4}$ || $\frac{23}{3} \times \frac{37}{4}$ || $\frac{92}{111}$ ||

---

a. 6 *corrected over* 9.
b. 30 *corrected over* 45.
c. *Repeated thus in MS.*

Divide for me $\frac{2}{3}$ by $\frac{3}{4}$. Say this: "2 times 4 makes 8, and 3 times 3 makes 9." Divide 8 by 9 and it will be $\frac{8}{9}$.

Divide for me $\frac{3}{5}$ by $\frac{5}{7}$. Say this: "3 times 7 makes 21, and 5 times 5 makes 25," and divide 21 by 25, and it will be $\frac{21}{25}$. And in this way you can do similar problems.

Divide for me 6 by $\frac{4}{5}$. You should put it as it is below and say: "5 times 6 makes 30, and 1 times 4 makes 4." Then divide 30 by 4. It will be $7\frac{1}{2}$.

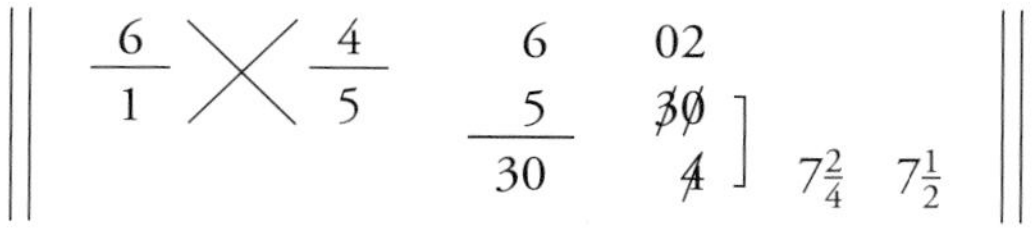

Divide for me $\frac{4}{5}$ by 9 and put it as it is below. And say: "5 times 9, 45, and 1 times 4 makes 4." Then divide 4 by 45 and it will make $\frac{4}{45}$.

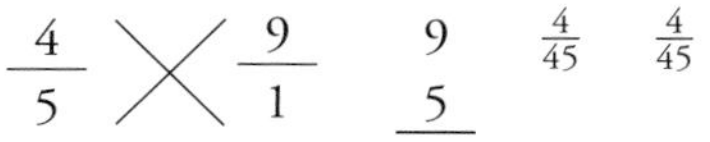

Divide for me $8\frac{2}{3}$ by 13. Say this: "3 times 8 makes 24 and the 2 above the line makes 26," and put $\frac{26}{3}$, and then multiply 1 times 26 makes 26, and divide this by 3 times 13, which will be $\frac{26}{39}$. And it's done. $8\frac{2}{3}$ by $\frac{13}{1}$ | $\frac{26}{3} \times \frac{13}{1}$. f. 7a

Divide for me 15 by $9\frac{6}{7}$. Put it as it will be below and say: "7 times 9 makes 63, and the 6 above the line makes $\frac{69}{7}$." And then put $\frac{15}{1}$. And then 7 times 15 makes 105 and 1 times 69 makes 69. Divide by 105 by 69 and you will get $1\frac{36}{69}$, which is factored to $\frac{12}{23}$. And so it is done. || $\frac{15}{1}$ by $9\frac{6}{7}$ || $\frac{15}{1} \times \frac{69}{7}$ ||

Divide for me $6\frac{5}{7}$ by $8\frac{7}{9}$. Say this: "6 times 7 makes 42 and 5 makes 47, which is $\frac{47}{7}$." And then say: "8 times 9 makes 72, and 7 makes 79." And put $\frac{79}{9}$. Then say: "7 times 79 makes 553," and then do 9 times 47 makes 423, and divide 423 by 553 and it will be $\frac{423}{553}$. || $6\frac{5}{7}$ by $8\frac{7}{9}$ || $\frac{47}{7} \times \frac{79}{9}$ ||

Let's suppose that we want to know what part $7\frac{2}{3}$ is of $9\frac{1}{4}$. This is the same problem as would be to say divide $7\frac{2}{3}$ by $9\frac{1}{4}$. Remember that we should say: "3 times 7 makes 21, and the 2 above the line makes 23," and put $\frac{23}{3}$, and then say: "4 times 9 is 36, and the 1 above the line makes 37," and put $\frac{37}{4}$. And then say: "4 times 23 makes 92," which you divide by $\frac{4}{3}$[1] times 37 which makes 111, which comes to $\frac{92}{111}$. And the problem is solved. || $7\frac{2}{3}$ by $9\frac{1}{4}$ || $\frac{23}{3} \times \frac{37}{4}$ || $\frac{92}{111}$ ||

1. This should be 3.

Partime $37\frac{2}{3}$ per $5\frac{3}{4}$. Si di' chosì: "$37\frac{2}{3}$ $5\frac{3}{4}$, 3 via 37 fa 111, e 2 ch'è de sovra la verga 113." E puo' metti $\frac{113}{3}$ e puo' di': "4 via 5 fa 20, e 3 ch'è de sovra fa $\frac{23}{4}$." E puo' moltipicha 113 per 4, che fa 452, li qual parti per 3 via 23, che fa 69, che ne vien $6\frac{38}{69}$. ||

Anchora per un'altra via e muodo. Truova un numero in che se truova $\frac{1}{3}\,\frac{1}{4}$, zoè che se truova li rutti da una parte e dal'altra, zioè $\frac{2}{3}$ fia $\frac{3}{4}$ che se truova 19 3 via 4, zioè in 12. Mo' mena zaschadon per 12 e di' in prima: "12 via 37 fa 444" e puo' die' tuor $\frac{2}{3}$ de 12, che son 8, i qual azonzi chon 444, serano
f. 7b 452. I quali salva, e | puo' moltipicha l'oltra parte per 12, ch'è 5 via 12 fa 60, e tuo' li $\frac{3}{4}$ de 12, che sono 9, i quali azonzi chon 60, che ne vien 69. Mo' parti 452 per 69, che ne ven $6\frac{38}{69}$. E se tu la vol provar fa' lo ritorno. Moltipicha $6\frac{38}{69}$ per $5\frac{3}{4}$, e fa' chosì: 6 via 69 fa 414 e zonzi 38, i qual è de sovra la verga, e farà 452, e metti $\frac{452}{69}$. E puo' fa' 4 via 5 fa 20, e 3 ch'è de sovra la verga fa $\frac{23}{4}$. Puo' moltipicha 452 via 23, fa 10396,[a] i quali parti per 4 via 69, ne vien $37\frac{2}{3}$, e chosì è fatta questa raxion.

Ponamo che noi voiamo partir 17 in $\frac{5}{7}$ e voiamo saver quanto vien per parte, questa è la so riegola, chomo ò ditto de sovra. E metti chomo sta qui de sutto, e puo' moltipicha 17 via 7, fa 119, i qual parti per 1 via 5, fa 5, che ne vien $23\frac{4}{5}$ via $\frac{5}{7}$, che ne vien a ponto 17. E per questo chotal muodo se può far tute le simille raxion mazuor e menor de queste. 17 per $\frac{5}{7}$ || $\frac{17}{1} \times \frac{5}{7}\ \frac{119}{5}$ ||

[TRAZER RUTTI DA RUTTI]

Inchina mo' noi abiamo mostrado chomo se die moltipicar e azonzer e partir rutti chon rutti e sany e rutti. Ora mostraremo chomo se die sutrar un rutto da un oltro rutto, senpre sapiando che 'l menor se debia trar dal mazuor, e che del menor non se può trazer el mazuor. E primo:

Trazi $\frac{1}{3}$ de $\frac{1}{2}$, roman $\frac{1}{6}$.
Trazi $\frac{1}{7}$ de $\frac{1}{5}$, roman $\frac{2}{35}$.
Trazi $\frac{1}{9}$ de $\frac{1}{7}$, roman $\frac{2}{63}$.
Trazi $\frac{3}{5}$ de $\frac{5}{7}$, roman $\frac{4}{35}$.
Trazi $\frac{5}{7}$ de $\frac{7}{9}$, roman $\frac{4}{63}$.

Se algon te dixese: "Trazi $\frac{1}{2}$ de $\frac{1}{3}$," questa non se può, perché $\frac{1}{2}\,\frac{1}{3}$ se truova in 6, el mezo de 6 sie 3 e lo terzo de 6 sie 2, adoncha non poria trar 3 da 2 che $\frac{1}{2}$ de $\frac{1}{3}$.

Se tu vuol trar $\frac{1}{3}$ de $\frac{1}{2}$ si di' chosì in croxe: "1 via 2 fa 2," el qual trazi de 1 via 3, fa 3. Adoncha roman 1, e quel salva.

f. 8a E puo' di': "3 via 2 fa 6," e metti sutto la verga, la qual sutto l'un serà 6, e serà chossì $\frac{1}{6}$.

---

a. 10396 *corrected over* 10366.

Divide for me $37\frac{2}{3}$ by $5\frac{3}{4}$. Say this: "$37\frac{2}{3}$ $5\frac{3}{4}$, 3 times 37 makes 111, and the 2 above the line makes 113." And then put $\frac{113}{3}$ and then say: "4 times 5 makes 20, and the 3 above makes $\frac{23}{4}$." And then multiply 113 by 4, which makes 452, which you divide by 3 times 23, which makes 69, which comes out $6\frac{38}{69}$. ||

Moreover, by another way and method. Find a number in which $\frac{1}{3}$ and $\frac{1}{4}$ are found, that is, the fractions are found on one side and the other, that is $\frac{2}{3}$ times $\frac{3}{4}$, which is found 19 3 times 4, that is in 12. Now reduce each by 12 and say first: "12 times 37 makes 444," and then you should take $\frac{2}{3}$ of 12, which is 8, which you add to 444, will be 452. Save these and | then multiply the other part f. 7b
by 12, which 5 times 12 makes 60, and take $\frac{3}{4}$ of 12, which is 9, which you add to 60, which makes 69. Now divide 452 by 69, which comes to $6\frac{38}{69}$. And if you want to prove it, do the converse. Multiply $6\frac{38}{69}$ by $5\frac{3}{4}$ and do this: 6 times 69 makes 414 and add 38, which is above the line, and it will make 452, and put $\frac{452}{69}$, and then do 4 times 5 makes 20, and the 3 above the line makes $\frac{23}{4}$. Then multiply 452 by 23, which makes 10,396, which you divide by 4 times 69, which comes to $37\frac{2}{3}$, and this problem is done.

Let's suppose that we want to divide 17 by $\frac{5}{7}$ and we want to know how much comes by division, this is the rule as I said above. And put as it is below, and then multiply 17 times 7, which makes 119, which you divide by 1 times 5, which makes 5, which comes to $23\frac{4}{5}$, which times $\frac{5}{7}$ comes exactly to 17. And by this same method you can do all of the similar problems, great and small, of this kind. 17 by $\frac{5}{7}$ // $\frac{17}{1} \times \frac{5}{7}$ $\frac{119}{5}$

To subtract fractions from fractions

Up to now we have shown how to multiply and add and divide fractions with fractions and integers and fractions. Now we shall show how one should subtract a fraction from another fraction, always knowing that the smaller should be taken from the larger and that the larger cannot be taken from the smaller. And first:

Take $\frac{1}{3}$ from $\frac{1}{2}$, the remainder is $\frac{1}{6}$.
Take $\frac{1}{7}$ from $\frac{1}{5}$, the remainder is $\frac{2}{35}$.
Take $\frac{1}{9}$ from $\frac{1}{7}$, the remainder is $\frac{2}{63}$.
Take $\frac{3}{5}$ from $\frac{5}{7}$, the remainder is $\frac{4}{35}$.
Take $\frac{5}{7}$ from $\frac{7}{9}$, the remainder is $\frac{4}{63}$.

If anyone says to you: "Take $\frac{1}{2}$ from $\frac{1}{3}$," this cannot be done because $\frac{1}{2}$ and $\frac{1}{3}$ are found in 6 and half of 6 is 3 and the third of 6 is 2, so as 3 cannot be taken from 2, so $\frac{1}{2}$ from $\frac{1}{3}$.

If you want to take $\frac{1}{3}$ from $\frac{1}{2}$, you say across: "1 times 2 makes 2," which you take from 1 times 3 makes 3. So 1 remains, and save this.

And then say: "3 times 2 makes 6," and put it below the line, what is under one will be 6, and so it f. 8a
will be $\frac{1}{6}$.

Anchora trazi $\frac{1}{7}$ de $\frac{1}{5}$. Si di' chosì: "1 via 5 fa 5, e 1 via 7 fa 7." Trazi 5 de 7, roman 2, e puo' di': "5 fia 7 35" e metti $\frac{2}{35}$. ||

Anchora trazi $\frac{1}{9}$ de $\frac{1}{7}$. Si di' chossì: "1 via 7 fa 7 e 1 via 9 fa 9." Trazi 7 de 9, roman 2. Puo' fa' 7 via 9 fa 63, e metti $\frac{2}{63}$.

Anchora trazi $\frac{3}{5}$ de $\frac{2}{3}$. Fa' 3 via 3 fa 9, el qual trazi de 2 via 5 fa 10, e roman 1, e puo' moltipicha le figore ch'è de sotto le verge, zoè 3 via 5 fa 15, e chosì roman $\frac{1}{15}$. ||

Anchora trazi $\frac{3}{5}$ de $\frac{5}{7}$. Fa' 3 via 7 fa 21, el qual trazi de 5 via 5 fa 25, e roman 4. Puoi moltipicha i nomeri de sutto, zoè 5 via 7 fa 35, e metti $\frac{4}{35}$. ||

Anchora trazi $\frac{5}{7}$ de $\frac{5}{9}$. Fa' 5 via 9 fa 45, el qual trazi de 5 via 7 35, e chosì roman 10. Puo' moltipicha i nomeri de sutto, zoè 7 via 9 fa 63, e metti chosì $\frac{10}{63}$. E chosì è fatta questa raxion. ||

Anchora trazi de[a] $8\frac{5}{7}$ de $13\frac{7}{9}$. Di' chosì: "8 de 13 roman 5," e metti chosì, e puo' trazi $\frac{5}{7}$ de $\frac{7}{9}$. Di' chosì: "5 via 9 fa 45," i qualli trazi de 7 via 7 fa 49, roman 4. Puo' moltipicha 7 via 9 fa 63, e metti $\frac{4}{63}$. ||

Se tu vollesi trar $\frac{7}{9}$ de $13\frac{3}{7}$ si fa chosì chomo è qui de sutto, e di': "7 via 13 fa 91 e 3 fa 94." E puo' moltipicha 7 via 7 in croxe fa 49, el qual trazi de 9 via 94 che fa 846, e roman 797, e te intrarà $12\frac{41}{63}$. || $\frac{7}{9}$ de $13\frac{3}{7}$ || $\frac{94}{7}$ || $\frac{7}{9}$ trazi $\frac{94}{7}$.

Anchora per lo simille muodo trazi $\frac{7}{9}$ de $13\frac{3}{7}$. Tu puo' far per un oltro muodo. Primo varda se tu po' trar $\frac{7}{9}$ de $\frac{3}{7}$, vedrai de non. Mo' tuo' 1 de 13 e sta 12 i qual[b] salva e de quello 1 fa $\frac{7}{7}$ e azonzi chon $\frac{3}{7}$.
f. 8b Serà $\frac{10}{7}$. Puo' trazi $\frac{7}{9}$ de $\frac{10}{7}$, fa' 7 via 7 fa 49, i qual trazi de 9 via 10 | che fa 90 e roman 41. Puo' moltipicha, resta $12\frac{41}{63}$.

Se tu volesti saver $7\frac{2}{3}$ che parte è de $9\frac{1}{4}$, questa è chotal raxion chomo serave a dir: "Partime $7\frac{2}{3}$ per $9\frac{1}{4}$." Rechordati che noi dovemo dir: "3 via 7 fa 21 e 2 ch'è de sovra fa 23." E metti $\frac{23}{3}$ e po' di': "4 via 9 fa 36, e 1 ch'è de sovra fa 37." E metti $\frac{37}{4}$ e puo' fa' 23 via 4 fa 92, e 3 via 37 fa 111. Mo' parti 92 in 111 e starà chosì $\frac{92}{111}$.

[Dar e tuor rutti]

Questo è l'amaistramento de saver dar e tuor quante parte sono i rutti che noi voiamo tuor over dar d'oni chossa in questo chotal muodo. Chonto te sia che 1ª parte sie $\frac{1}{2}$ de tutta la chosa intrega, e le 2 parte sie li $\frac{2}{3}$ de ttuta la chosa intrega.

---

a. *Sic in MS, for* trazi $8\frac{5}{7}$.

b. *From* tuo' *to* qual *added above the line with a reference mark.*

1. Here Michael proposes to take a larger number from a smaller, which he says above cannot be done. Presumably he actually intends to take $\frac{5}{9}$ from $\frac{5}{7}$.

Next take $\frac{1}{7}$ from $\frac{1}{5}$. You say like this: "1 times 5 makes 5, and 1 times 7 makes 7." Take 5 from 7, the remainder is 2, and then say: "5 times 7, 35," and put $\frac{2}{35}$.

Next take $\frac{1}{9}$ from $\frac{1}{7}$. You say like this: "1 times 7 makes 7 and 1 times 9 makes 9." Take 7 from 9, the remainder is 2. Then do 7 times 9 makes 63, and put $\frac{2}{63}$.

Next take $\frac{3}{5}$ from $\frac{2}{3}$. Do 3 times 3 makes 9, which you take from 2 times 5 makes 10, and the remainder is 1, and then multiply the figures that are below the line, that is 3 times 5 makes 15, and so $\frac{1}{15}$ remains.

Next take $\frac{3}{5}$ from $\frac{5}{7}$. Do 3 times 7 makes 21, which you take from 5 times 5 makes 25, and the remainder is 4. Then multiply the lower numbers, that is, 5 times 7 makes 35, and put $\frac{4}{35}$.

Next take $\frac{5}{7}$ from $\frac{5}{9}$.[1] Do 5 times 9 makes 45, which you take from 5 times 7, 35, and so the remainder is 10. Then multiply the lower numbers, that is 7 times 9 makes 63, and so put $\frac{10}{63}$. And that's how this problem is done.

Next take $8\frac{5}{7}$ from $13\frac{7}{9}$. Say this: "8 from 13 leaves 5," and put this, and then take $\frac{5}{7}$ from $\frac{7}{9}$. Say this: "5 times 9 makes 45," which you subtract from 7 times 7 makes 49, the remainder is 4, then multiply 7 times 9 makes 63, and put $\frac{4}{63}$.

If you would like to take $\frac{7}{9}$ from $13\frac{3}{7}$ you do it as shown below, and say: "7 times 13 makes 91 and 3 makes 94." And then multiply 7 times 7 across makes 49, which you take from 9 times 94 which makes 846, and 797 remains, and you will get $12\frac{41}{63}$. // $\frac{7}{9}$ from $13\frac{3}{7}$ // $\frac{94}{7}$ // $\frac{7}{9}$ subtracted from $\frac{94}{7}$.

Next in a similar way take $\frac{7}{9}$ from $13\frac{3}{7}$. You can do this another way. First note that if you try to take $\frac{7}{9}$ from $\frac{3}{7}$ you'll see that you can't. Now take 1 from 13 and it leaves 12; save this and from this 1 make $\frac{7}{7}$ and add to $\frac{3}{7}$. It will be $\frac{10}{7}$. Then subtract $\frac{7}{9}$ from $\frac{10}{7}$, do 7 times 7 makes 49, which you subtract from 9 times 10 | which makes 90 and 41 remains. Then multiply, which leaves $12\frac{41}{63}$. f. 8b

If you would like to know what part $7\frac{2}{3}$ is of $9\frac{1}{4}$, this is the same problem as it would be to say: "Divide for me $7\frac{2}{3}$ by $9\frac{1}{4}$." Remember that we should say: "3 times 7 makes 21 and the 2 which is above makes 23." And put $\frac{23}{3}$ and then say: "4 times 9 makes 36, and 1 which is above makes 37." And put $\frac{37}{4}$ and then do 23 times 4 makes 92 and 3 times 37 makes 111. Now divide 92 by 111 and it will be $\frac{92}{111}$.[2]

To give and take fractions

Here are the instructions on how to give and take however many parts the fractions are that we want to take or give from each thing in this way. Your calculation is that 1 part is $\frac{1}{2}$ of the whole thing and 2 parts are $\frac{2}{3}$ of the whole thing.

2. This problem is the same as one on fol. 7a, in slightly different wording.

E le 3 parte sie li $\frac{3}{4}$ de tuta la chosa intrega.
E le 4 parte sie li $\frac{4}{5}$ de tutta la chosa intrega.
E le 5 parte sie li $\frac{5}{6}$ de tutta la chosa intrega.
E le 6 parte sie li $\frac{6}{7}$ de tuta la chosa intrega.
E le 7 parte sie li $\frac{7}{8}$ de tuta la chosa intrega.
E le 8 parte sie li $\frac{8}{9}$ de tuta la chosa intrega.
E le 9 parte sie li $\frac{9}{10}$ de tuta la chosa intrega.

E chossì per ordene che senpre la chosa intrega è 1ª parte più che tuta la parte.

E se uno te dixese: "Dame una parte e $\frac{1}{2}$ de lire 7," sapi che $\frac{1}{2}$ parte sie $\frac{1}{2}$ de $\frac{1}{2}$, ch'è $\frac{1}{4}$. Azonzi insenbre una parte de lire 7 chol mezo d'una parte, zoè la mittà de 7 lire e la mittà dela mittà, zoè $\frac{1}{2}$ e $\frac{1}{4}$, sarà $\frac{3}{4}$. Adoncha tanto vien a dir: "Dame una parte e meza de lire 7," chomo: "Dame li $\frac{3}{4}$ de 7 lire." Fa' 3 via 7 fa 21, parti per 4 ne vien 5, zoè lire 5, soldi 5.

f. 9a Anchora dame 3 parte e $\frac{1}{3}$ de 9 lire. Rechordatti che li 3 parte son li $\frac{3}{4}$ e lo terzo de una parte sie el terzo de uno de queli quarti ch'è $\frac{1}{12}$. Adoncha 3 parte e $\frac{1}{3}$ de una parte sie $\frac{3}{4}$ e $\frac{1}{12}$, i quali azonti insenbre serano $\frac{5}{6}$ de 9 lire, e fa' 5 fia 9 fa 45, li quali parti per 6 che ne vien lire 7 soldi 10. E per questo chotal muodo farì dele oltre.

[L'altra guixa dela riegola del 3]

Questo sie l'amaistramento dela riegola del 3, per la qual riegola se può far tute le raxion de marchadantia et è partida in do parte e non in più, zoè in 2 guixe per le qual se può far le raxion de tutte le chose del mondo. E una guixa sie che son raxion senza nomen de alguna chosa d'altro numero che serà. Menzona 3 numeri deschanbiando l'uno dal'altro, e se chiama partidor lo primo numero che serà menzonado 2 volte, e lo segondo nomero die 'ser moltipichado per lo terzo, e quella moltipichazion se die partir in lo primo numero che serà menzonado 2 volte.

Per l'altra guixa dela riegola del 3, sapi che zascadona chosa che se menzona 3 chose per nome, per nomero 2 fiade. El primo nomero dela chosa che serà menzonado 2 volte die 'ser partidor, e li oltri 2 nomeri die 'ser moltipichadi l'un per l'oltrro e quel moltipicho se die partir in lo primo nomero che serà mentoado 2 volte per nomen e per nomero. Qui de sutto farò l'insenpio dela prima guixa dove se menzona numero senza nomen.

And 3 parts are $\frac{3}{4}$ of the whole thing.
And 4 parts are $\frac{4}{5}$ of the whole thing.
And 5 parts are $\frac{5}{6}$ of the whole thing.
And 6 parts are $\frac{6}{7}$ of the whole thing.
And 7 parts are $\frac{7}{8}$ of the whole thing.
And 8 parts are $\frac{8}{9}$ of the whole thing.
And 9 parts are $\frac{9}{10}$ of the whole thing.

And so on, in order that the whole thing is always 1 part greater than the whole part.

And if someone told me: "Give me one part and $\frac{1}{2}$ of 7 lire," know that $\frac{1}{2}$ part is $\frac{1}{2}$ of $\frac{1}{2}$, that is $\frac{1}{4}$. Add together one part of 7 lire, that is half of a part, that is half of 7 lire and half of the half, that is $\frac{1}{2}$ and $\frac{1}{4}$, which makes $\frac{3}{4}$. Then it is the same to say: "Give me a part and a half of 7 lire" as it is to say: "Give me $\frac{3}{4}$ of 7 lire." Do 3 times 7 makes 21, divide by 4 leaves 5, that is 5 lire and 5 soldi.[1]

Next give me 3 parts and $\frac{1}{3}$ of 9 lire. Remember that the 3 parts are $\frac{3}{4}$ and the third of a part is the third of one of those quarters, which is $\frac{1}{12}$. So 3 parts and $\frac{1}{3}$ of a part are $\frac{3}{4}$ and $\frac{1}{12}$, which added together will be $\frac{5}{6}$ of 9 lire, and do 5 times 9 makes 45, which divided by 6 comes to 7 lire 10 soldi. And in this way you will do the others. f. 9a

---

THE OTHER METHOD OF THE RULE OF THREE

Here are instructions on the rule of 3, by which rule can be done all of the problems of merchandise, and it's divided into two parts and no more, that is in 2 methods by which you can solve the problems of all the things in the world. And one method is that there are problems without the name of any thing that will be represented by a number. There are mentioned 3 numbers, changing one for the other, and the first number that will be mentioned 2 times is called the divisor, and the second number should be multiplied by the third, and that multiplication should be divided into the first number, which will be mentioned 2 times.

For the other method of the rule of 3, know that 3 things are each mentioned by name, 2 times by number. The first number of the thing that will be mentioned 2 times should be the divisor, and the other 2 numbers should be multiplied one by the other and this multiple should be divided by the first number that will be mentioned 2 times by name and by number. Below I will give an example of the first method where a number is mentioned without a name.

---

1. There were 20 soldi (shillings) to the lira (pound).

Se 3 fusse 4, che serave 5? Serave $6\frac{2}{3}$.
Se 5 fusse 7, che serave 9? Serave $12\frac{3}{5}$.
Se 13 fusse 15, che serave 17? Serave $19\frac{8}{13}$.
Se 17 fusse 23, che serave 32? Serave $43\frac{5}{17}$.

f. 10b[a] Fame questa raxion. Se $4\frac{5}{7}$ fusse $8\frac{3}{5}$, che seria $9\frac{5}{9}$? Tu die redur tuti in rutti i sany, over tuti i sany in rutti in questo muodo. E di': "4 via 7, azonto 5 ch'è de sovra la verga fa 33," e metti chossì $\frac{33}{7}$. E puo' va' al segondo e di': "5 via 8, azonto 3 ch'è de sovra la verga fa 43," e metti chosì $\frac{43}{5}$. E puo' va' al terzo numero e di': "9 via 9 fa 81, azonto 5 fa 86," e metti chosì $\frac{86}{9}$. E starà chosì una driedo al'oltra, cho[*mo*] vedirì qui de sotto.

$\frac{33}{7} \times \frac{43}{5} = \frac{86}{9}$ || 25886
1485 partidor ||

0
164
0427
11031
25886
14855 $17\frac{641}{1485}$
148

Finamo' t'ho mostrado chomo se die moltiplichar, azonzer e partir e sutrar roti da ruti e sany e ruti, e la riegola del 3 d'i ruti. E mo' te farò chognoser quanto è più un rutto d'un altro ruto.

[Trar rutti e poner rutti]

E digo qual è più $\frac{4}{5}$ o $\frac{3}{4}$, e quello ch'è più quanto quello ch'è men. Truova 1 numero che abia quarti e quinti, e questo se truova moltipichando le figure ch'è de sotto le verge, zoè 4 via 5 fa 20. Mo' dirai: "El quinto de 20 sie 4," adoncha $\frac{4}{5}$ sie 16, zoè $\frac{16}{20}$; el quarto de 20 sie 5, sie $\frac{3}{4}$ de 20, sie 15, zoè $\frac{15}{20}$. Po' po'[b] tu vederì che $\frac{4}{5}$ è $\frac{1}{20}$ più de $\frac{3}{4}$.

E puose far per un altro muodo. Moltipicha 19 in cruxe, là dove $\frac{4}{5}$ e $\frac{3}{4}$ de 4 via 4 fa 16, e puo' 3 via 5 fa 15. Ora trazi 15 de 16, roman 1, e puo' fa' 'na verga e metti uno de sovra, $\frac{1}{}$, e moltipicha quel de sutto, farà 20. Starà chusì $\frac{1}{20}$. È questo più de $\frac{3}{4}$, zoè $\frac{1}{20}$.

Anchora quanto è più $\frac{2}{3}$ e $\frac{2}{5}$ de $\frac{3}{4}$ e $\frac{1}{6}$. Questo è lo sso muodo chomo noi doveremo far. $\frac{1}{3}$, $\frac{1}{4}$, $\frac{1}{5}$, $\frac{1}{6}$ se truova in 60, sono 64, e li $\frac{3}{4}\,\frac{1}{6}$ sono 55. Mo' varda quanto è più 64 de 55, sono 9, el qual seria $\frac{9}{60}$. Adoncha è più $\frac{2}{3}$ e $\frac{2}{5}$ $\frac{9}{60}$ de $\frac{3}{4}\,\frac{1}{6}$. E per questa far le simille raxion.

---

a. *The sheet bearing pages 9b and 10a is lacking in the MS, having been torn out subsequent to binding.*
b. *Repeated thus in MS.*

If 3 were 4, what would 5 be? It would be $6\frac{2}{3}$.
If 5 were 7, what would 9 be? It would be $12\frac{3}{5}$.
If 13 were 15, what would 17 be? It would be $19\frac{8}{13}$.
If 17 were 23, what would 32 be? It would be $43\frac{5}{17}$.

Do this problem for me. If $4\frac{5}{7}$ were $8\frac{3}{5}$, what would $9\frac{5}{9}$ be? You should reduce all to fractions and integers, or all of the integers to fractions in this way. And say: "4 times 7, added to the 5 above the line makes 33," and so put $\frac{33}{7}$. And then go to the second and say: "5 times 8, added to the 3 above the line makes 43," and so put $\frac{43}{5}$. And then go to the third number and say: "9 times 9 makes 81, added to 5 makes 86," and so put $\frac{86}{9}$. And it will be one thing after the other, as you shall see below. f. 10b

$$\frac{33}{7} \times \frac{43}{5} = \frac{86}{9} \quad \Big\| \quad \begin{array}{l} 25886 \\ \quad 1485 \text{ divisor} \end{array} \quad \Big\|$$

```
   0
  164
 0427
 11031
 25886 ]
 14855 ]   17 641/1485
  148  ]
```

Up to now I have shown you how to multiply, add, and divide and subtract fractions from fractions and integers and fractions and the rule of 3 for fractions. And now I will make you know how much one fraction is larger than another.

To subtract fractions and combine fractions

And I say which is larger, $\frac{4}{5}$ or $\frac{3}{4}$, and whichever is larger, how much larger is it than the smaller? Find a number that has fourths and fifths, and this is found by multiplying the numbers that are under the line, that is 4 times 5 makes 20. Now you will say: "The fifth of 20 is 4," so $\frac{4}{5}$ is 16, that is $\frac{16}{20}$; the fourth of 20 is 5, that is $\frac{3}{4}$ of 20, it is 15, that is $\frac{15}{20}$. So then you will see that $\frac{4}{5}$ is $\frac{1}{20}$ more than $\frac{3}{4}$.

And it can be done by a different method. Multiply 19 crosswise, there where $\frac{4}{5}$ and $\frac{3}{4}$ of 4 times 4 makes 16 and then 3 times 5 makes 15. Then subtract 15 from 16, the remainder is 1, and then make a line and put one above, $\frac{1}{}$, and multiply what's below, it will make 20. So it will be $\frac{1}{20}$. And it is this much more than $\frac{3}{4}$, that is, $\frac{1}{20}$.

Next how much are $\frac{2}{3}$ and $\frac{2}{5}$ greater than $\frac{3}{4}$ and $\frac{1}{6}$? This is the way we should do it. $\frac{1}{3}$, $\frac{1}{4}$, $\frac{1}{5}$, $\frac{1}{6}$ is found in 60, they[1] make 64, and the $\frac{3}{4}$ and $\frac{1}{6}$ are 55. Now look how much 64 is larger than 55, it's 9, and that will be $\frac{9}{60}$. So $\frac{2}{3}$ and $\frac{2}{5}$ are $\frac{9}{60}$ larger than $\frac{3}{4}$ and $\frac{1}{6}$. And in this way similar problems are done.

---

1. I.e., $\frac{2}{3}$ and $\frac{2}{5}$.

E se tu vuol far per un oltro muodo fa' chosì. Azorzi[a] $\frac{2}{3}$ e $\frac{2}{5}$ a uno e serano $\frac{16}{15}$, e puo' azonzi $\frac{3}{4}\,\frac{1}{6}$,
f. 11a serano $\frac{22}{24}$. Mo' fa' per la regola del 3 dita[b] davanti, | zoè 16 via 24 fa 384, e puo' moltipicha 22 via 15 fa 330, e guarda quanto più 384 de 330, serano 54, e meti chosì $\frac{54}{\ }$. Puo' moltiplicha la figura de sotto, zoè 15 via 24 fa 360, e questa metti de sotto la verga, e starà chossì $\frac{54}{360}$. Schixa per 6, sarà chosì $\frac{9}{60}$, vedirì de sotto.

$\frac{16}{15}$	╳	$\frac{22}{24}$	fa 330 fa 384	384 330	roman $\frac{54}{360}$

Fame questa raxion de $\frac{5}{6}$ e $\frac{1}{3}$ levando $\frac{1}{2}$ e $\frac{1}{9}$, e di' quanto raman.[c] Questa sie la riegolla: $\frac{1}{6}\,\frac{1}{2}\,\frac{1}{3}\,\frac{1}{9}$ se truova per lo moltipicho delle figure de sotto l'una per l'oltra, che fa 324, e deli $\frac{5}{6}$ e $\frac{1}{3}$ de 324 sie 378, e lo $\frac{1}{2}$ e $\frac{1}{9}$ de 324 sie 198. Mo' lieva 198 de 378, roman 180, i quali sono $\frac{180}{324}$. Schixa per 36 e te roman $\frac{5}{9}$, e tanto te roman a tirar de $\frac{5}{6}\,\frac{1}{3}$ e $\frac{1}{2}$ e $\frac{1}{9}$.

Qui te dixese quanto è $\frac{3}{5}$ de $23\frac{1}{2}$ di' chossì: "Tuo' quel 3 ch'è de sovra la verga e chon esso moltiplicha $23\frac{1}{2}$, che fa $70\frac{1}{2}$, a partir per quel 5 ch'è di sotto la verga, che ne vien $14\frac{1}{10}$." E chosì è fatta.

Io voria saver $\frac{2}{3}$ de quanti quinti sono li $\frac{6}{7}$. Fa' chossì: parti $\frac{2}{3}$ in $\frac{6}{7}$, moltipicha 7 via $\frac{6}{7}$ fa 6 sany, e questi è lo partidor. E fa' 7 via $\frac{2}{3}$ fa $\frac{14}{3}$, a partir in 6 ne vien $\frac{7}{9}$ quinti. Ora di': "5 via 7 fa 35," parti in 9, ne vien $\frac{3}{5}$ e $\frac{8}{9}$ de quinto. È fatta.

Fame questa raxion. Se 4 fusse la mitade de 7, che parte seria 4 de 9? Varda quello ch'è la mitade de 7, zoè $3\frac{1}{2}$, e puo' varda che parte [*è*] 4 de 9. È lo $\frac{4}{9}$. Adoncha se 4 fose $3\frac{1}{2}$, che seria $\frac{4}{9}$? E seria $\frac{28}{72}$. Schixa, serano $\frac{7}{18}$.

f. 11b Perché avanti è ditto del trazer d'i ruti che non se può trazer el mazuor dal menor, per veder quanto resta se die intender in questo muodo. Ponyamo che per insenpio io voio trar $\frac{1}{3}$ da $\frac{1}{4}$. Pony che tu voi trar $\frac{1}{3}$ de 12 de 1 quarto de 12. Perché $\frac{1}{3}\,\frac{1}{4}$ se truova in 12, questo non se poria trar, perché el terzo di 12 sie 4, el quarto di 12 sie 3, sì che ben se vede chiaramente che 4 non se poria trar de 3.

Mo' se tu volesti trar $\frac{1}{3}$ de $\frac{1}{4}$, se poria intender a questo muodo, che tu volessi far de $\frac{1}{4}$ terzi, e quanto[d] de $\frac{1}{4}$ avesi fatto terzi trarne fori $\frac{1}{3}$. Adoncha romagneria $\frac{2}{3}$ de $\frac{1}{4}$, e questo seria de chosa intriega $\frac{1}{6}$, e tu averissi tratto $\frac{1}{3}$ de $\frac{1}{4}$, che seria de chosa intrega $\frac{1}{12}$. Sì che a questo muodo quanto[e] avesi a trar rutto de un altro rutto senpre trazi el menor dal mazuor, e quello che te ne venirà serà de quello che tu avera' lassado de quel rotto. E tuo' per esenplo chomo ò ditto de sovra e chomo te mostrarò.

---

a. *Sic in MS.*
b. dita *on top of* 3.
c. *Sic in MS.*
d. *Sic in MS.*
e. *Sic in MS.*

And if you want to do it another way, do this. Add $\frac{2}{3}$ and $\frac{2}{5}$ into one and it will be $\frac{16}{15}$, and then add $\frac{3}{4}$ and $\frac{1}{6}$, it will be $\frac{22}{24}$. Now do by the rule of 3 explained earlier, | that is, 16 times 24 makes 384 and f. 11a
then multiply 22 times 15 makes 330, and look how much larger 384 is than 330, which will be 54, and put it like this: $\frac{54}{\ }$. Then multiply the number below, that is, 15 times 24 makes 360, and put this below the line, and it will be $\frac{54}{360}$. Factored by 6, it will be $\frac{9}{60}$, which you will see below.

16	╳	22	makes 330	384	leaves $\frac{54}{360}$
15		24	makes 384	330	

Do me this problem of $\frac{5}{6}$ and $\frac{1}{3}$ taking off $\frac{1}{2}$ and $\frac{1}{9}$, and tell me how much remains. This is the rule: $\frac{1}{6}$ $\frac{1}{2}$ $\frac{1}{3}$ $\frac{1}{9}$ is found[1] by multiplying the numbers underneath by each other, which makes 324, and $\frac{5}{6}$ and $\frac{1}{3}$ of 324 is 378, and $\frac{1}{2}$ and $\frac{1}{9}$ of 324 is 198. Now take 198 from 378, the remainder is 180, which is $\frac{180}{324}$. Factor by 36 and you are left with $\frac{5}{9}$, and this is what's left from taking from $\frac{5}{6}$ and $\frac{1}{3}$ $\frac{1}{2}$ and $\frac{1}{9}$.

If someone said to you: "How much is $\frac{3}{5}$ of $23\frac{1}{2}$," say this: "Take the 3 that is above the line and multiply $23\frac{1}{2}$ by it, which makes $70\frac{1}{2}$, which is divided by the 5 below the line, which comes to $14\frac{1}{10}$." And that's how it's done.

I would like to know $\frac{2}{3}$ of how many fifths is $\frac{6}{7}$. Do it this way: divide $\frac{2}{3}$ by $\frac{6}{7}$,[2] multiply 7 times $\frac{6}{7}$, which makes 6 even, and this is the divisor. And do 7 times $\frac{2}{3}$ makes $\frac{14}{3}$ to divide by 6, which comes to $\frac{7}{9}$ fifths. Now say: "5 times 7 makes 35," divide by 9, it comes to $\frac{3}{5}$ and $\frac{8}{9}$ of a fifth. And it's done.

Do this problem for me. If 4 were the half of 7, what part would 4 be of 9? Look at what the half of 7 is, that is $3\frac{1}{2}$, and then look what part 4 is of 9. And it's $\frac{4}{9}$. So if 4 were $3\frac{1}{2}$, what would $\frac{4}{9}$ be? And it would be $\frac{28}{72}$. Factored, it would be $\frac{7}{18}$.

Because above it was said about subtracting that you can't take a larger one from a smaller one, to f. 11b
see how much remains you should understand it in this way. Let's say for example that I want to take $\frac{1}{3}$ from $\frac{1}{4}$. Suppose that you want to take $\frac{1}{3}$ of 12 from 1 quarter of 12. Because $\frac{1}{3}$ and $\frac{1}{4}$ are found in 12, this cannot be subtracted, because the third of 12 is 4 and the fourth of 12 is 3, so it can be easily seen that 4 cannot be taken from 3.

Now if you wished to take $\frac{1}{3}$ from $\frac{1}{4}$ it could be understood in this way, that you would like to make thirds of $\frac{1}{4}$ and when you had made thirds of $\frac{1}{4}$, take away $\frac{1}{3}$. Then there will remain $\frac{2}{3}$ of $\frac{1}{4}$, and this will be as a whole thing $\frac{1}{6}$, and you would have taken $\frac{1}{3}$ from $\frac{1}{4}$, which would be in all $\frac{1}{12}$. So that in this way when you would have to subtract a fraction from another fraction always subtract the smaller from the larger, and what results from this will be how much you will have left of this fraction. And take for an example as I have said above and as I will show you.

---

1. That is, their common denominator is found.
2. This should be "divide $\frac{6}{7}$ by $\frac{2}{3}$," and the following solution is thrown off.

Trazi $\frac{1}{3}$ de $\frac{1}{5}$. Fa' 1 via 5 fa 5, e puo' 1 via 3 fa 3. Tu vedi che 3 è menor de 5, adoncha trazi 3 de 5, roman 2, e puo' moltipicha le figure de sotto una per l'oltra e fara' 3 via 5 fa 15, e questo serà partidor de 2, e chonvien star in questo muodo $\frac{2}{15}$. E notta che tu a' ditto: "Trazi $\frac{1}{3}$ de $\frac{1}{5}$." E per intender che $\frac{1}{3}\,\frac{1}{5}$ se truova in 15, e questo $\frac{1}{5}$ el qual ay tratto el terzo sie de 15, nomeri 3, onde trazando el terzo di 3 manyfesto è che roman 2. Sì che ben sta $\frac{2}{15}$.

---

f. 12a [Chapitulli d'alzebra]

Quisti chapitoli li qual serà spezifichadi de sotto son chapitoli del'Alzibran, che fu uno saraxin lo qual chosì nomen avea, zoè alzebra, lo qual truova li ditti 6 e per li ditti chapitulli se può trovar zaschaduna raxion de radixe. E quando tu voi truovar alguna per li ditti 6 chapitulli e te bexogna chonponer quella o quele raxion che tu vuol truovar per tal magnera che la raxion vegna al muodo de uno de questi chapitulli, chomo ve serà dechiarado qui avanti.

Sapi che questi 6 chapitulli i primi, che son 3, sono dichonpoxitti e li oltri 3 sono chonpoxitti, e fase le sotil raxion maistrevele de nomeri per essi a chi se dotti a saverle ben adure, onde elli schomenza chossì.

Primo chapittolo	chosa ingual a numero	}
Segondo	zenso ingual a numero	}
Terzo	zenso ingual a chosa	}
Quarto chapitulo	zenso e chossa ingual a nomero	}
Quinto chapitolo	zenso e nomero ingual a chosa	}
Sextto chapitolo	chosa e numero ingual a zenso	}

Mo' nottarò la natura de ttuti 6 chapitolli qui chon esenpio, chomenzando dal primo el qual dixe "chosa ingual a numero." Sempre quando è fatta la edechazion e quel romagna chosa ingual a numero, se die partir lo numero per le chose e quelo che ne vien è numero e tanto val la chossa. Exenpio.

f. 12b Truovame 2 nomeri in proporzion che xe 1 da 2, che tanto faza el primo moltipichado per 7 quanto el segondo azonto chon 7. Pony che 'l primo fusse 1^{co}, chonvien che 'l segondo sia 2^{co}. Mo' moltipicha 1^{co} per 7, fa 7^{co}, e puo' azonzi 2^{co} chon 7^{n}, fa 2^{co} e 7^{n}, che sono inguali a 7^{co}. Mo' adequa le parte e trazi 2^{co} de 7^{co}, roman 5^{co}. E chosì tu ai da una parte 5^{co} e dal'altra 7^{n}. Parti li nomeri per le chose, vien $1\frac{2}{5}$, e tanto val la chosa. E tu ponesti el primo fuse 1^{co}, adoncha solo $1\frac{2}{5}$. El segondo tu ponesti 2^{co}, adoncha fulo[a] $2\frac{4}{5}$, e tanto el primo moltipichado per 7 quanto el segondo azonto 7.

---

a. *Sic in MS, evidently for* solo, *as for the previous problem.*

1. This is the beginning of a treatise on algebraic equations, which Michael follows other medieval authors in attributing to an Arabic scholar of that name.

Subtract $\frac{1}{3}$ from $\frac{1}{5}$. Do 1 times 5 makes 5, and then 1 times 3 makes 3. You see that 3 is less than 5, so subtract 3 from 5, the remainder is 2, and then multiply the numbers beneath by each other and it will be 3 times 5 makes 15, and this will be the divisor of 2, and it's usual to put it in this form: $\frac{2}{15}$. And note that you have said: "Take $\frac{1}{3}$ from $\frac{1}{5}$." And to understand that $\frac{1}{3}\,\frac{1}{5}$ are found in 15, and this $\frac{1}{5}$ from which you have taken a third is 3 numbers of 15, from which subtracting the third of 3 clearly leaves 2. So that it is really $\frac{2}{15}$.

The chapters of algebra

f. 12a

These chapters that will be specified below are chapters of Alzibran, who was a Saracen who had that name, that is, Algebra, who found these 6 chapters, and through these chapters you can find every problem involving roots.[1] And when you want to find anything by these 6 chapters, it is necessary for you to compose this or that problem that you want to find in such a way that the problem is done by the method of one of these chapters, as will be expressed henceforth.

Know that of these 6 chapters, the first, which are 3, are not composite and the other 3 are composite, and make subtle problems solvable with numbers by those who are brought to know well how to calculate them, whence they begin like this:[2]

First chapter	unknown equal to a number	}
Second	squared unknown equal to a number	
Third	squared unknown equal to an unknown	
Fourth chapter	squared unknown and unknown equal to a number	}
Fifth chapter	squared unknown and number equal to an unknown	
Sixth chapter	unknown and number equal to a squared unknown	

Now I will note here the nature of all of the 6 chapters with examples, beginning with the first, which is titled "unknown equal to a number." Whenever the equation is made and it works out to an unknown equal to a number, you should divide the number by the unknowns and what results is a number and that is the value of the unknown. Example:

Find me 2 numbers in proportion of 1 to 2, so that the first multiplied by seven is as much as the second added to 7. Put that the first is $1x$, and it follows that the second is $2x$. Now multiply $1x$ by 7, which makes $7x$, and then add $2x$ to 7, which makes $2x$ plus 7, which is equal to $7x$. Now equalize the sides and subtract $2x$ from $7x$, and the remainder is $5x$. And so you have on one side $5x$ and on the other side 7. Divide the numbers by the unknowns, it comes to $1\frac{2}{5}$, and that is the value of the unknown. And you put that the first was $1x$, so only $1\frac{2}{5}$. The second you put as $2x$, so it was $2\frac{4}{5}$, and the first multiplied by seven is as much as the second added to 7. f. 12b

---

2. For the mathematical notation of these equations, see Franci, "Mathematics in the Manuscript of Michael of Rhodes," vol. 3, pp. 126–128.

La natura del segondo chapitolo dove dixe "zenso ingual al numero" sie che se die partir li nomeri per li zenssi e quelo che ne vien la so radixe val la chosa.

Exempio. Trovame 1 nomero che tratone $\frac{1}{3}$ e $\frac{1}{4}$ el romagnente moltipichado in si medeximo fazia 12. Pony che 'l nomero fose 1^{co}, el $\frac{1}{3}$ el $\frac{1}{4}$ sie $\frac{7}{12}$. Tralo de 1^{co}, resta $\frac{5}{12}$. Ora moltipicha $\frac{5}{12}$ de chosa in si medeximo, fa $\frac{25}{144}$ de zensso, che sono inguali a 12. Mo' tu die partir 12 per $\frac{25}{144}$, zoè che tu parti el nomero per li zensi, viene $69\frac{3}{25}$, e la so radixe fo el nomero.

La natura del terzo chapitolo dove dize "zenso ingual a chosa." Dezi[a] partir le chose per li zensi e quelo che me venirà sie nomero. E tanto val la chosa.

Exenpio. Truovame 1 numero che moltipichado contra i so $\frac{2}{3}$ faza 3 chotanti del nomero trovado. Pony che 'l nomero sia 1^{co}. Mo' prendi li so $\frac{2}{3}$, che xe $\frac{2}{3}$ de chosa, e moltipicha $\frac{2}{3}$ de cosa via 1^{co}, fa $\frac{2}{3}$ de zensso, i qual son inguali a 3 tanto del nomero trovado, zoè a 3 chose. E chosì abiamo 3 chose
f. 13a ingual a $\frac{2}{3}$ | de zenso. Parti 3 per $\frac{2}{3}$, me vien $4\frac{1}{2}$, e tanto val la chosa. E tu ponesti 1^{co}. Adoncha fo el numero $4\frac{1}{2}$.

La natura del quarto chapitolo dove dize "zenso e chosa ingual a nomero." Se vuol partir tuta la edechazion per li zensi e poi demezar le chose e moltipichar in si medeximo, e quelo che ne vien azonzi sovra el nomero e la radixe dela suma men l'oltra mittade dele chose vien a valer la chosa.

Exenpio. Uno inprestò un oltro lire 20 per do ani a far chaico[b] d'ano. Quando vene in chavo de do ani el debitor rende el crededor tra pro' e chavedal lire 30. Domando a quanto fo inprestado la lira al mexe. Pony che fose inprestado a 1^{co} de danar, che val l'ano 12 chose de danar, che xe $\frac{1}{20}$ de lira, zoè 1^{co} de lira, sì che in lo primo ano tu ai tra pro' e chavedal lire 20 e 1^{co}. Ora fa per lo segondo ano e di': "Se 20 lire me dà lire 20 e 1^{co}, che me darà lire 20 e 1^{co}?" Moltipicha 20 e 1^{co} via 20 e 1^{co}, fa $1^{\square}$ e 40^{co} e 400^{n}, i qual partidi per 20 ne vien lire 20 e 2^{co} e $\frac{1}{20}$ de zenso, che sono inguali a 30 lire. Mo' se die adequar quanto è posibile. Trazi 20 de 30, roman 10, e mo' ai 10^{n} inguali a 20^{co} e $\frac{1}{20}$ de zenso. Fa' segondo la so natora che dixe "parti per li zensi," sì che adoncha parti 10 per $\frac{1}{20}$ de zenso, ne vien 200^{n}, e parti $\frac{1}{20}$ per $\frac{1}{20}$, ne vien $1^{\square}$, e puo' parti 2^{co} per $\frac{1}{20}$, fa 40^{co}. E chomo che tu hai 40^{co} e $1^{\square}$ inguali a 200^{n}, mo' se die tuor la mittà dele chose che sono 20^{co}, e moltipicha' in si fa 400. Azonzi chon lo nomero che son 200, farà 600, e radixe de 600 men el demezamento dele chose, che fo 20, val la chosa, e tanto fo inprestada la lira al mexe de danar.

---

a. *Sic in MS.*
b. *Sic in MS.*

The nature of the second chapter where it says "squared unknown equal to a number" is that you have to divide the numbers by the squared unknowns and the square root of the result is the unknown.

Example. Find me 1 number that when $\frac{1}{3}$ and $\frac{1}{4}$ are subtracted from it, the result multiplied by itself makes 12. Suppose that the number is $1x$, and $\frac{1}{3}$ and $\frac{1}{4}$ are $\frac{7}{12}$. Subtract it from $1x$; the remainder is $\frac{5}{12}$. Now multiply $\frac{5}{12}x$ by itself, which makes $\frac{25}{144}x^2$, which is equal to 12. Now you need to divide 12 by $\frac{25}{144}$, which is that you divide the number by the squared unknown, which comes to $69\frac{3}{25}$, and its square root is the number.

The nature of the third chapter where it says "squared unknown equal to an unknown." You should divide the unknowns by the squared unknowns and what results is the number. And that's the value of the unknown.

Example. Find me 1 number that when multiplied by $\frac{2}{3}$ of itself makes 3 times as much as the number found. Suppose that the number is $1x$. Now take $\frac{2}{3}$ of it, which is $\frac{2}{3}x$, and multiply $\frac{2}{3}x$ by $1x$, makes $\frac{2}{3}x^2$, which is equal to 3 times the number itself, that is $3x$. And so we have $3x$ equal to $\frac{2}{3} \mid x^2$. Divide 3 by $\frac{2}{3}$, I get $4\frac{1}{2}$, and that's the value of the unknown. And you put $1x$. So the number f. 13a
was $4\frac{1}{2}$.

The nature of the fourth chapter where it says "squared unknown and unknown equal to a number." It's necessary to divide the whole equation by the squared unknowns and then halve the unknowns and multiply them by themselves, and then add the result above the number, and the square root of the sum minus the other half of the unknowns comes to equal the unknown.

Example. One person loaned another 20 lire for 2 years to make a year's caique voyage.[1] When the two years are up, the debtor returns 30 lire to the creditor including interest and capital. I ask at what monthly rate per lira was the loan made. Say that it was loaned at $1x$ pennies,[2] which comes to $12x$ pennies per year, which is $\frac{1}{20}$ of a lira, that is $1x$ of a lira, so that in the first year you have between interest and capital 20 lire plus $1x$. Now do it for the second year and say: "If 20 lire gives me 20 lire plus $1x$, what will 20 lire plus $1x$ give me?" Multiply 20 and $1x$ by 20 and $1x$, it makes $1x^2$ and $40x$ and 400, which divided by 20 comes to 20 lire and $2x$ and $\frac{1}{20}$ of x^2, which is equal to 30 lire. Now we have to equalize as much as possible. Subtract 20 from 30, the remainder is 10, and now you have 10 equal to $20x$ and $\frac{1}{20}x^2$. Do according to the instruction that says "divide by the squared unknowns," so that you then divide 10 by $\frac{1}{20}$ of the squared unknown, which comes to 200, and divide $\frac{1}{20}$ by $\frac{1}{20}$ which comes to 1^2, and then divide $2x$ by $\frac{1}{20}$ and it makes $40x$. And as you have $40x$ and 1^2 equal to 200, now you can take half of the unknowns which is $20x$, and multiplied by itself makes 400. Add it to the number which is 200, it will make 600, and the square root of 600 minus half of the unknown, which was 20, is equal to the unknown, and that is the rate in pennies per month at which the loan was made.

1. Caique: a small trading vessel.
2. The lira (i.e., pound) is defined as equal to 240 pennies.

La natura del quinto chapitolo dove dize "zenso e nomero ingal a 1^{co}."

f. 13b El se vuol partir tuta la edechazion per li zensi e poi dimezar le chose e moltipicar in si medeximo e trarne lo nomero e la radixe de quelo trato del dimezamento dele chose, overo la radixe de quelo posto sul dimezamento dele chose val la chosa. E sapi che algone raxion te chonvenirà responder per lo primo muodo, e alchone per lo segondo, e alchone per intranbi li muodi.

Exenpio. Per un d'i muodi fame de 10 2 tal parte che moltipichada la diferenzia che xe dal'una al'altra in si medexima faza $20\frac{1}{4}$.

Pony che una dele parte fusse 1^{co} e l'altra roman 10 men 1^{co}. Mo' prendi la diferenzia ch'è dal'una al'altra, la qual vien a eser 2^{co} men [*10*].[a] Voyando redur a questo chapitulo che tu vuol saver chomo la diferenzia sia 2^{co} men 10, zonzi 2^{co} men 10 a 10 men 1^{co}, ch'è la segonda parte, et avera' 1^{co}. Adoncha vedi ben che a trar de 1^{co} 10 men 1^{co} te roman 2^{co} men 10, e tanto xe la diferenzia quanto una parte xe mazuor del'altra. La qual diferenzia, zoè 2^{co} men 10, moltipicha in si medexima, che monta 100^{n} e $4^{\square}$ men 40^{co}, che vien a eser inguali a $20\frac{1}{4}$. Mo' dovemo partir tuta la edechazion tanto chomo xe li zensi, zoè per 4, segondo la riegola, che ne vien $1^{\square}$ e 25^{n} men 10^{co} ingual a 5^{n} nomeri $\frac{1}{16}$. Mo' dovemo trar li men nomeri de zaschadona dele parte per adechar, zoè $5\frac{1}{16}$. A una dele parte romagnerà men numero, e del'altra romagnerà 19 e $\frac{15}{16}$ e $1^{\square}$ men 10^{co}. Mo' azonzi a zaschona dele parte 10^{co}, zoè tante chose quante à de men una dele parte, et avera' a una dele parte
f. 14a niuna chosa et al'altra parte 10^{co}. Adoncha serà romaxo | a una dele parte $1^{\square}$ 19^{co} e $\frac{15}{16}$ ingual a 10^{co}, zoè al'altra parte. Mo' te fazo asaver che lo prozeder che noi fesemo quanto[b] noi avesemo 100^{n} e $4^{\square}$ men 40^{co} ingual a $20\frac{1}{4}$ ave questo vero, che la edechazion non chonsona chon la riegola dada de sovra e la raxion par che sie questa, che non iera anchora tratto li men nomeri de zaschaduna dele parte né exiandio zonto le chose ben che la edechazion sia redutta a perfizion vera. Noi partisemo per tantto quanto fo li zensi avanti tenpo segondo chomo se demostra per riegola del ditto chapitolo, et dovemo tenir questo muodo che noi dovemo trar lo menor nomero, zoè $20\frac{1}{4}$ de zaschaduna dele parte. Romagnerà, dal'una dele parte nesun nomero et al'altra $79\frac{3}{4}$ e $4^{\square}$ men 40^{co}, e puo' se dovrà azonzer 40^{co}, che xe men dal'una dele parte, a zaschaduna parte, e romagnerà al'una dele parte $4^{\square}$ $79^{n}\frac{3}{4}$ de nomero et al'altra 40^{co}, che xe ben la chonsonanza ala riegola davanti, la qual dixe li zensi e li nomeri ingual ale chose, che partando mo' per li zensi ven ben tanto chomo vien per l'altro muodo, zoè $1^{\square}$ 19^{n} e $\frac{15}{16}$ ingual a 10^{co}. Mo' partimo le chose per 2 segando[c] la riegola, che ne vien 5, moltipicha' in si medeximo fa 25, la qual tranzinde el nomero, zoè $19\frac{15}{16}$, et averà che le chose vien a eser l'altra mittà dele chose, zoè 5 e più radixe de $5\frac{1}{16}$, la qual radixe xe $2\frac{1}{4}$. Adoncha vien a valer la chosa $7\frac{1}{4}$, la qual tu ponesti che fusse la prima parte, e l'altra parte vien a eser lo

a. 10 *lacking.*
b. *Sic in MS, evidently for* quando.
c. *Sic in MS, evidently for* segondo.
1. *10* lacking in MS.
2. Given incorrectly in MS as *19x*.

The nature of the fifth chapter, where it says "squared unknown and number equal to 1 unknown."

It is necessary to divide the whole equation by the squared unknowns and then halve the unknowns f. 13b and multiply them by themselves and take from them the number and the root of that taken from the halving of the unknowns, or rather the root of what is put over the half of the unknowns equal to the unknown. And know that for some problems you'll be better off using the first method, and for some the second, and for some both the methods.

Example. By one of the methods take 2 parts of 10 which when the difference of one from the other is multiplied by itself make $20\frac{1}{4}$.

Suppose that one of the parts was $1x$, and the other comes to $10 - 1x$. Now take the difference between one and the other, which comes to $2x$ minus 10.[1] Since in this chapter you want to convert what you want to know as the difference that there is of $2x$ minus 10, add $2x$ minus 10 to 10 minus $1x$, which is the second part, and you will have $1x$. Then you see well that when you subtract 10 minus $1x$ from $1x$, the remainder is $2x$ minus 10, and this is the difference of how much one part is larger than the other. Multiply this difference, that is, $2x$ minus 10, by itself, which comes to 100 and $4x^2$ minus $40x$, which is equal to $20\frac{1}{4}$. Now we should divide the whole equation by the amount of the squared unknowns, that is by 4, according to the rule, which comes to $1x^2$ and 25 minus $10x$, equal to $5\frac{1}{16}$. Now we must subtract the smaller numbers of each of the sides equally, that is $5\frac{1}{16}$. One of the sides will remain without numbers and on the other will remain 19 and $\frac{15}{16}$ and $1x^2$ minus $10x$. Now add $10x$ to both sides, that is, the amount of the unknown that is minus in one of the sides, and you will have on one side none of the unknown and on the other side $10x$. So there will be remaining | on one of the sides $1x^2$ 19[2] and $\frac{15}{16}$, equal to $10x$ which is on the other f. 14a side. Now I'll let you know that the procedure that we did when we had 100 and $4x^2$ minus $40x$ equal to $20\frac{1}{4}$ had this truth, that the equation is not consistent with the problem given above, and the reason why seems to be this, that the minus numbers were not subtracted from each side nor the unknowns added so that the equation could be brought to true perfection.[3] We divided by the amount that the squared unknowns were before the second time, as is shown in the rule for this chapter, and we should keep to this method of subtracting the lesser number, that is, $20\frac{1}{4}$ from each side. There will remain on one side no number and on the other $79\frac{3}{4}$ and $4x^2$ minus $40x$, and then $40x$, which is the amount by which one side is smaller than the other, should be added to each side, and there will remain on one of the sides $4x^2$ $79\frac{3}{4}$ and on the other $40x$, which is in accord with the rule put forth, which talks about squared unknowns and numbers equal to unknowns, which dividing now by the squared unknowns ends up pretty much like the other way, that is, $1x^2$ and $19\frac{15}{16}$ is equal to $10x$. Now we divide the unknowns by 2 following the rule, which comes to 5, multiplied by itself makes 25, which subtracts the number, that is $19\frac{15}{16}$, and the result will be that the unknown turns out to be the other half of the unknowns, that is, 5 plus the square root of $5\frac{1}{16}$ which is $2\frac{1}{4}$. So the value of the unknown is $7\frac{1}{4}$, which was what you put as the first part, and the other

3. This is one of the passages in which Michael appears to be "thinking out loud" about a problem and which indicate that the mathematics problems were not copied directly from another source but were worked out by Michael as he explained them. Cf. Franci, "Mathematics in the Manuscript of Michael of Rhodes," vol. 3, pp. 142–143.

romagnente infin a 10, zoè $2\frac{3}{4}$. E questa raxion sie resposta per lo primo muodo, zoè la mitade dele chose più radixe del dimezamento che fo ditto, e per altro muodo non se porà responder, siando mesa ala sovraditta edequazion toyando la diferenzia per lo muodo ch'è ditto. La raxion sie questa. f. 14b Se ttu avesi ditto la | diferenzia eser stada per oltro muodo perché la può eser, zoè 10 men 2^{co}, la raxion se porave aver resposta chomo ò dito per lo men, zoè per quello che vegnia a eser la chosa sì podeva eser la menor dele parte. La qual chosa non se puo' far siando la diferenzia 2^{co} men 10, perché 2^{co} chonvien eser più che 10, se diexe se die trar de quello che monta 2^{co}.

Echo l'asenpio. Se le chose vien a eser 5 e Rx[a] de $5\frac{1}{16}$, che siando dutto a numero vien a eser $7\frac{1}{4}$, tuta la parte 2^{co} vegnirà eser $\frac{2}{4}$, e la diferenzia se farà 2^{co} men 10. Adoncha trazi 10 de $14\frac{2}{4}$, roman $4\frac{1}{2}$, lo qual moltipicha' in si medemo fa $20\frac{1}{4}$. E se tu dixesi che la chosa fose 5 men Rx de $5\frac{1}{16}$, lo qual redutto a san vegnirave a eser $2\frac{3}{4}$, inposibile seria che de 2 sifatte cose 10 se podese trar, e perzò non se può responder fazando la diferenzia 2^{co} men 10 se non per lo più, segondo chomo se respoxe anchora.

Anchora chi mettesse una parte eser 1^{co} e 5 e l'altra partte 5 men 1^{co}, la diferenzia serave 2^{co}, e faravese per lo segondo chapitolo.

Fame de 10 2 tal parte che moltipichada la menor in si medexima e tratto quela moltipicazion dela moltipichazion dela diferenzia che xe dala mazuor ala menor, moltipicha' in si medexima, romagna 32. Domando: "Quanto serave zaschadona dele parte?"

f. 15a Pony che la menor parte sia 1^{co} e l'altra è romagnente inchina | 10, zoè 10 men 1^{co}. Mo' moltipicha' in si 1^{co} via 1^{co}, fa $1^{\square}$, puo' tuo' la diferenzia che'è dala mazuor ala menor, che xe 2^{co} men 10, e moltipicha' in si fa $4^{\square}$ e 100^{n} men 40^{co}. Mo' trazi la moltipichazion dela menor parte, zoè $1^{\square}$, dela moltipichazion dela diferenzia, zoè de $4^{\square}$ e 10^{n} men 40^{co}, roman $3^{\square}$ e 10^{n} men 4^{co}, lo qual xe ingual a 32^{n}, segondo la domandaxion de sovra. Adequa le parte e dà 40^{co} a zaschuna parte perché le mancha da una, et averemo $3^{\square}$ e 100^{n} inguālli a 32^{n} e 40^{co}. Adequa le parte e trazi 32^{n} de 100^{n}, et averemo de una parte $3^{\square}$ e 68^{n} inguali a 40^{co}. Mo' parti per li zensi et averemo $1^{\square}$ e $22\frac{2}{3}$ inguālli a $13^{co}\frac{1}{3}$. Mo' demeza le chose, fa $6\frac{2}{3}$, moltipicha' in si fa 44 e $\frac{4}{9}$. Trazi lo numero, zoè $22\frac{2}{3}$, roman $21\frac{7}{9}$, e la so Rx trazi del'oltra mittà dele chose, zoè de $6\frac{2}{3}$. Adoncha averemo che la menor parte sia $6\frac{2}{3}$ men Rx de $21\frac{7}{9}$, et redutta per lo segondo muodo che dixe lo quinto chapitollo, e del'altra parte die 'ser lo romagnente fin a 10, zoè $3\frac{1}{3}$ e Rx de $21\frac{7}{9}$. Mo' te fazo asaver che se tu avesti toltto la diferenzia per altro muodo, che la se puo' tuor segondo chomo xe demostrado davanti, el qual vien a eser 10 men 2^{co}, la resposta vegnirave simyle de questa che xe ditta de sovra. E per oltro

---

a. *In this MS,* Rx *appears to have been used as a mathematical symbol, equivalent to the modern* $\sqrt{\ }$, *rather than as an abbreviation for the word* radixe, *so it will not be expanded in this edition.*

part becomes the remainder from 10, that is, $2\frac{3}{4}$. And this problem is solved by the first method, that is, half of the unknowns plus the root of the aforementioned half, and you can't solve it the other way, putting it in the equation above subtracting the difference in the manner described. The problem is this. If you had said that the | difference should be done by the other way because it could be, that is 10 minus $2x$, the problem could have been solved as I said by the minus, that is by having that which became the unknown be the lesser of the two parts. This could not be done by taking the difference of $2x$ minus 10 because $2x$ works out to be more than 10; it has to be subtracted from what $2x$ turns out to be. f. 14b

Here is the example. If the unknowns came out to 5 and the square root of $5\frac{1}{16}$, which being turned into numbers becomes $7\frac{1}{4}$, all the side of $2x$ becomes $\frac{2}{4}$, and the difference will be $2x$ minus 10. Then take 10 from $14\frac{2}{4}$, and the remainder is $4\frac{1}{2}$, which when multiplied by itself makes $20\frac{1}{4}$. And if you said that the unknown was 5 minus the square root of $5\frac{1}{16}$, which when converted becomes $2\frac{3}{4}$, it is impossible that 10 could be subtracted from 2, and therefore you can't answer by making the difference $2x$ minus 10 except by plus, as in the response made earlier.

Again, whoever might hypothesize that one side is $1x$ and 5 and the other part 5 minus $1x$, the difference would be $2x$, and it would be done by the second chapter.

Make two parts from 10 which when the smaller is multiplied by itself and this multiplication is subtracted from the product of the difference between the larger and the smaller, multiplied by itself, the remainder is 32. I ask: "How much would each part be?"

Suppose that the smaller part is $1x$ and the other is the remainder from | 10, that is 10 minus $1x$. Now multiply by itself $1x$ by $1x$, it makes $1x^2$, then take the difference between the larger and the smaller, which is $2x$ minus 10, and multiply by itself, it makes $4x^2$ and 100 minus $40x$. Now subtract the multiplication of the smaller part, that is $1x^2$, from the multiplication of the difference, that is, from $4x^2$ and 100[1] minus $40x$, the remainder is $3x^2$ and 100[2] minus $40x$,[3] which is equal to 32 according to the question above. Equalize the sides and give $40x$ to each side because that much is lacking from one, and we will have on one side $3x^2$ and 100 equal to 32 and $40x$. Equalize the sides and subtract 32 from 100, and we will have on one side $3x^2$ and 68 equal to $40x$. Now divide by the squared unknowns and we will have $1x^2$ and $22\frac{2}{3}$ equal to $13\frac{1}{3}x$. Now halve the unknowns, it makes $6\frac{2}{3}$, multiply by itself, it makes 44 and $\frac{4}{9}$. Subtract the number, that is $22\frac{2}{3}$, the remainder is $21\frac{7}{9}$, and subtract its square root from the other half of the unknowns, that is from $6\frac{2}{3}$. So we will have that the smaller part is $6\frac{2}{3}$ minus the square root of $21\frac{7}{9}$, and converted by the second way that the fifth chapter gives, and on the other side should be the remainder from 10, that is, $3\frac{1}{3}$ and the square root of $21\frac{7}{9}$. Now I inform you that if you had taken the difference by the other way, the method I showed you before, which comes to 10 minus $2x$, the answer would be the same as the one that I've said above. And you couldn't answer by any other method, putting f. 15a

1. Given incorrectly in MS as *10.*
2. Given incorrectly in MS as *10.*
3. Given incorrectly in MS as *4x.*

muodo non se pottrà responder, metando che la menor parte fose 1^{co}. Vegnando adoncha la raxion
a questo chapitolo, la raxion non può venir altramente resposta se non numero men Rx, perché la
parte menor moltipicha' in si die fa[*r*] men che la diferenzia moltipicha' in si, da puo' che trazando
la moltipichazion dela menor dela moltipicazion dela diferenzia e die romagnir 32, da puo' che la
ditta raxion per oltro muodo non se può meter a questo chapitolo, se non è per li do muodi li qualli
f. 15b s'è | ditto de sovra, perché digando la chosa, el nomero e Rx la menor parte vegnirave a eser $6\frac{2}{3}$ e
radixe de 21 e $\frac{7}{9}$, la qual parte venirave a eser più cha intranbe le parte, le quale die 'ser 10, la qual
chosa inposibille che la parte sia mazuor che al tutto. E se tu vuol veder più chiaramente de zò,
prendi la Rx de $21\frac{7}{9}$, la qual è $4\frac{2}{3}$. Trazi de $6\frac{2}{3}$, resta 2, e tanto vien a eser la menor parte, e la
mazuor vien a eser el romagnente inchina 10, zoè 8, e la diferenzia vien a eser 2 tratto de 8. La
qual diferenzia sie 6, e moltipichada in ssi fa 36, ma moltipicha' la menor parte in si, zoè 2, fa 4,
e trazello de 36, e roman 32, po' la domandaxion.

Fame de 10 tal 2 parte che moltipichada l'una per l'oltra faza 21. Questa sie la riegolla. Pony che una dele parte sia 1^{co}, l'altra roman 10 men 1^{co}, e moltipicha' 1^{co} via 10 men 1^{co}, monta 10^{co} men $1^{\square}$, lo qual 10 men $1^{\square}$ sie ingual a 21. Mo' da' $1^{\square}$ a zaschadona dele parte, et avera' 10^{co} ingual a $1^{\square}$ e 21^{n}. Mo' dimeza le chosse, che ne vien 5 per parte, e moltipicha' in ssi fa 25. Mo' trazi lo nomero, zoè 21, da 25, resta 4 e radixe de questo 4 sono 2.[a] Zonzi, o vuol trar del'altra mitade dele chose, zoè di 5, et avera' 5, e radixe men[b] de 4, sono 2,[c] o vuol[d] de 5 men radixe de 4[e] resta 3, e tanto vallerà la chossa, e lo romagnente fina 10 valerà l'altra parte. Adoncha postu responder per lo più e per lo men quello che vien a valer la chosa? Mo' astu vezudo tutte le resposte le qual se può far per lo ditto quinto chapitollo? Se azonzi 2 a 5 serano 7, e l'oltra serà 3. Moltipicha' 3 via 7, fano 21. Anchor sutrazer 2 de 5 resta 3, e 3 via 7 fa 21, e questa è fata.[f]

La natura del sesto chapitolo dove dize: "Chosa e nomero ingual a zenso." El se die partir per li
f. 16a zenssi e puo' dimezar le chose e moltipicar | 1^{a} dela mittà in si medexima, e quela moltipichazion
zonzer sovra el nomero, e la radixe de quela suma più el dimezamento dele chose val la chosa.

Truovame un nomero che moltipichado per 10 e quela moltipichazion zonta a 39 faza tanto quanto el nomero moltipichado in ssi medeximo. Pony che 'l nomero sia 1^{co}, moltipicha' per 10 fa 10^{co}. Mo' azonzi a questa moltipichazion 39, e tu averai 10^{co} e 39^{n}, i quali die 'ser ingualli a 1^{co} moltipicha' in ssi, che monta $1^{\square}$. Mo' proziedi segondo la riegola sovra ditta che vien $1^{\square}$ ingual a 10^{co} e 39^{n}. Mo' dimeza le chose, che vien 5, e moltipicha' in si fa 25. La qual moltipicazion azonzi sovra el nomero, zoè 39^{n}, et avera' 64^{n} e Rx de 64^{n} più el dimezamento delle chose val la chosa, zoè lo

a. sono 2 *added above the line with a reference mark.*
b. men *added above the line with a reference mark.*
c. sono 2 *added above the line.*
d. radixe *crossed out with a horizontal line.*
e. resta 3 *added below the line with a reference mark.*
f. *From* Se azonzi *to* è fata *added between the lines in another ink.*

that the smaller part was $1x$. Coming then to the problem in this chapter, the problem can't be answered unless the number is less than the root, because the smaller part multiplied by itself must be smaller than the difference multiplied by itself, so that subtracting the multiplication from the lesser of the multiplication of the difference, there should remain 32; therefore this problem cannot be done by the method in this chapter, except by the two methods which are | discussed f. 15b
above, because saying the unknown, the number, and the root, the lesser part would be $6\frac{2}{3}$ and the root of $21\frac{7}{9}$, which part would be larger than the combination of the parts, which should be 10, which is impossible in that the part would be larger than the whole. And if you want to see this more clearly, take the root of $21\frac{7}{9}$, which is $4\frac{2}{3}$. Subtract it from $6\frac{2}{3}$, the remainder is 2, and that becomes the smaller part and the larger part becomes the remainder from 10, that is, 8, and the difference becomes 2 subtracted from 8. This difference is 6, and multiplied by itself makes 36, but multiply the smaller part by itself, that is 2, which makes 4 and subtract it from 36, the remainder is 32, as in the question.

From 10 make two parts which when multiplied one by the other make 21. This is the rule. Put that one of the parts is $1x$, the other remains 10 minus $1x$, and multiply $1x$ by 10 minus $1x$, it comes to $10x$ minus $1x^2$, which $10x$[1] minus $1x^2$ is equal to 21. Now add $1x^2$ to each side and you will have $10x$ equal to $1x^2$ and 21. Now halve the unknowns, which comes to 5 per side, and multiplied by itself makes 25. Now subtract the number, that is 21, from 25, the remainder is 4, and the root of this is 2. Add, or if you wish take from the other half of the unknown, that is from 5, and you will have 5, and less the root of 4, that is 2, or rather 5 minus the root of 4, leaves 3, and that is the value of the unknown, and the remainder from 10 will be the other part. So, can you answer by the higher and the lower which is the value of the unknown? Now have you seen all the answers that can be given for this fifth chapter? If you add 2 to 5 it will be 7, and the other will be 3. Multiply 3 times 7, which makes 21. Then subtract 2 from 5, which leaves 3, and 3 times 7 makes 21, and it's done.

The nature of the sixth chapter where it is said: "Unknown and number equal to squared unknown." It should be divided by the squared unknowns, and then halve the unknowns and multiply | one of the halves by itself, and add this multiplication to the number, and the square f. 16a
root of this sum plus the half of the unknowns is equal to the unknown.

Find me a number that multiplied by 10 and this multiplication added to 39 makes the same amount as the number multiplied by itself. Suppose that the number is $1x$, which multiplied by 10 makes $10x$. Now add to this multiplication 39, and you will have $10x$ plus 39, which should be equal to $1x$ multiplied by itself, which amounts to $1x^2$. Now proceed according to the rule given above which gives $1x^2$ equal to $10x$ plus 39. Now halve the unknown, which comes to 5, and multiply it by itself, which makes 25. Add this multiplication to the number, that is, 39, and you will have 64, and the square root of 64 plus half of the unknown equals the unknown, that is, the number. So you will have the result that the number should be 5 and the square root of 64 which is 8;

1. Given incorrectly in MS as *10*.

nomero. Adoncha averai che'l nomero sia 5 e Rx de 64^n sono 8. Azonti a 5 serano 13, moltipicha' in ssi fano 169, moltipicha' per 10 fano 130 e 39 fano 169.[a]

1°. Quando lo nomero è ingal ale chose dovemo partir lo numero per le chose, e quel che ne vien tanto val la chosa.

2°. Quando i zensi sono ingual al numero dovemo partir li nomeri per le zensi, a[b] la radixe de quel numero val la chosa.

3°. Quando le chose sono ingual a zensi dovemo partir le chose per li zensi, e quello numero tanto val la chosa.

4°. Quando lo numero è ingal ale chose e aali[c] zensi dovemo partir per li zensi, dimezar le chose e moltipicar in si medeximo, poner sovra el numero, e la radixe de quelo ne vien men lo dimezamento dele chose val la cosa.

f. 16b 5. Quando le chose sono ingual al nomero e ali zensi dovemo partir|ne i zensi, demezar le chose, moltipicar in si medeximo, trazene el numero, e la radixe de quelo tratto del dimezamento dele chose, overo la radixe de quelo posto sovra 'l dimezamento dele chose, val la chosa.

6. Quando i zensi sono ingual ale chose e al nomero dovemo partir per li zensi, dimezar le chose, moltipicar in si medeximo, poner sovra al numero, e la radixe de quelo più lo dimezamento dele chose val la chosa.

Quando i chubi sono ingual al numero dovemo partir lo numero per li chobi,[d] e quello ne vegnirà serà radixe chubicha, e tanto varà la chosa.

Quando li chubi sono ingual ale cose dovemo partir le chose per li chubi, e quelli che ne vegnirà serà Rx, e tanto val la chosa.

Quando i chubi sono ingual ai zensi dovemo partir li zensi per li chubi, e quel che ne vegnirà serà numero, e tanto val la chosa.

---

a. *From* sono 8 *to* fano 169 *added successivley, in the same hand but in a different ink, in the space at the end of the line and in the space immediately beneath it.*
b. *Sic in MS.*
c. *Sic in MS.*
d. *Sic in MS.*

added to 5 will be 13, multiplied by itself makes 169, multiplied by 10 makes 130, and that and 39 make 169.

---

1. When the number is equal to the unknowns, we should divide the number by the unknowns, and what results is equal to the unknown.[1]

2. When the squared unknowns are equal to the number, we should divide the numbers by the squared unknowns, and the square root of this number is the value of the unknown.

3. When the unknowns are equal to the squared unknowns, we should divide the unknowns by the squared unknowns, and this number is the value of the unknown.

4. When the number is equal to the unknowns plus the squared unknowns, we should divide by the squared unknowns, take half of the unknowns and multiply the result by itself, put above the number, and the square root of the result minus half of the unknowns is equal to the unknown.

5. When the unknowns are equal to the number plus the squared unknowns, we should divide | by f. 16b
the squared unknowns, halve the unknowns, multiply by itself, subtract from it the number, and the square root of that taken from the half of the unknowns, or rather the square root of that put above the half of the unknowns, is equal to the unknown.

6. When the squared unknowns are equal to the unknowns and the number, we should divide by the squared unknowns, halve the unknowns, multiply by itself, put above the number, and the square root of that plus the half of the unknowns is equal to the unknown.

When the cubes are equal to the number, we should divide the number by the cubes, and what results will be a cube root, and that will be the value of the unknown.

When the cubes are equal to the unknowns, we should divide the unknowns by the cubes, and what results will be a square root, and that is the value of the unknown.

When the cubes are equal to the squared unknowns, we should divide the squared unknowns by the cubes, and what results will be a number, and that is the value of the unknown.

---

1. For the mathematical notation of these equations and corrections of some errors, see Franci, "Mathematics in the Manuscript of Michael of Rhodes," vol. 3, p. 129.

Quando li zensi d'i zensi sono ingual al numero dovemo partir lo numero per li zensi d'i zensi, e quelo che ne vegnirà serà radixe di radixe, e tanto val la chosa.

Quando li zensi d'i zensi sono ingual ale chose dove[*mo*] partir le chose per li zensi d'i zensi, e quelo che ne vegnirà serà Rx chubicha, e tanto val la chosa.

Quando li zensi d'i zensi sono ingual ali zensi dovemo partir li zensi per li zensi d'i zensi, e quel che ne vien serà radixe, e tanto val la chosa.

Quando li zensi d'i zensi sono ingual a chubi dovemo partir li chobi per lli zensi d'i zensi, e quel che ne vegnirà serà numero, e tanto vual[a] la chosa.

f. 17a Quando lo zenso del chubo xe ingual al numero dovemo partir lo nomero | per lo zenso del chubo, e quelo che ne vien serà Rx relatta, e tanto vual la chosa.

8.[b] Quando lo zenso del chubo xe ingual ale chose dovemo partir le chose per lo zenso del chubo, e quello che ne vien serà radixe di radixe, e tanto val la chosa.

9. Quando lo zenso del chubo xe ingual al zenso dovemo partir lo zenso per lo zenso del chubo, e quel che ne vien serà Rx chubicha, e tanto val la chosa.

10. Quando lo zenso del chubo è ingual al chubo dovemo partir lo chubo per lo zenso del chubo, e quel che ne vien serà radixe, e tanto val la chosa.

11. Quando li zensi d'i chubi sono ingualli a zensi d'i zensi dovemo partir li zensi d'i zensi per li zensi d'i chubi, e quello che ne vien serà numero, e tanto val la chosa.

12. Quando i zensi sono inguali a zensi d'i zensi e a chubi dovemo partir per li zensi d'i zensi e demezar li chubi e moltipicar per si medeximo, poner sopra li zensi, e la Rx de quello meno lo dimezamento d'i chubi val la chosa.

13. Quando li chubi sono inguali a zensi d'i zenssi e a zenssi dovemo partir per li zensi d'i zensi e dimezar li chubi, moltipicar per si medeximo, trazende li zensi, e la radixe de quello posto sovra 'l dimezamento d'i chobi overo la Rx de quelo tratto del dimezamento d'i chubi val la chosa.

---

a. *Sic in MS (and likewise in the following sentence).*

b. *The marginal numbers on this page, in a different ink from the text, would appear to continue the paragraph listing on the previous page, but those are unnumbered, and there are 8 paragraphs there.*

When the squares of the squared unknowns are equal to the number, we should divide the number by the squares of the squares, and what results will be a square root of a square root, and that is the value of the unknown.

When the squares of the squared unknowns are equal to the unknowns, we should divide the unknowns by the squares of the squared unknowns, and what results will be a cube root, and that is the value of the unknown.

When the squares of the squared unknowns are equal to the squared unknowns, we should divide the squared unknowns by the squares of the squared unknowns, and what results will be square roots, and that is the value of the unknown.

When the squares of the squared unknowns are equal to cubes, we should divide the cubes by the squares of the squared unknowns, and what results will be a number, and that is the value of the unknown.

When the square of the cube is equal to the number, we should divide the number | by the square of the cube, and what results will be a square root, and that is the value of the unknown. f. 17a

8. When the square of the cube is equal to the unknowns, we should divide the unknowns by the square of the cube, and what results will be a square root of the square root, and that is the value of the unknown.

9. When the square of the cube is equal to the squared unknown, we should divide the squared unknown by the square of the cube, and what results will be a cube root, and that is the value of the unknown.

10. When the square of the cube is equal to the cube, we should divide the cube by the square of the cube, and what results will be a square root, and that is the value of the unknown.

11. When the squares of the cubes are equal to squares of the squared unknowns, we should divide the squares of the squared unknowns by the squares of the cubes, and what results will be a number, and that will be the value of the unknown.

12. When the squared unknowns are equal to squares of the squared unknowns plus cubes, we should divide by the squares of the squared unknowns and halve the cubes and multiply them by themselves, put above the squared unknowns, and the square root of this minus half of the cubes is equal to the unknown.

13. When the cubes are equal to squares of the squared unknowns plus squares, we should divide by the squares of the squared unknowns and halve the cubes, multiply by itself, subtract from the squared unknowns, and the square root of this put over the half of the cubes or rather the square root of this taken from the half of the cubes is equal to the unknown.

14. Quando li zensi d'i zensi sono ingual a chubi e a zensi dovemo partir per li zensi d'i zensi e dimezar li chubi, moltipicar in si medeximo, poner sopra[a] li zensi, e la radixe de quela più el
f. 17b dimezamen|to dei chubi val la chosa.

Quando zensi d'i zensi d'i zensi sono ingual al nomero se vuol partir nei zensi d'i zensi d'i zensi e dimezar li zensi, moltipichar per se medeximo e giunger chon lo numero, e la radixe dela giugione meno lo dimezamento d'i zensi val el zenso, doncha val la chosa, e la Rx dela diferenzia ch'è dal dimezamento d'i zensi infino ala Rx dela giuzione sono fate.

Quando la chosa, el zenso, el chubo sono inguali al nomero recha a uno chubo, poi parti i zensi per li chubi e quello che ne vien recha a Rx achubicha[b] e zonzi chol numero, e Rx chubicha di quelo che ne viene val la chosa.

Quando li zensi d'i zensi d'i zensi e zensi sono inguali al numero senpre debi rechar a uno zensso di zenso e dimezar[c] li zensi e moltipicar in si medeximo e zonzer sopra 'l numero, e di' che la chosa val la radixe dela diferenzia che dè dal dimezamento d'i zensi ala Rx di quello, e el zenso val la diferenzia che dè dal dimezamento d'i zensi ala radixe de quelo.

[CHOSE, ZENSI, CHUBI, RADIXE]

Notta che se tu vuol partir chosa i[*n*] numero si vien chosa e chosì a partir semegiantemente zenso e numero sì vien zenso, unde parti zò che tu voi in numero, ne vien quello medeximo ch'è di inanzi al numero, sichomo seria a dir parti $4^{\square}$ in 2^{n}, ne vien $2^{\square}$. E sapi che se vien a partir chosa in chosa ne vien numero, sichomo seria a dir parti 4^{co} in 2^{co} ne vien 2^{n}. La raxion sie questa che a moltipicar 2^{n} via 2^{co} fa 4^{co}. E a partir zenso in chosa ne vien chosa, chomo seria a dir partime $6^{\square}$ in 2^{co} ne
f. 18a vien 3^{co}. Anchora a partir | chubo in chosa ne vien zenso. Ponamo a partir 4 chubi in 2^{co}, sì vien $2^{\square}$. Anchora partir Rx de Rx in chosa vien chubo. Ponamo che tu voi partir Rx de Rx in 3^{co}, ne vien 2 chubi. Anchora sapi che a partir zenso in zenso ne vien chosa. Ponamo che tu voi partir $4^{\square}$ in $4^{\square}$ ne vien 1^{co}. Anchora a partir chubo in zenso ne vien chosa, chomo seria a dir 4 chubi in $2^{\square}$ ne vien

---

a. el nomero *crossed out with a horizontal line.*
b. *Sic in MS.*
c. le chose *crossed out with a horizontal line.*
1. The sense of this passage is unclear.
2. I.e., $4x$ cubed.
3. The expression used in the MS is *Rx*, which stands for *radixe* or root, but is here and elsewhere also used to mean square. This paragraph from here on is confused and has errors.
4. This is incorrect.

14. When the squares of the squared unknowns are equal to cubes plus squared unknowns, we should divide by the squares of the squared unknowns and halve the cubes, multiply by itself, put over the squared unknowns, and the square root of this plus the half | of the cubes is the value of the unknown. f. 17b

When the squares of the squares of the squared unknowns are equal to the number, one should divide into the squares of the squares of the squared unknowns and halve the squared unknowns, multiply by itself and add to the number, and the square root of the addition minus the half of the squared unknowns is equal to the squared unknown, so equal to the unknown, and the square root of the difference that there is from the half of the squared unknowns up to the square root of the addition are done.[1]

When the unknown, the square of the unknown, and the cube are equal to the number, bring to a cube, then divide the squared unknowns by the cubes and what results bring to the cube root and add to the number, and the cube root of the result is the value of the unknown.

When the squares of the squares of the squared numbers and the squared numbers are equal to a number, you should always bring to one square of a squared unknown and halve the squared unknowns and multiply by themselves and add to the number and say that the unknown has the value of the square root of the difference between the half of the squared unknowns and the square root of this, and the squared unknown is equal to the difference of the half of the squared unknown and its square root.

UNKNOWNS, SQUARED UNKNOWNS, CUBES, ROOTS

Note that if you want to divide an unknown by a number, the result is an unknown, and so to divide similarly a squared unknown by a number the result is a squared unknown, so divide what you want by the number, there results the same number which is equal to the number, as it would be to say divide $4x^2$ by 2, there results $2x^2$, and know that if it comes to dividing an unknown by an unknown the result is a number, as if it would be to say that divide $4x$ by $2x$ results in 2. The reason is that multiplying 2 by $2x$ makes $4x$. And to divide a squared unknown by an unknown results in an unknown, so this would be like saying divide for me $6x^2$ by $2x$, which comes to $3x$. Also to divide | a cube by an unknown results in a squared unknown. Let's try to divide 4 cubed[2] by $2x$; the result is $2x^2$. Then to divide a square[3] of a square by an unknown results in a cube. Let's say that you want to divide the square of the square by $3x$, the result is $2x$ cubed.[4] Also know that to divide a squared unknown by squared unknown results in an unknown.[5] Suppose that you want to divide $4x^2$ by $4x^2$, the result is $1x$.[6] Also to divide a cube by a squared unknown results in an unknown, as it would be to say $4x$ cubed divided by $2x^2$ results in $2x$. Also to divide the square of a f. 18a

5. This is incorrect.
6. The result is actually 1.

2^{co}. Anchora a partir Rx de Rx in zenso ne vien zenso, però che lo zenso sie la Rx de Rx de Rx. Mo' ponamo che tu voi partir Rx 4 de Rx in 2^{co}, ne vien $2^{\square}$. E queste sono le partizion che se fano nel'alzibra di sopra ditti nomi, com'è ditto per ordine.

[Chapitulli 18 d'alzibran]

Questi sono li 18 chapitoli trati de 6 chapitulli del'alzebran.

Primo chapitolo sie	Rx de Rx	ingual a chubi
segondo	Rx de Rx	ingual a zensi
terzo	Rx de Rx	ingual a chosa
quarto	Rx de Rx	ingual a numero
quinto	chubi	ingual a chosa
sesto	chubi	ingual a numero
settimo	chubi e zensi	ingual a chosa
octavo	chubi	ingual a zensi e chose
nono	chubi e chose	ingual a zensi
dezimo	Rx de Rx e zenso	ingual a chubo
undezimo	Rx de Rx e chubo	ingual a zenso
dodezimo	Rx de Rx e zensi	ingual a chubi
tredexeximo	Rx de zensi	ingual a chubi
f. 18b XIIII°	zensi	ingual a radixe di zensi
XV°	zensi	ingual a radixe de chose
XVI°	radixe de radixe e zensi	ingual a numero
XVII°	radixe de numero	ingual a chosa
XVIII°	zensi e numero	ingual a radixe de radixe

1. The text says "to divide the Rx of the Rx by a squared unknown results in a squared unknown, because the squared unknown is the Rx of the Rx of the Rx."
2. In the text the *4* is written above the *Rx*, perhaps as an exponent, but neither x^4 nor x^{-4} makes the problem work out.

squared unknown by a squared unknown results in a squared unknown, because the squared unknown is the root of the square of the square.[1] Now let's suppose that you want to divide the fourth root of the root by $2x$, the result is $2x^2$.[2] And these are the divisions that are made in algebra of the name given above, as set forth in order.

THE 18 CHAPTERS OF ALGEBRA

These are the 18 chapters taken from the 6 chapters of algebra.[3]

The first chapter is	squares of squared unknowns	equal to cubes	
second	squares of squared unknowns	equal to squared unknowns	
third	squares of squared unknowns	equal to an unknown	
fourth	squares of squared unknowns	equal to a number	
fifth	cubed unknowns	equal to an unknown	
sixth	cubed unknowns	equal to a number	
seventh	cubed unknowns and squared unknowns	equal to an unknown	
eighth	cubed unknowns	equal to squared unknowns and unknowns	
ninth	cubed unknowns and unknowns	equal to squared unknowns	
tenth	squares of squared unknowns and squared unknown	equal to a cube	
eleventh	squares of squared unknowns and cubed unknown	equal to squared unknown	
twelfth	squares of squared unknowns and squared unknowns	equal to cubes	
thirteenth	squares of squared unknowns	equal to cubes	
XIV	squared unknowns	equal to the squares of squared unknowns	f. 18b
XV	squared unknowns	equal to the squares of unknowns	
XVI	squares of squared unknowns and squared unknowns	equal to a number	
XVII	square of a number	equal to an unknown	
XVIII	squared unknowns and number	equal to the square of a squared unknown	

---

3. For the mathematical notation of these equations, see Franci, "Mathematics in the Manuscript of Michael of Rhodes," vol. 3, pp. 130–131.

[Veder zò che vuol dir chapitulli d'alzebran]

Dapoi che tu ai vezudo li sopraditti chapituli mo' te voio declarar chomo e zò che voglono dire. Sapi che in questi chapituli si dichono nomi chosì fatti, chose, zensi, chubi, numeri e radixe de radixe, perzò sapi che te scriverò da mo' avanti più breve in questi nomi, che io scriverò per figure de chosa una letera sì fatta "C," che vuol dir chosa. E per figure de zensi io scriverò una "Ç," che vuol dir zenso; e per figura de chubo scriverò "chu," che vuol dir chubo. E per figura de nomero scriverò "N," che vuol dir numero. E per radixe de radixe[a] scriverò "Rx," che vuol dir radixe. E quando io vorò dir "de radixe," io farò chosì "de Rx," che vuol dir de radixe, e si io vorò dir radixe de radixe, io scriverò "Rx de Rx." Le quali figure lievano tuti questi nomin,[b] inperzò tiente ben a mente.

Questo è lo muodo chomo se moltipicha questi nomi. Sapi che numero via chosa si fa chosa, e numero via zenso si fa zenso, e numero via chubo si fa chubo, sì che a moltipicar numero in che tu voi farà quello medeximo che tu moltipichi con lo numero.

Pony che tu voi moltipicar 3^{n} via 5^{co}, zoè $\frac{3}{N}$ via $\frac{5}{C}$. Moltipicha 3, ch'è de sovra lo numero, per 5, ch'è sovra C, monta $\frac{15}{C}$.

f. 19a E se tu voi moltipichar 4^{n} via $\frac{3}{C}$, si debi far chomo ò ditto de sora, che monta $\frac{12}{C}$, e chosì fa le simile raxion.

Mo' sapi che chosa via chosa fa zenso, sì che una chosa sie radixe de zensso, ancuor 1^{co} via zenso si fa chubo.

Chosa via chubo fa Rx de Rx. Ponamo che uno te dixesse: "3^{co} via 3^{co} che monta?" Tu die dir: "$\frac{9}{\text{Ç}}$." Anchora chi te dixesse: "$\frac{3}{C}$ via $\frac{4}{\text{Ç}}$ che monta?" Respondi: "El fa $\frac{12}{chu}$." Anchora chi te dixese: "$\frac{4}{C}$ via $\frac{5}{chu}$," fa 20 Rx de Rx chubicha. Mo' sapi che Ç via Ç fa Rx de Rx e zenso via numero fa Ç. Mo' tiente ben a mente che'l te serà molto utelle queste moltipichazion predditte.

Chi te dixese: "Che xe $\frac{5}{C}$ e $\frac{6}{N}$ moltipicha' via $\frac{5}{C}$ e $\frac{6}{N}$?" Tu die far chosì. Moltipicha in primo $\frac{5}{C}$ via $\frac{5}{C}$, che monta $\frac{25}{\text{Ç}}$, e puo' die far $\frac{5}{C}$ via $\frac{6}{N}$, che fa 30^{co}. E anchora diestu moltipichar da chavo $\frac{5}{C}$ via $\frac{6}{N}$, che monta 30^{co}, et averai 2 fiade 30^{co}, che fano 60^{co}, e puo' moltipicha $\frac{6}{N}$ via $\frac{6}{N}$, che faxe 36 numeri, sì che monta tutto $\frac{25}{\text{Ç}}\ \frac{60}{C}\ \frac{36}{N}$. E die far questa raxion e lo simille a questo per lo muodo che xe scritto qua de sovra e desegnado qua de ssotto.

---

a. *Thus repeated in MS.*
b. *Sic in MS.*

To see what the chapters of algebra mean

After you have seen the chapters above, now I want to demonstrate to you how and what they mean. Know that in these chapters are used the following words: unknowns, squared unknowns, cubes, numbers, and squares of squares, but know that to represent unknowns, I will write from now on the abbreviations of these words, that I will write a letter made like "C," and that indicates "unknown"; and for the form of squared unknowns I will write a "Ç," which means "squared unknown"; and for the figure of a cubed unknown I will write "chu," which means "cube"; and for the figure of a number, I will write "N," which means "number"; and for the square of a squared unknown, I will write "Rx," which means "root." And when I want to say "of the square," I will do this: "of Rx," which means "of the root"; and if I want to say the "square of a squared unknown," I will write "Rx of Rx." These figures bear all of these names; therefore keep them well in mind.

Here is the method to multiply these terms. Know that a number times an unknown makes an unknown, and a number times a squared unknown makes a squared unknown, and a number times a cubed unknown makes a cubed unknown, so that to multiply a number by whatever you want will be the same as when you multiply with the number.

Suppose that you want to multiply 3 by $5x$, that is 3^{n} by 5^{co}. Multiply 3, which has a superscript "n," by 5, which has a superscript "co," the result it 15^{co}.

And if you want to multiply 4 by $3x$, you should do as I have said above, which comes to $12x$, and so go the similar problems. f. 19a

Now know that an unknown times an unknown makes a squared unknown, as an unknown is the square root of a squared unknown; also an unknown times a squared unknown makes a cubed unknown.

An unknown times a cubed unknown makes the square of a squared unknown. Let's suppose that someone said to you: "$3x$ times $3x$, what does that amount to?" You should say $9x^2$. Again, if someone said to you: "$3x$ times $4x^2$, what does it amount to?" Respond: "It makes $12x^3$." Again, if someone said to you: "$4x$ times $5x^3$," it makes 20 times the square of the squared unknown cubic. Now know that x^2 times x^2 makes the square of a squared unknown, and a squared unknown times a number makes x^2. Now keep well in mind that these multiplications above will be very useful to you.

Someone asks you: "What is $5x$ plus 6 multiplied by $5x$ plus 6?" You should do it this way. First multiply $5x$ by $5x$, which makes $25x^2$, and then you should do $5x$ by 6, which makes $30x$. And then you should again multiply $5x$ times 6, which amounts to $30x$, and you will have 2 times $30x$ which makes $60x$, and then multiply 6 times 6, which makes 36, so that in all it comes to $25x^2$ $60x$ 36. And you should do this problem and ones similar to it by the method given above and put symbolically as below:

$\frac{5}{C}$ e $\frac{6}{N}$	$25^{\square}$	insuma $25^{\square}$ 60^{co} 36^{n}
	30^{co}	
$\frac{5}{C}$ e $\frac{6}{N}$	30^{co}	
	36^{n}	

Che fa $\frac{3}{C}$ e $\frac{4}{N}$ via $\frac{2}{C}$? Questo è lo muodo tu die moltipichar. In primo $\frac{2}{C}$ via $\frac{3}{C}$ fa $\frac{6}{Ç}$, e puoi moltipicar $\frac{2}{C}$ via $\frac{4}{N}$, fa $\frac{8}{C}$, sì che questa raxion monta $\frac{6}{Ç}$ e $\frac{8}{C}$.

Che fa $\frac{2}{C}$ men $\frac{4}{N}$ via $\frac{3}{C}$? Tu die in primo moltipichar $\frac{3}{C}$ via $\frac{2}{C}$, fa $\frac{6}{Ç}$, e poi moltipicha $\frac{3}{C}$ via $\frac{4}{N}$, fa $\frac{12}{C}$, e xe men, sì che tu dirai che fa $\frac{6}{Ç}$ men $\frac{12}{C}$. $\frac{2}{C}$ men $\frac{4}{N}$ via $\frac{3}{C}$ val $\frac{6}{Ç}$ men $\frac{12}{C}$.

f. 19b Che fa $\frac{4}{C}$ men $\frac{5}{N}$ via $\frac{2}{C}$ men $\frac{3}{N}$? Moltipicha in primo $\frac{4}{C}$ via $\frac{2}{C}$, fa $\frac{8}{Ç}$, e puo' moltipicha $\frac{3}{N}$ via $\frac{5}{N}$, fa $\frac{15}{N}$, e sono numeri plui, e puo' moltipicha $\frac{3}{N}$ via $\frac{4}{C}$, fa $\frac{12}{C}$, e xe men, moltipicha $\frac{5}{N}$ via $\frac{2}{C}$, fa $\frac{10}{C}$, e xe men, sì che monta tutto $\frac{8}{Ç}$ p[*i*]ù $\frac{15}{N}$ men $\frac{22}{C}$ segondo che par desegnado qui de sotto.

più 4^{co}	men	5^{n}	$8^{\square}$	zensi più	Son 8 zensi più 15 numeri men 22 chosse.
più 2^{co}	men	3^{n}	15^{n}	nomeri più	
			12^{co}	chose men	
			10^{co}	chose men	

[LIGAR ARZENTTI, FROMENTI, VINY]

E voio far una liga d'arzenti, volemo oro, o volemo biave, o volemo viny, o volemo ogi, che se posa ligar insieme. Ponamo per exenpio. Fa che tti aligi biave 5 a far farine. In prima el fromento val el ster soldi 58, l'orzo val el ster soldi 45, el meyo val el ster soldi 40, el surgo val el ster soldi 35 e la segala val el ster soldi 19. Voio tuor de queste a far farina che vagia el ster, mesidado tute insieme, el ster soldi 28. Quanto fromento, e quanto orzo, e quanto meyo, e quanto surgo, e quanta segalla voio de chadauna biava che mesidadi insieme vaia el ster soldi 28? Questo è lo muodo che tu debi far. In primo meterì in una posta soldi 58, e al'altra soldi 45, e al'oltra soldi 40, e al'oltra soldi 35, e al'oltra soldi 19, per lo muodo vedirì qui de sutto o indriedo, e chomenzarì a ligar 58 de fromento chon 19 che val la segalla, perché lo mazuor chon lo menor, e dirai: "Che diferenzia serà 28 de 58?" Serà 30, e quelli metti sovra de 19. E po' dirai: "Che diferenzia de 19 a 28?" Serà

$5x$ and 6	$25x^2$	in sum $25x^2$ $60x$ 36
	$30x$	
$5x$ and 6	$30x$	
	36	

What are $3x + 4$ times $2x$? Here is the method by which you should multiply. First, $2x$ times $3x$ makes $6x^2$, and then multiply $2x$ times 4 makes $8x$, so that this problem results in $6x^2 + 8x$.

What does $2x$ minus 4 times $3x$ make? You should first multiply $3x$ times $2x$, which makes $6x^2$, and then multiply $3x$ times 4, which makes $12x$, and it is minus, so that you will say that it makes $6x^2$ minus $12x$. $2x$ minus 4 times $3x$ equals $6x^2$ minus $12x$.

What does $4x$ minus 5 times $2x$ minus 3 make? Multiply first $4x$ times $2x$, it makes $8x^2$, and then f. 19b
multiply 3 times 5, it makes 15, and these are plus numbers, and then multiply 3 times $4x$, it makes $12x$, and it's minus, multiply 5 times $2x$, it makes $10x$, and it's minus, so it amounts in all to $8x^2$ plus 15 minus $22x$, as is shown below.

plus $4x$	╳	minus 5	$8x^2$	squared unknowns plus	It is 8 squared unknowns plus
plus $2x$		minus 3	15	numbers plus	15 numbers minus 22
			$12x$	unknowns minus	unknowns.
			$10x$	unknowns minus	

To mix[1] silver, grains, wines

And I want to make mixture of silver, we want gold, or we want grain, or we want wine, or we want oils that we want to mix together. Let's take an example. Suppose that you mix 5 grains to make flour. First, the wheat costs 58 soldi per bushel,[2] the barley is worth 45 soldi per bushel, the millet is worth 40 soldi per bushel, the sorghum is worth 35 soldi per bushel, and the rye is worth 19 soldi per bushel. I want to make from these a flour that is worth, mixing all together, 28 soldi per bushel. How much wheat, how much barley, how much millet, how much sorghum, and how much rye do I want of each grain so that when mixed together it is worth 28 soldi per bushel? This is the way that you should do it. First you will put in one place 58 soldi, and in the other 45 soldi, and in the other 40 soldi, and in the other 35 soldi, and in the other 19 soldi, as you will see below or opposite, and you will begin to mix 58 of wheat with the 19 which the rye is worth, because the most with the least, and you will say: "What difference will 28 be from 58?" It will be 30, and put this over 19. And then you will say: "What is the difference of 19 from 28?" It will be 9, and put this over 58.

1. The term used literally means "alloy."
2. The *ster* was about $2\frac{1}{3}$ modern bushels.

f. 20a 9, e quelli metti de sovra 58. E po' dirai: | "Che diferenzia da 45 a 28?" Serà 17, e questi metti sovra 19. E la diferenzia de 19 a 28 serà 9, e questi metti sovra 45 che val l'orzo. Po' dirai: "Che diferenzia da 40 a 28?" Serave 12, e questi metti sovra 19. E po' dirai: "Che diferenzia da 19 a 28?" Serave 9, e questi metti ssovra a 40 che val el meyo. Po' dirai: "Che diferenzia da 35 a 28?" Serà 7,[a] e questi metti sovra a 19.[b] Po' asuma tutto quello ch'è de sovra, che serà 66, e ttanti stera vuol de segalla, e quel ch'è de sovra 58, che son 9, e quel del'orzo ch'è de sovra, che xe 9, e quello ch'è de sovra lo meyo, che xe 9, e quello ch'è de sovra l'orzo, che xe 9, asumadi insenbre fano stera 102 del'una parte, e del'oltra, zoè stera 9 de fromenti, stera 9 d'orzo, stera 9 de meyo, stera 9 de surgo e stera 66 de segalla farà farina mesidada del valor de soldi 28 el ster, chomo avemo ditto de sovra. E per exenpio vui vedirì qui de ssutto desegnado e per questa farì ogni liga vollè far d'ogni raxion.

58	45	40	35	19				
12	fromenti	stera 9						7
17	orzo	stera 9						12
30	meyo	stera 9						17
7	surgo	stera 9	9	9	9	9		30
9	segala	stera 66	58	45	40	35		19
9								
9								
9								
suma stera 102					28			

58	45	40	35
9	9	9	9
522			
405	66		
360	19		
315	1254		
1254			
suma soldi 2856			

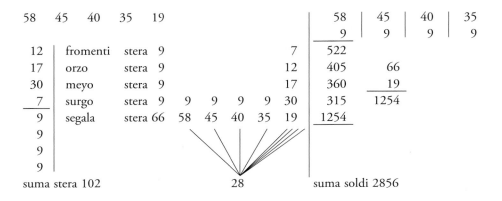

Stera[c] 102 per soldi 28 el ster die venir soldi 2856, e se non, serà falsa.

102	
28	soldi 2856
816	
204	
stera 102	28 |56

f. 20b Sì che adoncha stera 102 per soldi 28 el ster vien soldi 2856, e per lo simille mo|do segala stera 66 per soldi 19, e fromenti stera 9 per soldi 58, e orzo stera 9 per soldi 45, e meyo stera 9 per soldi 40, e sorgo stera 9 per soldi 35, monta in tutto soldi 2856. E questa raxion è fatta.

---

a. 7 *corrected above the line over* 12 *crossed out with a diagonal line.*

b. e da 19 a 28 serave 9, e questi metti sovra sovra 35 che val el sorgo. Po' dirai: "Che diferenzia da 30 a 28?" *crossed out with a horizontal line.*

c. Stera 108 per soldi 102 *crossed out with a horizontal line.*

And then you will say: | "What's the difference from 45 to 28?" It will be 17, and put this over 19. And the difference from 19 to 28 will be 9, and put this over 45, which is the value of the barley. Then you will say: "What's the difference from 40 to 28?" It will be 12, and put this over 19. And then you will say: "What's the difference from 19 to 28?" It will be 9, and put this over 40, which is the value of the millet. Then you will say: "What's the difference from 35 to 28?" It will be 7, and put this over 19.[1] Then add all that which is above, which will be 66, and that is the number of bushels of rye, and that which is above 58, which is 9, and that of the barley which is above, which is 9, and that which is above the millet, which is 9, and that which is above the barley, which is 9, added together make 102 bushels on one side, and on the other; that is, 9 bushels of wheat, 9 bushels of barley, 9 bushels of millet, 9 bushels of sorghum, and 66 bushels of rye will make a mixed flour worth 28 soldi per bushel, as we have said above. And as an example you will see this diagrammed below, and in this way you will make all mixtures that you wish in any problem. f. 20a

58 45 40 35 19

12	wheat	9 bushels
17	barley	9 bushels
30	millet	9 bushels
7	sorghum	9 bushels
9	rye	66 bushels
9		
9		
9		

sum 102 bushels

				7
				12
				17
9	9	9	9	30
58	45	40	35	19

28

58	45	40	35
9	9	9	9
522			
405	66		
360	19		
315	1254		
1254			

sum 2856 soldi

102 bushels for 28 soldi a bushel comes to 2,856 soldi, and if not, it will be wrong.

102	
28	2856 soldi
816	
204	

102 bushels 28 |56

So that 102 bushels for 28 soldi the bushel comes to 2,856 soldi, and in a similar way | 66 bushels of rye for 19 soldi, and 9 bushels of wheat for 58 soldi, and 9 bushels of barley for 45 soldi, and 9 bushels of millet for 40 soldi, and 9 bushels of sorghum for 35 soldi, it adds up in all to 2,856 soldi. And this problem is done. f. 20b

1. The following passage crossed out in MS: *And from 19 to 28 will be 9, and put this above 35 which is the value of the sorghum. Then you will say: "What is the difference from 30 to 28?"*

[Barattar merchadantie]

Son 2 homeny chi vuol baratar, l'uno a seda, chi val a chontadi la libra soldi $11\frac{3}{11}$, e mesela a baratto non so quanto, e vuol $\frac{1}{3}$ de deneri. L'oltro a pano, che val el brazo a chontadi soldi 60, e meselo a barato soldi 66. Adomando zò che la valse la seda a baratto. E prima per farla per la riegola del 3.

Per farla per la riegola del 3 tu die azonzer a quel che mese a barato quel del pano, che fu 66, la mittade, che serano 99, e la chaxion perché quel dela seda domanda $\frac{1}{3}$ in denari, perzò se azonze la mittade del 66, perché serano el terzo. E puo' azonzi al 60 33, serano 93. E diremo: "Se 93 torna 99, che tornerà $11\frac{3}{11}$ che xe lo prexio dela seda?" E chomenza a moltipicar, e diremo: "Se 93 me torna 99, che me tornerà $11\frac{3}{11}$?" Se moltipicho 11 via 11 fa 121, azonzer 3 fa 124 undexeximi, e diremo: "Se $\frac{93}{1}$ torna $\frac{99}{1}$, che tornerà $\frac{124}{11}$?" Serà a moltipichar 11 via 93, fa 1023, e questo è lo partidor, puo' moltipicha 99 via 124, fa 12276, e questi parti per 1023, insirà fuora soldi 12, e questo se die metter a baratto la libra de la seda, chomo vedirì qui de sutto per exenpio.

$\frac{93}{1}$	$\frac{99}{1}$	$11\frac{3}{11}$	11	$\frac{93}{1}$	$\frac{99}{1}$	$\frac{124}{11}$	93	1023	124
			11				11		99
							93		1116
			124				93		1116
			11						12276

0	1 | 0	5
000~~4~~0	5 | 5	5
~~12276~~		
~~10233~~	12	
~~102~~		

f. 21a El baratto indriedo fu fatta per la riegola del 3, e mo' te voio mostrar a farla per inpoxizion. Chomo dixe, la seda valse la libra soldi $11\frac{3}{11}$, e mesela a baratto non so quanto, el pano valse el brazo soldi 60, e mesela a bara' soldi 66, ma cholui dela seda vuol $\frac{1}{3}$ i denari.

Per farla per inpoxizion metemo aver messo 14. Diremo perché quel dela seda domandò el $\frac{1}{3}$ a cholui d'i pany, azonzemo la mittade de 66, serà 33, serano 99. Azonzi al 60 33, serano 93, e dixemo: "Se 93 torna 99, che serave $11\frac{3}{11}$?" Serave 12276. Parti per 11, serano 1116, e si moltipicho 14 che ò posto al'inpoxizion chon 93 serano 1302, e my voio che sia 1116. Adoncha sutrazi 1116 da 1302, serano più 186, e questi metti più ala prima inpoxizion. E per far la segonda inpoxizion

To barter merchandise

There are 2 men who want to barter, one has silk, which is worth $11\frac{3}{11}$ soldi in coins per pound, and I don't know how much he put into the barter, and he wants $\frac{1}{3}$ of the cash. The other has cloth, which is worth 60 soldi in coins per ell, and he put 66 soldi into the barter. I ask what the silk is worth in the barter. And first here's how to do it by the rule of 3.

To do it by the rule of 3, you should add to what the one with the cloth put into the barter, which is 66, its half, which will be 99, and the reason for this is because the one with the silk wants $\frac{1}{3}$ in coins, so you add the half of 66, which will be the third. And then add 33 to the 60, it will be 93. And we will say: "If 93 results in 99, what will $11\frac{3}{11}$, which is the price of the silk, result in?" And begin to multiply, and we will say: "If 93 brings me 99, what will $11\frac{3}{11}$ bring me?" If I multiply 11 by 11 it makes 121, add 3 makes 124 elevenths, and we will say: "If $\frac{93}{1}$ results in $\frac{99}{1}$, what will $\frac{124}{11}$ result in?" The next step will be to multiply 11 by 93, which makes 1,023, and this is the divisor, then multiply 99 by 124, it makes 12,276, and divide this by 1,023, it will result in 12 soldi, and this is what is put into barter for the pound of cloth, as you will see worked out below.

$\frac{93}{1}$ $\frac{99}{1}$ $11\frac{3}{11}$	11 11 ——— 124 ——— 11	$\frac{93}{1}$ $\frac{99}{1}$ $\frac{124}{11}$	93 11 ——— 93 93 ———	1023	124 99 ——— 1116 1116 ——— 12276

0 00040 12276 10233 102	12	1 | 0 / 5 | 5 5 / 5

The barter on the last page was done by the rule of 3, and now I want to show you how to do it by false position. As I have said, the silk was worth $11\frac{3}{11}$ soldi per pound, and I don't know how much is put into the barter, the cloth was worth 60 soldi per ell, and he put into the barter 66 soldi, but the one with the silk wants $\frac{1}{3}$ of the cash. f. 21a

To do this by false position we suppose that he put in 14. We will say because the one with the silk asks for $\frac{1}{3}$ from the one with the cloth, we add the half of 66, which will be 33, which will be 99. Add 33 to 60, it will be 93, and we say: "If 93 returns 99, what will $11\frac{3}{11}$ be?" It will be 12,276. Divide by 11, it will be 1,116, and if I multiply the 14 that I put in the false position with 93 it will be 1,302, and what I want is that it come out to 1,116. Then subtract 1,116 from 1,302, it will be plus 186, and put this as a plus at the first false position. And to do the second false position

pony che fuse 13, e diremo: "Se 93 fose 99, che serave $11\frac{3}{11}$?" Serano 1116. E puo' moltipicha 13 fia 93, fa 1209, et io voria che fusse 1116, adoncha è più 93. E se giremo l'inpoxizion e moltipichemo in croxe, e diremo: "186 via 13 fano 2418," puo' moltipicha 93 via 14, fano 1302, sotrazi 1302 de 2418, roman 1116. E per far partidor sutrazi 93 da 186, roman 93, parti 1116 per 93, insirano soldi 12, e tanto vuol eser messo la libra dela seda a baratto voiando $\frac{1}{3}$ in denari, chomo vedirì per exempio qui de sotto.

14 ╲ ╱ 13	93	186	2418	
più ╳ più	14	13	1302	
186 ╱ ╲ 93	372	558	1116	
93	93	186		
93 partidor	1302	2418		

0
1̸0
02̸8̸0
1̸1̸1̸6̸] 12
9̸3̸3̸
9̸

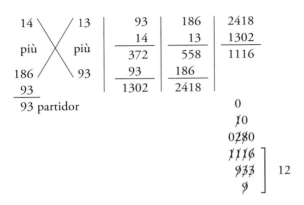

E questo de sovra è fatto per inpoxizion chomo ditto indriedo per la riegola del 3.

f. 21b E per far la ditta raxion per la chossa, zoè la chontrascritta per inpoxizion, el barattò l'uno a seda, e mesela a barato non so quanto, e vuol $\frac{1}{3}$ in denari, e val soldi $11\frac{3}{11}$ la libra, e l'oltro a pano, che val soldi 60 el brazo, e mesela a baratto soldi 66. Zò che se die meter la seda a baratto per la chosa.

Per far la ditta raxion per la chosa prima tu die trar de $11\frac{3}{11}$ che val a chontadi $\frac{1}{3}^{co}$. E perché tu à meso che sia $11\frac{3}{11}$ e 1^{co}, tratto de 1^{co} el $\frac{1}{3}$ te roman $\frac{2}{3}^{co}$, e tra' de $11\frac{3}{11}$ $\frac{1}{3}^{co}$, roman $11\frac{3}{11}$ men $\frac{1}{3}^{co}$. Tu die moltipicar per lo prexio che valse i pany in denari, che fu soldi 60 el brazo, e dir: "$\frac{2}{3}$ via 60 fa $\frac{120}{3}^{co}$." E se tu dixesti: "Se $11\frac{3}{11}$ fusse $\frac{2}{3}^{co}$, che serave 60?" noi savemo che a baratto fu messo 66, adoncha moltipicha 66 via $11\frac{3}{11}$, fa $\frac{8184}{11}$, e questo se die partir. Puo' moltipicha 66 via men $\frac{1}{3}^{co}$, fa 22^{co}. Adequa le parte, azonzi $\frac{120}{3}^{co}$ chon 22^{co}, serave 62^{co}, e questo è lo vostro partidor. Parti $\frac{8184}{11}$ per $\frac{62}{1}$, venyrà 12, e questo se die meter la seda a baratto, chomo vedirì qui de sutto per exenpio.

suppose that it was 13, and we will say: "If 93 was 99, what would $11\frac{3}{11}$ be?" It would be 1,116. And then multiply 13 by 93, which makes 1,209, and I would like it to have been 1,116, so then it's plus 93. And if we turn to the false position and multiply across, and we will say: "186 times 13 makes 2,418," then multiply 93 times 14, which makes 1,302, subtract 1,302 from 2,418, the remainder is 1,116. And to do the divisor, subtract 93 from 186, the remainder is 93, divide 1,116 by 93, it will result as 12 soldi, and that is the amount for which the pound of silk should be bartered, wanting $\frac{1}{3}$ in cash, as you will see in the example below.

14	13	93	186	2418
		14	13	1302
plus	plus	372	558	1116
186	93	93	186	
93		1302	2418	
93 divisor				

0
~~1~~0
0~~2~~~~8~~0
~~1~~~~1~~~~1~~~~6~~ | 12
~~9~~~~3~~~~3~~
~~9~~

And this above is done by false position, as was said on the previous page for the rule of 3.

And to do the same problem by an unknown, that is the one on the facing page by false position, one bartered silk, and put I don't know how much in the barter, and wants $\frac{1}{3}$ in cash, and it's worth $11\frac{3}{11}$ soldi per pound, and the other has cloth, which is worth 60 soldi per ell, and put into the barter 66 soldi. The amount of the silk should be put as the unknown. f. 21b

To do this problem by unknowns, first you should subtract $\frac{1}{3}x$ from $11\frac{3}{11}$ that it's worth in cash. And because you have said that it's $11\frac{3}{11}$ plus $1x$, subtract the $\frac{1}{3}$ from the $1x$ and $\frac{2}{3}x$ remains, and subtract $\frac{1}{3}x$ from $11\frac{3}{11}$, the remainder is $11\frac{3}{11}$ minus $\frac{1}{3}x$. You should multiply by the price that the cloths return in coins, which was 60 soldi per ell, and say: "$\frac{2}{3}$ times 60 makes $\frac{120}{3}x$," and if you said: "If $11\frac{3}{11}$ was $\frac{2}{3}x$, what would 60 be?" we know that 66 was put into the barter, so multiply 66 by $11\frac{3}{11}$, makes $\frac{8184}{11}$, and this should by divided. Then multiply 66 times minus $\frac{1}{3}x$, makes $22x$. Equalize the sides, add $\frac{120}{3}x$ to $22x$, it would be $62x$, and this is your divisor. Divide $\frac{8184}{11}$ by $\frac{62}{1}$, will result in 12, and this is how much silk should be put in the barter, as you will see in the example below.

$11\frac{3}{11}$	1^{co}	$11^{n}\frac{3}{11}$ men $\frac{1}{3}^{co}$	$\frac{2}{3}^{co}$	$\frac{60}{1}$	$\frac{120}{3}^{co}$	66
$\frac{124}{11}$	1^{co}	$\frac{66}{1}$	$\frac{8184}{11}$	$\frac{66}{3}^{co}$	$\frac{120}{3}$	serà 62^{co}

$$\frac{8184}{11} \times \frac{62}{1} \qquad \begin{array}{r} 62 \\ 11 \\ \hline 682 \end{array}$$

$$\begin{array}{r} 00 \\ 110 \\ 2360 \\ 8184 \\ 6822 \\ 68 \end{array} \Big] \; 12$$

E la sovraditta raxion è fatta per la chossa e vien ingual a quela inchontro fatta per inpoxizion, e per lo simille muodo vien a quela ch'è ala quarta indriedo al'inpoxizion fatta per la riegola del 3, e per simille farì tute le oltre raxion de simille muodo.

---

f. 22a Sono 2 homeny chi vuol baratar l'uno a pany, chi val el brazo soldi 60, e metela a barato soldi 66, e l'oltro a seda, che val la libra non so quanto, e mesella a baratto soldi 12 la libra, e vuol $\frac{1}{3}$ in denari. Adomando zò che valea la livra dela seda in denari. E primo per farla per la riegola del 3. Perché quel dela seda vuol $\frac{1}{3}$ di chontadi, adoncha azonzi a 66 la mittade, serano 99, azonzi a 60 33, serano 93. E dixemo: "Se 99 torna 93, che me tornerà 12?" Moltipicha 12 via 93, serano 1116, parti per 99, insirà del partidor $11\frac{3}{11}$, chomo vedirì qui de sotto.

$$\begin{array}{r} 66 \\ 33 \\ \hline 99 \end{array} \quad 99 \quad \begin{array}{r} 60 \\ 33 \\ \hline 93 \end{array} \quad 93 \quad 99 \quad 93 \quad 12 \quad \begin{array}{r} 93 \\ 12 \\ \hline 186 \\ 93 \\ \hline 1116 \end{array}$$

$$\begin{array}{r} 02 \\ 13 \\ 0227 \\ 1116 \\ 999 \\ 9 \end{array} \Big] \; 11\tfrac{27}{99} \quad \tfrac{3}{11}$$ Serano per la libra dela seda soldi $11\frac{3}{11}$.

El baratto de sovra è fato per la rigola del 3. Mo' te voio mostrar a farla per inpoxizion. Metti che la prima inpoxizion fuse 14, e perché quel dela seda [*vuol*] el terzo in denari, azonzi la mittade de 66, farano 99, et azonti 33 in 60 serano 93. E diremo: "Se 99 me torna 93, che serave 12?" Moltipichado 12 via 93 fano 1116, e se moltipicho quel 14 del'inpoxizion chon lo partidor, che sono 99, fano 1386, e mi voio che sia 1116. Adoncha sotrazi 1116 de 1386, roman eser più 270, e questi metti ala prima inpoxizion più. E per far la segonda inpoxizion metti che fusse 13, e diremo: "Se 99

$11\frac{3}{11}$	$1x$	$11\frac{3}{11}$ minus $\frac{1}{3}x$	$\frac{2}{3}x$	$\frac{60}{1}$	$\frac{120}{3}x$	66
$\frac{124}{11}$	$1x$	$\frac{66}{1}$	$\frac{8184}{11}$	$\frac{66}{3}x$	$\frac{120}{3}$	will be $62x$

$$\frac{8184}{11} \times \frac{62}{1} \qquad \begin{array}{r} 62 \\ 11 \\ \hline 682 \end{array}$$

00
110
2360
8184] 12
6822
68

And the problem above is done by unknowns and comes out the same as that done by false position, and in the same way comes to that which is on the page before the false position done by the rule of 3, and in a similar way you will do all the other problems.

There are two men who want to barter; one has cloth, which is worth 60 soldi per ell, and he puts 66 soldi into the barter, and the other has silk, which is worth I don't know how much per pound, and he put it into the barter at 12 soldi per pound, and he wants $\frac{1}{3}$ in cash. I ask how much the pound of silk was worth in cash. And to do this first by the rule of 3. Because the one with the silk wants $\frac{1}{3}$ in coins, then add to 66 its half, which will be 99, add 33 to 60, it will be 93. And we say: "If 99 returns 93, what will 12 return me?" Multiply 12 by 93, it will be 1,116, divide by 99, $11\frac{3}{11}$ will result from the division, as you will see here below. f. 22a

$$\begin{array}{r} 66 \\ 33 \\ \hline 99 \end{array} \quad 99 \quad \begin{array}{r} 60 \\ 33 \\ \hline 93 \end{array} \quad 93 \quad 99 \quad 93 \quad 12 \quad \begin{array}{r} 93 \\ 12 \\ \hline 186 \\ 93 \\ \hline 1116 \end{array}$$

02
13
0227
1116]
999] $11\frac{27}{99}$ $\frac{3}{11}$ It will be $11\frac{3}{11}$ soldi for the pound of silk.
9]

The barter above is done by the rule of 3. Now I want to show you how to do it by false position. Suppose that the first false position was 14, and because the one with silk wants one third in cash, add the half of 66, which will be 99, and add 33 to 60, which will be 93. And we will say: "If 99 gives me 93, what will 12 be?" 12 multiplied times 93 makes 1,116, and if I multiply the 14 of the false position with the divisor, which is 99, it makes 1,386, and I want it to be 1,116. So subtract 1,116 from 1,386, the remainder is plus 270, and put this at the first false position. And to do the

torna 93, che tornerà 12?" Moltipicha 12 via 93, fa 1116, e puo' moltipicha 13 via 99, fa 2287, e my voio che sia 1116. Adoncha sotrazi 1116 de 2287, roman 171 più, e questi metti ala segonda inpoxizion. Mo' moltipicha in croxe 14 via 171, fa 2394, e puo' moltipicha 13 via 270, fa 3510,
f. 22b sutrazi, roman 1116. Se trazi de 270 | 171 roman 99, parti 1116 per 99, insirà fuori soldi $11\frac{3}{11}$, e tanto valeva in denari la libra dela seda. E per mostrarvi insenpio, vedirì qui de sotto.

14		13	171	270	3510	270
più	X	più	14	13	2394	171
			684	810	1116	99
270		171	171	270		
			2394	3510		

Se die partir 1116 per 99.

02
13
0227
1116
999] $11\frac{27}{99}$ $\frac{3}{11}$
9

1	6	3
4	3	3

E per lo simille muodo farì ogn'altra inpoxizion de simille condizion.

E per voler far la ditta raxion per la chosa, pony che l'avese 1^{co} 12^{n}. Abatti de 12^{n} el terzo, 4, roman 8. Abatti de 1^{co} 4, roman 1^{co} men 4^{n}. E noi savemo che quel de 60 mese 66 el brazo, adoncha moltipicha 8 via 60, serà 480. Mo' moltipicha 66 via 1^{co}, fa 66^{co}, e moltipicha 66 via men 4, farà 264^{n}. Azonzi chon 480, fano 744. Parti per 66^{co}, insirà fora soldi $11\frac{3}{11}$. E muo' tu ay in primo la ditta raxion per la riegola del 3, e segondo per inpoxizion, e terza per la chosa, chomo per exenpio vedirì qui de sotto e farì ogn'altra simille raxion.

1^{co}	12^{n}	8	60	480	1^{co} men 4	66	66	66^{co}
	4		8			4	1^{co}	
			480			men 264		

66^{co} men 264^{n} più 480^{n} adequa

480
264
744 parti

1
020
188
744
666] $11\frac{18}{66}$ $\frac{3}{11}$
6

1. *2287* in MS.
2. *2287* in MS.

second false position suppose that it was 13, and we will say: "If 99 returns 93, what will 12 return?" Multiply 12 times 93, it makes 1,116, and then multiply 13 times 99, it makes 1,287,[1] and I want it to be 1,116. Then subtract 1,116 from 1,287,[2] the remainder is plus 171, and put this at the second false position. Now multiply across 14 times 171, it makes 2,394, and then multiply 13 times 270, it makes 3,510, subtract, the remainder is 1,116. If you subtract 171 from 270 | it leaves 99, divide 1,116 by 99, there will result $11\frac{3}{11}$ soldi, and that is how much the pound of silk was worth in cash. And to show you the example, you will see below. f. 22b

14	╲ ╱	13	171	270	3510	270
plus	╳	plus	14	13	2394	171
			684	810	1116	99
270	╱ ╲	171	171	270		
			2394	3510		

You should divide 1116 by 99.

02		1 | 6	3
~~1~~~~3~~		4 | 3	3
0~~2~~~~2~~7			
~~1~~~~1~~~~1~~~~6~~			
~~9~~~~9~~~~9~~	$11\frac{27}{99}$ $\frac{3}{11}$		
~~9~~			

And in the same way you will make any other false position for similar bartering.

And to do the problem by unknowns, put that you had $1x$ and 12. Subtract from 12 its third, 4, the remainder is 8. Subtract 4 from $1x$, the remainder is $1x$ minus 4. And we know that the one with 60 put 66 per ell, so multiply 8 times 60, it will be 480. Now multiply 66 by $1x$, it makes $66x$, and multiply 66 times minus 4, it will make 264. Add to 480, it makes 744. Divide by $66x$, it will result as $11\frac{3}{11}$ soldi. And now you have first the problem by the rule of 3, and second by false position, and third by unknowns, as you will see in the example below here, and you will do all other similar problems.

$1x$	12	8	60	480	$1x$ minus 4	66	66	$66x$
	4		8			4	$1x$	
			480			minus 264		

$66x$	minus	264	plus	480	equalize	480		1	
						264		0~~2~~0	
						744	divide	~~1~~~~8~~8	
								7~~4~~~~4~~	
								~~6~~~~6~~~~6~~	$11\frac{18}{66}$ $\frac{3}{11}$
								~~6~~	

f. 23a Sono 2 homeny chi vuol baratar l'uno a seda, che val in denar soldi $11\frac{3}{11}$ e mesela a baratto soldi 12, l'oltro a pany, che val el brazo a deneri non so quanti, e mesela a barato soldi 66. Domando zò che valeva el brazo de pano in denari.

E per far la ditta raxion per la riegola del 3 dixemo: "Se 12 ch'è lo barato dela seda me torna $11\frac{3}{11}$, che tornerà soldi 66 che val a baratto el brazo del pano?" Tu die in primo moltipichar in primo $\frac{124}{11}$ per $\frac{66}{1}$, insirà fuori 8184, e questi partidi per 11 fano 744, e questi partidi per lo tto partidor, che son 12, insirà fuor del partidor soldi 62. E questo valse el brazo del pano a chontadi, chomo vedirì qui de soto per exenpio.

12	$11\frac{3}{11}$	66
	$\frac{124}{3}$ $\frac{66}{1}$	

	000		00	
	~~1~~~~4~~~~4~~0		~~1~~~~2~~0	
124	~~8~~~~1~~~~8~~~~4~~		~~7~~~~4~~~~4~~	
66	~~1111~~	744	~~122~~	62
8184	~~11~~		~~1~~	

Sì che adoncha per la sovraditta raxion valleva el brazo del pano soldi 62.

E per far la ditta raxion per inpoxizion mettemo ch'ala prima inpoxizion fusse 14, e noi diremo: "Se 12 me torna $11\frac{3}{11}$, che tornerà 66 che xe lo baratto del brazo del pano?" Moltipicha $\frac{124}{11}$ per $\frac{66}{1}$, insirà 8184, a partir per 11 fano 744, e questi salva. Po' moltipicha 14 del'inpoxizion chon 12, che fu el partidor, serano 168, e questi sotrazi de 744, roman 576, e metti al'inpoxizion men. E per far la segonda inpoxizion ponamo che la fose 13. Moltipicha 13 via 12, fa 156, sutrazi de 744, roman 588, e metilli men ala segonda inpoxizion. Puo' moltipicha in croxe 14 via 588, fa 8232, e 13 via 576, fa 7488. Sotrazi de 8232, roman 744, e questi salva. Puo' sutrazi de 588 576 per far partidor, che son 12, adoncha parti 744 per 12, insirà soldi 62. E tanto valleva el brazo de pano in
f. 23b denari, | chomo vedirì qui de sotto per exenpio in figura.

14	13	576	588	8232	588
men	men	13	14	7488	576
576	588	1728	2352	744	012
		576	588	partir	partidor
		7488	8232		

00	
~~1~~~~2~~0	
~~7~~~~4~~~~4~~	
~~122~~	62 el brazo soldi 32 a denari
~~1~~	

1. *32* in MS.

There are 2 men who want to barter; one has silk which is worth in cash $11\frac{3}{11}$ soldi, and he put 12 soldi in the barter, the other has cloth, which is worth I don't know how much per ell, and put 66 soldi in the barter. I ask what the ell of cloth is worth in coins. f. 23a

And to do this problem by the rule of 3 we say: "If 12 which is the barter of the silk gives me $11\frac{3}{11}$, what will 66 soldi, which is the value in the barter of the ell of cloth, result in?" You should first multiply $\frac{124}{11}$ by $\frac{66}{1}$, it will come out to 8,184, and this divided by 11 makes 744, and this divided by your divisor, which is 12, will result in 62 soldi. And this is the value of the ell of cloth in coins, as you will see in the example below.

12	$11\frac{3}{11}$	66
	$\frac{124}{3}$ $\frac{66}{1}$	

	000		00	
124	1440		120	
66	8184		744	
8184	1111	744	122	62
	11		1	

So by this problem above the ell of cloth was worth 62 soldi.

And to do this problem by false position we suppose that 14 was at the first false position, and we will say: "If 12 gives me $11\frac{3}{11}$, what will 66, which is the barter of the ell of cloth, return?" Multiply $\frac{124}{11}$ by $\frac{66}{1}$, it will give 8,184, and this divided by 11 makes 744, and save this. Then multiply the 14 of the false position by 12, which was the divisor, which will be 168, and subtract this from 744, which leaves 576, and put it in the false position as minus. And to do the second false position, suppose that it was 13. Multiply 13 times 12, which makes 156, subtract this from 744, which leaves 588, and put it at the second false position as minus. Then multiply across 14 times 588, which makes 8,232, and 13 times 576, which makes 7,488. Subtract it from 8,232, which leaves 744, and save this. Then subtract 576 from 588 to get the divisor, which is 12, then divide 744 by 12, there will result 62 soldi. And this is the value of the ell of cloth in coins, | as you will see illustrated in the figure below. f. 23b

14		13	576	588	8232	588
	X		13	14	7488	576
minus		minus	1728	2352	744	012
576		588	576	588	to divide	divisor
			7488	8232		

00		
120		
744		
122	62 the ell	62[1] soldi in coins
1		

E per lo simille muodo farì ogn'altra raxion de baratti de simel muodo.

E per voler far la ditta raxion per la chosa, pony che fuse 1^{co} e 66^{n}. Moltipicha per $11\frac{3}{11}$, che fano $\frac{124}{11}$, per $\frac{66}{1}$, farano 8184. Parti per 11, fano 744. Moltipicha 1^{co} via fia[a] 12, fa 12^{co}. Parti 744^{n} per 12^{co}, serano soldi 62. E tanto valse el brazo del pano in denari, chomo per exenpio vedirì qui de sutto per figura.

1^{co} 66^{n} $11\frac{3}{11}{}^{n}$ $\frac{124}{11}{}^{n}$ $\frac{66}{1}$ 8184

11

000
1440
8184] 744
1111]
11]

1^{co}
12^{n}
fa 12^{co}

$\frac{12}{1}{}^{co}$ $\frac{744}{1}{}^{n}$

00
120
744] 62
122]
1]

E tanto vallse el brazo del pano in denari, zoè soldi 62. E questo è fatto.

Sono 2 chi vuol baratar l'uno a seda e l'oltro a pano. Quel dela seda val la libra soldi $11\frac{3}{11}$ in deneri, e mesela a barato soldi 12. Quel del pano val el brazo in denari soldi 62. Quando[b] die meter el brazo a baratto per defe[*n*]derse dal botto?

f. 24a E prima, per farla per la riegola del 3, se fa per questo muodo. "Se $11\frac{3}{11}$ | me torna 12, che me tornerà 62?" Moltipicha $\frac{62}{1}$ via $\frac{124}{11}$, serano 8184, e questi partidi per 124 serano[c] 66, e tanti se die m[*e*]tter a baratto, zoè el brazo soldi 66, chomo vedirì qui per exenpio.

E la raxion sovraditta non è ben posta, ma ben se die dir: "Se $11\frac{3}{11}$ torna 12 unittà, che tornerà 62?" Adoncha moltipicha 12 via 62, o per dir meyo 132 via 62, fa 8184, e questi partidi per 124 insirà soldi 66, e ttanto vuol eser messo el brazo del pano a baratto.

---

a. *Sic in MS.*
b. *Sic in MS, evidently for* quanto.
c. 744 *canceled with a horizontal line.*

And by the same method you will do every other problem of barter of a similar type.

And to do the same problem by unknowns, put that it was 1*x* plus 66. Multiply $11\frac{3}{11}$, which makes $\frac{124}{11}$, by $\frac{66}{1}$, it will make 8,184. Divide by 11, makes 744. Multiply 1*x* times 12, it makes 12*x*. Divide 744 by 12*x*, it will be 62 soldi. And that is the value of the ell of cloth in coins, as you will see worked out in the figure below.

1*x*   66 $11\frac{3}{11}$ $\frac{124}{11}$ $\frac{66}{1}$ 8184   000
11   1440   1*x*
8184   12
1111   744   makes 12*x*
11

$\frac{12}{1}x$ $\frac{744}{1}$ 00
120
744
122   62
1

And that is the value of the ell of cloth in coins, that is, 62 soldi. And it's done.

There are 2 who want to barter; one has silk and the other has cloth. The silk is worth $11\frac{3}{11}$ soldi per pound in coins, and he put 12 soldi in the barter. The cloth is worth 62 soldi per ell in coins. At how much should he put the ell in barter to avoid coming to blows?

And first, to do it by the rule of 3, it's done this way. "If $11\frac{3}{11}$ | gives me 12, what will 62 give me?" Multiply $\frac{62}{1}$ by $\frac{124}{11}$, it will be 8,184, and this divided by 124 will be 66, and so much should be put to barter, that is, 66 soldi per ell, as you will see worked out here. f. 24a

And the problem is not well stated above, but it should be put: "If $11\frac{3}{11}$ gives 12 units, what will 62 give?" Then multiply 12 times 62, or, to say it better, 132 times 62, makes 8,184, and this divided by 124 will result in 66 soldi, and for that much the ell of cloth should be put into the barter.

$11\frac{3}{11}$ 12 62 $\frac{124}{11}$ $\frac{12}{1}$ $\frac{62}{1}$

12	132
11	62
	8184

0
~~1~~0
07~~2~~
~~294~~0
~~8184~~] 66
~~1244~~
~~12~~

5	0
3	1

E per voler far la ditta raxion per inpoxizion mettemo che la prima inpoxizion fose 14 e la segonda inpoxizion fose 13. Noi moltipichemo 12 via 62, fano 744, e puo' moltipichemo 14 unyttà che xe dela prima inpoxizion con $11\frac{3}{11}$ che fo el partidor, fano $\frac{124}{11}$, a moltipicar con 14 fano 1736. E questi partidi per 11 fano $157\frac{9}{11}$, e mi vorave che fosse 744. Adoncha sutrazi $157\frac{9}{11}$ de 744, venyrà a manchar $586\frac{2}{11}$, e questi metti a manchar ala prima inpoxizion. E per lo simille a far la segonda moltipicha 13 via $\frac{124}{11}$, fano 1612, e questi partidi per 11 fano $146\frac{6}{11}$. E sutrati questi de 744 mancha $597\frac{5}{11}$, e questi metti a manchar ala segonda inpoxizion, chomo vedirì per figura.

14		13
men	X	men
$586\frac{2}{11}$		$597\frac{5}{11}$

$11\frac{3}{11}$ $\frac{12}{1}$ $\frac{62}{1}$

	62	$\frac{14}{1}$ $\frac{124}{11}$
	12	
744	744	1736
146	$157\frac{9}{11}$	
$597\frac{5}{11}$	$586\frac{2}{11}$	

00
~~11~~
0~~68~~9
~~1736~~]
~~1111~~] $157\frac{9}{11}$
~~11~~]

Moltipicha e parti, insirà fori del partidor el brazo soldi 66 chomo di sovra.

f. 24b E per far la ditta raxion per la chosa, quel che de $11\frac{3}{11}$ messe 12 e l'oltro chi val in deneri soldi 62 vuol saver zò che die meter el brazo a baratto. Pony che fose 1^{co} e 62^{n}. Moltipicha in primo 12 via 62, val 744, e questo salva. E moltipicha 1^{co} via $11\frac{3}{11}$, che fano $\frac{124}{11}$ de chossa, e parti chon 744, insirà fuor del partidor soldi 66. E tanto se vuol metter el brazo del pano a baratto, chomo vedirì qui per figura.

$11\frac{3}{11}$ 12   62 $\frac{124}{11}$ $\frac{12}{1}$ $\frac{62}{1}$

$$\begin{array}{r} 12 \\ 11 \\ \hline \end{array} \qquad \begin{array}{r} 132 \\ 62 \\ \hline 8184 \end{array}$$

0
10
072
2940
8184
1244 | 66
12

$$\begin{array}{c|c} 5 & 0 \\ \hline 3 & 1 \end{array}$$

And to do this problem by false position, we suppose that the first false position was 14 and the second false position was 13. We multiply 12 times 62, which makes 744, and then we multiply 14 units, which is the first false position, with $11\frac{3}{11}$ which was the divisor, which makes $\frac{124}{11}$, multiplied by 14 makes 1,736. And this divided by 11 makes $157\frac{9}{11}$, and I wanted it to be 744. Then subtract $157\frac{9}{11}$ from 744, it will result in $586\frac{2}{11}$ minus, and put this as a minus at the first false position. And to do the second by the same way, multiply 13 times $\frac{124}{11}$, which makes 1,612, and this divided by 11 makes $146\frac{6}{11}$. And when this is subtracted from 744 there lacks $597\frac{5}{11}$, and put this as a minus at the second false position, as you will see in the figure.

14 X 13
minus   minus
$586\frac{2}{11}$ $597\frac{5}{11}$

$11\frac{3}{11}$ $\frac{12}{1}$ $\frac{62}{1}$

$$\begin{array}{r} 744 \\ 146 \\ \hline 597\frac{5}{11} \end{array}$$ [1]

$$\begin{array}{r} 62 \\ 12 \\ \hline 744 \\ 157\frac{9}{11} \\ \hline 586\frac{2}{11} \end{array}$$

$\frac{14}{1}$ $\frac{124}{11}$
1736

00
11
0689
1736
1111 | $157\frac{9}{11}$
11

Multiply and divide, and the result of the division will be 66 soldi per ell as above.

And to do this problem by unknowns, the one that put in 12 of $11\frac{3}{11}$ and the other that is worth 62 soldi in coins wants to know at what sum he should put the ell into the barter. Put that it was 1*x* and 62. Multiply first 12 times 62, it is worth 744, and save this. And multiply 1*x* times $11\frac{3}{11}$, which makes $\frac{124}{11}$ of the unknown, and divide by 744, there will result from the division 66 soldi. And that is the price at which he should put the ell of cloth in the barter, as you will see in the figure here. f. 24b

1. I.e., $146\frac{6}{11}$.

1^{co} $11\frac{3}{11}$ 62^{n} 12^{n}

$$\begin{array}{r} 62 \\ \underline{12} \\ 124 \\ 62 \end{array} \qquad \frac{744}{1} \times \frac{124}{11} \qquad \begin{array}{r} 744 \\ \underline{11} \\ 744 \\ \underline{744} \\ 8184 \end{array}$$

0
10
072
2940
8184
1244 | 66
12

$$\begin{array}{c|c} 5 & 0 \\ \hline 3 & 1 \end{array} \qquad \begin{array}{c} 1 \\ \hline \hline 1 \\ \hline 1 \end{array}$$

Sono 2 chi vuol baratar l'uno a seda che val che val[a] a chontadi non so quanti, e mesela in baratto soldi 16, e vuol $\frac{1}{4}$ in denari, l'oltro a pano che val a chontadi soldi 8, e mesela a baratto soldi 10, e vuol $\frac{1}{3}$ in denari. Adomando zò che valeva la seda in denari, e per voler far la ditta raxion per la riegola del 3.

In primo choluy che de 8 fexe 10 e vuol $\frac{1}{3}$ in denari. Sotrazi de 10 el terzo, roman $6\frac{2}{3}$, e de 8 roman $4\frac{2}{3}$, e po' perché el primo domanda el quarto in denari adoncha azonzi sovra $6\frac{2}{3}$ el terzo, serano $8\frac{8}{9}$. Azonzi sovra $4\frac{2}{3}$ $2\frac{2}{9}$, farano $6\frac{8}{9}$, farano $\frac{62}{9}$, e l'altra parte serano $\frac{80}{9}$, e diremo: "Se $\frac{80}{9}$ torna $\frac{62}{9}$, che tornerà 16 che fo meso la seda a baratto?"

f. 25a Se moltipichi e parti serà el partidor 720 e quel che vorà eser partido serà 8928, insirà fuor del partidor $12\frac{288}{720}$, e per voler schizar serà $\frac{2}{5}$, zoè soldi $12\frac{2}{5}$. E talto[b] valse la seda in denari, chomo vedirì qui de sutto per exenpio e per figura.

$$\frac{80}{9} \times \frac{62}{9} \;—\; \frac{16}{1} \qquad \begin{array}{r} 80 \\ \underline{9} \\ 720 \end{array} \quad \begin{array}{r} 62 \\ \underline{9} \\ 558 \end{array} \quad \begin{array}{r} 558 \\ \underline{16} \\ 8928 \end{array}$$

2
03
178
8928
7200 | $12\frac{288}{720}$
72

E per voler schixar, schixa per 2 284, serà 144. Schixa per 2, serà 72. Schixa per 2, serà 36. Schixa 720 per 2, serà 360. Schixa per 2, serano 180. Schixa per 2, serà 90. E questi 90 schixa per 6, serano

---

a. *Repeated thus in the MS.*
b. *Sic in MS, evidently for* tanto.

1*x* $11\frac{3}{11}$ 62 12

62	$\frac{744}{1}$ ╳ $\frac{124}{11}$	744	
12		11	
124		744	
62		744	
		8184	

0
10
072
2940
8184
1244] 66
12

5	0
3	1

1
1
1

There are 2 who want to barter, one with silk which is worth I don't know how much in cash, and he puts into the barter 16 soldi, and wants $\frac{1}{4}$ in coins; the other has cloth that is worth in cash 8 soldi, and he puts 10 soldi into the barter, and wants $\frac{1}{3}$ in coins. I ask what the value of the silk in coins was, and wish to do the problem first by the rule of 3.

First, the one who from 8 makes 10 and wants $\frac{1}{3}$ in coins. Subtract from 10 its third, the remainder is $6\frac{2}{3}$, and from 8 remains $4\frac{2}{3}$, and then because the first wants a quarter in coins then add its third to $6\frac{2}{3}$, it will be $8\frac{8}{9}$. Add $2\frac{2}{9}$ to $4\frac{2}{3}$, it will make $6\frac{8}{9}$, which makes $\frac{62}{9}$, and the other part will be $\frac{80}{9}$, and we will say: "If $\frac{80}{9}$ gives $\frac{62}{9}$, what will the 16 bring that the silk was put in the barter at?"

If you multiply and divide, the divisor will be 720 and that which you want divided will be 8,928, the result of the division will be $12\frac{288}{720}$, and to factor, it will be $\frac{2}{5}$, that is, $12\frac{2}{5}$ soldi. And that is what the silk is worth in coins, as you will see worked out in the figure below. f. 25a

$\frac{80}{9}$ ╳ $\frac{62}{9}$ — $\frac{16}{1}$

80	62	558
9	9	16
720	558	8928

2
03
178
8928
7200] $12\frac{288}{720}$
72

And to factor, factor 288[1] by 2, it will be 144. Factor by 2, it will be 72. Factor by 2, it will be 36. Factor 720 by 2, it will be 360. Factor by 2, it will be 180. Factor by 2, it will be 90. And factor this

---

1. *284* in MS.

15. E questi 15 schixa per 3, serano 5. E schixar 36 per 6, serano 6. E questi schixadi per 3 serano 2. Adoncha serano $\frac{2}{5}$, fano in suma $12\frac{2}{5}$.

E per voler far la ditta raxion per inpoxizion metemo che la prima inpoxizion fuse 14 e la segonda fuse 13. Adoncha cholui che de 8 messe 10 e vuol $\frac{1}{3}$ in denari, sotrazi $3\frac{1}{3}$ de 10 e de 8, roman $6\frac{2}{3}$ e $4\frac{2}{3}$. E perché el primo domanda el quarto in denari, azonzemo el terzo de $6\frac{2}{3}$ a $6\frac{2}{3}$ e a $4\frac{2}{3}$, serano $8\frac{8}{9}$ e $6\frac{8}{9}$, fano $\frac{80}{9}$ $\frac{62}{9}$, e dixemo: "Se $\frac{80}{9}$ torna $\frac{62}{9}$, che serave 16 unyttà?" A moltipicar 16 62 fano 992, e questi partidi per 9 fano $110\frac{2}{9}$, e puo' moltipicha 14, che tu à posto ala prima inpoxizion, chon $\frac{80}{9}$, fano serano[a] 1120. A partir per 9 serano $124\frac{4}{9}$, e noi volemo che sia $110\frac{2}{9}$. Sotrazi, serano men $14\frac{2}{9}$, e chosì meteremo ala prima inpoxizion. E per lo simille moltipicha 13 via $\frac{80}{9}$, fano 1040, parti per 9, serave $115\frac{5}{9}$, e noi volemo che fusse $110\frac{2}{9}$. Adoncha è men $5\frac{3}{9}$, e meteremo ala segonda inpo-
f. 25b xizion, e puo' moltipicha in croxe 14 via $5\frac{3}{9}$, fano $74\frac{6}{9}$. E puo' moltipichemo 14 | via 13 chon $\frac{3}{9}$, serano $148\frac{8}{9}$. Sotrazi $74\frac{6}{9}$, roman[b] $110\frac{2}{9}$. Sotrazi $5\frac{3}{9}$ de $14\frac{2}{9}$, roman $8\frac{8}{9}$, serave $\frac{80}{9}$. A partir $110\frac{2}{9}$, serano $\frac{992}{9}$. Parti per 80, insirà $12\frac{2}{5}$, chomo vedirì qui per figora.

14 × 13
men men
$14\frac{2}{9}$ $5\frac{3}{9}$

$\frac{80}{9}$ $\frac{62}{9}$ $\frac{16}{1}$ 992
$110\frac{2}{9}$

80
14
1120
$124\frac{4}{9}$

$124\frac{4}{9}$
$110\frac{2}{9}$
men $14\frac{2}{9}$

$\frac{80}{9}$ $\frac{62}{9}$ $\frac{16}{1}$

80
13
1040
$115\frac{5}{9}$

$115\frac{5}{9}$
$110\frac{2}{9}$
men $5\frac{3}{9}$

$14\frac{2}{3}$
13

$184\frac{8}{9}$
$74\frac{6}{9}$
$110\frac{2}{9}$

14
$5\frac{3}{9}$
$76\frac{4}{9}$[c]

partir $\frac{992}{9}$ × $\frac{80}{9}$

0
13
992
800
8
$12\frac{32}{80}$ $\frac{2}{5}$

$14\frac{2}{5}$
$5\frac{3}{5}$
partidor $8\frac{8}{9}$

a. *Sic in MS.*
b. p *crossed out with a horizontal line.*
c. *This result, clearly erroneous, is crossed out with a diagonal line.*

90 by 6, it will be 15. And factor this 15 by 3, it will be 5. And to factor 36 by 6, it will be 6. And this factored by 3 will be 2. So it will be $\frac{2}{5}$, which makes in all $12\frac{2}{5}$.

And to do this problem by false position, we suppose that the first false position was 14 and the second was 13. Then for the one who puts 10 of 8 and wants $\frac{1}{3}$ in coins, subtract $3\frac{1}{3}$ from 10 and from 8, the remainder is $6\frac{2}{3}$ and $4\frac{2}{3}$. And because the first wants a quarter in coins, add a third of $6\frac{2}{3}$ to $6\frac{2}{3}$ and to $4\frac{2}{3}$, it will be $8\frac{8}{9}$ and $6\frac{8}{9}$, which makes $\frac{80}{9}$ and $\frac{62}{9}$, and we say: "If $\frac{80}{9}$ gives $\frac{62}{9}$, what would be 16 units?" To multiply 16 and 62 makes 992, and this divided by 9 makes $110\frac{2}{9}$, and then multiply 14, which you put at the first false position, with $\frac{80}{9}$, it will make 1,120. Divided by 9, it will be $124\frac{4}{9}$, and we wanted it to be $110\frac{2}{9}$. Subtract, it will be minus $14\frac{2}{9}$, and we'll put it as such at the first false position. And by the same way multiply 13 times $\frac{80}{9}$, it makes 1,040, divide by 9, it will be $115\frac{5}{9}$, and we wanted it to be $110\frac{2}{9}$. So it's minus $5\frac{3}{9}$, and we will put this at the second false position, and then multiply across 14 times $5\frac{3}{9}$ makes $74\frac{6}{9}$, and then we multiply 14 | times 13 with $\frac{3}{9}$, it will be $148\frac{8}{9}$.[1] Subtract $74\frac{6}{9}$, the remainder is $110\frac{2}{9}$. Subtract $5\frac{3}{9}$ from $14\frac{2}{9}$; the remainder is $8\frac{8}{9}$, which will be $\frac{80}{9}$. To divide $110\frac{2}{9}$, it will be $\frac{992}{9}$. Divide by 80, the result will be $12\frac{2}{5}$, as you will see in the figure below. f. 25b

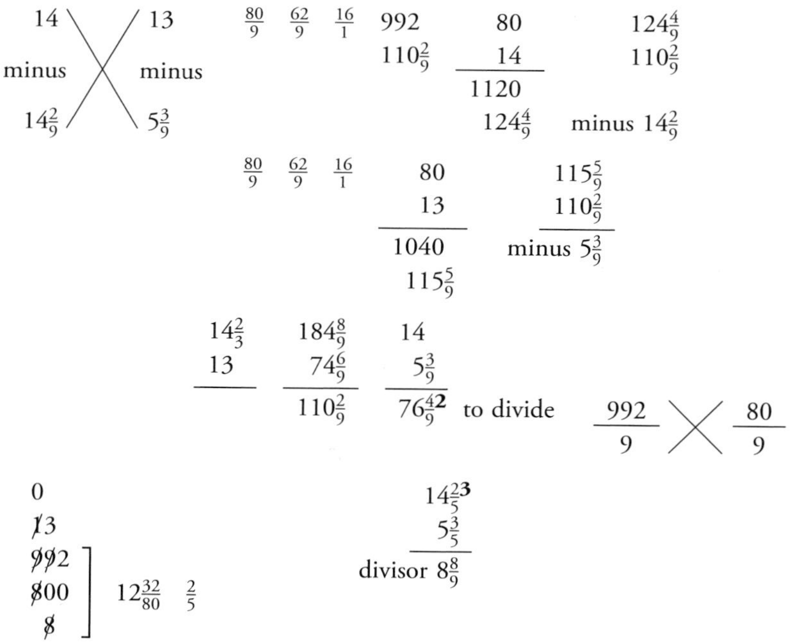

1. The correct multiplication would be 13 times $14\frac{2}{9}$ equals $184\frac{8}{9}$.
2. The problem is correctly worked out, except here, where the multiplication produces $76\frac{4}{9}$ rather than the correct $74\frac{6}{9}$; the incorrect answer is crossed out but not replaced by the correct one, which is used in the next stage (to the left).
3. These numbers should be $14\frac{2}{9}$ and $5\frac{3}{9}$.

E se la volesti far per la chosa pony che fuse 1^{co} e 16^{n}. Moltipicha per $\frac{62}{9}$, fano 992, a partir per 9 fano $110\frac{2}{9}$. Moltipicha 1^{co} via 80, fano $\frac{80^{co}}{9}$. Adoncha parti $110\frac{2}{9}$ per $\frac{80}{9}$. Insirà fuora $12\frac{2}{5}$.

1^{co} 16^{n} 62^{n} 00
16 992
999] $110\frac{2}{9}$ $\frac{992}{9}$ $\frac{80^{co}}{9}$

0
13
992
800] $12\frac{32}{80}$ $\frac{2}{5}$
8

---

f. 26a Sono 2 chi vuol baratar l'uno a lana, che val el miaro duchati 7, e metella in baratto duchati 8, e vuol $\frac{1}{4}$ in denari, e l'oltro a seda, che val a contadi el miaro duchati 12. Adomando: "Quando[a] la die meter a baratto voyando $\frac{1}{3}$ in denari?" Vegnirà a valler duchati $13\frac{1}{2}$. Quelo dela lana vuol baratar libre 1000 de lana, quanta seda ne averà? Averà libre $666\frac{2}{3}$. Quelo dela seda a libre $666\frac{2}{3}$ quanta lana averà? Averà libre miaro de lana, chomo vedirì qui de sotto per figura e per raxion per 3 muodi.

E primo per farla per la riegola del 3 abatemo de 8 el $\frac{1}{4}$, roman 6. Abatti 2 de 7, roman 5. Azonzi la mitade de 6, perché quel dela seda domandò $\frac{1}{3}$ in denari. Adoncha la mitade de 6 serà 3, asomadi serà 9. Azonzi a 5 3, serano 8. E diremo: "Se 8 torna 9, che serave 12?" Moltipicha 9 via 12, fa 108, e questo partido per 8 serano $13\frac{1}{2}$. E tanto die metter la seda in baratto dela lana.

Mo' volemo noi saver quanta seda averemo per libre 1000 de lana per li baratti, chomo vedirì qui de sotto per figora.

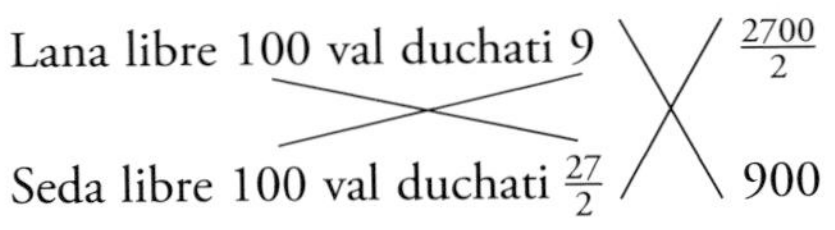

---

a. *Sic in MS, evidently for* quanto.

And if you wanted to do it by unknowns, put that it was $1x$ plus 16. Multiply by $\frac{62}{9}$, it makes 992, divided by 9 makes $110\frac{2}{9}$. Multiply $1x$ by 80, makes $\frac{80}{9}x$. Then divide $110\frac{2}{9}$ by $\frac{80}{9}$, the result will be $12\frac{2}{5}$.

$1x$ 16 62 00
16 992
999] $110\frac{2}{9}$ $\frac{992}{9}$ $\frac{80}{9}x$

0
~~1~~3
~~99~~2
~~8~~00] $12\frac{32}{80}$ $\frac{2}{5}$
~~8~~

There are two who want to barter; one has wool that's worth 7 ducats per thousandweight,[1] and he puts 8 ducats in the barter, and wants $\frac{1}{4}$ in coins, and the other has silk, which is worth in cash 12 ducats per thousandweight. I ask: "How much should he put in the barter if he wants $\frac{1}{3}$ in coins?" It will come to be worth $13\frac{1}{2}$ ducats. The one with the wool wants to barter 1,000 pounds of wool, how much silk will he have? He will have $666\frac{2}{3}$ pounds. The one with the silk has $666\frac{2}{3}$, how much wool will he have? He will have 1,000 pounds of wool, as you will see in the figure below and the problem solved 3 ways. f. 26a

And first, to do it by the rule of 3, we will subtract $\frac{1}{4}$ from 8, the remainder is 6. Subtract 2 from 7, which leaves 5. Add the half of 6, because the one with the silk wants $\frac{1}{3}$ in coins. Then the half of 6 will be 3, added together will be 9. Add 3 to 5, which will be 8. And we will say: "If 8 gives 9, what will 12 be?" Multiply 9 times 12, which makes 108, and this divided by 8 will be $13\frac{1}{2}$. And he should put that much silk into the barter of wool.

Now we want to know how much silk we will have per 1,000 pounds of wool by barter, as you will see in the figure below.

Wool, 100 pounds is worth 9 ducats $\frac{2700}{2}$

Silk, 100 pounds is worth $\frac{27}{2}$ ducats 900

1. In the problem that follows, these prices are calculated as being for 100 pounds rather than the 1,000 pounds specified in this introduction. The price of 12 ducats per hundred pounds is consistent with the value of silk at $11\frac{3}{11}$ soldi per pound, as the ducat was worth about 100 soldi in the first third of the fourteenth century; cf. Frederic C. Lane and Reinhold C. Mueller, *Money and Banking in Medieval and Renaissance Venice*, vol. 1, *Coins and Moneys of Account* (Baltimore: Johns Hopkins University Press, 1985), pp. 597–609.

E questo se fa: "Se $\frac{2700}{2}$ me tor[*n*]a $\frac{900}{1}$, che serave $\frac{1000}{1}$ libre de lana?" A moltipichar serà 1800000, e questo partido per 2700 insirà de fuor del partido libre $666\frac{2}{3}$ de seda. E per questa sifatta raxion tu puo' far ogni raxion simille.

f. 26b E per voler far la raxion inchontro per inpoxizion, ponamo che avesemo posto in la prima inpoxizion 600, e per questi dovesemo aver libre miaro de lana. Adoncha diremo: "Se 100 unittà val $\frac{27}{2}$, che valera 600?" Farà duchati 81. Abati el $\frac{1}{3}$, roman 54. Azonzi $\frac{1}{3}$, fa 72. Adoncha noi diremo: "Se 8 duchati me dà libre 100 de lana, che me darà duchati 72?" Moltipicha 100 via 72, serano 7200. E noi vosemo che fuse libre 1000 de lana. Adoncha venirà a manchar libre 100, e questi dovemo metter a manchar ala prima inpoxizion.

E per voler metera la segonda inpoxizion noi diremo che fose 900, e diremo: "Se 100 me torna $\frac{27}{2}$, che tornerà 900?" So che se moltipicho[a] e partto me die venir $121\frac{1}{2}$. Abati el $\frac{1}{3}$, resta 81. Azonzi el $\frac{1}{3}$, fa 108. E diremo: "Se duchati 8 che val el miaro de lana me torna 100, che me tornerà 108?" So che se moltipicho e parto me darà 1350, et io voria che fuse 1000. Adoncha ne vien più 350, e questi metimo ala segonda inpoxizion. Mo' moltipicha in croxe 600 via 350, fa 210000, e puo' moltipicha 900 via 100, vorano eser 90000. Adoncha più e men s'azonze. Adoncha zonti tuti fano 300000, e per far el partidor azonzi 350 chon 100, fano 450. Parti 300000, ne vien libre $666\frac{2}{3}$ de seda, chomo vedirì qui per figora.

600 \ / 900	900	90000	600	0		
men X più	100		350	0~~3~~3		
100 / \ 350			210000	~~366~~		
			90000	0~~6~~~~0~~~~0~~0		
				~~30000~~0		
				~~45~~000		$666\frac{2}{3}$
				~~455~~		
				~~4~~		

350
100
partidor 450

300000 die eser partido

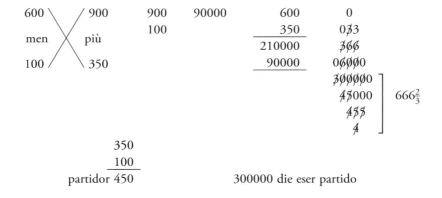

f. 27a E per far la raxion in driedo del baratto dela lana a seda, a volerla far per la chosa, pony che la seda vallese 1^{co}. Mo' fa chosì. Trazi de 8 el $\frac{1}{4}$, roman 6 e 5, puo' azonzi la mittade del 6 sovra 6 e sovra 5, serano 9 e 8, puo' di': "Se 8 fuse 9, che serave 12?" Tu sa' che die venir 1^{co}. Mo' se tu moltipicha 1^{co} via 8, fa 8^{co}, e moltipicha' 9 via 12, fa 108^{n}. Parti li nomeri per le chose, zoè 108^{n} per 8^{co}, vien $13\frac{1}{2}$. E tanto die valler la seda in baratto.

a. 100 via *crossed out with a horizontal line.*

And it's done this way: "If $\frac{2700}{2}$ gives me $\frac{900}{1}$, what will $\frac{1000}{1}$ pounds of wool be?" If it is multiplied, the result will be 1,800,000, and this divided by 2,700 will give a result of $666\frac{2}{3}$ pounds of silk. And with this calculation you can do all similar problems.

And to do the problem on the facing page by false position, let's suppose that we had put 600 in the first false position, and for this we would have had 1,000 pounds of wool. So we will say: "If 100 units is worth $\frac{27}{2}$, how much will 600 be worth?" It will be 81 ducats. If you subtract $\frac{1}{3}$, what remains is 54. Add $\frac{1}{3}$, which makes 72. So we will say: "If 8 ducats gives me 100 pounds of wool, what will 72 ducats give me?" Multiply 100 by 72, it will be 7,200. And we wanted it to be 1,000 pounds of wool. So there will come out to be minus 100 pounds, and we should put this as a minus at the first false position. f. 26b

And to do the second false position, we will say that it was 900, and we will say: "If 100 gives me $27\frac{1}{2}$, what will 900 give?" I know that if I multiply and divide I get $121\frac{1}{2}$. Subtract $\frac{1}{3}$, the remainder is 81. Add $\frac{1}{3}$, it makes 108. And we will say: "If 8 ducats which is what the thousandweight of wool is worth gives me 100, what will 108 give me?" I know that if I multiply and divide it will give me 1,350, and I wanted it to be 1,000. So it comes to plus 350, and we put this at the second false position. Now multiply across 600 times 350, which makes 210,000, and then multiply 900 times 100, and it will come out to 90,000. Then the plus and the minus are added together. When all are added together, they make 300,000, and to make the divisor add 350 to 100, which makes 450. Divide into 300,000, it comes to $666\frac{2}{3}$ pounds of silk, as you will see worked out here.

600 \	/ 900	900	90000	600	0	
minus X	plus	100		350	033	
				210000	366	
100 /	\ 350			90000	06000	
					300000	
					45000	$666\frac{2}{3}$
					455	
					4	

350
100
divisor 450

300000 should be divided

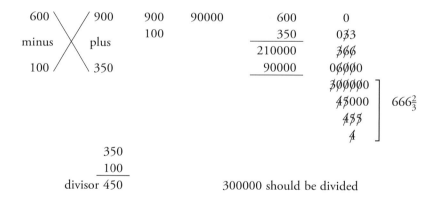

And to do the problem on the previous page of the barter of wool and silk; to do it by unknowns, put that the silk is worth $1x$. Now do this. Subtract from 8 its $\frac{1}{4}$, it leaves 6 and 5, then add the half of 6 to 6 and to 5, it will be 9 and 8, then say: "If 8 was 9, what would 12 be?" You know that it should come to $1x$. Now if you multiply $1x$ times 8, it makes $8x$, and multiply 9 times 12, it makes 108. Divide the numbers by the unknowns, that is 108 by $8x$, it comes to $13\frac{1}{2}$. And that is how much the silk in the barter should be worth. f. 27a

Mo' per saver quanta seda averemo per libre 1000 de lana questo è lo muodo. Pony che ne averemo 1^{co} de seda. Mo' savemo che per 1^{co} de seda averemo libre 1000 de lana, adoncha dixemo: "Se 100 libre de seda val duchati $13\frac{1}{2}$, che valerà 1^{co} de seda?" Fa 27^{co} da partir in 200, zoè $\frac{27}{200}^{co}$. Mo' abatti el $\frac{1}{3}$, roman $\frac{18}{200}$. Mo' azonzi el $\frac{1}{3}$, fa $\frac{24}{200}$. Mo' dovemo dir: "Se 8 me dà 100, che me darà $\frac{24}{200}$?" Moltipicha 100 via 24, fa 2400^{co}. Parti per 8 via 200, serano 1600, $\frac{2400}{1600}$, che sono ingual a 1000 libre de lana.

$\frac{2400}{1600}$ $\frac{1000}{1}$

0
0~~1~~1
~~1~~~~4~~~~4~~
0~~4~~~~6~~~~6~~6
~~1~~~~6~~~~0~~~~0~~~~0~~00
~~2~~~~4~~0000
2~~4~~00
~~2~~~~4~~

$666\frac{2}{3}$

E questa raxion è fatta per la chosa, che per libre 1000 de lana tu die aver libre $666\frac{2}{3}$ de seda. E per questo muodo farì ogn'altra simel raxion per la chosa.

f. 27b

Sono 2 chi vuol baratar, l'uno a seda, che val a chontadi non so quanti, e metela a baratto quel che la valeva e più la so radixe, e vuol $\frac{1}{4}$ in denari, l'oltro a pano, che val el brazo soldi 50, e meselo a baratto soldi 56. Adomando quanto valeva la seda a chontadi. E per voler far la ditta raxion per la chosa pony che la seda vallese $1^{\square}$, e si dixe che vuol più la so radixe. Adoncha fu $1^{\square}$ e 1^{co}. Puo' dixe che vuol el $\frac{1}{4}$ in denari. Adoncha sotrazemo el $\frac{1}{4}$ de $1^{\square}$, roman $\frac{3}{4}^{\square}$, e sutremo de 1^{co} $\frac{1}{4}$. Roman $\frac{3}{4}^{co}$. Adoncha sotremo de $1^{\square}$ $\frac{1}{4}$, roman $\frac{3}{4}^{\square}$, e sutremo de $1^{\square}$ $\frac{1}{4}^{co}$, roman $\frac{3}{4}^{co}$ men $\frac{1}{4}^{co}$. Mo' si moltipicho $\frac{3}{4}^{\square}$ men $\frac{1}{4}^{co}$ per 56 seria $\frac{168}{4}^{\square}$ men $\frac{56}{4}^{co}$, e puo' moltipichemo 50 via $\frac{3}{4}^{co}$, serano $\frac{150}{4}^{co}$, e $\frac{3}{4}^{\square}$ via 50 serà $150^{\square}$. Adoncha adequemo le parte et averemo men $\frac{56}{4}$ de chosa a $\frac{168}{4}^{\square}$, e non i mancharà niente. E demo $\frac{56}{4}^{co}$ a $\frac{150}{4}^{co}$, farano $\frac{206}{4}^{co}$. Mo' abatemo $\frac{150}{4}^{\square}$ de $\frac{168}{4}^{\square}$, roman $18^{\square}$. Partimo $\frac{206}{4}^{co}$ per $\frac{18}{4}^{\square}$, vegnirà $11\frac{4}{9}$, e tanto val la chosa. Moltipicha in ssi $11\frac{4}{9}$, fano $\frac{103}{4}$, e questi $\frac{103}{4}$ moltipichadi in si fano 10609. Moltipicha 9 via 9, fa 81, e questo è lo partidor. A partir 10609 insirà $130\frac{79}{81}$, e tanto valse la seda la libra a chontadi. Azonto più la so radixe, che sono $11\frac{4}{9}$, serano in soma $142\frac{34}{81}$, che fu el baratto dela seda, chomo vedirì per figora.

Now to know how much silk we will have for 1,000 pounds of wool, this is the way. Put that we will have $1x$ of silk. Now we know that for $1x$ of silk we will have 1,000 pounds of wool, so we say: "If 100 pounds of silk is worth $13\frac{1}{2}$ ducats, what will $1x$ of wool be worth?" It makes $27x$ to be divided by 200, that is, $\frac{27}{200}x$. Now subtract the $\frac{1}{3}$, the remainder is $\frac{18}{200}$. Now add the $\frac{1}{3}$, it makes $\frac{24}{200}$. Now we should say: "If 8 gives me 100, what will $\frac{24}{200}$ give me?" Multiply 100 times 24, it makes $2{,}400x$. Divide by 8 times 200, that is 1,600, $\frac{2400}{1600}$, which is equal to 1,000 pounds of wool.

$\frac{2400}{1600}$ $\frac{1000}{1}$

```
   0
  011
  144
 04666
1600000 ]
 240000 | 666 2/3
  2400  |
   24   ]
```

And this problem is done by unknowns, so for 1,000 pounds of wool you should have $666\frac{2}{3}$ pounds of silk. By this method you will do all other similar problems by unknowns.

---

f. 27b

There are 2 men who want to barter; one has silk, whose worth in cash I don't know, and he puts into trade its value plus the square root of its value, and wants $\frac{1}{4}$ in coins; the other has cloth, which is worth 50 soldi per ell, and he put into the barter 56 soldi. I ask how much the silk was worth in cash. And to do this problem by unknowns, put that the silk was worth $1x^2$, and it is said that it is worth that plus its square root. So it was $1x^2$ plus $1x$. Then it says that he wants $\frac{1}{4}$ in coins. So we subtract $\frac{1}{4}$ of $1x^2$, the remainder is $\frac{3}{4}x^2$; and we subtract $\frac{1}{4}$ from $1x$. The remainder is $\frac{3}{4}x$. Then we subtract $\frac{1}{4}$ from $1x^2$, the remainder is $\frac{3}{4}x^2$, and we subtract $\frac{1}{4}x$ from $1x^2$, the remainder is $\frac{3}{4}x^2$[1] minus $\frac{1}{4}x$. Now $\frac{3}{4}x^2$ minus $\frac{1}{4}x$ is multiplied by 56, it would be $\frac{168}{4}x^2$ minus $\frac{56}{4}x$, and then we multiply 50 by $\frac{3}{4}x$, it will be $\frac{150}{4}x$, and $\frac{3}{4}x^2$ times 50 will be $150x^2$. Then we equalize the sides and we will have minus $\frac{56}{4}x$ to $\frac{168}{4}x^2$, and nothing will be lacking. And we add $\frac{56}{4}x$ to $\frac{150}{4}x$, it makes $\frac{206}{4}x$. Now we subtract $\frac{150}{4}x^2$ from $\frac{168}{4}x^2$, the remainder is $18x^2$. We divide $\frac{206}{4}x$ by $\frac{18}{4}x^2$, there will come $11\frac{4}{9}$, and that is the value of the unknown. Multiply $11\frac{4}{9}$ by itself, it makes $\frac{103}{9}$,[2] and this $\frac{103}{9}$[3] multiplied by itself is 10,609, multiply 9 times 9, it makes 81, and this is the divisor. Divided into 10,609 will result in $130\frac{79}{81}$, and that was the value of the silk per pound in cash. Add to it its square root, which is $11\frac{4}{9}$, it will be in total $142\frac{34}{81}$, which was the barter of the silk, as you will see in the example.

---

1. $\frac{3}{4}x$ in MS.
2. $\frac{103}{4}$ in MS.
3. $\frac{103}{4}$ in MS.

$1^{\square}$ $1^{\square}$ 1^{co} $\frac{3^{\square}}{4}$ $\frac{3^{co}}{4}$ $\frac{3^{\square}}{4}$ men $\frac{1^{co}}{4}$ $\frac{3^{\square}}{4}$ $\frac{3^{co}}{4}$ $\frac{168^{\square}}{4}$ men $\frac{56^{co}}{4}$ $\frac{150^{\square}}{4}$ $\frac{150^{co}}{4}$

56^{co} $168^{\square}$ $\frac{206^{co}}{4}$ 206 42436 18 324

$150^{\square}$ 206 18

$18^{\square}$

03
014
1001
42436
32444
322
3

$130\frac{316}{324}$
$11\frac{4}{9}$
$142\frac{306}{729}$

E questo è fato per la chosa, e per far la prova vedirì in charte 63.

f. 28a

[Chonpanie de merchadantie]

Sono do homeny chi s'achonpagnia. L'uno mese in la chonpanya duchati 100 e l'oltro mese in la chonpanya piper sachi 3, e in chorto tenpo tra pro' e chavedal se truovò aver duchati 300. Quel che messe duchati 100 avè per dretta raxion duchati 90, e quel che messe sachi 3 de piper avè duchati 210. Adomando zò che fu messo el sacho de piper quanto[a] s'achonpanya.

In primo per farla per farla[b] per la riegola del 3 a saver zò che fu meso a valer el sacho de piper, noi diremo: "Se 90 eran 100, che iera 210?" Se moltipicho 100 via 210 fano 21000. Parti per 90, insirà $233\frac{3}{9}$, e perché i è sachi 3, parti per 3, venyrà duchati $77\frac{70}{90}$. E tanto fu messo a valer el sacho quando lor do s'achonpaniaron, chomo vedirì qui de sotto.

$\frac{90}{1} \times \frac{100}{1}$ — $\frac{210}{1}$ 21000

00
0333
21000
9000
99

$233\frac{30}{90}$ $\frac{3}{9}$

$\frac{21000}{90}$ $\frac{3}{1}$

3

270 partior

02
27
0711
a partir 21000
2700
27

$77\frac{210}{270}$ $\frac{7}{9}$

---

a. *Sic in MS, evidently for* quando.

b. *Thus repeated in MS.*

$1x^2 \quad 1x^2 \quad 1x \quad \frac{3}{4}x^2 \quad \frac{3}{4}x \quad \frac{3}{4}x^2 \quad \text{minus} \quad \frac{1}{4}x \quad \frac{3}{4}x^2 \quad \frac{3}{4}x \quad \frac{168}{4}x^2 \quad \text{minus} \quad \frac{56}{4}x \quad \frac{150}{4}x^2 \quad \frac{150}{4}x$

$56x \quad 168x^2 \quad \frac{206}{4}x \quad 206 \quad 42436 \quad 18 \quad 324$

$150x^2 \quad 206 \quad 18$

$18x^2$

03
014
1001
42436
32444 | $130\frac{316}{324}$
322 | $11\frac{4}{9}$
3 | $142\frac{306}{729}$

And this is done by unknowns, and to see the proof, look on folio 63.

f. 28a

MERCHANT COMPANIES

There are two men who form a company. One put 100 ducats into the company and the other put three sacks of pepper into the company, and in short time between profit and capital it was found to have 300 ducats. The one who put 100 ducats had by correct accounting 90 ducats, and the one who put in 3 sacks of pepper had 210. I ask what value was attributed to the sack of pepper when the company was formed.

First to do it by the rule of 3, to know at what value the sack of pepper was put, we will say: "If 90 was 100, what will 210 be?" If I multiply 100 by 210, it makes 21,000. Divide by 90, it will result in $233\frac{3}{9}$, and because it's three sacks, divide by 3; it will come out $77\frac{70}{90}$ ducats. And that's what the value of the sack was put at when the two of them formed the company, as you will see here below.

$\frac{90}{1} \times \frac{100}{1} — \frac{210}{1} \quad 21000$

00
0333
21000
9000 | $233\frac{30}{90}$ $\frac{3}{9}$
99

$\frac{21000}{90} \quad \frac{3}{1}$

3

270 divisor

02
27
0711
to divide 21000
2700 | $77\frac{210}{270}$ $\frac{7}{9}$
27

E per far la ditta raxion per inpoxizion metti che ala prima inpoxizion fuse 150 per sachi 3, el segondo chonpagno mese duchati 100, serano 250. Puo' di': "Se 250 me dà tra pro' e chavedal duchati 300, che me darà 150 che fu messo al'inpoxizion?" Adoncha moltipicha 300 via 150,
f. 28b fano 45000. Puo' moltipicha 210 | via 250 fano 52500, sotrazi, serano più 7500. Puo' moltipicha 300 via 300, fa 90000. Moltipicha 400 via 210, fano 84000. Sotrazi, serano men 6000. Puo' moltipicha in croxe 150 via 6000, fano 2250000. Puo' moltipicha 300 via 7500, fano 900000. Azonzi, serano 3150000. Parti, insirà fuor $233\frac{4500}{13500}$, chomo vedirì qui de sotto per figora.

150 ╲ ╱ 300
più ╳ men
7500 ╱ ╲ 6000

$\frac{250}{1} \times \frac{300}{1}$ — $\frac{150}{1}$

300	250
150	210
000	000
1500	250
300	500
45000	52500

52500
45000
più 7500

$\frac{250}{1} \times \frac{300}{1}$ — $\frac{300}{1}$ 90000

6000
150 | 2250000 | 400 | 90000
900000 | 210 | 84000
a partir 3150000 | | men 6000

E azonto a 300 dela segonda 100 fano 400. Moltipicha 400 fia 210 fano 84000.

7500
6000
13500

0
0~~1~~
~~1~~~~4~~4
0~~4~~~~0~~~~0~~
~~1~~~~5~~~~5~~~~5~~5
~~3~~~~1~~~~5~~~~0~~~~0~~00
~~1~~~~3~~~~5~~0000
~~1~~~~3~~~~5~~00
~~1~~~~3~~~~5~~

$233\frac{4500}{13500}$ | $\frac{3}{9}$

~~23~~[a] $\frac{3150000}{13500} \times \frac{3}{1}$
3
40500

031
0~~3~~~~1~~~~5~~5
~~3~~~~1~~~~5~~~~0~~~~0~~00
~~4~~0~~5~~000
~~4~~0~~5~~0

$77\frac{31500}{40500}$ | $\frac{7}{9}$

a. *Number crossed out with diagonal line.*

And to do this problem by false position, suppose that for the first false position it was 150 for 3 sacks, the second partner put 100 ducats, it will be 250. Then say: "If 250 gives me 300 between profit and capital, what will the 150 give me that was put in the false position?" Then multiply 300 by 150, which makes 45,000. Then multiply 210 | times 250, which makes 52,500, subtract, it will f. 28b be plus 7,500. Then multiply 300 by 300, makes 90,000. Multiply 400 times 210, which makes 84,000. Subtract, it will be minus 6,000. Then multiply across 150 times 6,000, which makes 2,250,000. Then multiply 300 times 7,500, which makes 900,000.[1] Add, it will be 3,150,000. Divide, the result will be $233\frac{4500}{13500}$, as you will see worked out below.

150 ╲ ╱ 300
plus ╳ minus
7500 ╱ ╲ 6000

$\frac{250}{1}$ ╳ $\frac{300}{1}$ — $\frac{150}{1}$

300
150
000
1500
300
45000

250
210
000
250
500
52500

52500
45000
plus 7500

$\frac{250}{1}$ ╳ $\frac{300}{1}$ — $\frac{300}{1}$ 90000

6000
150
2250000
900000
to divide 3150000

400
210

90000
84000
minus 6000

And 100 added to 300 of the second makes 400. Multiply 400 by 210 makes 84000.

7500
6000
13500

0
01
144
0400
15555
3150000
1350000
13500
135

$233\frac{4500}{13500}$ | $\frac{3}{9}$

23 $\frac{3150000}{13500}$ ╳ $\frac{3}{1}$
3
40500

031
03155
3150000
405000
4050

$77\frac{31500}{40500}$ | $\frac{7}{9}$

1. Michael gives these two results in the wrong order (the first should be 900,000, the second 2,250,000), though it doesn't affect what follows.

E per far la ditta raxion per inpoxizion pony che la prima inpoxizion fusse 2, e tu debi dir: "Se 100
f. 29a torna 90, che serà 2?", che tu ha posto ala prima inpoxizion. Se moltipi[*cho*] | 90**a** via 2 fa 180, e questi partidi per 100 serano $1\frac{8}{10}$, e my voio che sia 210. Adoncha serano men $208\frac{2}{10}$. E la segonda inpoxizion metemo che la fose 3, e noi diremo: "Se 100 torna 90, che tornerà 3?" Se moltipicho 3 via 90 fa 270, e questi partidi per 100 farano $2\frac{7}{10}$, e noi volemo che fose 210. Adoncha abatello de 210, roman $207\frac{3}{10}$. E questi metti men ala segonda inpoxizion. Moltipicha in croxe e parti chomo dele oltre avemo fatto, inssirà chomo indriedo avemo fatto. E per questa farì ogn'altra raxion.

Anchora per la ditta raxion moltipicha chomo de sovra avemo ditto, chomo vedirì qui de sotto per figora.

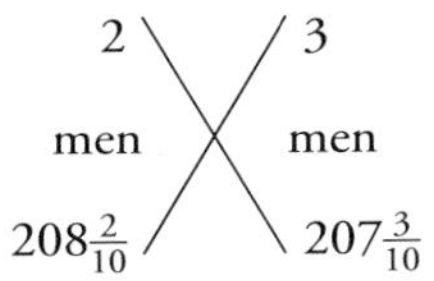

Anchora se la ditta raxion fose posto 1 sacho, moltipicha 3 volte l'inpoxizion per veder zò che val el sacho, chomo vedirì qui de sotto per figora.

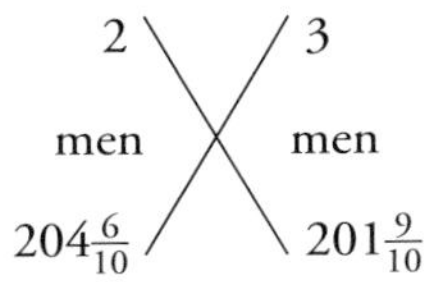

100 90 6

100 90 9

f. 29b E per far la ditta raxion per la chosa, pony che fose 100ⁿ e 3ᶜᵒ chi vadagnase tra pro' e chavedal duchati 300. Moltipichemo 100 via 300, fa 30000, puo' moltipiche[*mo*] 3ᶜᵒ via 90, fa 270ᶜᵒ, puo' moltipichemo 100 via 90, fa 9000, che son ingual a 30000. Sotrazi 9000, resta 21000. Parti per 270ᶜᵒ, insirà $77\frac{210}{270}\,\frac{7}{9}$. E tanto valse el sacho, chomo vedirì qui de sotto per figora.

1^{co}	100^{n}	100^{n}	3^{co}	300	02	
		90	90	100	27	
		9000	270^{co}		0711	
				30000	21000	
				9000	2700	$77\frac{210}{270}$**b** $\frac{7}{9}$
				21000	27	

a. 100 via 2 fano 200, se parto per 90 insirà fuora $2\frac{2}{9}$, e my voio che ssia 210. Adoncha abatti $2\frac{2}{9}$ de 210, roman *crossed out.*
b. *An extra* 0 *crossed out.*

And to do the same problem by false position, suppose that the first false position was 2, and you should say: "If 100 gives 90, what will 2 be?", which you have put at the first false position. If I multiply | 90[1] by 2 it makes 180, and this divided by 100 will be $1\frac{8}{10}$, and I want it to be 210. So it will be minus $208\frac{2}{10}$. And for the second false position, we suppose that it was 3, and we will say: "If 100 returns 90, what will 3 return?" If I multiply 3 times 90 it makes 270, and this divided by 100 will make $2\frac{7}{10}$, and we wanted it to be 210. So subtract it from 210, the remainder is $207\frac{3}{10}$. And this we put as minus at the second false position. Multiply across and divide as in the others that we've done, it will come out as we've done before. And in this way you will do all other problems. f. 29a

Then for this problem it's multiplied as we have said above, as you will see worked out below.

2	3
minus	minus
$208\frac{2}{10}$	$207\frac{3}{10}$

Then if the problem was put as 1 sack, multiply the false position by three to see the value of the sack, as you will see worked out below.

2	3	100	90	6			
					100	90	9
minus	minus						
$204\frac{6}{10}$	$201\frac{9}{10}$						

And to do this problem by an unknown, put that it was 100 plus $3x$ which was worth 300 ducats between profit and capital. We multiply 100 by 300, which makes 30,000, then we multiply $3x$ by 90, which makes $270x$, then we multiply 100 by 90, which makes 9,000, which is equal to 30,000. Subtract 9,000, the remainder is 21,000. Divide by $270x$, there will result $77\frac{210}{270}\ \frac{7}{9}$. And that was the value of the sack, as you will see worked out here below. f. 29b

$1x$	100	100	$3x$	300	02	
		90	90	100	27	
		9000	$270x$		0711[2]	
				30000	21000	
				9000	2700	$77\frac{210}{270}\ \frac{7}{9}$
				21000	27	

1. The following crossed out in MS: *100 times 2 makes 200, divided by 90 results in $2\frac{2}{9}$, and I wanted it to be 210, so subtract $\frac{2}{9}$ from 210, the remainder is*

2. Michael sometimes neglects to cross out numbers in *galera* division problems; in this line, the first 1 should properly be crossed out.

Sono 2 chonpagni chi s'achonpagna. L'uno messe in la chonpagna duchati 300 e l'oltro messe piper chargi 8, e stete in la chonpagna ano 1, e in questo tenpo se truovò aver tra pro' e chavedal duchati 600. Quel che mese duchati 300 avè per dreta raxion duchati 250, e quel che mese piper chargi 8 avè per dretta raxion duchati 350. V'adomando zò che valse el chargo de piper quanto[a] fu fatta la dita chonpania e che valse tuto el piper, che fu chargi 8.

In primo per far la ditta raxion per la riegola del 3 tu farà: "Se 250 m'eran 300, che tornerà 350?" Adoncha moltipicha 300 via 350, fano 105000, e questi partidi per 250 insirà duchati 420. E tanto fu messo a valer tuto el piper quanto[b] fu fatta la ditta chonpanya. E voiando veder zò che valse el chargo, parti 420 per 8, insirà duchati $52\frac{1}{2}$. E tanto valse el chargo.

f. 30a E per far la ditta raxion posto indriedo per inpoxizion, pony che ala prima inpoxizion fusse messo 2. Adoncha dovemo dir: "Se 300 che messe el primo torna 250, che serave 2 ch'ò posto ala prima inpoxizion?" Se moltipicho 2 via 250 fano 500, e questi partidi per 300 insirà fuor $1\frac{2}{3}$, e noi volemo che fuse 350, che avè cholui che mese el piper. Adoncha sutrazi de $1\frac{2}{3}$ de 350, vegnirà a romanir $348\frac{1}{3}$, e questi pony che manchase ala prima inpoxizion. Poi faremo per la segonda inpoxizion che fusse 3, e diremo: "Se 300 erano 250, che serave 3?" Se moltipicho 3 via 250 fano 750, e questi partidi per 300 serano $2\frac{1}{2}$, e noi vossemo che fose 350. Adoncha abatti $2\frac{1}{2}$ de 350, roman $347\frac{1}{2}$. Poi moltipicha in croxe 2 via $347\frac{1}{2}$, fano 695, e puo' moltipicha 3 via $348\frac{1}{3}$, fano 1045, e questo sutrato el menor dal mazuor roman 350. E puo' per far el partior sutrazi de $348\frac{1}{3}$ $347\frac{1}{2}$, roman $\frac{5}{6}$. Parti $\frac{350}{1}$ per $\frac{5}{6}$, insirà fuor 420, e questi valleva el piper quanto[c] fu fatta la chonpanya. E per veder zò che fu messo el chargo parti 420 per 8, insirà duchati $52\frac{1}{2}$, chomo vedirì qui per figura.

a. *Sic in MS, evidently for* quando.
b. *Sic in MS, evidently for* quando.
c. *Sic in MS, evidently for* quando.

There are two partners who make a company. One put 300 ducats into the company and the other put 8 loads of pepper, and stayed in the company 1 year, and in this time it was found that there was between profit and capital 600 ducats. The one who put 300 ducats had by correct accounting 250 ducats, and the one who put 8 loads of pepper had by correct accounting 350 ducats. I ask you how much the load of pepper was worth when the company was founded and what all the pepper was worth, which was 8 loads.

First to do the problem by the rule of 3, you will do: "If 250 was 300, what would 350 give?" Then multiply 300 by 350, makes 105,000, and this divided by 250 will give 420 ducats. And that is the value of all of the pepper when the company was made. And wanting to see how much a load was worth, divide 420 by 8, it will give $52\frac{1}{2}$ ducats. And that is the value of the load.

And to do the problem set out on the back of this page by false position, suppose that at the first false position a 2 was put. Then we should say: "If 300 that the first one put gives 250, what will the 2 give that I have put at the first false position?" If I multiply 2 times 250, it makes 500, and this divided by 300 will give $1\frac{2}{3}$, and we wanted it to be the 350 that the one who put in the pepper had. So subtract $1\frac{2}{3}$ from 350, there will remain $348\frac{1}{3}$, and put that this much was minus at the first false position. Then we will suppose for the second false position that it was 3, and we will say: "If 300 was 250, what would 3 be?" If I multiply 3 by 250 it makes 750, and this divided by 300 will be $2\frac{1}{2}$, and we wanted it to be 350. So, subtract $2\frac{1}{2}$ from 350, the remainder is $347\frac{1}{2}$. Then multiply across 2 times $347\frac{1}{2}$, makes 695, and then multiply 3 times $348\frac{1}{3}$, makes 1045, and here subtract the lesser from the larger, which leaves 350. And to make the divisor, subtract $347\frac{1}{2}$ from $348\frac{1}{3}$, the remainder is $\frac{5}{6}$. Divide $\frac{350}{1}$ by $\frac{5}{6}$, it will result in 420, and this was the value of the pepper when the company was made. And to see at what the load was set, divide 420 by 8, there will result $52\frac{1}{2}$ ducats, as you will see worked out here. f. 30a

2		3	300	250	2	250	500	2		
men		men				2		500		
								300	$1\frac{200}{300}$	$\frac{2}{3}$
$348\frac{1}{3}$		$347\frac{1}{2}$	300	250	3	250	750			
						3		1		
								750		
								300	$2\frac{150}{300}$	$\frac{1}{2}$

	350	350	$\frac{350}{1}$	$\frac{5}{6}$
	$1\frac{2}{3}$	$2\frac{1}{2}$		
	men $348\frac{1}{3}$	men $347\frac{1}{2}$		

$348\frac{1}{3}$	$348\frac{1}{3}$	$347\frac{1}{2}$	1045	0	
$347\frac{1}{2}$	3	2	695	0~~1~~0	
partior $\frac{5}{6}$	1045	695	a partir 350	~~2~~~~1~~~~0~~0	
				~~5~~~~5~~5	420

Fu messo el piper duchati 420, e questi partidi per 8.

0	
0~~2~~4	
~~4~~~~2~~~~0~~	
~~8~~~~8~~	$52\frac{4}{8}$ $\frac{2}{4}$ $\frac{1}{2}$

f. 30b E per far la ditta raxion per la chosa tu ssa ben che 'l primo mese duchati 300, adoncha dixe cholui che messe el piper: "Fu messo 1 ^{co}, zoè 300 e 1 ^{co}, e si avè tra pro' e chavedal duchati 600." Adoncha noi faremo: "Se 300 e 1 ^{co} me dà 600, che me darà 300?" Farano 180000, e se partimo per 250 venirà 250. Adoncha moltipichemo 250 via 300, fano 75000, e puo' moltipichemo 250 via 1 ^{co}, fano 250 ^{co}, che son ingual a questi nomeri. Adoncha dovemo sotrar de 180000 75000, venirà a romanir 105000, e questi partidi per 250 ^{co} insirà 420 duchati. E tanto valse tuto el piper quanto[a] fu fatta la chonpanya. E se volesi veder zò che valse el chargo, parti 420 per 8 chargi, insirà fuori duchati $52\frac{1}{2}$. E tanto fu meso a valler el chargo, chomo vedirì qui de sotto per figura.

a. *Sic in MS, evidently for* quando.

2	3	300 250 2	250 500 2
minus	minus		2
$348\frac{1}{3}$	$347\frac{1}{2}$	300 250 3	250 750
			3

~~5~~00
~~3~~00] $1\frac{200}{300}$ | $\frac{2}{3}$

1
~~7~~50
~~3~~00] $2\frac{150}{300}$ | $\frac{1}{2}$

350	350	$\frac{350}{1}$ $\frac{5}{6}$
$1\frac{2}{3}$	$2\frac{1}{2}$	
minus $348\frac{1}{3}$	minus $347\frac{1}{2}$	

$348\frac{1}{3}$	$348\frac{1}{3}$	$347\frac{1}{2}$	1045	0
$347\frac{1}{2}$	3	2	695	0~~1~~0
divisor $\frac{5}{6}$	1045	695	to divide 350	~~2~~~~1~~~~0~~0
				~~5~~~~5~~5] 420

The pepper was set at 420, and this divided by 8.

0
0~~2~~4
~~4~~~~2~~~~0~~
~~8~~~~8~~] $52\frac{4}{8}$ $\frac{2}{4}$ $\frac{1}{2}$

And to do this problem by unknowns, you know well that the first one put in 300 ducats, so the one who put in the pepper said: "1*x* was put in, that is 300 plus 1*x*, and there was a total of 600 ducats between profit and capital." So we will do the following: "If 300 plus 1*x* gives me 600, what will 300 give me?"[1] It will be 180,000, and if we divide by 250 it will come out to 250. Then we multiply 250 by 300, which makes 75,000, and then we multiply 250 times 1*x*, which makes 250*x*, which is equal to these numbers. Then we should subtract 75,000 from 180,000, and the remainder will be 105,000, and this divided by 250*x* will give 420 ducats. And that was the value of all the pepper when the company was made. And if you would like to see what the worth of the load of pepper was, divide 420 by 8 loads, the result will be $52\frac{1}{2}$ ducats. And that is what the load was worth, as you will see worked out below. f. 30b

1. The solution becomes confused at this point.

1^{co}	300^{n} e 1^{co}	300^{n}	600^{n}	300^{n}	600	180000
					300	75000
					000	105000
					000	
					1800	

300	250^{co}	0		0		
250		0~~1~~		0~~2~~4		
000		0~~25~~0		~~4~~~~2~~~~0~~		
1500		~~1~~~~0~~~~5~~~~0~~00		~~8~~~~8~~	$52\frac{4}{8}$	$\frac{1}{2}$
600		~~2~~~~5~~000	420			
75000		~~2~~~~5~~5				
		2				

[Pagar nuolli in gallia o in nave]

Sono 2 chi vien del viazo de Soria e portò in nave piper sachi 5 e l'oltro portò in nave sachi 9. E vegnando in luogo dove i marchadanti vuolse li so sachi, el patron i domandò el nolo, e de achordo f. 31a el patron tolse de chadaun de loro sacho 1 de piper. El patron vende li 2 sachi de piper | e a choluy che avea sachi 9 i dè per so resto duchati 20, e a choluy che avea sachi 5 i dè per so resto duchati 32, e dise elo era pagado del so nuolo, e v'adomando zò che paga de nolo per sacho e zò che fu vendudo el sacho.

E voyando far la ditta raxion per inpoxizion pony che ala prima inpoxizion metemo che pagasse duchato 1. Adoncha per sachi 9 vorave de nollo duchati 9. El patron i dè per resto duchati 20, che val 29. Adoncha el segondo se 'l paga per sacho 1 duchato 1, per sachi 5 serà duchati 5. El patron i dè per so resto duchati 32, val 37, e non vuol salvo 29. Adoncha avè de più duchati 8. E chosì meteremo al'inpoxizion per 1 che m'ò poxo più 8. Mo' femo la segonda inpoxizion. Ponamo che pagasse duchati 2 per sacho. Adoncha 2 fia 9 fa 18, e 20 avè del patron val 38. El segondo 2 fia 5 val 10, e 32 avè del patron fa 42, e non die aver più cha 38, adoncha avè più 4. E chosì metemo al'inpoxizion per 2 che m'ò puxo più 4. E moltipicha in croxe un fia 4 fa 4, e 2 fia 8 fa 16. Più e più, sutrazi el men del più, roman 12. Sotrazi per far partidor 4 de 8, roman 4, e questo è lo partidor. Partir 12 per 4 serano 3, e tanti duchati paga de nolo per sacho. E per veder quanti duchati fu vendudo el sacho, el primo 3 fia 9 fa 27, e 20 avè del patron fano 47. El segondo 3 fia 5 fa 15, e 32 avè del patron fa 47. Adoncha el sacho fu vendudo per duchati 47, e pagallo per sacho de nollo f. 31b duchati 3, cho[*mo*] vedirì qui inchontro | per figura d'inpoxizion, e per questa farì tute le oltre simille raxion.

$1x$	300 and $1x$	300	600	300	600	180000	
					300	75000	
					000	105000	
					000		
					1800		

300	$250x$	0		0		
250		01		024		
000		0250		420		
1500		105000		88	$52\frac{4}{8}$	$\frac{1}{2}$
600		25000	420			
75000		255				
		2				

To pay freight charges for a galley or ship

There are two men who return from the voyage to Syria, and one carried in the ship 5 sacks of pepper, and the other carried in the ship 9 sacks. And coming to the place where the merchants wanted their sacks, the captain asked them for the freight charge, and by agreement the captain took from each of them 1 sack of pepper. The captain sold the 2 sacks of pepper | and to the one who had 9 sacks he gave 20 ducats for his change, and to the one who had 5 sacks he gave 32 ducats for his change, and he said that he had been paid for his freight, and I ask you what was the freight charge per sack and at what price the sack was sold. f. 31a

And wanting to do this problem by false position, suppose at the first false position that we put that he paid 1 ducat. So for 9 sacks the freight would have been 9 ducats. The captain gave him 20 ducats change, which is worth 29 ducats. Then if the second one paid 1 ducat per sack, for 5 sacks it would be 5 ducats. The captain gives him for his change 32 ducats, which is a value of 37 ducats, and we only want 29 ducats. So he had plus 8 ducats. And so we will put at the false position for 1 that he had given me plus 8 ducats. Now let's make the second false position. Let's suppose that he paid 2 ducats per sack. So 2 times 9 makes 18, and the 20 he had from the captain makes 38. The second 2 times 5 makes 10, and the 32 he had from the captain makes 42, and he shouldn't have more than 38, so he had plus 4. And so we put for the false position for 2 that now I've put plus 4. And multiply across, one times 4 makes 4, and 2 times 8 makes 16. Plus and plus, subtract the lesser from the greater, the remainder is 12. To make the divisor, subtract 4 from 8, the remainder is 4, and this is the divisor. Divide 12 by 4, it will be 3, and that many ducats per sack is the freight charge. And to see how many ducats the sack was sold for, first 3 times 9 makes 27, and the 20 that the captain had makes 47; for the second 3 times 5 makes 15, and the 32 the captain had makes 47. So the sack was sold for 47 ducats, and each sack paid 3 ducats for freight, as you shall see opposite here | worked out by false position, and in this way you will do all other similar problems. f. 31b

1		2	9	5	37	18	10	42
			20	32	29	20	32	38
più	X	più	29	37	più 8	38	42	più 04
8		4						

16	4	16	12	8	4	00	
		4		4		12	
						4	3

9		10	5
3			3
27	47		15
20	47		32

E per far la sovrascritta raxion per la chosa pony che 'l primo, che avè sachi 9, pagasse 1^{co} per sacho. Adoncha per 9 sachi pagaria 9^{co}. El segondo pagaria per sachi 5 5^{co}. E puo' el patron dè per so resto al primo 20^{n}, e quel del segondo i dè per so resto 32^{n}. Seria 9^{co}, 20^{n}, 5^{co}, 32^{n}. Adequa le parte. Più 9^{co} più 5^{co}, sotrazi l'ono del'oltro, roman 4^{co}, più 20^{n}, più 32^{n}. Sutrazi el menor del mazuor, romagnerà 12^{n}. A partir in 4^{co} adoncha serano 3, e tanti pagarà per nolo per sacho. E per veder zò che valse el sacho, 3 fia 9 fa 27, e 20 per resto fa 47. E 3 fia 5 fa 15, e 32 per so resto fa ben 47. Sì che paga el sacho per nolo duchati 3, e sì fo vendudo el sacho per duchati 47, chomo vedirì qui de sutto per figura.

1^{co}	9^{co}	20^{n}	5^{co}	32^{n}	9^{co}	4^{co}	32^{n}	12^{n}	4^{co}
					5^{co}		20^{n}		

00		9		5
12		3		3
4	3	27	47	15
		20	47	32

f. 32a Sono 2 chi vien de viazio. El primo porta sachi 40 de lana, el segondo porta sachi 30 de lana, e vuol pagar el nolo. I merchadanti dè al patron el primo sachi 2, el segondo sacho 1. El patron vende i sachi, e al primo i dè per so resto duchati 10, al segondo dise mancharli duchati 7. Adomando zò che paga de nolo per sacho e zò che fu vendudo el sacho.

E prima per far la ditta raxion per inpoxizion pony che 'l primo avesse pagado duchato 1 per sacho, per 40 sachi pagerà duchati 40, e 10 avè per resto fano 50, e fu sachi 2. Adoncha vene el sacho duchati 25. El segondo duchati per sacho voralo duchati 30, e sì disse chi manchasse duchati 7. Adoncha non die aver salvo duchati 25 et alo duchati 30, avanza duchati 5. Mo' dixe llo patron

1	2	9	5	37	18	10	42	
plus	plus	20	32	29	20	32	38	
		29	37	plus 8	38	42	plus 04	
8	4							

16	4	16	12	8	4	00	
		4		4		~~12~~	
						~~4~~	3

9		~~10~~	5
3			3
27	47		15
20	47		32

And to do the problem above by unknowns, put that the first one, who had 9 sacks, paid 1*x* per sack. Then for 9 sacks he would pay 9*x*. The second would pay 5*x* for 5 sacks. And then the captain gave as change to the first one 20, and to the second he gives as change 32. It would be 9*x*, 20, 5*x*, 32. Equalize the sides. Plus 9*x* and plus 5*x*, subtract one from the other, the remainder is 4*x*, plus 20 and plus 32. Subtract the lesser from the greater, there will remain 12. Divided by 4*x* will be 3, and that's how much the freight will cost per sack. And to see what the sack was worth, 3 times 9 makes 27, and 20 as change makes 47. And 3 times 5 makes 15, and 32 for his change makes 47. So each sack pays 3 ducats freight, and the sack was sold for 47 ducats, as you will see worked out below.

1*x*	9*x*	20	5*x*	32	9*x*	4*x*	32	12	4*x*
					5*x*		20		

00		9		5
~~12~~		3		3
~~4~~	3	27	47	15
		20	47	32

There are 2 men who come on a voyage. The first carries 40 sacks of wool, the second carries 30 sacks of wool, and wants to pay the freight. The first merchant gives to the captain 2 sacks and the second 1 sack. The captain sells the sacks, and he gives to the first man 10 ducats change and to the second he says that he's 7 ducats short. I ask what the freight charge per sack is and what the sack was sold for. f. 32a

And first to do this problem by false position, suppose that the first had paid 1 ducat per sack, so he will pay 40 ducats for 40 sacks, and the 10 he had as change makes 50, and it was 2 sacks. So the sack went for 25 ducats. The second would have had 30 ducats per sack, and it's said that 7 ducats were lacking. So he was supposed to have only 25 ducats per sack and he has 30, so it's 5 ducats

mancharlli 7, adoncha sotrazi 5 de 7, i mancherà duchati 2. Adoncha per 1 che m'ò poxo mancha 2. E chosì metti al'inpoxizion prima. E per far la segonda inpoxizion pony che fusse 2. Adoncha 2 fia 40 fa 80, e 10 val 90 per 2 sachi. Venyrà a valer el sacho duchati 45. El segondo, 2 fia 30 fano 60, e non vuol valer più cha 45. Sutrazi 45 de 60, roman più 15. El patron dise averli manchado duchati 7, sutrazi de 15 roman più 8, e questi metti più ala segonda inpoxizion per 2 che m'ò poxo più 8. Moltipicha in croxe e dirai: "1 fia 8 fa 8 e 2 fia 2 fa 4," men e più, azonzi 8 e 4 fa 12, azonzi 2 chon 8, è questo lo partidor. Parti 12 per 10, serano duchati $1\frac{1}{5}$, e tanto paga per sacho de nollo. E
f. 32b per veder quanti duchati fo vendudo el sacho tu fara' $1\frac{1}{5}$ | fia 40 fa 48, e 10 avè dal patron fano duchati 58, a partir in sachi 2 fano duchati 29. E tantto fu ve[*n*]dudo el sacho. El segondo, a moltipichar $1\frac{1}{5}$ fia 30 fa 36. Mo' dise el patron che i mancha duchati 7. Sutrazi 7 de 36, roman ben 9, chomo vedirì qui de sotto per figora.

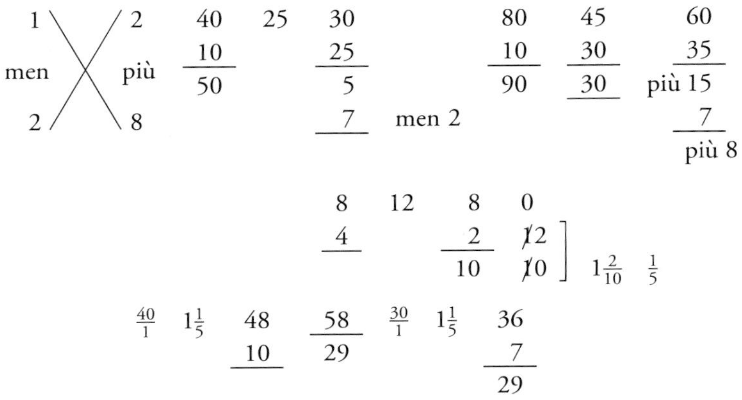

E per far la ditta raxion per la chosa pony che 'l primo avese pagado 1^{co} per sacho. Adoncha per sachi 40 pagaria 40^{co} e 10^{n}, a partir per 2 sachi seria 20^{co} più 5^{n}. El segondo die pagar 1^{co} per sacho. Adoncha averà 30^{co} men 7^{n}. Adequa le parte. 30^{co} più 20^{co} più più[a] 5^{n} men 7^{n}, più e più chosse, sotrazi el menor dal mazuor, 20^{co} de 30^{co}, roman 10^{co}. Azonzi più 5^{n} men 7^{n}, fano 12^{n}, a partir per 10^{co}. Parti, e partido averì duchati $1\frac{1}{5}$, e tanto paga el sacho per nolo. E per veder zò che fu vendudo el sacho, $1\frac{1}{5}$ fia 40 val 48, e 10 per resto fa 58. Parti per 2 sachi, fano 29, e moltipicha $1\frac{1}{5}$ via 30, fa 36. Abatti 7 che mancha, serano 29.

---

a. *Repeated thus in MS.*

over. Now since the captain said that he was 7 short, then subtract 5 from 7, there will be 2 ducats short. So for 1 that I put, it's minus 2. And we put this for the first false position. And to do the second false position, suppose that it was 2. Then 2 times 40 makes 80, plus 10 makes 90 for 2 sacks. So the sack will come out to be worth 45 ducats. The second man, 2 times 30 makes 60, and it should only be worth 45. Subtract 45 from 60, the remainder is 15. The captain said that he was 7 ducats short, subtract from 15, the remainder is plus 8, and put this as a plus for the second false position for 2 that I've put, plus 8. Multiply across and you will say: "1 times 8 makes 8 and 2 times 2 makes 4," plus and minus, add 8 and 4 makes 12, add 2 to 8, and this is the divisor. Divide 12 by 10, it will be $1\frac{1}{5}$ ducats, and this was the freight charge per sack. And to see for how many ducats the sack was sold, you will do $1\frac{1}{5}$ | times 40 makes 48, and the 10 that the captain had makes 58 ducats, divided into 2 sacks makes 29 ducats. And that's what the sack was sold for. The second, multiply $1\frac{1}{5}$ by 30 makes 36. Now the captain says that it was 7 ducats short. Subtract 7 from 36; the remainder is 29,[1] as you will see worked out below. f. 32b

1	2	40	25	30		80	45	60
minus	plus	10		25		10	30	35
		50		5		90	30	plus 15
2	8			7	minus 2			7
								plus 8

8	12	8	0		
4		2	~~1~~2		
		10	~~1~~0	$1\frac{2}{10}$	$\frac{1}{5}$

$\frac{40}{1}$	$1\frac{1}{5}$	48	58	$\frac{30}{1}$	$1\frac{1}{5}$	36
		10	29			7
						29

And to do this problem with unknowns, put that the first had paid $1x$ per sack. So for 40 sacks he would pay $40x$ and 10, divided by 2 sacks it would be $20x$ plus 5. The second should pay $1x$ per sack. So he will have $30x$ minus 7. Equalize the sides. $30x$ plus $20x$, plus 5 minus 7, positive and positive unknowns, subtract the lesser from the larger, $20x$ from $30x$, the remainder is $10x$. Add plus 5 minus 7, which makes 12, divide by $10x$. Divide, and divided you will have $1\frac{1}{5}$ ducats, and that is the cost per sack of the freight. And to see how much the sack was sold for, $1\frac{1}{5}$ times 40 is 48, and 10 for change makes 58. Divide by 2 sacks, it makes 29, and multiply $1\frac{1}{5}$ by 30, which makes 36. Subtract the 7 that's lacking, which will be 29.

1. *9* in MS.

1^{co} 40^{co} più 10^{n} più 20^{co} più 5^{n} più 30^{co} men 7^{n} più 30^{co} più 5^{n}

più 20^{co} men 7

10^{co} 12^{n}

$1\frac{1}{5}$ fia 40 fa 48 0

10 $\not{1}2^{n}$

58 $\not{1}0^{co}$ $1\frac{2}{10}$ $\frac{1}{5}$

duchati 29

$1\frac{1}{5}$ 30 fa 36

7

duchati 29

f. 33a

[ZUGAR DADI IN BARATARIA]

Sono 3 chi vuol zugar[a] ai dadi. El primo vadagna al segondo el terzo d'i so denari, el segondo vadagna al terzo el quarto d'i so denari. Quanto[b] se levano dal zuogo se truovò che chadaun di loro avea in man duchati 8, ma el terzo vadagnò al primo el quinto d'i so denari. Adomando: "Quanti avea zaschon di loro quanto[c] se meseno al zuogo che levadi se truovò per zaschon duchati 8?"

E per voler far la ditta raxion per inpoxizion pony ala prima inpoxizion che fusse 8, ala segonda inpoxizion fosse 6. Mo' chonvien che ttu metti de sotto 8 sifatto nomero che dagando el terzo al primo possa aver per dar el quintto al terzo, e quel segondo romagna chon 8, abiando dal terzo el quartto. Adoncha mettemo sotto 8 6, perché toyando el primo el terzo, serano 2, averà 10, ben porà dar el quinto al terzo. El primo romagnerà chon 8. El segondo, dagando $\frac{1}{3}$ al primo d'i 6, roman chon 4, che dovemo meter al terzo, che dagando el quarto el segondo debia romanir chon 8. Adoncha mettemo sotto 6 16. Se de 16 tuo' el quartto, el segondo roman chon 8, el terzo roman chon 12. Mo' el terzo vuol el quinto dal primo, che son 10, 2. Roman el primo chon 8, el segondo chon 8, el terzo chon 14, e noi vosemo che l'avesse 8. Ado[*ncha*] è più 6, e meteremo al'inpoxizion
f. 33b prima 8 e 6 e 16 più 6. E per far la segonda inpoxizion chomo è de|chiaratto ala prima, mettemo 6, e de sotto 6 12, e de sotto 0. Adoncha el primo tuo' dal segondo el terzo de 12, fano 4, azonti chon 6 fano 10, per dar el quinto al terzo. Mo' el segondo roman chon 8, el terzo sta chon 0 e die aver dal primo el quinto, che son 2. Adoncha el terzo averà 2, e noi volemo che l'abia 8. Adoncha mancha 6. Mo' moltipicha in croxe e fa 6 fia 8 fa 48, e 6 fia 6 fa 36. Men e più, azonzi 48 chon 36, fano 84, azonzi de sotto per far partidor. 6 e 6 fa 12, parti 84 per 12, insirà duchati 7, e tanti avea el primo chonpagno. E per far per lo segondo, 6 fia 6 fa 36, e 6 fia 12 fa 72, azonti insieme fano 108, e questi partidi per 12 fano 9, e tanti avè el segondo chonpagno. E per far el terzo moltipicha 6 fia 16, fa 96, e questi partidi per 12 fa 8, e tanti avè el terzo chonpagno. Mo' aprova la ditta raxion. El primo 7, el segondo 9. Tuo' el terzo che son 3. Averà el primo 10, el segondo roman chon 6. Tuo' el quarto dal terzo, che son 8, 2, averà el segondo 8. El terzo roman chon 6. Toyando el quinto dal primo, cha à 10, sono 2 e 6, val ben 8, chomo vedirì qui de sotto per figora.

---

a. zugar *added above the line.*
b. *Sic in MS, evidently for* quando.
c. *Sic in MS, evidently for* quando.

1x	40x	plus 10	plus 20x	plus 5	plus 30x minus 7	plus 30x	plus 5
						plus 20x	minus 7
						10x	12

$1\frac{1}{5}$ times 40 makes 48 / 10 / 58 / 29 ducats — 0 / ~~1~~2 / ~~1~~0x] $1\frac{2}{10}$ $\frac{1}{5}$ — $1\frac{1}{5}$ 30 makes 36 / 7 / 29 ducats

---

f. 33a

To play dice in barter

There are three who want to play dice. The first wins from the second a third of his money, the second wins from the third a quarter of his money. When they got up from the game, it turned out that each of them had in his hand 8 ducats, but the third won from the first a fifth of his money. I ask you: "How much did each one have when they started to play, so that when they got up it turned out that each had 8 ducats?"

And to do the problem by false position, suppose at the first false position that it was 8, at the second false position that it was 6. Now you should put below the 8 a number such that when one third of it is given to the first, enough will remain to give a fifth to the third, and that the second is left with 8, having a quarter from the third. So we put 6 under 8, because the first taking a third, it will be 2, he will have 10, so he can well give a fifth to the third. The first will remain with 8. The second, giving $\frac{1}{3}$ of the 6 to the first, remains with 4, which we should put at the third, which giving a quarter [to] the second should remain with 8. So we put 16 beneath 6. If he takes a quarter of 16, the second remains with 8, the third remains with 12. Now the third wants a fifth of the first which is 10, and that is 2. The first remains with 8, the second with 8, the third with 14, and we wanted him to have 8. So it's plus 6, and we will put at the first false position 8 and 6 and 16 plus 6. And to do the second false position, as it is | declared at the first, we put 6, and 12 below 6, and 0 below. f. 33b
So the first took from the second a third of 12, which makes 4, add to 6 makes 10, to give a fifth to the third. Now the second remains with 8, the third stays with 0 and should have a fifth of the first, which is 2. So the third will have 2, and we wanted him to have 8. So it is minus 6. Now multiply across and it makes 6 times 8 makes 48, and 6 times 6 makes 36. Plus and minus, add 48 to 36, makes 84, add below to make the divisor. 6 and 6 makes 12, divide 84 by 12, the result will be 7 ducats, and that's what the first companion had. And to do the second, 6 times 6 makes 36, and 6 times 12 makes 72, added together makes 108, and this divided by 12 makes 9, and that's what the second companion had. And to do the third, multiply 6 times 16, makes 96, and this divided by 12 makes 8, and that's what the third companion had. Now prove the problem. The first 7, the second 9. Take a third, which is three. The first will have 10, the second remains with 6. Take a quarter of the third, which is 8, 2, the second will have 8. The third remains with 6. Taking a fifth from the first, who has 10, are 2 and 6, which is worth 8, as you will see worked out below.

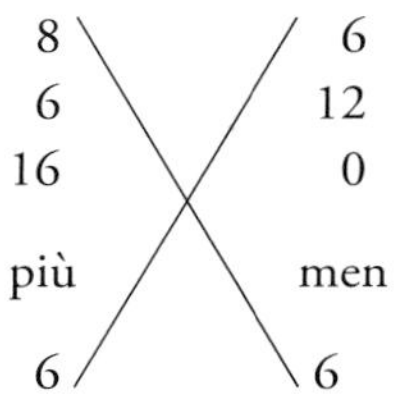

48	0		72	0	
36	~~1~~0		36	0~~1~~0	
84	~~84~~		108	~~108~~	
	~~12~~	7		~~12~~	9

96		
6	0	
	~~1~~0	
	~~96~~	
	~~12~~	8

primo 8
segondo 8
terzo 8

primo 7
segondo 9
terzo 8

10 primo
6
2 segondo 8

terzo 6
2

f. 34a E per voler far la sovrascritta raxion per la chossa pony che 'l | primo chonpagno avesse più 1^co^, el segondo chonpagno per forza die aver più 30^n^ men 3^co^, che dagando al primo el terzo d'i so denari romagna chon 10^n^. Adoncha daremo al primo più 10^n^ e men 1^co^, averallo 10^n^ più 1^co^ men 1^co^, roman 0. El segondo roman chon più 20^n^ men 2^co^. El terzo chonvien aver tanto che dagando el so quarto al segondo romagna chon 8^n^. Adoncha vorallo aver men 48^n^ più 8^co^. A dar al segondo men 12^n^ più 2^co^, abatti men 12^n^ da più 20^n^, vegnirà a romagnir chon 8^n^. E darlli $\frac{1}{4}$ de più 8^co^ serano 2^co^ più. A sbater 2^co^ men serà romaxo 0. El terzo roman aver men 36^n^ più 6^co^. Mo' el terzo vuol dal primo el quintto d'i so, che sono più 2^n^. Asbatti de 36^n^ men, roman 34^n^ 6^co^ ingual a più 8^n^. Adequa le parte. Men 34^n^ più 8^n^ fano 42, a partir in 6^co^, 6 7 fa 42, e tanti avè el primo chonpagno. El segondo vuol dar $\frac{1}{3}$ al primo, adoncha serà 9. Dandoi 3 serà 10, el segondo roman chon 6. El segondo[a] vuol el quarto dal terzo per aver 8^n^, adoncha avelo 8^n^. Dando al segondo 2^n^ roman chon 8^n^, el terzo roman chon 6^n^. El primo, dagando al terzo el so quinto, sono 2. Adoncha el terzo arà ben 8, 8 primo, 8 segondo, 8 el terzo, chomo vedirì qui de sotto per figora.

più 1^co^	più 30^n^	men 3^co^	men 48^n^	più 8^co^	
	più 20^n^	men 2^co^	men 36^n^	più 6^co^	
	men 12	più 2^co^	più 2		
	più 8	0^co^	men 34^n^	più 6^co^	ingual più 8

adequa men 34 7^n^ 9^n^ 8^n^ Serà primo 8, segondo 8, terzo 8.
più 8 42^n^
6^co^

a. segondo *corrected above the line over* terzo *crossed out with a horizontal line.*

8	6	48	0		72	0		
6	12	36	10		36	010		
16	0	84	84		108	108		
plus	minus		12	7		12	9	
6	6	96						
		6	0					
			10					
			96					
			12	8				

first 8
second 8
third 8

first 7
second 9
third 8

10 first
6
2 second 8

third 6
2

And to do this problem by unknowns, put that the | first companion had plus 1*x*, the second companion then had to have had plus 30 minus 3*x*, so that giving to the first a third of his money he remains with 10, so that, after giving a third of his money to the first one, he has 10 left. Then we will give to the first plus 10 and minus 1*x*, he will have 10 plus 1*x* minus 1*x*, the remainder is 0. The second remains with plus 20 minus 2*x*. The third should have enough so that giving a quarter of his to the second he remains with 8. So he will have minus 48 plus 8*x*. To give the second minus 12 plus 2*x*, subtract minus 12 from plus 20, he will come out with 8. And to give him $\frac{1}{4}$ of plus 8*x* will be plus 2*x*. To subtract minus 2*x*, there will remain 0. The third remains with minus 36 plus 6*x*. Now the third wants a fifth from the first, which is plus 2. Subtract from minus 36, the remainder is 34 and 6*x*, equal to plus 8. Equalize the sides. Minus 34 plus 8 makes 42, to divide by 6*x*, 6 7 makes 42, and that's what the first companion had. The second wants to give $\frac{1}{3}$ to the first, so it will be 9. Giving 3 will be 10, the second remains with 6. The second wants a quarter of the third to have 8, so he will have 8. Giving the second 2, he remains with 8, the third remains with 6. The first, giving the third his fifth, is 2. So the third will have 8, the first 8, the second 8, the third 8, as you will see worked out below. f. 34a

plus 1*x*	plus 30	minus 3*x*	minus 48	plus 8*x*	
	plus 20	minus 2*x*	minus 36	plus 6*x*	
	minus 12	plus 2*x*	plus 2		
	plus 8	0*x*	minus 34	plus 6*x*	equals 8

equalize minus 34   7 9 8
plus 8   42
6*x*

It will be for the first 8, for the second 8, for the third 8.

f. 34b Sono 4 chi vuol zugar ai dadi. El primo vadagnò al segondo el $\frac{1}{3}$ d'i so denari. El segondo vadagnò al terzo el $\frac{1}{4}$ d'i so denari. El $\frac{1}{3}$ vadagnò al quarto el quinto d'i so denari. El quartto vadagnò al primo el sesto d'i so denari. Levadi che sono dal zuogo, chadaun di loro s'ano trovado in man duchati 10. Adomando: "Quanti ne fo li denari che avè chadaun di loro quanto[a] i chomenzò a zugar?"

E per far la ditta raxion per inpoxizion pony che ala prima inpoxizion fuse 10, e ala segonda fuse 6, e per lo terzo fusse 24, e al quarto fusse men 40. La chaxion sie questa che 'l primo che à 10 e vuol dar al quarto el sesto. Adoncha chonvien che 'l segondo abia 6, perché el primo tuo' el terzo de 6, sono 2, el primo roman chon 12, el segondo roman chon 4. Mo' vuol el quarto dal terzo chonpagno, che 'l segondo romagna chon 10. Adoncha vorallo che 'l terzo avesse 24. Dagando el quarto, che son 6, el segondo averà 10, el terzo roman chon 18. Mo' voralo dal quartto el quintto. Adoncha vorave lo quartto men 40, che dagando men $\frac{1}{5}$, che son 8, a più 18 roman chon 10. El quarto roman chon men 32, e vorallo dal primo el sesto, che son 2 più. Adoncha sotrazi da men 32 più 2, roman men 30, e noi vossemo che fusse più 10, adoncha roman men 40.

E per far la segonda inpoxizion pony el primo 8, el segondo 12, el terzo 8, el quarto 20. El primo 8, el terzo de 12, 4, azonti chon 8 fa 12. El segondo roman chon 8. El terzo 8, dar 2 alo 8 fa 10. El quarto à 20, tuò el quinto, son 4, azonti a 6, che roman al terzo, serano 10. El quarto 16, e tuo' el sesto dal primo che son 2, arà 18, e noi volemo 10, a più 8.

f. 35a Puo' moltipicha in croxe e fara' 8 fia 10 val 80, 8 fia 40 fa 320, azonto 80 fano 400, perché più e men s'azonze. Puo' azonzi men 40 più 8 fa 48, parti 400 per 48, insirà fuor del partidor duchati $8\frac{1}{3}$, e tanto avè el primo chonpagno quando se meseno a zuogar. Segui l'inpoxizion.

10	8
6	12
24	8
men 40	20
men 40	più 8

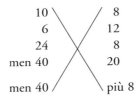

Prima, moltipichado in croxe 80 e 320 fa azonti insieme 400. E questi partidi per 48 serà $8\frac{1}{3}$. Puo' el segondo, 6 fia 8 48, 40 fia 12 480, azonti sono 528. E questi partidi per 48, el segondo averà 11. Per lo terzo, 8 fia 24 fa 192, 8 fia 40 320, azonti insieme sono 512. E questo partido per 48, arà el terzo $10\frac{2}{3}$. Per lo quarto, 8 fia men 40 fa 320, 40 fia 20 fa 800, tratto 320 roman 480. Partidi per 48, el quarto chonpagno averà 10.

E per far la sovrascritta raxion per la chosa pony che 'l primo avese 1^{co}, chonvien che 'l segondo abia men 36^{n} men 3^{co}. E perché el primo vuol $\frac{1}{3}$ de 36, serà più 12^{n} men 1^{co}. Adoncha 1^{co} più 1^{co} men abatti, roman 0. Mo' daremo 12^{n} più. Averà el primo 12^{n}, el segondo roman più 24^{n} men 2^{co}. Mo' volemo del terzo chonpagno el quarto. Adoncha volemo che abia men 56^{n}, che dagando

a. *Sic in MS, evidently for* quando.

There are 4 who want to play dice. The first won from the second $\frac{1}{3}$ of his money. The second won from the third $\frac{1}{4}$ of his money. The third won from the fourth one fifth of his money. The fourth won from the first one sixth of his money. Getting up from the game, each of them found that he had 10 ducats in his hand. I ask: "How many ducats did each have when they began to play?" f. 34b

And to do the problem by false position, suppose that at the first false position it was 10, and for the second it was 6 and for the third it was 24 and for the fourth it was minus 40.[1] The situation is that the first who had 10 wants to give a sixth to the fourth. So it works out that the second had 6, because the first took a third of 6, which is 2, the first remains with 12, the second remains with 4. Now he wants a quarter from the third companion, so the second remains with 10. So it will work out that the third had 24. Giving a quarter, which is 6, the second will have 10, the third remains with 18. Now, from the fourth we need a fifth. So the fourth will have minus 40, which giving minus $\frac{1}{5}$, which is 8, to plus 18 remains with 10. The fourth remains with minus 32, and should have a sixth from the first, which is plus 2. So subtract plus 2 from minus 32, the remainder is minus 30, and we wanted it to be 10, so the remainder is minus 40.

And to do the second false position, suppose that the first has 8, the second 12, the third 8, the fourth 20. The first 8, a third of 12, 4, added to 8 makes 12. The second remains with 8. The third 8, to give 2 to the 8 makes 10. The fourth has 20, take away a fifth, which is 4, added to 6, which remains to the third, will be 10. The fourth 16, and take a sixth from the first which is 2, will have 18, and we wanted 10, so it is plus 8.

Then multiply across and it will be 8 times 10 makes 80, 8 times 40 makes 320, added to 80 makes 400, since it's plus and minus it's added. Then add minus 40 plus 8 makes 48, divide 400 by 48, there will result from the division $8\frac{1}{3}$ ducats, and that's how much the first companion had when they started to play. The false position follows. f. 35a

10	8
6	12
24	8
minus 40	20
minus 40	plus 8

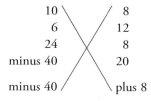

First, multiplied across 80 and 320, added together makes 400. And this divided by 48 will be $8\frac{1}{3}$. Then the second, 6 times 8 48, 40 times 12 480, added together is 528. And this divided by 48, the second will have 11. For the third, 8 times 24 makes 192, 8 times 40 320, added together is 512. And this divided by 48, the third will have $10\frac{2}{3}$. For the fourth, 8 times minus 40 makes 320, 40 times 20 makes 800, 320 subtracted leaves 480. Divided by 48, the fourth companion will have 10.

And to do this problem by unknowns, put that the first had $1x$, it works out that the second had minus 36 minus $3x$. And because the first wants $\frac{1}{3}$ of 36, it will be plus 12 minus $1x$. So plus $1x$ and minus $1x$ subtracted leaves 0. Now we will give plus 12. The first will have 12, the second gets 24 minus $2x$. Now we want a quarter from the third companion. So we want him to have minus 56, of

1. The meaning is: Put 10 at the first false position for what the first player had, then the second would have had 6, the third would have had 24, and the fourth would have had minus 40.

14^n men, che son el quarto, romagna el segondo chon 10^n. E si volemo che abia più 8^{co}, che dagando più 2^{co} a men 2^{co} roman 0, el terzo romagnerà men 42^n più 6^{co}. Mo' volemo del quarto chonpagno el quinto. Adoncha 5 fia 52 fa 260^n più, abatti più 52 de più 260, roman chon 208^n. Abatti men 42 da più 42, roman 0, e darli 10 più. Adoncha roman più 10. Puo' demo lo quinto
f. 35b delle chose, | che son 6^{co}, e faremo 5 fia 6 fa 30^{co} men. Si li demo el quinto de 30^{co} romagnerà 0. El quarto roman chon 24^{co} men dado dal primo, che son 2 a 208^n serano 210^n, e 24^{co} men ingual a 80^n.[a] Adequa le parte. Demo demo[b] più 10 a più 210, zoè sotrarlo perché più e più, roman 200^n partir per 24^{co}, e questo partido insirà fuora duchati $8\frac{1}{3}$. E tanto avè el primo, el segondo 11, el terzo $10\frac{2}{3}$, el quarto 8. E per questo fa le oltre simille.

più[c]	1^{co}	più 36	men 3^{co}		più 24^n	men 2^{co}	56[d] men	più 8^{co}	più 260^n	30^{co}
		12	1		men 14	più 2^{co}	14	più 2^{co}		
	12	24	2^{co}		10	0^{co}	men 42	più 6	208	24^{co}
			0				10	più 5	2	
			0~~4~~8				5		210	24^{co} 10
			~~200~~						10	
			~~24~~	$8\frac{8}{24}$ $\frac{1}{3}$ el primo					200	24^{co}

[Chonprar 3 1ª zoya]

Sono 3 chi vuol chonprar una zoya la qual se domanda ducati 12. Dixe el primo ai altri 2: "Se vuy me dè la mitade d'i vostri apresso i mie io averò ducati 12." Dixe el segondo ai altri 2: "Se me dè el terzo d'i vostri averò ducati 12." Dixe el terzo ai altri 2: "Se me dè el quarto d'i vostri denari apreso i mie averò ducati 12." Io adomando quanti ne avea zascadun di loro in burssa.

E per far la ditta raxion per la riegola del 3, in che se truova $\frac{1}{2}$, $\frac{1}{3}$, $\frac{1}{4}$? Se truova in 12. Serano la mittà de 24, 12 serà $\frac{2}{3}$ de 18, 12 serà $\frac{3}{4}$ de 16. E per dir meyo, se tu vuol mezo del 12 vuol altri tanti,
f. 36a serano 24. E perché l'oltro domanda el terzo de 12, darallo la mità | de 12, serano 18. E perché l'oltro domanda el quarto de 12, seria[e] el terzo 4, serano 16. Mo' azonzi 24, 18, 16, fano 58, abatti un homo, roman 2. Parti 58 per 2, serano 29. Mo' faremo quanto è più 29 de 24, serano 5, e quanti è più 29 de 18, sono 11, e quanti è più 29 de 16, sono 13. Suma 5, 11, 13, fa 29. Adoncha el primo 5, el segondo 11, el terzo 13. Se 'l primo tuo' la mitade de 11 e 13 son 12, e 5 serano 17,[f] adoncha diremo: "Se[g] 17 val 12, che serave 5?" Venirà, moltipichada e partida, $3\frac{9}{17}$. E per lo simille, se 17 val 12, che serave 11? Serà $7\frac{13}{17}$. E per lo terzo, se 17 val 12, che serave 13? Serà $9\frac{3}{17}$.

a. 24 *crossed out with a diagonal line.*
b. *Thus repeated in MS.*
c. men *crossed out with a diagonal line.*
d. 56 *corrected over* 42.
e. el quarto *crossed out with a diagonal line.*
f. 17 *corrected over* 18.
g. 17 *crossed out with a diagonal line.*

which giving minus 14, which is the quarter, leaves the second with 10. And we wanted him to have plus 8*x*, which giving plus 2*x* to minus 2*x* leaves 0, the third will remain with minus 42 plus 6*x*. Now we want a fifth from the fourth companion. So 5 times 52 makes plus 260, subtract plus 52 from plus 260, leaves 208. Subtract minus 42 from plus 42, the remainder is 0, and give him plus 10. So the remainder is plus 10. Then we give the fifth of the unknown, | which is 6*x*, and we will do 5 times 6 makes minus 30*x*. If we give him the fifth of 30*x* there will remain 0. The fourth remains with minus 24*x* given by the first, which is 2 plus 208, will be 210, and minus 24*x* equals 80. Equalize the sides. We give plus 10 to plus 210, that is subtract since it's plus and plus, the remainder is 200 to divide by 24*x*, and this divided will result in $8\frac{1}{3}$ ducats. And that's how much the first had, the second 11, the third $10\frac{2}{3}$, the fourth 8. And in this way do other similar problems. f. 35b

plus 1*x* plus 36 minus 3*x* plus 24 minus 2*x* 56 minus plus 8*x* plus 260 30*x*
12 1 minus 14 plus 2*x* 14 plus 2*x*
12 24 2*x* 10 0*x* minus 42 plus 6 208 24*x*
0 10 plus 5 2
0~~4~~8 5 210 24*x* 10
~~2~~0~~0~~ 10
~~2~~~~4~~ $8\frac{8}{24}$ $\frac{1}{3}$ the first 200 24*x*

Three men to buy a jewel

There are 3 who want to buy a jewel which takes 12 ducats. The first says to the other 2: "If you give me a half of yours, together with mine I will have 12 ducats." The second says to the other 2: "If you give me a third of yours, I will have 12 ducats." The third says to the other 2: "If you give me a quarter of your money, together with mine I will have 12 ducats." I ask how much each of them had in his purse.

And to do the problem by the rule of 3, in what are $\frac{1}{2}$, $\frac{1}{3}$, and $\frac{1}{4}$ found? They are found in 12. It will be half of 24, 12 will be $\frac{2}{3}$ of 18, 12 will be $\frac{3}{4}$ of 16. And to say better, if you want half of the 12 that all want, it will be 24. And because the other wants a third of 12, he will give half | of 12, it is 18. f. 36a
And because the other asks for a quarter of 12, a third would be 4, it will be 16. Now add 24, 18, 16, makes 58, subtract one man, there remain 2. Divide 58 by 2, it will be 29. Now we do how much 29 is greater than 24, it will be 5, and how much 29 is greater than 18, it is 11, and how much 29 is greater than 16, it is 13. Add 5, 11, 13, makes 29. So the first has 5, the second 11, the third 13. If the first takes half of 11 and 13 it's 12, and 5 will be 17, so we will say: "If 17 is worth 12, what would 5 be?" Multiplied and divided, it will come to $3\frac{9}{17}$. And in the same way, if 17 is worth 12, what would 11 be? It will be $7\frac{13}{17}$. And for the third, if 17 is worth 12, what would 13 be? It will be $9\frac{3}{17}$.

E per far la ditta per inpoxizion pony che la prima inpoxizion fuse 6 e l'oltra fose 3. E per saver meter per 2 chonpagny, tu ssa che li 2 vorave aver tanti che si ve dà la mittade tu vuol aver 12. Adoncha per forza chonvien che l'abia quey 2 12. Mo' diremo noi: "Fame de 12 2 parte che dagando el terzo del'una parte al'oltra faza faza[a] 10." Adoncha serave la prima 9 e la segonda[b] 3. Adoncha el primo domanda ai do la mittade che ano 12, serano 6, e luy 6 che fa ben 12. Dixe el segondo, che à 9: "Dame el terzo d'i vostri. Averò 12." L'uno à 6 e l'oltro 3. El terzo serà 3 e 9 fa 12. Dixe el terzo: "Dame el quarto de 6 e 9 e averò 12." El quarto de 15 sono $3\frac{3}{4}$, e luy à 3, sono

f. 36b $6\frac{3}{4}$. Adoncha i mancha $5\frac{1}{4}$, e chosì me|teremo al'inpoxizion. E per meter la segonda inpoxizion serà 3, e li oltri per forza chonvien aver 18, sì che se li dà la mittà, che sono 9, averà 12. Adoncha fame de 18 2 parte che dagando el terzo del'una parte al'oltra faza 11. Adoncha venyrà $7\frac{1}{2}$ e $10\frac{1}{2}$. El primo, che à 3, domanda li altri 2 la mittade, che sono 18, son 9 e 3 val ben 12. Dixe el segondo, che à $7\frac{1}{2}$, ai altri 2 che ano $13\frac{1}{2}$: "Dame el terzo." Serano $4\frac{1}{2}$, e luy $7\frac{1}{2}$, ben averà 12. Dixe el terzo, che à $10\frac{1}{2}$, ai altri 2: "Dame el quarto d'i vostri," che sono $10\frac{1}{2}$, adoncha chon li so averalo $13\frac{1}{8}$. Adoncha à più $1\frac{1}{8}$, e mettila chosì al'inpoxizion. Moltipicha in croxe e parti, insirà la prima $3\frac{54}{102}$ e la segonda $7\frac{78}{102}$, e la terza $9\frac{18}{102}$, chomo vedirì qui de sotto per figora.

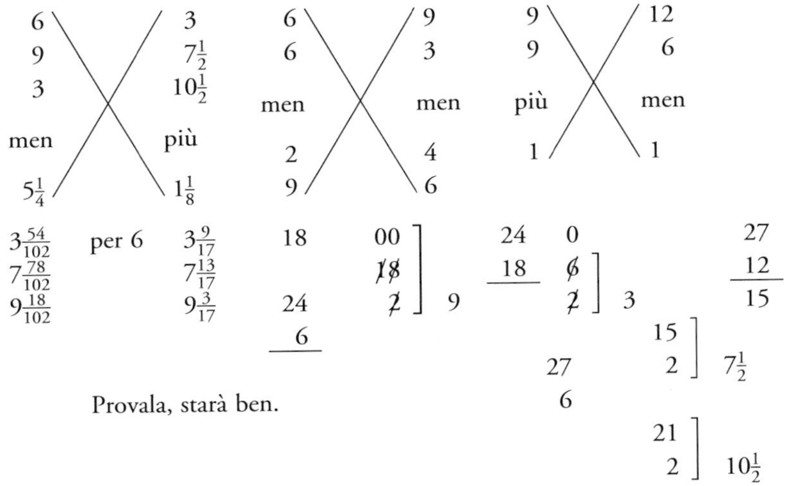

Provala, starà ben.

f. 37a E per far la ditta raxion per la chosa pony che 'l primo avese 1^{co}. | Chonvien che 'l segondo el terzo per forza abia più 24 men 2^{co}, che 'l primo domanda la mitade de 24, sono 12, e la mitade de men 2^{co} men 1^{co}. Adoncha 1^{co} più da 1^{co} men roman 0, e roman chon 12, sì che roman aver el segondo el terzo 24 men 1^{co}. Adoncha che parte poremo tuor de 24 men 1^{co} che dagando el terzo delo romagnente abia 12? Serà 6 de 24, resta 18. Tuor el terzo de 18 serano 6. Adoncha ben à el segondo 12. Mo' dixe el segondo, che à 6: "Truovame un numero che abatudo el terzo romagna 1." Serà $1\frac{1}{2}$. Tuo' el terzo, serà $\frac{1}{2}^{co}$. Adoncha averà el segondo 6^{n} e $\frac{1}{2}^{co}$. E del segondo e del primo tuo' el quarto che $1\frac{1}{2}$ per numero più $\frac{3}{8}^{co}$, fin 12 mancha $10\frac{1}{2}$ men $\frac{3}{8}^{co}$. Suma el primo, el segondo, el terzo, ave-

a. *Thus repeated in the text.*
b. *6 crossed out with a diagonal line.*

And to do this by false position, suppose that the first position was 6 and the other was 3. And to know how to do it for 2 companions, you know that the 2 would have so much that if they gave you half you would have 12. So, by necessity it works out that these two have 12. Now we will say: "Make me 2 parts of 12 so that giving the third of one part to the other makes 10." So the first would be 9 and the second 3. So the first asks the two for half of the 12 that they have, it will be 6, and he has 6 which makes 12. The second, who has 9, says: "Give me the third of yours. I will have 12." One has 6 and the other 3. The third will be 3 plus 9 makes 12. The third says: "Give me a quarter of 6 and 9 and I will have 12." A quarter of 15 is $3\frac{3}{4}$, and he has 3, it's $6\frac{3}{4}$. So, it's minus $5\frac{1}{4}$, and this we | will put at the false position. And to do the second false position it will be 3, and the f. 36b
others will then work out to have 18, so that if he gives them half, which will be 9, he will have 12. So take 2 parts of 18 so that giving the third one part, makes 11 for the other. So it will come to $7\frac{1}{2}$ and $10\frac{1}{2}$. The first, who has 3, asks for the half of the other 2, which is 18, it's 9 plus 3 makes 12. The second, who has $7\frac{1}{2}$, says to the other two, who have $13\frac{1}{2}$: "Give me a third." It will be $4\frac{1}{2}$, and he has $7\frac{1}{2}$, so he will have 12. The third, who has $10\frac{1}{2}$, says to the other two: "Give me a quarter of yours," which is $10\frac{1}{2}$, so with his own he will have $13\frac{1}{8}$. So he has plus $1\frac{1}{8}$, and put it this way in the false position. Multiply across and divide, the first will come out to $3\frac{54}{102}$ and the second $7\frac{78}{102}$ and the third $9\frac{18}{102}$, as you will see worked out below.

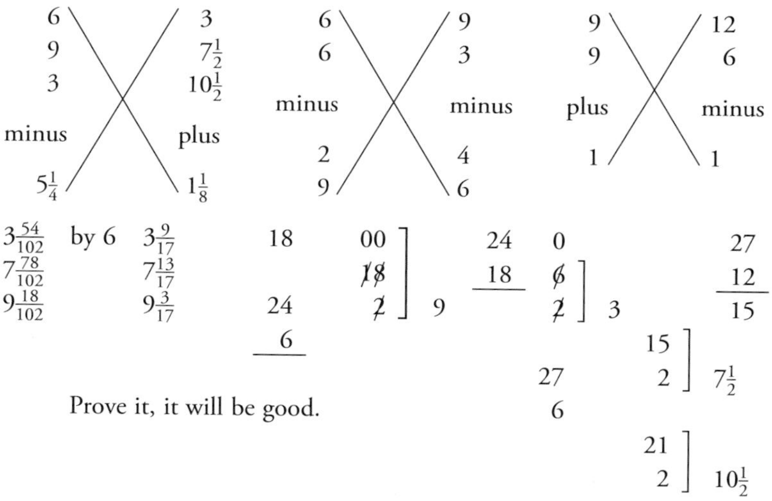

And to do this problem by unknowns, put that the first had $1x$. | It works out that the second and f. 37a
the third have to have plus 24 minus $2x$, of which the first asks for half of the 24, it's 12, and the half of the minus $2x$ is minus $1x$. So plus $1x$ from minus $1x$ gives 0, and he remains with 12, so the second and the third remain with 24 minus $1x$. So what part can we take from 24 minus $1x$ that giving the third of what remains he will have 12? It will be 6 from 24, the remainder is 18. Take a third of 18, it will be 6. So the second has 12. Now the second, who has 6, says: "Find me a number from which subtracted a third the remainder is 1." It will be $1\frac{1}{2}$. Take a third, it will be $\frac{1}{2}x$. So the second will have 6 and $\frac{1}{2}x$. Now take from the first and the second a quarter, which is $1\frac{1}{2}$ in number plus $\frac{3}{8}x$, to 12 there lacks $10\frac{1}{2}$ minus $\frac{3}{8}x$. Add the first, the second, and the third, it will

rano $16\frac{1}{2}$ per numero. $1\frac{1}{8}^{co}$ sono inguali 24^{n} men 1^{co}. Adequa et averemo $2\frac{1}{8}^{co}$. Sutrazi de 24 $16\frac{1}{2}$, serano $7\frac{1}{2}$. Parti li nomeri per le chose, serano $3\frac{18}{34}$, el segondo $7\frac{26}{34}$, el terzo $9\frac{6}{34}$, chomo vedirì qui de sotto, per esenpio. E perché dixe qui de sotto $1\frac{1}{8}$, sie che asumado 1^{co} del primo a meza del segondo, che sono $\frac{3}{2}^{co}$, dando el quarto, serano $\frac{3}{8}^{co}$, serano men, e sotrato $\frac{3}{8}$ de $\frac{3}{2}$ resta aver $1\frac{1}{8}^{co}$ e nome $10\frac{1}{2}$.[a]

1^{co}	24^{n}	men 2^{co}	24				1	
			6				$0\not{3}8$	
			18	men 1^{co}		$1^{co}\frac{1}{2}$	$\not{1}\not{2}\not{0}$	]
12	più 1^{co}						$\not{3}\not{4}$	$3\frac{18}{34}$
	men 1^{co}	0						
			$1\frac{1}{2}$	$1\frac{1}{2}^{co}$				
$6\frac{1}{2}^{co}$			$1\frac{1}{2}$		$\frac{3}{2}^{co}$	$\frac{1}{4}$	$\frac{3}{8}^{co}$	
$10\frac{1}{2}$	men $\frac{3}{8}^{co}$		$1\frac{1}{2}$					
$16\frac{1}{2}$	$1\frac{1}{8}^{co}$	ingual 1^{co}		24		$7\frac{1}{2}$	$\frac{15}{2}$	$\frac{17}{8}$
	$2^{co}\frac{1}{8}$			$16\frac{1}{2}$				

---

f. 37b

[Uno che fa testamento a una dona graveda]

El'è uno homo che fa testamento e lasa una so muyer graveda lire 7 de grossi chon questa chondizion che se la faxe figlo maschio lire 4 sia dela mare e lire 3 sia del figlo, e se la faxe fia femena vuol che per so maridar abia lire 4 e la madre abia lire 3. Ven caxo che morto el marido, fexe 1° fio e 1ª fia. Adomando che vien a chadaun de ttuti 3. Fa chosì. Se 'l figlo avese 3, la madre die aver 4, e la fia die aver el terzo de 4 de più ch'à la madre, che xe 1 e $\frac{1}{3}$. Zonzi a lire 4, fa $5\frac{1}{3}$. Mo' dirai: "Sono 3 chi s'achonpagna. El primo messe 3, el segondo messe 4, el terzo messe $5\frac{1}{3}$, et ano vadagnado lire 7. Che die aver zaschadun di loro?" Asuma 3, 4, $5\frac{1}{3}$, fa 12 e $\frac{1}{3}$, e dirai: "Se $12\frac{1}{3}$, che son $\frac{37}{3}$, me rende lire 7, che tornerà 3?" $1\frac{26}{37}$. E per lo segondo: "Se $\frac{37}{3}$ torna 7, che serà 4?" Serà $2\frac{10}{37}$. E per lo terzo: "Se $\frac{37}{3}$ torna 7, che serà $5\frac{1}{3}$?" Serà $3\frac{3}{111}$. Asuma insieme, serà ben $1\frac{26}{37}$, $2\frac{10}{37}$, $3\frac{3}{111}$, serano ben 7. E chossì fa ttute le oltre simille raxion.

[Chortelli e vazine]

Fame questa raxion: 5 chortelli e 4 vazine val soldi 13; 6 chortelli 4 vazine val soldi 22. Che valse el chortello e che valse la vazina?

---

a. *From* E perché dixe *to* nome $10\frac{1}{2}$ *added in the left margin in a different ink.*

be $16\frac{1}{2}$ in number. $1\frac{1}{8}x$ is equal to 24 minus $1x$. Equalize and we will have $2\frac{1}{8}x$. Subtract $16\frac{1}{2}$ from 24, it will be $7\frac{1}{2}$. Divide the numbers by the unknowns, it will be $3\frac{18}{34}$, the second $7\frac{26}{34}$, the third $9\frac{6}{34}$, as you will see worked out below. And because it says below here $1\frac{1}{8}$, it is the $1x$ of the first added to half of the second, which is $\frac{3}{2}x$, giving a quarter it will be $\frac{3}{8}x$, it will be minus, and $\frac{3}{8}$ subtracted from $\frac{3}{2}$ will leave $1\frac{1}{8}$ and not at all $10\frac{1}{2}$.[1]

$1x$	24	minus $2x$	24			1		
			6			0~~3~~8		
			18	minus $1x$	$1\frac{1}{2}x$	~~120~~		
12	plus $1x$					~~34~~	$3\frac{18}{34}$	
	minus $1x$	0						
			$1\frac{1}{2}$		$1\frac{1}{2}x$			
$6\frac{1}{2}x$			$1\frac{1}{2}$			$\frac{3}{2}x$	$\frac{1}{4}$	$\frac{3}{8}x$
$10\frac{1}{2}$	minus $\frac{3}{8}x$		$1\frac{1}{2}$					
$16\frac{1}{2}$	$1\frac{1}{8}x$	equals $1x$		24	$7\frac{1}{2}$	$\frac{15}{2}$	$\frac{17}{8}$	
	$2\frac{1}{8}x$			$16\frac{1}{2}$				

A MAN WHO LEAVES A WILL FOR A PREGNANT WIFE

f. 37b

There is a man who makes a will and leaves to his pregnant wife 7 lire di grossi with the condition that if she had a son 4 lire go to the mother and 3 lire go to the son, and if she has a daughter he wants her to have 4 lire for her dowry and the mother would have 3 lire. It happens that the husband died, and she had one son and one daughter. I ask you what came to each of them. Do it this way. If the son had 3, the mother would have 4, and the daughter should have a third of 4 more than the mother, which is $1\frac{1}{3}$. Added to 4 lire, it makes $5\frac{1}{3}$. Now you will say: "There are 3 who form a company. The first puts in 3, the second puts in 4, the third puts in $5\frac{1}{3}$, and they have a profit of 7 lire. What should each of them get?" Add 3, 4, $5\frac{1}{3}$, it makes $12\frac{1}{3}$, and you will say: "If $12\frac{1}{3}$, which is $\frac{37}{3}$, gives me 7 lire, what will 3 give me?" $1\frac{26}{37}$. And for the second: "If $\frac{37}{3}$ gives 7, what will 4 be?" It will be $2\frac{10}{37}$. And for the third: "If $\frac{37}{3}$ gives 7, what will $5\frac{1}{3}$ be?" It will be $3\frac{3}{111}$. Add all together, it will be $1\frac{26}{37}$, $2\frac{10}{37}$, $3\frac{3}{111}$, it comes to 7. And do all similar problems this way.

KNIVES AND SHEATHS

Do this problem for me. 5 knives and 4 sheaths are worth 13 soldi; 6 knives and 4 sheaths are worth 22 soldi. What is the knife worth and what is the sheath worth?

---

1. This sentence is added in the left margin in a different ink. I thank Linda Carroll for the suggestion that *nome* here is a contraction of *no meno* meaning "not at all."

E per far la ditta per inpoxizion pony ala prima che 'l chortelo valese soldo 1, chortelli 5 valeria soldi 5, inchina soldi 13 roman soldi 8. E le vazine son 4, die valler soldi 2 la vazina. El segondo, f. 38a 6 chortelli valeria valeria[a] soldi 6 e 4 | vazine valeria soldi 8. Adoncha 6 e 8 fa 14, e noi volemo che sia 22. Adoncha men 8, e pony questo ala prima inpoxizion. E per far la segonda inpoxizion pony che valese el chortello soldi 2. Adoncha 5 chortelli valeria soldi 10. Inquina soldi 13 avanza soldi 3, e questi partidi per 4 serano $\frac{3}{4}$ per vazina. Adoncha el segondo per chortelli 6 serano soldi 12 e $\frac{3}{4}$, per 4 vazine seria soldi 15, e noi volemo che sia soldi 22. Adoncha mancha 7, e questo metti ala segonda inpoxizion. Ala prima diremo: "Per 1 che m'ò posto men 8." Ala segonda: "Per 2 che m'ò posto mancha 7." Moltipicha in croxe e parti. Valerà el chortello 9 e la vazina 8. Moltipicha 5 fia 9, 45; 4 fia 8, 32. Sutrazi 32 de 45, roman 13. Moltipicha 6 fia 9, 54, e 4 fia 8, 32. Sotrazi 32 de 54, roman 22. E questo è fatto, chomo vedirì qui de sotto per figura.

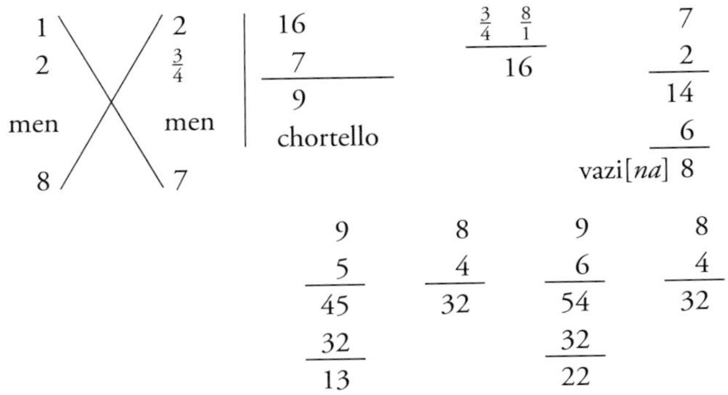

Adoncha valse el chortello valse soldi 8 e la vazine men 8.

f. 38b E per far la ditta raxion per la chosa pony che 'l chortelo valese 1^{co}. Chonvien che 4 vazine vallesse 13^{n} men 1^{co}, 5 chortili[b] valeria 5^{co}, e per lo simille 6 chortelli 6^{co}, 4 vazina val più 13^{n}[c] 5^{co}, che son ingual a 22^{n}. Adequa le parte, più 22^{n} più 13^{n}, trazi el menor dal mazuor, roman 9, e tanto val el chortello. Adoncha 5 fia 9 val 45, trazi 13, roman 32, e tanto valse men 4 vazine. Adoncha la vazina valse men 8. Moltipicha 6 fia 9 val 54, abatti per vazine 4 32, roman 22, chomo vedirì qui de sotto per figora.

a. *Thus repeated in the MS.*
b. chortili *corrected over* chortilo.
c. più *crossed out with a diagonal line.*

And to do it by false position, suppose at the first that the knife was worth 1 soldo, 5 knives would be worth 5 soldi, from 13 soldi there remain 8 soldi. And there are 4 sheaths, so each sheath must be worth 2 soldi. The second, 6 knives would be worth 6 soldi and 4 | sheaths would be worth 8 f. 38a
soldi. So 6 and 8 make 14, and we wanted it to be 22. So minus 8, and put this at the first false position. And to do the second false position suppose that the knife was worth 2 soldi. So 5 knives would be worth 10 soldi. From 13 soldi there are 3 soldi left, and this divided by 4 will be $\frac{3}{4}$ per sheath. So the second for 6 knives will be $12\frac{3}{4}$ soldi and for 4 sheaths will be 15 soldi, and we wanted it to be 22 soldi. So it is minus 7, and put this at the second false position. At the first we will say: "For 1 that I've put, minus 8." At the second: "For 2 that I've put, minus 7." Multiply across and divide. The knife will be worth 9 and the sheath 8.[1] Multiply 5 times 9, 45; 4 times 8, 32. Subtract 32 from 45, the remainder is 13. Multiply 6 times 9, 54, and 4 times 8, 32. Subtract 32 from 54, the remainder is 22. And this is done, as you will see worked out below.

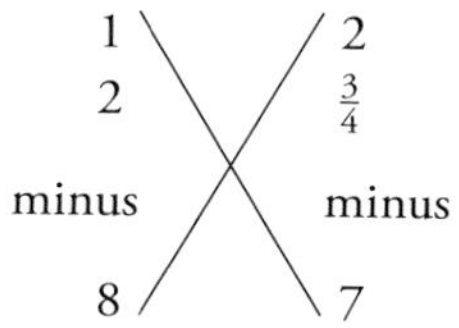

16
7
———
9
knife

$\frac{3}{4}$ $\frac{8}{1}$
———
16

7
2
———
14
6
———
sheath 8

9
5
———
45
32
———
13

8
4
———
32

9
6
———
54
32
———
22

8
4
———
32

So the knife is worth 9[2] soldi and the sheath minus 8.

And to do the problem by unknowns, put that the knife was worth $1x$. It works out that 4 sheaths f. 38b
were worth 13 minus $1x$, 5 knives would be worth $5x$, and in the same way 6 knives would be worth $6x$, 4 sheaths are worth plus 13 and $5x$, which is equal to 22. Equalize the sides, plus 22 plus 13, subtract the smaller from the larger, the remainder is 9, and this is the value of the knife. Then 5 times 9 makes 45, subtract 13, the remainder is 32, and that is worth minus 4 sheaths. So the sheath is worth minus 8. Multiply 6 times 9 is 54, subtract 32 for 4 sheaths, the remainder is 22, as you will see worked out below.

---

1. Actually the sheath is worth minus 8 soldi.
2. *8* in MS.

[Chavezo de pano]

1^{co} 13^{n} men 1^{co} | 5^{co} 6^{co} più 13^{n} 5^{co} ingual 22^{n}

22	45[a]	8	45	00	
13	32	4	13	~~32~~	8
9	13			~~4~~	

9
6
54 22
32
22

---

f. 39a

[Chavezo de pano]

E voyo chonprar un chavezo de pano. Se li don lire 3 del brazo me mancharà soldi 4, e se li don soldi 32 del brazo m'avanza soldi 44. Quanti braza era el chavazo e quanti erano i denari in borsa?

E per far la ditta raxion per inpoxizion pony che valese lire 3. Adoncha moltipicha 3 fia 3 fa 9 lire, abatti soldi 4 che mancha, roman lire 8 soldi 16. E puo' fa 3 fia 32 fa 96, sutrazi de lire 8 soldi 16 soldi 96, roman lire 4 che son soldi 80, e de questi sotrazi soldi 44, roman soldi 36, e questi metti che ssia più al'inpoxizion. Puo' faremo la segonda inpoxizion vallesse lire 4. Adoncha moltipicha 3 fia 4 fa 12, sutrazi 4, roman lire 11 soldi 16, e puo' moltipicha 4 fia 32 fa 128, sserano lire 6 soldi 8, e questo sotrazi de lire 11 soldi 16, roman lire 5 soldi 8. E noi volemo che sia soldi 44. Adoncha sotrazi 44 de soldi 108, roman più 64. E diremo per 4 che m'ò puxo più 64, e moltipicha in croxe 3 fia 64 fa 192, e 4 fia 36 fa 144, più e più, sotrazi la menor dela mazuor, roman soldi 48. Poi per far partidor quel de sotto sutrazi 36 de 64, roman 28, e questo è lo partidor. A partir 48 insirà del partidor[b] $1\frac{5}{7}$ de brazo de pano. Adoncha per aprovar la ditta raxion faremo: "Se 1 brazo me dà soldi 60, che me darà brazo $1\frac{5}{7}$?" Daratte soldi $102\frac{6}{7}$. E perché dise che manchase soldi 4 abattillo, roman
f. 39b soldi $98\frac{6}{7}$. E se'l volemo dar del brazo soldi 32, monta soldi $54\frac{6}{7}$, sotrazillo | de soldi $98\frac{6}{7}$ $54\frac{6}{7}$, roman soldi 44, che ben a' ditto [*e*]ra verittade, chomo vedirì qui de sotto per figura.

---

a. *Operation crossed out with a diagonal line.*
b. soldi *crossed out with a diagonal line.*

1*x* 13 minus 1*x* | 5*x* 6*x* plus 13 5*x* equals 22

22	45[1]	8	45	00	
13	32	4	13	32 (crossed out)	
9	13			4 (crossed out)	8

9
6
54 22
32
22

A CLOTH REMNANT

f. 39a

And I want to buy a remnant of cloth. If it goes for 3 lire per ell, I will be 4 soldi[2] short, and if it goes for 32 soldi per ell, I will be 44 soldi ahead. How many ells was the cloth and how much money was in the purse?

And to do this problem by false position, suppose that it was worth 3 lire.[3] Then multiply 3 times 3 makes 9 lire, subtract the 4 soldi that are lacking, and it remains 8 lire 16 soldi. And then do 3 times 32 makes 96, subtract 96 soldi from 8 lire 16 soldi, it leaves 4 lire which is 80 soldi, and subtract from this 44 soldi, the remainder is 36 soldi, and put this as a plus at the false position. Then we will suppose at the second false position that it was worth 4 lire.[4] So multiply 3 times 4 makes 12, subtract 4, the remainder is 11 lire 16 soldi, and then multiply 4 times 32 makes 128, it will be 6 lire 8 soldi, and subtract this from 11 lire 16 soldi, the remainder is 5 lire 8 soldi. And we want it to be 44 soldi. So subtract 44 from 108 soldi, the remainder is plus 64. And we will say for 4 that I have put plus 64, and multiply across 3 times 64 makes 192 and 4 times 36 makes 144, plus and plus, subtract the lesser from the greater, the remainder is 48 soldi. Then to make the divisor, that below, subtract 36 from 64, the remainder is 28, and this is the divisor. Dividing 48, the result of the division will be $1\frac{5}{7}$ ells of cloth. Then to prove this problem we will do: "If 1 ell gives me 60 soldi, what will $1\frac{5}{7}$ give me?" It will give you $102\frac{6}{7}$ soldi. And because it said that 4 soldi were lacking, subtract it, the remainder is $98\frac{6}{7}$ soldi. And if we wanted to give 32 soldi per ell, it comes to $54\frac{6}{7}$, subtract | $54\frac{6}{7}$ from $98\frac{6}{7}$ soldi, the remainder is 44 soldi, which is correct, as you will see f. 39b
worked out below.

---

1. Operation crossed out with a diagonal line.
2. The lira was worth 20 soldi.
3. This should probably read "suppose that it was 3 ells."
4. This should probably read "that it was 4 ells."

3 \ / 4 | 3 lire 9 serà lire 8 soldi 16 | roman lire 4
più X più | 3
36 / \ 64 |

32 serà 96 lire 4 16 | serà soldi 80 più 36
3
44
36

4
3 lire 12

lire 11 soldi 16
lire 6 soldi 8
lire 5 soldi 8

3[a] 32 |
2 4 |
128 / lire 6 soldi 8

108
44
più 64

64 36
3 4
192
144
48 a partir

64
36
28 partidor

20
48
28] $1\frac{20}{28}$ $\frac{5}{7}$

Se $\frac{1}{1}$ $\frac{60}{1}$ $\frac{12}{7}$

006
720]
777] $102\frac{6}{7}$ $98\frac{6}{7}$
4

Se $\frac{1}{7}$ $\frac{32}{1}$ $\frac{12}{7}$

0 $54\frac{6}{7}$
036
384]
77] $54\frac{6}{7}$ soldi 44

E per far la ditta raxion per la chosa pony che 'l pano vallesse 1^{co} de brazo. Chonvien che abia a darli soldi 60 del brazo. Serave 60^{co} de soldo men 4. E de questi denari medeximi se i don 32 soldi serà più 44, zoè i don 32^{co}. Avanza 44^{n}, che son ingual a 60^{co} men 4. Adequa le parte et averemo 28^{co} ingual a 48^{n}. Parti per le chosse, venirà pano brazo $1\frac{5}{7}$, et averà in borsa soldi $98\frac{6}{7}$.[b]

---

f. 40a **[Numeri in proporzion]**

Truovame 2 nomeri in proporzion chomo è 2 a 5, e trato 7 dela mazuor parte faza tanto quanto azonto 3 sula menor. Pony che fusse 2^{co} e 5^{co}. Per voler trar 7 dela mazuor roman 5^{co} men 7^{n}, e

a. *Numbers crossed out with a diagonal line.*
b. $\frac{6}{7}$ *corrected over* $\frac{7}{6}$.

3 ╲ ╱ 4 | 3 lire 9 it will be 8 lire 16 soldi | there remain 4 lire
plus ╳ plus | 3
36 ╱ ╲ 64 |

32 it will be 96 4 lire 16 | it will be 80 soldi plus 36
3
44
36

4
3 lire 12
11 lire 16 soldi
6 lire 8 soldi
5 lire 8 soldi

$\frac{3^1}{2}$ 32 | 108
4 | 44
128 ⁄ 6 lire 8 soldi plus 64

64 36
3 4
192
144
48 to divide

64
36
28 divisor

20
~~48~~ $1\frac{20}{28}$ $\frac{5}{7}$
~~28~~

If $\frac{1}{1}$ $\frac{60}{1}$ $\frac{12}{7}$ 006
~~720~~
~~777~~ $102\frac{6}{7}$ $98\frac{6}{7}$
4

If $\frac{1}{7}$ $\frac{32}{1}$ $\frac{12}{7}$ 0
~~036~~ $54\frac{6}{7}$
~~384~~
77 $54\frac{6}{7}$ 44 soldi

And to do this problem by unknowns, put that the cloth was worth 1*x* per ell. It works out that it was necessary to give him 60 soldi per ell. It will be 60*x* soldi minus 4. And from the same money, if it's at 32 soldi it will be plus 44, that is, it goes for 32*x*. So there is plus 44, which is equal to 60*x* minus 4. Equalize the sides and we will have 28*x* equal to 48. Divide by the unknowns, the cloth will be worth $1\frac{5}{7}$ per ell and there will be $98\frac{6}{7}$ soldi in the purse.

Numbers in proportion

f. 40a

Find me 2 numbers in the proportion of 2 to 5, and for which 7 subtracted from the larger makes as much as 3 added to the smaller. Put that it was 2*x* and 5*x*. To take 7 from the larger, the remainder

1. Numbers crossed out with a diagonal line.

per voler azonzer 3 sula menor serà 2^{co} e 3^{n}, che son ingual. Adoncha azonti più 3^{n} men 7^{n} fano 10^{n}. Sutrazi 2^{co} de 5^{co}, roman 3^{co}, a partir 10^{n} serà $3\frac{1}{3}$, e tanta val la chossa. E noi ponesemo 2^{co}, adoncha serà $6\frac{2}{3}$. E ponessemo 5^{co}, adoncha 5 fia $3\frac{1}{3}$ fa $16\frac{1}{3}$. Abati 7 de $16\frac{1}{3}$, roman $9\frac{1}{3}$. Azonzi 3 su el $6\frac{2}{3}$, roman $9\frac{2}{3}$, chomo vedirì qui sutta[a] per figora.

2^{co}	5^{co}	2^{co} e 3^{n}	5^{co} men 7^{n}	7^{n}	01	
				3^{n}	~~10~~	
$16\frac{2}{3}$[b]					~~3~~	$3\frac{1}{3}$
7						$3\frac{1}{3}$
$9\frac{1}{3}$						$6\frac{2}{3}$

E per far la sovrascritta raxion per inpoxizion pony che 'l primo fuse 10, el segondo fusse 25, azonto 3 al 10 fano 13, e trato 7 de 25 roman 18, e noi volemo 13, adoncha più 5. E la segonda 12 e 30. Trato de 30 7 roman 23; azonto a 12 3 serano 15, e noi volemo che ssia[c] 15, adoncha più 8. Moltipicha e parti, insirà fuori 20 la menor, 50 la mazuor. Parti 50 per 3, serà $16\frac{2}{3}$. Parti 20 per 3, serano $6\frac{2}{3}$. Azonti 3 a $6\frac{2}{3}$, serano $9\frac{2}{3}$. Trato de $16\frac{2}{3}$ 7, roman $9\frac{2}{3}$.

10	12	80	200	25	10	12	30
25	30	60	150	7	3	3	7
		20	50	18	13	15	23
più	più			13			15
5	8			5 più			8 più

8	02		02		$6\frac{2}{3}$	$16\frac{2}{3}$
5	~~50~~		~~20~~		3	7
3	~~3~~	$16\frac{2}{3}$	~~3~~	$6\frac{2}{3}$	$9\frac{2}{3}$	$9\frac{2}{3}$

f. 40b Truovame 2 numeri in proporzion chomo è 2 da 5 che tanto faza azontto l'un chon l'altro quanto moltipichado l'uno per l'altro. Adoncha pony che fusse 2^{co} e 5^{co}. Mo' moltipicha 2^{co} fia 5^{co} fa $10^{\square}$, a zonzer 2^{co} e 5^{co} fa 7^{co}, che son $\frac{7}{10}$, e tanto val la chossa. E noi ponessemo 2^{co}. Adoncha 2 fia 7 fa 14. A partir per 10 serà $1\frac{4}{10}$, serano $1\frac{2}{5}$. E l'oltro ponessi 5. Adoncha 5 fia 7 fa 35 dezimi, serano $3\frac{1}{2}$. Moltipicha, serano $\frac{49}{10}$. Azonzi, serano $\frac{49}{10}$, chomo vedirì qui de sotto per exenpio.

---

a. *Sic in MS.*

b. 1 *written over* 2.

c. 23 *crossed out with a diagonal line.*

is $5x$ minus 7, and to add 3 to the smaller will be $2x$ plus 3, which are equal. So add plus 3 minus 7 makes 10. Subtract $2x$ from $5x$, leaves $3x$, to divide into 10 will be $3\frac{1}{3}$, and that is the value of the unknown. And we put $2x$, so it will be $6\frac{2}{3}$. And we put $5x$, so 5 times $3\frac{1}{3}$ makes $16\frac{2}{3}$. Subtract 7 from $16\frac{2}{3}$, it leaves $9\frac{2}{3}$.[1] Add 3 to $6\frac{2}{3}$, the remainder is $9\frac{2}{3}$, as you will see worked out below.

$2x$	$5x$	$2x$ and 3	$5x$ minus 7	7	01	
				3	~~10~~	
$16\frac{2}{3}$[2]					~~3~~	$3\frac{1}{3}$
7						$3\frac{1}{3}$
$9\frac{2}{3}$[3]						$6\frac{2}{3}$

And to do this problem by false position, suppose that the first was 10, the second was 25, 3 added to 10 makes 13, and 7 subtracted from 25 leaves 18, and we wanted 13, so plus 5. And for the second false position it is 12 and 30. 7 subtracted from 30 leaves 23; 3 added to 12 will be 15, and we want it to be 15, so plus 8. Multiply and divide, the lower will be 20 and the higher 50. Divide 50 by 3, it will be $16\frac{2}{3}$. Divide 20 by 3, it will be $6\frac{2}{3}$. Add 3 to $6\frac{2}{3}$, it will be $9\frac{2}{3}$. 7 subtracted from $16\frac{2}{3}$ leaves $9\frac{2}{3}$.

10	12	80	200	25	10	12	30
25	30	60	150	7	3	3	7
plus	plus	20	50	18	13	15	23
				13			15
5	8			plus 5			plus 8

8	02		02		$6\frac{2}{3}$	$16\frac{2}{3}$
5	~~50~~		~~20~~		3	7
3	~~3~~	$16\frac{2}{3}$	~~3~~	$6\frac{2}{3}$	$9\frac{2}{3}$	$9\frac{2}{3}$

Find me 2 numbers in the proportion of 2 to 5 such that one added to the other is equal to one f. 40b
multiplied by the other. So put that it was $2x$ and $5x$. Now multiply $2x$ times $5x$, it makes $10x^2$, add $2x$ and $5x$, it makes $7x$, which is $\frac{7}{10}$, and that is the value of the unknown. And we put $2x$. So 2 times 7 makes 14, divided by 10 will be $1\frac{4}{10}$, which will be $1\frac{2}{5}$. And for the other you have put 5. So 5 times 7 makes 35 tenths, it will be $3\frac{1}{2}$. Multiply, it will be $\frac{49}{10}$. Added, it will be $\frac{49}{10}$, as you will see worked out below.

---

1. MS reads *5 times $3\frac{1}{3}$ makes $16\frac{1}{3}$. Subtract 7 from $16\frac{1}{3}$, it leaves $9\frac{1}{3}$.*
2. *$16\frac{2}{3}$* incorrectly changed to *$16\frac{1}{3}$* in MS.
3. *$9\frac{1}{3}$* in MS.

$\frac{7}{5}$ $\frac{49}{10}$ 14 $\frac{49}{10}$ 2^{co} 5^{co} 7^{co} 10□ $\frac{14}{10}$ $1\frac{2}{5}$ $3\frac{1}{2}$ $\frac{7}{2}$
35

f. 41a Truovame 2 nomeri in proporzion chomo è 4 a 7, che trato la mittà dela menor dala mazuor e lo romagnente partido per l'oltra mitade men 7 faza 10. Pony che fusse 4^{co} e 7^{co}, a trar 2^{co} de 7^{co} roman 5^{co}. Mo' si parto 5^{co} per 2^{co} men 7^{n} die far 10^{n}. Adoncha moltipicha 2^{co} fia 10, fa 20^{co}, e 7 fia 10, fa 70^{n}. Adequa, sutrazi 5^{co} de 20^{co}, roman 15^{co} e 70^{n}. Parti 70^{n} per per[a] 15^{co}, venirà $4\frac{2}{3}$, e tanto val la chosa. E noi ponesemo 4^{co}. Adoncha moltipicha 4 fia $4\frac{2}{3}$, fano $18\frac{2}{3}$. E per que[b] ancuor ponesemo 7^{co}, moltipicha 7 fia $4\frac{2}{3}$, fano $32\frac{2}{3}$. Adoncha sutrazi la mittà dela menor, serà $9\frac{1}{3}$, da $32\frac{2}{3}$, roman $23\frac{1}{3}$. Parti per l'oltra mitade, che son $9\frac{1}{3}$ men 7, serano $2\frac{1}{3}$, a partir $23\frac{1}{3}$ serano $\frac{70}{3}$ $\frac{7}{3}$.

0 4^{co} e 7^{co} 5^{co} 2^{co} men 7^{n} 20^{co} 70^{n}
70 5
77] 10 partior 15^{co} 70 a partir

1
30
70
15] $4\frac{2}{3}$ | $18\frac{2}{3}$ $4\frac{2}{3}$ $32\frac{2}{3}$
4 7

$32\frac{2}{3}$ $23\frac{1}{3}$ $9\frac{1}{3}$ 00
$9\frac{1}{3}$ 7 70
$2\frac{1}{3}$ 77] 10
$\frac{7}{3}$ $\frac{70}{3}$

f. 41b Truovame un numero che tratto $\frac{1}{3}$ e $\frac{1}{4}$ e lo romagnente moltipichado per 10 faza 100. Pony che fusse 1^{co}. $\frac{1}{3}\frac{1}{4}$ sono de $12\frac{7}{12}$. Sutrazi 1^{co}, roman $\frac{5}{12}$, che son ingual a 100^{n}. Moltipicha 12 fia 100, fa 1200, a partir per 5 serano 240. Trazi el quarto de 240^{n}, serano 60. Trazi el terzo 240, serà 80. Ben serà lo resto 100. Vedirì qui de sotto.

1^{co} $\frac{1}{3}$ $\frac{1}{4}$ $\frac{7}{12}$ 1^{co} $\frac{5}{12}$ $\frac{100}{1}$ 020 240
1200 140
555] 240 100
80
60
100^{n}

a. *Thus repeated in MS.*
b. *Sic in MS.*

$\frac{7}{5}$ $\frac{49}{10}$ 14 $\frac{49}{10}$ $2x$ $5x$ $7x$ $10x^2$ $\frac{14}{10}$ $1\frac{2}{5}$ $3\frac{1}{2}$ $\frac{7}{2}$
35

Find me 2 numbers in the proportion of 4 to 7, so that half of the smaller subtracted from the larger and the remainder divided by the other half minus 7 makes 10. Put that it was $4x$ and $7x$, to subtract $2x$ from $7x$ leaves $5x$. Now if I divide $5x$ by $2x$ minus 7 it should give 10. So multiply $2x$ times 10, it makes $20x$, and 7 times 10, it makes 70. Equalize, subtract $5x$ from $20x$, it leaves $15x$ and 70. Divide 70 by $15x$, it will come to $4\frac{2}{3}$, and that is the value of the unknown. And we put $4x$. So multiply 4 times $4\frac{2}{3}$, it makes $18\frac{2}{3}$. And because we also put $7x$, multiply 7 times $4\frac{2}{3}$, it makes $32\frac{2}{3}$. So subtract half of the smaller, it will be $9\frac{1}{3}$, from $32\frac{2}{3}$, the remainder is $23\frac{1}{3}$. Divide by the other half, which is $9\frac{1}{3}$, minus 7, it will be $2\frac{1}{3}$, to divide into $23\frac{1}{3}$ will be $\frac{70}{3}$ $\frac{7}{3}$. f. 41a

0 $4x$ and $7x$ $5x$ $2x$ minus 7 $20x$ $70x$
70 5
77] 10 divisor $15x$ 70 to divide

1
30
70
15] $4\frac{2}{3}$ | $18\frac{2}{3}$ $4\frac{2}{3}$ $32\frac{2}{3}$
4 7

$32\frac{2}{3}$ $23\frac{1}{3}$ $9\frac{1}{3}$ 00
$9\frac{1}{3}$ 7 70
$2\frac{1}{3}$ 77] 10
$\frac{7}{3}$ $\frac{70}{3}$

Find me a number such that after subtracting $\frac{1}{3}$ and $\frac{1}{4}$ from it, the remainder multiplied by 10 makes 100. Put that it was $1x$. $\frac{1}{3}$ and $\frac{1}{4}$ are $\frac{7}{12}$ in twelfths. Subtract $1x$, the remainder is $\frac{5}{12}$, which is equal to 100. Multiply 12 times 100, it makes 1,200, divided by 5 it will be 240. Subtract a quarter of 240, it will be 60. Subtract the third from 240, it will be 80. And the remainder is indeed 100. You will see it here below. f. 41b

$1x$ $\frac{1}{3}$ $\frac{1}{4}$ $\frac{7}{12}$ $1x$ $\frac{5}{12}$ $\frac{100}{1}$ 020 240
1200 140
555] 240 100
80
60
100

E per far la ditta raxion per inpoxizion pony che fusse ala prima 12, ala segonda 24. El terzo de 12 serà 4, el quarto serà 3, sono 7, roman 5. 5 fia 19 fa 95, e mi voio che sia 100. Adoncha è men 5. E la segonda 24. El $\frac{1}{3}$ el $\frac{1}{4}$ sono 14, inchina 24 resta 10. Moltipicha per 19, fano 190, e my voio che sia 100. Adoncha più 90. E moltipicha in croxe e parti, insirano 240. Dame el terzo, serano 80. Deme[a] el quarto, serano 60. Sutrazi 140 de 240, roman ben 100, chomo avemo ditto de sovra.

12		24
men	X	più
5		90

12
3
4
5
19
95

100
95
5 men
9
5

24
8
6
10
19

190
100
90 più

95 partior

90
12
1080
120
partir 1200
95

24
5
120

19
1

240
80 el terzo
60 el quarto
240
140
100

100

00
32
0480
22800
9555
99
240

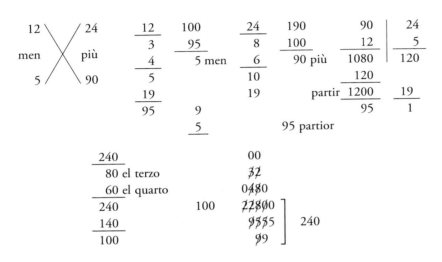

f. 42a Truovame un numero che tanto faza el so terzo moltipichado per 19 quanto el so quarto moltipichado in si medeximo. Pony che fusse 1^{co}, el terzo serave $\frac{1}{3}^{co}$. Moltipicha' per 19 fa 19 terzi de chossa, e tolto el so quarto serave $\frac{1}{4}^{co}$. Moltipichado in si fano $\frac{1}{16}^{\square}$. Adoncha parti le chose per li zenssi. Moltipicha 19 fia 16, fano 304. Parti per 3, serano $101\frac{1}{3}$. E questo val la chossa. Mo' tuo' el so terzo, serano $33\frac{7}{9}$, serano $\frac{304}{9}$. Moltipicha' per 19 farano $\frac{5776}{9}$. Mo' tuo' el so quarto, serà $25\frac{1}{3}$, serano $\frac{76}{3}$. Moltipicha' 76 fia 76 fa $\frac{5776}{9}$. E questo è fatto.

1^{co} $\frac{1}{3}^{co}$ $\frac{19}{1}^{co}$ $\frac{19}{3}^{co}$ $\frac{1}{4}$ $\frac{1}{4}$ $\frac{1}{16}^{\square}$ $\frac{19}{3}$

304
333 $101\frac{1}{3}$
$33\frac{7}{9}$

$\frac{304}{9}$ $\frac{19}{1}$ $\frac{5776}{9}$

76
76

5776
5776

0
021
101
44 $25\frac{1}{4}$

a. *Sic in MS.*

And to do this problem by false position, suppose 12 for the first and 24 for the second. A third of 12 will be 4, a quarter will be 3, they make 7, the remainder is 5. 5 times 19 makes 95, and I want it to be 100. So it is minus 5. And the second is 24. $\frac{1}{3}$ and $\frac{1}{4}$ of it makes 14, up to 24 makes 10. Multiply by 19, makes 190, and I wanted it to be 100. So it's plus 90. And multiply across and divide, the result is 240. Give me a third, it will be 80. Give me a quarter, it will be 60. Subtract 140 from 240 the remainder is 100, as we said above.

12 ╲ ╱ 24	12	100	24	190	90	24
minus ╳ plus	3	95	8	100	12	5
	4	5 minus	6	90 plus	1080	120
5 ╱ ╲ 90	5		10		120	
	19		19	to divide	1200	19
	95	9			95	1
		5		95 divisor		

240		00	
80 a third		32	
60 a fourth		0480	
240	100	22800	
140		9555	240
100		99	

Find me a number such that a third of it multiplied by 19 is equal to a quarter of it multiplied by itself. Put that it was $1x$, and the third would be $\frac{1}{3}x$. Multiplied by 19, the result is 19 thirds of the unknown, and its quarter would be $\frac{1}{4}x$. Multiplied by itself, the result will be $\frac{1}{16}x^2$. Then divide the unknowns by the squared unknowns. Multiply 19 by 16, and that makes 304. Divide by 3, which will be $101\frac{1}{3}$. And this is the value of the unknown. Now take its third, which will be $33\frac{7}{9}$, or $\frac{304}{9}$. Multiplied by 19 it will be $\frac{5776}{9}$. Now take its quarter, it will be $25\frac{1}{3}$, which will be $\frac{76}{3}$. Multiply 76 times 76 it makes $\frac{5776}{9}$. And it's done. f. 42a

$1x$ $\frac{1}{3}x$ $\frac{19}{1}x$ $\frac{19}{3}x$ $\frac{1}{4}$ $\frac{1}{4}$ $\frac{1}{16}x^2$ $\frac{19}{3}$

304	
333	$101\frac{1}{3}$
	$33\frac{7}{9}$

$\frac{304}{9}$ $\frac{19}{1}$ $\frac{5776}{9}$

76	5776
76	5776

0	
021	
101	
44	$25\frac{1}{4}$

f. 42b Truova un numero che, abatudo $\frac{1}{3}$, el romagnente moltipichado per 16 faza tanto quanto moltipichado el so terzo del dito nomero per lo quinto del ditto numero. Pony che fusse 1^{co}, abati 1 terzo, roman $\frac{2^{co}}{3}$, e questi $\frac{2^{co}}{3}$ moltipichadi per 16 fano $\frac{32}{3}$. Puo' $\frac{1^{co}}{3}$ moltipichado per $\frac{1}{5}$ de chosa fa $\frac{1^{\square}}{15}$. Moltipicha 15 fia 32, fa 480, e questi parti per 3, serano 160. Adoncha tanto val la chossa, e questo fu el nomero. Mo' trazi el terzo de 160, serano $53\frac{1}{3}$, e lo romagnente serano $106\frac{2}{3}$. Moltipicha per 16, fano 5120, e puo' el terzo $53\frac{1}{3}$ e lo quinto de 160 serano 32. Adoncha moltipicha $53\frac{1}{3}$ per 32.

5120 1^{co} $\frac{2^{co}}{3}$ 16 32^{co} $\frac{1^{co}}{3}$ $\frac{1}{5}$ $\frac{1^{\square}}{15}$ $\frac{32}{3}$ 480 / 333] 160

$106\frac{2}{3}$ / $53\frac{1}{3}$ $106\frac{2}{3}$ $\frac{16}{1}$ 5120 $53\frac{1}{3}$ $\frac{32}{1}$

---

f. 43a [Chonpanie per 4 chonpagni]

Quatro chonpagni fano 1ª conpagna. El primo mese a primo dì de marzo duchati 15. El segondo messe al primo dì de mazio duchati 25. El terzo messe al primo dì d'octobrio duchati 40. El quarto messe al primo dì novenbrio duchati 80. E si ano tegnodo questa chonpanya inchina dì ultimo fevrer proximo et ano vadagnado duchati 100. V'adomando zò che die aver zaschaduno di loro. E per voler far la ditta raxion per la riegola del 3 fa che tutty 3 avesse la suma de tuti 4. Moltipicha li sso denar per li mexi, che xe lo primo 180, el segondo 250, el terzo 200, el quinto[a] 320. Asumadi tuti fano 950, e diremo: "Se 950 me dà 100, che me darà 180 che fu del primo?" E questi partidi per lo partidor serà duchati $18\frac{90}{95}$. Per le segondo: "Se 950 fose 100, che serave 250?" Moltipichar e partir, serà duchati $26\frac{300}{950}$. El terzo, moltipichado e partido, serà duchati $21\frac{50}{950}$. El quarto, moltipichado e partido, serà duchati $33\frac{650}{950}$. Asuma, serà in tutto duchati 100.

E per far la ditta raxion per inpoxizion metti che moltipichado el primo li so duchati 15 che messe in la chonpaggnia[b] chon lo so tenpo, che fu mexi 12, serave 180. E questo adopia' farà 360. E per lo simille fara' al segondo. Moltipichadi i sso duchati 25 per mexi 10 fano 250, e questi adopiadi fano 500. E per lo terzo che mese duchati 40, moltipicha' per mexi 5 farà 200, e questi adopiadi farano 400. E metti in la inpoxizion prima el quarto. Moltipicha i so denari, che fu duchati 80, per
f. 43b mexi 4, fano 320, e questi adopia, fano 640. E me[*ti*] | al primo 360, el segondo 500, el terzo 400, el quarto 640. Mo' avemo chonpida la prima inpoxizion. E faremo la segonda inpoxizion. 3 fia 180 fa

---

a. *Sic in MS, evidently in plave of* quarto.
b. *Sic in MS.*

Find a number such that, with $\frac{1}{3}$ subtracted, the remainder multiplied by 16 makes as much as a third of the number multiplied by a fifth of the number. Put that it was $1x$, subtract 1 third, the remainder is $\frac{2}{3}x$, and this $\frac{2}{3}x$ multiplied by 16 makes $\frac{32}{3}$. Then $\frac{1}{3}x$ multiplied by $\frac{1}{5}x$ makes $\frac{1}{15}x^2$. Multiply 15 times 32, which makes 480, and this divided by 3 will be 160. So this is the value of the unknown, and that was the number. Now subtract a third of 160, it will be $53\frac{1}{3}$, and the remainder will be $106\frac{2}{3}$. Multiply it by 16, which makes 5,120, and then a third is $53\frac{1}{3}$ and a fifth of 160 will be 32. Then multiply $53\frac{1}{3}$ by 32. f. 42b

$$5120 \quad 1x \quad \tfrac{2}{3}x \quad 16 \quad 32x \quad \tfrac{1}{3}x \quad \tfrac{1}{5} \quad \tfrac{1}{15}x^2 \quad \tfrac{32}{3} \quad \left.\begin{matrix}480\\333\end{matrix}\right] \; 160$$

$$\begin{matrix}106\frac{2}{3}\\53\frac{1}{3}\end{matrix} \quad 106\tfrac{2}{3} \quad \tfrac{16}{1} \quad 5120 \quad 53\tfrac{1}{3} \quad \tfrac{32}{1}$$

COMPANIES FOR 4 PARTNERS

f. 43a

Four partners make a company. The first one put in 15 ducats on the first day of March.[1] The second put in 25 ducats on the first day of May. The third put in 40 ducats on the first day of October. The fourth put in 80 ducats on the first day of November. And so they have held this company until the last day of the following February and have made a profit of 100 ducats. I ask you what each of them should have. And to do this problem by the rule of three, put that all three had the sum of all four. Multiply their money by the months, so that the first has 180, the second 250, the third 200, the fourth[2] 320. Added all together it makes 950, and we will say: "If 950 gives me 100, what will 180 give me, which was the amount of the first?" And this divided by the divisor will be $18\frac{90}{95}$ ducats. For the second: "If 950 was 100, what would 250 be?" Multiply and divide, it will be $26\frac{300}{950}$. The third, multiplied and divided, will be $21\frac{50}{950}$. The fourth, multiplied and divided, will be $33\frac{650}{950}$ ducats. Add together, it will be in all 100 ducats.

And to do this problem by false position, suppose that for the first the 15 ducats that he put in the company multiplied by his time, which was 12 months, would be 180. And this doubled will be 360. And the second will be done in a similar way. His 25 ducats multiplied by 10 makes 250, and this doubled makes 500. And for the third, the 40 ducats he put in multiplied by 5 months will be 200, and this doubled will make 400. And put the fourth at the first false position. Multiply his money, which was 80 ducats, by 4 months, it makes 320, and double this, which makes 640. And put | for the first 360, the second 500, the third 400, the fourth 640. Now we have completed the first false position. And we will make the second false position. 3 times 180 makes 540, and put f. 43b

---

1. The Venetian year ran from March 1 through the end of February.
2. *fifth* in MS.

540, e questi metti de sovra al'inpo[*xi*]zion. El segondo, 3 fia 250[a] fa 750, e questa metti ale[b] segonda. El terzo, 3 fia 200 fa 600, e questo metti al terzo. El quarto, 3 volte 320 fa 960, e questo metti al quarto. E moltipicha in croxe. 360 fia tutta la suma d'i deneri moltipichadi chon lo tenpo, che sono 2750, fano 990000. Puo' moltipicha 1800 fia 540, fano 972000, e questa sutratto da 990000 roman 180000. Puo' sotratto per far partidor da 2750 1800, roman 950, e questo è lo partidor. Parti 18000 per 950, insirà fuor duchati $18\frac{900}{950}$. Avanti che tu abi moltipichado in cruxe, asuma el primo, el segondo, el terzo, el quarto, che sono 1900, e noi volemo che ssia duchati 100. Adoncha è più 1800. E per lo simille la segonda. El primo, el segondo, el terzo, el quarto asumadi sono 2850, e noi volemo che sia duchati 100. Adoncha è più 2750. Segui chomo è ditto de sovra, in croxe moltipicha e parti, insirà el primo $18\frac{900}{950}$, el segondo $26\frac{300}{950}$, el terzo $21\frac{50}{950}$, el quarto $33\frac{650}{950}$. Asumadi fano duchati 100, chomo vedirì qui de sotto per figora.

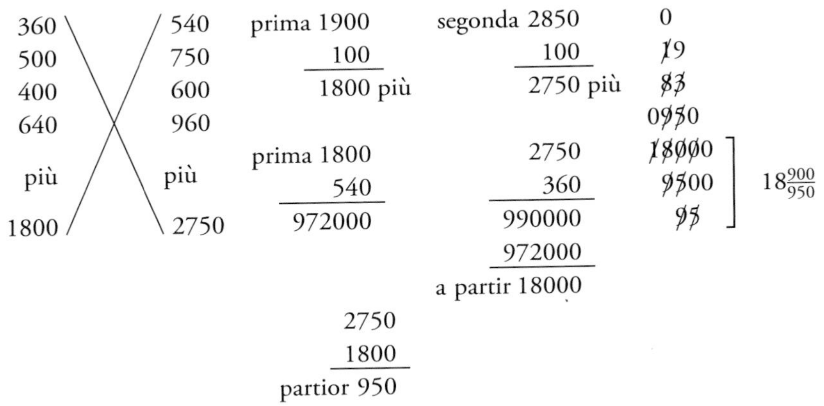

f. 44a E se la volesti far la raxion fata indriedo per chonpania de 4 chonpagni per la chossa, moltipichado li denari chon el tenpo serave, el primo chonpagno serave 180, el segondo, moltipichado i deneri chon el tenpo, fa 250. El terzo, moltipichado chon el tenpo, fano 250.[c] El terzo, moltipichado i denari chon el tenpo, fano 200. El quarto, moltipichado i denari chon el tenpo, fano 320. El primo, 180^{n}, asuma' el segondo, el terzo, el quarto, fano 770^{n}. Adoncha se 180 me dà 1^{co}, che me darà 770^{n}? Serà 770^{co}. Se parto 770^{co} per 180 dieme venir 1^{co}. Ponamo che 'l primo avese 1^{co}, el segondo, el terzo, el quarto per forza chonvien che l'abia 100^{n} men 1^{co}. Adoncha se parto 770^{co} per 180^{n} die venir 100 men 1^{co}. Adoncha moltipicha 180 fia 100, fano 18000^{n}, e puo' moltipicha 180 fia 1^{co}, fano 180^{co}. Adequa le parte, azonzi 180 a 770, serano 950^{co}. Parti 18000^{n}, insirà fuor per lo partior per lo primo chonpagno duchati $18\frac{90}{95}$, e per lo simille farì el segondo, el terzo, el quarto. Azonzi uno chon l'altro, averano tuti 4 duchati 100, chomo vedirì qui de sotto per figura.

a. 250 *corrected over* 350.
b. Sic in MS.
c. *Sic in MS, evidently as an error of copying. The correct statement is the following one.*

this above for the false position. The second, 3 times 250[1] makes 750, and put this for the second. The third, 3 times 200 makes 600, and put this for the third. The fourth, 3 times 320 makes 960, and put this for the fourth. And multiply across. 360 times the sum of the money multiplied by the time, which is 2,750, makes 990,000. Then multiply 1,800 by 540, which makes 972,000, and this subtracted from 990,000 leaves 18,000.[2] Then I subtract 1,800 from 2,750 to make the divisor, it leaves 950, and this is the divisor. Divide 18,000 by 950, which will result in $18\frac{900}{950}$ ducats. Before you have multiplied across, add the first, the second, the third, and the fourth, which is 1,900, and we wanted it to be 100 ducats. So it is plus 1,800. And do the second false position the same way. The first, the second, the third, and the fourth added are 2,850, and we wanted them to be 100 ducats. So it is plus 2,750. Follow the instructions given above, multiply across and divide, and the result will be for the first $18\frac{900}{950}$, the second $26\frac{300}{950}$, the third $21\frac{50}{950}$, the fourth $33\frac{650}{950}$. Added they make 100 ducats, as you will see worked out below.

360	540	first 1900	second 2850	0	
500	750	100	100	19	
400	600	1800 plus	2750 plus	83	
640	960			0950	
		first 1800	2750	18000	
plus	plus	540	360	9500	$18\frac{900}{950}$
1800	2750	972000	990000	95	
			972000		
			to divide 18000		
		2750			
		1800			
		divisor 950			

And if you wanted to do the problem of the company of 4 partners on the preceding page by unknowns, the sum of the first companion, with the money multiplied by the time, would be f. 44a
180; for the second, money multiplied by time would be 250. For the third, money multiplied by time makes 200.[3] For the fourth, the money multiplied by the time makes 320. The first, 180, add the second, the third, the fourth, makes 770. So if 180 gives me $1x$, what will 770 give me? It will be $770x$. If I divide $770x$ by 180 I should get $1x$. Let's suppose that the first had $1x$, the second, the third, the fourth then would have had to have 100 minus $1x$. So if I divide $770x$ by 180 I should get 100 minus $1x$. Then multiply 180 times 100, which makes 18,000, and then multiply 180 by $1x$, which makes $180x$. Equalize the sides, add 180 to 770, it will be $950x$. Divide 18,000, there will result from the division for the first companion $18\frac{90}{95}$ ducats, and in the same way you will do the second, the third, the fourth. Add one with the other, all 4 will have 100 ducats, as you will see worked out below.

1. *350* in MS.
2. *180000* in MS.
3. The manuscript first says *the third multiplied by the time makes 250* and then repeats the sentence correctly as here.

180	770	|	180	1^{co}	770	|	770^{co}	|	180^{co}	1^{co}	770^{co} 100^{n}
men 1^{co}			770	18000^{n}		|	180^{n} più 1^{co}		770^{co}	100^{n} men 1^{co}	

$$\begin{array}{r} 770 \\ \underline{180} \\ \text{partior } 950^{co} \end{array}$$

$$\begin{array}{r} 180^{co} \\ \underline{770} \\ 950^{co} \end{array} \qquad 18000^{n}$$

$$\left.\begin{array}{r} \not{0}9 \\ \not{8}\not{3} \\ 0\not{9}\not{5}0 \\ \not{1}\not{8}\not{0}\not{0}0 \\ \not{9}\not{5}00 \\ \not{9}\not{5} \end{array}\right] \; 18\tfrac{900}{950}$$

---

f. 44b

[Zenzer, chanella, zafaran]

E voyo chonprar zenzer e chanela, zafaran, per duchati 100, e voio che 'l zenzer vaya el 100 duchati 16, la chanella el 100 duchati 26, el zafaran el 100 duchati 80. E voio a pexo do tanto chanella cha cha[a] zenzer, e sì voyo a pexo 3 tanto zafaran cha chanella. E per voler far la ditta per inpoxizion, vedirì qui de sotto, pony che metesse ala prima inpoxizion che chonprasse 50 libre[b] de zenzer, e vorave de chanella libre 100, e vorave de zafaran libre 300. E chosì metti ala prima inpoxizion 50, 100, 300. Poi diremo: "Libre 50 de zenzer varia duchati 8, libre 100 chanela varia duchati 26, libre 300 de zafaran varia duchati 240, asumadi fano 274, e my voio che sia 100." Adoncha serà più 174, adoncha diremo: "Più 174." E per la segonda meteremo 100, 200, 600. Libre 100 de zenzer varia duchati 16, libre 200 chanela varia duchati 52, libre 600 de zafaran varia duchati 480, asumadi fano 548, e my voyo che sia 100. Adoncha pony più 448 e moltipicha in croxe e parti, insirà $18\frac{68}{274}$.

---

a. *Thus repeated in MS.*
b. chanella *crossed out with a horizontal line.*

180 770 | 180 1x 770 | 770x | 180x 1x 770x 100

minus 1x 770 18000 | 180 plus 1x 770x 100 minus 1x

770
180
divisor 950x

180x 18000
770
950x

~~0~~9
~~8~~~~3~~
0~~9~~~~5~~0
~~1~~~~8~~~~0~~~~0~~0
~~9~~~~5~~00 $18\frac{900}{950}$
~~9~~~~5~~

Ginger, cinnamon, saffron

f. 44b

And I want to buy ginger and cinnamon, and saffron, for 100 ducats, and I want the ginger to go for 16 ducats per 100, the cinnamon for 26 ducats the 100, the saffron for 80 ducats the 100. And I want twice as much cinnamon by weight as ginger, and I also want 3 times as much saffron by weight as cinnamon. And to do this problem by false position, as you will see below, suppose that I put at the first false position that I bought 50 pounds of ginger, and I wanted 100 pounds of cinnamon, and I wanted 300 pounds of saffron. And so put 50, 100, and 300 at the first false position. Then we will say: "50 pounds of ginger would be worth 8 ducats, 100 pounds of cinnamon would be worth 26 ducats, 300 pounds of saffron would be worth 240 ducats, added together makes 274, and I wanted it to be 100." So it will be plus 174, so we will say: "plus 174." And for the second we will suppose 100, 200, 600. 100 pounds of ginger would be worth 16 ducats, 200 pounds of cinnamon would be worth 52 ducats, 600 pounds of saffron would be worth 480 ducats, added together makes 548, and I wanted it to be 100. So put plus 448 and multiply across and divide, the result will be $18\frac{68}{274}$.

50	100	8	16	448	174	
100	200	26	52	50	100	
300	600	240	480	22400	17400	
		274	548	17400		
più	più	100	100	5000 a partir	448	
174	448	174 più	448 più		174	
					partior 274	

5000

0
066
220
3368
5000
2744
27

$18\frac{68}{274}$	$\frac{136}{548}$	zenzer
$36\frac{272}{548}$	———	chanela
$109\frac{268}{548}$	———	zafaran

f. 45a E per voler far la raxion indriedo de zenzer, chanela, zafaran per la chosa pony che tuolesse 1^{co} de zenzer. Adoncha 1^{co} de zenzer a duchati 16 el 100 monta $\frac{16}{100}{}^{co}$. E puo' 2 tanto de chanella serà 2^{co} de chanella. A duchati 26 el 100 mo[*nta*] $\frac{52}{100}{}^{co}$. E puo' 3 tantto zafaran, a duchati 80 el 100, monta $\frac{480}{100}{}^{co}$. Suma tuti et avera' $\frac{548}{100}$, che sono ingualli a 100 duchati. Adoncha parti 100 duchati per le chose, vegnirà $18\frac{136}{548}$. E tanto val la chosa. E la chanella 2 volte tanto serà $36\frac{272}{548}$, el zafaran 3 tanto serà $109\frac{268}{548}$. E tante libre vuol de zaschaduna chosa, chomo vedirì per figora.

50	100	8	16	448	174
100	200	26	52	50	100
300	600	240	480	22400	17400
		274	548	17400	
plus	plus	100	100	5000 to divide	448
174	448	174 plus	448 plus		174
					divisor 274

5000

0
0~~0~~6
~~220~~
~~330~~8
~~5000~~
~~2744~~
~~2~~7

$18\frac{68}{274}$	$\frac{136}{548}$	ginger
$36\frac{272}{548}$	———	cinnamon
$109\frac{268}{548}$	———	saffron

And to do the problem of the ginger, cinnamon, and saffron on the previous page by unknowns, put that he took 1x of ginger. So 1x of ginger at 16 ducats per 100 comes to $\frac{16}{100}x$. And then 2 times as much cinnamon will be 2x of cinnamon. At 26 ducats per 100, it comes to $\frac{52}{100}x$. And then 3 times as much saffron, at 80 ducats the 100, comes to $\frac{480}{100}x$. Add it all and he will have $\frac{548}{100}$, which is equal to 100 ducats. Then divide 100 ducats by the unknowns; it will come to $18\frac{136}{548}$. And that is the value of the unknown. And the cinnamon twice as much will be $36\frac{272}{548}$, the saffron 3 times as much will be $109\frac{268}{548}$. And there will be that many pounds of each thing, as you will see worked out. f. 45a

1^{co}	$\frac{16}{100}^{co}$	$\frac{52}{100}^{co}$	$\frac{480}{100}^{co}$	1600	
2^{co}				5200	
6^{co}				6800	480
				10000	100

$$\begin{array}{r} 4800000 \\ 680000 \\ \hline 5480000 \end{array} \qquad \frac{548|0|0|0|0}{100|0|0|0|0}$$

$$\frac{548}{100} \qquad \frac{100}{1}$$

1
2
057
450
05626
10000
5488
54

$18\frac{136}{548}$ zenzer
$36\frac{272}{548}$ chanela
$109\frac{268}{548}$ zafaran

E questo è fatto per la chossa.

---

f. 45b

[Partir 4 un sachetto denari]

Sono 4 chi vuol partir 1 sachetto de duchati. El primo averse e tolse fuora $\frac{1}{5}$ men 3 de tutta la suma. El segondo mese man e tolse fuora $\frac{1}{6}$ più 4. El terzo mese la man e tuolse $\frac{1}{7}$ men 5. El quarto tuolse $\frac{1}{10}$ più 8. Romaxe in sachetto duchati 20. Quanti fu la suma del sachetto? E per far la ditta raxion per la chosa e voiando far la ditta raxion fa suma. El men fu 8, el più fu 12. Sotrazi 8 de 12, roman 4. Puo' azonzi $\frac{1}{5}\,\frac{1}{6}$, fa $\frac{11}{30}$; azonzi $\frac{1}{7}\,\frac{1}{10}$, fano $\frac{17}{70}$; e questi azonti insieme fano $\frac{1280}{2100}$. Schixa, serano $\frac{64}{105}$ più 4^{n}. Sutrazi 1^{co} de $\frac{64}{105}$, roman $\frac{41}{105}^{co}$, che son ingual men 4^{n} ingual a 20^{n}. Adequa le parte, più 20^{n} men 4^{n}, azonti serano 24^{n}. Parti per $\frac{41}{105}^{co}$, e voler moltipichar 24 fia 105 fano 2520. Parti per $\frac{41}{105}^{co}$, insirà del partidor $61\frac{19}{105}$. E tanto val la chossa. E per aprovar la ditta, dame el quinto de $61\frac{19}{105}$, seràno $12\frac{124}{125}$. Dame el sesto, serà $10\frac{124}{630}$. El settimo serà $8\frac{544}{735}$. El dezimo serà $6\frac{124}{1050}$. Aprovalo, starà ben.

$1x \quad \frac{16}{100}x \quad \frac{52}{100}x \quad \frac{480}{100}x$

$2x$

$6x$

$$\begin{array}{r} 1600 \\ 5200 \\ \hline 6800 \\ \hline 10000 \end{array} \qquad \frac{480}{100}$$

$$\begin{array}{r} 4800000 \\ 680000 \\ \hline 5480000 \end{array}$$

$$\frac{548|0|0|0|0}{100|0|0|0|0}$$

$\frac{548}{100} \quad \frac{100}{1}$

1
2
057[1]
450
05626
10000
5488
54

$18\frac{136}{548}$ ginger
$36\frac{272}{548}$ cinnamon
$109\frac{268}{548}$ saffron

And this is done by unknowns.

FOUR DIVIDE A SACK OF COINS

f. 45b

There are 4 men who want to divide a sack of ducats. The first opened it and took out $\frac{1}{5}$ minus 3 of the total. The second put his hand in and took out $\frac{1}{6}$ plus 4. The third put his hand in and took out $\frac{1}{7}$ minus 5. The fourth took out $\frac{1}{10}$ plus 8. There remained 20 ducats in the sack. How much was the total of the sack? And to do this problem by unknowns, to do the problem take the sum. The minus was 8, the plus was 12. Subtract 8 from 12, which leaves 4. Then add $\frac{1}{5}\,\frac{1}{6}$, which makes $\frac{11}{30}$; add $\frac{1}{7}\,\frac{1}{10}$, which makes $\frac{17}{70}$; and these added together make $\frac{1280}{2100}$. Factor; it will be $\frac{64}{105}$ plus 4. Subtract $1x$ from $\frac{64}{105}$;[2] there remains $\frac{41}{105}x$, which is equal to minus 4, equal to 20. Equalize the sides, plus 20 minus 4, added they will be 24. Divide by $\frac{41}{105}x$, and multiplying 24 times 105 makes 2,520. Divided by $\frac{41}{105}x$, the result of the division will be $61\frac{19}{105}$.[3] And that is the value of the unknown. And to prove this, give me a fifth of $61\frac{19}{105}$, which will be $12\frac{124}{125}$. Give me a sixth, which will be $10\frac{124}{630}$. A seventh will be $8\frac{544}{735}$. A tenth will be $6\frac{124}{1050}$. Prove it; it will be fine.

---

1. This numeral should be 3.
2. Correctly this should be "Subtract $\frac{64}{105}$ from $1x$."
3. This should be $61\frac{19}{41}$, and the following calculations are thrown off.

$$\begin{array}{llllllll}
\frac{1}{5} & \text{men} & 3^{n} & \text{men } 8^{n} & 12 & \frac{1}{5}\ \ \frac{1}{6}\ \ \frac{11}{30}\ \ \frac{1}{7}\ \ \frac{1}{10} \\
\frac{1}{6} & \text{più} & 4 & & 8 & \\
\frac{1}{7} & \text{men} & 5 & \text{più } 12^{n} & \overline{4^{n}} & \\
\frac{1}{10} & \text{più} & 8 & & & \frac{17}{70}\quad \frac{11}{30}\quad \frac{1280}{2100} \\
 & & & \frac{24^{n}}{1}\ \ \frac{41}{105}^{co}\ \ \frac{64}{105} & \text{men} & 4^{n}\ \ 20^{n}\ \ \text{più}\ \ 61\frac{19}{105}
\end{array}$$

f. 46a [Truovar numeri]

Truovame un numero che abatudo $\frac{1}{3}$ el romagnente moltipichado per 4 faza 4 più cha la mittà del nomero moltipicha' per 4. Pony che fusse 1^{co}, abati $\frac{1}{3}$, roman $\frac{2}{3}^{co}$. Moltipicha per 4, serano $\frac{8}{3}^{co}$, e puo' tuo' la mitade del nomero che è $\frac{1}{2}^{co}$. Moltipicha $\frac{4}{2}^{co}$ men 4^{n}, e questo è inngal a $\frac{8}{3}^{co}$ più 4^{n}. Adequa le parte. $\frac{8}{3}^{co}$ sono $2\frac{2}{3}^{co}$; $\frac{4}{2}^{co}$ sono 2^{co}. Adoncha sotrazi 2^{co} de $2\frac{2}{3}^{co}$, roman $\frac{2}{3}^{co}$ a partir $\frac{4}{1}^{n}$. Adoncha 3 fia 4 fa 12. Parti per 2^{co}, serà 6. Aprovalla. Tuo' el terzo de 6, serà 2, roman 4. Moltipicha 4 fia 4, fa 16, e puo' la mitade del nomero ch'è 6, serano 3. Moltipicha 3 fia 4, 12, che ben fu 4 più.

$$\begin{array}{llllllll}
1^{co} & \frac{2}{3}^{co} & \frac{8}{3}^{co} & \frac{1}{2}^{co} & \frac{4}{2}^{co} & \text{men } 4^{n} & \frac{8}{3}^{co} & \text{più } 4^{n} \\
\frac{4}{2}^{co} & \frac{2}{3}^{co} & \frac{4}{1} & 12 & & 4 & 16^{n} & \frac{6}{3} \\
 & & & 2 & 6 & 4 & 12\quad \text{più } 4 & 4
\end{array}$$

E per far la ditta per inpoxizion pony che tu avese posto ala prima inpoxizion 12, poristi[a] che tratto el terzo de 12 roman 8. Moltipicha' per 4, fa 32. Puo' abatti la mitade de 12, roman 6. 4 fia 6 fa 24, e my voria che fusse più 4. Adoncha sotrazi del 32, roman più 4. Puo' ala segonda pony che fose 9, el terzo serave 3, roman 6. E quello moltipicha' per 4, fa 24. Tuo' la mitade de 9, serà $4\frac{1}{2}$. Molti-
f. 46b picha' per 4, fano 18, e my voio che sia 22. Adoncha è più 2. Moltipicha e trazi, serano 12. A par|tir in 2 fano 6, chomo vedirì qui de sotto per figora.

$$\begin{array}{lllllllll}
12 \diagdown\ \diagup 9 & 36 & 00 & & 12 & 32 & & 4 \\
\text{più} \times \text{più} & 24 & \not{1}\not{2} & & 4 & 24 & & 2 \\
4 \diagup\ \diagdown 2 & \overline{12} & \not{2} & 6 & \overline{8} & \overline{4} & & \\
 & & & & 4 & & 9 & 6 \\
 & & & 24 & & & & 4 \\
 & & 12 & 18 & & & & \\
 & & 6 & \overline{2} & & & &
\end{array}$$

a. *Sic in MS.*

$\frac{1}{5}$	minus	3	minus 8		12	$\frac{1}{5}$	$\frac{1}{6}$	$\frac{11}{30}$	$\frac{1}{7}$	$\frac{1}{10}$	
$\frac{1}{6}$	plus	4			8						
$\frac{1}{7}$	minus	5	plus	12	4						
$\frac{1}{10}$	plus	8				$\frac{17}{70}$		$\frac{11}{30}$	$\frac{1280}{2100}$		

$\frac{24}{1}$ $\frac{41}{105}x$ $\frac{64}{105}$ minus 4 20 plus $61\frac{19}{105}$

FIND NUMBERS

f. 46a

Find me a number such that with $\frac{1}{3}$ subtracted, the remainder multiplied by 4 makes 4 more than half of the number multiplied by 4. Put that it was $1x$; subtract $\frac{1}{3}$, which leaves $\frac{2}{3}x$. Multiplied by 4, it will be $\frac{8}{3}x$, and then take a half of the number, which is $\frac{1}{2}x$. Multiply, $\frac{4}{2}x$ minus 4, and this is equal to $\frac{8}{3}x$ plus 4. Equalize the sides. $\frac{8}{3}x$ is $2\frac{2}{3}x$; $\frac{4}{2}x$ is $2x$. Then subtract $2x$ from $2\frac{2}{3}x$, which leaves $\frac{2}{3}x$ to divide into $\frac{4}{1}$. Then 3 times 4 makes 12. Divide by $2x$; it will be 6. Prove it. Take a third from 6, which will be 2; it leaves 4. Multiply 4 times 4, which makes 16, and then half of the number which is 6, will be 3. Multiply 3 times 4, 12, which indeed was 4 more.

$1x$	$\frac{2}{3}x$	$\frac{8}{3}x$	$\frac{1}{2}x$		$\frac{4}{2}x$	minus 4		$\frac{8}{3}x$	plus 4	
$\frac{4}{2}x$	$\frac{2}{3}x$	$\frac{4}{1}$	12		4	16			$\frac{6}{3}$	
			2	6	4	12	plus 4		4	

And to do this by false position, suppose that you had put 12 at the first false position; you would put that having taken a third from 12, leaves 8. Multiplied by 4, it makes 32. Then subtract half of 12, which leaves 6. 4 times 6 makes 24, and I would like it to be plus 4. So subtract from 32, which leaves plus 4. Then at the second suppose that it was 9; a third would be 3, which leaves 6. And that multiplied by 4 makes 24. Take a half of 9, which will be $4\frac{1}{2}$. Multiplied by 4, it makes 18, and I want it to be 22. So it's plus 2. Multiply and subtract, it will be 12. To divide | into 2 makes 6, as you will see worked out below. f. 46b

12	9	36	00			12	32		4
plus	plus	24	~~12~~			4	24		2
		12	~~2~~		6	8	4		
4	2					4		9	6
				24					4
			12	18					
			6	2					

Truovame un numero che abatudo i sso $\frac{2}{3}$ e sul romagnente azonto 7 faza tanto quanto moltipichado i so $\frac{2}{3}$ per 4. Pony che fusse 1^{co}, a tuor $\frac{2}{3}^{co}$ romagnerà $\frac{1}{3}^{co}$ e 7^{n} ingual a $\frac{8}{3}^{co}$, che fu moltipichadi $\frac{2}{3}^{co}$ per 4. Abatti $\frac{1}{3}^{co}$ de $\frac{8}{3}^{co}$, roman $\frac{7}{3}^{co}$. A partir 7^{n} insirà 3.

1^{co} $\quad \frac{1}{3}^{co}$ e 7^{n} $\quad \frac{2}{3}^{co}$ $\quad \frac{4}{1}$ $\quad \frac{8}{3}^{co}$ $\quad \frac{1}{3}^{co}$ $\quad$ roman $\quad \frac{7}{3}$ $\quad \frac{7}{1}$

00
21 (struck through)
7] 3

E per farla per inpoxizion pony ala prima 9, ala segonda 6.

9 × 6
più più
14 × 7

84
63
21
7] 3

3
7
10

6
4
24
10
più 14

2
7
9

4
4
16
9
più 7

9
7
più 63

14
6
più 84

84
63
21

00
21 (struck through)
7 (struck through)] 3

1
7
fa 8

4
2
8 fa

---

f. 47a [Martoloyo de navichar a mente]

Questa[a] raxion chiamada martoloyo per navichar a mente, chomo di sutto pono per singolo.

Prima sie alargar, segonda sie avanzar, terza sie ritorno, quarta sie avanzo di ritorno, e chadauna de queste à fiolle 8. E prima:

Alargar 8, zoè 20, 38, 55, 71, 83, 92, 98, 100.

Avanzar 8, zoè 98, 92, 83, 71, 55, 38, 20, 0.

a. *Above, in the left margin, in the same ink, the sign of a Greek cross, of uncertain significance.*

Find me a number such that $\frac{2}{3}$ from it and 7 added to the remainder makes as much as $\frac{2}{3}$ of it multiplied by 4. Put that it was $1x$; subtracting $\frac{2}{3}x$ will leave $\frac{1}{3}x$ and 7 equal to $\frac{8}{3}x$, which was $\frac{2}{3}x$ multiplied by 4. Subtract $\frac{1}{3}x$ from $\frac{8}{3}x$; it leaves $\frac{7}{3}x$. To divide by 7 will result in 3.

$1x$ $\quad \frac{1}{3}x$ and 7 $\quad \frac{2}{3}x \quad \frac{4}{1} \quad \frac{8}{3}x \quad \frac{1}{3}x$ $\quad$ leaves $\quad \frac{7}{3} \quad \frac{7}{1}$

00
~~21~~]
7] 3

And to do this by false position, put 9 at the first, 6 at the second.

9 ╲ ╱ 6   84   3   6   4
plus ╳ plus   63   7   4   4
14 ╱ ╲ 7   21]   10   24   2   16
7] 3   10   7   9
plus 14   9   plus 7

9   14
7   6   84   00
plus 63   plus 84   63   ~~21~~]   1
21   7] 3   7
makes 8

4
2
8 makes

MARTELOIO—TO NAVIGATE MENTALLY

f. 47a

This method is called the "marteloio" for navigating mentally, as I set forth systematically below.[1]

First is the distance off course; second is the advance; third is the return; fourth the advance on the return, and each of these has 8 daughters. And first:

The distance off course 8, that is, 20, 38, 55, 71, 83, 92, 98, 100.

The advance 8, that is, 98, 92, 83, 71, 55, 38, 20, 0.

1. For the *marteloio*, see Franci, "Mathematics in the Manuscript of Michael of Rhodes," vol. 3, pp. 140–142, and Piero Falchetta, "The Portolan of Michael of Rhodes," vol. 3, pp. 207–209.

Ritorno 8, zoè 51, 26, 18, 14, 12, 11, $10\frac{1}{5}$, 10.

Avanzo di ritorno, zoè 50, 24, 15, 10, $6\frac{1}{2}$, 4, $2\frac{1}{5}$, 0.

E per exenpio dela ditta riegolla ponamo che una tera te stia per levante mia 100, quanto voio andar intro levante e sirocho, che son quarte 2, che la tera me stia quarta de tramontana al griego, che son quarte 7, e che seroyo alargo dal ditto luogo? E questo sie lo muodo sutto scritto per raxion sovra ditta.

Prendi l'alargar de quel che tu domandi che tu vuol che romagna, che son quarte 7, che son $\frac{98}{10}$. Poi zonzeremo $\frac{2}{4}$ che tu va e 7 che tu domandi, serano 9. Adoncha chì è lo ritorno de quarte 9, sono $10\frac{1}{5}$. Moltipichemo e dixemo in questo muodo $\frac{98}{10}\ \frac{51}{5}$, e moltipichado insirà, moltipichado e partido, mia $99\frac{48}{50}$.[a]

f. 47b Poy volemo dir quanti mia seremo alargi dal ditto luogo. E voyando far la ditta raxion, prendi el chontrario de quel che tu à fatto e dirai che xe l'alargar de quarte due, sono $\frac{38}{10}$, e lo rittorno e lo rittorno[b] de quarte 9, che son $10\frac{1}{5}$, e moltipicha $\frac{38}{10}$ fia $\frac{51}{5}$, moltipicha e parti, insirà mia $38\frac{38}{50}$. E tanto seremo alargi dal ditto luogo. E per lo simille modo fara' ogn'altra raxion simille.

E per voler archizar, zoè voltezar, saver tornar in croxe e veder l'avanzar arai fatto, e per lo simille de avanzar, chomo vedirì qui de sotto per singolo.

E per esenpio la mia via sie per levantte e non de poso andar, e von mia 100 per la quarta de sirocho al'ostro. Quanti mia voyo andar ala quarta de griego alo levante che io vegna ala mia croxe, e quanto arò avanzado? Questo è lo muodo, che xe l'alargar de quarte 5, son $\frac{83}{10}$, e lo ritorno de quarte 3, sono 18. Moltipicha $\frac{83}{10}$ fia $\frac{18}{1}$, fano 1494, e questi partidi per 10 serano mya $149\frac{4}{10}$, e tanti mya voio chaminar che sia ala croxe.

E chi te domandasse che averemo avanzado, questo è lo muodo. Prendi l'avanzo di ritorno de quarte 3, che son 15, e l'alargar de quarte 5, che son $\frac{83}{10}$, moltipichadi e partidi serano $124\frac{5}{10}$, e puo' azonzi l'avanzo del'alargar de quarte 5, sono 55, azonti fano mia $179\frac{5}{10}$. Tanto avra' avanzado.

f. 48a E per un'altra raxion la mia via è per ponente e non di posso andar, e von quarta de ponente al garbin mia 100. El vento va avanti, e von intro ponente garbin mia 100. El vento va avanti, e von ala quarta de garbin inver lo ponente. El vento va avanti, e von per garbin mia 100. Adomando quanty mia voyo venir per maistro ch'io vegna ala mia croxe, e questo è lo muodo, e quanto averò

---

a. e tanto seramo alargi dal ditto luogo *crossed out with a wavy line.*

b. *Thus repeated in MS.*

1. MS has *terra,* which could also mean "city."

2. Michael calls these units, which are one-eighth of a 90° arc or one-thirty-second of a circumference, *quarte* (quarters), but the modern term in navigation is "points."

3. For medieval navigational and wind directions, see p. 347, note 8, below.

The return 8, that is, 51, 26, 18, 14, 12, 11, $10\frac{1}{5}$, 10.

The advance on the return, that is, 50, 24, 15, 10, $6\frac{1}{2}$, 4, $2\frac{1}{5}$, 0.

And for an example of this rule, let's say that a land[1] is 100 miles to the east of you, how much do I want to go east-southeast, which is 2 points,[2] so that the land will be north by east of me, which is 7 points, and how wide would I be from that place?[3] And below is written the way according to this technique.

Take the distance off course that you want to be left with, which is 7 points, which is $\frac{98}{10}$. Then we will add the 2 points that you go and the 7 that you seek, which will be 9. So here is the return of 9 points, which is $10\frac{1}{5}$. We multiply and we say in this way $\frac{98}{10}$ $\frac{51}{5}$, and multiplied together, multiplied and divided, $99\frac{48}{50}$ miles.[4]

Then we want to say how many miles we will be away from that place. And wanting to make this f. 47b
calculation, do the opposite of what you have done and you will say that the distance off course of two points is $\frac{38}{10}$, and the return[5] of 9 points, which is $10\frac{1}{5}$, and multiply $\frac{38}{10}$ times $\frac{51}{5}$, multiply and divide, will make $38\frac{38}{50}$ miles. And we will be this far from that place. And by the same method you will do all other similar calculations.

And if you want to tack, that is to go about, to know how to return to the course and see the advance you have made, it is done in the same way as the advance, as you will see systematically below.

And for example, my course is to the east and I cannot go that way, and we go 100 miles southeast by south. How many miles do I want to go northeast by east so that I come to my course, and how much will I have advanced? This is the method, which is the distance off course of 5 points, which is $\frac{83}{10}$, and the return of 3 points, which is 18. Multiply $\frac{83}{10}$ by $\frac{18}{1}$, which makes 1,494, and this divided by 10 will be $149\frac{4}{10}$ miles, and we want to travel this many miles to our course.

And if someone asked you how much we will have advanced, here is the way. Take the advance on the return of 3 points, which is 15, and the distance off course of 5 points, which is $\frac{83}{10}$, multiplied and divided will be $124\frac{5}{10}$, and then add the advance on the distance off course of 5 points, which is 55, added makes $179\frac{5}{10}$ miles. You will have advanced that much.

And by another calculation, my course is to the west and I cannot go that way, and we go 100 miles f. 48a
west by south. The wind draws ahead, and we go toward west-southwest 100 miles. The wind goes on, and we go southwest by west. The wind goes on, and we go southwest 100 miles. I ask how many miles I want to go northwest so I come to my course, and this is the way, and how much I will have advanced. And we will say that the distance off course of one point will be 20, and of 2

---

4. *And you will be that far wide of the place* crossed out in MS.
5. *and the return* repeated in MS.

avanzado. E diremo che xe l'alargar de quarta una, serà 20, e de $\frac{2}{4}$[a] 38, e de $\frac{3}{4}$ 55, e de $\frac{4}{4}$ 71. Azonti tutti insieme serano 184. E puo' diremo che lo ritorno de $\frac{4}{4}$ sono 14. Moltipicha' $\frac{184}{10}$ chon $\frac{14}{1}$ fano 2576, partidi per 10 fano mya $257\frac{6}{10}$. E tanti mya averì chaminado per venir ala croxe.

E che averò avanzado che xe l'avanzo di ritorno de quarte 4? Sono 10, moltipichadi per $\frac{184}{10}$ fano 1840, partidi per 10 serano mia 184. E po' diremo che xe l'avanzo di largar. De quarta 1ª sono 98, e $\frac{2}{4}$ 92, e $\frac{3}{4}$ 83, e $\frac{4}{4}$ 71. Asumadi fano 344, azonzi chon 184, fano 528. E tanto averai avanzado.

E chi tte domandase: "Per[b] un'oltra raxion una tera me sta per ponente la sera, e non dicho quanto mia la notte von. La notte von per maistro mia $41\frac{8}{10}$ e la maitina me sta quel teren intro ponente garbin. Quanti mie son alargo mo' che me sta el teren intro ponente garbin, e quanti mia era alargo la sera quando me steva per ponente?" Tu die far per questo muodo e dir che m'alargo per $\frac{2}{4}$ $\frac{38}{10}$, e lo
f. 48b ritorno de $\frac{6}{4}$ son 11. Adoncha moltipicha 38 fia | 11 fano 418, parti per 10, serano mia $41\frac{8}{10}$ tu avera' chaminado.

E che serastu alargo quando te sta intro ponente garbin, che xe l'alargar de quatro quarte sono $\frac{71}{10}$ e lo ritorno de $\frac{6}{4}$ son 11? Moltipicha $\frac{71}{10}$ fia $\frac{11}{1}$, fano 781, parti per 10, serano mia $78\frac{1}{10}$. Tanto serastu alargo.

E per saver quel che tu ieri alargo la sera quando te steva per ponente, torna per lo contrario del tuo andar, e va per sirocho quel che te romagna per ponente, quel che te steva intro ponente garbin tu anderà mia $53\frac{1}{5}$. Abatti el quarto, roman mia $39\frac{9}{10}$, e seratu alargo mia $126\frac{4}{5}$. Abati el quarto, ne roman mia $96\frac{2}{5}$. E tanto eristu alargo la sera.

1 3°

alargar			avanzar			ritorno			avanzo de ritorno		
1	quarta	20	1	quarta	98	1	quarta	51	1	quarta	50
2	—	38	2	—	92	2	—	26	2	—	24
3	—	55	3	—	83	3	—	18	3	—	15
4	—	71	4	—	71	4	—	14	4	—	10
5	—	83	5	—	55	5	—	12	5	—	$6\frac{1}{2}$
6	—	92	6	—	38	6	—	11	6	—	4
7	—	98	7	—	20	7	—	$10\frac{1}{5}$	7	—	$2\frac{1}{5}$
8	—	100	8	—	0	8	—	10	8	—	0

a. *Though written as fractions, this and the following are really a succession of navigational "quarters" (points on the compass).*
b. Che averò *crossed out with a horizontal line.*

points[1] 38, and of 3 points 55, and of 4 points 71. Added all together it will be 184. And then we will say that the return of 4 points is 14. Multiply $\frac{184}{10}$ with $\frac{14}{1}$ makes 2,576, divided by 10 makes $257\frac{6}{10}$ miles. And that is how many miles you will have traveled to get to the course.

And what will I have advanced on the return of 4 points? It is 10, multiplied by $\frac{184}{10}$ makes 1,840, divided by 10 will be 184 miles. And then we will say what the advance is on the distance off course. From one point it is 98, and 2 points 92, and 3 points 83, and 4 points 71. Added together they make 344; added to 184, it makes 528. And you will have advanced this much.

And if someone asks you: "As another calculation, a land is west of me in the evening and I can't say how many miles we go in the night. In the night we go northwest $41\frac{8}{10}$ miles and in the morning that land is toward the west-southwest. How many miles is the distance now that the land is toward the west-southwest, and how many miles was it distant in the evening when it was to my west?" You should do it this way and say that my distance off course for 2 points is $\frac{38}{10}$ and the return of 6 points is 11. Therefore multiply 38 by | 11, which makes 418; divided by 10 it will be $41\frac{8}{10}$, the f. 48b
number of miles that you will have traveled.

And how far away will you be when your location is west-southwest, of which the distance off course of 4 points is $\frac{71}{10}$ and the return of 6 points is 11? Multiply $\frac{71}{10}$ by $\frac{11}{1}$, which makes 781, divide by 10; it will be $78\frac{1}{10}$ miles. You will be that far away.

And to know how far away you were yesterday when you were to the west, go back the opposite of your course, and go as far southeast as you had gone to the west, such that you are at west-southwest; you will go $53\frac{1}{5}$ miles. Reduce by a quarter and $39\frac{9}{10}$ miles remains, and you will be $126\frac{4}{5}$ miles away. Reduce by a quarter; there remains $96\frac{2}{5}$ miles. And you were that far away in the evening.

1 3°[2]

distance off course			advance			return			advance on return		
1	point	20	1	point	98	1	point	51	1	point	50
2	—	38	2	—	92	2	—	26	2	—	24
3	—	55	3	—	83	3	—	18	3	—	15
4	—	71	4	—	71	4	—	14	4	—	10
5	—	83	5	—	55	5	—	12	5	—	$6\frac{1}{2}$
6	—	92	6	—	38	6	—	11	6	—	4
7	—	98	7	—	20	7	—	$10\frac{1}{5}$	7	—	$2\frac{1}{5}$
8	—	100	8	—	0	8	—	10	8	—	0

1. The navigational term is rendered by a fraction: $\frac{2}{4}$, and so on for the other quarters (points) below.
2. Uncertain significance.

f. 49a Sono 3 chonpagni chi fano una chonpanya. El primo mese duchati 20 e stete in la chonpanya mexi 2. El segondo messe in la chonpanya duchati 10 e stette un tenpo non so quanto. El terzo mese in la chonpanya marche 10 d'oro e stette mexi 5. E in questo tenpo ano vadagnado tuti 3 duchati 20. Al primo ducha de vadagno duchati 7, al secondo duchati 5, al terzo duchati 8. Adomando che tenpo stette el primo[a] e zò che valse 10 marche del'oro.

E per far la ditta raxion tti fara' moltipicha del primo i so denari e dirai: "2 fiade 20 fa 40." E questo è lo primo. E per far lo segondo tu dirai: "Se 7 me dà 40, che me darà 5 che avè de vadagno el segondo?" Moltipicha e parti, serano $28\frac{4}{7}$, e questi partidi per i so duchati 10 che mese in la chonpanya serano $2\frac{6}{7}$. E questi mexi stette el segondo. E per lo simille faremo per lo terzo. Se 7 me dà 40, che me darà 8? Moltipicha e parti, serano duchati $45\frac{5}{7}$. E questi partidi per lo so tenpo, che fu mexi 5, serano duchati $9\frac{1}{7}$. E tanti duchati valse 10 marche d'oro.

Mo' assuma duchati 40 del primo, e del segondo $28\frac{4}{7}$, e del terzo $45\frac{5}{7}$, fano 800 settimy, e diremo: "Se $\frac{800}{7}$ me dà duchati 20, che me darà 40?" Moltipicha e parti, serano duchati 7. E per lo simille muodo: "Se $\frac{800}{7}$ me dà 20, che me darà $28\frac{4}{7}$?" Serano 5. E per lo simille el terzo: "Se $\frac{800}{7}$ me dà 20, che serave $45\frac{5}{7}$?" Insirà fora duchati 8. E questa è fatta.

f. 50a[b]

[GALINE E GALLI]

Fame questa raxion: 3 galine men 1 gallo val soldi 40 e 5 galli men 1ª gallina val soldi 30. Che valse el gallo e che valse la galina? E per farla per inpoxizion vedirì qui de sotto.

E se la volesti far per inpoxizion pony che la galina valese soldi 15. Ado[*n*]cha 3 galine valleria soldi 45, e noi non avemo ezetto soldi 40. Adoncha sotrazi 40 de 45, roman 5 men valleria el gallo, sì che meteremo ala prima inpoxizion de sovra 15 e de sotto men 5. E la segonda inpoxizion metti che la galina vallese soldi 20, 3 galline valleria soldi 60, e noi non avemo che valesse se non soldi 40. Adoncha sutrazi 40 de 60, roman 20 men valse el gallo. E chosì metti de sora 20 e de sotto men 20. E po' faremo 5 fia 5 galli fa 25 galli, abatti 1ª galina che val 15, roman 10, e noi volemo che ssia 30. Adoncha è men 20. Po' diremo: "5 fia 20 fa 100." Abatti 1ª galina che val 20, roman 80, e noi vollemo che sia 30. Adoncha serà più 50. E chosì de sotto, moltipicha e parti, chomo vedirì per figora.

---

a. *Sic in MS, evidently in place of* segondo.

b. *Folio 49b is blank except for the usual* Ihesus *at the top and a rope border at the bottom.*

There are three partners who make a company. The first one put in 20 ducats and stayed in the company 2 months. The second one put 10 ducats in the company and I don't know how long he stayed. The third one put 10 marks of gold in the company and stayed 5 months. And in this period all three of them have made a profit of 20 ducats. The share of profits for the first one is 7 ducats, for the second one 5 ducats, for the third one 8 ducats. I ask how much time the second[1] one stayed in and what 10 marks of gold was worth. f. 49a

And to do this problem, you will multiply the first one by his money and you will say: "2 times 20 makes 40." And this is the first one. And to do the second one, you will say: "If 7 gives me 40, what will 5 give me which was the profit of the second one?" Multiply and divide, it will be $28\frac{4}{7}$, and this divided by his 10 ducats that he put in the company will be $2\frac{6}{7}$. And the second one stayed in this many months. And we will do the third one in the same way. If 7 gives me 40, what will 8 give me? Multiply and divide, it will be $45\frac{5}{7}$ ducats. And this divided by his time, which was 5 months, will be $9\frac{1}{7}$ ducats. And that's how much 10 marks of gold was worth.[2]

Now add the 40 ducats of the first one, and $28\frac{4}{7}$ of the second one, and $45\frac{5}{7}$ of the third one, which makes 800 sevenths, and we will say: "If $\frac{800}{7}$ gives me 20 ducats, what will 40 give me?" Multiply and divide, it will be 7 ducats. And by the same way: "If $\frac{800}{7}$ gives me 20, what will $28\frac{4}{7}$ give me?" It will be 5. And the third one in the same way: "If $\frac{800}{7}$ gives me 20, what will $45\frac{5}{7}$ be?" The result will be 8 ducats. And it's done.

Hens and roosters

f. 50a

Do this problem for me: 3 hens minus 1 rooster are worth 40 soldi and 5 roosters minus 1 hen are worth 30 soldi. What was the rooster worth and what was the hen worth? And to do this by false position, you will see here below.

And if you would like to do it by false position, suppose that the hen was worth 15 soldi. Then 3 hens would be worth 45 soldi, and we wanted only 40 soldi. So subtract 40 from 45, there remains minus 5 as what the rooster was worth, so we put for the first false position 15 on top and minus 5 below. And for the second false position put that the hen was worth 20 soldi, 3 hens would be worth 60 soldi, and we had that they were worth only 40 soldi. So subtract 40 from 60, which leaves minus 20, which was the value of the rooster. And so put 20 above and minus 20 below. And then we will do 5 times 5 roosters makes 25 roosters.[3] Subtract 1 hen which is worth 15, which leaves 10, and we wanted it to be 30. So it's minus 20. Then we will say: "5 times 20 makes 100." Subtract 1 hen which is worth 20, which leaves 80, and we wanted it to be 30. So it will be plus 50. And so below, multiplied and divided, as you will see worked out.

---

1. MS has *first.*
2. In reality, a mark of pure gold produced 67 ducats: Alan M. Stahl, *Zecca: The Mint of Venice in the Middle Ages* (Baltimore: Johns Hopkins University Press, 2000), p. 31.
3. This should be 25 soldi.

15			20
5			20
men			più
20			50

50
15
250
50
750
400
1150 a partir

50
20
partior 70

0
043
1150
700
7

$16\frac{3}{7}$ la galina
3
$49\frac{2}{7}$
40
el galo soldi $9\frac{2}{7}$

$9\frac{2}{7}$
5
$46\frac{3}{7}$
30
$16\frac{3}{7}$

f. 50b E per voler far la ditta raxion per la chossa pony che la galina vale|se 1^{co}, 3 galine valeria 3^{co}. Chonvien che 'l galo vaya tanto che abatando de 3^{co} romagna 40^{n}. Per forza chonvien valer 3^{co} men 40^{n}, sì che abatti 3^{co} men 40^{n} de 3^{co}, roman 40^{n}. Poi varda che val 5 galli. 5 fia 3^{co} val 15^{co}, e puo' moltipicha 5 fia 40^{n}, fa men 200^{n}, che son inguali a 30^{n}. Adequa le parte, abatti de 15^{co} 1^{co}, roman 14^{co}. Numeri, zoè tu abatti 1^{co} de 15^{co} che valse la galina. Azonzi 200^{n} a 30^{n}, fano 230^{n}, parti per 14^{co}, insirà soldi $16\frac{3}{7}$ val la galina. El gallo valse tanto che abatudo de 3 galine roman 40^{n}. Adoncha 3 galline valse soldi $49\frac{2}{7}$, sì che 'l gallo valse men soldi $9\frac{2}{7}$. E provala, tu truovera' che starà ben cho[*mo*] farì moltipicha' 5 galli fia soldi $9\frac{2}{7}$, fano soldi $46\frac{3}{7}$. Sutrazi soldi 30 che val i galli, roman el valor dela galina che manchava i gali, che sono soldi $16\frac{3}{7}$, chomo vedirì qui de sotto per figora.

1^{co}	3^{co} men 40^{n}	5	15^{co}	roman 14^{co}	40	200^{n}	230^{n}
		3	1^{co}		5	30	

$49\frac{2}{7}$
40
$9\frac{2}{7}$ e tanto val el gallo
5
$46\frac{3}{7}$

$46\frac{3}{7}$
30
$16\frac{3}{7}$

0
03
196
230
144
1

$16\frac{3}{7}$ e tanto val la galina

$46\frac{3}{7}$
$16\frac{3}{7}$
che valse i gali 30^{n}

$49\frac{2}{7}$
$9\frac{2}{7}$ galli
40^{n} che valse le galine

15   20
5   20
minus   plus
20   50

50
15
250
50
750
400
1150 to divide

50
20
divisor 70

0
0$\not{4}$3
$\not{1}\not{1}\not{5}$0
$\not{7}$00
$\not{7}$

16$\frac{3}{7}$ the hen
3
49$\frac{2}{7}$
40
the rooster soldi 9$\frac{2}{7}$

9$\frac{2}{7}$
5
46$\frac{3}{7}$
30
16$\frac{3}{7}$

And to do this problem by unknowns, put that the hen was worth | $1x$, 3 hens would be worth $3x$. It works out that the rooster is worth what subtracted from $3x$ leaves 40. By necessity it's worth $3x$ minus 40, so that you subtract $3x$ minus 40 from $3x$; there remains 40. Then note what 5 roosters are worth. 5 times $3x$ is worth $15x$, and then multiply 5 times 40, which makes minus 200, which is equal to 30. Equalize the sides, subtract $1x$ from $15x$, there remains $14x$. For the numbers, that is you subtract $1x$ from the $15x$ that the hen was worth. Add 200 and 30, which makes 230, divide by $14x$, the result is 16$\frac{3}{7}$ soldi for the value of the hen. The rooster was worth the amount that, when 3 hens are subtracted from it, leaves 40. So 3 hens were worth 49$\frac{2}{7}$, then the rooster was worth less, 9$\frac{2}{7}$ soldi. And prove it; you will find that it works as well as when you multiply 5 hens times 9$\frac{2}{7}$ soldi, which makes 46$\frac{3}{7}$ soldi, and subtract the 30 soldi that the roosters are worth, which leaves the value of the hen which the roosters were lacking, which is 16$\frac{3}{7}$ soldi, as you will see worked out below. f. 50b

$1x$ $3x$ minus 40   5 / 3 $15x$ / $1x$ leaves 14   40 / 5   200 / 30   230

49$\frac{2}{7}$
40
9$\frac{2}{7}$ the value of the rooster
5
46$\frac{3}{7}$

46$\frac{3}{7}$
30
16$\frac{3}{7}$

0
0$\not{3}$
$\not{1}\not{9}$6
$\not{2}\not{3}$0
$\not{1}\not{4}\not{4}$
$\not{1}$

16$\frac{3}{7}$ and the hen was worth that much

46$\frac{3}{7}$
16$\frac{3}{7}$
what the roosters were worth 30

49$\frac{2}{7}$
9$\frac{2}{7}$ roosters
40 what the hens were worth

f. 51a Fame questa raxion. 5 galine più un gallo val soldi 40, 4 galli più 1ª galina val soldi 32. Che valse el galo e che valse la galina, per voler far per inpoxizion?

E per far la ditta per inpoxizion pony che la galina valese soldi 7. Adoncha adoncha[a] 5 galine valea soldi 35. Adoncha el galo valea soldi 5, ch'è lo romagnente de soldi 40. Poi diremo: "4 gali val soldi 20 e 1ª galina soldi 7 val soldi 27," e noi volemo che sia soldi 32. Adoncha è men 5, e per questa fa la segonda.

7	6	98	1				1	
5	10	30	0~~6~~4		$\frac{70}{19}$	$\frac{5}{1}$	~~4~~3	
		128	~~12~~~~8~~				7~~0~~	
men	più		~~1~~~~9~~	$6\frac{14}{19}$			~~1~~~~9~~	$3\frac{13}{19}$
5	14							

gali $25\frac{5}{19}$
$6\frac{14}{19}$
soldi 32

galine $33\frac{13}{19}$
$6\frac{6}{19}$
soma soldi 40ⁿ

E per far la ditta raxion per la chosa pony che la galina vallese 1^{co}. Chonvien che 1 gallo vaya zoè soldi 40^{n} men 5^{co}, e noi savemo che un gallo val 40^{n} men 5^{co}, e sì volemo saver zò che die vallia[b] 4 galli, che fa 160^{n} men 20^{co}. E poi volemo saver quanto val 1ª gallina, e zà avemo sapudo che val 1^{co}. Sì che adoncha azonzi insenbre 1^{co} chon 160^{n} men 20^{co}. Averemo 160^{n} men 19^{co}. Parti li nomeri per le chose, ne vien $6\frac{14}{19}$, e tanto val la chosa. E noi metesemo che la galina valese 1^{co},
f. 51b adoncha valse soldi $6\frac{14}{19}$. Adoncha | moltipicha per 5 galine, serano $33\frac{13}{19}$. El gallo valse lo resto inchina soldi 40, che sono $6\frac{6}{19}$. Adoncha moltipicha 4 galli fia $6\frac{6}{19}$, monta $25\frac{5}{19}$, azonta 1ª galina che val $6\frac{14}{19}$, ben serano soldi 32 chomo ò ditto de sovra. Vedirì per figora.

1^{co}	5^{co}	40^{n}	men	5^{co}	160^{n}	men	19^{co}	32^{n}	1	
160^{n} men 20^{co}					32				0~~6~~4	
					128		19^{co}		~~12~~~~8~~	
									~~1~~~~9~~	$6\frac{14}{19}$
										$33\frac{13}{19}$ suma 40^{n}
					$25\frac{5}{19}$					$6\frac{6}{19}$
					$6\frac{14}{19}$					
					suma 32^{n}					

a. *Thus repeated in MS.*
b. *Thus in MS.*

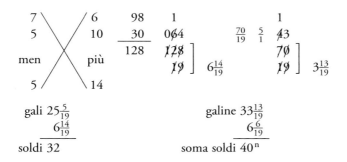

Do this problem for me. 5 hens plus a rooster are worth 40 soldi, 4 roosters plus a hen are worth 32 soldi. What was the rooster worth, and what was the hen worth, doing this problem by false position? f. 51a

And to do it by false position, suppose that the hen was worth 7 soldi. So 5 hens would be worth 35 soldi. Then the rooster would be worth 5 soldi, which is the remainder of 40 soldi. Then we will say: "4 roosters are worth 20 soldi and 1 hen 7 soldi, it is worth 27 soldi," and we wanted it to be 32 soldi. So it is minus 5, and do the second method for this one.

7 \ / 6
5 X 10
minus plus
5 / \ 14

98
30
———
128

1
0~~6~~4
~~1~~~~2~~~~8~~]
~~1~~~~9~~] $6\frac{14}{19}$

$\frac{70}{19}$ $\frac{5}{1}$

1
~~4~~3
7~~0~~]
~~1~~~~9~~] $3\frac{13}{19}$

roosters $25\frac{5}{19}$
$6\frac{14}{19}$
———
soldi 32

hens $33\frac{13}{19}$
$6\frac{6}{19}$
———
sum soldi 40

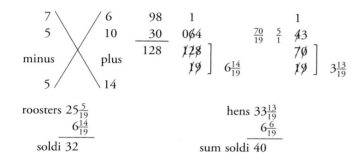

And to do this problem by unknowns, put that the hen was worth 1*x*. It works out that 1 rooster is worth 40 soldi minus 5*x*, and we know that a rooster is worth 40 minus 5*x*, and if we want to know what 4 roosters are worth, this would be 160 minus 20*x*. And then we want to know how much 1 hen is worth, and we have already known that it's worth 1*x*. So then add together 1*x* with 160 minus 20*x*. We will have 160 minus 19*x*. Divide the numbers by the unknowns, which comes to $6\frac{14}{19}$, and that is the value of the unknown. And we put that the hen was worth 1*x*, so it's worth $6\frac{14}{19}$ soldi. Then | multiply by 5 hens; it will be $33\frac{13}{19}$. The rooster was worth the rest up to 40 soldi, which is $6\frac{6}{19}$. Then multiply 4 roosters by $6\frac{6}{19}$, which comes to $25\frac{5}{19}$, add 1 hen which is worth $6\frac{14}{19}$; it will then come out to 32, as I've said above. You will see it worked out. f. 51b

1*x* 5*x* 40 minus 5*x*
160 minus 20*x*

160 minus 19*x* 32
32
———
128 19*x*

1
0~~6~~4
~~1~~~~2~~~~8~~]
~~1~~~~9~~] $6\frac{14}{19}$
$33\frac{13}{19}$ sum 40
$6\frac{6}{19}$

$25\frac{5}{19}$
$6\frac{14}{19}$
———
sum 32

Fame questa raxion. 7 galine 3 galli val soldi 54. 4 galli 5 galine val soldi 32. Che val el gallo e che val la galina? Chomo vedirì qui de sotto.

E per far la ditta raxion per inpoxizion ponamo che la galina valese soldi 6. Adoncha 7 galine valeria soldi 42. Inchina la suma di soldi 54 serano men soldi 12. Adoncha 3 gali valse soldi 12, che vien al gallo soldi 4. E chosi mettemo 6, e sotto 6 4, e diremo: "4 galli val soldi 16, 5 galine val soldi 30, e 16 val 46," e noi volemo che sia soldi 32, adoncha è più soldi 14. E per far la segonda pony che fusse fuse[a] 5.[b] 5 fia 7 fa 35, inchina 54 seria 19. E questi partidi in 3 galli seria per gallo soldi $6\frac{1}{3}$. E chosì mettemo sutto 5 al'inpoxizion che vaia el gallo. Adoncha galli 4 valeria soldi $25\frac{1}{3}$,
f. 52a e per galine 5 seria soldi 25. Azonti serano | soldi $50\frac{1}{3}$, e mi voio che sia soldi 32. Adoncha è più soldi $18\frac{1}{3}$. E chosì metti al'inpoxizion e moltipicha in croxe e parti chomo tu à fatto dele oltre. E per figora vui vedirì qui de sotto.

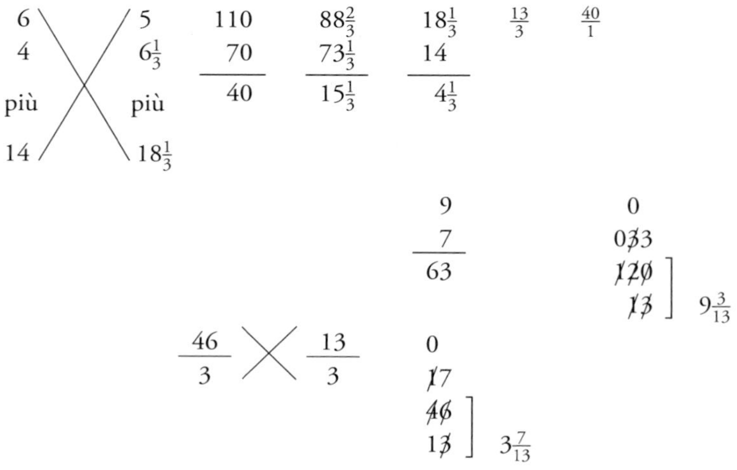

La galina valse soldi $9\frac{3}{13}$, el galo valse men soldi $3\frac{7}{13}$, $64\frac{8}{13}$.

E per far la ditta raxion per la chosa poni che la galina valese 1^{co}, adoncha 7 galine valse 7^{co}, chonvien per forza che 3 galli vaya soldi 54 men 7^{co}. E poi dixe: "4 galli e 5 galine val soldi 32," e noi savemo che 3 galli val soldi 54 men 7^{co}. Adoncha 4 galli val el terzo più de 3 galli, che monta soldi 72 men. Azonzi el terzo a 7^{co}, serà $9^{co}\frac{1}{3}$. E 5 galline savemo che val 5^{co}, sì che 4 gali e 5 galine val soldi 72 men $9^{co}\frac{1}{3}$ e 5^{co} più, che sono ingali a soldi 32. Adequemo le parte et averemo 40^{n} ingual a $4^{co}\frac{1}{3}$, perché l'adequar che noi avemo fatto sì fu che noi avemo sotratto de soldi[c] 72 32, però

a. *Thus repeated in MS.*
b. *Corrected over* 15.
c. 32 *crossed out with diagonal line.*

Do this problem for me. 7 hens and 3 roosters are worth 54 soldi. 4 hens and 5 roosters are worth 32 soldi. What is a rooster worth and what is a hen worth? You will see how below.

And to do the problem by false position, we suppose that the hen was worth 6 soldi. Then 7 hens would be worth 42 soldi. Up to the sum of 54 soldi would be minus 12 soldi. So 3 roosters were worth 12 soldi, which comes to 4 soldi per rooster. And so we put 6, and 4 below 6, and we will say: "4 roosters are worth 16 soldi, 5 hens are worth 30 soldi, and 16 makes 46," and we wanted it to be 32 soldi, so it is plus 14 soldi. And to do the second false position, suppose that it was 5. 5 times 7 makes 35, from there to 54 would be 19. And this divided by 3 roosters would be $6\frac{1}{3}$ soldi per rooster. And we put this in the false position below 5, as what the rooster is worth. So 4 roosters would be worth $25\frac{1}{3}$ soldi, and it would be 25 soldi for 5 hens. Added together it will be | $50\frac{1}{3}$ soldi, f. 52a
and I want it to be 32 soldi. So it is plus $18\frac{1}{3}$ soldi. And put this at the false position and multiply across and divide as you have done for the others. And you will see it worked out below here.

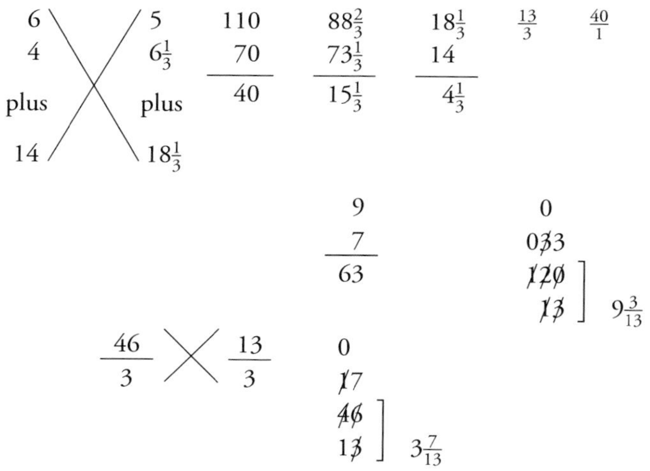

The hen was worth $9\frac{3}{13}$ soldi, the rooster was worth minus $3\frac{7}{13}$ soldi, $64\frac{8}{13}$.

And to do this problem by unknowns, put that the hen was worth $1x$, then 7 hens were worth $7x$, it follows by necessity that 3 roosters are worth 54 soldi minus $7x$. And then say: "4 roosters and 5 hens are worth 32 soldi," and we know that 3 roosters are worth 54 soldi minus $7x$. So 4 roosters are worth one third more than 3 roosters, which amounts to 72 soldi minus something. Add a third to $7x$, it will be $9\frac{1}{3}x$. And we know that 5 hens are worth $5x$, so 4 roosters and 5 hens are worth 72 soldi minus $9\frac{1}{3}x$ and plus $5x$, which is equal to 32 soldi. Equalize the sides and we will have 40 equal to $4\frac{1}{3}x$, because the equalization that we have done was that we have subtracted 32 from 72

romaxe soldi 40, e po' noi avemo sotratto 5^{co} de $9^{co}\frac{1}{3}$, però roman $4^{co}\frac{1}{3}$, che son per partir 40^{n} per $4^{co}\frac{1}{3}$, e vegnirà partido soldi $9\frac{3}{13}$. E tanto val la chosa, zoè 1ª galina. El galo chonvien a valer per
f. 52b forza | men $3\frac{7}{13}$, e la raxion sie che moltipichade 7 galine per $9\frac{3}{13}$ serà soldi $64\frac{8}{13}$, e lo spender fu soldi 54. Adoncha quel che fu spexo de più vien a valer men li galli. Adoncha, sotratto de $64\frac{8}{13}$ 54 resta soldi $10\frac{8}{13}$, e questi partidi per 3 galli vien per gallo men soldi $3\frac{7}{13}$.

1^{co} 7^{co} 54^{n} 7^{co} $9\frac{1}{3}$ soldi 54 men 7^{co}

18 $2\frac{1}{3}$

72

men 72^{n} 40^{n} $9^{co}\frac{1}{3}$ $4^{co}\frac{1}{3}$ 40^{n}

32 5

$\frac{40^{n}}{1}$ $\frac{13^{co}}{3}$ 0 $64\frac{8}{13}$

033 54 el galo men $3\frac{7}{13}$

120 $10\frac{8}{13}$

la galina soldi $9\frac{3}{13}$ 13 $9\frac{3}{13}$ 3

7

[Partir 3 chonpagni duchati 100]

Fame questa raxion. Sono 3 chi volono partir duchati 100. El primo vuol una parte de 100 e un più. El segondo vuol 2 parte chomo el primo e 2 più. El terzo vuol 3 parte chomo el segondo e 3 più. E t'adomando: "Quanti n'avè il primo, e quanti avè el segondo, e quanti avè el terzo?"

E se la volesti far per inpoxizion pony che 'l primo avese 1, e più 1 val 2. El segondo el segondo[a] 2 tanti, e 2 più che sono 6. El terzo vuol 3 tanti chomo el segondo, adoncha averalo 21. Asumadi serano 29, e noi vossemo che fuse 100. Adoncha serà men 71, e chosì metti al'inpoxizion. E per
f. 53a far la ditta segonda inpoxizion pony che fose 2 e 1 de | più val 3. El segondo vuol 2 tanti, e do più, serano 8. El terzo vorà 3 tanti, e 3 più, serano 21. E sumadi insieme fano 38, e noi vosemo che fuse duchati 100, adoncha è men 62. A chossì fato muodo metti ala segonda inpoxizion, moltipicha 12 in croxe, parti, sutrazi. El primo averà duchati $9\frac{8}{9}$, el segondo duchati $20\frac{16}{9}$, el terzo duchati $63\frac{48}{9}$, e sumadi i rutti chon li sany serano duchati 100, chomo vedirì qui de sotto per figura.

a. *Thus repeated in MS.*

soldi, but 40 soldi remained, and then we have subtracted $5x$ from $9\frac{1}{3}x$, but $4\frac{1}{3}x$ remain, which makes 40 to be divided by $4\frac{1}{3}x$, and divided it will come to $9\frac{3}{13}$ soldi. And that is the value of the unknown, that is, one hen. The rooster then by necessity works out to be worth | minus $3\frac{7}{13}$, and f. 52b
the solution is that 7 hens multiplied by $9\frac{3}{13}$ will be $64\frac{8}{13}$ soldi, and the outlay was 54 soldi. So that which was spent as a positive is worth the value of the roosters as a negative. So, subtract 54 from $64\frac{8}{13}$, which leaves $10\frac{8}{13}$ soldi, and this divided by 3 roosters leaves minus $3\frac{7}{13}$ soldi per rooster.

$1x$ $7x$ 54 $7x$ $9\frac{1}{3}$ 54 soldi minus $7x$

18 $2\frac{1}{3}$

72

minus 72 40 $9\frac{1}{3}x$ $4\frac{1}{3}x$ 40

32 5

$\frac{40}{1}$ $\frac{13}{3}x$ $\not{0}$ $64\frac{8}{13}$

$0\not{3}3$ 54 the rooster minus $3\frac{7}{13}$

$\not{1}\not{2}\not{0}$ $10\frac{8}{13}$

the hen $9\frac{3}{13}$ soldi $\not{1}\not{3}$ $9\frac{3}{13}$ 3

7

Three partners divide 100 ducats

Do this problem for me. There are 3 who want to divide 100 ducats. The first one wants a part of the 100 and one more. The second one wants 2 parts like the first one and 2 more. The third one wants 3 parts like the second one and 3 more. And I ask you: "How much did the first one have, and how much did the second one have, and how much did the third one have?"

And if you wanted to do it by false position, suppose that the first one had 1, and plus 1 is 2. The second one had 2 times as much, and plus 2, which is 6. The third one wants 3 times as much as the second one,[1] so he will have 21. Added together it will be 29, and we wanted it to be 100. So it will be minus 71, and put this in the false position. And to do the second false position, suppose that it was 2 and 1 | more makes 3. The second wants 2 times, and two more, which will be 8. The f. 53a
third one will want 3 times as much, and 3 more, which will be 21.[2] And added together they make 38, and we wanted it to be 100 ducats, so it's minus 62. So having done it this way, put it at the second false position, multiply 12 across, divide, subtract. The first one will have $9\frac{8}{9}$ ducats, the second one $20\frac{16}{9}$ ducats, the third one $63\frac{48}{9}$ ducats, and when the fractions are added to the integers, it will come to 100 ducats, as you see worked out below.

1. *and plus 3* omitted.
2. This should be 27 (as shown correctly in the calculations that follow).

2	╲ ╱	3	2	100		3	100	
			6	29		8	38	
men	╳	men	21	71 men		27	62 men	
71	╱ ╲	62	29			38		

62	71	213	71
2	3	124	62
124	213	partir 89	9 partior

08
$\not{8}\not{9}$
$\not{9}$] $9\frac{8}{9}$ primo suma duchati 100
$20\frac{16}{9}$ segondo
$63\frac{48}{9}$ terzo

E se la volesti far per la chosa, pony che 'l primo avese 1^{co} e un più, el segondo 2 tanto più 2, serano 2^{co} 4^{n}, el terzo vuol 6^{co} 15^{n}. Suma' serano 9^{co} 20^{n}, ingual a duchati 100. Sutrazi 20 de 100, roman 80^{n}. A partir per 9^{co} serano $8\frac{8}{9}$, e tanto val la chosa, e un de più serano $9\frac{8}{9}$. El segondo averà $21\frac{7}{9}$, el terzo $68\frac{3}{9}$, azontti insieme serano duchati 100.

f. 53b 1^{co} 1 più 2^{co} 4^{n} 6^{co} 15^{n} 9^{co} 20^{n} 100^{n} 08
20 $8\not{0}^{n}$] $8\frac{8}{9}$
$\not{9}$

$9\frac{8}{9}$ primo $9\frac{8}{9}$ segondo $21\frac{7}{9}$ terzo $68\frac{3}{9}$ / 100

[CHONPRAR SCINALLI]

Fame questa raxion. El è un chi chonpra al marchado schinali 4. El primo pexa libre 5. El segondo pexa libre 4, e valse più cha 'l primo per bontade la libra pizulli 3. El terzo pexa libre 3, e valse per bontade la libra più cha 'l primo pizulli[a] 4. El quarto pexa libre 2, e valse per bontade più cha 'l primo la libra pizolli 6. V'adomando, aver spexo in questi 4 schinalli soldi 20, zò che valse la libra de zascadon scinal.

---

a. p *crossed out with a diagonal line.*

2 \ / 3 — minus X minus — 71 / \ 62

2	100		3	100	
6	29		8	38	
21	71 minus		27	62 minus	
29			38		

62	71	213	71
2	3	124	62
124	213	to divide 89	9 divisor

08
8̸9̸]
9̸] $9\frac{8}{9}$ first   total 100 ducats
$20\frac{16}{9}$ second
$63\frac{48}{9}$ third

And if you wanted to do it by unknowns, put that the first one had $1x$ plus one, the second one twice as much plus 2, it will be $2x$ 4, the third one wants $6x$ 15. Add; it will be $9x$ 20, equal to 100 ducats. Subtract 20 from 100, there remains 80. Divided by $9x$ it will be $8\frac{8}{9}$, and that is the value of the unknown, and one more will be $9\frac{8}{9}$. The second one will have $21\frac{7}{9}$, the third one $68\frac{3}{9}$, added together it will be 100 ducats.

$1x$ 1 plus $2x$ 4 $6x$ 15 $9x$ 20 100 08]   f. 53b
20 80̸] $8\frac{8}{9}$
9̸]

$9\frac{8}{9}$ first $9\frac{8}{9}$ second $21\frac{7}{9}$ third $68\frac{3}{9}$ / 100

To buy sturgeon spines

Do this problem for me. There is a man who buys 4 sturgeon spines[1] in the market. The first weighs 5 pounds. The second weighs 4 pounds and because of its quality was worth 3 piccoli[2] per pound more than the first. The third weighs 3 pounds and because of its quality was worth 4 piccoli per pound more than the first. The fourth weighs 2 pounds, and because of its quality was worth 6 piccoli more per pound than the first. I ask you, having spent 20 soldi on these 4 spines, what each spine was worth per pound.

1. Dried sturgeon spines were a delicacy imported to Venice from ports on the Black Sea; see Angéliki Tzavara, "À propos du commerce vénitien des 'scienali' (schinalia) (première moitié du XVe siècle)," in Damien Coulon, Catherine Otten-Froux, Paule Pagès, and Dominique Valérian, eds., *Chemins d'Outre Mer: Études d'histoire sur la Méditerranée médiévale offertes à Michel Balard* (Paris: Publications de la Sorbonne, 2004), 2: 813–826.
2. There were 12 piccoli (pennies) to the soldo (shilling).

E per far la ditta raxion per inpoxizion pony ala prima inpoxizion che la libra de scinal vallese pizulli 10. Adoncha 5 libre valeria pizulli 50, e 4 libre pizulli 40, e 3 libre pizulli 30, e 2 libre pizulli 20. Puo' diremo: "3 fia 4, 12." Puo' diremo: "Libre 3 3[a] fia 4, 12." Puo' diremo: "2 6 fa 12." Asumandi insieme fano 176, e noi volemo che sia soldi 20, che son pizulli 240. Adoncha serà men pizulli 64. E chosì farì ala prima inpoxizion. E per far la segonda pono che fusse 12. 5 fia 12 fa 60, 4 fia 12 fa 48, 3 fia 12 fa 36, 2 fia 12 fa 24, e 3 fia 12, che val pizulli de più, fano 36. Azontti insieme fano 204, e noi volemo che sia 240. Adoncha è men 36. E chossì metti ala segonda inpoxizion.

f. 54a

10 \ / 12	50	20	240	[c]60	240
men X men	40	12	176	48	204
64 / \ 36	30	240	[b]64 men	36	men 36
	20		64	24	
	36		36	36	
	176		partior 28	204	

36 360	64	768	01	
10	12	360	14	
	128	408 partir	226	
	64		408	
$14\frac{4}{7}$			288	$14\frac{16}{28}$
			2	

$72\frac{6}{7}$
$70\frac{2}{7}$
$55\frac{5}{7}$
$41\frac{1}{7}$
} suma 240^{n}

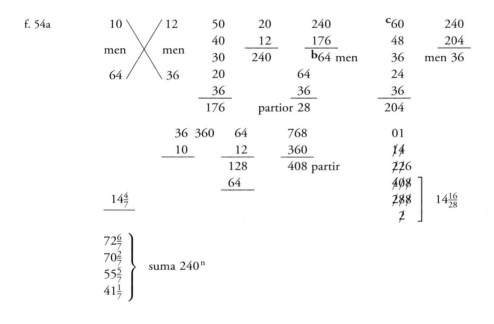

E per far la ditta raxion per la chosa, pony che la libra de scinal vallese 1^{co}. Adoncha libre 5 de schinal valeria 5^{co}. E per lo simille libre 4 de scinal valeria 4^{co} più 12^{n}. E per lo simille libre 3 de scinal valeri[*a*] 3^{co} più 12^{n}. E per lo simile 2 libre de scinal valeria 2^{co} più 12^{n}, che son ingal a 20 soldi, che son 240 pizulli. Azonzi le chosse, serano 14^{co}. Azonzi più el valor d'i schinali, che son 36, et averemo 14^{co} 36^{n} ingual a 240^{n}. Sutrazi 36 de 240^{n}, roman 204^{n}, e questi partidi per le chose serano pizulli $14\frac{4}{7}$. E tanto valse la libra del primo scinal. El segondo valse la libra pizulli $17\frac{4}{7}$. El terzo valse la libra $18\frac{4}{7}$. El quarto valse la libra $20\frac{4}{7}$. Sumadi serano pizulli 240, soldi 20.

---

a. *Thus repeated in MS.*
b. *2 crossed out with diagonal line.*
c. *2 crossed out with diagonal line.*

And to do this problem by false position, suppose at the first false position that the pound of sturgeon spines was worth 10 piccoli. Then 5 pounds would be worth 50 piccoli, and 4 pounds 40 piccoli, and 3 pounds 30 piccoli, and 2 pounds 20 piccoli. So we will say: "3 times 4, 12." Then we will say: "3 pounds times 4, 12." Then we will say: "2 6 makes 12." Added together they make 176, and we wanted it to be 20 soldi, which is 240 piccoli. So it will be minus 64 piccoli. And you will put this at the first false position. And to do the second, I suppose that it was 12. 5 times 12 makes 60, 4 times 12 makes 48, 3 times 12 makes 36, 2 times 12 makes 24, and 3 times 12, which is the value of the extra piccoli, is 36. Added together they make 204, and we wanted it to be 240; so it's minus 36. And put this at the second false position.

f. 54a

10 ╲ ╱ 12	50	20	240	60	240	
minus ╳ minus	40	12	176	48	204	
64 ╱ ╲ 36	30	240	64 minus	36	minus 36	
	20		64	24		
	36		36	36		
	176		divisor 28	204		

36 360	64	768	01	
10	12	360	14	
	128	408 to divide	226	
	64		408	
$14\frac{4}{7}$			288	$14\frac{16}{28}$
			2	

$72\frac{6}{7}$
$70\frac{2}{7}$
$55\frac{5}{7}$
$41\frac{1}{7}$
} total 240

And to do the problem by unknowns, put that the pound of sturgeon spines was worth $1x$. So 5 pounds of sturgeon spines would be worth $5x$. And in the same way 4 pounds of sturgeon spines would be worth $4x$ plus 12. And in the same way 3 pounds of sturgeon spines would be worth $3x$ plus 12. And in the same way 2 pounds of sturgeon spines would be worth $2x$ plus 12, which is equal to 20 soldi, which is 240 piccoli. Add the unknowns, it will be $14x$. Add the extra value of the spines, which is 36, and we will have $14x$ 36 equal to 240. Subtract 36 from 240, there remains 204, and this divided by the unknowns will be $14\frac{4}{7}$ piccoli. And that was the value of the pound of the first sturgeon spine. The second was worth $17\frac{4}{7}$ piccoli per pound. The third was worth $18\frac{4}{7}$ per pound. The fourth was worth $20\frac{4}{7}$ per pound. Added together they will be 240 piccoli, 20 soldi.

[Far de 12 2 parte e de 20 2 parti]

f. 54b

1^{co}	5^{co}	4^{co} più 12^{n}	3^{co} più 12^{n}	2^{co} più 2^{n}	5^{co}	14^{co}	36^{n}	240^{n}

240^{n}	5^{co}	0
36	4^{co}	0~~2~~
204^{n}	3^{co}	~~1~~~~0~~8
	2^{co}	~~2~~~~0~~~~4~~
	14	~~1~~~~4~~~~4~~
		~~1~~

] $14\frac{8}{14}$

$14\frac{4}{7}$

$17\frac{4}{7}$

$18\frac{4}{7}$

$20\frac{4}{7}$

$72\frac{6}{7}$
$70\frac{2}{7}$
$55\frac{5}{7}$
$41\frac{1}{7}$
} suma 240^{n}

[Far de 12 2 parte e de 20 2 parti]

Fame de 12 do parte che moltipichada l'una per 7 e l'oltra per 11 faza 100. Pony che fusse 1^{co} e 12^{n} men 1^{co}. Moltipicha più 1^{co} fia 7, fa più 7^{co}, e moltipicha 11 fia 12, fa 132 men 11^{co}, che son ingual a 100^{n}. Adequa. Demo 11^{co} men a 132^{n}, e non i mancharà niente; e demoli a 100^{n}, serano più 11^{co}. Sutrazi 7^{co} più da 11^{co} più, roman 4^{co}. So[*trazi*] 100 de 132, roman 32, e questi partidi per 4^{co} serano 8^{n}, e l'oltra parte inchina 12 serano 4. Adoncha moltipicha 7 fia 8, fa 56^{n}, e 4 fia 11 fa 44^{n}. Azonti sono ben 100.

1^{co} 12^{n} men 1^{co}

7^{co} 132^{n} / 100; $\frac{32^{n}}{1} \times \frac{4^{co}}{1}$; 100; 11^{co} / 7 / 4^{co}; 00 / ~~3~~~~2~~ / ~~4~~]

8	4
7	11
56	44
44	
100	

f. 55a Fame de 12 2 parte che moltipichada l'una per 7 e l'oltra per 11 | faza 10. Pony che fusse 1^{co} l'una parte e l'oltra fusse 12^{n} men 1^{co}. Moltipicha 1^{co} fia 7, fa 7^{co}, e 12 fia 11, fa 132^{n}, che son ingual a soldi 10^{n}. Adequa, tu avera' a trar de 11^{co} 7^{co}, roman 4^{co} a partir 122. Parti, serano $30\frac{1}{2}$, inchina

1*x*	5*x*	4*x* plus 12	3*x* plus 12	2*x* plus 2	5*x*	14*x*	36	240	f. 54b
					4*x*				
			240		3*x*		0		
			36		2*x*		02		
			204		14		168		
							204		
							144	$14\frac{8}{14}$	
							1		

$14\frac{4}{7}$
$17\frac{4}{7}$
$18\frac{4}{7}$
$20\frac{4}{7}$

$72\frac{6}{7}$
$70\frac{2}{7}$
$55\frac{5}{7}$
$41\frac{1}{7}$
total 240

DIVIDE 12 IN 2 PARTS AND 20 IN 2 PARTS

Find me two parts of 12 such that one multiplied by 7 and the other by 11 make 100. Put that they were 1*x* and 12 minus 1*x*. Multiply plus 1*x* by 7, which makes plus 7*x*, and multiply 11 times 12, which makes 132, minus 11*x*, which is equal to 100. Equalize the sides. Let's give minus 11*x* to 132, and nothing will be lacking; and let's give it to 100, which will be plus 11*x*. Subtract plus 7*x* from plus 11*x*, there remains 4*x*. Subtract 100 from 132, there remains 32, and this divided by 4*x* will be 8, and the other part up to 12 will be 4. Then multiply 7 times 8, which makes 56, and 4 times 11, which makes 44; added they are indeed 100.

1*x*	12 minus 1*x*							
7*x*	132	$\frac{32}{1}$ × $\frac{4}{1}x$	100	11*x*	00			
	100			7	32			
				4*x*	4	8	4	
						7	11	
						56	44	
						44		
						100		

Find me 2 parts of 12 such that one multiplied by 7 and the other by 11 | make 10. Put that one part was 1*x* and the other was 12 minus 1*x*. Multiply 1*x* times 7, which makes 7*x*, and 12 times 11, which makes 132, which is equal to 10 soldi. Equalize, you will have to subtract 7*x* from 11*x*, which leaves 4*x*, to divide into 122. Divide, it will be $30\frac{1}{2}$, up to 12 there is $18\frac{1}{2}$. If I multiply 7 f. 55a

12 mancha $18\frac{1}{2}$. Se moltipicho 7 fia $30\frac{1}{2}$ fa $213\frac{1}{2}$, e si moltipicho 11 fia $18\frac{1}{2}$ serano $203\frac{1}{2}$. Adoncha sutrazi de $113\frac{1}{2}$ $103\frac{1}{2}$, resta ben 10.

1^{co}	12^{n} men 1^{co}			
7^{co}	132^{n} / 10 / partir 122	men 11^{co}	primo 11^{co} / 7^{co} / partior 4^{co}	00 / ~~1~~~~2~~2 / ~~4~~4] $30\frac{2}{4}$ $\frac{1}{2}$
men $18\frac{1}{2}$ / 7	$103\frac{1}{2}$	$113\frac{1}{2}$ / $103\frac{1}{2}$ / 10		E questo è fatto.

Truovame 1 numero che abatudo $\frac{1}{3}$ el romagnente moltipichado per 4 faza 4 più che la mittà del nomero moltipicha' per 4. E per farla per la chossa pony che fuse 1^{co}. Abatti el terzo, roman $\frac{2}{3}^{co}$, moltipicha' per 4 fa $\frac{8}{3}^{co}$. E puo' tuo' la mità del numero ch'è $\frac{1}{2}^{co}$. Moltipicha' per 4 fa $\frac{4}{2}^{co}$. E questo è ingual a $\frac{8}{3}^{co}$ più 4^{n}. Adequa.

1^{co} $\quad \frac{2}{3}^{co}$ $\quad \frac{8}{3}^{co}$ $\quad \frac{1}{2}^{co}$ $\quad \frac{4}{2}^{co}$ $\quad$ per 4 $\quad \frac{8}{3}^{co}$ $\quad \frac{4}{1}$ $\quad \frac{4}{6}$

$\frac{4}{2}^{co}$ ingual $\frac{8}{3}^{co}$ per[a] 4 ingual $\frac{4}{6}^{co}$ per $\frac{4}{1}$ $\quad$ 00 / ~~2~~~~4~~ / ~~4~~] 6^{n}

E per farla per inpoxizion vedirì qui de sotto per figora.

12		9	36 / 24 / 12	00 / ~~1~~~~2~~ / ~~2~~] 6		8	32 / 24 / 8 / 4 / più 4	24 / 18 / 6 / 14 / più 2
più	X	più						
4		2						
			4 / 2 / 2	3 / 4 / 12	4 / 4	16 / 12 / più 4^{n}		

a. *Corrected over* $\frac{1}{2}^{co}$.

by $30\frac{1}{2}$ it makes $213\frac{1}{2}$, and if I multiply 11 by $18\frac{1}{2}$ it will be $203\frac{1}{2}$. Then subtract $203\frac{1}{2}$[1] from $213\frac{1}{2}$;[2] the remainder will be 10.

1x 12 minus 1x

7x 132 minus 11x first 11x 00

10 7x 122

to divide 122 divisor 4x 44] $30\frac{2}{4}$ $\frac{1}{2}$

minus $18\frac{1}{2}$ $203\frac{1}{2}$[3] $213\frac{1}{2}$[4]

7 $203\frac{1}{2}$[5]

10 And it's done.

Find me a number such that when $\frac{1}{3}$ is subtracted the remainder multiplied by 4 makes 4 more than half of the number multiplied by 4. And to do this by unknowns, put that it was 1x. Subtract a third, which leaves $\frac{2}{3}x$; multiplied by 4 makes $\frac{8}{3}x$. And then subtract half of the number, which is $\frac{1}{2}x$. Multiplied by 4 makes $\frac{4}{2}x$; and this is equal to $\frac{8}{3}x$ plus 4. Equalize.

1x $\frac{2}{3}x$ $\frac{8}{3}x$ $\frac{1}{2}x$ $\frac{4}{2}x$ by 4 $\frac{8}{3}x$ $\frac{4}{1}$ $\frac{4}{6}$

$\frac{4}{2}x$ equals $\frac{8}{3}x$ by 4 equals $\frac{4}{6}x$ by $\frac{4}{1}$ 00

24

4] 6

And to do it by false position, you will see it worked out here.

12 9 36 00 8 32 24

plus plus 24 12 24 18

12 2] 6 8 6

4 2 4 14

4 16 plus 4 plus 2

2 3 4 12

2 4 4 plus 4

12

1. *$103\frac{1}{2}$* in MS.
2. *$113\frac{1}{2}$* in MS.
3. *$103\frac{1}{2}$* in MS.
4. *$113\frac{1}{2}$* in MS.
5. *$103\frac{1}{2}$* in MS.

f. 55b Fame de 20 2 parte che moltipichada zaschaduna in ssi medexima e azonte le moltipichazion insenbre faza 208. Pony che fusse 1^{co} e 20^{n} men 1^{co}. Moltipichadi in si fano più $1^{\square}$ e men $1^{\square}$ 40^{co} 400^{n}. Azonti insieme fano $2^{\square}$ 40^{co} ingual 208^{n}. Adequa le parte. Averemo a un d'i ladi 208^{n} e 40^{co}, e del'altra parte averemo $2^{\square}$ e 400^{n}. Adoncha sutrazi 208^{n} de 400^{n}, roman 192^{n}. Parti per $2^{\square}$, serano 96^{n}. Parti 40^{co} per $2^{\square}$, serano 20^{co}. Demeza le chose, serano 10^{co}, moltipicha' in ssi fano 100^{n}. Adoncha de 100^{n} 96^{n} roman 4 più lo demezamento dele chose, che la mitade de 20^{co} son 10^{co}, e la radixe de 4 sono 2, che val 12. E tanto val la chosa. E l'altra parte chonvien eser 8. Mo' moltipicha 8 fia 8 fa 64, e 12 fia 12 fa 144. Asumadi insieme fano ben 208.

1^{co}	20^{n} men 1^{co}		$2^{\square}$	400^{n}	40^{co}	208	40^{co}	
1^{co}	20^{n} men 1^{co}			208			20^{co}	
$2^{\square}$	20^{co}			192				
	20^{co}			96	10		100	
$2^{\square}$	400^{n} 40^{co}	più $1^{\square}$			10		96	
							4	
			le chose 10				2	2
				2	12		radixe	

12	8	144	
12	8	64	
144	64	208	Fa ben 208^{n}.

f. 56a Truovame 1 numero che abatudo el so terzo e quel terzo | moltipichado per 5 e lo romagnente del nomero, che son $\frac{2}{3}$, moltipichado per lo so terzo e azonte insieme quele moltipicazion faza 33. Pony che fuse 1^{co}, tuo' $\frac{1}{3}$, roman $\frac{2}{3}^{co}$, moltipicha' per 5 farà $\frac{5}{3}^{co}$. Azonte le moltipicazion insenbre farà $\frac{5}{3}^{co}$. Mo' moltipicha $\frac{2}{3}^{co}$ chon $\frac{1}{3}^{co}$, farano $\frac{2}{9}^{\square}$ $\frac{5}{3}^{co}$ ingual a 33^{n}. Parti $\frac{5}{3}^{co}$ per $\frac{2}{9}^{\square}$, venirave $7^{co}\frac{1}{2}$. Demeza le chose, tuo' la mitade de $7^{co}\frac{1}{2}$, che so[*n*] $3^{co}\frac{3}{4}$, parti 33 per $\frac{2}{9}^{\square}$, serano $148\frac{1}{2}$. Puo' moltipicha el demezamento in si. Sono $\frac{15}{4}$. 15 fia 15 fa 225. A partir per 4 fia 4, fa 16, insirà fuora $14\frac{1}{16}$. E questi azonti chon $148\frac{1}{2}$ serano $162\frac{9}{16}$ men $3\frac{3}{4}$, e tanto val la chosa. Mo' questo nomero à radixe. La so radixe sie $12\frac{3}{4}$, e per aprovarla la radixe si xe vero moltipicha $12\frac{3}{4}$ in si, farà ben $16\frac{9}{16}$. Abatti el dimezamento, che son $3\frac{3}{4}$, da $12\frac{3}{4}$, roman 9. Aprovala: tuo' el $\frac{1}{3}$ de 9, serà 3; moltipichà per 5, serà 15. Tuo' $\frac{2}{3}$ de 9, serano 6. A moltipichar per 3 serano 18. Adoncha azonzi l'una chon l'oltra. Serano ben 33, chomo vedirì per figora.

Find me 2 parts of 20 such that when each one is multiplied by itself and the products are added together it makes 208. Put that it was $1x$ and 20 minus $1x$. Multiplied by itself it makes plus $1x$ and minus $1x^2$ $40x$ 400. Added together it makes $2x^2$ $40x$ equal to 208. Equalize the sides. We will have on one of the sides 208 and $40x$ and on the other side we will have $2x^2$ and 400. Then subtract 208 from 400, which leaves 192. Divide by $2x^2$; it will be 96. $40x$ divided by $2x^2$ will be $20x$. Halve the unknowns, it will be $10x$; multiplied by itself it makes 100. Then 96 from 100 leaves 4 plus the half of the unknowns, which is half of $20x$, which is $10x$, and the square root of 4 is 2, which is equal to 12. And that is the value of the unknown. And the other part works out to be 8. Now multiply 8 times 8, which makes 64, and 12 times 12, which makes 144. Added together they indeed make 208. f. 55b

$1x$	20 minus $1x$	$2x^2$	400	$40x$	208	$40x$	
$1x$	20 minus $1x$		208			$20x$	
$2x^2$	$20x$		192				
	$20x$		96	10		100	
$2x^2$	400 $40x$ | plus $1x^2$			10		96	
						4	
		the unknowns 10				2	2
			2	| 12 |		root	

12	8	144	
12	8	64	
| 144	| 64 |	208	It makes 208.

Find me 1 number such that when its third is subtracted, and that third | is multiplied by 5 and the remainder of the number, which is $\frac{2}{3}$, is multiplied by its third and added together to this product, it makes 33. Put that it was $1x$, subtract $\frac{1}{3}$, there remains $\frac{2}{3}x$. Multiplied by 5 will make $\frac{5}{3}x$. The products added together will make $\frac{5}{3}x$. Now multiply $\frac{2}{3}x$ by $\frac{1}{3}x$, it will make $\frac{2}{9}x^2$ $\frac{5}{3}x$ equal to 33. Divide $\frac{5}{3}x$ into $\frac{2}{9}x^2$; it will come to $7\frac{1}{2}x$. Halve the unknowns, take the half of $7\frac{1}{2}x$, which is $3\frac{3}{4}x$, divide 33 by $\frac{2}{9}x^2$, it will be $148\frac{1}{2}$. Then multiply the half by itself. It is $\frac{15}{4}$. 15 times 15 makes 225. To divide by 4 times 4, which makes 16, the result will be $14\frac{1}{16}$. And this added to $148\frac{1}{2}$ will be $162\frac{9}{16}$, minus $3\frac{3}{4}$, and that is the value of the unknown. Now this number has a square root. Its square root is $12\frac{3}{4}$, and to prove that the root is right, multiply $12\frac{3}{4}$ by itself, it will make $162\frac{9}{16}$.[1] Subtract the half, which is $3\frac{3}{4}$, from $12\frac{3}{4}$, there remains 9. Prove it: take $\frac{1}{3}$ of 9, it will be 3; multiply by 5, it will be 15. Take $\frac{2}{3}$ of 9, it will be 6. To multiply by 3 will be 18. Then add one to the other. It will be 33, as you will see worked out. f. 56a

1. *$16\frac{9}{16}$* in MS.

1^{co}	$\frac{2^{co}}{3}$	$\frac{1^{co}}{3}$	$\frac{5}{1}$	$\frac{5^{co}}{3}$	$\frac{2^{\square}}{9}$	$\frac{5^{co}}{3}$	$\frac{33}{1}$	03		
								~~45~~]		
								~~6~~]	$7\frac{3}{6}$	$\frac{1}{2}$

$\frac{2^{\square}}{9}$	$\frac{33}{1}$	297	$\frac{9}{6}$	3	fa 15	
		$148\frac{1}{2}$		5		
		$14\frac{1}{16}$	3		fa 18	$162\frac{9}{16}$ men $3\frac{3}{4}$
		$12\frac{3}{4}$			33	
$12\frac{3}{4}$		$3\frac{3}{4}$				0
$12\frac{3}{4}$						$\frac{15}{4}$ ~~02~~
$162\frac{9}{16}$						~~1~~~~0~~1
						~~225~~]
						~~166~~] $14\frac{1}{16}$
						~~1~~]

f. 56b [Numeri quadratti]

Quisti sono chiamati nomeri chadratti che azonto su al nomero e trato un oltro nomero faza quadratto. Farai per questa rigolla. Prendi 2 nomeri al to piaxer, ponamo che sia 1 e 2. Moltipicha' in si, 1 fia 1 fa 1, 2 fia 2 fa 4; azonti insieme fa 5. E puo' truova 1 numero che azonto e trato de 5 faza quadratto. Questo nomero serà 4. Azonto al 5 fa 9, l'è ben quadratto perché la so radixe de 9 son 3. Sutrazi de 5 4, roman 1. Adoncha 1 fia 1 fa 1, e ben è quadratto. Adoncha la radixe de questo sie 3. Moltipicha questo 3 chon lo nomero trovasti, che fu 4, fa 12. Adopiallo, serà 24, e questo salva. Puo' moltipicha 5 in si, fa 25. Mo' azonzi 24 chon 25, fa 49, ch'è ben quadratto. Anche tratto 24 de 25 è ben 1, ch'è quadrato. Mo' volemo noi veder quanti nomeri trovaremo in questo quadratto, zoè 24, el qual vene per lo dopio de 12: in primo 1, 4, 9, 16, e più non. Mo' chomenzemo da una. Chomenzemo da 16, e si parto 24 per 16 vien $1\frac{1}{2}$. È questo nomero da trar e meter. E puo' parti 25 per 16, vien $1\frac{9}{16}$. È questo lo nomero trovado che è ben quadratto, che la sso radixe sie $1\frac{1}{4}$. Mo' azonzi $1\frac{1}{2}$ su $1\frac{9}{16}$, farà $3\frac{1}{16}$, et anche questo è nomero quadratto, che la so radixe sie $1\frac{3}{4}$. Mo' trazi $1\frac{1}{2}$ da $1\frac{9}{16}$, roman $\frac{1}{16}$, e questo è nomero quadrato che la so radixe sie $\frac{1}{4}$. E chosì da tuti li altri nomeri farì, zoè 9, 4, 1, e si più ne voi più ne truoverai.

f. 57a E se alguno te disise: "Truovame 1 numero quadratto che azonto e tratto $1\frac{1}{2}$ ssia quadratto," respondi che ssia $1\frac{9}{16}$.

E se 'l dixese $2\frac{6}{9}$, seria $2\frac{7}{9}$.

E se 'l dixese 6, seria $6\frac{1}{4}$.

E se 'l dixese 24, seria 25.

$1x$ $\frac{2}{3}x$ $\frac{1}{3}x$ $\frac{5}{1}$ $\frac{5}{3}x$ $\frac{2}{9}x^2$ $\frac{5}{3}x$ $\frac{33}{1}$ 03

$\not{4}\not{5}$]

$\not{6}$] $7\frac{3}{6}$ $\frac{1}{2}$

$\frac{2}{9}x^2$ $\frac{33}{1}$ 297 $\frac{9}{6}$ 3 makes 15

$148\frac{1}{2}$ 5

$14\frac{1}{16}$ 3 makes 18 $162\frac{9}{16}$ minus $3\frac{3}{4}$

$12\frac{3}{4}$ 33

$12\frac{3}{4}$ $3\frac{3}{4}$

$12\frac{3}{4}$

$162\frac{9}{16}$

0

$\frac{15}{4}$ $\not{0}\not{2}$

$\not{1}\not{0}1$

$\not{2}\not{2}\not{5}$]

$\not{1}\not{6}\not{6}$] $14\frac{1}{16}$

$\not{1}$]

SQUARE NUMBERS

f. 56b

We call square numbers those numbers which added to a number and subtracted from another number make a square. You will do it by this rule. Take 2 numbers of your choice, put that they are 1 and 2. Multiply by themselves, 1 times 1 makes 1, 2 times 2 makes 4; added together they make 5. And then find 1 number which added to and subtracted from 5 makes a square. This number will be 4. Added to 5 it makes 9, which is indeed a square because the square root of 9 is 3. Subtract 4 from 5, there remains 1. Then 1 times 1 makes 1, and it's indeed a square. So the square root of this is 3. Multiply this 3 with the number you found, which was 4, it makes 12. Double it, it will be 24, and save this. Then multiply 5 by itself, which makes 25. Now add 24 to 25; it makes 49, which is indeed a square. Also subtract 24 from 25 and it's 1, which is a square. Now we want to see how many numbers we will find in this square, that is 24, which comes from the double of 12: at first 1, 4, 9, 16 and no more. Now let us begin from 1. Let us begin from 16, and if I divide 24 by 16 it comes to $1\frac{1}{2}$. This is the number to subtract and put. And then divide 25 by 16, which comes to $1\frac{9}{16}$. This is the number that has been found that is indeed a square, whose root is $1\frac{1}{4}$. Now add $1\frac{1}{2}$ to $1\frac{9}{16}$, which will make $3\frac{1}{16}$, and this number is also a square, whose root is $1\frac{3}{4}$. Now subtract $1\frac{1}{2}$ from $1\frac{9}{16}$, which leaves $\frac{1}{16}$, and this is a square number whose root is $\frac{1}{4}$. And so you will do for all the other numbers, that is 9, 4, 1, and if you want more of them you will find more.

And if someone said to you: "Find me 1 square number such that $1\frac{1}{2}$ added to it and subtracted from it will be a square," respond that it is $1\frac{9}{16}$. f. 57a

And if he said $2\frac{6}{9}$, it would be $2\frac{7}{9}$.

And if he said 6, it would be $6\frac{1}{4}$.

And if he said 24, it would be 25.

E perché i'ò ditto de sovra numero quadratto che chosa se chiama, sapi che tuti li nomeri che hano radixe son quadrati, chomo dirò alguni qui. Primo, 1 sie quadratto perché moltipichado in si fa 1. Segondo, 4 è quadratto perché la so radixe 2; 9 perché la so radixe 3; 16 perché la so radixe 4; 25 perché la so radixe 5. E chossì fara' tuti li oltri, chomo vedirì qui de sotto per figora.

1 2

1

4

5 fia 5 fa 25 | suma $\frac{49}{1}$

4

fa 9 radixe 3 12, adopia serà 24

$\frac{1}{16}$ 4 9 16

$\frac{1}{4}$ $\frac{1}{4}$ | $\frac{1}{16}$ 0 0 tuor e meter $1\frac{8}{16}$

18 19 $1\frac{9}{16}$

24 25 $3\frac{1}{16}$

16 $1\frac{8}{16}$ 16 $1\frac{9}{16}$

$1\frac{3}{4}$

---

f. 57b [Chonprar pano 2 persone]

Fame questa raxion. 2 homeny chonpra pano e à tanti denari l'un chomo l'altro. El primo chonpra braza 3 de pano e avanzalli soldi 4; l'oltro chonpra braza 7 e manchalli soldi 7. Adomando: "Quanti denari avea chadaun di loro? e quanti soldi valeva el brazo?"

E per far la ditta raxion per inpoxizion, pony che gostasse soldi 3 el brazo. Adoncha a soldi 3 el brazo veniria soldi 9, e puo' soldi 4 che 'l avanza sono, azonti, 13. E questo fu el primo chonpagno. Mo' diremo per lo segondo chonpagno che 'l chonpra braza 7. Adoncha ala raxion del primo, die gostar braza 7 a soldi 3 el brazo, soldi 21. Mo' dixe che roman debitto soldi 7. Adoncha sotrazi 7 de 21, roman 14, e noi volemo che ssia chomo à el primo, che son soldi 13. Adoncha avanza de più soldo 1, e questo metti più ala prima inpoxizion, e diremo per 3 che m'ò puxo più 1. E per far la segonda inpoxizion pony che gostasse el brazo soldi 4. 3 braza valeria soldi 12 e 4 che avanza al primo val 16. El segondo, braza 7 per 4 soldi el brazo monta soldi 28. Mancha 7, abatti 7 de 28, roman 21, e noi volemo che ssia soldi 16 chomo el primo chonpagno. Adoncha, sutrato 16 de 21, roman più 5. E diremo per 4 ch'è posto più 5 ala segonda inpoxizion. Moltipicha in croxe e diremo: "3 fia 5 val 15" e puo' "Un fia 4 fa 4." Più e più, sotrazi el men del più. Adoncha, sutratto 4 de 15, roman 11. E questi die 'ser partidi.

f. 58a E per far partior dela ditta raxion sutrazi 1 de 5, roman 4, e questo è lo partior. Parti 11 per 4, vien $2\frac{3}{4}$, e tanti soldi valse el brazo del pano. Aprovalla e diremo: "3 braza fia soldi $2\frac{3}{4}$ fa soldi $8\frac{3}{4}$, e più

And because I said above what is called a square number, know that all the numbers that have square roots are squares, as I will name some here. First, 1 is a square because multiplied by itself makes 1. Second, 4 is a square because its root is 2; 9 because its root is 3; 16 because its root is 4; 25 because its root is 5. And so you will do all the others, as you will see worked out below.

1 2

1

4

5 times 5 makes 25

4

makes 9 root 3 12, doubled will be 24 | sum $\frac{49}{1}$

$\frac{1}{16}$ 4 9 16

$\frac{1}{4}$ $\frac{1}{4}$ | $\frac{1}{16}$ 0 0 Subtract and put $1\frac{8}{16}$

18 19 $1\frac{9}{16}$

24 25 $3\frac{1}{16}$

16] $1\frac{8}{16}$ 16] $1\frac{9}{16}$

$1\frac{3}{4}$

Two people buy cloth

f. 57b

Do this problem for me. 2 men buy cloth, and one has as much money as the other. And the first one buys 3 ells of cloth and has 4 soldi extra; the other one buys 7 ells and is 7 soldi short. I ask: "How much money did each of them have? And how many soldi was one ell worth?"

And to do this problem by false position, suppose that it cost 3 soldi per ell. Then at 3 soldi per ell it would come to 9 soldi, and then the 4 soldi that are extra, added in, make 13. And this was the first partner. Now we will say for the second partner that he buys 7 ells. Then at the rate of the first, 7 ells at 3 soldi per ell should cost 21 soldi. Now say that this leaves a debt of 7 soldi. So subtract 7 from 21, which leaves 14, and we wanted it to be like the first, which is 13 soldi. So it is plus 1 soldo extra, and put this as a plus at the first false position, and we will say for 3 that we have put plus 1. And to do the second false position, suppose that the ell cost 4 soldi. 3 ells would be worth 12 soldi and the 4 extra for the first one is worth 16. The second one, 7 ells for 4 soldi per ell amounts to 28 soldi. There lacks 7, subtract 7 from 28, it leaves 21, and we wanted it to be 16 soldi like the first partner. So subtract 16 from 21, which leaves plus 5. And we will say for 4 that plus 5 is put at the second false position. Multiply across and we will say: "3 times 5 is worth 15," and then: "one times 4 makes 4." Plus and plus, subtract the lesser from the greater. Then subtract 4 from 15, which leaves 11. And these should be divided.

And to make the divisor for this problem subtract 1 from 5, which leaves 4, and this is the divisor. f. 58a
Divide 11 by 4, which comes to $2\frac{3}{4}$, and this is the number of soldi that the ell of cloth is worth.

soldi 4 azonti serano soldi $12\frac{3}{4}$. E tanti soldi avea chadaun di loro in borssa. Adoncha el sego[*ndo*] 7 braza moltipicha' per $2\frac{3}{4}$ serano soldi $19\frac{1}{4}$. Mo' i mancha soldi 7. Adoncha, sotratto 7 de 19 roman al segondo soldi $12\frac{1}{4}$ in borssa sì al'uno chomo al'oltro, cho[*mo*] vedirì qui de sotto per figura.

3 \ / 4 — 3 — fa 13 — primo
più X più — 3 — 14 — el segondo
1 / \ 5 — 9 — 7 — più 1 — ala prima
4 — 3
21
7

4 — fa 16 — primo — 5 — 4
3 — 24 — 3 — 1
12 — 7 — più 5 — fa 15 — 4
4 — 4 — 4
28
03 — 7 — a partir 11 — 5
~~11~~ — 1
~~4~~] $2\frac{3}{4}$ — partior 4
3 — $2\frac{3}{4}$
$8\frac{1}{4}$ — primo $12\frac{1}{4}$ — 7
4 — $19\frac{1}{4}$ / $12\frac{1}{4}$ / el segondo
7

E per voler far la ditta raxion per la chosa, pony che 'l brazo vallesse 1^{co}. Adoncha 3 braza valleria 3^{co} più 4^{n}. El segondo die valler 7 braza 7^{co}, chomo vallea el primo, men 7^{n}. Adoncha più 3^{co} e
f. 58b più 4 numeri e più chosse 7^{co} men 7^{n}. Adoncha adequemo le parte. | Sutrezemo più 3^{co} de più 7^{co}, roman 4^{co} più 4^{n} men 7^{n}. Azonti el men chon el più, serano 11^{n} ingual a 4^{co}. Se die partir li nomeri per le chosse, insirà $2\frac{3}{4}$. E tanti soldi valse el brazo de pano. Aprovallo, starà ben, vedirì qui de sotto.

Prove it, and we will say: "3 ells times $2\frac{3}{4}$ soldi makes $8\frac{1}{4}$[1] soldi, and then the 4 soldi added will be $12\frac{1}{4}$[2] soldi." And each of them had that many soldi in his purse. Then the second one: 7 ells multiplied by $2\frac{3}{4}$ will be $19\frac{1}{4}$ soldi. Now 7 soldi are lacking. So, subtract 7 from $19\frac{1}{4}$,[3] there remains for the second one $12\frac{1}{4}$ soldi in his purse as for the other, as you will see worked out below.

3 4 3 makes 13 first
plus plus 3 14 the second
9 7 plus 1 at the first
1 5 4 3
21
7

4 makes 16 first 5 4
3 21[4] 3 1
12 7 plus 5 makes 15 4
4 4 4
28
03 7 to divide 11 5
~~11~~ 1
~~4~~] $2\frac{3}{4}$ divisor 4
3 $2\frac{3}{4}$
$8\frac{1}{4}$ first $12\frac{1}{4}$ 7
4 $19\frac{1}{4}$ / $12\frac{1}{4}$ / the second
7

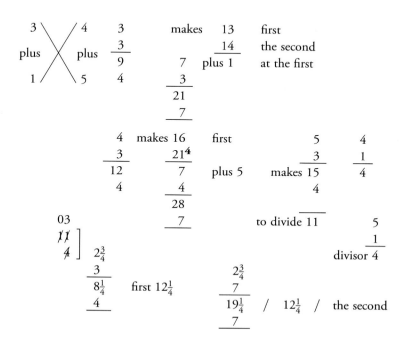

And to do this problem by unknowns, put that the ell was worth $1x$. Then 3 ells would be worth $3x$, plus 4. The second one should be worth 7 ells, $7x$, as the first one, minus 7. Then plus $3x$ and plus 4 is plus $7x$ minus 7. Then we equalize the sides. | We subtract plus $3x$ from plus $7x$, which leaves $4x$, plus 4 minus 7. Add the minus to the plus; it will be 11 equal to $4x$. The numbers should be divided by the unknowns, which will give $2\frac{3}{4}$. And that's how many soldi the ell of cloth was worth. Prove it, it will come out well, as you will see below. f. 58b

1. *$8\frac{3}{4}$* in MS.
2. *$12\frac{3}{4}$* in MS.
3. *19* in MS.
4. *24* in MS.

[Chonprar 4 1ª peza de pano]

1^{co}	3^{co} più 4^{n}	7^{co} men 7^{n}	7^{n} 11^{n} / 4^{n}	7^{co} / 3^{co} / 4^{co}

$2\frac{3}{4}$
3
soldi $8\frac{1}{4}$
4
soldi $12\frac{1}{4}$

03^{n}
11
4^{co}] $2\frac{3}{4}$

$2\frac{3}{4}$
7
$19\frac{1}{4}$
7
soldi $12\frac{1}{4}$

[Chonprar 4 1ª peza de pano]

Fame questa raxion. 4 homeny chonpra 1ª peza de pano. L'un tuolse la mitade, l'altro $\frac{1}{3}$, l'oltro $\frac{1}{5}$, l'oltro $\frac{1}{6}$, e romaxe braza 20. Adomando quanto fu longa ttutta la peza.

E per far la ditta raxion per la chosa pony che fusse 1^{co}, el vuol $\frac{1}{2}^{co}$, $\frac{1}{3}^{co}$, $\frac{1}{5}^{co}$, $\frac{1}{6}^{co}$, e questi azonti serano $\frac{216}{180}$. Adoncha vuolemo che ssia 1^{co}, adoncha abatti 1^{co}, serano $\frac{36}{180}^{co}$, che son ingual a 20 braza. Moltipicha 20 fia 180, fa 3600, e questo partido per 36 inssirà fuor braza 100. E si voio tuor el mezo de 100 serà 50, el terzo de 100 serano $33\frac{1}{3}$, el quinto de 100 serano 20, el sesto de 100 serano $16\frac{2}{3}$, asumadi serano braza 120. Adoncha la peza è me[*n*] el 100 sutratto de 120, roman 20. E questo fu el pano che fu longa.

f. 59a 1^{co} $\frac{1}{2}^{co}$ $\frac{1}{3}^{co}$ $\frac{1}{5}^{co}$ $\frac{1}{6}^{co}$ | $\frac{5}{6}^{co}$ $\frac{1}{5}$ $\frac{1}{6}$ $\frac{11}{30}^{co}$ $\frac{5}{6}^{co}$ $\frac{216}{180}^{co}$ 1^{co} $\frac{36}{180}^{co}$ $\frac{20}{1}$

180
20
3600

1
2
048
1544
3600
2166
21] $16\frac{144}{216}$

1^{co}

$\frac{36}{180}$ $\frac{20}{1}$

0000
3600
3666
33] 100

100
50
$33\frac{1}{3}$
20
$16\frac{2}{3}$
120

120
100
20 e questo fa le brazi di pano perché el 100 fu men

1x 3x plus 4 | 7x minus 7 7 11 / 4 7x / 3x / 4x

$2\frac{3}{4}$
3
$8\frac{1}{4}$ soldi
4
$12\frac{1}{4}$ soldi

03
~~11~~]
~~4~~x] $2\frac{3}{4}$

$2\frac{3}{4}$
7
$19\frac{1}{4}$
7
$12\frac{1}{4}$ soldi

FOUR PEOPLE BUY A PIECE OF CLOTH

Do this problem for me. 4 men buy 1 piece of cloth. One took half for himself, the other $\frac{1}{3}$, the other $\frac{1}{5}$, the other $\frac{1}{6}$, and there remained 20 ells. I ask how long the whole piece was.

And to do this problem with unknowns, put that it was 1x, which calls for $\frac{1}{2}x$, $\frac{1}{3}x$, $\frac{1}{5}x$, $\frac{1}{6}x$, and these added together will be $\frac{216}{180}$. And we want it to be 1x, so subtract 1x; it will be $\frac{36}{180}x$, which is equal to 20 ells. Multiply 20 times 180, which makes 3,600, and this divided by 36 will result in 100 ells. And if you want to take half of 100 it will be 50, a third of 100 will be $33\frac{1}{3}$, a fifth of 100 will be 20, a sixth of 100 will be $16\frac{2}{3}$, added together it will be 120 ells. Then the piece is the 100 subtracted from 120, which leaves 20. And this was how long the cloth was.

1x $\frac{1}{2}x$ $\frac{1}{3}x$ $\frac{1}{5}x$ $\frac{1}{6}x$ | $\frac{5}{6}x$ $\frac{1}{5}$ $\frac{1}{6}$ $\frac{11}{30}x$ $\frac{5}{6}x$ $\frac{216}{180}x$ 1x $\frac{36}{180}x$ $\frac{20}{1}$ f. 59a

180
20
3600

1
~~2~~4[1]
0~~4~~~~8~~
~~1~~~~5~~~~4~~4
~~3~~~~6~~~~0~~~~0~~]
~~2~~~~1~~~~6~~~~6~~] $16\frac{144}{216}$
~~2~~~~1~~]

$\frac{36}{180}$ $\frac{20}{1}$

1x

0000
~~3~~~~6~~~~0~~~~0~~]
~~3~~~~6~~~~6~~~~6~~] 100
~~3~~~~3~~]

100
50
$33\frac{1}{3}$
20
$16\frac{2}{3}$
120

120
100
20 and the ells of cloth make this, because the 100 was minus

1. *4* lacking in MS.

E per far la ditta raxion per la riegola del 3 truova un numero che abia $\frac{1}{2}$, $\frac{1}{3}$, $\frac{1}{5}$, $\frac{1}{6}$, e questo se truova in 60. La mitade de 60 son 30, el terzo de 60 sun 20, el quinto de 60 sun 12, el sesto de 60 son 10. Asumandi in tutto fano 72. Adoncha è più cha 60 12, li qualli serano men. E noi faremo: "Se 12 me fa 60 men, che serave 20?" Serano a moltipichar 1200 men. E questi partidi per 12 serano 100. Adoncha dame la mitade de 100, serano 50, el terzo de 100 serano $33\frac{1}{3}$, el quinto de 100 serano 20, el sesto de 100 serano $16\frac{2}{3}$. Asumadi serano 120, e noi vossemo men 100. Adoncha la peza era men 100 e più 20 braza. Sutratto de più 120 men 100, resta ben 20 braza.

60	72	12 men	60	20	0000		100
30	60		20		1200		50
20	men 12		00		1222	100 men	$33\frac{1}{3}$
12			120		11		20
10							$16\frac{2}{3}$
72							più 120

più 120
men 100
20 braza

f. 59b E per far la ditta raxion per inpoxizion pony che vallese 60. Adoncha la mitade de 60 sono 30, el terzo de 60 sono 20, che val 50, quinto de 60 sono 12, che val 62, el sesto de 60, 10, val 72, e noi vossemo che fosse 20, e 72 val 92. Adoncha è più cha 60 32, e questo metti ala prima inpoxizion, per 60 che m'ò puxo più 32. E per far la segonda inpoxizion pony che fusse 120. La mitade son 60, el terzo 20, el quinto 24, el sesto 20. Asumadi sono 144, e 20 che son avanzadi val 164, e noi vosemo che 'l fose 120, adoncha più 44. E chossì dirai: "Per 120 che m'ò puxo più 44." Moltipicha e party, insirà fuor 100 men. Adoncha la peza non fu più cho[*mo*] vedirì qui de sotto per figura.

And to do this problem by the rule of 3, find a number that has $\frac{1}{2}$, $\frac{1}{3}$, $\frac{1}{5}$, $\frac{1}{6}$, and this is found in 60. Half of 60 is 30, a third of 60 is 20, a fifth of 60 is 12, a sixth of 60 is 10. Added all together they make 72. So it's 12 more than 60, which will be minus. And we will say: "If 12 gives me minus 60, what would 20 be?" The number to multiply will be minus 1,200. And this divided by 12 will be 100. So give me a half of 100, which will be 50, a third of 100 will be $33\frac{1}{3}$, a fifth of 100 will be 20, a sixth of 100 will be $16\frac{2}{3}$. Added together it will be 120, and we wanted minus 100. So the piece will be minus 100, and plus 20 ells. Minus 100 subtracted from plus 120 leaves 20 ells.

60
30
20
12
10
72

72
60
minus 12

minus 12

60 20
20
00
120

0000
1200
1222
11
minus 100

100
50
$33\frac{1}{3}$
20
$16\frac{2}{3}$
plus 120

plus 120
minus 100
20 ells

And to do this problem by false position, suppose that it was worth 60. So half of 60 is 30, a third f. 59b
of 60 is 20, which is worth 50, a fifth of 60 is 12, which is worth 62, a sixth of 50, 10, which is worth 72, and we wanted it to be worth 20, and 72 makes 92.[1] So it is 32 more than 60, and put this at the first false position, for 60 that I have put plus 32. And to do the second false position suppose that it was 120. Half is 60, a third 40,[2] 1 fifth 24, a sixth 20. Added together it is 144, and the 20 that is extra is worth 164, and we wanted it to be 120, so it is plus 44. And so you will say: "For 120 that I have, I have put plus 44." Multiply and divide; the result will be minus 100. So the piece was not more, as you will see worked out below.

1. This seems to be an error for "and the 20 that is extra is worth 92, and we wanted it to be 60."
2. *20* in MS.

60 \	/ 120	60	120	60	120	44
		30	60	44	32	32
più	più	20	40	240	240	partior 12
32 /	\ 44	12	24	240	360	
		10	20	2640	3840	
		72	144		2640	
		20	20		a partir 1200	
		92	164			
		60	120			
		più 32	più 44			

sutrazi 120	fu braza 20	00		100
100		1200		50
20		1222	100	$33\frac{1}{3}$
		11		20
				$16\frac{2}{3}$
				suma 120

E questo è fatto per inpoxizion. E per lo simille farì ogn'altra raxion.

---

f. 60a

[Tre chonpagni ala mitade di so denari 2 più]

Sono 3 chonpagni. Dixe el primo ai altri 2: "I mie denari sono la mitade d'i vostri e 2 più." El segondo dixe ai altri 2: "Li mie denari sono el terzo d'i vostri e 3 più." Dixe el terzo ai altri 2: "Li mie denari sono el quarto d'i vostri e 4 più." Adomando quanti ne avè zaschadun di loro, e se la volesti far per inpoxizion.

E per far la ditta raxion per inpoxizion pony che tuti 3 aveseno soldi 11. Adoncha el primo domanda la mittade e 2 più. Adoncha daremo al primo el terzo più 2. Abatti de 11 2, roman 9. El terzo de 9, 3, serano 3 e do 5. Adoncha el primo à 5. El segondo dise voler el terzo più 3. Adoncha sutra' 3 de 11, roman 8. Mo' demo al segondo el quarto de 8, sono 2, e 3 val 5. E per voler far del terzo che vuol el quarto più 4, sutrazi 4 de 11, roman 7, e questi parti per 5. Serà $1\frac{2}{5}$, e questi serano del terzo. Adoncha el primo 5, el segondo 5, el terzo $5\frac{2}{5}$, fano $15\frac{2}{5}$, e noi vosemo che fusse 11. Adoncha serà più $4\frac{2}{5}$, e questo pony ala prima inpoxizion. E per far la segonda inpoxizion pony che vallese soldi 14. Abatti 2, roman 12, el terzo de 12 serà 4, e 2 val 6. E questo serà del primo. El segondo abatti 3 de 14, roman 11, e questo partido per 4 serano azonti $5\frac{3}{4}$. E per lo terzo abatti 4 de 14, roman 10. El quinto de 10 sono 2, val 6. El primo 6, el segondo $5\frac{3}{4}$, el terzo 6. Asumadi

60 ╲ ╱ 120	60	120	60		120	44
plus ╳ plus	30	60	44		32	32
32 ╱ ╲ 44	20	40	240		240	divisor 12
	12	24	240		360	
	10	20	2640		3840	
	72	144			2640	
	20	20			to divide 1200	
	92	164				
	60	120				
	plus 32	plus 44				
	subtract 120	was 20 ells	00		100	
	100		1200		50	
	20		1222	100	$33\frac{1}{3}$	
			11		20	
					$16\frac{2}{3}$	
					sum 120	

And this was done by false position. And you will do all other problems the same way.

THREE PARTNERS—TO HALF OF HIS MONEY TWO MORE

f. 60a

There are 3 partners. The first says to the other two: "My coins are half of yours and 2 more." The second one says to the other two: "My coins are a third of yours and 3 more." The third one says to the other 2: "My coins are a quarter of yours and 4 more." I ask how many each had, and if you would like to do it by false position.

And to do this problem by false position, suppose that all three had 11 soldi. Then the first one asks for half and 2 more.[1] Then we will give to the first one a third plus 2. Subtract 2 from 11, which leaves 9. A third of 9 is 3, it will be 3 and two 5. So the first one has 5. The second one said he wanted a third plus 3. So subtract 3 from 11, which leaves 8. Now we give to the second one a quarter of 8, which is 2, and 3 makes 5. And to make for the third one the fourth plus 4 that he wants, subtract 4 from 11, which leaves 7, and divide this by 5. It will be $1\frac{2}{5}$, and this will be that of the third one. So the first one 5, the second one 5, the third one $5\frac{2}{5}$, which makes $15\frac{2}{5}$, and we wanted it to be 11. So it will be plus $4\frac{2}{5}$, and put this at the first false position. And to do the second false position suppose that it was worth 14 soldi. Subtract 2, which leaves 12, a third of 12 will be 4, and 2 makes 6. And this will be that of the first one. For the second one subtract 3 from 14, which leaves 11, and this divided by 4 will be $5\frac{3}{4}$ added. And for the third one subtract 4 from 14, which leaves 10. A fifth of 10 is 2, which makes 6. The first one has 6, the second one $5\frac{3}{4}$, the third

1. The nature of the problem changes at this point from how much each man had to what each asked from the others.

fano $17\frac{3}{4}$, e noi vosemo che 'l fose 14. Adoncha più $3\frac{3}{4}$, e chossì metti ala segonda inpoxizion. Moltipicha e parti chomo vedirì qui de sotto per figora.

11 ╲ ╱ 14	14	11	$4\frac{2}{5}$	00
più ╳ più	$4\frac{2}{5}$	$3\frac{3}{4}$	$3\frac{3}{4}$	~~1~~~~1~~4
$4\frac{2}{5}$ ╱ ╲ $3\frac{3}{4}$	$61\frac{3}{5}$	$41\frac{1}{4}$		~~4~~~~0~~~~7~~] $31\frac{4}{13}$
	$41\frac{1}{4}$			~~1~~~~3~~~~3~~
	$20\frac{7}{20}$			~~1~~

$\frac{13}{20}$	$\frac{407}{20}$	primo $11\frac{10}{13}$
		segondo $10\frac{1}{13}$
	suma $31\frac{4}{13}$	terzo $9\frac{6}{13}$

f. 60b E per far la chontrascritta raxion per la chosa pony che tuti 3 avesse 1^{co} men 2^{n}. Perché el primo domanda ai altri 2 aver la mitade d'i so denari più 2, adoncha el primo vuol la mittade de 1^{co} men 2^{n}. Se 'l domanda la mitade se die dar el terzo, adoncha el terzo de 1^{co} serà $\frac{1}{3}^{co}$. Muo' vuol el terzo men de 2^{n}, che son men $\frac{2}{3}^{n}$, e lui vuol aver più 2^{n}. Sutra' $\frac{2}{3}^{n}$ de 2^{n}, roman $1\frac{1}{3}^{n}$. Adoncha el primo averà $\frac{1}{3}^{co}$ più $1\frac{1}{3}^{n}$. Mo' volemo far per lo segondo e diremo che 'l fuse 1^{co} men 3^{n}. Adoncha chostui domanda el terzo. Se vuol dar el quarto. Daremoli de 1^{co} $\frac{1}{4}^{co}$, e de men 3^{n} el quarto serà $\frac{3}{4}^{n}$ men, e luy vorave più 3. Sutratto $\frac{3}{4}^{n}$ de 3 roman più $2\frac{1}{4}^{n}$. Mo' faremo per lo terzo che vuol el quarto più 4. Adoncha diremo che fusse 1^{co} men 4^{n}. Demoi el quinto de 1^{co}, serà $\frac{1}{5}^{co}$. El quinto de 4^{n} serano $\frac{4}{5}^{n}$, e lui vuol più 4. Sutratto de 4 roman $3\frac{1}{5}^{n}$. Adoncha el primo $\frac{1}{3}^{co}$ più $1\frac{1}{3}^{n}$; el segondo $\frac{1}{4}^{co}$ più $2\frac{1}{4}^{n}$; el terzo $\frac{1}{5}^{co}$ più $3\frac{1}{5}^{n}$. Mo' moltipicha over azonzi $1\frac{1}{3}$, $2\frac{1}{4}$, $3\frac{1}{5}$ fano $\frac{407}{60}^{n}$. Mo' azonzi $\frac{1}{3}^{co}$ $\frac{1}{4}^{co}$ $\frac{1}{5}^{co}$, fano $\frac{47}{60}^{co}$, che sono ingualli a 1^{co}. Adequa le parte. Trazi $\frac{47}{60}$ de una chossa, roman $\frac{13}{60}^{co}$ a partir $\frac{407}{60}^{n}$. Parti 407^{n} per 13^{co}, insirà fuor del partidor 31 e $\frac{4}{13}$, e tanti denari aveano tuti 3. El primo avea soldi $11\frac{10}{13}$, el segondo soldi $10\frac{1}{13}$, el terzo soldi $9\frac{6}{13}$. Asumadi insieme fano soldi $31\frac{4}{13}$, chomo vedirì qui de sutto.

1^{co}	1^{co} men 2^{n}	primo	$\frac{1}{3}^{co}$ più $1\frac{1}{3}^{n}$	$1\frac{1}{3}$	$2\frac{1}{4}$	$\frac{1}{3}^{co}$	$\frac{1}{4}^{co}$	primo è $11\frac{10}{13}$
	1^{co} men 3^{n}	segondo	$\frac{1}{4}^{co}$ più $2\frac{1}{4}^{n}$	$\frac{4}{3}$	$\frac{9}{4}$	$\frac{7}{12}$	$\frac{1}{5}$	segondo $10\frac{1}{13}$
	1^{co} men 4^{n}	terzo	$\frac{1}{5}^{co}$ più $3\frac{1}{5}^{n}$	$\frac{43}{12}$	$\frac{16}{5}$		$\frac{47}{60}^{co}$	terzo $9\frac{6}{13}$
								suma $31\frac{4}{13}$

$\frac{47}{60}^{co}$	$\frac{407}{60}^{n}$ ingual 1^{co}	$\frac{13}{60}^{co}$ partior	215	00
	$\frac{47}{60}$		192	~~1~~~~1~~4
			a partir 407	~~4~~~~0~~~~7~~] $31\frac{4}{13}$
				~~1~~~~3~~~~3~~
				~~1~~

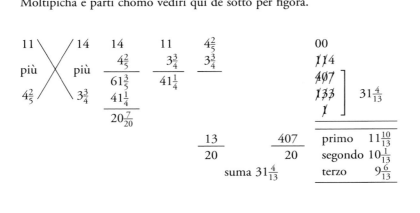

one 6. And they add up to $17\frac{3}{4}$, and we wanted it to be 14. So it is plus $3\frac{3}{4}$, and put this at the second false position. Multiply and divide as you will see worked out below.

11 ╲ ╱ 14	14	11	$4\frac{2}{5}$		00	
plus ╳ plus	$4\frac{2}{5}$	$3\frac{3}{4}$	$3\frac{3}{4}$		~~1~~~~1~~4	
	$61\frac{3}{5}$	$41\frac{1}{4}$			~~4~~~~0~~7	
$4\frac{2}{5}$ ╱ ╲ $3\frac{3}{4}$	$41\frac{1}{4}$				~~1~~~~3~~~~3~~	$31\frac{4}{13}$
	$20\frac{7}{20}$				~~1~~	
			$\frac{13}{20}$	$\frac{407}{20}$	first $11\frac{10}{13}$	
					second $10\frac{1}{13}$	
			sum $31\frac{4}{13}$		third $9\frac{6}{13}$	

And to do the problem written opposite by unknowns, put that all three had $1x$ minus 2. Because the first one asks the other 2 to have half of his coins plus 2, then the first one wants half of $1x$ minus 2. If he asks for half, one must put a third, so a third of $1x$ will be $\frac{1}{3}x$. Now the third one wants 2 less, which is minus $\frac{2}{3}$, and he wants plus 2. Subtract $\frac{2}{3}$ from 2, which leaves $1\frac{1}{3}$. So the first one will have $\frac{1}{3}x$ plus $1\frac{1}{3}$. Now we want to do this for the second one, and we will say that is was $1x$ minus 3. And this one calls for the third. So a quarter should be given to him. We will give him $\frac{1}{4}x$ from $1x$, and a quarter of minus 3 will be minus $\frac{3}{4}$, and he wanted plus 3. $\frac{3}{4}$ subtracted from 3 leaves plus $2\frac{1}{4}$. Now we will do for the third one who wants a quarter plus 4. Then we will say that it was $1x$ minus 4. We give him a fifth of $1x$, which will be $\frac{1}{5}x$. A fifth of 4 will be $\frac{4}{5}$, and he wanted plus 4. Subtracted from 4 it leaves $3\frac{1}{5}$. So the first one has $\frac{1}{3}x$ plus $1\frac{1}{3}$; the second one $\frac{1}{4}x$ plus $2\frac{1}{4}$; the third one $\frac{1}{5}x$ plus $3\frac{1}{5}$. Now multiply or rather add $1\frac{1}{3}$, $2\frac{1}{4}$, and $3\frac{1}{5}$, which makes $\frac{407}{60}$. Now add $\frac{1}{3}x$, $\frac{1}{4}x$, and $\frac{1}{5}x$, which makes $\frac{47}{60}x$, which is equal to $1x$. Equalize the sides. Subtract $\frac{47}{60}$ from one x, which leaves $\frac{13}{60}x$ to divide into $\frac{407}{60}$. Divide 407 by $13x$, the result of the division will be 31 and $\frac{4}{13}$, and that is the number of the coins that all 3 had. The first had $11\frac{10}{13}$ soldi,[1] the second $10\frac{1}{13}$ soldi, the third $9\frac{6}{13}$ soldi. Added together they make $31\frac{4}{13}$ soldi, as you will see below. f. 60b

$1x$	$1x$ minus 2	first	$\frac{1}{3}x$ plus $1\frac{1}{3}$	$1\frac{1}{3}$ $\quad$ $2\frac{1}{4}$	$\frac{1}{3}x$ $\quad$ $\frac{1}{4}x$	first is $11\frac{10}{13}$
	$1x$ minus 3	second	$\frac{1}{4}x$ plus $2\frac{1}{4}$	$\frac{4}{3}$ $\quad$ $\frac{9}{4}$	$\frac{7}{12}$ $\quad$ $\frac{1}{5}$	second $10\frac{1}{13}$
	$1x$ minus 4	third	$\frac{1}{5}x$ plus $3\frac{1}{5}$	$\frac{43}{12}$ $\quad$ $\frac{16}{5}$	$\frac{47}{60}x$	third $9\frac{6}{13}$
						sum $31\frac{4}{13}$

$\frac{47}{60}x$	$\frac{407}{60}$ equals $1x$	$\frac{13}{60}x$ divisor	215	00	
	$\frac{47}{60}$		192	~~1~~~~1~~4	
			to divide 407	~~4~~~~0~~7	
				~~1~~~~3~~~~3~~	$31\frac{4}{13}$
				~~1~~	

1. Up to now, the problem had been in terms of denari (coins or pennies); now it is specifically soldi (shillings or soldino coins worth 12 pennies).

---

f. 61a **[FAR 3 UNA CHAXA]**

Sono 3 chonpagni e si tuol a far 1ª chaxa. L'uno la vuol far in 6 dì, l'altro la vuol far in 7 dì, l'oltro la vuol far in 26 dì. Adomando in quanti dì la farà tuti 3.

E per voler far la dita raxion per inpoxizion pony che ala prima inpoxizion fusse posto 3. Adoncha 3 maistri faria la ditta chaxa in $\frac{3}{6}$, e 3 maistri la farà in $\frac{3}{7}$, e 3 maistri la farà in $\frac{3}{26}$. Asumadi insieme $\frac{3}{6}$, $\frac{3}{7}$, $\frac{3}{26}$ fano 1140, e noi vosemo che fuse 1092. Adoncha sutrazi l'un per l'oltro, roman men 48. E questo die 'ser posto ala prima inpoxizion, per 3 che m'ò poxo men 48. E per voler far la segonda inpoxizion pony che fuse 4. Fazo che 3 maystri la farà in $\frac{4}{6}$, $\frac{4}{7}$, $\frac{4}{26}$, e questi asumadi fano 1520. Tratti 1092, resta men 428, e metera' ala segonda inpoxizion, per 4 che m'ò puxo, men 428. Moltipicha in croxe e parti, insirà fuor zorny $2\frac{332}{380}$, chomo vedirì per figora.

3 ╲ ╱ 4 $\frac{3}{6}$ $\frac{3}{7}$ $\frac{3}{26}$ 1140 $\frac{4}{6}$ $\frac{4}{7}$ $\frac{4}{26}$ 1520
men ╳ men 1092 1092
men 48 men 428
48 ╱ ╲ 428

428 428 3
48 3 0~~4~~3
più 380 1284 partir ~~1~~~~0~~~~9~~2]
192 ~~3~~~~8~~0] $2\frac{332}{380}$

E per far la ditta per la chossa pony che fusse $\frac{1}{6}^{co}$, $\frac{1}{7}^{co}$, $\frac{1}{26}^{co}$, serave $\frac{380}{1092}$, e noi faremo: "Se $\frac{380}{1092}$ me dà $\frac{1}{1}$, che me darà $\frac{1}{1}$ chaxa?" E questo moltipichado e partido serave zorny $2\frac{332}{1380}$, chomo vedirì qui de sotto per esenpio per figora.

1^{co} $\frac{1}{6}^{co}$ $\frac{1}{7}^{co}$ $\frac{1}{26}^{co}$ $\frac{13}{42}^{co}$ $\frac{1}{26}^{co}$ $\frac{380^{co}}{1092}$ ╳ $\frac{1}{1}$ dì = $\frac{1}{1}$ caxa

3
0~~4~~3
~~1~~~~0~~~~9~~2]
~~3~~~~8~~0] $2\frac{332}{380}$

E 3 maistri farà la chaxa in zorny $2\frac{332}{380}$.

f. 61b **[FAR DE 10 2 PARTE]**

Fame de 10 2 parte che moltipichad'una in si medexima, e tratto la menor dela mazuor, romagna 10.

THREE MEN BUILD A HOUSE

f. 61a

There are three partners and they undertake to build a house. One wants to do it in 6 days, the other wants to do it in 7 days, the other wants to do it in 26 days. I ask in how many days all three will do it.

And to do this problem by false position, suppose that 3 was put at the first false position. So three masters would do the house in $\frac{3}{6}$, and 3 masters would do the house in $\frac{3}{7}$, and 3 masters would do it in $\frac{3}{26}$. Added together, $\frac{3}{6}$, $\frac{3}{7}$, and $\frac{3}{26}$ make 1,140, and we wanted it to be 1,092. So subtract one from the other, which leaves minus 48. And this should be put at the first false position, 48 for the 3 that I have put. And to do the second false position, suppose that it was 4. I put that 3 masters will do it in $\frac{4}{6}$, $\frac{4}{7}$, and $\frac{4}{26}$, and these added together make 1,520. Subtracted from 1,092, there remains minus 428, and this will go at the second false position, minus 428 for the 4 I have put. Multiply across and divide, the result will be $2\frac{332}{380}$ days, as you will see worked out.

3 ╲╱ 4
minus ╳ minus
48 ╱╲ 428

$\frac{3}{6}$ $\frac{3}{7}$ $\frac{3}{26}$

1140
1092
minus 48

$\frac{4}{6}$ $\frac{4}{7}$ $\frac{4}{26}$

1520
1092
minus 428

428
48
plus 380

428
3
1284
192

3
043
to divide 1092]
380] $2\frac{332}{380}$

And to do it by unknowns, put that it was $\frac{1}{6}x$, $\frac{1}{7}x$, and $\frac{1}{26}x$, which will be $\frac{380}{1092}$, and we will say: "If $\frac{380}{1092}$ gives me $\frac{1}{1}$, what will $\frac{1}{1}$ house give me?" And this multiplied and divided will be $2\frac{332}{380}$ days, as you will see worked out in the figure below.

$1x$ $\frac{1}{6}x$ $\frac{1}{7}x$ $\frac{1}{26}x$ $\frac{13}{42}x$ $\frac{1}{26}x$

$\frac{380x}{1092}$ ╳ $\frac{1}{1}$ day = $\frac{1}{1}$ house

3
043
1092]
380] $2\frac{332}{380}$

And 3 masters will make the house in $2\frac{332}{380}$ days.

MAKE 2 PARTS OF 10

f. 61b

Find me 2 parts of 10 such that each multiplied by itself, and the lesser subtracted from the greater, leaves 10.

E per far la ditta raxion per la chosa pony che fusse 1^{co} 10^{n} men 1^{co}, e questi moltipichadi in ssi fano questo scriverò avanti. Men 1^{co} via men 1^{co} fa più $1^{\square}$, e puo' men 1^{co} via più 10^{n} fa 10^{co}, e men via 1^{co}[a] più 10^{n} fa 10^{co}, azonti insieme serano men 20^{co}. E puo' 10^{n} via 10^{n} fa più 100^{n}. Et averemo più $1^{\square}$, men 20^{co}, più 100^{n}. Adequemo le parte. Più 100^{n}, più 10^{n}, sutratto 10^{n} de 100^{n}, roman 90^{n}, ingual a 20^{co}. Mo' parti li nomeri per le chosse, parti 90^{n} per 20^{co}, serano $4\frac{1}{2}$. E questo è l'una parte. E l'oltra vol eser $5\frac{1}{2}$ per servir a 10^{n}. Mo' faremo noi in questo muodo. 2 via $4\frac{1}{2}$ fa $\frac{9}{2}$, e do via $5\frac{1}{2}$ fa $\frac{11}{2}$, e questi moltipicha' in ssi, $\frac{9}{2}$ via $\frac{9}{2}$ fa $\frac{81}{2}$, e questo partido per 4 fano $20\frac{1}{4}$. E puo' 11 via 11 mezi fano $\frac{121}{2}$, partido per 4 serano $30\frac{1}{4}$, e tratto la menor dela mazuor, trane $20\frac{1}{4}$ de $30\frac{1}{4}$. Ben roman 10^{n}, chomo vedirì qui de sotto per figura.

1^{co}	10^{n} men 1^{co}	10 men 1^{co}	più	$1^{\square}$	più 100^{n} men 20^{co} più 10^{n}
		10 men 1^{co}	men	20^{co}	10
		100^{n} 20^{co} $1^{\square}$	più	100^{n}	90^{n} men 20^{co}
					20
					$4\frac{1}{2}$ $5\frac{1}{2}$

01		00		$\frac{9}{2}$	$\frac{9}{2}$	$\frac{11}{2}$	$\frac{11}{2}$	$30\frac{1}{4}$
81		121						$20\frac{1}{4}$
44	$20\frac{1}{4}$	44	$30\frac{1}{4}$					10^{n}

E se la volesti far per inpoxizion, pony che fusse 6, 4, e la segonda fusse 7, 3. E chomenzemo a moltipichar in si, 6 via 6 fa 36, e 4 via 4 fa 16. Sutratto 16 de 36, resta 20, e noi vose[*mo*] che fusse 10. Adoncha più 10 ala prima inpoxizion. E la segonda, 7 via 7 fa 49, 3 via 3 fa 9. Sutratto 9 de 49,
f. 62a roman 40, e noi vossemo che fuse 10, adoncha serà più 30. | Ora moltipicha in cruxe. 6 via 30 fa 180^{n} e 7 via 10 fa 70. Sutrazi 70 de 180, roman 110, e puo' moltipicha 4 via 30, fa 120, e 3 via 10, fa 30. Sutratto 30 de 120 serano 90, sutratto de 30 10, roman 20. E questo è lo partidor. Parti 110 per 20, serano $5\frac{1}{2}$, e parti 90 per 20, serano $4\frac{1}{2}$. 2 fia $4\frac{1}{2}$ fa $\frac{9}{2}$, moltipichado in ssi fano 81, e questo partido per 4 fa $20\frac{1}{4}$. Puo' 2 $5\frac{1}{2}$ fa $\frac{11}{2}$, moltipichadi in si fano 121. E questo partido per 4 fano $30\frac{1}{4}$. Sutratto la menor, ch'è $20\frac{1}{4}$, ben roman 10, chomo vedirì qui de sotto per figura.

---

a. *Sic in MS, for* men 1^{co} via.

And to do this problem by unknowns, put that it was $1x$ and 10 minus $1x$, and these muliplied by themselves make this which I shall write further. Minus $1x$ times minus $1x$ makes plus $1x^2$, and then minus $1x$ times plus 10 makes $10x$, and minus $1x$ times plus 10 makes $10x$, added together they will be minus $20x$. And then 10 times 10 makes plus 100. And we will have plus $1x^2$, minus $20x$, plus 100. Equalize the sides. Plus 100, plus 10, 10 subtracted from 100, which leaves 90, equal to $20x$. Now divide the numbers by the unknowns, divide 90 by $20x$, it will be $4\frac{1}{2}$. And this is one side. And the other would be $5\frac{1}{2}$ to get to 10. Now we will do it in this way. 2 times $4\frac{1}{2}$ makes $\frac{9}{2}$, and two times $5\frac{1}{2}$ makes $\frac{11}{2}$, and this multiplied by itself, $\frac{9}{2}$ times $\frac{9}{2}$ makes $\frac{81}{4}$,[1] and this divided by 4 makes $20\frac{1}{4}$. And then 11 times 11 halves makes $\frac{121}{4}$,[2] divided by 4 will be $30\frac{1}{4}$, and the lesser subtracted from the greater, take $20\frac{1}{4}$ from $30\frac{1}{4}$. The result is 10, as you will see worked out below.

$1x$	10 minus $1x$	10 minus $1x$	plus	$1x^2$	plus 100 minus $20x$ plus 10	
		10 minus $1x$	minus	$20x$	10	
		100 $20x$ $1x^2$	plus	100	90 minus $20x$	
					20	
					$4\frac{1}{2}$	$5\frac{1}{2}$

01		001[3]		$\frac{9}{2}$	$\frac{9}{2}$	$\frac{11}{2}$	$\frac{11}{2}$	$30\frac{1}{4}$
~~81~~		~~121~~						$20\frac{1}{4}$
~~44~~	$20\frac{1}{4}$	~~44~~	$30\frac{1}{4}$					10

And if you wanted to do it by false position, suppose that it was 6, 4, and the second was 7, 3. And we begin by multiplying by itself, 6 times 6 makes 36, and 4 times 4 makes 16. Subtract 16 from 36, which leaves 20, and we wanted it to be 10. So it's plus 10 at the first false position. And the second, 7 times 7 makes 49, 3 times 3 makes 9. 9 subtracted from 49, which leaves 40, and we wanted it to be 10, so it will be plus 30. | Now multiply across. 6 times 30 makes 180 and 7 times f. 62a
10 makes 70. Subtract 70 from 180, which leaves 110, and then multiply 4 times 30, which makes 120, and 3 times 10, which makes 30. 30 subtracted from 120 will be 90, 10 subtracted from 30 leaves 20. And this is the divisor. Divide 110 by 20, which will be $5\frac{1}{2}$, and divide 90 by 20, which will be $4\frac{1}{2}$. 2 times $4\frac{1}{2}$ makes $\frac{9}{2}$, multiplied by itself makes 81, and this divided by 4 makes $20\frac{1}{4}$. Then 2 $5\frac{1}{2}$ makes $\frac{11}{2}$, multiplied by itself makes 121. And this divided by 4 makes $30\frac{1}{4}$. Subtracting the lesser, which is $20\frac{1}{4}$, leaves 10, as you will see worked out below.

1. $\frac{81}{2}$ in MS.
2. $\frac{121}{2}$ in MS.
3. *1* lacking in MS.

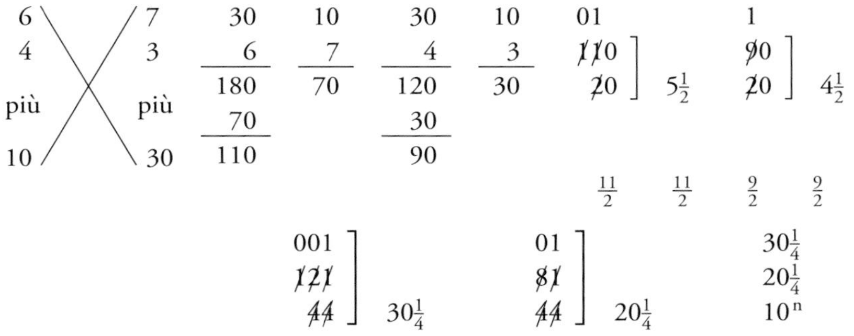

6 4 più 10
7 3 più 30
30 6 180 70 110
10 7 70
30 4 120 30 90
10 3 30
01 110 20 5½
1 90 20 4½
11/2 11/2 9/2 9/2
001 121 44 30¼
01 81 44 20¼
30¼ 20¼ 10ⁿ

Fame de 10 2 parte che moltipichada una per 7 faza tanto quanto moltipichada l'oltra per 9.

E per far la ditta raxion per inpoxizion pony che fusse 6, 4. Se moltipicho 6 fia 7 fa 42, e se moltipicho 4 fia 9 fa 36, e noi vollemo che sia in tutto 36. Abatti 36 de 42, roman più 6, e questi metti ala prima inpoxizion per 6 e 4 che m'ò posto più 6. E per far la segonda, metto 5[a] e 5,[b] e sì moltipicho 5 fia 7 fa 35, e 5 fia 9 fa 45, e mi voio che sia 35, adoncha più 10, e metti ala segonda
f. 62b inpoxizion per 5 e 5 che m'ò posto più 10. Moltipicha 6 via 10, fa | 60 e 6 via 5 fa 30, azonti fa 90, e 4 via 10 fa 40, e 6 via 5 fa 30, azonti fano 70. Azontto 10 chon 6 fano 16, e questo è lo partior. Parti 90 per 16, insirà fuor $5\frac{5}{8}$, e parti 70 per 16, insirà $4\frac{3}{8}$. Adoncha fa 5 via 8 fa 40, e 5 de sovra la verga fa 45 octavy, e 4 via 8 fa 32. Azonto 3 ch'è sovra la verga fa 35[c] octavy. Se moltipicho 7 via 45 fa 315, e se moltipicha 9 via 35 fa 315, e tanto val l'una moltipichada per 7 quanto l'oltra moltipichada per 9, chomo vedirì qui de sotto per figura.

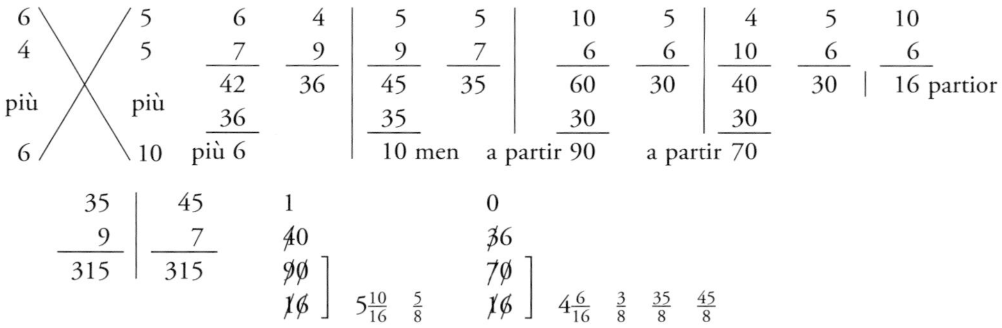

6 4 più 6
5 5 più 10
6 7 42 36 più 6
4 9 36
5 9 45 35 10 men
5 7 35
10 6 60 30 a partir 90
5 6 30
4 10 40 30 a partir 70
5 6 30
10 6 16 partior
35 9 315
45 7 315
1 40 90 16 5 10/16 5/8
0 36 70 16 4 6/16 3/8 35/8 45/8

---

a. *Corrected over* 7.
b. *Corrected over* 3.
c. 35 *corrected over* 38.

6 ╲ ╱ 7
4 ╳ 3
plus / plus
10 ╱ ╲ 30

30 × 6 = 180; 70; 110
10 × 7 = 70
30 × 4 = 120; 30; 90
10 × 3 = 30

01 / 110 (struck) / 20 (struck)] $5\frac{1}{2}$
1 / 90 (struck) / 20 (struck)] $4\frac{1}{2}$

$\frac{11}{2}$ $\frac{11}{2}$ $\frac{9}{2}$ $\frac{9}{2}$

001 / 121 (struck) / 44 (struck)] $30\frac{1}{4}$
01 / 81 (struck) / 44 (struck)] $20\frac{1}{4}$
$30\frac{1}{4}$ / $20\frac{1}{4}$ / 10

Find me 2 parts of 10 such that one multiplied by 7 makes as much as the other multiplied by 9.

And to do this problem by false position, suppose that it was 6 and 4. If I multiply 6 times 7 it makes 42, and if I multiply 4 times 9 it makes 36, and we wanted it to be 36 in all. Subtract 36 from 42, which leaves plus 6, and put this plus 6 at the first false position: for 6 and 4 that I have put, plus 6. And to do the second, suppose 5 and 5, and if I multiply 5 times 7 it makes 35, and 5 times 9 makes 45, and I want it to be 35, so it's plus 10, and put plus 10 at the second false position: for 5 and 5 that I have placed, plus 10. Multiply 6 times 10, it makes | 60, and 6 times 5 makes 30, added makes 90, and 4 times 10 makes 40, and 6 times 5 makes 30, added makes 70. 10 added with 6 makes 16, and this is the divisor. Divide 90 by 16, the result will be $5\frac{5}{8}$, and divide 70 by 16, the result will be $4\frac{3}{8}$. Then do 5 times 8 makes 40 and 5 above the line makes 45 eighths, and 4 times 8 makes 32. The 3 that's above the line added in makes 35 eighths. If I multiply 7 times 45 it makes 315, and if 9 is multiplied times 35 it makes 315, and one multiplied by 7 is the same as the other multiplied by 9, as you will see worked out below. f. 62b

6 ╲ ╱ 5
4 ╳ 5
plus / plus
6 ╱ ╲ 10

6 × 7 = 42; 36; plus 6
4 × 9 = 36
5 × 9 = 45; 35; 10 minus
5 × 7 = 35
10 × 6 = 60; 30; to divide 90
5 × 6 = 30
4 × 10 = 40; 30; to divide 70
5 × 6 = 30 | 10 + 6 = 16 divisor

35 × 9 = 315 | 45 × 7 = 315

1 / 40 (struck) / 90 (struck) / 16 (struck)] $5\frac{10}{16}$ $\frac{5}{8}$
0 / 36 (struck) / 70 (struck) / 16 (struck)] $4\frac{6}{16}$ $\frac{3}{8}$ $\frac{35}{8}$ $\frac{45}{8}$

E se la volesti far per la chossa pony che fusse 1^{co} 10^{n} men 1^{co}. Se moltipicho 10^{n} men 1^{co} che son dal'un ladi per 9, so che die far 90^{n} men 9^{co}. E se moltipicho l'oltra parte, ch'è più 1^{co}, chon 7, fano più 7^{co}, et averemo più 7^{co} 90^{n} men 9^{co}. Adequemo le parte e daremo men 9^{co} a 90^{n}, e non i mancharà niente. Mo' demo 9^{co} men a 7^{co} più. Azontto el più chon lo men serano 16^{co} ingual a 90^{n}, e questi partidi per 16 insirà fuo[*ra*] $5\frac{10}{16}\frac{5}{8}$, e lo resto inchina 10 serano $4\frac{3}{8}$. E questi, moltipichada l'una per 7 e l'oltra 9, serà ingual chomo a' ditto de sovra.

1^{co} $\quad$ 10^{n} men 1^{co} $\quad$ 90^{n} men 9^{co} $\quad$ 7^{co} più $\quad$ più 7^{co} $\quad$ 90^{n} $\quad$ 9^{co} men

45	35	$4\frac{3}{8}$	9	
5	9		16^{co}	1
ingal 315	315 ingal			~~4~~0
				~~90~~
				~~16~~ $5\frac{10}{16}$ $\frac{5}{8}$

f. 63a [Truovar un numero che moltipichado per la so mittade]

Truovame un numero che moltipichado per la so mitade, e quela moltipichazion partida per lo so terzo del nomero truovado, vegna 20.

E se la volesti far per inpoxizio[*n*] pony che ala prima inpoxizion fusse 12. Moltipichada per la so mitade seria 72, e partida per lo so terzo, che sono 4, seria 18, e noi volemo che sia 20. Adoncha è men 2. E chosì metti ala prima per 12 che m'ò posto men 2. E per far la segonda pony che fusse 24. Moltipicha 12 via 24, che xa[a] la mittade del 24, serano 288, e questi partidi per lo so terzo, che sono 8 de 24, insirà 36, e noi volemo che sia 20. Adoncha è più 16. E chosì metemo ala segonda inpoxizio[*n*] per 24 che m'ò posto più 16. Mo' moltipicha 12 via 16 fa 192, e 2 via 24 fa 48, azonti insieme fano 240. E per far partidor, più 16 men 2 azonti fano 18. Die 'ser partidi 240 per 18. Insirà fuora $13\frac{6}{18}$, $\frac{3}{9}$, $\frac{1}{3}$, che moltipichado $13\frac{1}{3}$, che serà el numero truovado per la so mitade, e quela moltipichazion partida per lo so terzo vegnirà 20.

---

a. *Sic in MS.*

And if you wanted to do it by unknowns, put that it was 1*x*, 10 minus 1*x*. If I multiply 10 minus 1*x* which is on one side by 9, I know that it should give 90 minus 9*x*. And if I multiply the other side, that is, plus 1*x*, by 7, it makes plus 7*x*, and we will have plus 7*x*, 90 minus 9*x*. We equalize the sides and we will give minus 9*x* to 90, and nothing will be lacking. Now we give minus 9*x* to plus 7*x*. Adding the plus with the minus will be 16*x* equal to 90, and these divided by 16 will result in $5\frac{10}{16}\frac{5}{8}$, and the rest up to 10 will be $4\frac{3}{8}$. And these, the one multiplied by 7 and the other by 9, will be equal, as said above.

1*x*	10 minus 1*x*	90 minus 9*x*	7*x* plus	plus 7*x*	90	9*x* minus
				9		
	45	35	$4\frac{3}{8}$	16*x*	1	
	5	9			~~4~~0	
	equals 315	315 equals			~~90~~	
					~~16~~	$5\frac{10}{16}$ $\frac{5}{8}$

FIND A NUMBER WHICH MULTIPLIED BY ITS HALF

f. 63a

Find me a number which when multiplied by its half, and that product divided by a third of the found number, comes to 20.

And if you want to do it by false position, suppose that it was 12 at the first false position. Multiplied by its half would be 72, and divided by its third, which is 4, would be 18, and we want it to be 20. So it is minus 2. And so put minus 2 at the first false position for 12 that we have put. And to do the second suppose that it was 24. Multiply 12 times 24, which is half of 24, the result will be 288, and this divided by its third, which for 24 is 8, will result in 36, and we wanted it to be 20. So it is plus 16. And so we put 16 at the second false position for the 24 that we've put. Now multiply 12 times 16 makes 192, and 2 times 24 makes 48, added together they make 240. And to make the divisor, plus 16 minus 2 added together make 18. 240 should be divided by 18. The result will be $13\frac{6}{18}$, $\frac{3}{9}$, $\frac{1}{3}$, $13\frac{1}{3}$, which multiplied by its half and that product divided by its third will come to 20.

12 ╲ ╱ 24
men ╳ più
2 ╱ ╲ 16

192 16
48 2
a partir 240 18 partior

0
0~~3~~
~~1~~~~0~~6
~~2~~~~4~~~~0~~
~~1~~~~8~~~~8~~ | $13\frac{6}{18}$ $\frac{3}{9}$ $\frac{1}{3}$ el numero
~~1~~

$13\frac{1}{3}$ $6\frac{2}{3}$ 40 0
$\frac{40}{3}$ $\frac{20}{3}$ 20 0~~8~~8
$\frac{800}{9}$ ╳ $\frac{40}{9}$ 00 ~~8~~~~0~~~~0~~
80 ~~9~~~~9~~ | $88\frac{8}{9}$

$13\frac{1}{3}$ 0
$4\frac{4}{9}$ ~~8~~00
$\frac{40}{9}$ ~~4~~00 | 20
4

El numero fu $13\frac{1}{3}$. Moltipichado per la so mitade fu $\frac{800}{9}$. Partido per lo so terzo, che fu $\frac{40}{9}$, è ben insiu 20, chomo dixe de sovra.

f. 63b E per far la chontrascritta raxion per la chosa pony che 'l numero fusse 1^{co}, e questo numero moltipichado per la so mitade sia $\frac{1}{2}^{co}$, e puo' el terzo del numero che xe $\frac{1}{3}^{co}$ partido $\frac{1}{2}^{co}$ insirà fuor $\frac{2}{3}^{co}$, che son ingual a 20. Adoncha moltipicha $\frac{2}{3}$ via $\frac{20}{1}$, serave 40^{n}, a partir per 3^{co} insirà fuora de partir $13\frac{1}{3}$. Aprovemo che questo sia lo nomero. La mitade de $13\frac{1}{3}$ serave $6\frac{2}{3}$, serano $\frac{20}{3}$, serano $\frac{40}{3}$. Mo' volemo moltipichar 20 fia 40, fano 800, e diese partir per 3 via 3 fa 9. Adoncha partido 800 per 9 insirà $88\frac{8}{9}$, e questi serano $\frac{800}{9}$. Mo' volemo el terzo del numero che xe $13\frac{1}{3}$. El so terzo serave $\frac{40}{9}$. Adoncha parti $\frac{800}{9}$ per $\frac{40}{9}$ e die insir 20 chomo fu la domanda de sovra. Vedirì qui de sotto per figora.

12 minus 2 × 24 plus 16

192, 48 — to divide 240

16, 2 — 18 divisor

0
03
166
240
188
1

$13\frac{6}{18}$ $\frac{3}{9}$ $\frac{1}{3}$ the number

$13\frac{1}{3}$ $\frac{40}{3}$ $6\frac{2}{3}$ $\frac{20}{3}$

40
20
00
80

0
088
800
99

$88\frac{8}{9}$

$\frac{800}{9}$ × $\frac{40}{9}$

$13\frac{1}{3}$
$4\frac{4}{9}$
$\frac{40}{9}$

0
800
400
4

20

The number was $13\frac{1}{3}$. Multiplied by its half it was $\frac{800}{9}$. Divided by its third, which was $\frac{40}{9}$, it comes to 20, as was said above.

And to do the problem written opposite by unknowns, put that the number was $1x$, and this number multiplied by its half or $\frac{1}{2}x$, and then the third of the number, which is $\frac{1}{3}x$, divided by $\frac{1}{2}x$ will result in $\frac{2}{3}x$, which is equal to 20. Then multiply $\frac{2}{3}$ times $\frac{20}{1}$, which will be 40, to divide by $3x$; $13\frac{1}{3}$ will result from the division. Let's prove that this is the number. Half of $13\frac{1}{3}$ will be $6\frac{2}{3}$, it will be $\frac{20}{3}$, it will be $\frac{40}{3}$. Now we want to multiply 20 times 40, which makes 800, and this should be divided by 3 times 3, which makes 9. Then 800 divided by 9 will result in $88\frac{8}{9}$, and this will be $\frac{800}{9}$. Now we want a third of this number, which is $13\frac{1}{3}$. Its third will be $\frac{40}{9}$. Then divide $\frac{800}{9}$ by $\frac{40}{9}$ and 20 should result as was in the question above. You will see it worked out below. f. 63b

1^{co} $\frac{1}{2}^{co}$ $\frac{1}{3}^{co}$ $\frac{2}{3}^{co}$ ingal a $\frac{20}{1}^{n}$

20	0		
2	~~1~~1		
40	~~40~~	$13\frac{1}{3}$	$\frac{40}{3}$ $\frac{20}{3}$
	~~33~~	$6\frac{2}{3}$	$\frac{800}{9}$

$\frac{800}{9}$ $\frac{40}{9}$

0			
0~~8~~8		$13\frac{1}{3}$	
~~800~~	$88\frac{8}{9}$	$4\frac{4}{9}$	$\frac{40}{9}$
~~99~~			
00			
~~8~~~~0~~0			
~~4~~00	20		
~~4~~			

E per far una prova d'una raxion fatta qui driedo a charte 27 chomo dize: "2 homeny baratta l'uno a seda che val a chontadi non sa quanti, e metela a baratto quel che la valeva, e vuol $\frac{1}{4}$ in denari e più la so radixe, e l'oltro a pany chi val soldi 50 el brazo, e mesela a baratto soldi 56, adomanda zò che la valsse la seda a chonttadi, la seda valse a chontadi soldi $142\frac{34}{81}$." Per far la prova senza la radixe fu $130\frac{79}{81}$, e la radixe fu $11\frac{4}{9}$, che sono $\frac{36}{81}$. Adoncha se moltipicho 130 via 81, azonti $\frac{79}{81}$, fano $\frac{10609}{81}$, e se moltipicho 142 via 81, azonti 34, fano 11536, e de questi sutratto el quarto, che sono 2884, resta
f. 64a netto 8652, | e sutratto de 10609 2884 roman 7725. E voiando aprovar la ditta raxion, tu fara': "Se 7725 me torna 8652, che serave 50 che valse a chontadi el pano?" Tu trovera' che serà 56, chomo vedirì qui de sotto per figora, e per questa far ogn'altra raxion simille.

$1x$ $\frac{1}{2}x$ $\frac{1}{3}x$ $\frac{2}{3}x$ equal to $\frac{20}{1}$

20
2
40

0
1̸1
4̸0̸
3̸3̸] $13\frac{1}{3}$

$6\frac{2}{3}$

$\frac{40}{3}$ $\frac{20}{3}$
$\frac{800}{9}$

$\frac{800}{9}$ $\frac{40}{9}$

0
08̸8
8̸0̸0̸
9̸9̸] $88\frac{8}{9}$

$13\frac{1}{3}$
$4\frac{4}{9}$

$\frac{40}{9}$

00
8̸0̸0
4̸00] 20
4̸

And to do the proof of a problem done back on folio 27[1] where it said: "2 men barter, one has silk whose worth in money he does not know, and puts it into the barter at its worth, and wants $\frac{1}{4}$ in coins plus its square root, and the other has cloth that is worth 50 soldi per ell, and puts it into the barter for 56 soldi, it asks how much the silk was worth in money, the silk was worth in money $142\frac{34}{81}$ soldi." To do the proof, without the root it was $130\frac{79}{81}$, and the root was $11\frac{4}{9}$, which is $\frac{36}{81}$. So 130 is multiplied by 81, add $\frac{79}{81}$, which makes $\frac{10609}{81}$, and if I multiply 142 by 81, and add 34, it makes 11,536, and one quarter subtracted from this, which is 2,884, leaves 8,652, | and 2,884 f. 64a
subtracted from 10,609 leaves 7,725. And if you wish to prove the problem, you will do: "If 7,725 gives me 8,652, what would the result be if the value of the cloth was 50?" You will find that it will be 56, as you will see worked out here below, and in this way you will do all other similar problems.

1. That is, fol. 27b, above.

$1^{\square}$ $1^{\square}$ 1^{co} $\frac{3}{4}^{\square}$ $\frac{3}{4}^{co}$ $\frac{50}{1}^{n}$ $\frac{150}{4}^{\square}$ $\frac{150}{4}^{co}$ $\frac{168}{4}^{\square}$ men $\frac{56}{4}^{co}$

56 $150^{\square}$

$1^{\square}$ $\frac{3}{4}^{\square}$ men $\frac{1}{4}^{co}$ a partir 206 partior $\frac{18}{4}^{\square}$

56 56 $\frac{18}{4}^{\square}$ $\frac{206}{4}^{co}$

0
01
128
206
188
1] $11\frac{8}{18}$ $\frac{4}{9}$ $\frac{36}{81}$

$11\frac{4}{9}$ 103 309 10609 00
$11\frac{4}{9}$ 103 000 81 21
103 0257
10609
8111 $130\frac{79}{81}$
88 $11\frac{36}{81}$
$142\frac{34}{81}$

10609 11536
81 7725 2884
2884 8652

e diremo 7725 8652 50 00
50 0410
432600 4633
087650
432600
77255] 56
772

e per lo simille suma 8652 7725 56 000
56 03210
432600
86522] 50
865

e per lo simille suma 50 56 8652
e per lo simille 56 50 7725

E per questa farì ogn'altra raxion simille chomo vui vedì. Adoncha se parto 8652 per 81 venirà soldi $106\frac{66}{81}$, e tanto valse la seda a barato.[a] E se parto 7725 per 81 venirà soldi $95\frac{30}{81}$, e tanto valse la seda a chontadi la livra. E se la volé aprovar aprovela che starà ben.

a. A barato *corrected above the line over* a chontadi *crossed out with a horizontal line.*

$1x^2$ $1x^2$ $1x$ $\frac{3}{4}x^2$ $\frac{3}{4}x$ $\frac{50}{1}$ $\frac{150}{4}x^2$ $\frac{150}{4}x$ $\frac{168}{4}x^2$ minus $\frac{56}{4}x$

56 $150x^2$

$1x^2$ $\frac{3}{4}x^2$ minus $\frac{1}{4}x$ to divide 206 divisor $\frac{18}{4}x^2$

56 56 $\frac{18}{4}x^2$ $\frac{206}{4}x$

0
01
128
206
188 | $11\frac{8}{18}$ $\frac{4}{9}$ $\frac{36}{81}$
1

$11\frac{4}{9}$ 103 309 10609 00
$11\frac{4}{9}$ 103 000 81 21
103 0257
10609
8111 | $130\frac{79}{81}$
88 | $11\frac{36}{81}$
$142\frac{34}{81}$

10609 11536
81 7725 2884
2884 8652

And we will say 7725 8652 50 00
50 0410
432600 4633
087650
432600
77255 | 56
772

And in the same way sum 8652 7725 56 000
56 03210
432600
86522 | 50
865

And in the same way sum 50 56 8652
And in the same way 56 50 7725

And in this way you will do any other similar problem, as you will see. So 8,652 divided by 81 will come to $106\frac{66}{81}$ soldi, and this is the value of the silk in barter. And if I divide 7,725 by 81 it will come to $95\frac{30}{81}$ soldi, and this is the value of the silk per pound in cash. And if you want to prove it, prove it, because it will come out right.

f. 64b

[Piper per 100 libre per impoxizion]

El zentener de piper val duchati 16, che valeria libre 200 per inpoxizion? E noi meteremo ala prima inpoxizion che vallese 20, e sì diremo: "Se livre 200 val duchati 20 del'inpoxizion, che valeria libre 100?" Moltipicha e parti, insirà fuor duchati 10, e noi vossemo che fuse 16, serà men 6, e diremo: "Per 20 che m'ò puxo serano men 6." E per voler far la segonda inpoxizion metemo che vallese 10, e noi diremo: "Se 200 val 10, che serà 100?" Moltipicha e parti, insirà duchati 5, e noi vosemo che 'l fose lo prexio, zoè 16. Adoncha serà men 11. E chosì meterì ala segonda inpoxizion. Moltipicha in croxe e parti, insirà 32, chomo vedirì qui de sotto per figura.

20 ╲ ╱ 10
men ╳ men
6 ╱ ╲ 11

200 20 | 100 / 20 / 000 / 200 | 00 / 2000 / 2000 / 20] 10 | 16 / 10 / men 6 | 20 / 11 / 220 / 60 / a partir 160

200 10 | 100 / 10 | 00 / 1000 / 200] 5 | 16 / 5 / 11 men

0 / 010 / 160 / 55] 32

E per la riegola del 3, se 100 me dà 16 ch'è 200?

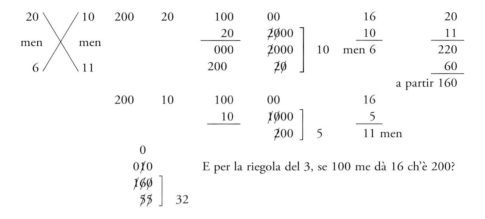

E per far la ditta raxion per la chosa, se 100 me torna 16, ch'è 200? Se moltipicho 16 fia 200 me die venir 1^{co}. Adoncha moltipicha 1^{co} via 100^{co}. A partir insirà duchati 32, chomo vedirì per figora.

100^{n} / 1^{co} | 16^{n} | 200^{n} / 16 | 1^{co} | 3200^{n} | 100^{co} | 00 / 3200 / 1000 / 10] 32

100 16 200 1^{co}

PEPPER FOR 100 POUNDS BY FALSE POSITION

f. 64b

The hundredweight of pepper is worth 16 ducats; what would 200 pounds be worth by false position? And we will suppose at the first false position that it was worth 20, and so we will say: "If 200 pounds is worth 20 ducats by false position, what would 100 pounds be worth?" Multiply and divide, it will result in 10 ducats, and we wanted it to be 16, it will be minus 6, and we will say: "For the 20 that I have put it will be minus 6." And to do the second false position we suppose that it was worth 10, and we will say: "If 200 is worth 10, what will 100 be?" Multiply and divide, the result will be 5 ducats, and we wanted it to be the price, that is, 16. So it will be minus 11. And you will put this at the second false position. Multiply across and divide, the result will be 32, as you will see worked out below.

20 \ / 10
minus X minus
6 / \ 11

200 20 | 100, 20, 000, 200 | 00, 2000, 2000, 20] 10 | 16, 10, minus 6 | 20, 11, 220, 60, to divide 160

200 10 | 100, 10 | 00, 1000, 200] 5 | 16, 5, 11 minus

0
010
160
55] 32

And by the rule of 3, if 100 gives me 16, what is 200?

And to do this problem by unknowns, if 100 gives me 16, what is 200? If I multiply 16 by 200 the result is 1*x*. So multiply 1*x* by 100.[1] Dividing will result in 32 ducats, as you will see worked out below.

100 1*x* | 16 | 200 16 | 1*x* | 3200 | 100*x* | 00, 3200, 1000, 10] 32 | 100 16 200 1*x*

1. *100x* in MS.

f. 65a

[PIPER INVESTIDA]

E voio spender duchati 32 in piper che 'l zentener vaia duchati 16. Mettemo ala prima inpoxizion che 'l fusse libre 100, e per libre 100 vosemo duchati 16, e noi volemo che sia per duchati 32. Adon[*cha*] serà men duchati 16, e chosì meteremo al'inpoxizion. E per far la segonda mettemo che fuse libre 50, valeria duchati 8, e noi vosemo spender duchati 32. Adoncha serà men 24. Moltipicha e parti, insirà del partidor duchati over piper libre 200.

100	50	100	50	800	00	
		24	16		~~1~~~~6~~00	a partir
men	men	2400	300	24	~~888~~	200
16	24	800	50	16	E per la regola del 3, se	
		1600		partior 8	16 me torna 100, ch'è 32?	

E per far la ditta raxion per la chosa, se 16 me torna 100, che serà 32? Se moltipicho 100 via 32 farà 3200, e se partto per 16^{co} serano libre 32.

16 100 200 1^{co}

[BRAZA DE PANO E TELA E SOLDI E GROSI MESIDADI]

Fame questa raxion. Braza 3 de pano val livre 8 de zera; livre 5 de zera val braza 9 de tella; braza 7 de tela val soldi 54; soldi 25 val soldi[a] 6. Adomando per grossi 32 quante braze de pano averò. E per far la ditta, moltipicha 3 via 5 fa 15; 15 via 7 fa 105; e 25 via 105 fa 2625. Questi sono braza de pano. Puo' 8 fia 9 fa 72, e 51 via 72 fa 3888, e 6 via 3888 fa 23325. Questi sono grosi. E se volesti veder per grosi 32 quante braze averò, fa' chossì: "Se 23325 me torna 2625, che serave grosi 32?" A moltipichar 2625.[b]

---

a. *Sic in MS.*

b. *Corrected over* 23325 *crossed out with a diagonal line; the problem is continued on the bottom of the facing page.*

1. MS has *soldi* here, but the context of later calculations makes it clear that grossi are meant. In this period, the grosso coin was worth 4 soldini; what must be meant here is the grosso of account, defined as $\frac{1}{24}$ of the gold ducat. 25 soldi to 6 grossi would then imply a rate of 100 soldi to the ducat; this was the commercial rate from 1411 to 1425, but is cited as an official rate as late as 1433: cf. Lane and Mueller, *Money and Banking in Medieval and Renaissance Venice*, 1: 602–609.

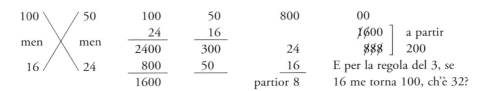

Pepper invested

f. 65a

And I want to spend 32 ducats on pepper of which the hundredweight goes for 16 ducats. We suppose at the first false position that it was 100 pounds, and for 100 pounds we spent 16 ducats, and we want it to be for 32 ducats. So it will be minus 16 ducats, and this is what we put at the false position. And to do the second, we suppose that it was 50 pounds, which would be worth 8 ducats, and that we wanted to spend 32 ducats. So it will be minus 24. Multiply and divide, the result of the division will be 200 ducats or rather pounds of pepper.

100	╲ ╱	50	100	50	800	00	
minus	╳	minus	24	16		~~1~~~~6~~00	to divide
			2400	300	24	~~888~~	200
16	╱ ╲	24	800	50	16	And by the rule of 3, if 16	
			1600		divisor 8	gives me 100, what is 32?	

And to do the problem by unknowns, if 16 gives me 100, what will 32 be? If I multiply 100 times 32 it will make 3,200, and if I divide by 16*x* it will be 32 pounds.

16 100 200 1*x*

Ells of cloth and canvas and soldi and grossi mixed up

Do this problem for me. 3 ells of cloth are worth 8 pounds of wax; 5 pounds of wax are worth 9 ells of canvas; 7 ells of canvas are worth 54 soldi; 25 soldi are worth 6 grossi.[1] I ask how much cloth I will have for 32 grossi. And to do this, multiply 3 times 5, which makes 15; 15 times 7, which makes 105; and 25 times 105, which makes 2,625. These are ells of cloth. Then 8 times 9, which makes 72; and 54[2] times 72, which makes 3,888; and 6 times 3,888, which makes 23,325.[3] These are grossi. And if you want to see how many ells I will have for 32 grossi, do this: "If 23,325 gives me 2,625, what will 32 grossi be?" To multiply 2,625.[4]

---

2. *51* in MS.
3. This should be 23,328; and the rest of the problem is thrown off accordingly.
4. The problem is continued on the bottom of the facing page.

f. 65b [Zoya a vadagnar e perder]

E ò chonprado una zoya per duchati 25, e voio vadagnar 7 per 100. Adomando quanto voio vender la ditta. Per inpoxizion pony che la vendese 30, e noi diremo: "Se 25 me torna 5 che son men cha 30, che me darà 100?" Moltipicha e parti, insirà 20, e noi vosemo vadagnar 7. Adoncha è men 13. E chosì pony ala prima.

30 \ / 28[a]
più X più
13 / \ 5

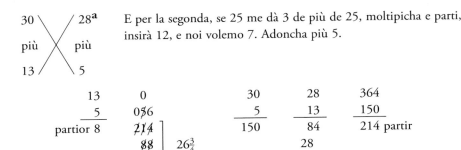

E per la segonda, se 25 me dà 3 de più de 25, moltipicha e parti, insirà 12, e noi volemo 7. Adoncha più 5.

13	0		30	28	364
5	0~~5~~6		5	13	150
partior 8	~~2~~~~1~~~~4~~		150	84	214 partir
	~~8~~~~8~~	$26\frac{3}{4}$		28	

E per la riegola del 3, se 100 me torna 107 ch'è 25?

E per far la ditta raxion per la chosa, se 100 me torna 107, ch'è 25? Se moltipicho 107 via 25 me die venir 1^{co}. Adoncha moltipicha 100 via 1^{co} e parti, insirà $26\frac{3}{4}$.

100^{n} 107^{n} 25^{n} 1^{co}

---

per[b] 32, fa 84000. Parti per 23325, insirà fuor braza $3\frac{14025}{23325}$, e se vosè saver per braza 25 quanti grosi averì, questo è lo muodo: "Se 2625, che son d'i brazi, me torna 23325, che son d'i grosi, che serave braza 25?" Moltipicha 25 fia 23325 fano 583125, e questo parti per 2625, insirà del partior $222\frac{375}{2625}$, e questi serano grosi. E per lo simille farì ogn'altra raxion sì fatta.

f. 66a [Zoya a chonprar e vender che vadagna per 100 e che perde per 100]

E ò chonprado una zoya per duchati 38, e voio perder duchati 8 per 100. Adomando quanto la voio vender. Per inpoxizion pony che la vendese 30, e noi diremo: "Se 38 me dà 8 che son più de

---

a. *Corrected over* 32 *crossed out with a diagonal line.*

b. *Continued from fol. 65a.*

PROFIT AND LOSS ON A JEWEL

f. 65b

And I have bought a jewel for 25 ducats, and I want to make a profit of 7 percent. I ask for how much I want to sell it. By false position suppose that I sold it for 30, and we will say: "If 25 gives me 5, what is less than 30 that will give me 100?" Multiply and divide, the result will be 20, and we wanted a profit of 7. So it is plus[1] 13, and put this at the first.

30		28
plus	X	plus
13		5

And for the second, if 25 gives me 3 more than 25, multiply and divide, the result will be 12, and we wanted 7. So, plus 5.

13	0		30	28	364
5	0~~5~~6		5	13	150
divisor 8	~~214~~		150	84	214 to divide
	~~88~~	$26\frac{3}{4}$		28	

And by the rule of 3, if 100 gives me 107, what is 25?

And to do this problem by unknowns, if 100 gives me 107, what is 25? If I multiply 107 times 25, $1x$ should come to me. So multiply 100 by $1x$ and divide, the result will be $26\frac{3}{4}$.

100 107 25 $1x$

---

by[2] 32, which makes 84,000. Divide by 23,325, the result will be $3\frac{14025}{23325}$ ells, and if I wanted to know how many grossi I would have for 25 ells, this is the way: "If 2,625, which are the ells, gives me 23,325, which are the grossi, what would 25 ells be?" Multiply 25 times 23,325, which makes 583,125, and divide this by 2,625, the result of the division will be $222\frac{375}{2625}$, and these will be grossi. And you will do all other problems the same way.

TO BUY AND SELL A JEWEL, THE PERCENT GAINED AND LOST

f. 66a

And I have bought a jewel for 38 ducats, and I want to lose 8 percent. I ask for how much I want to sell it. By false position suppose that I sold it for 30, and we will say: "If 38 gives me 8 that is more

---

1. *minus* in MS.
2. Continued from fol. 65a.

30, che me darà 100?" Tu sa che die venir 800, e questi partidi per 38 insirà $21\frac{1}{19}$, e noi vosemo perder 8. Adoncha è men $13\frac{1}{19}$, e chosì mettemo ala prima. E la segonda mettemo che fose 28, e diremo: "Se 38 fusse men del 28 al 38 che son 10, che seria 100?" Tu sa che die venir 1000, e questi partidi insirà fuor $26\frac{6}{19}$, e noi vosemo perder 8. Adoncha è men $18\frac{6}{19}$. E moltipicha e parti, insirà $34\frac{96}{100}$, chomo vedirì per figora.

30 men $13\frac{1}{19}$ — 28 men $18\frac{6}{19}$

30	28	$18\frac{6}{19}$
$18\frac{6}{19}$	$13\frac{1}{19}$	$13\frac{1}{19}$
$549\frac{9}{19}$	$365\frac{9}{19}$	$5\frac{5}{19}$
$365\frac{9}{19}$	184 a partir	

E per la riegola del 3, se 100 me torna 92, che serà 38?

184
19
3496

$\frac{184}{1}$ $\frac{100}{19}$ partior $\frac{100}{19}$

00
~~3~~~~4~~96
~~1~~000
~~1~~0

$34\frac{96}{100}$

E per far la ditta raxion per la chosa, se 100 me torna 92, che serave 38? Moltipicha 92 fia 38, e die venir 1^{co}, e moltipicha una chosa via 100, fa 100^{co}, a partir $34\frac{96}{100}$.

100 92 38 1^{co}

---

f. 66b E ò chonprado una zoya per duchati 42 e si l'ò venduda per duchati 51 adomando zò che ò vadagnado per 100. E per farla per inpoxizion pony che fusse 10, e noi diremo: "Se 100 torna 110, che serave 42?" Moltipicha e parti, insirà $46\frac{1}{5}$, e noi vosemo che fuse 51. Adoncha serà men $4\frac{4}{5}$, e chosì metti ala prima. E per la segonda metti che fuse 12, e noi diremo: "Se 100 fuse 112, che serave 42?" Moltipicha e parti, insirà $47\frac{1}{25}$, e noi vosemo che fuse 51. Adoncha serà men $3\frac{24}{25}$, e chosì metti ala segonda inpoxizion.

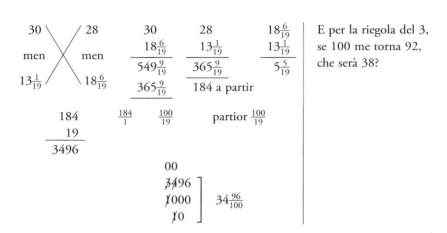

than 30, what will 100 give me?" You know that 800 should come, and this divided by 38 will result in $21\frac{1}{19}$, and we wanted to lose 8. So it is minus $13\frac{1}{19}$, and we put this at the first false position. And for the second we suppose that it was 28, and we will say: "If 38 was less than 28 to 38, which is 10, what would 100 be?" You know that 1,000 should come, and this divided will result in $26\frac{6}{19}$, and we wanted to lose 8. So it is minus $18\frac{6}{19}$. And multiply and divide, the result will be $34\frac{96}{100}$, as you will see worked out.

30	28	30	28	$18\frac{6}{19}$	And by the rule of 3,
minus	minus	$18\frac{6}{19}$	$13\frac{1}{19}$	$13\frac{1}{19}$	if 100 gives me 92,
$13\frac{1}{19}$	$18\frac{6}{19}$	$549\frac{9}{19}$	$365\frac{9}{19}$	$5\frac{5}{19}$	what will 38 be?
		$365\frac{9}{19}$	184 to divide		

184
19
3496

$\frac{184}{1}$ $\frac{100}{19}$ divisor $\frac{100}{19}$

00
3496
1000
10] $34\frac{96}{100}$

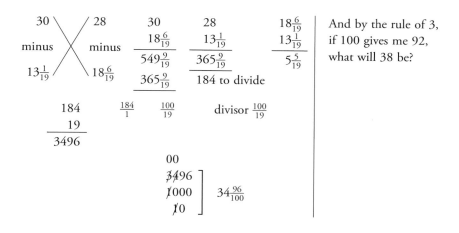

And to do this problem by unknowns, if 100 gives me 92, what would 38 be? Multiply 92 by 38, and it should come to $1x$, and multiply one unknown by 100, it makes $100x$, to divide $34\frac{96}{100}$.

100 92 38 $1x$

And I've bought a jewel for 42 ducats and I've sold it for 51 ducats, I ask what percent I've gained. f. 66b
And to do this by false position, suppose that it was 10, and we will say: "If 100 gives 110, what would 42 be?" Multiply and divide, the result will be $46\frac{1}{5}$, and we wanted it to be 51. So it will be minus $4\frac{4}{5}$, and put this at the first. And for the second, suppose that it was 12, and we will say: "If 100 was 112, what would 42 be?" Multiply and divide, the result will be $47\frac{1}{25}$, and we wanted it to be 51. So it will be minus $3\frac{24}{25}$, and put this at the second false position.

[Partir 4, 3, roman 1, qual fu el partior]

10	12	10	12	$57\frac{3}{5}$	$4\frac{4}{5}$	
men	men	$3\frac{24}{25}$	$4\frac{4}{5}$	$39\frac{3}{5}$	$3\frac{24}{25}$	
$4\frac{4}{5}$	$3\frac{24}{25}$	$39\frac{3}{5}$	$57\frac{3}{5}$	a partir 18	partior $\frac{105}{125}$	$\frac{18}{1}$

E per la regola del 3 se 42 me dà 51 che serave 100?

04	
0155	
2250	$21\frac{45}{105}$
1055	
10	

E per far la ditta raxion per la chossa, se 42 me torna 51, che serave 100? So che me die venir 1^{co}. Se moltipicho 1^{co} via 100 fa 100^{co}. A partir insirà $21\frac{45}{105}$.

42 51 100 1^{co}

[Partir 4, 3, roman 1, qual fu el partior]

El è un chi domanda: "In 4 3 roman 1, qual fu el mio partior?" El partior è più qual romagnente. Adoncha se fuse 2, 2 fia 3 fa 6, e un 7, non può eser partido per 4. E se 'l fuse 3, 3 fia 3 9, e un 10, non può partir per 4. Se 'l fuse 4, 3 fia 4 12, e un 13, non se può partir per 4. E se 'l fose 5, 3 fia 5 fa 15, e un val 16, ben se può partir per 4, perché insirà 4. Adoncha seria $\frac{5}{4}$. Adoncha parti $\frac{4}{1}$ per $\frac{5}{4}$, insirà dela gallia 3, roman 1. Adoncha $\frac{5}{4}$ fu el partior.

---

f. 67a E ò chonprado una zoya per duchati 60, e si l'ò venduda per duchati 50 adomando zò che perdo per 100. Per farla per inpoxizion pony che 'l fusse persso 10 per 100. Adoncha diremo: "Se 100 torna 90, che tornerà 60?" Moltipicha e parti, insirà 54, e noi vosemo che 'l fusse 50, adoncha è più 4. E per la segonda pony che fosse 12. Diremo: "Se 100 fuse 88, che serave 60?" Moltipicha e parti, serà $52\frac{4}{5}$, e noi vosemo quel fose 50, adoncha è più $2\frac{4}{5}$, e metti ala segonda. Segui.

10	12	10	12	48	4	$\frac{6}{5}$	$\frac{20}{1}$	0	
più	più	$2\frac{4}{5}$	4	28	$2\frac{4}{5}$			044	
4	$2\frac{4}{5}$	28	48	20	$1\frac{1}{5}$			100	$16\frac{2}{3}$
								66	

E per la riegola del 3, se 60 torna 50, che serave 100?

---

1. This is presumably a reference to the *galera* method of division: see Franci, "Mathematics in the Manuscript of Michael of Rhodes," vol. 3, p. 119.

10		12
minus	X	minus
$4\frac{4}{5}$		$3\frac{24}{25}$

10	12	$57\frac{3}{5}$	$4\frac{4}{5}$
$3\frac{24}{25}$	$4\frac{4}{5}$	$39\frac{3}{5}$	$3\frac{24}{25}$
$39\frac{3}{5}$	$57\frac{3}{5}$	to divide 18	divisor $\frac{105}{125}$ $\frac{18}{1}$

And by the rule of 3, if 42 gives me 51, what will 100 be?

04
01̸5̸5
2̸2̸5̸0̸
1̸05̸5̸
1̸0

$21\frac{45}{105}$

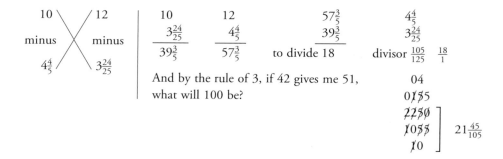

And to do this problem by unknowns, if 42 gives 51, what will 100 be? I know that it should come to 1*x*. If I multiply 1*x* times 100 it makes 100*x*. By dividing, the result will be $21\frac{45}{105}$.

42 51 100 1*x*

Divide 4 and 3 with a remainder of 1, what was the divisor?

There is someone who asks: "In 4 and 3 there remains 1, what was my divisor?" The divisor is more than the remainder. So if it was 2, 2 times 3 makes 6, and 1 is 7; it can't be divided by 4. And if it was 3, 3 times 3 is 9, and one is 10; it can't be divided by 4. If it was 4, 3 times 4 is 12, and 1 is 13; it can't be divided by 4. And if it was 5, 3 times 5 makes 15, and one makes 16; it can indeed be divided by 4, because the result will be 4. So it will be $\frac{5}{4}$. So divide $\frac{4}{1}$ by $\frac{5}{4}$, the result of the galley[1] will be 3 with a remainder of 1. So $\frac{5}{4}$ was the divisor.

And I have bought a jewel for 60 ducats, and if I have sold it for 50 ducats, I ask what percent I lose. To do this by false position, suppose that it was for 10 percent. So we will say: "If 100 returns 90, what will 60 return?" Multiply and divide, the result will be 54, and we wanted it to be 50, so it is plus 4. And for the second suppose that it was 12. We will say: "If 100 was 88, what would 60 be?" Multiply and divide, it will be $52\frac{4}{5}$, and we wanted it to be 50, so it is plus $2\frac{4}{5}$, and put this at the second. Continue. f. 67a

10		12
plus	X	plus
4		$2\frac{4}{5}$

10	12	48	4	$\frac{6}{5}$	$\frac{20}{1}$	0
$2\frac{4}{5}$	4	28	$2\frac{4}{5}$			04̸4
28	48	20	$1\frac{1}{5}$			1̸0̸0̸
						6̸6̸

$16\frac{2}{3}$

And by the rule of 3, if 60 results in 50, what will 100 be?

E per far la ditta raxion per la chossa, se 60 torna 50, che serave 100? Se moltipicho 60 via 100 so che me die venir 1^{co}, adoncha moltipicha 1^{co} via 100, fa 100^{co}. Parti, insirà $16\frac{2}{3}$.

60 50 100 1^{co}

[Moltipichar per radixe]

Per moltipichar per radixe per inpoxizion, più più, men men, sutrazi; più men, azonzi. E chosì el partior, men via più fa men, e più via men fa men, men via men fa più, più via più fa più.

più via	17	men	4
più via	5	men	3
men	20	plus	12
men	51	plus	85

più 12	men 51
più 85	men 20
più 97	men 71

97
71
26
E questo è fatto.

E per un altro muodo più 17 men 4 roman 13; più 5 men 3 roman 2. Moltipicha 2 fia 13, fa 26, e tantto vien a far[a] più via 17 men 4, più 5 men 3. E questo è fatto.

f. 67b [Zoy a vender e perder quanto fu el chavedal]

Ò vendudo 1ª zoya per duchati 25 e ò perso 7 per 100. Quanto fu el mio chavedal? Pony che fuse perso 10. Noi diremo: "Se 25 torna 10, che serave 93 che son mancho 7?" Moltipicha e parti, serano $37\frac{1}{5}$. Noi vosemo che fuse 100, adoncha è men $62\frac{4}{5}$ ala prima inpoxizion. E per la segonda metti che fusse perso 12, e diremo: "Se 25 torna 12, che tornarà 93?" Moltipicha e parti, serano $44\frac{16}{25}$, e noi vosemo che fuse 100. Adoncha men $55\frac{9}{25}$. E chosì metti ala segonda inpoxizion.

---

a. *Evident traces of erased writing between the lines.*

And to do this problem by unknowns, if 60 results in 50, what will 100 be? If I multiply 60 times 100, I know that 1*x* should come to me, then multiply 1*x* times 100, it makes 100*x*. Divide, the result will be $16\frac{2}{3}$.

60 50 100 1*x*

Multiplying by roots

To multiply by roots by false position: plus plus, minus minus, subtract; plus minus, add. And so for the divisor, minus times plus makes minus, and plus times minus makes minus, minus times minus makes plus, plus times plus makes plus.

plus times	17	minus	4	plus 12	minus 51	97
plus times	5	minus	3	plus 85	minus 20	71
minus	20		plus 12	plus 97	minus 71	26
minus	51		plus 85			And it's done.

And by another method, plus 17 minus 4 leaves 13; plus 5 minus 3 leaves 2. Multiply 2 times 13, which makes 26, and this is the same as plus 17 minus 4 times plus 5 minus 3. And it's done.

To sell a jewel and lose, how much was the capital? f. 67b

I have sold 1 jewel for 25 ducats and I have lost 7 percent. What was my capital? Suppose that 10 was lost. We will say: "If 25 results in 10, what will 93 which is 7 short be?" Multiply and divide, it will be $37\frac{1}{5}$. We wanted it to be 100, so it is minus $62\frac{4}{5}$ at the first false position. And for the second suppose that 12 was lost, and we will say: "If 25 results in 12, what will 93 result in?" Multiply and divide, it will be $44\frac{16}{25}$, and we wanted it to be 100. So it is minus $55\frac{9}{25}$. And put this at the second false position.

10 — 12   25   10   93   0   25   12   93

men — men   10   13   12

$62\frac{4}{5}$ — $55\frac{9}{25}$ 930   303   186

930   93

255    $37\frac{1}{5}$ 1116

2

100   01   100

$37\frac{1}{5}$ 13 $44\frac{16}{25}$

men $62\frac{4}{5}$ 0316   men $55\frac{9}{25}$

1116

255    $44\frac{16}{25}$

2

$55\frac{9}{25}$ $553\frac{3}{5}$ $62\frac{4}{5}$ $753\frac{3}{5}$

10   12 $553\frac{3}{5}$

$753\frac{3}{5}$ a parti[*r*] 200

$62\frac{4}{5}$ $7\frac{11}{25}$ [1]

$55\frac{9}{25}$ 7   2

06

126

3484

5000

1866 $26\frac{164}{186}$ $\frac{82}{93}$

$\frac{186}{25}$ $\frac{200}{1}$ 18

E per far la ditta raxion per la riegola del 3. Se 93 eran 100 che serave 25? Seguilla.

E per far la ditta raxion per la chosa, se 93 100 25. So che se moltipicho 25 fia 100 me die venir 1 co. Adoncha moltipicha 25 fia 1 co, fa 25 co. Parti lo moltipicho per 25, insirà $26\frac{82}{93}$.

93   100   25   1 co

10 ╲ ╱ 12   25   10

minus ╳ minus

$62\frac{4}{5}$ ╱ ╲ $55\frac{9}{25}$

93
10
———
930

0
13
303[1]
930
255
2
] $37\frac{1}{5}$

25   12

93
12
———
186
93
———
1116

100
$37\frac{1}{5}$
———
minus $62\frac{4}{5}$

01
13
0316
1116
255
2
] $44\frac{16}{25}$

100
$44\frac{16}{25}$
———
minus $55\frac{9}{25}$

$55\frac{9}{25}$
10
———

$553\frac{3}{5}$

$62\frac{4}{5}$
12
———
$753\frac{3}{5}$

$753\frac{3}{5}$
$553\frac{3}{5}$
———
to divide 200

$62\frac{4}{5}$
$55\frac{9}{25}$
———

$7\frac{11}{25}$
7
———

$\frac{186}{25}$ $\frac{200}{1}$

1[2]
2
06
126
3484
5000
1866
18
] $26\frac{164}{186}$ $\frac{82}{93}$

And to do the problem by the rule of 3. If 93 was 100, what would 25 be? Continue it.

And to do the problem by unknowns, it's 93 100 25. I know that if I multiply 25 by 100, 1*x* should come to me. So multiply 25 by 1*x*, which makes 25*x*. Divide the product by 25, the result will be $26\frac{82}{93}$.

93 100 25 1*x*

---

1. This and the line above seem to be in error.
2. Incomplete in MS.

f. 68a

[Piper a vender e vadagnar quanto fu el chavedal]

Ò vendudo un sacho de piper per duchati 25. Ò vadagnado 7 per 100. Adomando quanto fu el mio chavedal. E per far la dita raxion per inpoxizion pony che vadagnase 10. Adoncha noi diremo: "Se 25 eran 10 del'inpoxizion, che serave 107?" Moltipicha e parti, insirà $42\frac{4}{5}$, e noi vosemo che fusse 100. Adoncha è men $57\frac{1}{5}$, e questo metti ala prima inpoxizion. E per far la segonda noi diremo che vadagnase 20. Se 25 torna 20, che serave 107? Moltipicha e parti, insirà $85\frac{3}{5}$, e noi vosemo che fuse 100. Adoncha è men $14\frac{2}{5}$. Moltipicha e parti, insirà $23\frac{39}{107}$, chomo vedirì qui de sotto per figora.

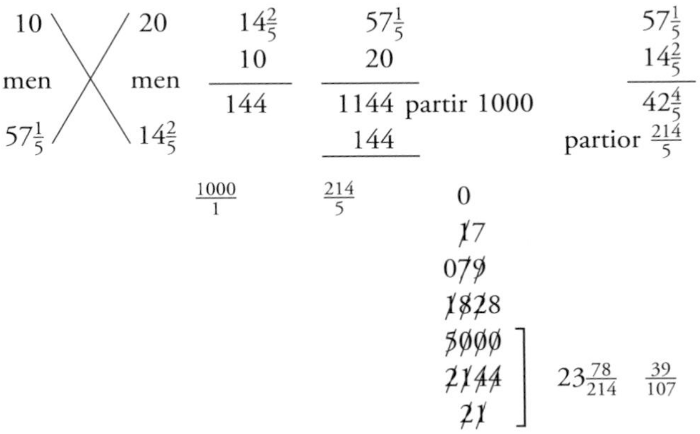

E per farla per la riegola del 3, se 107 eran 100, ch'è 25?

E per far la ditta raxion per la chosa, se 107, 100, 25. Tu sa che se moltipicho 25 via 100 die far 1^{co}. Adoncha moltipicha 1^{co} via 107^{n}, fa 107^{co}. A partir el moltipicho insirà $23\frac{39}{107}$. 107^{n} 100^{n} 25^{n} 1^{co} 107^{co} 5000.

---

f. 68b

[Piper in Alesandria vadagnar e perder, chondur in Venyexia]

El piper in Alesandria per quanto chonprerò che chondutto in Veniexia, là che val duchati 75 el chargo, vadagna 35 per 100? E per far la ditta raxion per inpoxizion pony che fusse chonprado per 50, e noi diremo: "Se 75 torna 50, che serave 135?" Moltipicha 50 via 135 e parti, insirà 90, e noi vosemo che fuse 100. Adoncha serà men 10, e questi metti ala prima inpoxizion. E per far la segonda noi diremo che fuse chonprado per 40, e sì diremo: "Se 75 era 40, che serave 135?" Moltipicha 40 via 135; moltipicha e parti, insirà 72, e noi vosemo che fusse 100. Adoncha serà men 28, e chosì metermo ala segonda inpoxizion. Moltipicha e parti chomo avè fatto dele oltre, insirà $55\frac{5}{9}$.

PEPPER TO SELL AND PROFIT, HOW MUCH WAS THE CAPITAL?

f. 68a

I have sold a bag of pepper for 25 ducats. I have made a profit of 7 percent. I ask how much my capital was. And to do this problem by false position, suppose that 10 was gained. Then we will say: "If 25 was 10 of the false position, what would 107 be?" Multiply and divide, the result will be $42\frac{4}{5}$, and we wanted it to be 100. So it is minus $57\frac{1}{5}$, and put this at the first false position. And to do the second, we will say that 20 was gained. If 25 results in 20, what would 107 be? Multiply and divide, the result will be $85\frac{3}{5}$, and we wanted it to be 100. So it is minus $14\frac{2}{5}$. Multiply and divide, the result will be $23\frac{39}{107}$, as you will see worked out below.

10		20	$14\frac{2}{5}$	$57\frac{1}{5}$	$57\frac{1}{5}$
minus	X	minus	10	20	$14\frac{2}{5}$
$57\frac{1}{5}$		$14\frac{2}{5}$	144	1144 to divide 1000	$42\frac{4}{5}$
				144	divisor $\frac{214}{5}$

$\frac{1000}{1}$ $\frac{214}{5}$

0
17
079
1828
5000
2144] $23\frac{78}{214}$ $\frac{39}{107}$
21

And to do it by the rule of 3, if 107 was 100, what is 25?

And to do this problem by unknowns, it is 107, 100, 25. You know that if I multiply 25 by 100 it should be $1x$. So multiply $1x$ by 107, which makes $107x$. To divide the product will result in $23\frac{39}{107}$.
107 100 25 $1x$ $107x$ 5000.

PROFIT AND LOSS ON PEPPER IN ALEXANDRIA, TO SHIP TO VENICE

f. 68b

For how much will I buy pepper in Alexandria so that brought to Venice, where it is worth 75 ducats per load, it will make a profit of 35 percent? And to do this problem by false position, suppose that it was bought for 50, and we will say: "If 75 results in 50, what would 135 be?" Multiply 50 times 135 and divide, the result will be 90, and we wanted it to be 100. So it will be minus 10, and put this at the first false position. And to do the second we will say that it was bought for 40, and so we will say: "If 75 was 40, what would 135 be?" Multiply 40 by 135; multiply and divide, the result will be 72, and we wanted it to be 100. So it will be minus 28, and we will put this at the second false position. Multiply and divide as you have done for the others, the result will be $55\frac{5}{9}$.

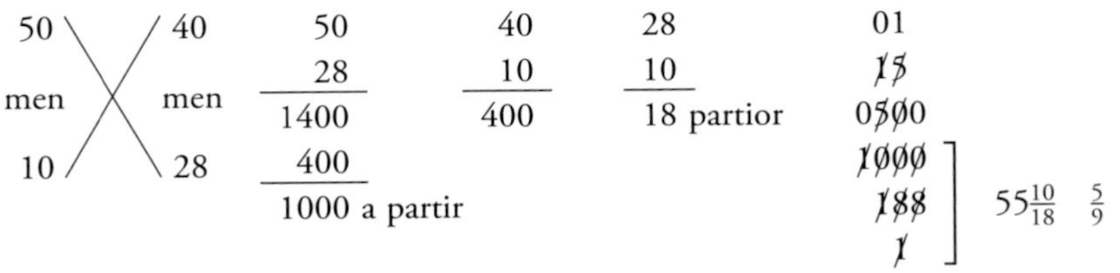

E per farla per la riegola del 3, se 135, 100, 75, insirà $55\frac{5}{9}$.

E per far la ditta raxion per la chosa, se 135, 100, 75. Tu sa che se moltipicho 75 via 100 me die venir 1^{co}. Adoncha moltipicha 1^{co} via 135^{n} fa 135^{co}. A partir lo moltipicho die venir $55\frac{5}{9}$.

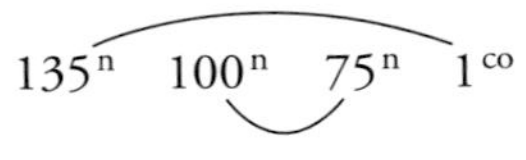

---

f. 69a El piper in Allesandria per quanto chonprerò che chondutto in Veniexia, là che val el chargo duchati 75, perda 35 per 100? E per far la ditta per inpoxizion pony che gostase 80, e noi diremo: "Se 75 torna 80, che serave 65?" Moltipicha e parti, insirà $69\frac{1}{3}$, e noi vosemo che fuse 100, adoncha men $30\frac{2}{3}$. E chosì meti ala prima inpoxizion. E per far la segonda ponamo che valese duchati 90, e noi diremo: "Se 75 torna 90, che serave 65?" Moltipicha e parti, insirà 78, e noi vosemo che fuse 100. Adoncha serà men 22, e chosì metti ala segonda inpoxizion. Moltipicha e parti, insirà duchati $115\frac{5}{13}$.

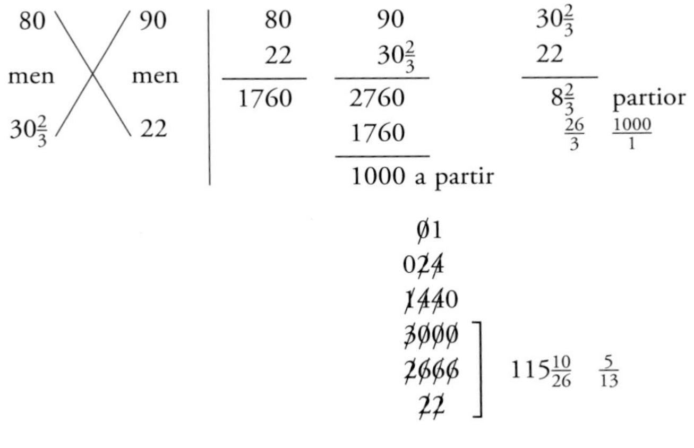

E per la riegolla del 3, se 65 torna 100, che serà 75?

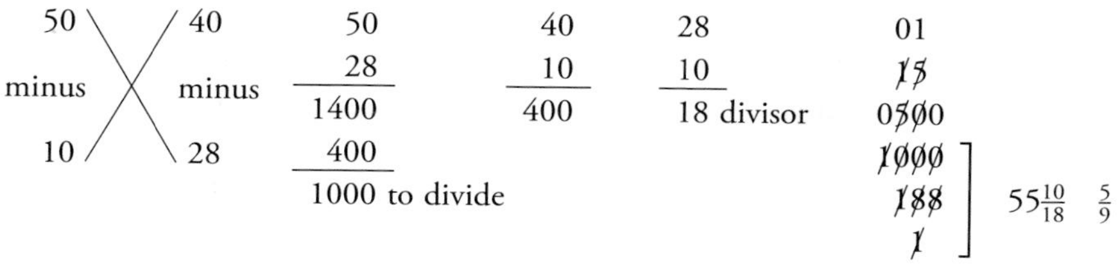

And to do it by the rule of 3, it is 135, 100, 75, the result will be $55\frac{5}{9}$.

And to do this problem by unknowns, it is 135, 100, 75. You know that if I multiply 75 by 100 I should get 1*x*. So multiply 1*x* by 135, it makes 135*x*. Dividing what's multiplied should result in $55\frac{5}{9}$.

For how much will I buy pepper in Alexandria so that brought to Venice, where it is worth 75 ducats per load, it will lose 35 percent? And to do this problem by false position, suppose that it cost 80, and we will say: "If 75 results in 80, what would 65 be?" Multiply and divide, the result will be $69\frac{1}{3}$, and we wanted it to be 100, so it is minus $30\frac{2}{3}$. And put this at the first false position. And to do the second, let's suppose that it was worth 90 ducats, and we will say: "If 75 results in 90, what would 65 be?" Multiply and divide, the result will be 78, and we wanted it to be 100. So it will be minus 22, and put this at the second false position. Multiply and divide, the result will be $115\frac{5}{13}$ ducats. f. 69a

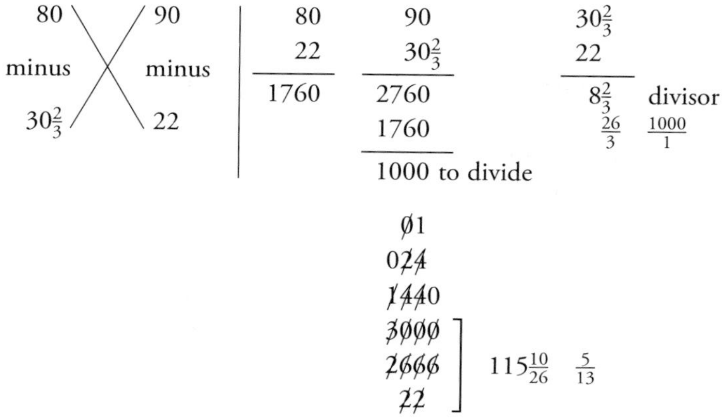

And by the rule of 3, if 65 results in 100, what will 75 be?

E per la chosa, se 65 torna 100, che serave 75? Tu sa che se moltipicho 75 via 100 me die venir 1^{co}, adoncha moltipicha 1^{co} via 65^{n}, fa 65^{co}. A partir el moltipicho insirà duchati $115\frac{5}{13}$. Tanto vuol chonprar in Allesandria per perder in Venyexia.

---

f. 69b [Chonprar braza 5 de fustagno per duchato e dar la mitade braza 4 per duchato e l'oltra mittade braza 7 per 2 duchati e vadagnar 1 duchato, quanti fu le braza e quanto el chavedal]

Uno homo chonpra braza 5 de fustagno per duchato, e puo' vende la mitade del ditto fustagno che lo à chonprado braza 4 per duchato e l'oltra mitade vende braza 7 per 2 duchati, e quando l'avè vendudo se truova aver vadagnado duchato 1. Adoncha quante braza ne chonpra e quanto fu el chavedal? Ponamo che 'l chonprase braza 56 a 5 braza per duchato, serave duchati $11\frac{1}{5}$, e questi metti de parte per chavedal. E puo' partiremo 56 in 2 parte, serà per parte 28, e chomenzaremo a vender 4 braza per duchato; chonvien che da braza 28 aver duchati 7. Puo' diremo per l'altra mitade, che son braza 28, a vender 7 braza per duchati 2, adoncha per braza 28 noi averemo duchati 8. E sì faremo 7 e 8 fa 15 duchati, e noi volemo che sia tra vadagno e chavedal $12\frac{1}{5}$. Adoncha sutratto $12\frac{1}{5}$ de 15 roman più $2\frac{4}{5}$, e chosì metemo ala prima inpoxizion. E per far la segonda inpoxizion pony che fuse braza 48. Per braza 5 per duchato seria duchati $9\frac{3}{5}$, e questi salva per chavedal. Mo' partimo 48 per mitade, serave 24 per parte, e venderemo 1ª parte 4 braza per duchato, noi averemo duchati 6. Puo' venderemo braza 7 per duchati 2, noi averemo duchati $6\frac{6}{7}$. E questi duchati asumadi fano duchati $12\frac{6}{7}$, e noi vosemo che fuse tra vadagno e chavedal duchati $10\frac{3}{5}$. Adoncha, sutratti de $12\frac{6}{7}$ $10\frac{3}{5}$, resta più $2\frac{9}{35}$, e questi metti ala segonda inpoxizion. Puo' moltipicha in croxe 56 via $2\frac{9}{35}$, fano $126\frac{2}{5}$, e 48 via $2\frac{4}{5}$, fa $134\frac{2}{5}$; e de questi, sutratto la menor dela mazuor, roman 8, e questo se die partir. E per far partior, sutratto de $2\frac{4}{5}$ $2\frac{9}{35}$, roman $\frac{95}{175}$, e questo è partior.

f. 70a Adoncha partimo 8 per $\frac{95}{175}$. Devemo moltipichar 8 fia 175, fano 1400, e questi partidi per 95, insirà braza $14\frac{70}{95}$, chomo vedirì qui de sotto per figora deponer la inpoxizion.

And by unknowns, if 65 results in 100, what would 75 be? You know that if I multiply 75 by 100, I should get 1*x*, so multiply 1*x* by 65, it makes 65*x*. Dividing what's multiplied will result in $115\frac{5}{13}$ ducats. That's how much you should pay in Alexandria to lose in Venice.

BUYING FUSTIAN AT 5 ELLS PER DUCAT AND SELLING HALF AT 4 ELLS PER DUCAT AND THE OTHER HALF AT 7 ELLS PER 2 DUCATS AND PROFITING 1 DUCAT, HOW MANY ELLS WERE THERE AND WHAT WAS THE CAPITAL? f. 69b

A man buys 5 ells of fustian[1] per ducat, and then sells half of this fustian that he has bought at 4 ells per ducat and he sells the other half at 7 ells for 2 ducats, and when he has sold it he finds that he has made a profit of 1 ducat. So how many ells did he buy and how much was the capital? Let's suppose that he bought 56 ells at 5 ells per ducat, it would be $11\frac{1}{5}$ ducats, and put this in the place for capital. And then we will divide 56 in 2 parts, which will be 28 per part, and we will begin to sell 4 ells per ducat, it works out from 28 ells to have 7 ducats. Then we will say for the other half that it is 28 ells to sell 7 ells for 2 ducats, so for 28 ells we will have 8 ducats. And so we will have 7 and 8, which makes 15 ducats, and we wanted it to be $12\frac{1}{5}$ including profit and capital. So $12\frac{1}{5}$ subtracted from 15 leaves plus $2\frac{4}{5}$, and put this at the first false position. And to do the second false position, suppose that it was 48 ells. At 5 ducats per ell it would be $9\frac{3}{5}$ ducats, and save this for the capital. Now we divide 48 in half, which will be 24 per half, and we will sell 1 part at 4 ells per ducat; we will have 6 ducats. Then we will sell 7 ells for 2 ducats; we will have $6\frac{6}{7}$ ducats. And these ducats added together make $12\frac{6}{7}$ ducats, and we wanted there to be $10\frac{3}{5}$ ducats between profit and capital. So, subtract $10\frac{3}{5}$ from $12\frac{6}{7}$, which leaves plus $2\frac{9}{35}$, and put this at the second false position. Then multiply across 56 times $2\frac{9}{35}$, which makes $126\frac{2}{5}$, and 48 by $2\frac{4}{5}$, which makes $134\frac{2}{5}$; and of this subtract the lesser from the greater, which leaves 8, and this should be divided. And to make the divisor, subtract $2\frac{9}{35}$ from $2\frac{4}{5}$, which leaves $\frac{95}{175}$, and this is the divisor.

So we divide 8 by $\frac{95}{175}$. We should multiply 8 times 175, which makes 1,400, and this divided by 95 will result in $14\frac{70}{95}$ ells, as you will see worked out below showing the false position. f. 70a

---

1. Fustian is a fabric containing cotton and linen.

56 ╲╱ 48
più ╳ più
$2\frac{4}{5}$ ╱╲ $2\frac{9}{35}$

28 4 per duchato serà duchati 7
28 7 per duchato serà duchati 8 ⟩ 15 $\frac{56}{55}$] $11\frac{1}{5}$
15
$12\frac{1}{5}$
più $2\frac{4}{5}$

chavedal $11\frac{1}{5}$
vadagno 1
serano $12\frac{1}{5}$

$12\frac{6}{7}$
$10\frac{3}{5}$
più $2\frac{9}{35}$

24 4 per duchato serà duchati 6
24 7 per 2 duchati serà duchati $6\frac{6}{7}$ ⟩ $12\frac{6}{7}$

03
~~48~~
~~5~~] $9\frac{3}{5}$ vadagno 1
fano $10\frac{3}{5}$

56
$2\frac{9}{35}$
$126\frac{2}{5}$

48
$2\frac{4}{5}$
$134\frac{2}{5}$

$134\frac{2}{5}$
$126\frac{2}{5}$
8 a partir

$2\frac{4}{5}$
$2\frac{9}{35}$
$\frac{95}{175}$ partior

$\frac{95}{175}$ $\frac{8}{1}$
8

07
~~49~~
0~~55~~0
~~1400~~
~~955~~
~~9~~] $14\frac{70}{95}$

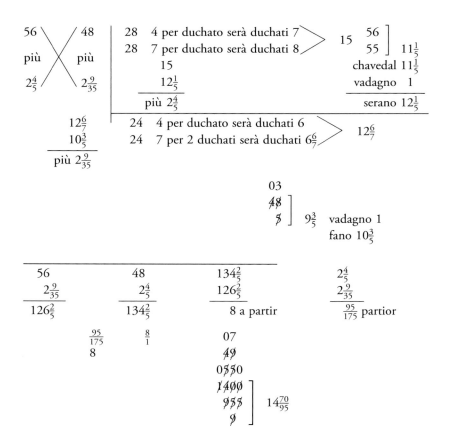

E per far la ditta raxion per la chosa pony che fusse chonprado 1^{co} de fostagno. Adoncha si se chonpra 5 per duchato, doverave eser $\frac{1}{5}^{co}$, e questo salva per chavedal. Puo' dixe voler vender la mitade per braza 4 per duchato. Adoncha parti 1^{co} per mitade, seria $\frac{1}{2}^{co}$. E se braza 4 me dà un duchato, che me darà $\frac{1}{2}^{co}$? Me darà $\frac{1}{8}^{co}$. Puo' per l'altra mitade diremo: "Se 7 braza me dà duchati 2, che me darà $\frac{1}{2}^{co}$?" Me darà $\frac{1}{7}^{co}$, e questo azontto fano $\frac{15}{56}^{co}$, e questo sotratto de $\frac{1}{5}^{co}$ serano $\frac{19}{280}^{co}$. E
f. 70b diremo: "Se $\frac{19}{280}^{co}$ me dà $\frac{1}{1}^{n}$, che | e[a] questo fatto in croxe serà partior 19^{co}, a partir 280^{n}, insirà da moltipichar e partir braza $14\frac{14}{19}$, chomo vedirì qui de sotto per figora.

a. *This page is not continuous with the preceding, facing, one.*

56 plus $2\frac{4}{5}$ ╳ 48 plus $2\frac{9}{35}$

28 4 per ducat will be 7 ducats
28 7 per ducat will be 8 ducats
15
$12\frac{1}{5}$
plus $2\frac{4}{5}$

15 56
55] $11\frac{1}{5}$
capital $11\frac{1}{5}$
profit 1
will be $12\frac{1}{5}$

$12\frac{6}{7}$
$10\frac{3}{5}$
plus $2\frac{9}{35}$

24 4 per ducat will be 6 ducats
24 7 per 2 ducats will be $6\frac{6}{7}$ ducats
$12\frac{6}{7}$

03
48
5] $9\frac{3}{5}$ profit 1
makes $10\frac{3}{5}$

56
$2\frac{9}{35}$
$126\frac{2}{5}$

48
$2\frac{4}{5}$
$134\frac{2}{5}$

$134\frac{2}{5}$
$126\frac{2}{5}$
8 to divide

$2\frac{4}{5}$
$2\frac{9}{35}$
$\frac{95}{175}$ divisor

$\frac{95}{175}$ $\frac{8}{1}$
8

07
49
0550
1400
955
9] $14\frac{70}{95}$

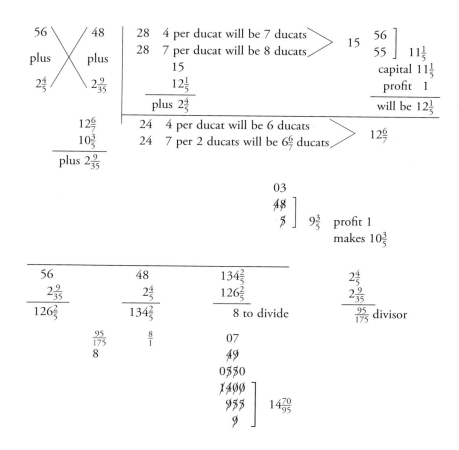

And to do this problem by unknowns, put that $1x$ of fustian was bought. So if it is bought at 5 per ducat, it would be $\frac{1}{5}x$, and save this for the capital. Then it says that he wants to sell half for 4 ells per ducat. So divide $1x$ in half, which would be $\frac{1}{2}x$. And if 4 ells give me 1 ducat, what will $\frac{1}{2}x$ give me? It would give me $\frac{1}{8}x$. Then for the other half we will say: "If 7 ells gives me 2 ducats, what will $\frac{1}{2}x$ give me?" It will give me $\frac{1}{7}x$, and this added together makes $\frac{15}{56}x$, and this subtracted from $\frac{1}{5}x$ will be $\frac{19}{280}x$. And we will say: "If $\frac{19}{280}x$ gives me $\frac{1}{1}$, what | and[1] this done across, the divisor will be $19x$, to divide into 280, the result of multiplying and dividing will be $14\frac{14}{19}$ ells, as you will see worked out below. f. 70b

1. This text is not continuous with that of the facing page, but it seems to continue the same calculations.

1^{co} $\frac{1}{5}^{co}$ de duchato $\frac{1}{2}^{co}$ $\frac{4}{1}$ $\frac{1}{1}$ $\frac{1}{2}^{co}$ $\frac{1}{8}^{co}$

$\frac{7}{1}$ $\frac{2}{1}$ $\frac{1}{2}^{co}$ $\frac{2}{14}$ $\frac{1}{7}^{co}$ | $\frac{1}{7}^{co}$ $\frac{1}{8}^{co}$ — $\frac{15}{56}^{co}$ | $\frac{15}{56}^{co}$ $\frac{1}{5}^{co}$

75
56
—
19 $\frac{1}{1}$ $\frac{1}{1}$
—
280

1
0~~5~~
~~1~~~~9~~4
~~2~~~~8~~~~0~~
~~1~~~~9~~~~9~~
~~1~~] $14\frac{14}{19}$ de brazo

E per far la prova dela ditta raxion noi diremo $14\frac{14}{19}$ zò fu vuove. A chonprarlli 5 a soldo venirà de chavedal soldi $2\frac{90}{95}$, e vender la mittade 4 a soldo. La mitade de $\frac{280}{19}$ serano $\frac{140}{19}$, e questi dadi 4 per soldo insirà soldi $1\frac{16}{19}$. E l'altra mitade, dadi 7 per 2 soldi, insirà soldi $2\frac{14}{133}$.[a] Mo' adoncha azonzi $1\frac{16}{19}$ con 2 e $\frac{14}{133}$, insirà $3\frac{2394}{2527}$. E questi sun ingual a $\frac{90}{95}$. Adoncha, sutratto de $3\frac{2394}{2527}$, roman 1. E questo fu el vadagno, chomo dixe de sovra.

$14\frac{14}{19}$ — $\frac{280}{19}$ la mità $7\frac{7}{19}$ — $\frac{140}{19}$ $\frac{4}{1}$ $\frac{1}{1}$ $\frac{140}{19}$

6
0~~7~~4
~~1~~~~4~~~~0~~
~~7~~~~6~~] $1\frac{64}{76}$ primo

$\frac{7}{1}$ $\frac{2}{1}$ $\frac{140}{19}$

1
0~~2~~4
~~2~~~~8~~~~0~~
~~1~~~~3~~~~3~~] $2\frac{14}{133}$ segondo

$1\frac{16}{19}$
$1\frac{14}{133}$

$2\frac{90}{95}$
$3\frac{2394}{2527}$ $\frac{90}{95}$
90
—
0000
22743
—
227430

2394
95
—
11970
21546
—
227430
227430
—
| 1 | 0 E questo è fatto per raxion.

$3\frac{2394}{2527}$
$2\frac{90}{95}$
—
resta 1 de vadagno

---

a. $\frac{14}{133}$ *in place of* $\frac{90}{95}$ *crossed out with diagonal lines.*

$1x$ $\quad \frac{1}{5}x$ of a ducat $\quad \frac{1}{2}x \quad \frac{4}{1} \quad \frac{1}{1} \quad \frac{1}{2}x \quad \frac{1}{8}x$

$\frac{7}{1} \quad \frac{2}{1} \quad \frac{1}{2}x \quad \frac{2}{14} \quad \frac{1}{7}x$ | $\frac{1}{7}x \ \frac{1}{8}x$ / $\frac{15}{56}x$ $\quad \frac{15}{56}x \quad \frac{1}{5}x$

75
56
———
19 $\quad \frac{1}{1} \quad \frac{1}{1}$
———
280

1
05
194
280
199 $\quad 14\frac{14}{19}$ ells
1

And to do the proof of this problem, we will say that it was $14\frac{14}{19}$ eggs.[1] To buy 5 for a soldo the capital will come to $2\frac{90}{95}$ soldi, and to sell half for 4 per soldo. Half of $\frac{280}{19}$ will be $\frac{140}{19}$, and this given at 4 per soldo will result in $1\frac{16}{19}$ soldi. And the other half, given at 7 for 2 soldi, will result in $2\frac{14}{133}$ soldi. So now add $1\frac{16}{19}$ to 2 and $\frac{14}{133}$; the result will be $3\frac{2394}{2527}$. And this is equal to $\frac{90}{95}$. So subtracted from $3\frac{2394}{2527}$, there remains 1. And that was the profit, as was said above.

$14\frac{14}{19}$ / $\frac{280}{19}$ $\quad$ half $\quad 7\frac{7}{19}$ / $\frac{140}{19}$ $\quad \frac{4}{1} \quad \frac{1}{1} \quad \frac{140}{19}$

6
074
140
76 $\quad 1\frac{64}{76}$ first

$\frac{7}{1} \quad \frac{2}{1} \quad \frac{140}{19}$

1
024
280
133 $\quad 2\frac{14}{133}$ second

$1\frac{16}{19}$
$1\frac{14}{133}$

$2\frac{90}{95}$
$3\frac{2394}{2527} \quad \frac{90}{95}$
90
———
0000
22743
———
227430

2394
95
———
11970
21546
———
227430
227430
| 1 | 0 $\quad$ And this was done in the problem.

$3\frac{2394}{2527}$
$2\frac{90}{95}$
———
leaves 1 of profit

1. The problem began with ells of fustian and prices in ducats; it has become eggs and prices in soldi here.

f. 71a Uno homo à vuove in 2 zeste, tante in l'uno quante in l'altro, e vende l'un zesto soldi 5 e l'oltro vende soldi 9, e dele 7 men a soldo che non se de' el primo zesto. Adomando quante fu le vuove e quanti ne de' a soldo.

E per far la ditta raxion per inpoxizion pony che fuse dadi 14 a soldo, e noi diremo 5 via 14 fa 70. Mo' el segondo vuol men 7. Adoncha 9 via 7 fa 63, e noi vosemo che fose 70, adoncha è men 7, per 14 chom'ò poxo men 7. E per far la segonda inpoxizion pony che fuse 15, moltipicha 5 via 15 fa 75, men 7 fu 8, 8 via 9 fa 72, e noi vosemo che fuse 75, adoncha è men 3. Segui l'inpoxizion, moltipicha e parti, insirà vuove $15\frac{3}{4}$ per soldo, chomo vedirì qui de sutto per figora. E la soma dele vuove serano per zesto vuove $78\frac{3}{4}$.

14 ╲ ╱ 15
men ╳ men
7 ╱ ╲ 3

14 · 5 · 70 · 63 · men 7 | 7 · 9 · 63

15 · 5 · 75 · 72 · men 3 | 8 · 9 · 72

14 · 3 · 42 | 15 · 7 · 105 · 42 · a partir 63

7
3
4 partior

0
~~7~~3
~~63~~
~~44~~] $15\frac{3}{4}$
5 5
le vuove $78\frac{3}{4}$ suma

$15\frac{3}{4}$
9 9
8
72
6

$78\frac{3}{4}$ suma

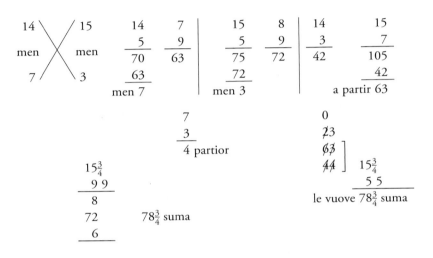

E se la volesti far per la chosa pony che fuse 1^{co} a soldo, seria 5^{co} più 35^{n}.

~~1~~
5^{co}
9^{co}

4^{co} più 35^{n}

03
~~35~~]
~~4~~] $8\frac{3}{4}$
9 9
tanti fo $78\frac{3}{4}$ per zesto

8
$7\frac{3}{4}$
$15\frac{3}{4}$
5
tanti fu $78\frac{3}{4}$ al'oltro zesto

f. 71b

Se algun volese chognoser el baratar, qual è avantazado e qual el chontrario, vedirì qui de sotto. Sono do chi vuol baratar, lo primo quel che val duchati 8 mete 10, l'oltro quel che val 20 vuol 24. Chomo se chognose, vui chonzerè al muodo qui de sotto. Moltipicha in croxe 8 fia 24 fa 192

A man has eggs in 2 baskets, as many in one as in the other, and he sells one basket for 5 soldi and sells the other for 9 soldi, and gives them for 7 less per soldo than he gives those in the first basket. I ask how many eggs there were and how much they were per soldo. f. 71a

And to do the problem by false position, suppose that 14 were given per soldo, and we will say 5 times 14 makes 70. Now the second is worth 7 less. So 9 times 7 makes 63, and we wanted it to be 70, so it's minus 7; for the 14 that I've put it's minus 7. And to do the second false position, suppose that it was 15, multiply 5 times 15 makes 75, minus 7 was 8, 8 times 9 makes 72, and we wanted it to be 75, so it is minus 3. Continue the false position, multiply and divide, the result will be $15\frac{3}{4}$ eggs per soldo, as you will see worked out below. And the sum of the eggs will be $78\frac{3}{4}$ eggs per basket.

14	╲ ╱	15	14	7	15	8	14	15
minus	╳	minus	5	9	5	9	3	7
7	╱ ╲	3	70	63	75	72	42	105
			63		72			42
			minus 7		minus 3			to divide 63

$15\frac{3}{4}$	7	0
9 9	3	13
8	4 divisor	63] $15\frac{3}{4}$
72	$78\frac{3}{4}$ sum	44] 5 5
6		the eggs $78\frac{3}{4}$ sum

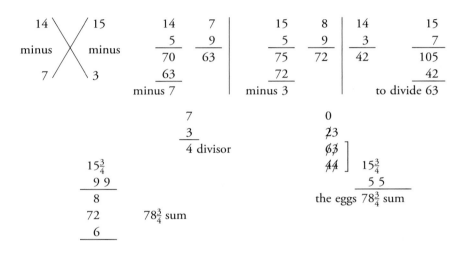

And if you wanted to do it by unknowns, put that it was $1x$ per soldo, it would be $5x$ plus 35.

1	$4x$ plus 35	03	8
$5x$		35]	$7\frac{3}{4}$
$9x$		4] $8\frac{3}{4}$	$15\frac{3}{4}$
		9 9	5
		it was this much $78\frac{3}{4}$ per basket	it was this much $78\frac{3}{4}$ in the other basket

f. 71b

If anyone wants to know about barter, what is profit and what is the contrary, you will see below. There are two who want to barter, the first puts 10 ducats for what is worth 8, the other asks 24 for what is worth 20. You will know below how it is known. Multiply across 8 times 24, which makes

e 10 via 20 fa 200. A sutrar 192 de 200 romagnerà 8, che è 'l meio quel che de 8 fa 10. Adoncha abatti 8 de 10, roman 2. Parti 8 per 2, serà 4. E puo' dirai: "Qui è desavantazado quel che de 20 fa 24." Parti 4 per 24, serà $\frac{1}{6}$. Adoncha voravello in chontadi $\frac{1}{6}$ per non aver el botto, chomo vedirì qui de sutto per figora.

8	X	10	200	200	10	0		$\frac{4}{24}$	$\frac{1}{6}$
20		24	192	192	8	~~8~~			
				8	2	~~2~~	4		

E per veder la verittà trazemo el sesto de 24, roma[*n*] 20,[a] e trazi 4 de 20, roman 16. Vui vedirì a moltipichar in croxe tanto l'una parte chomo l'oltra.

20	24	16	X	20	160	
4	4	8		10	160	E questo è baratto chomun.
16	20					

Sono due chi vuol baratar l'uno a pany e l'oltro a seda. Quel del pano, quel che val a chontadi 18 ne vuol a baratto 24, e vuol $\frac{1}{4}$ di chontadi. E quel dela seda val a chontadi la livra 24. Adomando zò che la die meter a baratto, voyando $\frac{1}{3}$ in denari. Vui vedirì qui de sotto per figora.

21	27	24	01		18	24
		27	~~6~~~~4~~8		6	6
		648	~~2~~~~1~~1	$30\frac{6}{7}$	12	18
			2		9	9
					21	27

Tantto vuol meter la seda per defenderse dal botto del pano.

f. 72a Do homeny baratta l'uno a pany e l'oltro a seda. Dize cholui del pano: "5 braza del mio pano val duchati 15." Dize cholui dela seda: "8 livre dela mia seda val duchati 16." Adomando: "Per livre 15 de seda quanti braza de pano averò?" E questo se chiama svariado.

braza	5	X	15	80	braza de pano
livre	8		16	120	livre de seda

a. *Corrected over* 18 *crossed out with a horizontal line.*

192, and 10 times 20, which makes 200. Subtracting 192 from 200 leaves 8, which is the better of what 8 from 10 makes. So subtract 8 from 10, it leaves 2. Divide 8 by 2, it will be 4. And then you will say: "Here is the deficit that 24 makes from 20." Divide 4 by 24; it will be $\frac{1}{6}$. So, there will have to be $\frac{1}{6}$ in cash to not come to blows, as you will see worked out below.

8	×	10	200	200	10	0		$\frac{4}{24}$	$\frac{1}{6}$
20		24	192	192	8	$\not{8}$			
				8	2	$\not{2}$	4		

And to see the truth, we subtract a sixth from 24, which leaves 20, and subtract 4 from 20, which leaves 16. You will see it multiplied across, one side as much as the other.

20	24	16	×	20	160	
4	4	8		10	160	And this is common barter.
16	20					

There are two who want to barter, one has cloth, the other has silk. The one who has cloth, for that which is worth 18 in cash he wants 24 in barter, and he wants $\frac{1}{4}$ in cash. And the one with the silk is worth 24 in cash per pound. I ask what he should put in barter, wanting $\frac{1}{3}$ in coins. You will see it worked out below.

21	27	24	01		18	24
		27	$\not{6}\not{4}8$		6	6
		648	$\not{2}\not{1}1$	$30\frac{6}{7}$	12	18
			2		9	9
					21	27

This is the amount of silk to put in to avoid taking a hit[1] from the cloth.

Two men barter; one has cloth and the other has silk. The one with cloth says: "5 ells of my cloth are worth 15 ducats." The one with silk says: "8 pounds of my silk are worth 16 ducats." I ask: "For 15 pounds of silk, how many ells of cloth will I have?" And this is called diversified barter. f. 72a

ell	5	×	15	80	ell of cloth
pound	8		16	120	pound of silk

1. The Venetian idiom has the same meaning as the equivalent in English—i.e., to lose money.

120 livre de seda me dà 80 braza de pano, che me darà livre 15 de seda? Moltipicha e parti, insirà braza 10 de pano. E se vuolesti far de braza 10, quanta seda? Se 80 me torna 120, che serave 10? Insirà livre 15.

Sono 2 chi vuol baratar, l'uno a pany e l'oltro a seda. Dize choluy del pano: "El mio pano val 6, e a baratto dela vostra seda voio 8 e voio $\frac{1}{4}$ di chontadi, e sì voio baratar 10 braza de pano." Dize cholui dela seda: "La livra dela mia seda val 10." Voio saver zò che die meter in baratto, voiando $\frac{1}{5}$ in denari.

6	8	4	6	$\frac{11}{2}$	$\frac{15}{2}$	$\frac{10}{1}$	
2	2	$5\frac{1}{2}$	$1\frac{1}{2}$				
4	6		$7\frac{1}{2}$		22 partior		30
							10
							300

1
0$\not{2}$
$\not{1}\not{8}$4
$\not{3}\not{0}\not{0}$
$\not{2}\not{2}\not{2}$] $13\frac{14}{22}$ $\frac{7}{11}$
$\not{2}$

Per aprovarla, 8 fia 10 val 80, abatti $\frac{1}{4}$ roman 60, azonti el quarto serano 75, e puo' moltipicha. Se $13\frac{7}{11}$ fuse livra 1, ch'è 75? Moltipicha e parti, insirà livra $5\frac{5}{10}$. E per lo simille fa tutto le oltre.

---

f. 72b

[VEDER PERCHÉ SE SEDEMEZA LE CHOSSE E PARTIR PER ZENSSI]

Volemo noi dimostrar per che raxion dimeza le chosse. Nela riegola d'alzebran sichomo noi diremo zenso e chossa ingual al nomero, el qual sie nel quarto chapittollo dove dize: "A partir tutta la edechazion per li zensi e poi dimezar le chosse e moltipichar in si medeximo, e quelo che ne vien azonzi sopra el nomero, e la radixe dela suma men l'oltra mittade dele chosse vien a valer la chosa." E per mostrar la ditta la chonvien seguir per questo chotal muodo. Sichomo noi diremo $1^{\square}$ 10^{co} sono ingual a 39^{n}, e questo principio del demostramento serà chossì chomo serà un zenso, el qual serà desegnatto □, el qual zenso non sapiamo quanto el sia. Or per saver quanto è, noi dimezemo le 10^{co}, e serano 5^{co}, lo qual noi azonzeremo al ditto zensso in questo muodo:

120 pounds of silk gives me 80 ells of cloth, what will 15 pounds of silk give me? Multiply and divide, the result will be 10 ells of cloth. And if you wanted to do 10 ells, how much silk would it take? If 80 gives me 120, what will 10 be? The result will be 15 pounds.

There are 2 who want to barter, one has cloth and the other has silk. The one with cloth says: "My cloth is worth 6, and in trade I want 8 of your silk and I want $\frac{1}{4}$ in cash, and also I want to barter 10 ells of cloth." The one with silk says: "A pound of my silk is worth 10." I want to know what should be put into trade, wanting $\frac{1}{4}$[1] in coins.

6	8	4	6	$\frac{11}{2}$	$\frac{15}{2}$	$\frac{10}{1}$	
2	2	$5\frac{1}{2}$	$1\frac{1}{2}$				
4	6		$7\frac{1}{2}$	22 divisor			30
							10
							300

1
0~~2~~
~~18~~4
~~300~~
~~222~~
~~2~~ | $13\frac{14}{22}$ $\frac{7}{11}$

To prove it, 8 times 10 is worth 80, subtracting $\frac{1}{4}$ leaves 60, adding a quarter will be 75, and then you multiply. If $13\frac{7}{11}$ was 1 pound, what is 75? Multiply and divide, the result will be $5\frac{5}{10}$ pounds. And do all the others the same way.

To see how unknowns are halved and to divide by squared unknowns f. 72b

We want to show by what process unknowns are halved. In the rule of algebra, as we will explain, a squared unknown and an unknown are equal to a number, which is in the fourth chapter where it says: "Divide the whole equation by the squared unknowns and then halve the unknowns and multiply by themselves, and add what results to the number, and the root of the sum minus the other half of the unknowns is going to be equal to the unknown." And to show this, it will be useful to proceed in this way. So we will say $1x^2$ and $10x$ are equal to 39, and this beginning of the demonstration will be as the equivalent of a squared unknown, which will be designated □, the squared number whose value we don't know. Now, to find out how much it is, we halve the $10x$, and it will be $5x$, which we will add to the squared unknown in this way.

---

1. $\frac{1}{5}$ in MS.

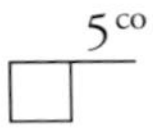

E le oltre 5^{co} azonzeremo al'altro latto del zensso, e serà chosì:

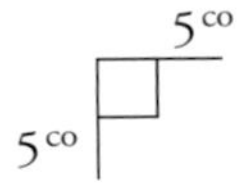

Dapoi moltipicheremo le ditte 5^{co} in si medexime, e serano chosì:

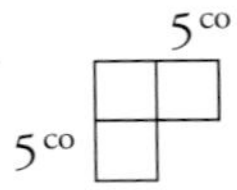

Abiamo adoncha che $1^{\square}$ e queste 10^{co} demezade e moltipichade in si sono ingual a 39^{n}. E per voler saver quanto sie lo zensso noi chonpiremo el quadrillatto, e serave chossì:

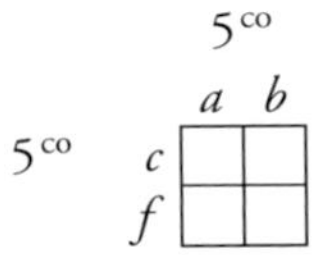

Sapiamo adoncha che latto *a b* sono 5^{co}, e latto *c f* sono 5^{co}, che sono ingual *a b*. Adoncha per ogni faza sono moltipichado in ssi 5 fa 25, e tanto tien el quadrilatto. El romagnente de ttuto è 39^{n}, azonti insieme sono 64^{n}, e la radixe de questi sono 8. E tanto è zascona faza del quadrilatto. Adoncha savemo che una parte del quadro son 8 in tutto, e *a b* xe 5. Trande[a] 5 de 8, roman 3; moltipicha 3 via 3, fa 9, e 9 fu el zensso. E tanto valse la chossa, zoè 3.

5^{co}

		a	*b*
5^{co}	*c*	zenso	moltipichado in si
	f	moltipichado in si	39

a. *Sic in MS, perhaps for* trai, *or* sottrai, *as could be supposed from the context.*

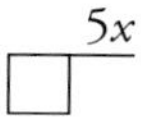

And the other $5x$ we will add to the other side of the squared unknown, and it will be like this:

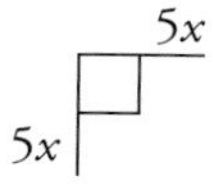

Then we will multiply these $5x$'s by themselves, and it will be like this:

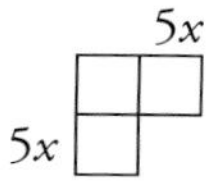

We have then that $1x^2$ and this $10x$ halved and multiplied by itself are equal to 39. And to find out how much the squared unknown is, we will complete the quadrilateral, and it will be like this:

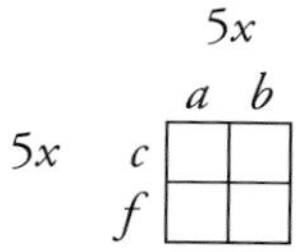

We know then that side *ab* is $5x$, and side *cf* is $5x$, which is equal to *ab*. So for each side, 5 is multiplied by itself, which makes 25, and this is the size of the quadrilateral. The remainder of all of this is 39, which added together is 64, and the root of this is 8. And each side of the quadrilateral is that big. So we know that one part of the square is 8 in all and *ab* is 5. Subtract 5 from 8, which leaves 3; multiply 3 times 3, which makes 9, and 9 was the square. And that is the value of the unknown, that is, 3.

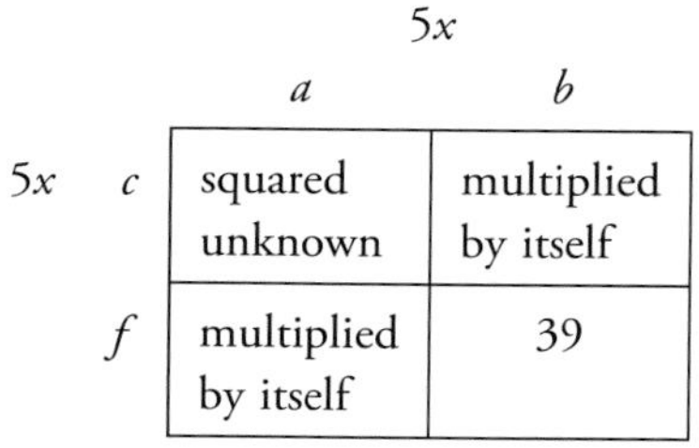

f. 73a

[Veder 1 marchadante a far 2 viazi e non saver chon quanti denari, quanto fu el chavedal]

Uno marchadante va in viazio e porta merchadantia per denari non sa quanti, e va in questo viazio e vadagna per 100 non sa quanti. Mo' truovasse aver tra pro' e chavedal duchati 150. Mo' fallo el segondo viazio chon questi duchati 150 in merchadantia, e vadagna duchati 50 più per 100 ch'al primo viazio, e truovase aver tanti denari che de un duchato à fato 3. Adomando: "Quanto fu el chavedal e quanti vadagna in prima e puo' per cento?"

E se la volesti far per la chosa, questa intra nel sesto chapitolo dove dize: "Chosa e numero ingual a zenso." Se die partir le chose e li nomeri per li zensi e demezar le chose e moltipichar in si e azonzer sovra el nomero, e la radize de quelo più el demezamento delle chosse val la chossa.

Pony che 'l portasse al viazio 1^{co} e truovase aver tra pro' e chavedal duchati 150. Al segondo viazio chon questi 150 vadagnò duchati 50 più che non fe' al primo viazio, et à fatto de 1 duchato 3. Adoncha moltipicha 150 via 150, fano 22500, chomo è a dir: "Se 1^{co} me dà duchati 150, che serave 150?" Fa' lo numero ditto de sovra. E puo' dixe aver vadagnado duchati 50 più ch'al primo, che porta 1^{co}. Adoncha vuol vadagnar $\frac{1}{2}^{co}$, e questa $\frac{1}{2}^{co}$ moltipichada per 150 fano 75^{co}. E perché à fatto de 1 3, moltipichemo 1^{co} fi[*a*] 3^{n} fa 3^{co}, e questo moltipichado in si fa $3^{\square}$. Puoi faremo chomo dixe la riegola, partiremo le chosse per li zensi, e partide le chose roman 25^{co}, e partidi li nomeri roman 7500^{n}. Mo' averemo $3^{\square}$, 25^{co}, 7500^{n}. Mo' volemo demezar le chose, serano $12\frac{1}{2}^{co}$.
f. 73b Moltipicha $12\frac{1}{2}$ via $12\frac{1}{2}$, fano $156\frac{1}{4}$, azonti a 7500 fano $7656\frac{1}{4}$. Mo' tore|mo la radixe de questo, che son $87\frac{1}{2}$, azonto el dimezamento, $12\frac{1}{2}$, fano 100^{n}, e tanto fu el chavedal. Adoncha se 'l chave[*dal*] fu 100 al primo viazio, vadagnando 50 per 100, ben fu duchati 150 el segondo viazio 50 più ch'al primo. Adoncha noi faremo: "Se 100 fu 150, che serave 200?" Fano 30000. E questi partidi per 100 fano 300. Adoncha de 100 duchati fe' 300, ben fexe de 1 3, chomo vedirì qui de sotto.

To see a merchant make two voyages and not know with how much money, how much was the capital?

f. 73a

A merchant goes on a voyage and carries merchandise whose worth is unknown, and he goes on this voyage and earns an unknown percent. Now he has found that he has 150 ducats between profit and capital. Now he makes a second voyage with these 150 ducats in merchandise, and he gains 50 ducats more per 100 than on the first voyage, and he finds that he has as much money as if from one ducat he had made 3. I ask: "How much was his capital and how much did he gain on the first and then what percent?"

And if you wanted to do it by unknowns, this falls into the sixth chapter where it says: "Unknown and number equal to a square." The unknown and the number should be divided by the squared unknowns, and halve the unknowns, and multiply by themselves and add to the number, and the root of this plus half of the unknowns is the value of the unknown.

Put that he took $1x$ on the voyage and found that he had between profit and capital 150 ducats. On the second voyage with this 150 he gained 50 more than he had on the first voyage, and had gained 3 ducats from 1. So multiply 150 by 150, which makes 22,500, as if to say: "If $1x$ gives me 150 ducats, what will 150 be?" Put this number on top. And then it says that he earned 50 ducats more than on the first, when he carried $1x$. So his profit is $\frac{1}{2}x$, and this $\frac{1}{2}x$ multiplied by 150 makes $75x$. And because he made 3 from 1, we multiply $1x$ by 3, which makes $3x$, and this multiplied by itself makes $3x^2$. Then we will do as the rule says, we will divide the unknowns by the squared unknowns, and when the unknowns are divided it leaves $25x$, and when the numbers are divided it leaves 7,500. Now we will have $3x^2$, $25x$, 7,500. Now we want to halve the unknowns, which will be $12\frac{1}{2}x$. Multiply $12\frac{1}{2}$ by $12\frac{1}{2}$, which makes $156\frac{1}{4}$, added to 7,500 makes $7{,}656\frac{1}{4}$. Now, we will take | the root of this, which is $87\frac{1}{2}$, added to the half, $12\frac{1}{2}$, which makes 100, and this much was the capital. So if the capital was 100 on the first voyage, gaining 50 per 100, the second voyage's 150 ducats was indeed 50 more than on the first. So we will say: "If 100 was 150, what will 200 be?" It makes 30,000. And this divided by 100 makes 300. So from 100 ducats he made 300, that is, he made 3 from 1, as you will see below. f. 73b

1^{co} 150^{n} 150^{n} 22500^{n} 1^{co} $\frac{1}{2}^{co}$ $\frac{150}{1}^{n}$ 75^{co} 1^{co} 3^{n} 3^{co} $1^{\square}$

$3^{\square}$ 75^{co} 25^{co} | 22500^{n} $12\frac{1}{2}$ $12\frac{1}{2}$ | 75^{co} $3^{\square}$ $156\frac{1}{4}$ | 25^{co} | 22500 $3^{\square}3^{\square}3^{\square}3^{\square}$ | 7500^{n} 156 7656

100 150 200

0
30000
10000
100
1 | 300

8
058
1237
7656
867
6 | $87\frac{87}{174}$

radixe $87\frac{1}{2}$
$12\frac{1}{2}$
100

Uno merchadante se parte e va in viazio e porta marchadantia per denari non sa quanti e vadagna duchati non sa quanti, e truovase aver tra pro' e chavedal duchati 150. E puo' fa un oltro viazio e vadagna duchati 50 per 100 più ch'al primo viazio, e truovase in Venyexia duchati 300. Adomando chon quanti duchati andò ai ditti viazi. E per far la ditta raxion pony che portasse al primo viazio 1^{co}, e fatto quello se truovase 150^{n}, et al segondo viazio 50 più ch'al primo. Adoncha fu vadagna' $\frac{1}{2}^{co}$. Perzò diremo: "Se 1^{co} me dà 150^{n}, che serave 150^{n} e $\frac{1}{2}^{co}$?" Serave 22500, e meza chossa via 150^{n} fano 75^{co}. Mo' dovemo partir per 1^{co}. So che me die venir 300^{n}. Adoncha moltipicha 1^{co} via
f. 74a 300^{n}, fa 300^{co} che son ingual a 22500^{n} e 75^{co}. Mo' adequemo le parte e traremo | 75^{co} de 300^{co}, roman 225^{co}, e questo è lo partidor, e partido insirà 100^{n}. E questi fu li denari che mese in marchadantia quando se mese al primo viazio. E per aprovarlo, se 100 torna 150, che serave 150? e più 50 val 200, serave 30000, e questi partidi per 100 serano 300. È per esenpio fatto qui de sotto.

1^{co} 150^{n} 150^{n} $\frac{1}{2}^{co}$ 150^{n} 150^{n} 22500^{n} $\frac{1}{2}^{co}$ $\frac{150}{1}^{n}$ 75^{co} 1^{co} 300^{n} 300^{co} ingual

300^{co} 225^{co}
75

000
22500
22555
222
2 | 100

Un merchadante va in viazio e porta in merchadantia per denari non sa quanti e vadagna non sa quanti, e truovase aver duchati 120. Puo' fa un oltro viazio chon questi duchati 120 e vadagna 20 per cento più ch'al primo viazio, e trovasse tra pro' e chavedal duchati 200. V'adomando quanti fu el chavedal. E per farla per la chossa pony che 'l portasse al primo viazio 1^{co}, e truovasse aver du-

$1x$ 150 150 22500 $1x$ $\frac{1}{2}x$ $\frac{150}{1}$ $75x$ $1x$ 3 $3x$

$1x^2$

$3x^2$	$75x$	22500	$75x$	$25x$	22500	
	$25x$	$12\frac{1}{2}$	$3x^2$		$3x^2 3x^2 3x^2 3x^2$	7500
		$12\frac{1}{2}$	$156\frac{1}{4}$			156
						7656

100 150 200

8
058
1237
7656 | $87\frac{87}{174}$
867
6

root $87\frac{1}{2}$
$12\frac{1}{2}$
100

0
30000
10000 | 300
100
1

A merchant takes off and goes on a voyage and takes merchandise of unknown value and gains an unknown amount of ducats, and finds that he has 150 ducats between profit and capital. And then he takes another voyage and gains 50 ducats per 100 more than on the first voyage, and finds that he has 300 ducats in Venice. I ask with how many ducats he went on these trips. And to do this problem, put that he carried $1x$ on the first voyage, and having done this he found that he had 150, and on the second voyage 50 more than on the first. So it was a profit of $\frac{1}{2}x$. For this we will say: "If $1x$ gives me 150, what will 150 and $\frac{1}{2}x$ be?" It will be 22,500, and half the unknown times 150 makes $75x$. Now we should divide by $1x$. I know that the result should be 300. So multiply $1x$ times 300, which makes $300x$, which is equal to 22,500 and $75x$. Now we equalize the sides and we will take | $75x$ from $300x$, which leaves $225x$, and this is the divisor, and divided the result will be 100. And this was the amount of money that he put into merchandise when he took the first trip. And to prove it, if 100 results in 150, what will 150 be? And plus 50 it's worth 200, it will be 30,000, and this divided by 100 will be 300. It's worked out below as an example. f. 74a

$1x$ 150 150 $\frac{1}{2}x$ 150 22500 $\frac{1}{2}x$ $\frac{150}{1}$ $75x$ $1x$ 300 $300x$ equals

150

$300x$ $225x$

75

000
22500
22555 | 100
222
2

A merchant goes on a voyage and takes merchandise of an unknown value, and he gains an unknown amount and finds that he has 120 ducats. Then he takes another voyage with these 120 ducats and earns 20 percent more than on the first voyage, and finds that he has 200 ducats between profit and capital. I ask you how much his capital was. And to do this by unknowns, put

chati 120. El se die far: "Se 1^{co} me dà 120^{n}, che serave 120^{n}?" Seria 14400^{n}, puo' per 20 che vadagna serà $\frac{1}{5}^{co}$. Moltipicha $\frac{1}{5}^{co}$ via 120^{n}, fano 24^{co}, e se parto per 1^{co} so che me die venir duchati 200. Adoncha 200 via 1^{co} fa 200^{co}, che son ingual a 14400^{n} e 24^{co}. Adequa le parte. Tra' 24^{co} de 200^{co}, roman 176^{co}; parti 14400^{n} per 176^{co} e tu averai $81\frac{144}{176}\frac{9}{11}$. E tanti funo li denari posti ali ditti viazi. E per aprovarlla farì: "Se $81\frac{9}{11}$ torna $38\frac{2}{11}$, che seria 100?" Seria $46\frac{2}{3}$, e questi fu vadagnadi al primo viazio, e questi azonti a 120 fano $166\frac{2}{3}$. E per vederla dirai: "Se 100 torna $166\frac{2}{3}$, che serave 120?" Moltipichado e partido serano duchati 200, i qualli se truovò a chonpimento del viazio. È per esenpio qui de ssotto.

1^{co} 120^{n} 120^{n} 1^{co} $\frac{1}{5}^{co}$ 120^{n} 14400^{n} $\frac{1}{5}^{co}$ $\frac{12}{1}^{n}$ 24^{co} 1^{co} 200^{n} 200^{co}
120^{n} 24

176^{co} 14400 / 176] $81\frac{9}{11}$ $81\frac{9}{11}$ $38\frac{2}{11}$ 100^{n} $46\frac{2}{3}$ | 120 / $46\frac{2}{3}$ | $166\frac{2}{3}$

100^{n} $166\frac{2}{3}$ 120^{n} Serano 200 duchati.

f. 74b

[Truovar numeri]

Truovame 3 numeri che zaschadun moltipichado in si e azonto le moltipichazion insenbre faza 100. Prima truova 3 nomeri che zascon moltipichado in ssi e azonto le moltipichazion insenbre faza numero quadratto. I qual numeri son questi: 2, 3, 6, che moltipichadi zascadun in ssi e azonto le moltipichazion insenbre fano 49, che serà numero quadratto, et io voria fusse 100. Adoncha 100 via 49 fa 4900. Ora truova 3 nomeri quadratti che azonti insieme faza 4900, i qual serano questi: 4356, la so Rx serà 66; l'oltro 400, la so Rx serà 20; l'oltro 144, la so radixe 12. E zascuna de queste radixe partti per 7 ch'è la radixe de 49. La prima $9\frac{3}{7}$, la segonda $2\frac{6}{7}$, la terza $1\frac{5}{7}$, che moltipichadi questi 3 nomeri in si, el primo fa $88\frac{44}{49}$, el segondo fa $8\frac{8}{49}$, el terzo fa $2\frac{46}{49}$. Asumadi insenbre fano 100. Et è fatta questa raxion.

[Partir ducati 54 alguny e domandar uno la radixe de ttuti]

Sono alguny chonpagni chi parteno duchati 54, e vien 1 el qual domanda la radixe dela parte che aspetta a chadaun di loro. E puo', partidi li ditti denari, tocha a choluy per radixe da cholor ch'ano partido duchati 14. V'adomando: "Quanti fu li chonpagny e quanti denari thocha a chadaun di loro per parte sì che possa eser spexi duchati 54, e tolti le radixe funo 14?" Et per voler far la ditta raxion pony che 1 chonpagno avesse duchati 16, e la radixe fu 4, el segondo avese duchati 9, la

that he took $1x$ on the first voyage and found that he had 120 ducats. It should go: "If $1x$ gives me 120, what would 120 be?" It would be 14,400, then for the 20 it gains it will be $\frac{1}{5}x$. Multiply $\frac{1}{5}x$ times 120, which makes 24, and if I divide by $1x$, I know that it should give me 200 ducats. So 200 times $1x$ makes $200x$, which is equal to 14,400 and $24x$. Equalize the sides. Subtract $24x$ from $200x$, which leaves $176x$, divide 14,400 by $176x$, and you will have $81\frac{144}{176}\ \frac{9}{11}$. And this was the amount of money taken on the voyages. And to prove it, you will do: "If $81\frac{9}{11}$ results in $38\frac{2}{11}$, what would 100 be?" It would be $46\frac{2}{3}$, and this was gained on the first voyage, and this added to 120 makes $166\frac{2}{3}$. And to see it, you will say: "If 100 results in $166\frac{2}{3}$, what will 120 be?" Multiplied and divided it will be 200 ducats which he found he had at the end of the voyage. It's shown below as an example.

$1x$ 120 120 $1x$ $\frac{1}{5}x$ 120 14400 $\frac{1}{5}x$ $\frac{120}{1}$ $24x$ $1x$ $200x$ $200x$
120 24

$176x$ 14400] $81\frac{9}{11}$ $38\frac{2}{11}$ 100 $46\frac{2}{3}$ | 120 | $166\frac{2}{3}$
176] $81\frac{9}{11}$ $46\frac{2}{3}$

100 $166\frac{2}{3}$ 120 It will be 200 ducats.

f. 74b

To find numbers

Find me 3 numbers such that each multiplied by itself and the products added together make 100. First find 3 numbers such that each multiplied by itself and the product added together make a square number. These numbers are: 2, 3, 6, which each multiplied by itself and the product added together make 49, which will be a squared number, and I wanted it to be 100. So 100 times 49 makes 4,900. Now find 3 squared numbers that added together make 4,900, which will be these: 4,356, its root will be 66; the other 400, its root will be 20; the other 144, its root 12. And each of these roots is divided by 7, which is the root of 49. The first is $9\frac{3}{7}$, the second $2\frac{6}{7}$, the third $1\frac{5}{7}$; if these 3 numbers are multiplied by themselves the first makes $88\frac{44}{49}$, the second makes $8\frac{8}{49}$, and the third makes $2\frac{46}{49}$; added together they make 100. And this problem is done.

Some divide 54 ducats and one asks for the square root of all

There are some partners who divide 54 ducats, and 1 comes who asks for the root of the part that belongs to each of them. And then, after the money was divided, he gets 14 ducats as the root of what they divided. I ask you: "How many partners were there, and how much goes to each of them as their share if 54 ducats could be spent, and when the roots were taken they were 14?" And to do this problem, suppose that 1 partner had 16 ducats, and the root was 4; the second had 9 ducats,

radixe fu 3. Adoncha inchina 54 resta 29 e la radixe fu.[a] 4 e 3 fa 7, inchina 14 mancha 7, e noi
f. 75a faremo: "Fame de 7 2 parte[b] | che moltipichada zaschuna in si et azonte le moltipichazion insenbre faza 29." Pony che 2 parte fusse questo: 1^{co} 7^{n} men 1^{co}, e moltipichadi questi in si fano $2^{\square}$ 49^{n} 14^{co}, ingual a 29^{n}. Adequa le parte et averai $2^{\square}$ 14^{co} 20^{n}. Parti per li zensi, averai 7^{co} 10^{n}. Demeza le chosse et averai $3^{co}\frac{1}{2}$. Moltipicha in si, fa $12\frac{1}{4}$. Trane el nomero ch'è 10, roman $2\frac{1}{4}$, e la radixe de questo, $1\frac{1}{2}$, azonto al dimezamento, che xe $3\frac{1}{2}$, fa 5. Inchina 7 mancha 2. A moltipichar in si 2 fia 2 fa 4, e 5 via 5 fa 25. Azonti serà ben 29, chomo vedirì qui de sotto per figora.

16
9
25
4
duchati 54

4
3
5
2
duchati 14

1^{co} 7^{n} men 1^{co}
1^{co} 7^{n} men 1^{co}
$1^{\square}$ 49^{n} 14^{co} $1^{\square}$

$2^{\square}$ 49^{n} 14^{co} 29^{n}
29
$2^{\square}$ 20^{n} 14^{co}
10^{n} 7^{co}

$3^{co}\frac{1}{2}$ $12\frac{1}{4}$
$3\frac{1}{2}$ 10
$2\frac{1}{4}$

29 / 7

$3\frac{1}{2}$ $1\frac{1}{2}$ 9
5 2 9 81
5 2
25 4

Adoncha fu 4 chonpagny. El primo avè duchati 16, e dè per radixe duchati 4. El segondo duchati 9, per radixe 3. El terzo 25, per radixe 5. El quartto 4 duchati, e per radixe 2. Adoncha asuma le radixe, serà duchatai 14, e azonti i denari serano duchati 54. Et è fatta questa raxion.

---

f. 75b

[Far de 14 3 parte in proporzion che moltipichada l'una per 2, l'oltra per 3, l'oltra per 4, e zonto le moltipichazion faza 48]

V chapitolo[c]

Fame de 14 3 parte in proporzion che moltipichada l'una per 2 e l'oltra per 3 e l'oltra per 4, e azonte le moltipichazion insenbre faza 48. Pony che la prima parte fusse 1^{co}, le oltre 2 parte sono

---

a. *Sic in MS, with no space left.*
b. V chapitolo *centered at bottom of page.*
c. *Centered at top of age, in position usually occupied by* Ihesus.

the root was 3. So up to 54 there remains 29 and the root was.[1] 4 and 3 make 7, up to 14 there is 7 lacking, and we will say: "Make me 2 parts of 7[2] | which when each is multiplied by itself and the products added together make 29." Put that 2 parts were this: $1x$ 7 minus $1x$, and this multiplied by itself makes $2x^2$ 49 $14x$, equal to 29. Equalize the sides and you will have $2x^2$ $14x$ 20. Divide by the squared numbers; you will have $7x$ 10. Halve the unknowns and you will have $3\frac{1}{2}x$. Multiply by itself; it makes $12\frac{1}{4}$. Subtract from it the number, which is 10, it leaves $2\frac{1}{4}$, and the root of this, $1\frac{1}{2}$, added to the half, which is $3\frac{1}{2}$, makes 5. Up to 7 there is 2 lacking. To multiply by itself, 2 times 2 makes 4, and 5 times 5 makes 25. Added together they will be 29, as you will see worked out below. f. 75a

16	4
9	3
25	5
4	2
ducats 54	ducats 14

$1x$	7	minus	$1x$
$1x$	7	minus	$1x$
$1x^2$	49	$14x$	$1x^2$

$2x^2$	49	$14x$	29
	29		
$2x^2$	20	$14x$	
	10	$7x$	

$3\frac{1}{2}x$	$12\frac{1}{4}$
$3\frac{1}{2}$	10
	$2\frac{1}{4}$

29 / 7

$3\frac{1}{2}$	$1\frac{1}{2}$	9	
5	2	9	81
5	2		
25	4		

So there were 4 partners. The first one had 16 ducats, and gave as the root 4 ducats. The second one 9 ducats, as a root 3. The third one 25, as a root 5. The fourth one 4 ducats, and as a root 2. So add together the roots, it will be 14 ducats, and the money added together will be 54 ducats. And this problem is done.

Divide 14 in 3 parts so that when one is multiplied by 2, the other by 3, and the other by 4, the products summed make 48 f. 75b

Chapter V[3]

Find for me 3 parts of 14 in such a proportion that when one is multiplied by 2, and the other by 3, and the other by 4, the products added together make 48. Put that the first part was $1x$, the other

1. Not completed in the manuscript, with no space left. At this point the method of solution switches from double false position to unknowns.

2. *Chapter V* centered at bottom of page.

3. Centered at top of page, in position usually occupied by *Jesus*.

14^{n} men 1^{co}. Ora moltipicha la prima, zoè 1^{co}, per 2^{n}, fa 2^{co}, resta fina 48 48^{n} men 2^{co}. Ora bixogna far de 14 men 1^{co}[a] 2[b] parte che l'una moltipichada per 3 e l'oltra per 4, e azonte le moltipichazion insenbre, faza 48 men 2^{co}, e fane mo' chosì: moltipicha 14^{n} men 1^{co} per 3, fa 42 men 3^{co}. Sutrazi de 48 men 2^{co}, mancha 6 e 1^{co}. E questa serà la parte mazuor, e quella de mezo serà el romagnente fina 14 men 1^{co}, zoè 8 men 2^{co}. Siché avemo la prima 1^{co}, la segonda 8 men 2^{co}, la terza 6 e 1^{co}, le quale azonte fa ben 14. E moltipicha la prima per 2, la segonda per 3, la terza per 4, fa aponto 48. Mo' volemo noi saver quando[c] sono 3 nomeri proporcionadi che tanto fa el primo moltipichado per lo terzo quanto el segondo moltipichado in si medeximo. Però moltipicha 1^{co} via 6^{n} e 1^{co}, fa 6^{co} e 1^{co}, e puo' moltipicha in si 8^{n} men 2^{co}, fa 64^{n} men 32^{co} e più $4^{\square}$, che sono ingualli a 6^{co} e $1^{\square}$. Adequa le parte et averemo 38^{co} ingualli a $3^{\square}$ e 64^{n}. Parti per li zenssi, averemo $12\frac{2}{3}^{co}$ ingualli a $1^{\square}$ e $21\frac{1}{3}$. Dimeza le chosse, averemo $6\frac{1}{3}$. Moltipicha' in si fa $40\frac{1}{9}$. Trazene el nomero, zoè $21\frac{1}{3}$, resta $18\frac{7}{9}$, e la radixe de questo sie $4\frac{1}{3}$. Trazilo dela mitta dele chosse, tanto val la chossa che è $6\frac{1}{3}$, resta 2. E tanto fu el primo nomero. El segondo, che fo 8 men 2^{co}, resta 4, perché la chossa valse men 2. Abatti de 8 4, roman 4. E la terza che fu 6 e 1^{co}, vien a eser 8, perché la chossa valse più 2, e 6 fa ben 8.

f. 76a Notta che se tu volesi far de alchuna quantità 2 parte che moltipichando l'una per zerta quantittà, e per lo simille l'oltra per zerta quantittà, e puo' sumar le moltipichazion e quela suma fese alchuna quantittà, esempio: "E voio far de 12 do parte che moltipichando l'una per 2 e l'oltra per 4 faza 40," moltipicha 12 per 2, fa 24, varda quanto mancha a 40, el mancha 16, partillo per la diferenzia che xe da 2 a 4, la qual è 2, che vien 8. E tanto l'una parte; l'oltra, el romagnente fin a 12, che xe 2. E questo ò demostrado per dechiarazion d'un passo dela sovraditta raxion.

[TRUOVAR 2 HOMENY A DENARI; DIXE EL PRIMO: "SE TU ME DÀ $\frac{1}{3}$ D'I TTO AVERÒ 24"; EL SEGONDO: "SE TU ME DÀ $\frac{1}{5}$ D'I TTO ARÒ 56"]

IIII chapitollo[d]

Do homeny à denari. Dixe el primo al segondo: "Se tu me dà el terzo d'i tto denari, ne averò 24." Dixe el segondo al primo: "E se tu me dè el quinto d'i to, cho[*n*] i mie ne averò 56 e più la radixe de quello che te romagnerà." Domando: "Quanti n'avea zaschadun?"

Qui te chonvien tal chosa che abia el primo que[e] quando l'averà dado al segondo el quinto i romagna tal chossa che abia radixe. Per la qual chossa meterò quel[f] primo avesse $1^{\square}$ e $\frac{1}{4}$ de zensso. Chon-

---

a. men 1^{co} *added in the right margin.*
b. men 1^{co} *added between the lines with a reference mark, an addition rendered superfluous by the preceding marginal annotation.*
c. *Sic in MS.*
d. *Centered.*
e. *Sic in MS, evidently for* che.
f. *Sic in MS, evidently for* che 'l.

2 parts are 14 minus $1x$. Now multiply the first, which is $1x$, by 2, which makes $2x$. There remains to 48, 48 minus $2x$. Now it is necessary to make 2 parts of 14 minus $1x$ such that one multiplied by 3 and the other by 4, and the products added together, makes 48 minus $2x$, and we will do it this way: multiply 14 minus $1x$ by 3, which makes 42 minus $3x$. Subtract it from 48 minus $2x$; there is 6 and $1x$ lacking. And this will be the greater part, and the half will be the remainder up to 14 minus $1x$, that is, 8 minus $2x$. Thus we have that the first is $1x$, the second 8 minus $2x$, the third 6 and $1x$, which added together make 14. And multiply the first by 2, the second by 3, the third by 4, which makes 48. Now we want to know how much are 3 numbers in proportion so that the first multiplied by the third is the same as the second multiplied by itself. So multiply $1x$ by 6 and $1x$, which makes $6x$ and $1x^2$,[1] and then multiply 8 minus $2x$ by itself, which makes 64 minus $32x$ and plus $4x^2$, which is equal to $6x$ and $1x^2$. Equalize the sides and we will have $38x$ equal to $3x^2$ and 64. Divide by the squared numbers; we will have $12\frac{2}{3}x$ equal to $1x^2$ and $21\frac{1}{3}$. Halve the unknowns; we will have $6\frac{1}{3}$. Multiply by itself; it makes $40\frac{1}{9}$. Subtract the number, that is, $21\frac{1}{3}$, which leaves $18\frac{7}{9}$, and the root of this is $4\frac{1}{3}$. Take it from half of the unknowns; it gives the value of the unknown which is $6\frac{1}{3}$, with 2 left over. And this was the first number. The second, which was 8 minus $2x$, leaves 4, because the unknown was worth minus 2. Subtract 4 from 8, which leaves 4. And the third, which was 6 and $1x$, works out to be 8, because the unknown was worth plus 2 and 6, which makes 8.

Note that if you wanted to find two parts of any quantity which when multiplying one by a certain quantity, and similarly the other by a certain quantity, and then adding the products and this sum making another quantity, for example: "And I want to make 2 parts of 12 such that multiplying one by two and the other by 4 makes 40," multiply 12 by 2, which makes 24, note how much is lacking to make 40, 16 is lacking, divide it by the difference between 2 and 4, which is 2; it comes to 8. And that is the value of one part; the other is the remainder to 12, which is 4.[2] And this I have demonstrated through explanation in the course of the problem above. f. 76a

---

FIND 2 MEN WITH MONEY; THE FIRST ONE SAYS: "IF YOU GIVE ME $\frac{1}{3}$, I WILL HAVE 24"; THE SECOND ONE: "IF YOU GIVE ME $\frac{1}{5}$, I WILL HAVE 56"

Chapter IIII

Two men have coins. The first one says to the second one: "If you give me a third of your coins, I will have 24." The second one says to the first one: "And if you give me a fifth of yours, with my own I will have 56 plus the square root of what you have left." I ask: "How many did each have?"

Here you need to make the unknown what the first one had, of which when he will have given a fifth to the second one, the remainder is the square root of the unknown. With this as the

1. *1x* in MS.
2. *2* in MS.

vien el segondo aver 72 men $3\frac{3}{4}^{\square}$. E se 'l segondo dà al primo el terzo d'i soi, i darà 24 men $1\frac{1}{4}$. |
f. 76b Averà ben el primo 24. Mo' el primo che à $1\frac{1}{4}$ de zensso darà el quinto al segondo, che serà $\frac{1}{4}$, e roman chon $1^{\square}$. El segondo vegnirà aver 72 men $3^{\square}\frac{1}{2}$, che sono ingualli a 56 e 1^{co}, perché 1^{co} sì è radixe de $1^{\square}$. Mo' adequa le parte et averai 16^{n} ingualli a $3^{\square}\frac{1}{2}$ e 1^{co}. Parti per li zensi, averemo $4\frac{4}{7}$ de nomero, ingualli a $1^{\square}$ e $\frac{2}{3}^{co}$. Dimeza le chosse, avera' $\frac{1}{7}^{co}$. Moltipicha in ssi, fa $\frac{1}{49}$. Azonzillo al nomero, averemo $4\frac{29}{49}$. E chosì la radixe de $4\frac{29}{49}$ men la mittà dele chosse, che xe $\frac{1}{7}$, val la chossa. E tu ponesti che 'l primo avese $1\frac{1}{4}^{\square}$. Adoncha moltipicha Rx $4\frac{29}{49}$ men $\frac{1}{7}$ in ssi medesimo e metti chomo sta qui de sotto e moltipicha men $\frac{1}{7}$ via men $\frac{1}{7}$ fa $\frac{1}{49}$, e puo' moltipicha $\frac{1}{7}$ men via $\frac{225}{49}$. Tu die redur $\frac{1}{7}$ a Rx, e farà $\frac{1}{49}$. E moltipicha $\frac{1}{49}$ via $\frac{225}{49}$, fa $\frac{225}{2401}$, e puo' anche el simille fa $\frac{225}{2401}$. Azonzilo insenbre, fa men Rx de $\frac{900}{2401}$. E puo' moltipicha $\frac{225}{49}$ via Rx, fa $\frac{225}{49}$. Siché per questo moltipichamento tu à più $\frac{1}{49}$ e più $\frac{225}{49}$, che xe $4\frac{29}{49}$ e men Rx de $\frac{900}{2401}$, zoè la qual Rx sie $\frac{30}{49}$. E se ttu azonzi insenbre $\frac{1}{49}$ chon $4\frac{29}{49}$, tu avera' 4 e $\frac{30}{49}$. E questo è quello che val el zensso, zoè $4\frac{30}{49}$ men $\frac{30}{49}$, che resta aponto 4. E noi ponesemo $1\frac{1}{4}$. Adoncha se 'l zenso val 4, $1\frac{1}{4}$ valeria 5, e tanto avea el primo. El segondo chonvien aver 57.

Pruovala. Se 'l segondo dà el terzo d'i soi al primo, i darà 19, e 5 da ssi val 24, e se 'l primo dà al segondo $\frac{1}{5}$ d'i soi, i darà 1, siché el segondo averà 58. E ben se vede che 'l primo romaxe chon 4 e quel[a] segondo avè 58, siché el segondo avè ben 56 e la Rx de 4, che è 2, che azonto a 56 fa ben 58. E per lo simille fare' dele oltre sì fatte raxion.

f. 77a

[Truovar numeri quadrati de radixe d'ogni numero chubicho]

Tratado de truovar radixe quadratta de ogni numero.

Qua se chomenza lo tratado dele radixe quadratte e chubiche de nomeri che abia radixe dischrette. Onde è da saver che zaschadon numero per si medeximo moltipichado sie radixe dela sua moltipichazion, zoè radixe quadratta. In questo muodo 4 è radixe de 16, però che 4 via 4 fa 16, e 5 è radixe de 25, però 5 via 5 fa 25, e chossì de ogni numero moltipichado in si medeximo. E radixe chubiche è dita quella quando lo numero è moltipichado in si medeximo e quela multipichazion vien moltipichada poi chon raxon[b] quadratta dela moltipichazione. In questo muodo 3 moltipichado in si fa 9, e questo 9 moltipichado per la so radixe quara, zoè per 3, fa 27, el qual vien ditto che 3 è radixe chubicha de 27, e 4 è radixe chubicha de 64, e 5 è radixe chubicha de 125.

---

a. *Sic in MS, evidently for* che 'l.
b. *Sic in MS.*

unknown, I will put that the first one had $1x^2$ and $\frac{1}{4}$ of the squared unknown. Then it follows that the second one had 72 minus $3\frac{3}{4}x^2$. And if the second one gives to the first one a third of his, he will give 24 minus $1\frac{1}{4}x^2$.**1** | The first one will then have 24. Now the first one, who has $1\frac{1}{4}$ of the squared unknown, will give a fifth to the second one, which will be $\frac{1}{4}$, and he remains with $1x^2$. The second one will come to have 72 minus $3\frac{1}{2}x^2$, which is equal to 56 and $1x$, because $1x$ is the square root of $1x^2$. Now equalize the sides and you will have 16 equal to $3\frac{1}{2}x^2$ and $1x$. Divide by the squared unknowns, we will have $4\frac{4}{7}$ as a number, equal to $1x^2$ and $\frac{2}{7}x$.**2** Halve the unknowns, you will have $\frac{1}{7}x$. Multiplied by itself, it makes $\frac{1}{49}$. Add it to the number; we will have $4\frac{29}{49}$. And so the square root of $4\frac{29}{49}$ minus half of the unknown, which is $\frac{1}{7}$, is equal to the unknown. And you put that the first one had $1\frac{1}{4}x^2$. So multiply the root of $4\frac{29}{49}$ minus $\frac{1}{7}$ by itself and put as it is below, and then multiply minus $\frac{1}{7}$ times minus $\frac{1}{7}$, which makes $\frac{1}{49}$, and then multiply minus $\frac{1}{7}$ times $\frac{225}{49}$. You should reduce $\frac{1}{7}$ to the root and it will make $\frac{1}{49}$. And multiply $\frac{1}{49}$ by $\frac{225}{49}$, which makes $\frac{225}{2401}$, and then similarly it also makes $\frac{225}{2401}$. Add it together, it makes minus the root of $\frac{900}{2401}$. And then multiply $\frac{225}{49}$ times the root, which makes $\frac{225}{49}$. So from this multiplication you have plus $\frac{1}{49}$ and plus $\frac{225}{49}$, which is $4\frac{29}{49}$ and minus the root of $\frac{900}{2401}$, whose root is $\frac{30}{49}$. And if you add $\frac{1}{49}$ together with $4\frac{29}{49}$, you will have 4 and $\frac{30}{49}$. And this and the value of the squared unknown, that is, $4\frac{30}{49}$ minus $\frac{30}{49}$, leaves exactly 4. And we put $1\frac{1}{4}$. So if the squared unknown is worth 4, $1\frac{1}{4}$ would be worth 5, and that is what the first one had. The second one works out to have 57. f. 76b

Prove it. If the second one gives a third of his to the first one he will give 19, and 5 of his own is worth 24, and if the first one gives to the second one $\frac{1}{5}$ of his, he will give him 1, so that the second one will have 58. And it is clearly seen that the first one remains with 4 and that the second one had 58, so that the second one had 56 and the square root of 4, which is 2, which added to 56 indeed makes 58. And you will do the other problems the same way.

---

f. 77a

To find square and cube roots of every number

A treatise on finding the square root of any number.

Here begins the treatise of square and cube roots of the numbers that have discrete roots. One should know that each number multiplied by itself is the root of its product, that is, the square root. In this way 4 is the root of 16, because 4 times 4 makes 16, and 5 is the root of 25 because 5 times 5 make 25, and so for any number multiplied by itself. And cube roots refer to that thing when the number is multiplied by itself and that product is then multiplied with the square root**3** of the product. In this way 3 multiplied by itself makes 9, and this 9 multiplied by its square root, that is by 3, makes 27, by which it is said that 3 is the cube root of 27, and 4 is the cube root of 64, and 5 is the cube root of 125.

---

1. x^2 lacking in MS.
2. $\frac{2}{3}x$ in MS.
3. MS has *raxon* (reason or problem) in place of *radixe* (root).

E prima te ò mostrado la riegola de truovar la radixe quadrata de zaschadun numero che tte avegnisse lo qual avese radixe descretta, in questo muodo. Tu die adesso numerar le figure le qual chontien lo numero del qual tu voi truovar la radixe quadratta, e se le figure fusse despare tu die meter
f. 77b sotto la prima figura, la qual è inver | la man senestra, zoè ala destra dela nota, el numero in tal figura la qual moltipichada in ssi medexima monta tanto intregamente quanto la dita figura de sovra, overo più apresso che puol, zoè che moltipichada la dita figura di sutto in ssi medexima, tratta la ditta moltipichazion dela quantità dela figura di sovra, romagna niente, overo che romagna men che se puol. E poi redupia quela figura la qual fu metuda de sotto, e meti quelle adopiazion 1ª figura più avanti inver la man senestra dela notta e depena la prima figura, la qual moltipichada in ssi medexima overo fu adoblada. Poi metti una figura tal più avanti inver man destra, zoè ala sinistra dela adopiasion, che moltipichada per la figura la qual fu doplada, deffaza le figure che li sono di sopra insieme chon quello che avanza, zoè dinanzi. Se alguna chosa fusse avanzado dela figura ditta dananzi, chonsiderando lo ditto avanzo romagnise per numero, zoè dezene tante quante lo ditto avanzo romagnise per numero, trazando la prima moltipicazion dela prima figura, e anchora moltipicada la dita figura metuda per la qual al presente noi parlemo in si medexima, deffaza la figura che li xe de sovra et eziando quelo che fu avanzado. Se alguna chosa avanzase trazando la moltipichazion dela prima figura moltipichada per l'adoplaxion dela prima figura fuor de quello xe ditto, zoè delo residuo di sovra lo qual fuse avanzado, chonsiderando lo ditto residuo eser desene chomo fu chonsiderato lo primo residuo, intendando che la prima figura la qual moltipichi per l'adopiasion sia talle che moltipichada per la ditta moltipicazion deffaza la figura che li xe de sovra
f. 78a e lo residuo | segondo chomo xe ditto, in tal muodo che moltipichada per la figura in si medema, la qual è moltipichada per la adopiaxion, si posa dala figura che li xe de sovra, overo dala figura de sovra cho[*n*]zonta chon alguno residuo che li fuse avanzado davanti, intendando senpre queli residii per dezene, zetto che la ultima moltipichazione, trazandola che alguna quantitade delo residuo che roman non se die intender altro che propriamente lo ditto residuo, e la radixe descretta serave tornada de quela quela[a] quantitade la qual romagnise trazando l'ultimo residuo de ttuta la quantitade del numero, partando per mitade ziaschaduna figura che fose adoplada in lo prozeder dela ditta riegola, zoè tolendo la mitade de tuto quello che montase le figure adoplade, salvando senp[*r*]e intrega la ultima figura del prozeso dela sovraditta riegola. E simelmente die prozeder in zaschaduna quantità del numero, destruzando tute le figure per el sovraditto muodo, andando senpre adoblando la figura la qual xe adoblada solamente una fiada fina l'ultima. Ma se le figure fusseno pare, debi chomenzar dala segonda figura che xe a man senestra, e sutto quella metti 1ª tal ffigura che moltipichada in si medexima deffaza la figura che li xe de sovra e quela che i xe davanti intregamente a più presso che se puol, sichomo t'ò amaistrado in lo numero che fusse disparo, e poi debì a operar e prozedere chomo t'ò demostrado in questo chapitulo che xe de sovra, e questa sie la pridicha.

f. 78b Ponamo che tu voi truovar la radixe de queste figure, zoè 6964321, zoè quadratta. Fa chosì: metti sotto lo primo 6 che xe da man senestra 1ª figura, tal che moltiplichada per si e la moltipichazion tratta dal ditto 6 faza lo 6 al più che 'l può, la qual figura vien ad eser 2, lo qual moltipicha' in si fa 4, lo qual tratto del 6 roman 2. Da' de pena al 6 e schrivy al 2 sovra al 6, puo' redopla lo 2, serano 4.

a. *Thus repeated in MS.*

And I have shown you first the rule for finding the square root of every number that should come to you that has a discrete root, in this way.[1] For this you should number the figures which contain the number of which you want to find the square root, and if there is an odd number of figures, you should put under the first figure, the one that is toward | the left hand, that is, to the right of the note, the number in such a figure which when multiplied by itself goes wholly into the value of this figure above, or as closely as possible, that is, that when this figure below is multiplied by itself, and this product is subtracted from the value of the figure above, there remains nothing, or what remains is as small as can be. And then double that figure that was put below, and put this doubling one figure forward toward the left hand of the note and cross out the first figure, the one that was multiplied by itself or was doubled. Then put a figure enough forward toward the right, that is, to the left of the doubling, which multiplied by the figure that was doubled, goes into the figures that are together above as well as the one that comes forward, that is, the one in front. If anything was brought forward from the figure said to be in front, considering that this number brought forward remained for the number, that is, so many tenths as this amount brought forward remained for the number, subtracting the first multiplication from the first figure, and also having multiplied this figure about which we are speaking at present by itself, it goes into the figure that is above it and also the one that was brought forward. If anything is brought forward subtracting the multiplication of the first figure multiplied by the doubling of the first figure beyond what is said, that is, of the residue above what was brought forward, considering this residue to be drawn as the first residue was considered, intending that the first figure which you multiply by the doubling be such that multiplied by this multiplication it goes into the figure that is above and the residue | as is said, in such a way that multiplied by the figure itself, that which is multiplied by the doubling is put by the figure that is above it, or by the figure above added to any residue that may have been brought forward in front, always understanding these residues as tenths, so that the last multiplication, subtracting any quantity of the residue that remains, not be understood as anything other than this very residue, and the discrete root will result from that quantity that remains from subtracting the final residue from the entire quantity of the number, dividing in half each figure that was doubled in the process of this problem, that is, taking half of everything that the doubled figures amount to, always keeping whole the last figure of the process of this problem. And in this way you should proceed with each value of the number, crossing out all the figures in this way, always doubling the figure which is doubled only once until the end. But if the figures were even, you should begin with the second figure on the left, and under this put first one figure which when multiplied by itself goes into the figure that is above and that which is forward wholly as closely as possible, as I have taught you in the case of the number that was uneven, and then operate and proceed as I have shown you in this paragraph above and that which precedes it.

f. 77b

f. 78a

Let's suppose that you want to find the root of this figure: 6,964,321, that is, the square root. Do this: put under the first 6 that is on the left hand one figure which multiplied by itself and the product subtracted from 6 makes 6 as closely as it can, which figure works out to be 2, which multiplied by itself makes 4, which subtracted from 6 leaves 2. Give a pen mark on the 6 and write the f. 78b

1. These instructions are long, complicated, and obscure; the translation is an attempt to render the instructions as given rather than to make full sense of them. The procedure is more clearly understood from the example that follows.

E da' de pena al 2, e va chon el 4 1ª figura più avanti, in tal muodo che 'l 4 serà sotto al 9 che ven driedo al 6. E muo' truova 1ª tal figura che moltiplichata per lo 4 ch'è sotto lo 9 deffaza le 2 figure che sono de sovra, che reprexentta 29, in tal muodo che, moltipichada la figura che mettesti, la sia moltiplichada in si medexima, deffaxa lo romagnente al più che 'l può. La qual figura vien a eser 6, lo qual moltipicha' per 4 faza 24. Lo qual trazi dal 29 che xe sovrà, roman 5. E puo' moltipicha 6 per si fa 36, lo qual trazi de 56, roman 20. E chossì schriverai e darai de pena chomo se faxe al partir dela gallia. E poi redopla lo 6 che metesti sotto lo 6, fa 12, e chon 4 che li è dananzi fa 52, chon 'l qual va 1ª figura più avanti, in tal muodo che lo 5 vegna sotto el 6 e lo 2 sotto el 4. Metti sotto lo 3 1ª tal figura più avanti che moltiplichada per 52 deffaza le 3 figure che reprexentta 204 al più che 'l se può. E quela medema figura moltiplichada in si deffaza lo residuo che roman fin'al 3, la qual figura vien a eser 3. Lo qual moltipichado per 52 fa 156, lo qual trazi de 204, roman 48. Puo' moltiplicha 3 in si, fa 9, lo qual trazzi de 483, roman 474. Puo' se die da chonsiderar le figure che
f. 79a son scritte sotto reprexenta 523, e se tu azonzi el 3 fa 526, | lo qual porta una figura più avanti in tal muodo che la vegia sutto el 2, e lo 2 sotto el 3, e lo 5 sotto el 2. E puo' metti una tal figura sotto l'uno che xe ala man destra, che moltipichada per 526 deffaza lo 4742 che xe de sopra al più che se pò, che vien eser 9. Lo qual moltipichado per 526 faxe 4734, lo qual tratto de 4742 roman 8. Siché tu ai di sopra per figura intriega 81, lo qual disfarai per moltipichar 9 in si medemo, fa 81. Abatti 81 de 81, e roman niente. Mo' arai tu de sutto per figure intriege quelo che reprexenta 5269. Salva lo 9, a partir per 2 lo 5260, che ne vien 2630, azonto 9 faxe 2639. Questa è la radixe quadra dele figure lo chui numero fuse despare, chomo vedirì per figura.

  00
 0~~4~~~~2~~0
0~~2~~~~5~~7~~6~~0
~~2~~~~5~~~~0~~~~8~~~~4~~~~8~~0
~~6~~~~9~~~~6~~~~4~~~~3~~~~2~~~~1~~]
~~2~~~~4~~~~6~~~~2~~~~3~~~~6~~~~9~~]   2639
  ~~5~~~~5~~~~2~~]

E per lo simile muodo tu poi truovar la radixe quadratta de quelli numeri che fu separadi, chomenzando dala figura che xe inver man senestra andar inver man destra, fazando chomo t'ò mostrado di sopra. Ponamo che noi voiamo truovar la radixe de 1225. Se chomenza dala figura segonda, che reprexenta 200, e metila sotto lo 3, e moltipicha per si, fa 9, lo qual trazi de 12 che xe sopra, roman
f. 79b 3, e puo' redopia lo 3 | che metesti sotto el primo 2 e prozedi chomo t'ò ditto, zoè mostrado, zoè lo 3 adoplado fa 6, e quelo metti sotto lo segondo 2. E puo' truova un numero, lo qual metti sotto al 5, per lo qual moltipichado lo 6 che xe sotto le 2 figure che xe sotto el 32 deffaza lo 32, overo più apresso che se può. Lo qual serà 5, moltipichado per 6 fa 30, abatutto 30 de 32 roman 2, lo qual 2

2 above the 6, then double the 2, which will be 4. And give a pen mark to the 2, and go one figure forward with the 4, in such a way that the 4 will be below the 9 which comes after the 6. And now find one figure which when multiplied by the 4 that is below the 9 goes into the 2 figures that are above, which is represented by 29, in such a way that the figure that you have put multiplied by itself goes into the remainder as much as it can. This figure works out to be 6, which multiplied by 4 makes 24. When this is subtracted from the 29 which is above, there remains 5. And then multiply 6 by itself, it makes 36, which you subtract from 56, which leaves 20. And this you will write and give a pen mark as you do in division by the galley method. And then double the 6 that you have put below the 6, which makes 12, and with the 4 that is in front of it makes 52, with what goes one figure forward, in such a way that the 5 goes below the 6 and the 2 below the 4. Put beneath the 3 one such figure forward that when multiplied by 52 goes into the 3 figures that represent 204 as closely as possible. And that same figure multiplied by itself goes into the residue that remains up to 3, which figure turns out to be 3. This multiplied by 52 makes 156, which you subtract from 204, which leaves 48. Then multiply 3 by itself, which makes 9, which you subtract from 483, which leaves 474. Then you should consider the figures that are written below represent 523, and if you add the 3 it makes 526, | which carries one figure forward in such a way that it comes below the 2, and the 2 below the 3, and the 5 below the 2. And then put a figure below the one that is on the right hand, such that multiplied by 526 it goes into the 4,742 that is above as much as possible, which turns out to be 9, which multiplied by 526 makes 4,734, which when subtracted from 4,742 leaves 8. So you have above in the figure the whole number 81, which you will go into by multiplying 9 by itself, which makes 81. Subtract 81 from 81, and nothing remains. Now you will have below whole figures which represent 5,269. Save the 9, divide the 5,260 by 2, which comes to 2,630, adding the 9 makes 2,639. This is the square root of the figures, which number was odd, as you will see worked out. f. 79a

```
   00
  0420
 025760
2508480
6964321 ]
2462369 ]   2639
  552   ]
```

And in a similar manner you can find the square root of those numbers that were separated, beginning from the figure that is toward the left going toward the right hand, doing as I have shown above. Let's suppose that we want to find the root of 1,225. If you begin with the second figure, which represents 200, and put below it a 3, and multiply it by itself, it makes 9, which you subtract from the 12 that is above, which leaves 3, and then double the 3 | which you put under the first 2 and proceed as I told you, as is shown, that is, 3 doubled makes 6, and put this under the second 2. And then find a number, which you put under the 5, which when multiplied by the 6 that is under the 2 figures that are below the 32 goes into 32, or as closely as possible. This will be 5, which multiplied by 6 makes 30, 30 subtracted from 32 leaves 2, which 2 in front of 5 represents 25, f. 79b

dananzi a 5 reprexenta 25, e puo' moltipicha' per si fano 25, abatudo de 25 e roman niente. E chusì tu vedi che radixe quadratta de 1225 sie 35. Ed è fatta, vedirì per figura.

```
  00
0320
1225 ]
 365 ]  35
```

Se tu volesti truovar le radixe chubicha de zaschadun nu[*m*]ero, o grande o pizollo che 'l se ssia, el se vuol far chussì. In prima truovar un numero lo qual sia da 9 in zuxo, lo qual numero vien chiamado per l'aitta del'aresmetticha digitto, lo qual digitto tu die meter sutto la figura la qual in luogo del'ultimo numero inverso la man senestra. E ssia tal lo ditto numero sutto meso ala ditta figura che proditto el ditto numero a radixe chuba, quela suma che ne vegnirà se puosa trazer de tute le figure
f. 80a le qual fusse davanti la ditta figura, la qual è in luogo del mellesimo, chon | tantto la ditta figura propria milesimo chon le oltre figure le qual fusse dananti la ditta figura, ho veramente quela fusse sola senza aver oltre figure davanti, pur sia tal lo ditto numero ditto digitto che, dutto a radixe chuba, elo chonsuma tutte le figure ditte dananzi o più apreso che se possa. E zà fatto, moltiplicha lo ditto numero per 3, la qual moltipichazion metti sutto ala terza figura de man destra chomenzando dal ditto mezo avanti ditto. E fatta la ditta moltipichazion, metti lo ditto digitto in primo truovado sotto la ditta moltipichazion, e puo' moltipicha quelo numero sottomeso al so triplado, sorameso lo so triplado chon lo so triplado, reputando tutta fiada tute le figure del triplado eser dexene. E quelo che ne ven salva, questo siando fatto.

È de chatar anchora uno numero lo qual tu à [*a*] meter sotto la prosima figura la qual driedo la prima sutto la qual tu metesti lo triplado chon lo so sottotriplado, tuta fiada andando ala man destra. Lo qual digitto, o ver numero, sia sì fatto che moltiplichado chon lo triplado solo qual è de sovra lo sottotriplado, chontando lo ditto triplado in questa parte eser pur numero o numero con dexene, figure chon figure, chomo elo ve[*n*] per singolo. La qual moltipichazion zonta chon la moltipichazion che tu festi de triplado chon lo sottotriplado, e quela suma che te vegnirà de queste due moltipichazion zonte l'una chon l'altra.

Anchora moltiplichade chon lo numero che mo' da driedo è truovado, sia tal suma quela destruza
f. 80b tute le figure | le qual te è romaxe infina al triplado o più apreso che se può. Questo fatto, lo numero driedo truovado redu' a chubo, e quello che te vegnirà trazillo de tutte le figure che è davanti sovra avanzade. E questo fatto, o el te avanzerà alchuna chossa o niente. E se alchuna chossa te avanza, trala da tutte le figure le qual tu volesti trar over truovar radixe chuba. E quelo che te romagnirà serà lo numero che avea la radixe chuba, serà li 2 numeri li qual truovasti metandoli a sequenzia l'un del'altro chomo tu li truovasti. E se 'l te avanzase, quele figure seria tutte quele che averia chubo, e

and then multiplied by itself makes 25, and subtracted from 25 leaves nothing. And so you see that the square root of 1,225 is 35. And it's done, as you will see in the figure.

```
 00
0̸3̸2̸0
1̸2̸2̸5̸ ]
 3̸6̸5̸ ]  35
```

If you want to find the cube root of every number, no matter how large or small it may be, it should be done this way.[1] First find a number which is 9 or lower, which number is called by help from arithmetic a digit, which digit you should put beneath the figure which is in the place of the last number to the left. And this number put under the figure is such that, this number put to the cube,[2] the resulting sum can be subtracted from all the figures that were in front of this figure, which is in the thousands place, with | the amount of this figure's own thousand with the other f. 80a
figures that were in front of this figure, or truly that one was alone without having other figures before, so that this number called a digit, when reduced to its cube, uses up all of the figures before it or as closely as is possible. And having done this, multiply this number by 3, and put this product under the third figure to the right beginning with that middle specified above. And having done this multiplication, put that digit found first under this product, and then multiply the number put under by its triple, having put its triple above with its triple, considering each time all the figures of the triple to be tens, and save what remains when this is done.

And it's necessary to find another number which you have to put under the closest figure behind the first under which you put the triple with its under-triple, each time going to the right. This digit, or number, may be made so that, multiplied with its triple only which is over the under-triple, counting this triple in this part to be a simple number or a number with tens, figures with figures, as they come one by one, you add this product with the product that you made of the triple with the under-triple, and add together the sums that you will get from these two products.

Then multiplied with the number that is now found behind, there may be such a sum that goes into all the figures | which have remained to you up to the triple or as close as is possible. Having f. 80b
done this, reduce the number found behind to a cube, subtract the result from all the figures that have been put forward above. And having done this, either you will have something as a remainder or nothing. And if you have a remainder, subtract it from all the figures from which you wanted to extract or find the cube root. And what remains to you will be the number that had the cube root; it will be the 2 numbers that you found, putting them in sequence one with the other as you found them. And if you have a remainder, those figures would be all that would have a cube, and their

---

1. This procedure is better understood from the example that follows.
2. The manuscript has *radixe chuba*, which is sometimes used for "cube" as well as "cube root."

la sso radixe chuba seria queli numeri li quali tu averai truovadi, chomo t'ò ditto davanti. E se 'l numero del qual tu voi truovar radixe chuba fuse de più figure, zoè che l'avesse 2 figure o più messe in luogo de million, senpre è da fare chomo t'ò ditto davanti. Mo' per chonpir le oltre figure, tu die metter queste 2 figure truovade, zoè questi 2 numeri, lasando l'ultimo in logo preditto secondo che tu lo metesti, e l'altro lo qual tu truovasti in prima lassallo davanti da quello inverso la man senestra. Poi, triplado tuti questi 2 numeri, zoè moltipicha' per 3, e quella moltipichazion redu' avanti inver la man destra, metando tuta fiada la terza figura ultima dela ditta moltipichazion sotto la terza figura dele figure de sovra e le oltre figure del ditto triplado ch'è inver la man senestra chossì per ordene chossì chomo ele va. E sotto lo ditto triplado metti lo so sottotriplado, zoè le 2 figure che noi avemo ditto davanti. E questo fatto, tuo' 1 digitto over 1 numero lo qual metti ala prima
f. 81a figura | la qual è sequente a quela andando inver la man destra, e se[*n*]pre va fazando chomo t'ò ditto davanti senpre.

Tute le figure che tu truoverai serà 'l Rx chuba del numero che tu vorai truovar, metandole a sequenzia l'una driedo al'altra chome tu le truoverai, e per questo muodo postu truovar ogni radixe chuba de algun numero de quante figure vuol eser se sia e se lo averà Rx chuba lo numero. E se lo numero non averà Rx chuba, quelo che tu truoverai serà Rx chuba la più prossima che tu porai. Favelando a numero son a quelo che te avanzarà, trazilo dal numero che tu voi truovar Rx chuba, e quelo che te romagnerà averà Rx chuba, e la sso Rx chuba serà i numeri che tu avera' truovada.

Se noi volemo saver la radixe chuba de 123456 la più prosima que[a] se posa qua[*n*]to in le ditte figure, faremo per la dita riegola de sovra scritta, che in prima è da metter 1 numero sotto el 3, lo qual è da man dretta, e sia fatto che reduto a Rx chuba destruza più che 'l pò quele tre figure, zoè 1, 2, 3, che representa 123, lo qual numero sie 4. Lo qual dutto a Rx chuba monta 64. Lo qual 64 trazi de 123, roman 59. Poi tripla quello 4, zoè moltiplicha per 3, fa 12. Lo qual 12 metti sotto alo 5, e lo 1 chon 12 sotto lo 4, ch'è la terza figura dala man destra dele figure. E sotto lo ditto 12 metti
f. 81b lo so sotto triplado, zoè lo 4, e puo' metti che quello 12 sia 12 | dexene, che dira' 120, lo qual moltiplicha per lo sso triplado, zoè per 4, e averai 480, lo qual salva. Puo' truova 1 numero, lo qual mettilo so[*tto el*] 6, zoè sotto l'ultima figura destra, lo qual sia sì fatto che moltiplicha via lo triplado, zoè con 12, e quela moltipichazion zonta con 480, e la suma che ne vegnirà sia fata che moltiplichada chon lo ditto numero, quela suma che n[*e*] devegnirà che suma tutte quelle figure ch'è al ditto 5, zoè ala segonda figura de man destra, lo qual numero sie 9. Lo qual moltipicha' chon 12, fa 108. Lo qual azonzi a 480, monta 588. Lo qual 588 moltiplicha via lo ditto numero truovado, zoè per 9, e averai 5292. La qual moltipichazion trazi de 5945, roman 653. Azonzi l'ultima figura, che sono 6, a 653, serano 6536. Trazi lo chubicho, che sono 729, resta 5807.[b] Poi redu' quelo 9 a Rx chuba, che ne vien 729, li qual trazi de 6536, roman 5807. E questa suma che te roman sie quelo che te avanza del ditto numero del qual tu voi truovar radixe chuba. Mo' trazi 5807 de 123456, roman 117649, e questo sie quel numero che à radixe chuba. Ala so Rx chuba sie li 2 numeri, zoe 49, che tu truovasti, e se tu la vuol truovar moltiplicha 49 via 49, monta ben 117649, el qual sie numero chubo lo qual ò ditto. Et è fatta.

---

a. *Sic in MS.*

b. *From* Azonzi l'ultima figura *to* 5807 *added in the right margin with reference mark.*

cube root would be those numbers which you would have found, as I said to you before. And if the number of which you want to find the cube root had more figures, that is, if it had 2 figures or more put in the millions place, it is always necessary to do it as I have told you before. Now to complete the other figures, you should put these 2 found figures, that is, these 2 numbers, leaving the last in the place where you put it, and the other which you found first leave it in front of that one toward the left. Then these 2 numbers are tripled, that is multiplied by 3, and this product put forward toward the right, always putting the third last figure of this product under the third figure of the figures above and the other figures of this triple that are toward the left in the same order as they go. And under this triple put its under-triple, that is, the 2 figures that we have said before. And having done this, take 1 digit or 1 number, which you put at the first figure | which follows going to the right, and always doing as I have always told you before. f. 81a

All the figures that you will find will be the cube root of the number that you would like to find, putting them in sequence one after the other as you find them, and in this way you can find every cube root of any number, no matter how many figures it has, if the number has a cube root. And if the number does not have a cube root, what you find will be the nearest cube root that you can find. Speaking of the number that remains, subtract it from the number of which you want to find the cube root, and what remains will have a cube root, and its cube root will be the number that you have found.

If we want to know the cube root of 123,456 as closely as can be done for these figures, we will do it by the rule written above, of which the first is to put 1 number under the 3 which is on the right side, and it should be done so that reduced to its cube it goes into those three figures as much as possible, that is, 1, 2, 3, which represent 123, which number is 4, which when reduced to a cube amounts to 64. Subtract this 64 from 123, which leaves 59. Then triple this 4, that is, multiply it by 3, which makes 12. Put this 12 under the 5, and the 1 with 12 under the 4, which is the third figure on the right of the figures. And under this 12 put its under-triple, that is, the 4, and then put that this 12 is 12 | tens, which is to say 120, which you multiply by its triple, that is by 4, and you will have 480, which you save. Then find 1 number which you put under the 6, that is, under the last figure on the right, which should be such that you multiply by the triple, that is by 12, and you add this product to 480, and the sum that results will be such that multiplied with this number, that sum that results sums all those figures which are at that 5, that is, at the second figure from the right; which number is 9. Multiply this by 12, it makes 108. Add it to 480, it amounts to 588. Multiply this 588 by the number found, that is by 9, and you will have 5,292. Subtract this product from 5,945; it leaves 653. Add the last figure, which is 6, to 653; it will be 6,536. Subtract the cube, which is 729; it leaves 5,807. Then reduce that 9 to its cube, which comes to 729, subtract it from 6,536, which leaves 5,807. And this sum that remains is the remainder of the number of which you want to find the cube root. Now subtract 5,807 from 123,456, which leaves 117,649, and this is the cube number that has the cube root. At its cube root are 2 numbers, that is 49, which you found, and if you want to find it, multiply 49 by 49,[1] it indeed comes to 117,649, which is the cube that I said. And it is done. f. 81b

1. I.e., 49 by 49 by 49.

E se tu vuol truovar la radixe chuba de questo 123456789 è da far in prima tutto questo che tu ài fatto, e zò faro,[a] le tto figure dela qual tu ài voiudo truovar la Rx chuba serà sì termenade che te romagnerà pur in 5807789. Puo' diestu meter quelo 49 sotto la quarta figora che è da man destra, zoè lo 9 del 49 diestu meter sotto lo 7 e lo 4 del 49 diestu meter sotto la nulla. Puo' diesti triplar 49
f. 82a per 3, | che te ne vegnirà 147. Lo qual 147 metti sotto la segonda figura da man destra, zoè lo 7 mezo lo 8 e lo 4 per mezo lo 7 seguende[b] inver la senestra, e lo 1 per mezo l'oltro 7, e lo 49 sottometti alo 147. E poi moltiplicha 49 via 147 dexene, che te ne vegnirà 72030, lo qual salva. Poi truova 1 numero lo qual sottometti al'ultima figura, zoè al 9, e quelo numero moltiplicha chon 147, e quela suma azonta a 72030, e quelo che te vien tal suma che destruza tutte le figure ch'è infina a chavo del triplado, la qual figura sie 7. Moltiplicha 7 via 147, monta 1029. Lo qual 1029 azonzi chon 7230 et averai 73059. La qual suma anchora moltiplicha chon quel numero truovado, zoè chon 7, e monta 511413. Lo qual trazi de 580778, roman 693659. Puo' redu' questo a Rx chuba, che vien 343. La qual trazi de 693659, roman 693316, e questo sie quelo che te avanza de 123456789. E quelo avanzo tratto de questo 123456789 roman 122763473, e questo numero sie quello che à Rx chuba. E la so Rx chuba sie tuti quelli numeri che tu truovasti mesi a sequenzia l'uno al'oltro, chusì per ordene chomo tu li truovasti, li qual chussì mesi vien 497. Mo' moltiplicha 497 via 497, monta 247009. Anchora moltiplichallo via 497, monta ben 122763473, e per questo muodo puostu truovar tute Rx chube, over più apreso che se puol de zaschadun numero chomo grande el se sia, intendando a responder per numero senza algun numero. E questa raxion sie fatta.

f. 82b

[Moltipicar radixe chon numero o radixe de zò che tu vuol; a partir radixe chon numero; a zonzer, a sutrar numeri per radixe]

Quy apresso se chomenza algun introditorio del'alzebra sovra la radixe, chomo se moltipicha, partir e zonzer e trazer, inperzò che a voler inprender aresmettecha el bixogna la praticha de moltipichar, partir e zonzer e trazer d'ogni numero e quantitade ananzi che noi tratiamo del'alzebran e dele sue chostione. Sì diremo del'opera dele radixe quelo che sia nezesario, chon zò sia chosa che le questione asolutte per l'alzebran non posano eser ben fatte senza la doltrina dele ditte radixe. È da sapere in prima che chossa è radixe,[c] onde dicho che radixe de numero sie numero in si moltipichado fa quelo medeximo numero, chomo dirò. 3 è radixe de 9, 4 è Rx de 16, 5 è Rx de 25, 6 è Rx de 36. E sono numeri che ano Rx e alguny numeri che non ano Rx, le chue Rx son ditte Rx sorde, chon zò sia chosa che son inposibille quello nel truovare, ma in che muodo più apreso che possiamo segondo el numero, e quele vegnirà dimostreremo.

Ponyamo che noi voiamo truovar la Rx de 10 prosimamente, fara' chussì: truoverai in prima nel numero la più prosima Rx che ssia in 10, lo qual serà 3, che è Rx de 9, lo qual chontra 10 roman

a. *Sic in MS, probably for* fato.
b. *Sic in MS.*
c. xa ditto *added in the left margin by another hand, most probably at a later time.*
1. Each of the "betweens" in this sentence seems to be a mistake for "below."
2. *7230* in MS.
3. Michael has omitted a step in his explanation: the subtraction leaves 69365, and then you have to "add the last figure" (i.e., tack on the last digit of the original number)—thus, put 9 at the end to get 693659.

And if you want to find the cube root of this 123,456,789, it is to be done exactly as you have done, and when it is done, your figures of which you wanted to find the cube root will have turned out so that you will remain with 5,807,789. Then you should put that 49 under the fourth figure on the right, that is, you should put the 9 of the 49 under the 7 and you should put the 4 of the 49 under the naught. Then you should triple the 49 by 3, | which results in 147. Put this 147 under the second figure to the right, that is, the 7 between[1] the 8 and 4 between the 7 following toward the left, and the 1 between the other 7, and put the 49 below the 147. And then multiply 49 by 147 tens, which will come to 72,030, which you save. Then find 1 number which when you put it below the last figure, that is, the 9, and you multiply this number with 147, and you add that sum to 72,030, what results goes into all the figures which are up to the beginning of the triple, which figure is 7. Multiply 7 times 147, it comes to 1,029. Add this 1,029 with the 72,030[2] and you will have 73,059. This sum you multiply again with the number you have found, that is, with 7, and it comes to 511,413, which you subtract from 580,778, which leaves 693,659.[3] Then reduce this[4] to a cube, which comes to 343. Subtract this from 693,659, which leaves 693,316, and that is the remainder from 123,456,789. And this remainder subtracted from 123,456,789 leaves 122,763,473, and this number is the one that has a cube root. And its cube root is all those numbers that you found put in a sequence one with the other, in the order in which you found them, which put this way comes to 497. Now multiply 497 by 497, it comes to 247,009. Multiply it again by 497, it indeed comes to 122,763,473, and in this way you can find all of the cube roots, or as close as you can to any number however large it may be, trying to answer by a number without any number.[5] And this problem is done. f. 82a

To multiply a root with a number or with whatever you wish, to divide a root by a number, to add, to subtract numbers by a root f. 82b

Here begins an introduction to algebra concerning the root, how it is multiplied, divided, and added, and subtracted, because whoever wants to learn arithmetic needs the knowledge of multiplying, dividing, and adding and subtracting every number and quantity before we treat algebra and its questions. So we will say about the operation of roots only what is necessary, as it is the case that the questions solved by algebra cannot be well done without the knowledge of these roots. And it's necessary to know first what a root is, whence I say that the root of a number is that number which when multiplied by itself makes that same number, as I shall say. 3 is the root of 9, 4 is the root of 16, 5 is the root of 25, 6 is the root of 36. And there are numbers that have roots and some numbers that don't have roots, which roots are called deaf roots, that is, that they are impossible to find, except in the closest fashion which we can according to the number, and we will demonstrate these results.

Let's suppose that we want to find the root of 10 approximately; you will do it this way: you will first find the closest root that there is in 10, which will be 3, which is the root of 9, which up to 10

4. That is, take the cube of 7.
5. This probably means "without any remainder."

1, lo qual parti per lo dopio dela radixe, zoè per 6. Ne vien $\frac{1}{6}$, lo qual azonzi chon 3, et averai $3\frac{1}{6}$ per la radixe de 10, non a ponto ma apreso.

E per truovar la radixe de 40 la più apresso radixe serà 6, el qual è Rx de 36, lo qual trazi de 40. Roman 4, lo qual parti per lo dopio de sie, che son 12, el qual $\frac{1}{3}$ azonzi chon 6. Avera' $6\frac{1}{3}$ per Rx de
f. 83a 40 prosimamente. | E per questo muodo porai truovar Rx de ogno numero segondo l'abacho. Ma segondo mexura la radixe d'ogni numero se può aver a ponto per geometria.

Seguitta el moltipicar el partir dele Rx.[a]

Se tu volesi multipichar radize de numero per Rx de numero, over partir, multipicha l'una chontra l'altra, e la radixe de quella moltiplichazione fare.

E se vui volesi partir l'una per l'altra, parti Rx per radixe e la radixe de quelo avenimento ne verà.

Ponyamo che avesti a moltipichar Rx de 7 in Rx de 8. Moltipicha 7 via 8, fa 56, e la radixe de 56 farà la multipichazione.

E se tu volesti moltipicar 3 in radixe de 8 farai chussì: recherai a radixe quadratta, ch'è 9, e puoi moltipicha 8 via 9, fa 72. Del qual 72 pigla la Rx per la adomandatta multipichazione.

Anchora. Se ttu volesti moltipicar radize de $2\frac{1}{3}$ per radixe de 2, moltipicha $2\frac{1}{3}$ via 2, monta 49, e la radixe de quello 49, ch'è 7, ne vien per la ditta moltipichazione.

Anchora. Se ttu volesti moltipicar over pligrare[b] li $\frac{3}{4}$ de radixe de 32, redu' $\frac{3}{4}$ a radize, ch'è $\frac{9}{16}$, e poi pigla li $\frac{9}{16}$ de 32, ch'è 18, e la Rx de 18 serà li $\frac{3}{4}$ de 32.

f. 83b Se tu volesti partir radixe de 20 in radixe de 5, parti 20 per 5, che ne vien 4, e la radixe de 4, che è 2, vien del ditto partimento.

Anchora. Se volesti partir Rx de 40 per 2, recha 2 a radixe, che è 4. Ora parti 40 per 4, son 10, e Rx de 10 vien delo adomandatto partimentto.

E se avesi a zonzer Rx de 7 chon Rx de 8, e puo' moltipichar in si medeximo, farai chossì: dela multipichazione de radixe de 7 ne vien 7 in si medeximo, e dela moltipichazione de radixe de 8 in si medemo ne vien 8. 8 chon 7 fa 15, e poi moltiplichare radixe de 7 chon Rx de 8 e radopialo, che moltipichado Rx de 7 in Rx de 8 fa Rx de 56. Lo qual radopiado fa radixe de 224, inperzò che

---

a. *Centered on the page.*
b. *Sic in MS, probably for* piglare.
1. Centered on page.
2. This problem is erroneous as written; seemingly it was intended to read "multiply the root of $2\frac{1}{3}$ by the root of 21," for which Michael's solution would be correct.

leaves 1, which you divide by the double of the root, that is, by 6. It comes to $\frac{1}{6}$, which you add to 3, and you will have $3\frac{1}{6}$ for the root of 10, not exactly but near.

And to find the root of 40, the closest root will be 6, which is the root of 36, which you subtract from 40. There remains 4, which you divide by the double of six, which is 12, which $\frac{1}{3}$ you add to 6. You will have $6\frac{1}{3}$ as the root of 40 approximately. | And in this way you will be able to find the f. 83a
root of any number according to the abbacus. But you can find the exact root of every number according to a measurement by geometry.

There follows multiplying and dividing of roots.[1]

If you would like to multiply the root of a number by the root of a number, or divide, multiply one by another and take the root of this product.

And if you would like to divide one by the other, divide the root by the root and the root of that result will come.

Let's suppose that you had to multiply the root of 7 by the root of 8. Multiply 7 by 8, which makes 56, and the root of 56 will be the product.

And if you wanted to multiply 3 times the root of 8 you will do this: you will bring it to the squared root, that is 9, and then multiply 8 times 9, which makes 72. Take the root of that 72 for the required product.

Again. If you wanted to multiply the root of $2\frac{1}{3}$ by the root of 2,[2] multiply $2\frac{1}{3}$ times 2, it comes to 49, and the root of that 49, which is 7, comes from this multiplication.

Again. If you wanted to multiply or take $\frac{3}{4}$ of the root of 32, reduce $\frac{3}{4}$ to its root,[3] that is, $\frac{9}{16}$, and then take $\frac{9}{16}$ of 32, which is 18, and the root of 18 will be $\frac{3}{4}$ of that of 32.

If you wanted to divide the root of 20 by the root of 5, divide 20 by 5, which comes to 4, and the f. 83b
root of 4, which is 2, comes from this division.

Again. If you wanted to divide the root of 40 by 2, take 2 to its root,[4] which is 4. Now divide 40 by 4, which is 10, and the root of 10 comes from this division.

And if you had to add the root of 7 with the root of 8 and then multiply it by itself, you would do it this way: from the multiplication of the root of 7 by itself comes 7, and from the multiplication of the root of 8 by itself comes 8. 8 with 7 makes 15, and then multiply the root of 7 with the root of 8 and double it, which root of 7 multiplied by the root of 8 makes the root of 56. Which

3. That is, its square.
4. That is, its square.

tanto vien a dire quanto a moltipichare per 2. E a redopiare Rx de 56 per 2 fa Rx de 224. E chosì averai per la demandatta moltipichazione 15 a radixe de 224.

E se ttu avesi a moltipichar 7 e Rx de 8 in 7 men Rx de 8, faresti in prima 7 via 7 fa 49, e dela moltipichazione de Rx de in meno Rx de 8 sì ne perviene meno 8. Tralo de 49, roman 41, e poi dela moltipichazione de 7 più Rx de 8 e in 7 meno Rx de 8. Azonzi insieme, ne viene niente. Adoncha, moltipichado 7 e Rx de 8 in 7 meno Rx de 8 sì ne pervien 41.

E se avesti a moltipichar 10[a] e radixe de 20 via 12 e Rx de 20, moltipicha in prima 10 in 12, sì fa
f. 84a 120, e poi radixe de 20 | in radixe de 20, fano 20, et averai in tutto 140. Ora ài a moltipichar 10 in Rx de 20 e 12 in Rx de 20. Adoncha, sichomo a moltipichare 22 in radixe de 20 e se vuole moltipichare 22 in Rx de 20, reccha 22 a Rx, che è 484. Ora ài a moltiplichare 484 via 20, fa 9680, e la Rx de 9680 fa.[b] El qual azonzi a 140, et avera' per la domandata moltiplichazione 140 e Rx de 9680.

Se vuolessi multipichare 4 e radixe de 9 via 5 men radixe de 4, fa' chossì per chaxella e di': "5 via 4 fa 20," e puo' moltipicha Rx de 9 men Rx de 4, e fa' 4 via 9 fa 36 men Rx de 36. Tralo de 20, roman 14. Poi moltiplicha in croxe redutto a Rx via Rx de 9, fa Rx 225, dela qual trazi 4. Redutto a Rx, moltipichado per men Rx de 4 fa men Rx de 64. Sì che trata questa Rx de 8 de quela de 225, ch'è 15, roman 7, el qual azonzi a 14. Serano 21, chomo vedirì qui de sotto per figura.

4	più	Rx	de	9		4	5
						4	5
						16	25
5	men	Rx	de	4		4	9
						64	225 Rx più 15

più 20	men 8	64 Rx men 8
più 15	men 6	
più 35	14 men	35
		14
		21

Moltiplichado per radixe el sovrascritto chapittollo serà 21. Et è fatta.

---

a. 10 *added above the line with reference mark.*
b. *Number lacking in MS, with no space left for it.*

doubled makes the root of 224, because that's what it amounts to from multiplying by 2.[1] And to double the root of 56 by 2 makes the root of 224. And so you will have for the required multiplication 15 and[2] the root of 224.

And if you had to multiply 7 and the root of 8 by 7 minus the root of 8, you would first do 7 times 7 makes 49, and from the multiplication of the root by the minus root of 8 comes minus 8. Subtract it from 49, which leaves 41, and then from the multiplication of 7 plus the root of 8 by 7 minus the root of 8, add together, it comes to nothing. Then 7 and the root of 8 multiplied by 7 minus the root of 8 comes to 41.

And if you had to multiply 10 and the root of 20 by 12 and the root of 20, first multiply 10 by 12, which makes 120, and then the root of 20 | by the root of 20, which makes 20, and you will have f. 84a
in all 140. Now you have to multiply 10 by the root of 20 and 12 by the root of 20. So, as it is necessary to multiply 22 by the root of 20 and if you want to multiply 22 by the root of 20, take 22 to its root, which is 484. Now you have to multiply 484 by 20, which makes 9,680, and the root of 9,680 makes.[3] Add this to 140, and you will have for the required multiplication 140 and the root of 9,680.

If you wanted to multiply 4 and the root of 9 by 5 minus the root of 4, do as in the box and say "5 times 4 makes 20," and then multiply the root of 9 minus the root of 4, and do 4 times 9 makes 36, minus the root of 36. Subtract it from 20, which leaves 14. Then multiply across 5 reduced to its root by the root of 9, which makes the root of 225, from which you subtract 4. Reduced to its root, multiplied by minus the root of 4, it makes minus the root of 64. Then take this root of 8 from that of 225, which is 15, which leaves 7, which you add to 14. It will be 21, as you will see worked out below.

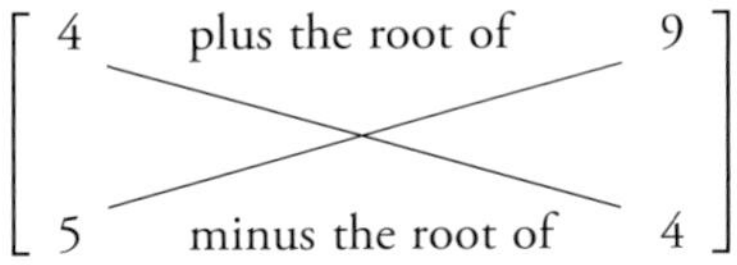

4	5
4	5
16	25
4	9
64	225 root plus 15

plus 20	minus 8	64 root minus 8
plus 15	minus 6	
plus 35	14 minus	35
		14
		21

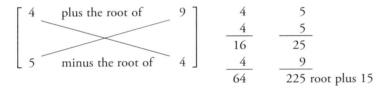

Multiplied by the root, the paragraph above will be 21. And it's done.

1. That is, 2 squared.
2. MS has *a*, which means "to."
3. Number lacking in MS, with no space left for it.

f. 84b Quando volessi rezonzer over trazer Rx de numero chon Rx de numero, dei moltiplichare l'una per l'oltra e la Rx del moltipichamentto radopiare, lo qual redopiatto rezonzer chon l'una e chon l'altra Rx che tu voi rezonzer, e la Rx del rezozamento fara' chomo tu volessi razonzer la Rx de 8[a] chon la Rx del 18. Dicho che dei moltiplichare 8 via 18, fa 144, piglare la radixe de 12, la quale redopia', fa 24, lo qual zonzi chon 18 e chon 8, zoè chon 26, fa 50. E chossì fara' el ditto razozamento, zoè Rx de 50.

E se tu volesti trazer radize de 8 de radize de 18, lo dopio dela Rx dela moltiplichazione de 8 in 18, el qual è 24, trazi delo azozamentto de 8 in 18, zoè de 26, roman 2, e chosì averai Rx de 8 de Rx de 18, roman Rx de 2. E questo medemo rezozamentto e sotramento tuto in altro modo se può fare. Zoè che se tu vuol razonzer radize de 8 chon quela de 18, parti la Rx de 18 per la Rx de 8, zoè senpre la mazuor Rx dala menor Rx, che ne vien Rx de $2\frac{1}{4}$, la chui Rx è $1\frac{1}{2}$, e truovase chossì: recha $2\frac{1}{4}$ a quarti, che son $\frac{9}{4}$, la ssoa Rx sie $\frac{3}{2}$, zoè $1\frac{1}{2}$, al qual azonzi senpre uno, farà $2\frac{1}{2}$. Lo qual $2\frac{1}{2}$ moltiplicha in quela in quela[b] Rx che fu partida, zoè in Rx de 8. Adoncha ài a moltiplichar $2\frac{1}{2}$ via Rx de 8, e quela farà Rx de 50 per lo adomandatto azonzamentto.

E se a questo medeximo muodo vorai trazer Rx de 8 chon Rx de 18, parti per 8, che ne vien Rx de
f. 85a $2\frac{1}{4}$, el quale è $1\frac{1}{2}$, del qualle trazi sempre 1, roman $\frac{1}{2}$. Lo qual $\frac{1}{2}$ mol|tiplicha per quela Rx che fu partida, zoè per Rx de 8, el quale fa Rx de 2 per lo romagnente, tratola de 8 dela Rx de 18. E chussì farai quando avessi a rezonzer ho trare de Rx for de de[c] numeri. Se[d] inperò è da sapere che se la moltipichazione overo partimento del'una Rx nel'altra serà numero lo quale non abia perixamente radixe, che le ditte Rx non si possano rezonzer nì trare, sichomo manifestamente apare nel prexente demostramento. E dotene esenpio qui apresso nel seguente chapittullo.

Io voi azonzer Rx de 7 chon Rx de 8. Tu die moltipichar 7 via 8, fano 56, e de questo se vuol pligiare la Rx de 56. Non à Rx, ma la Rx de 56 se vuol redopiare, zoè 4 via 56 fa 224. Ora se vuol razonzer 7 e 8, fano 15, e azonzi la Rx de 224, che fa 15, che è Rx de 224. Abiamo che rezonti inssieme Rx de 7 chon Rx de 8 fano presso la Rx de 224 azonto sovra 15, la Rx dela suma per l'adomandatto agiugiamentto.

E se tu volesti moltipichar Rx chuba de 5 chon Rx chuba de 10, moltipicha 5 via 10, fa 50, e la Rx chuba de 50 fa la ditta moltipichazione. E se tu vuolesti moltiplichar 3 in Rx chuba de 10, redu' 3 a Rx chuba, ch'è de 9. 3 via 3 fa 9 e 3 9 via fa 27. Ora moltipicha 27 via 10, fa 270, e chussì ài per la moltipichazione de 3 in Rx de 10, zoè chuba, Rx de 270, zoè chuba.

---

a. 8 *corrected over* 18.
b. *Thus repeated in MS.*
c. *Thus repeated in MS.*
d. *Thus in MS, possibly for* et.

When you would like to add together or subtract the root of a number with the root of a number, you must multiply one by the other and double the root of the product, which when doubled you add with one and the other root that you want to add, and the root of the sum will be as if you wanted to add the root of 8 with the root of 18. I say that you should multiply 8 times 18, which makes 144, take the root of 12, which you double, which makes 24, which you add to 18 and to 8, that is to 26, which makes 50. And so you will do the addition, that is, the root of 50. f. 84b

And if you wanted to subtract the root of 8 from the root of 18, the double of the root of the product of 8 by 18, which is 24, subtract it from the sum of 8 and 18, that is, 26, which leaves 2, and so you will have the root of 8 from the root of 18, which leaves the root of 2. And this same addition and subtraction can be done in a completely different way. That is, that if you want to add the root of 8 with that of 18, divide the root of 18 by the root of 8, that is, always the greater by the lesser root, which comes to the root of $2\frac{1}{4}$, whose root is $1\frac{1}{2}$, and it is found this way: put $2\frac{1}{4}$ in quarters, which is $\frac{9}{4}$, its root is $\frac{3}{2}$, that is $1\frac{1}{2}$, to which you always add 1, which will make $2\frac{1}{2}$. This $2\frac{1}{2}$ you multiply by that root that was divided, that is, by the root of 8. Then you have to multiply $2\frac{1}{2}$ by the root of 8, and this will make the root of 50 for the requested addition.

And if in this same way you would like to subtract the root of 8 from the root of 18, divide by 8, which comes to the root of $2\frac{1}{4}$, which is $1\frac{1}{2}$, from which you always subtract 1, which leaves $\frac{1}{2}$. Multiply this $\frac{1}{2}$ | by that root that was divided, that is, by the root of 8, which makes the root of 2 as the remainder; subtract it from 8 from the root of 18.[1] And so you will do when you have to add or subtract the roots of numbers. But on the other hand it's necessary to know that if the multiplication or the division of one root with the other will be a number that doesn't have an even root, that these roots cannot be added nor subtracted, as clearly appears in the present demonstration. And I give you an example of it near here in the following paragraph. f. 85a

I want to add the root of 7 with the root of 8. You should multiply 7 by 8, which makes 56, and from this it's necessary to take the root of 56. It doesn't have a root, but the root of 56 should be doubled, that is, 4 times 56 makes 224. Now if you want to add 7 and 8, it makes 15, and add the root of 224, which makes 15 and the root of 224.[2] We have that added together, the root of 7 with the root of 8 makes next to the root of 224 added to 15, the root of the sum for the requested addition.[3]

And if you wanted to multiply the cube root of 5 with the cube root of 10, multiply 5 times 10, which makes 50, and the cube root of 50 makes this multiplication. And if you wanted to multiply 3 by the cube root of 10, reduce 3 to its cube, which is of 9. 3 times 3 makes 9 and 3 times 9 makes 27. Now multiply 27 by 10, which makes 270, and so you have for the multiplication of 3 by the root of 10, that is, the cube root, the root of 270, that is, the cube root.

1. I.e., "when the root of 8 is subtracted from the root of 18." The required solution has already been given ("the root of 2"); the last phrase seems to be a restatement of the problem.
2. *15 which is the root of 224* in MS.
3. That is, the solution to this problem is $\sqrt{7} + \sqrt{8} = \sqrt{15 + \sqrt{224}}$.

f. 85b E se vuolesti partir Rx chuba de 50 in Rx chuba de 10, | moltipicha 5 via 10, fa 50, e la Rx chuba de 50 fa la ditta moltipichazione. E se vuolesti moltipichar 3 in Rx chuba de 10, redu' 3 a Rx chuba ch'è de 9, zoè 3 via 3 fa 9 e 3 via 9 fa 27. Ora moltipicha 27 via 10, fano 270, e chussì ài per moltipichazione de 3 in Rx de 10, zoè chuba, Rx de 270, zoè chuba.

E se vuolesti partir Rx chuba de 50 in Rx chuba de 10, parti 50 per 10, ne vien 5. E la Rx chuba de 5 ne vien partando Rx chuba de 50 in Rx chuba de 10.

E se tu volesti partir 10 per la Rx chuba de 5, recha 10 a Rx chuba in questo muodo: 10 via 10 fano 100, e 100 via 10 fano 1000. Ora ài rechato 10 a radize chuba ch'è 1000. Ora ài a partir 1000 per 5, ne vien 200. E chussì averai che partando 10 per radixe chuba de 5 sì ne vien radixe chuba de 200.

E se vuolesti partir radixe chuba de 50 per 2, recha 2 a radize chuba, ch'è 8, zoè 2 via 2 fa 4 e 2 via 4 fa 8. Ora parti 50 per 8, ne vien $6\frac{1}{4}$. E chussì ài per lo ditto partimentto Rx chuba de $6\frac{1}{4}$.

E se volesti multiplichar Rx chuba de 10 per Rx quadratto[a] de 6, farai chussì: recha 10 a Rx quadratta, che è 100. Recha 6 a Rx chuba, ch'è 216, zoè 6 via 6 fa 36, e 6 via 36 fa 216. Ora moltipicha 100 via 216, fa 21600, o volettu dir la Rx chuba da la Rx quadratta de 21600. Simiglantemente se tu volesti partir Rx chuba de 10 per Rx quadratta de 6, parti 100 per 216, ne vien $\frac{25}{54}$, averai al partimento ne verà Rx de Rx chuba de $\frac{25}{54}$.

f. 86a E se vuolesi partir Rx quadratta de 10 per la Rx chuba de 6, parti 216 per 100, che ne vien $2\frac{4}{25}$, e la Rx dela radize chuba de $2\frac{4}{25}$ che ne vien dal ditto partimento. E sapi che bene che non se puossa responder per numero pezisamente,[b] alguna volta se può a queste chontra tali respondere più brevemente per la tal chossa. [*In*]tendo per exenpio Rx de Rx chubicha de 100 sie chomo pura Rx chuba de 10, inperzò che la Rx de 100 sie 10, e anche Rx de Rx chubicha de 125 sie chomo pura Rx quadratta de 5, inperzò che la Rx de 125 sie 5.

E se vuolesti azonzer Rx chubicha chon Rx chubicha, over trazer Rx chubicha de Rx chubicha, debi partir la mazuor dala menor de quelo che te vien. Pigla la Rx overo trare 1° se tu à 1 a trare, e de quello che tu averai dopo la zonta hovero da puo' la menovatta moltiplicha chontra ala menor Rx chubicha dividesti. Chomo se tu volesti razonzer radize chubicha de 12 chon le radize chubiche de 96, parti 96 per 12, che ne vien 8, del qua[*l*] 8 pigla Rx chuba, ch'è 2, el qual azonzi a uno, farà 3, el qual 3 moltipicha in Rx chubicha de 12, fano Rx chubicha de 324, inperzò che 3 se vuol rechar a radixe chubicha, ch'è 27. Ora se vuol moltiplichar 12 via 27, fano 324. E chossì ài che zezonzento radixe Rx de 12 fano chon Rx chubicha de 96, farà Rx chubicha 324.

---

a. *Sic in MS.*

b. *Sic in MS, probably in place of* prezisamente.

1. From this point, the text of this paragraph repeats that of the previous problem, which was about multiplication, not division.

And if you wanted to divide the cube root of 50 by the cube root of 10, | multiply[1] 5 times 10, which makes 50, and the cube root of 50 completes this multiplication. And if you wanted to multiply 3 by the cube root of 10, reduce 3 to the cube which is of 9, that is, 3 times 3 makes 9 and 3 times 9 makes 27. Now multiply 27 by 10, which makes 270, and so you have for the multiplication of 3 by the root of 10, that is, the cube root, the root of 270, that is, the cube root. f. 85b

And if you wanted to divide the cube root of 50 by the cube root of 10, divide 50 by 10, which comes to 5. And the cube root of 5 comes from dividing the cube root of 50 by the cube root of 10.

And if you wanted to divide 10 by the cube root of 5, take 10 to its cube in this way: 10 times 10 makes 100, and 100 times 10 makes 1,000. Now you have taken 10 to its cube, which is 1,000. Now you have to divide 1,000 by 5, which comes to 200. And so you will have that dividing 10 by the cube root of 5 comes to the cube root of 200.

And if you wanted to divide the cube root of 50 by 2, take 2 to its cube which is 8, that is 2 times 2 makes 4 and 2 times 4 makes 8. Now divide 50 by 8, which comes to $6\frac{1}{4}$. And so you have for this division the cube root of $6\frac{1}{4}$.

And if you wanted to multiply the cube root of 10 by the square root of 6, you will do it this way: take 10 to its square, which is 100. Take 6 to its cube, which is 216; that is, 6 times 6 makes 36, and 6 times 36 makes 216. Now multiply 100 times 216, which makes 21,600, or rather the cube root of the square root of 21,600. Similarly if you wanted to divide the cube root of 10 by the square root of 6, divide 100 by 216, which comes to $\frac{25}{54}$; you will get from the division the root of the cube root of $\frac{25}{54}$.

And if you wanted to divide the square root of 10 by the cube root of 6, divide 216 by 100,[2] which comes to $2\frac{4}{25}$, and the root of the cube root of $2\frac{4}{25}$ which results from this division. And know that when one cannot answer with a precise answer, sometimes it's possible to answer such questions more briefly for such a thing. I mean for example the root of the cube root of 100 is like the simple cube root of 10, because the root of 100 is 10, and also the root of the cube root of 125 is like the simple square root of 5, since the root of 125 is 5. f. 86a

And if you wanted to add a cube root to a cube root, or subtract a cube root from a cube root, you must divide the greater by the lesser of what results. Take the root or rather subtract one if you have one to subtract, and from that which you will have after the addition or rather after the subtraction multiply against the lesser cube root you have divided. As if you wanted to add the cube root of 12 with the cube root of 96, divide 96 by 12, which comes to 8, from which 8 take the cube root, which is 2, which you add to one, which will make 3, and that 3 you multiply by the cube root of 12, which makes the cube root of 324, because 3 is to be reduced to its cube, which is 27. Now it is necessary to multiply 12 by 27, which makes 324. And so you have that the cube root of 12 added to the cube root of 96 will make the cube root of 324.

---

2. This is the solution for dividing the square root of 6 by the cube root of 10 (rather than the square root of 10 by the cube root of 6, as stated).

f. 86b E se tu avessi a moltiplichar Rx chuba de 6 e radixe | chubicha de 48 in Rx chubicha del 48 meno Rx chubicha del 6, moltiplicha Rx chubicha de 48 via Rx chubicha de 48, fa Rx chubicha de 2304. Ora ài a moltiplichare Rx chuba de 48 Rx chuba de 6 e poi Rx chubicha de 48 via Rx chuba de 6 meno, dela qual moltiplichazione averai niente. E poi a moltiplicar Rx chuba de 6 via Rx chubicha de 6 meno fa meno Rx chubicha de 36, et ài Rx chubicha de 2304 meno Rx chubicha de 36. Adonque se se[a] può eser, trai Rx chuba de 36 de Rx chubicha de 2304 e parti 2304 per 36, che 'n dè 64. Piane la Rx chubicha ch'è 4 e trar 1, roman 3. Ora moltipicha 3 via Rx chubicha de 36, recha 3 a chobicha Rx, ch'è 27. Ora moltiplicha 27 via 36, fano 972, e chossì ài che trando Rx chubicha de 36 de Rx chubicha de 2304 roman Rx chubicha di 972. E chussì tu ài che moltipichando Rx chubicha de 6 e Rx chobicha de 48 via Rx chubicha de 48 meno Rx chubicha de 6 fa Rx chubicha de 972.

Se avesi a moltiplicar Rx de Rx de 10 per Rx de Rx de 20, moltiplicha 10 via 20, fano 200, et averai per la domandatta moltiplichazione Rx de Rx de 200. E se vuolesi moltiplichar 12 Rx de Rx de 10, recha 12 in radixe de Rx e di': "12 via 12 fa 144, e 144 via 144 fa 20736." Ora moltiplicha 10 via 20736, fa 207360. E la Rx dela Rx de 207360 averai per la dimandatta moltiplichazione.

f. 87a Se tu volesti partir 12 per Rx de Rx de 10, partira' 2736 in 10, che ne vien $2073\frac{3}{5}$, e la ssoa Rx dela Rx ne vien dal ditto partimento.

Se vuolesti moltiplichar Rx de Rx de 20 per Rx de 6, fa chosì: recha Rx de 6 a Rx de Rx, ch'è 36, e poi moltiplicha 20 via 36, monta 720. E se volesti partir Rx de Rx de 20 per Rx de 6, parti 20 in 36, che monta $\frac{5}{9}$, e la Rx de Rx de $\frac{5}{9}$ ne vien dal ditto partimento.

E chomo se die moltiplichar Rx de Rx de 10 men Rx de 2 via altrotanto? Moltiplicha in prima Rx de Rx de 10, fa Rx de 10. Ora moltiplicha Rx de 2 meno via Rx de 2 meno, fa 2 più. Poi ài a moltiplichare Rx de Rx de 10 via Rx de 2, fa Rx de Rx de 40. Lo qualle se vuole redopiare, zoè a moltipicare per 2, fano Rx de Rx de 640. Et averai per la dimandatta moltipichazione Rx de 10 più 2 e meno Rx de Rx de 640.

E se avessi a moltiplichar Rx de Rx de 20 e Rx de 3 meno 2 via Rx de Rx de 20 e Rx de 3 meno 2, fa chossì: moltiplicha in prima Rx de Rx de Rx de 20 e Rx de 3 per si medeximo, zoè Rx de Rx de 20 in ssi fa 20, e Rx de 3 via Rx de 3 fa 3, e cho[*n*] Rx de 20 più 3. Or ài a moltiplichar Rx de Rx de
f. 87b 20 via Rx de 3, fano Rx de 180, el qual se vuol redopiare, farà Rx de Rx | de 2880, et averai per la

a. *Thus repeated in MS.*
1. *2736* in MS.
2. There is one "root of" too many here.

And if you had to multiply the cube root of 6 and the | cube root of 48 by the cube root of 48 f. 86b
minus the cube root of 6, multiply the cube root of 48 by the cube root of 48, which makes the cube root of 2,304. Now you have to multiply the cube root of 48 by the cube root of 6 and then the cube root of 48 by the cube root of minus 6, from which multiplication you will have nothing. And then multiplying the cube root of 6 by minus the cube root of 6 makes minus the cube root of 36, and you have the cube root of 2,304 minus the cube root of 36. So if it is possible, subtract the cube root of 36 from the cube root of 2,304 and divide 2,304 by 36, which is 64. Take the cube root, which is 4, and subtract 1, which leaves 3. Now multiply 3 times the cube root of 36, take 3 to the cube, which is 27. Now multiply 27 by 36, which makes 972, and so you have that subtracting the cube root of 36 from the cube root of 2,304 leaves the cube root of 972. And so you have that multiplying the cube root of 6 and the cube root of 48 times the cube root of 48 minus the cube root of 6 makes the cube root of 972.

If you had to multiply the root of the root of 10 by the root of the root of 20, multiply 10 times 20, which makes 200, and you will have for the requested product the root of the root of 200. And if you wanted to multiply 12 by the root of the root of 10, take 12 to the root of the root and say: "12 times 12 makes 144, and 144 times 144 makes 20,736." Now multiply 10 times 20,736, which makes 207,360. And you will have the root of the root of 207,360 for the requested product.

If you wanted to divide 12 by the root of the root of 10, you will divide 20,736[1] by 10, which f. 87a
comes to $2{,}073\frac{3}{5}$, and its root of the root comes from this division.

If you wanted to multiply the root of the root of 20 by the root of 6, do this: take the root of 6 to the root of the root, which is 36, and then multiply 20 times 36, which comes to 720. And if you wanted to divide the root of the root of 20 by the root of 6, divide 20 by 36, which comes to $\frac{5}{9}$, and the root of the root of $\frac{5}{9}$ is the result of this division.

And how does one multiply the root of the root of 10 minus the root of 2 by as much? First multiply the root of the root of 10, which makes the root of 10. Now multiply minus the root of 2 by minus the root of 2, which makes plus 2. Then you have to multiply the root of the root of 10 by the root of 2, which makes the root of the root of 40. This has to be doubled, that is, multiplied by 2, which makes the root of the root of 640. And you will have for the requested multiplication the root of 10 plus 2 and minus the root of the root of 640.

And if you had to multiply the root of the root of 20 and the root of 3 minus 2 times the root of the root of 20 and the root of 3 minus 2, do it this way: first multiply the root of the root of the root of[2] 20 and the root of 3 by themselves, that is, the root of the root of 20 by itself makes 20,[3] and the root of 3 times the root of 3 makes 3. And with the root of 20 plus 3, you now have to multiply the root of the root of 20 by the root of 3, which makes the root of[4] 180, which must be doubled. It will make the root of the root | of 2,880, and you will have for the product of the root f. 87b

3. Correctly, "makes the root of 20."
4. Correctly, "the root of the root of."

moltiplichazione de Rx de Rx de 20 e radixe de 3 in si medesimo Rx de Rx de 2880 e Rx de 20[a] più 3. Apreso ài a moltiplichare Rx de Rx de 20 e Rx de 3 via meno 2, fano Rx de Rx de 320 e Rx de 12 meno. I quali se vuol redopiare, fano Rx de Rx de 5120 e Rx de 48, et anchor queli altri numeri Rx de Rx de 2880 et Rx de 20 plui 3 meno Rx de Rx de 5120 e meno Rx de 48. Ora vi zonzi 2 via meno 2 fa plui 4, e cho' insuma Rx de Rx de 2880 e Rx de 20 plui 7 meno Rx de Rx de 5120, et anche meno Rx de 48. Ma quanto[b] avera' piena dotrina de razonzer e trar d'ogni maniera de Rx, se potrai razonzer e trare che là dove è de bexogno che sia possibille.

Quando tu volesti razonzer Rx de Rx de 20 chon la Rx de Rx de 320, partirai 320 per 20, che 'n devien 16, dela qual pia la Rx de Rx, ch'è 2, e azonzi 1, farà 3. El qual 3 moltiplicha in Rx dela Rx de 20, fano Rx de Rx de 1620, e chosì ài che zonzendo insieme Rx de Rx de 20 chon la Rx dela Rx de 320 fano Rx de Rx de 1620.

E se vorai trare Rx de Rx de 20 dela Rx de Rx de 320, parti 320 per 20, ne vien 16, el quale pia la Rx dela Rx, ch'è 2, del qual senpre ne trai 1. Ne roman 1. Apresso moltiplicha 1 via Rx de Rx de
f. 88a 20, fano Rx de Rx de 20, de Rx de Rx de 320. Sì romane Rx de Rx de 20. E anche | sapi che tutte le radixe dele Rx non si possano zonzere né trazere se non quele che 'l partimento del'una nel'altra avese Rx de Rx. E chosì anche ti ramento dele chubiche, che se le divixione non aveseno radixe chubiche, quelle non se possano razonzer nì trare.

Quando volesemo partire algun numero overo alguna Rx o Rx de Rx, overo numero a Rx per alguno binomio, debia in prima moltiplichare lo binomio chon lo residuo. E quela moltiplichazione sempre viene numero e partidore per lo residuo de quelo binomio quando se moltiplicha per lo residuo. E quelo che ne vien è quelo ch'è domandado.

E se volesti partir alguna quantità delo residuo die moltiplichar quelo che noi vogliamo partire per lo binomio de quelo residuo. E quela moltiplichazione dibiamo[c] partire per lo ditto partidore che se truovò quando che se moltiplichò di binomio per lo suo residuo. E dotti per esenpio questo:

Parti 60 per 4 e Rx de 12. Adomando quelo che ne vien. Tu die in prima truovar el partidor in quisto muodo: die moltiplichare 4 e Rx de 12 via 4 meno Rx de 12 in questo muodo: 4 via 4 fa 16. E puo' tu die molti[*pli*]char 4 via Rx de 12 e 4 via meno Rx de 12 e azonzi chon 16. Farà pur
f. 88b 16, inperzò che de quele moltipichazion non de pervene niente. E poi moltiplicha Rx de 12 | via meno Rx de 12, azonzi chon 16, farà pur 4. Doncha moltiplichando lo binomio per lo suo residuo viene per quela moltiplichazione 4. E quisto è lo partidore. Poi se vole moltiplichare quela quanti-

---

a. 20 *corrected over* 3.
b. *Sic in MS.*
c. *Sic in MS, very probably for* dobiamo.

of the root of 20 and the root of 3 by itself the root of the root of 2,880 and the root of 20 plus 3. Next you have to multiply the root of the root of 20 and the root of 3 by minus 2, which makes the root of the root of 320 and the root of 12 minus.[1] This must be doubled, which makes the root of the root of 5,120 and the root of 48, and also those other numbers: the root of the root of 2,880 and the root of 20 plus 3 minus the root of the root of 5,120 and minus the root of 48. Now you add 2 times minus 2, which makes plus 4, and together with the root of the root of 2,880 and the root of 20 plus 7 minus the root of the root of 5,120 and also minus the root of 48. Now when you have a full knowledge of adding and subtracting all kinds of roots, you will be able to add and subtract where it is necessary and it is possible.

When you would like to add the root of the root of 20 with the root of the root of 320, you will divide 320 by 20, which comes to 16, from which you take the root of the root, which is 2, and add 1, which will make 3. Multiply this 3 by the root of the root of 20, which makes the root of the root of 1,620, and so you have that adding together the root of the root of 20 with the root of the root of 320 makes the root of the root of 1,620.

And if you would like to subtract the root of the root of 20 from the root of the root of 320, you divide 320 by 20, which comes to 16, from which you take the root of the root, which is 2, from which you always subtract 1. There remains 1. Now multiply 1 times the root of the root of 20, which makes the root of the root of 20, from the root of the root of 320 there remains the root of the root of 20. And also | know that all of the roots of the roots cannot be added nor subtracted f. 88a
except those whose division one from the other has a root of a root. And I remind you that it's the same for the cubes, that if the divisions do not have a cube root, they cannot be added or subtracted.

When we wanted to divide any number or rather any root or root of a root, or rather number with root by any binomial, first it's necessary to multiply the binomial with the remainder. And that multiplication always comes with the number and divisor for the remainder of that binomial when it is multiplied by the remainder. And the result is that which was sought.

And if you wanted to divide any quantity of the remainder, you must multiply that which we want to divide by the binomial of that remainder. And we should divide that product by that divisor that was found when the binomial was multiplied by its remainder. And I give you this as an example:

Divide 60 by 4 and the root of 12. I ask you what comes from it. You should first find the divisor in this way: you should multiply 4 and the root of 12 by 4 minus the root of 12 in this way: 4 times 4 makes 16. And then you should multiply 4 times the root of 12 and 4 times minus the root of 12 and add to 16. It will make simply 16, because from that multiplication came nothing. And then multiply the root of 12 | times minus the root of 12, add to it 16, which will make simply 4. Then f. 88b
multiplying the binomial by its remainder gives 4 by that multiplication. And this is the divisor. Then that quantity that we want to divide should be multiplied, that is 60, by the remainder of

1. I.e., minus both these numbers.

tade che noi voyamo partire, zoè 60, per lo residuo de quel binomio, zoè per 4 meno Rx de 12. E puo' moltiplicha 60 via 4 meno Rx de 12, fano 240 meno Rx de 43200. E quele se vuol partir per 4 che fo partidore, viene 60 meno Rx de 2700. E quisto è quello ch'è adomandatto, zoè 60 meno Rx de 2700.

E se volesti partir 40 in Rx de 12 e per Rx de 18, die chavare 12 da 18, roman 6, e moltiplicha 6 6, fa 36, e questo serà partidore. Ora se vuol rechare 40 a Rx, che è 1600. Ora moltipicha 18 in 1600, fano 28800, e partir in 36, che 'n devien 800. Poi se vuole moltiplichare 12 via 1600, monta 19200, a partir per 36 che 'n devien $533\frac{1}{3}$. E diremo che partido 40 per Rx de 12 e Rx de 18 ne vien radixe de 800 meno Rx de $533\frac{1}{3}$.

E se volesti partir 20 per Rx de 5 e Rx de 7, la riegola sie questa: trazi 5 de 7, roman 2, e questo è partidore. Ora die moltiplichare 20 via Rx de 7, fano Rx de 2800, inperò che se vuole rechar 20 a Rx, fano 400. Ora se vuol moltiplichare 7 via 400, fano 2800. Apresso[a] se vuol partir Rx de 2800 in 2, ne vien Rx de 700. Puoi se vuole moltiplichare 20 via Rx de 5, fano Rx de 2000, inperò che se
f. 89a vuol | rechar 20 a Rx, che è de 400, e moltiplichar 5 via 400, fano 2000. Apreso se vuol partir le 2000 in 2, ne vien Rx de 500. E cossì responderai che partando 20 per Rx de 7 e radixe de 5 sì ne viene Rx de 500. E se dizesse che io volesse partire 20 in Rx de 7 meno Rx de 5, chossì faresti chomo ài fatto ora. Mo' responderesti che 'l ditto partimento ne verebe Rx de 700 e Rx de 500.

E se avessi a partire 20 per 3 e Rx de 5 farai chusì: recha 3 a radixe, ch'è 9. Hora è sichomo dizessi: "Io voio partir 20 per Rx de 9 et e[b] Rx de 5," che sai che se vuole chavar 5 de 9 roman 4. E questo è lo partidore. Poi se vuol moltiplichare 20 via Rx de 9, zoè 20 via 3, fa 60, a partir per 4 ne vien 15. Hora se vuol moltiplichare 20 via Rx de 5, fa Rx de 2000. A partir per 4 ne vien Rx de 125. E chossì abiamo che partando 20 per 3 e Rx de 5 sì ne vien 15 meno Rx de 125. Mo' partendo 20 per 3 meno Rx 5 sì ne vien 15 e radize de 125. E chosì farai quando averai a partir numero, over Rx de numero per do Rx. E sapi che quele 2 Rx nel qualle apartire se potrano azonzer insieme a partir pur per una radize solla.

E se tu volesti apartir 50 e Rx de 145 per 6 e Rx de 20, sì partiremo 50 per 6 et Rx de 20. E poi sì
f. 89b partiremo Rx | de 145 per 6 e Rx de 20. Ma prima partiremo 50 per 6 e Rx de 20. Truova' el partidore, die far 6 via 6, fano 36. Chavane 20 de 36, roman 16, e poi moltiplicha 50 via 6, fano 300. A partir per 16 ne vien $18\frac{3}{4}$. Poi moltiplicha 50 via 50, fano 2500, e poi 36 via 2500, fano

---

a. in 2 ne vien *crossed out with diagonal lines.*
b. *Thus repeated in MS.*

that binomial, that is by 4 minus the root of 12. And then multiply 60 by 4 minus the root of 12, it makes 240 minus the root of 43,200. And that should be divided by 4 which was the divisor, which comes to 60 minus the root of 2,700. And this is that which was sought, that is, 60 minus the root of 2,700.

And if you wanted to divide 40 by the root of 12 and the root of 18, you should take 12 from 18, which leaves 6, and multiply 6 by 6, which makes 36, and this will be the divisor. Now it is necessary to take 40 to its square, which is 1,600. Now multiply 18 by 1,600, which makes 28,800, and divide by 36, which comes to 800. Then 12 should be multiplied by 1,600, which comes to 19,200, to divide by 36, from which comes $533\frac{1}{3}$. And we will say that 40 divided by the root of 12 and the root of 18 comes to the root of 800 minus the root of $533\frac{1}{3}$.

And if you wanted to divide 20 by the root of 5 and the root of 7, the rule is this: subtract 5 from 7, which leaves 2, and this is the divisor. Now you should multiply 20 times the root of 7, which makes the root of 2,800, but it's necessary to take 20 to its root, which makes 400. Now it's necessary to multiply 7 by 400, which makes 2,800. Next it's necessary to divide the root of 2,800 by 2, which comes to the root of 700. Then 20 should be multiplied by the root of 5, which makes the root of 2,000, but it's necessary | to take 20 to its root, which is 400, and multiply 5 by 400, which makes 2,000. Now it's necessary to divide the 2,000 by 2, which comes to the root of 500. And so you will answer that dividing 20 by the root of 7 and the root of 5 comes to the root of 500.[1] And if I said that I wanted to divide 20 by the root of 7 minus the root of 5, you would do as you have done now. Now you would answer that this division would come to the root of 700 and the root of 500. f. 89a

And if you had to divide 20 by 3 and the root of 5 you would do it this way: take 3 to its root, which is 9. Now it is as if you said: "I want to divide 20 by the root of 9 and the root of 5," because you know that if it's necessary to take 5 from 9, the remainder is 4. And this is the divisor. Now it's necessary to multiply 20 times the root of 9, that is, 20 times 3, which makes 60, and that divided by 4 comes to 15. Now it's necessary to multiply 20 times the root of 5, which makes the root of 2,000. Divided by 4, that comes to the root of 125. And so we have that dividing 20 by 3 and the root of 5 comes to 15 minus the root of 125. Now dividing 20 by 3 minus the root of 5 comes to 15 and the root of 125. And you will do it this way when you have to divide a number, or the root of a number, by two roots. And know that those two roots in which you want to divide can be added together to divide simply by only a single root.

And if you wanted to divide 50 and the root of 145 by 6 and the root of 20, we will then divide 50 by 6 and the root of 20. And then we will divide the root | of 145 by 6 and the root of 20. But first we will divide 50 by 6 and the root of 20. Find the divisor; you should do 6 times 6, which makes 36. Take 20 from 36, which leaves 16. And then multiply 50 times 6, which makes 300. To divide by 16 comes to $18\frac{3}{4}$. Then multiply 50 times 50, which makes 2,500, and then 36[2] times 2,500, f. 89b

---

1. Michael has dropped a term in his summation; it should come to "the root of 700 minus the root of 500."
2. This number should be 20, which throws this piece of his solution off.

90000. A partir in 256, zoè nel quadratto del 16, viene $351\frac{9}{16}$. E chussì ài che partando 50 per 6 e Rx de 20 vien $18\frac{3}{4}$ meno Rx de $351\frac{9}{16}$. E questo se vuol salvare. Ora se vuol partir Rx de 145 per 6 e Rx de 20. Moltiplicha 6 via 6, fa 36, e moltiplicha 36 via 145, fano 5220. A partir in 256 ne vien $20\frac{25}{64}$. Ora moltiplicha 20 via 145, fano 2900. A partir in 256 ne vien $11\frac{21}{64}$. E chussì abiamo che partendo 145 per 6 e Rx de 20 ne vien $18\frac{3}{4}$ e Rx de $20\frac{25}{64}$ meno Rx de $351\frac{9}{16}$ e anche meno Rx de $11\frac{21}{64}$. E chusì farai le semeyante raxion. E questo abasti oramai per le raxion dele radixe che asai abiamo.

Anchora te voio amaistrar che senpre tutte le questione te vegnise dade pony che senpre sia 1^{co} o zensso, ma la più fiade pony che fusse 1^{co}, e la non te vegnise ben. Pony che la questione fusse $1^{\square}$, e azò che tu sapi la magnera e delo zenso e dela chosa, che chossa via chossa fa un zenso. E zenso via zenso fano zenso de zenso. E chosa via zenso fa chubo. E chosa via chubo zenso de zenso. E chossa via numero fa tante chosse quando[a] xe li numeri. E zenso via numero tanti zensi quanto è lo numero, salvo che chossa via 4 numeri fa 4^{co}. E notta che si te vegnise zenso rotto ingual al numero
f. 90-1a over la chosa, debilo redur a sano. E per quel medeximo che | tu moltiplichi lo zenso, chossì farai el numero over la chosa, sichomo seria a dir $\frac{1}{2}$ zenso a redur a san, che serà ingual a 6^{n}. Noi faremo 2 via mezo serà 1 san, e chosì moltiplicheremo 2 via 6^{n} fano 12. E questo se fa per achordar el zenso chon lo numero. E chossì farai se 'l te avegnyse zenso ingual a numero over a chossa, rotto o chossa rotta ingual a numero.

E notta che se te avegnise 1^{chu} e $1^{\square}$ de zenso ingual a un zenso, debi produr ogni chossa per zenso. E sapi che 'l se te avegnisse ingual al numero e non se poria aprodur, ch'è la raxion perchè 'l muda spezia. Ma in tute le oltre guixe che te avegnise, tu lo pora' produr ezetto per lo sopradditto chapitullo, zoè numero. Verbi grazia se 'l te avegnise uno chubo et un zenso di zenso ingual a $2^{\square}$, debi produr ogni chosa per zenso. Unde a produr chubicho in zenso vien 1^{co}, e la raxion perché chossa via zensso fa 1 chubo, e perzò aprodur chubo in zenso fa chossa. Mo' debi produr zenso di zensso per zensso; deven zensso, e la raxion perché zenso fia zensso fa zenso di zensso, inperzò che aprodur zenso di zenso per zenso fa zenso. Mo' debi produr 2 zensi per zenso e vero 2 numeri, e la raxion che ssia questa che 2 via zenso fa $2^{\square}$, inperzò aprodur 2 zensi per zenso li 'ndevien 2, onde debi dir che tanto val 1 chubo e 1 zenso di zenso ingual a $2^{\square}$ quando $1^{\square}$ 1^{co} ingual a 2 drame. E per questa via debi produr ogny inguazion.

---

a. *Sic in MS, evidently for* quanto.

which makes 90,000. Divided by 256, that is, the square of 16, it comes to $351\frac{9}{16}$. And so you have that dividing 50 by 6 and the root of 20 comes to $18\frac{3}{4}$ minus the root of $351\frac{9}{16}$. And this should be saved. Now it's necessary to divide the root of 145 by 6 and the root of 20. Multiply 6 times 6, which makes 36, and multiply 36 times 145, which makes 5,220. Divided by 256, it comes to $20\frac{25}{64}$. Now multiply 20 by 145, which makes 2,900. Divided by 256, it comes to $11\frac{21}{64}$. And so we have that dividing 145 by 6 and the root of 20 comes to $18\frac{3}{4}$ and the root of $20\frac{25}{64}$ minus the root of $351\frac{9}{16}$ and also minus the root of $11\frac{21}{64}$. And you will do similar problems this way. And at this point you have finished with the problems of the roots, which we have had enough of.

Now I want to teach you that for all the questions that come to you, always put that there may always be 1 unknown or squared unknown, but more often put that it was 1 unknown, and that it didn't come out right for you. Put that the question was one squared, and so that you know the form of the squared unknown and of the unknown, that an unknown times an unknown makes a squared unknown. And a squared unknown times a squared unknown makes the squared unknown of a squared unknown. And an unknown times a squared unknown makes a cube. And an unknown times a cube makes the squared unknown of a squared unknown. And an unknown times a number makes as many unknowns as the number is. And a squared unknown times a number as many squared unknowns as the number is, except that an unknown times 4 numbers makes $4x$. And note that if you came up with a fractional squared unknown equal to the number, or to the unknown, you should reduce it to an integer. And by the same way as | you multiply the squared unknown, so you will do for the number or the unknown, which would be to say that $\frac{1}{2}$ of a squared unknown is to be reduced to a whole, which would be equal to 6. We will do 2 times half, which will be 1 whole, and so we will multiply 2 times 6 makes 12. And this is done to bring the squared unknown into accord with the number. And you will do it this way if you had a squared unknown equal to a number or an unknown, fraction or fractional unknown equal to a number. f. 90-1a

And notice that if you had $1x^3$ and $1x^2$ of a squared unknown equal to a squared unknown, you must factor everything by a squared unknown. And know that if you had it equal to a number and it couldn't be factored, it's because the problem is the duty on spices. But in all the other forms that it might come to you, you could factor it except for this chapter, that is, to a number. As an example, if you had a cube and the square of a square equal to $2x^2$, you should factor everything by a squared unknown. Hence to factor a cube by a squared unknown comes to $1x$, which is the case because an unknown times a squared unknown makes 1 cube, and because to factor a cube by a squared unknown makes an unknown. Now you must factor the square of a square by a square, which becomes a square, and it's because a squared unknown times a squared unknown makes the squared unknown of a squared unknown, because factoring the squared unknown of the squared unknown by a squared unknown makes a squared unknown. Now you must factor 2 squared unknowns by a squared unknown and 2 numbers, and that's because 2 times a squared unknown makes $2x^2$, so to factor 2 squared unknowns by a squared unknown comes to 2, from which you should say that 1 cube and 1 squared unknown of a squared unknown is equal to $2x^2$ when $1x^2$ $1x$ is equal to 2. And in this way you should factor every equation.

f. 90-1b

[Trar el dado, a saver quanti ponti; a penssar, a tuor denari e saver quanti]

Per trar el dado e saver zo che ponto averà trato o un dado o 2 o 3, chisto serà el muodo. Pony che un dado fusse 6, e l'oltro fusse 5, e l'oltro fusse 4. Adopia quel che tu à getado che fu 6, serano 12; azonzi 5, serano 17. Moltipicha per 5, serano 85. Azonzi lo segondo dado che fu 5, serano 90. E questo tuto moltipicha per 10, serano 900. Azonto el terzo dado che fu 4, serano 904. E de questa suma sutratto 250, roman 654. Adoncha el primo fu 6, el segondo fu 5, el terzo fu 4, chomo vedirì qui de sotto.

6	5	4	adopia	6	serà	12			
				6		5			
				5		17	85	900	900
						5	5	10	

904
250
6 · 5 · 4 adoncha fu 6 / 5 / 4 /

Se vuolesti pensar de saver de denari, pony tu à in man duchati 6. Per saver quanti averai questo è lo muodo. Fali azonzer la mitade de quelo, serano 9, e domandarai se 'l ge n'è rutto; tien un e dì quel faza ssano. Zoè, se l'è $4\frac{1}{2}$ fa che ssia 5, e se non serà rotto non tenir niente. Puo' dirai la segonda vuolta vuolta:[a] "Azonzì la mittade," e domandelli se 'l ge n'è schavazado. Se 'l ge dixe sì, teny 2, el schavazado fallo sano, e domandarai quanti[b] vuolte la so suma. D'ogny 9 arai 4, e de questo 9 azonti $4\frac{1}{2}$ serano $13\frac{1}{2}$. Falo sano, serà 14; abati 9, serano 4. El segondo, schavazado 2, serano 6. Adoncha fu tanti.

f. 91-1a

[Eser 15 christiany e 15 zudie in un ballo, a nonbrar e getar ogny 9 un e gettar fuora i zudiei]

Sono 15 cristiany e 15 zudei e per fortuna se vuolse guitar d'ogny 9 un in aqua. Che modo se può far che i zudei vagano in aqua e i christiany romagna? Se die chonzar in un ballo e meter per lo muodo dirò qui de sotto.

---

a. *Thus repeated in MS.*
b. *Thus in MS.*
1. This refers to a traditional Venetian method of divination using dice, adding the points together to tell to whom a certain thing will happen: cf. Giuseppe Boerio, *Dizionario del dialetto veneziano*, 2nd ed. (Venice, 1856; reprinted, Milan: A. Martello, 1971), *s.v.* trar.

To cast dice, to know how many points; to think, to take money and know how much f. 90-1b

To throw dice and know what point one die or 2 or 3 will have thrown, here is the method.[1] Put that one die was 6, and the other was 5, and the other was 4. Double the one that you threw that was 6, which will be 12; add 5, which will be 17. Multiply by 5, which will be 85. Add the second die that was 5, which will be 90. And all of this multiplied by 10 will be 900. Add the third die that was 4, which will be 904, and from this sum subtract 250, which leaves 654. So the first was 6, the second was 5, the third was 4, as you will see below:

6	5	4	double	6	it will be	12			
				6		5			
				5		17	85	900[2]	900
						5	5	10	

904	
250	
6 · 5 · 4	so it was 6 / 5 / 4 /

If you would like to think about knowing about money, put that you had in hand 6 ducats. This is the way to know how much you would have. Add on the half of it, which will be 9, and you will ask if there is a fraction; hold on to one and say that it was a whole number. That is, if it is $4\frac{1}{2}$ put that it is 5, and if it is not a fraction don't hold on to anything. Then you will say the second time: "Add the half," and ask if there is any remainder. If there is, say yes, hold on to 2, and make the remainder whole, and you will ask how many times its sum is. From every 9 you will have 4, and of this 9 add $4\frac{1}{2}$, which will be $13\frac{1}{2}$. Make it whole, which will be 14, subtract 9, which will be 5.[3] The second, with 2 remaining, will be 6. So it was that much.

Fifteen Christians and 15 Jews are in a circle;[4] count and throw out each ninth one and throw out the Jews f. 91-1a

There are 15 Christians and 15 Jews and by chance it is necessary to throw each 9th one in the water. By what method can it be done so that the Jews go in the water and the Christians remain? You should arrange in a circle and put it in the way I will tell you to below.

2. This should be 90.

3. *4* in MS.

4. The term is *ballo*, which may represent the *ballo tondo*, a traditional Venetian circle dance; cf. Boerio, *Dizionario, s.v.* balo, or it may be cognate to the modern Italian *palo*, meaning a pole, and denote a raft, as suggested by the reference in the problem to throwing in water.

Quatuor cristiany, quinque zudei, duo cristiany, unus zudeo, tres cristiany, unus zudeo et uno cristiano, bis duo zudei e do cristiany, tres zud'i, unus cristiano, do zudie, do cristiany, sofizite unus zudeo.

Quatuor, quinque, duo, unus, tres, unus et unus, bis duo, tres, unus, bis duo, sofizit unus.

E per vuoler dequiarar el ditto chapitullo sie per questi 5 nomy per letre vochabille:

Populeam virgam mater regina tenebatt.

a e i o sun quatuor, signyficha po a e i o u sun quinque, significha pu a e sun duo, significha le a est unus, signyficha am	isti sunt quatuor, quinque, duo, unus

a e i sun tres, signyficat vir a est unus, signyficat gam a est uno, significat ma	tres, unus et uno

f. 91-1b a e est duo, significat ter a e est duo, significat re a e i est tres, significat gi a est unus, significat na[a] a e est duo, signyficat te a e est duo, signyficat ne a est unus, signyficat bat	bis duo, tres, unus,[b] bis duo, suficiit unus

[Chavar denari d'una quantità e romagnir in man alguny, a saver che fu el chavedal]

E son alguny chi vuol chavar d'una quantità de denari $\frac{1}{3}$, $\frac{1}{4}$, $\frac{1}{5}$, e romaxelli duchati 20. Adomando zò che fu el chavedal. E per voler far la ditta raxion per la riegola del 3 truovame 1 numero ch'abia $\frac{1}{3}$, $\frac{1}{4}$, $\frac{1}{5}$. 60 sia questo numero. El terzo de 60 20, el quarto serave 15, el quinto 12, fano 47. È men de 60 13, el qual achonzerà in questo muodo, chomo vedirì per esenpio.

---

a. na *corrected over* nam.
b. et uno *crossed out with a horizontal line.*

Four Christians, five Jews, two Christians, one Jew, three Christians, one Jew and one Christian, twice 2 Jews and two Christians,[1] three Jews, one Christian, two Jews, two Christians, one Jew finishes.

Four, five, two, one, three, one and one, twice two, three, one, twice two, one finishes.[2]

And the way to declare this paragraph is by these 5 words by vowel letters:

"Populeam virgam mater regina tenebat."[3]

a e i o are quatuor, it refers to po[4] a e i o u are quinque, it refers to pu a e are duo, it refers to le a is unus, it refers to am	These are quatuor, quinque, duo, unus

---

a e i are tres, it refers to vir a is unus, it refers to gam a is uno, it refers to ma	tres, unus et uno

f. 91-1b

a e is duo, it refers to ter a e is duo, it refers to re a e i is tres, it refers to gi a is unus, it refers to na a e is duo, it refers to te a e is duo, it refers to ne a is unus, it refers to bat	bis duo, tres, unus, bis duo, suficiit unus

---

TO EXTRACT COINS FROM A QUANTITY AND SOME REMAIN IN THE HAND; FIGURE OUT WHAT THE CAPITAL HAD BEEN

And there are some who want to take from a quantity of coins $\frac{1}{3}$, $\frac{1}{4}$, $\frac{1}{5}$, and 20 ducats are left. I ask you what the capital was. And to do this problem by the rule of 3, find me 1 number that has $\frac{1}{3}$, $\frac{1}{4}$, $\frac{1}{5}$. 60 is this number. A third of 60 20, a fourth will be 15, a fifth 12, which makes 47. It's 13 less than 60, which will fit in this way, as you will see worked out.

1. I.e., two of each.
2. This section is in Latin.
3. "The queen mother held a poplar branch."
4. That is, *o* is the fourth vowel.

60	60	13	20	60	0	
20	47	60	1	20	01	
15	13			00	0334	
12				120	1200	
47					133	$92\frac{4}{13}$
					1	

E per far la ditta raxion per inpoxizion pony che ala prima fusse 60. E noi diremo: "El terzo de 60 son 20, el quarto 15, fa 35, el quinto 12, fano 47, e 20 fano 67." E noi vossemo che fuse 60. f. 90-2a Adoncha serà più 7. E questo serà dela prima inpoxizion. E per far la se|gonda inpoxizion pony che fusse 120, lo terzo seria 40, el quarto seria 30, el quinto seria 24, fano in suma 94, e 20 serano 114, e noi vossemo che fusse 120. Adoncha è men 6, e questi metti ala segonda inpoxizion e moltipicha in croxe e parti. Insirà $92\frac{4}{13}$, e tanto fu el chavedal. Se tu chavy el terzo, el quarto, el quinto, resterà 20, chomo ò ditto de sovra, e vedirì qui per figura.

60	120	60	67	120	120	
		20	60	40	114	
plù	men	15	plù 7	30	men 6	
7	6	12		24		
6		47		94	0	
13		20		20	01	
		67		114	0334	
	60	120	840		1200	
	6	7	360		133	$92\frac{4}{13}$
	360	840	1200		1	
				$92\frac{4}{13}$		$30\frac{10}{13}$
				$72\frac{4}{13}$		$23\frac{1}{13}$
				20		$18\frac{6}{13}$
						$92\frac{4}{13}$

E per far la ditta raxion per la chossa pony che fusse $\frac{1}{3}^{co}$, $\frac{1}{4}^{co}$, $\frac{1}{5}^{co}$; azonti insieme serano $\frac{47}{60}$. Adoncha se trazo 1^{co} de 60, zoè 47 de 60, roman 13 sesantexima, che son ingual a 20^{n}. Adoncha moltipicha 20 fia 60, fano 1200, e questo partido fano $92\frac{4}{13}$. Vedirì qui de sotto.

60	60	13	20	60	0	
20	47	60	1	20	0~~1~~	
15	13			00	0~~33~~4	
12				120	~~1200~~	
47					~~133~~	$92\frac{4}{13}$
					~~1~~	

And to do this problem by false position, suppose that it was 60 at the first. And we will say: "A third of 60 is 20, a fourth 15, which makes 35, a fifth 12, which makes 47, and 20 makes 67." And we wanted it to be 60. So it will be plus 7. And this will be at the first false position. And to make the | second false position, suppose that it was 120, the third would be 40, the fourth would be 30, the fifth would be 24, which makes in all 94, and 20 would be 114, and we wanted it to be 120. So it is minus 6, and put this at the second false position, and multiply across and divide. The result will be $92\frac{4}{13}$, and that was the amount of the capital. If you subtract the third, the fourth, and the fifth the remainder will be 20, as I said above, and as you will see worked out here. f. 90-2a

60	120		60	67	120	120	
			20	60	40	114	
plus	minus		15	plus 7	30	minus 6	
7	6		12		24		
6			47		94	0	
13			20		20	0~~1~~	
			67		114	0~~33~~4	
		60	120	840		~~1200~~	
		6	7	360		~~133~~	$92\frac{4}{13}$
		360	840	1200		~~1~~	
					$92\frac{4}{13}$		$30\frac{10}{13}$
					$72\frac{4}{13}$		$23\frac{1}{13}$
					20		$18\frac{6}{13}$
							$92\frac{4}{13}$

And to do this problem by unknowns, put that it was $\frac{1}{3}x$, $\frac{1}{4}x$, $\frac{1}{5}x$; added together they are $\frac{47}{60}$. So if I subtract 1x from 60, that is 47 of 60, it leaves 13 sixtieths, which is equal to 20. So multiply 20 times 60, which makes 1,200, and this divided makes $92\frac{4}{13}$. You will see it below.

$\frac{1}{3}^{co}$ $\frac{1}{4}$ $\frac{7}{12}$ $\frac{1}{5}$ $\frac{47}{60}$

$$\begin{array}{r} 60 \\ \underline{47} \\ 13 \end{array} \qquad \frac{13}{60} \times \frac{20}{1} \qquad \begin{array}{r} 60 \\ \underline{20} \\ 1200 \end{array} \qquad \left.\begin{array}{r} 1 \\ 0\not{3}3 \\ \not{1}\not{2}\not{0}0 \\ \not{1}\not{3}3 \\ 1 \end{array}\right] 92\tfrac{4}{13}$$

$$\begin{array}{r} \underline{92\tfrac{4}{13}} \\ 30\tfrac{10}{13} \\ 23\tfrac{1}{4} \\ \underline{18\tfrac{6}{13}} \end{array} \qquad \text{summa } 72\tfrac{4}{13} \qquad \begin{array}{r} 92\tfrac{4}{13} \\ \underline{72\tfrac{4}{13}} \\ 20 \end{array}$$

f. 90-2b [Saver che tenpo veny in Veniexia e chon chi son stado in viazio]

Al nomen di Dio. Qui de sotto scriverò mi Michalli da Ruodo el tenpo veny in Veniexia. Zò fu 1401 adì 5 zugno.

E primo m'achordiè in Manfredonya per homo da remo chon el nobille homo miser Piero Loredan fu de miser Alvixe Loredan, siando chapetagno l'egregio miser Andrea Benbo, in la varda, el mio chomitto ser Michiel Russo e paron zurado ser Antuonyo Rizo.

M'achordiè per homo da remo del 1402 in la varda chon el nobile homo miser Marcho Grimani, siando in la varda chapetagno l'egregio miser Marin Charavelo, el mio chomitto ser Michiel di Verisielli, el mio paron ser Zorzi Giallinà de Chandia.

M'achordiè per homo da remo del 1403 in la varda chon el nobile homo miser Andrea da Molin, siando chapetagno l'egregio meser Charlo Zen, el mio chomitto ser Nicholetto Testa, el mio paron ser Manolli Filaretto. Fu al tenpo che ronpessemo a Modon el Pozichardo.

M'achordiè per homo da remo del 1404 in lo viazio de Fiandria chon el nobille homo miser Marcho Chorer, siando chapetagno l'egregio miser Fantin Michiel, el mio chomitto ser Alvixe de Boninsegna, el mio paron ser Boldisera Rocho.

M'achordiè per proder del 1405 in lo viazio de la varda chon el nobile homo miser Piero Miany, siando chapetagno l'egregio miser Marin Charavello, el mio chomitto ser Michiel di Verissieli, el mio paron ser Cristofallo Mezany. Et in questo fu levado nochiero forsi per la mittà del viazio, e questa fu la prima honor che avy.

1. Michael seems to have skipped the last step of this *galera*; it should look identical to the one just above.

2. This should be $23\frac{1}{13}$.

3. For the context of this autobiographical section, see Alan M. Stahl, "Michael of Rhodes: Mariner in Service to Venice," vol. 3, pp. 79–92.

$\frac{1}{3}x \quad \frac{1}{4} \quad \frac{7}{12} \quad \frac{1}{5} \quad \frac{47}{60}$

$$\begin{array}{r} 60 \\ \underline{47} \\ 13 \end{array} \qquad \frac{13}{60} \times \frac{20}{1} \qquad \begin{array}{r} 60 \\ \underline{20} \\ 1200 \end{array} \qquad \left.\begin{array}{r} 1 \\ 0\not{3}3 \\ \not{1}\not{2}\not{0}0 \\ \not{1}\not{3}3 \\ 1 \end{array}\right] \; 92\tfrac{4}{13}$$

[1]

$$\begin{array}{r} 92\tfrac{4}{13} \\ \hline 30\tfrac{10}{13} \\ 23\tfrac{1}{4} \\ 18\tfrac{6}{13} \\ \hline \end{array} \qquad \text{sum } 72\tfrac{4}{13} \qquad \begin{array}{r} 92\tfrac{4}{13} \\ 72\tfrac{4}{13} \\ \hline 20 \end{array}$$

[2]

KNOW WHEN I CAME TO VENICE, WITH WHOM I HAVE VOYAGED

f. 90-2b

In the name of God. I, Michael of Rhodes, shall write below about the time I came to Venice.[3] It was on June 5, 1401.

And first, I signed on in Manfredonia[4] as an oarsman with the nobleman Pietro Loredan, son of the late Alvise Loredan, the captain of the guard fleet being the distinguished Andrea Bembo, my *comito*[5] Michele Russo and the *paron zurado* Antonio Rizzo.

I signed on as oarsman in 1402 in the guard fleet with the nobleman Marco Grimani, the captain of the guard fleet being the distinguished Marino Caravello, my *comito* Michele di Verisielli, my *paron* Giorgio Giallinà of Candia.

I signed on as oarsman in 1403 in the guard fleet with the nobleman Andrea da Molin, the captain being the distinguished Carlo Zeno, my *comito* Nicoletto Testa, my *paron* Manolli Filaretto. It was at this time that we routed Boucicaut at Modon.[6]

I signed on as oarsman in 1404 on the voyage to Flanders with the nobleman Marco Correr, the captain being the distinguished Fantino Michiel, my *comito* Alvise de Boninsegna, my *paron* Baldassarre Rocco.

I signed on as *proder* in 1405 in the guard voyage with the nobleman Pietro Miani, the captain being the distinguished Marino Caravello, my *comito* Michele di Verisielli, my *paron* Cristoforo Mezani. And in this voyage I was promoted to *nochier* for about half of the voyage, and this was the first honor that I received.

---

4. On the coast of Apulia.
5. For the duties and remuneration of this and other shipboard offices that Michael held, see Stahl, "Michael of Rhodes," vol. 3, pp. 40–74.
6. For this incident, see Stahl, "Michael of Rhodes," vol. 3, pp. 51–54.

f. 91-2a M'achordiè per proder in Londra del 1406 chon el nobile homo miser Marcho Chorer, siando chapetagno l'egregio miser Fantin Michiel, el mio chomitto ser Alessio Chonzanave, el mio paron ser Boldisera Rocho.

M'achordiè per nochiero in la varda del 1407 chon el nobille homo miser Bertuzi Diedo, siando chapetanyo l'egregio miser Fantin Michiel, mio chomitto ser Piero Gatta, el mio paron ser Cristofaro d'Anchona.

M'achordiè per nochiero in la varda del 1408 chon el nobile homo miser Franzescho Baxadona, siando chapetagno l'egregio miser Piero Zivrany, el mio chomitto ser Michiel di Virissielli, el mio paron ser Nichuola Biancho.

M'achordiè per nochiero in la varda del 1409 chon l'egregio chapetagno miser Nichuolò Fuschullo, el mio armiraio ser Iachomello di Rinoldo, el mio chomitto el mio chomitto[a] ser Marcho Bochetta, el mio paron ser Bortholamio Durazier.

M'achordiè per nochiero in la varda del 1410 chon l'egregio miser Piero Zivrany, armiraio ser Iachomello di Rinoldo, el mio chomitto ser Michiel di Verissielli, el mio paron ser Zuan Tardo.

M'achordiè per nochiero in lo viazio de Fiandria del 1411 chon l'egregio miser Bernardo Pasqualigo, abiando in la ditta per chapetagno l'egregio miser Lunardo Mazanigo, el mio armiraio ser Marcho de Malamocho, el mio chomitto ser Iachomello di Rinoldo, el mio paron ser Zuan Tardo.

f. 91-2b M'achordiè per nochiero al viazio de Fiandria del 1412 chon l'egregio miser Vido da Chanal "el grasso," abiando in galia per chapetagno l'egregio miser Marcho Zustignan, el mio armiraio ser Michalletto de Benedetto, el mio chomitto ser Michalletto Giallitti, el mio paron ser Antuonyo da Chorfu.

M'achordiè per nochiero al viazio de Fiandria del 1413 chon el nobille homo miser Piero Marzello, abiando in la ditta per chapetagno l'egregio miser Marcho Lunbardo, el mio armiraio ser Alvixe de Bonainsegna, el mio chomitto ser Michiel di Verissielli, el mio paron ser Pasqualin d'Ogniben.

M'achordiè per paron in la varda chon lo nobille homo miser Iachomo Barbarigo del 1414, el nostro chapetagno l'egregio miser Piero Zivrany, el nostro chomitto ser Michiel di Verissielli.

M'achordiè in la varda chon el nobille homo miser Nicholò di Priolli del 1415, el nostro chapetagno miser Nicholò Fuschullo, el mio chomitto ser Zanin de Lanzillotto, e mi so paron. E in questo troviè mia moier Dorattia morta.[b]

---

a. *Repeated in MS.*

b. *From* E in questo *to* morta *added in the same hand, almost certainly at a later time, in another ink (seppia rather than dark brown).*

I signed on as *proder* in London[1] in 1406 with the nobleman Marco Correr, the captain being the distinguished Fantino Michiel, my *comito* Alessio Conzanave, my *paron* Baldassarre Rocco. f. 91-2a

I signed on as *nochier* in the guard fleet of 1407 with the nobleman Bertuccio Diedo, the captain being the distinguished Fantino Michiel, my *comito* Pietro Gatta, my *paron* Cristoforo d'Ancona.

I signed on as *nochier* in the guard fleet of 1408 with the nobleman Francesco Basadonna, the captain being the distinguished Pietro Civran, my *comito* Michele di Verisielli, my *paron* Nicolò Bianco.

I signed on as *nochier* in the guard fleet of 1409 with the distinguished captain Nicolò Foscolo, my *armiraio* Jacomello di Rinaldo, my *comito* Marco Bochetta, my *paron* Bartolomeo Durazir.

I signed on as *nochier* in the guard fleet of 1410 with the distinguished Pietro Civran, *armiraio* Jacomello di Rinaldo, *comito* Michele di Verisielli, my *paron* Giovanni Tardo.

I signed on as *nochier* on the voyage to Flanders of 1411 with the distinguished Bernardo Pasqualigo, having there as captain the distinguished Leonardo Mocenigo, my *armiraio* Marco de Malamocco, my *comito* Jacomello di Rinaldo, my *paron* Giovanni Tardo.

I signed on as *nochier* on the voyage to Flanders of 1412 with the distinguished Guido da Canal "the fat," having in the galley as captain the distinguished Marco Giustinian, my *armiraio* Micheletto de Benedetto, my *comito* Micheletto Giallitti, my *paron* Antonio da Corfu. f. 91-2b

I signed on as *nochier* on the voyage to Flanders of 1413 with the nobleman Pietro Marcello, having there as captain the distinguished Marco Lombardo, my *armiraio* Alvise de Boninsegna, my *comito* Michele di Verisielli, my *paron* Pasquale Dognibene.

I signed on as *paron* in the guard fleet with the nobleman Giacomo Barbarigo in 1414, our captain the distinguished Pietro Civran, our *comito* Michele di Verisielli.

I signed on in the guard fleet with the nobleman Nicolò di Priuli in 1415, our captain Nicolò Foscolo, my *comito* Giovanni di Lanzillotto, and I was *paron*. And at this point I found that my wife Dorotea had died.[2]

1. Presumably on the Flanders fleet, whose destination was London.

2. Sentence added by Michael in a later hand.

M'achordiè per paron in la varda del 1416 chon el nobile homo miser Iachomo Barbarigo, el nostro chapetagno l'egregio miser Piero Loredan, el mio chomitto ser Thomao Pissatto. Ala vituoria d'i Turchi.

M'achordiè per homo de chonseyo al viazio de Fiandria del 1417 chon el nobelle homo miser Zuane Malipiero, el mio chomitto ser Antuonio Chopo, el mio paron ser Puolo Negro, siando chapetagno in le ditte miser Franzescho Pixany.

f. 92a M'achordiè per paron in la varda del 1418 chon el nobile homo miser Brancha Loredan, siando chapetagno l'egregio miser Nicholò Chapello, el mio chomitto ser Thomao Pissatto.

M'achordiè patron d'una galliotta in la varda d'Albanya del 1419, siando chapetagno l'egregio miser Franzescho Benbo.

M'achordiè paron in lo viazio de Fiandria del 1420 chon el nobille homo miser Andria Queriny, siando in la ditta per chapetagno l'egregio miser Zuan Diedo, morì in Fiandria, el mio armiraio ser Alovixe de Bonainsegna, el mio chomitto ser Nichuola da Molin.

M'achordiè per chomitto in lo viazio de la Tana del 1421 chon el nobile hom[*o*] miser Benedetto Michiel, abiando per chapetagno in la ditta l'egregio miser Vido da Chanal "el Bivilaqua," el mio armiraio ser Michalletto de Benedetto, el mio paron ser Zanin Parixotto.

M'achordiè per armiraio in la varda del 1422 chon l'egregio miser Nichollò Chapello, el mio chomitto ser Zanin de San Rafael, el mio paron ser Pasqualin Marango[*n*]. Et in questo viazio morì mio fio Thodorin.[a]

M'achordiè per chomitto in llo viazio d'Alessandria chon el nobile homo miser Felipo da Chanal, el mio paron ser Bortholamio Negro, siando chapetagno l'egregio miser Thomao Duodo. 1423.

M'achordiè per chomitto in la varda del 1424 chon el nobile homo miser Polo Pasqualigo, mio paron Pasquali, chapetagno miser Piero Loredan, ai fatti de Galipoli.

f. 92b M'achordiè per chomitto in la varda del 1425 chon el nobille homo miser Alovixe Loredan, siando chapetagno l'egregio miser Fantin Michiel, el mio paron Antuonyo Negro.

M'achordiè per chomitto in lo viazio de Trabexonda del 1426 chon el nobelle homo miser Felipo da Chanal, el mio paron ser Bortholamio Negro, siando chapetagno l'egregio miser Puolo Pasqualigo.

---

a. *From* Et in questo viazio *to* Thodorin *added in the same hand, almost certainly at a later time, in another ink (seppia rather than dark brown).*

1. For this incident, see Stahl, "Michael of Rhodes," vol. 3, pp. 63–64.

2. On the river Don near the Sea of Azov, in Crimea.

I signed on as *paron* in the guard fleet of 1416 with the nobleman Giacomo Barbarigo, our captain the distinguished Pietro Loredan, my *comito* Tommaso Pissato. [We were present] at the victory over the Turks.[1]

I signed on as *homo de conseio* on the voyage to Flanders of 1417 with the nobleman Giovanni Malipiero, my *comito* Antonio Coppo, my *paron* Paolo Negro, the captain for this being Francesco Pisani.

I signed on as *paron* in the guard fleet of 1418 with the nobleman Branca Loredan, the captain being the distinguished Nicolò Capello, my *comito* Tommaso Pissato. f. 92a

I signed on as *paron* of a small galley in the guard fleet of Albania in 1419, the captain being the distinguished Francesco Bembo.

I signed on as *paron* on the voyage to Flanders of 1420 with the nobleman Andrea Querini, the captain of this being the distinguished Giovanni Diedo, who died in Flanders. My *armiraio* was Alvise de Boninsegna, my *comito* Nicolò da Molin.

I signed on as *comito* in the voyage to Tana[2] of 1421 with the nobleman Benedetto Michiel, having for captain the distinguished Guido da Canal "Bevilaqua,"[3] my *armiraio* Micheletto de Benedetto, my *paron* Giovanni Parisotto.

I signed on as *armiraio* in the guard fleet of 1422 with the distinguished Nicolò Capello, my *comito* Giovanni de San Rafaele, my *paron* Pasqualino Marangon.[4] And my son Teodorino died on this voyage.[5]

I signed on as *comito* on the voyage to Alexandria with the nobleman Filippo da Canal, my *paron* Bartolomeo Negro, the captain being the distinguished Tommaso Duodo. 1423.

I signed on as *comito* in the guard fleet of 1424 with the nobleman Paolo Pasqualigo, my *paron* Pasquale,[6] captain Pietro Loredan, for the events at Gallipoli.[7]

I signed on as *comito* in the guard fleet of 1425 with the nobleman Alvise Loredan, the captain being the distinguished Fantino Michiel, my *paron* Antonio Negro. f. 92b

I signed on as *comito* on the voyage to Trebizond of 1426 with the nobleman Filippo da Canal, my *paron* Bartolomeo Negro, the captain being the distinguished Paolo Pasqualigo.

---

3. I.e., teetotaler.
4. As *marangon* means ship's carpenter, it might not be used as a family name here.
5. Sentence added by Michael in a later hand. It is not clear whether Teodorino was on the voyage or just died during it. See Stahl, "Michael of Rhodes," vol. 3, p. 72.
6. No other name given.
7. For this incident, see Stahl, "Michael of Rhodes," vol. 3, p. 73.

Et in questo ano andiè patron d'un arssil in Puya per chargar chavalli a Trany e a Bexeye.

M'achordiè per chomitto al viazio dela Tana del 1427 chon el nobille homo miser Nicholò Manolessio, el mio paron ser Marcho Marinzi, siando chapetagno l'egregio miser Troilo Malipiero.

M'achordiè per armiraio in la varda chon l'egregio miser Andria Mozenygo, el mio chomitto ser Antuonyo de Rizardo, el mio paron ser Nicholò Venier, del 1428.

M'achordiè per homo de chonseyo in lo viazio de Fiandria chon l'egregio miser Nicholò Michiel, el mio chomitto ser Greguol d'Antuonyo, el mio paron ser Bernardo, siando chapetagno l'egregio miser Frederigo Chontariny, del 1430.

M'achordiè in la varda per chomitto chon el nobile homo miser Puolo Pasqualigo, el mio paron ser Bortholamio Dobra, siando chapetagno l'egregio miser Piero Loredan. Ala Riviera avesemo vituoria d'i Zenovexi e my veny per tera ferido e vasto, del 1431.

f. 93a M'achordiè per chomitto in la varda del 1432 chon el nobille homo miser Vidal Miany, provedeor chon miser Piero Loredan, el mio paron ser Nichola de Chandia, el nostro chapetagno miser Piero Loredan.

M'achordiè per chomitto in Alessandria chon el nobile homo miser Zuan Loredan, mio paron ser Iachomello Rizo, siando chapetagno l'egregio miser Lorenzo Chapello, del 1433.

M'achordiè per chomitto al viazio dele Aque Morte del 1434 chon el nobile homo miser Zuan da Molin, el mio paron ser Polo Manega.

M'achordiè per homo de chonseio chon el nobille homo miser Zacharia Donado al viazio de Monchastro, siando chapetagno el spectabelle homo miser Alvixe Loredan, mio chomitto Lazaro Parixotto e mio paron Pollo Negro, del 1435.

M'achordiè per armiragio chon el spectabille homo miser Franzescho Chapello, el mio chomitto Lazaro Parixotto, paron Chorzulla, del 1436 al viazio de Fiandria. Et in questo viazio truovyè mia moier Chataruza morta.[a]

M'achordiè per chomitto chon el nobille homo miser Alvixe Benbo chon le gallie del papa in Chostantinopolli per l'inperador, del 1437. Mio chapetagno l'egregio miser Antonio Chondolmer, mio paron Nichola de Chandia.

---

a. *From* Et in questo viazio *to* morta *added in the same hand, almost certainly at a later time, in another ink (seppia rather than dark brown).*

1. The *arsile* appears to have been the unrigged hulk of an old galley, used as a prison ship or horse transport; cf. A. Jal, *Glossaire nautique* (Paris, 1848), s.v. *arsile,* quoting Dandolo, 1468: "arsili . . . sono corpi di galee disforniti."

2. It was probably in connection with this voyage that Michael received the copy of the fleet orders of Mocenigo which he included later in the manuscript; see below, pp. 324–345.

And in this year I went as patron of an arsella[1] in Apulia to load horses in Trani and Bisceglie.

I signed on as *comito* on the voyage to Tana of 1427 with the nobleman Nicolò Manolesso, my *paron* Marco Marinzi, the captain being the distinguished Troilo Malipiero.

I signed on as *armiraio* in the guard fleet with the distinguished Andrea Mocenigo, my *comito* Antonio de Rizardo, my *paron* Nicolò Venier, in 1428.[2]

I signed on as *homo de conseio* on the voyage to Flanders with the distinguished Nicolò Michiel, my *comito* Gregorio d'Antonio, my *paron* Bernardo,[3] the captain being the distinguished Federico Contarini, in 1430.

I signed on in the guard fleet as *comito* with the nobleman Paolo Pasqualigo, my *paron* Bartolomeo Dobra, the captain being the distinguished Pietro Loredan. We had a victory along the coast over the Genoese and I went home over land, wounded and broken, in 1431.[4]

I signed on as *comito* in the guard fleet of 1432 with the nobleman Vitale Miani provisor to Pietro Loredan, my *paron* Nicolò da Candia, our captain Pietro Loredan. f. 93a

I signed on as *comito* to Alexandria with the noblemen Giovanni Loredan, my *paron* Jacomello Rizzo, the captain being the distinguished Lorenzo Capello, in 1433.

I signed on as *comito* on the voyage to Aigues Mortes in 1434 with the nobleman Giovanni da Molin, my *paron* Paolo Manega.

I signed on as *homo de conseio* with the nobleman Zaccaria Donà on the voyage to Moncastro,[5] the captain being the noteworthy Alvise Loredan, my *comito* Lazaro Parisotto and my *paron* Paolo Negro, in 1435.

I signed on as *armiraio* with the noteworthy Francesco Capello, my *comito* Lazaro Parisotto, *paron* Corzulla,[6] in 1436 on the voyage to Flanders. And on this voyage I found my wife Cataruccia dead.[7]

I signed on as *comito* with the nobleman Alvise Bembo with the papal galleys to Constantinople for the emperor,[8] in 1437, my captain the distinguished Antonio Condulmer, my *paron* Nicolò da Candia.

---

3. No other name given.
4. For this incident, see Stahl, "Michael of Rhodes," vol. 3, pp. 77–78.
5. Akkerman, at the mouth of the Dniester on the Black Sea.
6. No other name given.
7. Sentence added by Michael in a later hand.
8. For this voyage, see Stahl, "Michael of Rhodes," vol. 3, p. 92.

M'achordiè per hom[*o*] de chonseio del 1438 chon el nobille homo miser Batista Chonttariny[a] al viazio de Londra, chapetagno l'egregio miser Antonio Diedo, mio chomytto ser Nychollò de la Zudecha, paron Antonio Chalafao.

f. 93b 1439

M'achordiè in lo viazio del papa in Romanya. Portassimo l'imperador in Chostantinopolli, chapetagno dela ditta l'egregio miser Antuonyo Chondolmer, abiando la gallia del nobille homo miser Andrea Gritti, abiando patron in lla ditta miser Nichollò Gritti, ssiando in chonpanya gallie 2 dela Tana, chapetagno dele ditte miser Marcho Zacor,[b] patrony miser Andrea Contariny e miser Franzescho Manollessio, armyrayo delle ditte Nichollò Dellegende, mio armyrayo ser Benedetto d'Ardovin, mio paron zurado Antuonio Parexin.

1440

M'achordiè in lo viazio de Zipri per portar la raina,[c] per armyrayo del nobille homo miser Benedetto Dandullo, andassemo a Zerine, a Famagosta, a Zafo, a Barutto, a Zafo, a Famagosta e tornasemo in Venyexia. Mio chomitto ser Bortholomyo Fiorian, paron Zan Bafo. Tornassemo in Veniexia. Gallia bella.

M'achordiè in lo viazio de Fiandria chon el nobille hom[*o*] miser Alvixe Diedo del 1441. Nostro chapetanyo miser Lorenzo Minyo, gallie 4. Mio chomito Nycholò dela Zudecha, paron Dagnel Furlan.

1442[d]

M'achordiè per hom[*o*] de chonseyo chon el nobille hom[*o*] miser Piero Orio in lo viazio d'Alessandria, chapetagno miser Marin da Molin, mio chomitto Daniel Forlan.

M'achordiè del 1443 hom[*o*] de chonseyo chon el nobille hom[*o*] miser Bartholomeo Pixany al viazio de Londra, chapetagno miser Zorzi Valaresso, mio chomitto ser Nichollò dela Zudecha, el tuxo. E del 1444 avy la stadera, adì 28 zener.[e]

---

a. *At this point there appear small ink marks above the line which seem to represent the blotting of the ink from the corresponding line (22) of the facing page (fol. 93b).*

b. *Sic in MS, though not an attested Venetian noble family name.*

c. *Sic in MS, evidently for* reina.

d. *Above the line.*

e. *From* E del 1444 *to* 28 zener *added in the same hand, almost certainly at a later time, in another ink (seppia rather than dark brown).*

I signed on as *homo de conseio* in 1438 with the nobleman Battista Contarini on the voyage to London, the captain the distinguished Antonio Diedo, my *comito* Nicolò dalla Giudecca, *paron* Antonio Calafano.[1]

1439[2]

f. 93b

I signed up on the papal voyage to Romania.[3] We carried the emperor to Constantinople.[4] Captain in this the distinguished Antonio Condulmer, having the galley of the nobleman Andrea Gritti, whose patron was Nicolò Gritti, with two Tana galleys accompanying. Their captain was Marco Zaco[5] and patrons Andrea Contarini and Francesco Manolesso, *armiraio* of them Nicolò Dellegende, my *armiraio* Benedetto Dardoin, my *paron zurado* Antonio Paresin.

1440

I signed up on the voyage to Cyprus to carry the queen.[6] For *armiraio*, the nobleman Benedetto Dandolo. We went to Zerine,[7] to Famagusta, to Jaffa, to Beirut, to Jaffa, to Famagusta, and we returned to Venice. My *comito* Bartolomeo Florian, *paron* Giovanni Baffo. We returned to Venice. A beautiful galley.

I signed up on the voyage to Flanders with the nobleman Alvise Diedo in 1441. Our captain Lorenzo Minio, 4 galleys. My *comito* Nicolò dalla Giudecca, *paron* Daniele Furlan.

1442

I signed on as *homo de conseio* with the nobleman Pietro Orio on the voyage to Alexandria, the captain Marino da Molin, my *comito* Daniele Furlan.

I signed on in 1443 as *homo de conseio* with the nobleman Bartolomeo Pisani on the voyage to London, the captain Giorgio Valaresso, my *comito* Nicolò dalla Giudecca, the bald.[8] And on January 28, 144[5],[9] I had the steelyard.[10]

---

1. As *calafao* means caulker, it may not be used here as a family name.
2. At this point the writing, which has been continuous except for noted additions, changes with each annual entry, but appears to remain in the hand of Michael of Rhodes.
3. That is, to the Byzantine Empire.
4. On return from Ferrara, having signed a document proclaiming the union of the Churches.
5. This family name does not occur among those of members of the nobility of Venice in the period, who would usually have been the men selected as captains of a merchant galley. However, members of the Zaco family appear in lists of the members of the Great Council of Crete of the fourteenth century: see Sally McKee, *Uncommon Dominion: Venetian Crete and the Myth of Ethnic Purity* (Philadelphia: University of Pennsylvania Press, 2000), p. 41.
6. Medea Paleologo, daughter of the marquis of Monferrato, married in 1440 to John Lusignan, king of Cyprus.
7. Probably either Cerigo or Sira, in the Cyclades.
8. Literally, tonsured.
9. As the Venetian calendar began in March, the MS gives the year as 1444.
10. The official weighing station of Venice.

f. 95a[a]

[Veder chalandario]

Marzo, adì 16 ore 12, la notta[b] ore 12, el zorno ore 12, à dì 31.

1. S. Erchuliani, S. Alban veschovo. *
2. S. Primittivi marturi.
3. S. Trezenti chonfesori.
4. S. Fellizitti monazi, S. Fuscha.
5. S. Iulliany epischopi, S. Lizio veschovo.
6. S. Felizitatis marturi.
7. S. Perpetue virgine, S. Felizitta virgine. *
8. S. Pontu marturi, S. Honorati marturi.
9. S. Actalle abatte, S. 40 marturi.
10. S. Ieronymo marturi.
11. S. 40 marturi.
12. S. Gregorii pape e 7 chonfessori.
13. S. Mazedony presbiteri. *
14. S. Insursii marturi. E se la luna faxe in questo dì, el sol oschura.
15. S. Longiny marturi. * + El sol va in Aries.
16. S. Ectene.
17. S. Greduetes marturi. * +
18. S. [. . .].[c] * +
19. S. Olutrui marturi. +
20. S. Chonpuritti marturi epischopi. +

f. 95b 21. S. Benedetti abate confesori.

---

a. *Fols. 94a and 94b are blank; fol. 94a has the usual* Ihesus *at the top.*

b. *Sic in MS.*

c. *Not specified in MS.*

1. For the chronological information in this calendar, see Faith Wallis, "Michael of Rhodes and Time Recknoning: Calendar, Almanac, Prognostication," vol. 3, pp. 289–292.
2. Bishop of Perugia and martyr, d. 549.
3. Bishop of Angers, d. 560.
4. The asterisks in this section of the manuscript indicate the days of the rising of those stars which were thought to foretell changes in the weather, while the crosses indicate those days thought to be perilous; cf. Wallis, "Michael of Rhodes and Time Reckoning," vol. 3, p. 292.
5. Roman martyr, first century, also known as Primitiva.
6. Deacon, Faenza.
7. Third-century martyr of Ravenna.
8. Third-century bishop of Rome.
9. Bishop of Rome, martyred 257.
10. Martyred at Carthage, 203.
11. Martyred together at Carthage, third century.

f. 95a

See the calendar

March.[1] On the 16th, at 12 o'clock, the night has 12 hours and the day 12 hours; it has 31 days.

1. St. Herculanus,[2] St. Albinus bishop.[3] *[4]
2. St. Primitivus martyr.[5]
3. St. Terentius confessor.[6]
4. St. Felix monk, St. Fosca.[7]
5. St. Julian bishop,[8] St. Lucius bishop.[9]
6. St. Felicity martyr.[10]
7. St. Perpetua virgin, St. Felicity virgin.[11] *
8. St. Pontianus martyr,[12] St. Honoratus martyr.
9. St. Attala,[13] forty holy martyrs.[14]
10. St. Jerome martyr.
11. Forty holy martyrs.[15]
12. St. Gregory pope[16] and seven confessors.
13. St. Macedonius priest.[17] *
14. St. Innocent martyr. And if the moon is new on this day, the sun obscures it.[18]
15. St. Longinus martyr.[19] + The sun enters Aries.
16. St. Adelaide.[20]
17. St. Gertrude martyr.[21] * +
18. St. .[22] * +
19. St. Leontius martyr.[23] +
20. St. Gundebert martyr and archbishop.[24] +
21. St. Benedict abbot and confessor.[25] f. 95b

12. Martyred at Spoleto, 169.
13. Abbot of Bobbio, d. 627.
14. Martyred in Cappadocia, 320.
15. Repeated from March 9.
16. Gregory I, the Great, pope 590–604.
17. The fourth-century bishop Macedonius was considered a heretic by most Christians.
18. Cf. Wallis, "Michael of Rhodes and Time Reckoning," vol. 3, p. 292.
19. Centurion, martyred in the first century.
20. Uncertain reading.
21. Abbess of Nivelles, d. 659.
22. Blank in MS; given as "S. nulla" in Pietro di Versi, *Raxion de' marineri: Taccuino nautico del XV secolo*, ed. Annalisa Conterio (Venice: Comitato per la Pubblicazione delle Fonti relative alla Storia di Venezia, 1991), and in Giorgetta Bonfiglio Dosio, ed., *Ragioni antique spettanti all'arte del mare et fabriche de vasselli: Manoscritto nautico del sec. XV* (Venice: Comitato per la Pubblicazione delle Fonti relative alla Storia di Venezia, 1987).
23. Uncertain reading.
24. Uncertain reading.
25. Of Nursia, founder of the abbey of Monte Cassino and author of monastic rule, died 543.

22. S. Poli epischopi.
23. S. Felizis, 20 marturi.
24. S. Quiriny marturi.
25. S. anonziaxon beate Marie.
26. S. Montanyny presbitero.
27. S. Chastulli marturi.
28. S. Terpotis marturi.
29. S. Marzi e Victoriny.
30. S. Segondo marturo.
31. S. Victonis abatte.

Avril, adì 5 ore 19, el zorno ore 13 e la notte ore 11, à dì 30.

1. S. Agabundi.
2. S. Vizendi et Abundi.
3. S. Bugiandi virgine.
4. S. Ambroxii epischopi.
5. S. Erenis virgine. *
6. S. Zelestiny pape. +
7. S. Chopie virgine.
8. S. Marine virgine.
9. S. Apolony epischopi.
10. S. Zachaelis pape, S. Zacharia.

f. 96a

11. S. Leonis pape.
12. S. Iuliny pape, S. Rafael. *
13. S. Efimie virgine. Se la volta la luna lo sol oschura.
14. S. Tiburtii, Valeriany marturi.
15. S. Elene regine. El sol in signo de Tauro.

---

1. Bishop of Narbonne, third century.
2. Martyred at Rome, first century.
3. Of Tarsus.
4. Martyred at Pisa, first century.
5. Of Arethusa on Mount Lebanon, martyred fourth century.
6. Of Carthage, martyred 484.
7. Martyred at Asti, 119.
8. Bishop of Como, d. 468.
9. Vincent Ferrer (?): given as such by Contario in her edition of the *Raxion de' marineri* and Bonfiglio Dosio in her edition of the *Ragioni antique*, but he was not canonized until 1455.
10. Same as April 1.
11. Bishop of Milan, d. 397.
12. Martyred at Thessalonica, 304.

22. St. Paul bishop.[1]
23. St. Felix, twenty martyrs.
24. St. Quirinus martyr.[2]
25. Holy Annunciation of the Blessed Virgin Mary.
26. St. Montanino priest.
27. St. Castore martyr.[3]
28. St. Torpezio martyr.[4]
29. SS. Mark[5] and Victorinus.[6]
30. St. Secondus martyr.[7]
31. St. Vitus abbot.

April, on the 5th day at the 19th hour, the day has 13 hours and the night 11 hours; it has 30 days.

1. St. Abundius.[8]
2. St. Vincent[9] and St. Abundius.[10]
3. St. Burgunda virgin.
4. St. Ambrose bishop.[11]
5. St. Irene virgin.[12] *
6. St. Celestine pope.[13] +
7. St. Sophia virgin.[14]
8. St. Marina virgin.[15]
9. St. Apollonius bishop.
10. St. Zachary pope.[16] St. Zachary.[17]
11. St. Leo pope.[18] f. 96a
12. St. Julius pope,[19] St. Raphael.[20] *
13. St. Euphemia virgin.[21] If the moon is turning,[22] the sun will obscure it.
14. SS. Tiburtius and Valerian martyrs.[23]
15. St. Helena queen.[24] The sun in the sign of Taurus.

---

13. Pope 423–432.
14. Martyred at Rome, first century.
15. Martyred in Egypt, ca. 750, relics translated to Venice.
16. Pope 741–742.
17. Zechariah, father of John the Baptist.
18. Leo I, the Great, pope 440–461.
19. Julius I, pope 341–352.
20. Archangel.
21. Martyred in Chalcedon, 307.
22. I.e., new.
23. Martyred with Maximus, third century.
24. Mother of Emperor Constantine, d. 328.

16. S. Fansti marturi, S. Sidero.
17. S. Inozenzii.
18. S. Querine virgine.
19. S. Simeonis epiper,[a] pape Leo.
20. S. Motutto et Ingredi marturi. *
21. S. M[*a*]ximiany marturi.
22. S. Savi pape, Georgio marturo.[b]
23. S. Georgii marturi.
24. S. Liberalis chonfessori.
25. S. Marzi apostolli evangeliste. A ore 2 el zorno 14.
26. S. Marzelli, S. Anastaxia.
27. S. Athanasi pape.
28. S. Victalis marture.
29. S. Petri marturi.
30. S. Queriny epischopi.

Mazio zorny 31, adì 16 ore 9, el dì ore 15 e la notte ore 9.

1. S. apostolorum Felipi e Iachobi. *
2. S. Athanasi papa, S. Ieremia papa.

f. 96b

3. S. inventi sancta cruxe.
4. S. Fluriany, S. Gutaldo.
5. S. Gostine virgine.
6. S. Iohane ante Porta Latina. +
7. S. Mutralis Fame, S. Anzollo.
8. S. Michaelis. +
9. S. Felizis pape, S. Agioppo dali vermy.[c]

---

a. *Sic in MS, probably in place of* epischopi.
b. Georgio marturo *added later, in another hand with other ink.*
c. S. Agioppo dali vermy *added later, in another hand with other ink (seppia instead of dark brown).*
1. Same as April 11?
2. Warrior martyred at Lydda in Palestine ca. 303; name added in different hand. Lacking in Pietro di Versi, *Raxion de' marineri*; present in Bonfiglio Dosio, ed., *Ragioni antique*.
3. Same as April 22? In Pietro di Versi, *Raxion de' marineri*, and Bonfiglio Dosio, ed., *Ragioni antique*, given as St. Gregory martyr.
4. Martyred in 68; patron saint of Venice.
5. Anastasia martyred in 304. St. Marcellus lacking in Pietro di Versi, *Raxion de' marineri*, and Bonfiglio Dosio, ed., *Ragioni antique*.
6. No pope of the name; perhaps Athanasius, patriarch of Alexandria, d. 373.
7. Martyred at Ravenna, second century.
8. Dominican preacher, assassinated by Cathars 1252.

16. St. Faustus martyr, St. Isidore.
17. St. Innocent.
18. St. Quirina virgin.
19. St. Simon bishop, Pope Leo.[1]
20. SS. Theotimus and Ingrid martyrs. *
21. St. Maximianus martyr.
22. SS. Sabba pope, George martyr.[2]
23. St. George martyr.[3]
24. St. Liberale confessor.
25. St. Mark apostle and evangelist.[4] At the second hour, the day has 14 hours.
26. St. Marcellus, St. Anastasia.[5]
27. St. Athanasius pope.[6]
28. St. Vitalis martyr.[7]
29. St. Peter martyr.[8]
30. St. Quirinus bishop.

May has 31 days; on the 16th day at the 9th hour, the day has 15 hours and the night 9 hours.

1. Holy apostles Philip and James.[9] *
2. St. Athanasius[10] pope, St. Geremia pope.[11]
3. Finding of the Holy Cross. f. 96b
4. St. Florian,[12] St. Gotthard.[13]
5. St. Augustine virgin.[14]
6. St. John before the Latin Gate.[15]
7. St. Marziale (?),[16] St. Angelo.[17]
8. St. Michael.[18] +
9. St. Felix pope, St. Hazop of the worms.[19]

---

9. James the Minor, martyred 62.
10. Probably Athanasius, patriarch of Alexandria, whose feast day in the Catholic Church is May 2.
11. No pope of this name. Given as St. Geremia in Pietro di Versi, *Raxion de' marineri*, and Bonfiglio Dosio, ed., *Ragioni antique*. Geremia (Jeremiah) was an Old Testament prophet venerated in Venice as a saint; Old Testament figures were venerated in the Eastern tradition and in those parts of Italy that had been under Byzantine rule.
12. Martyred at Lorch, ca. 304.
13. Bishop of Hildesheim, d. 1038.
14. St. Augustine, bishop of Hippo, d. 430, conversion commemorated May 5, *virgine* apparent miscopying of "conversione."
15. John the Evangelist, d. 101.
16. Third-century bishop of Limoges.
17. Possibly a reference to the eve of St. Michael; see May 8.
18. The Archangel, day of celebration of his apparition or revelation.
19. "St. Hazop of the worms" added in different ink, possibly in another hand; it does not appear in Pietro di Versi, *Raxion de' marineri*, or Bonfiglio Dosio, ed., *Ragioni antique*.

10. S. Pinachi marturi, S. Veneral virgine.
11. S. Mametti epischopi.
12. S. Arpuliny e Pancrati, S. Leo archileo. *
13. S. Sernati epischopi.
14. S. Benvegnati chonfessori, S. Zuane Grixostomo.
15. S. Valentiny chonfessori, S. Illaris chonfesor. Se 'l fa la luna el sol oschura, et in questi dì el sol in Gemini.
16. S. Isidoris marturi.
17. S. Peregrany marturi. +
18. S. nullo.
19. S. Prudenziany virgine. +
20. S. Iustu et Ustadii, S. Potenziane virgine.
21. S. Elene virgine, S. Tadeo Stade marturi.
22. S. Chastu Emilei marturi.
23. S. Dixiderii epischopi, S. Elena.
24. S. Servile e Satrivime marturi.
25. S. translatto Franzeschi, S. Bullo marturo.
26. S. Lutrii Erchudulli.
f. 97a 27. S. Iohanes papa marturo.
28. S. Trentris epischopi.
29. S. Maximi[a] epischopi.
30. S. translatto S. Nicholao.
31. S. Petronille, S. Chanzian.

Zugno à zorny 30, adì 5 ore 16, el dì ore 16 e la notte ore 8.

1. S. Floriani e Segondi.
2. S. Marzelli e Petri.

---

a. *The* a *in* Maximi *added above the line.*
1. Martyred at Alexandria, 250.
2. Martyred at Rome under Antoninus Pius.
3. Bishop of Vienne, d. ca. 476.
4. Martyred at Rome, first century.
5. Martyred at Rome, 304.
6. "St. Leo archileo" lacking in Pietro di Versi, *Raxion de' marineri*; present in Bonfiglio Dosio, ed., *Ragioni antique*.
7. Bishop of Tongres, d. 384.
8. Perhaps St. Benvenuto Scotivoli, d. 1282, bishop and Franciscan.
9. Bishop of Constantinople, d. 407.
10. Bishop of Passau, d. 440.
11. Bishop of Pavia, d. 376.
12. The plural in the MS is perhaps in error, as the transit always occurs on one specific day.
13. Perhaps Isidore the farmer, d. 1130.

10. St. Epimachus martyr,[1] St. Veneranda virgin.[2]
11. St. Mamertus bishop.[3]
12. SS. Apuleius[4] and Pancras,[5] St. Leo archileo.[6] *
13. St. Servatius bishop.[7]
14. St. Benvenutus confessor,[8] St. John Chrysostom.[9]
15. St. Valentine confessor,[10] St. Hilary confessor.[11] If the moon is new the sun obscures it, and in these days[12] the sun is in Gemini.
16. St. Isidore martyr.[13]
17. St. Peregrinus martyr.[14] +
18. St. nothing.
19. St. Prudence virgin.[15] +
20. SS. Justus and Eustace, St. Potentia virgin.
21. St. Helen virgin, SS. Thaddaeus and Eustace martyrs.[16]
22. SS. Castus and Emilius martyrs.[17]
23. St. Desiderius bishop,[18] St. Helen.
24. SS. Servilius and Saturninus martyrs.
25. Translation of St. Francis,[19] St. Servulus martyr.
26. St. Eleutherius.[20]
27. St. John pope martyr.[21] f. 97a
28. St. Terentius bishop.[22]
29. St. Maximinus bishop.[23]
30. Holy translation of St. Nicholas.[24]
31. St. Petronilla,[25] St. Cassian.

June has 30 days; on the 5th at 16 hours, the day has 16 hours and the night 8.

1. SS. Florian and Secundus.
2. SS. Marcellinus and Peter.[26]

---

14. Bishop of Auxerre, martyred 304.
15. Pietro di Versi, *Raxion de' marineri*, has also "S. Iohannes marturi, S. Daniel"; Bonfiglio Dosio, ed., *Ragioni antique*, has also "S. Iohannes marturo."
16. Martyred at Rome, 118.
17. Martyred together in Africa, 250.
18. Bishop of Vienne, d. 608.
19. Of Assisi, d. 1226.
20. Pope 182–193.
21. John I, pope 523–526.
22. Martyred near Pesaro in 251.
23. Archbishop of Trier, d. 349.
24. Of Bari, bishop of Myra, d. fourth century.
25. Reputed daughter of Peter.
26. Exorcists, martyred together at Rome 304.

3. S. Genisini, Erasmi, S. Prabino.
4. S. Quirizi marturi.
5. S. Bonofazii e Quiriny marturi.
6. S. Amatii chonfessori, S. Bonifazio.
7. S. Lunziany marturi. + *
8. S. Primi e Naboris marturi.
9. S. Primi Felizial marturi.
10. S. vigillia.
11. S. Bernaba apostolo.
12. S. Baxilidis marturi. *
13. S. Antuonyo prosbitero.[a]
14. S. Elutri pape, S. Alesio chonfesor.
15. S. Vin, Modesti e Senzie, S. Vido. In questo dì el sol in Chanzer.
16. S. Feriore marturi.
17. S. vizillia. Se vuolta la luna el sol oschura.

f. 97b

18. S. Marzi e Marzeliany marturi.
19. S. Gervaxii e Portaxii marturi.
20. S. Selverine epischopi.
21. S. Ruiny abatte, S. Elbany marturi.
22. S. Paulli, S. Iulliany marturi.
23. S. vigillia nativitatis S. Iohanis.
24. S. nativitas Iohanis. A ore 23, el dì ore 16 e la note ore 8.
25. S. aparitti beati Marzi.
26. S. Iohanes e Paulli.
27. S. 7 dormienti.
28. S. Leonis pape vigillia.
29. S. apostolorum Petri e Pauli.
30. S. Negotta.

---

a. *Sic in MS.*
1. Mime, martyred at Rome 286.
2. Bishop of Formies, martyred fourth century.
3. Reading uncertain.
4. Three-year-old, martyred at Rome 304.
5. Anglo-Saxon missionary to Europe, d. 754.
6. Bishop of Siscia, martyred 309.
7. Martyred at Rome, ca. 287.
8. Martyred at Milan, ca. 304.
9. Brothers, martyred together at Rome, ca. 287.
10. Martyred on Cyprus, first century.
11. Soldier, martyred with companions, third or fourth century.
12. Of Padua, Franciscan, d. 1231.
13. Pope 182–193, martyred at Rome.

3. SS. Genesius[1] and Erasmus,[2] St. Pergentinus.[3]
4. St. Quiricus martyr.[4]
5. SS. Boniface[5] and Quirinus martyrs.[6]
6. St. Amatius confessor, St. Boniface.
7. St. Lucian martyr. + *
8. SS. Primus[7] and Nabor[8] martyrs.
9. SS. Primus, Felitianus martyrs.[9]
10. Holy vigil.
11. St. Barnabas apostle.[10]
12. St. Basilides martyr.[11] *
13. St. Anthony priest.[12]
14. St. Eleutherius pope,[13] St. Alexis confessor.[14]
15. SS. Vitus, Modestus, and Crescenza,[15] St. Vitus. On this day the sun is in Cancer.
16. St. Ferreolus martyr.[16]
17. Holy vigil. If the moon is turning, the sun obscures it.
18. SS. Mark and Marcellinus martyrs.[17] f. 97b
19. SS. Gervasius and Protasius martyrs.[18]
20. St. Severinus bishop.[19]
21. St. Rufinus abbot, St. Alban martyr.[20]
22. SS. Paul and Julian martyrs.[21]
23. Holy vigil of the birth of St. John.
24. Birth of St. John.[22] At 23 hours, the day has 16 hours and the night 8.
25. Holy apparition of St. Mark.
26. SS. John and Paul.[23]
27. The seven sleeping saints.[24]
28. St. Leo pope, vigil.
29. SS. Peter[25] and Paul[26] apostles.
30. St. nothing.[27]

---

14. Of Edessa, d. 417.
15. Martyred together in Lucania, fourth century.
16. Martyred at Besançon, 212.
17. Martyred at Rome, ca. 287.
18. Martyred together, first century.
19. Bishop of Cologne, d. ca. 403.
20. Martyred at Mainz, 400.
21. Martyred at Rimini, third century.
22. The Baptist, d. ca. 27.
23. Brothers, martyred together at Rome, 362 or 363.
24. Martyrs in Ephesus, third century.
25. Martyred at Rome, 67.
26. Martyred at Rome, 67.
27. The word *negotta* is characteristic of the dialect of the region of Bergamo; I thank Linda Carroll for this observation.

Luyo zorny 31, adì 16 ore 6, lo dì ore 15 e la note ore 9.

1. S. Marzeliani marturi vizillia.
2. S. Pruzesii marturi, vixatazio S. Marie.
3. S. Mustiole virgine.
4. S. Dixiderii epischopi.
5. S. Liodori epischopi. *
6. S. translatto S. Thomao, S. Chabriel.
7. S. Rofiny e Segondi, Roxina Segonda.
8. S. Tiliani marturi.
9. S. Athanasii marturi.
10. S. Unfonsse e Pangrati, S. Patriniano.

f. 98a

11. S. translatto S. Benedetti.
12. S. Pii et Ermachulle, S. Marchuola.
13. S. ditazio beate Marie, S. Alessio chonfesor.
14. S. Amuetti pape, Querenziny e Rillitta.
15. S. Iustiny e Rufiny epischopi. + Et in questo dì el sol in Leo.
16. S. Quirinzi et Inlerte, S. Mar[*i*]ne virgine.
17. S. Alesii chonfesori, S. Marine virgine. + La luna fa el sol oscura.
18. S. negutta. *
19. S. Arseny.
20. S. Malgaritta virgine.
21. S. Paresedis virgine.
22. S. Marie Madelene.
23. S. Apolenaris epischopi.
24. S. Cristine virgine, vigilia.
25. S. Iachobi e Christofori.

---

1. Exorcist, martyred at Rome, third or fourth century.
2. Martyred in Poitou, ca. 613.
3. Visit of Mary to St. Elizabeth, mother of John the Baptist, while both were pregnant; observation instituted 1378.
4. Martyred in Tuscany, 273.
5. Bishop of Vienne, d. 608.
6. Bishop of Altino, d. ca. 407.
7. Archbishop of Canterbury, martyred 1170, translation of relics July 7, 1222.
8. Archangel.
9. Names given twice, second time corrupted.
10. Reading uncertain.
11. Bishop of Bologna, d. ca. 450.
12. Of Nursia, founder of Monte Cassino.
13. Pope 158–167.
14. Name given in two different dialect versions.
15. Reading uncertain.

July has 31 days; on the 16th day at the 6th hour, the day has 15 hours and the night has 9 hours.

1. St. Marcellinus martyr,[1] vigil.
2. St. Prudentius martyr,[2] Visitation of the Blessed Virgin Mary.[3]
3. St. Mustiola virgin.[4]
4. St. Desiderius bishop.[5]
5. St. Heliodorus bishop.[6] *
6. Holy translation of St. Thomas,[7] St. Gabriel.[8]
7. SS. Rufinus and Secundus.[9]
8. St. Vitalianus martyr.
9. St. Athanasius martyr.
10. SS. Alphonse[10] and Pancras, St. Paternian.[11]
11. Translation of St. Benedict.[12] f. 98a
12. SS. Pius[13] and Ermagora.[14]
13. Detail of the Blessed Virgin Mary;[15] St. Alexis confessor.[16]
14. SS. Anacletus pope,[17] Quirinzini, and Relitta.
15. SS. Justin and Rufinus bishops. And on this day the sun enters Leo.
16. SS. Quirinzi and Hilary,[18] St. Marina virgin.[19]
17. St. Alexis confessor,[20] St. Marina virgin.[21] + The moon is new, the sun obscures it.
18. St. nothing. *
19. St. Arsenio.[22]
20. St. Margaret virgin.[23]
21. St. Praxedis virgin.[24]
22. St. Mary Magdalene.[25]
23. St. Apollinaris bishop.[26]
24. St. Christina virgin,[27] vigil.
25. SS. James[28] and Christopher.[29]

---

16. Confessor at Edessa, died at Rome 417.
17. Pope 100–112, martyred at Rome.
18. Reading uncertain.
19. Died in Egypt, ca. 750; relics translated to Venice 17 July.
20. Same as July 13, above.
21. Same as July 16, above.
22. Anchorite of Sceté, died ca. 449.
23. Martyred at Antioch, third century.
24. Died at Rome, third century.
25. Died ca. 66.
26. Of Ravenna, d. ca. 78.
27. Martyred in Lake Bolsena, third or fourth century.
28. The Major (of Compostella), martyred 44.
29. Martyred in Syria, third century.

26. S. Pastoris gardenalis, S. Ana.
27. S. Almorò marturo.
28. S. Pantalon marturo.
29. S. Fasuny marturo.
30. S. Abdon e Sodon marturi.
31. S. Fantiny chonfesori.

Avosto zorny 31, adì 5 ore 13, el dì ore 14 e la note ore 10.

1. S. vinchula S. Petri. *
2. S. Fanstin pape.
3. S. inventi[a] S. Stefany.
4. S. Iustine pape.
5. S. Dominizi e Dominy chonfesorii.

f. 98b

6. S. Sisti e Salvatoris.
7. S. Donati epischopi.
8. S. Quiriachi marturi.
9. S. Romany, vigillia.
10. S. Laurenzii marturi.
11. S. Ziberti marturi.
12. S. Elana[b] virgine, S. Chanzian.
13. S. Ipolitti e Chanziane marture.
14. S. Chanzian, S. Felixe, vigilia. *
15. S. asension beate Marie. + El sol in segno de Virgo.
16. S. Arnulfi epischopi.
17. S. Mametis marturi. Se la luna fa el sol oschura.
18. S. Margii marturi.
19. S. Bernarde chonfesore, S. Samoel.

---

a. *Corrected over* Ius.
b. *Sic in MS for* Elena.
1. Of Rome, martyred ca. 160; though brother to Pius I, he was not a cardinal.
2. Mother of Mary.
3. Martyred at Nicomedia, 303.
4. Martyred at Nicomedia, 303.
5. Persians, martyred at Rome, 250.
6. Apostle, martyred 67.
7. No pope of this name.
8. Protomartyr, d. 33; body found 415.
9. No pope of this name.
10. Founder of the Dominican order, died August 6, 1221.
11. Reading uncertain.

26. St. Pastor cardinal,[1] St. Anne.[2]
27. St. Hermolaus martyr.[3]
28. St. Pantalon martyr.[4]
29. St. Faustinus martyr.
30. St. Abdon and St. Sennen martyrs.[5]
31. St. Fantinus confessor.

August has 31 days; on the fifth day at the 13th hour, the day has 14 hours and the night 10 hours.

1. St. Peter in chains.[6] *
2. St. Faustinus pope.[7]
3. Finding of St. Stephen.[8]
4. St. Justin pope.[9]
5. SS. Dominic[10] and Dominus[11] confessors.
6. SS. Sixtus[12] and Saviour.[13] f. 98b
7. St. Donatus bishop.[14]
8. St. Cyriacus martyr.[15]
9. St. Romanus,[16] vigil.
10. St. Lawrence martyr.[17]
11. St. Gilbert martyr.
12. St. Helen virgin, St. Cassian.
13. SS. Hypolitus and Cassian martyrs.[18]
14. St. Cassian, St. Felix,[19] vigil. *
15. Assumption of the Blessed Virgin Mary.[20] + The sun enters the sign of Virgo.
16. St. Arnulf bishop.[21]
17. St. Mansuetus martyr.[22] If the moon is new, the sun obscures it.
18. St. Mark martyr.
19. St. Bernard confessor,[23] St. Samuel.[24]

---

12. Pope Sixtus II, 260–261.
13. Transfiguration of Jesus Christ.
14. Bishop of Arezzo, martyred 362.
15. Bishop of Ancona, martyred in the Holy Land, 133.
16. Martyred at Rome, fourth century.
17. Soldier, martyred at Rome, 258.
18. Martyred at Rome, 258.
19. Martyred together at Rome, 259.
20. Venetian usage has the same term for the Assumption of Mary to Heaven as for the Ascension of Christ.
21. Of Metz, d. 640.
22. In Egypt, fifth century.
23. Of Clairvaux, d. 1153.
24. Old Testament prophet, venerated in Venice.

20. S. Donati marturi.
21. S. ecta[a] beate Marie.
22. S. Furtunati, vigillia. *
23. S. nullo, vigillia.
24. S. Genyso marturi, S. Bortholomeo.
25. S. Alesandrini et Anastaxia. Ore 20, el dì ore 13 e la note ore 11.
26. S. nullo.
27. S. negotta.
28. S. Pauliny epischopi, S. Agostin.
29. S. Felizis et Audattii. Digolazio S. Iohannis.
30. S. nullo.
31. S. negotta.

Setenbrio zo[*r*]ny 30, adì 15, lo zorno ore 12 e la notte ore 12.

f. 99a

1. S. Egidii et Rinziany.
2. S. Antuonii marturi.
3. S. Fumie virgine.
4. S. Marzelli marturi, s. Moixe.
5. S. Erchulliany marturi.
6. S. Zacharia.
7. S. Andriany marturi.
8. S. Nativitas beate Marie.
9. S. Gregorii marturi.
10. S. Ilarii pape.
11. S. Pronti e Iazenti.
12. S. Sirii chonfessori.
13. S. Ligorii marturi.
14. S. exaltade S. Cruzis.

---

a. *Probably for* ottava *here and also elsewhere.*
1. Same as August 7?
2. Martyred at Arles, 303.
3. Apostle, martyred at Rome, 47.
4. Soldier of the Theban legion, martyred 288.
5. Bishop of Hippo, d. 430.
6. Martyred together at Rome, fourth century.
7. The Baptist.
8. Abbot in Languedoc, died 726.

20. St. Donatus martyr.[1]
21. Octave of the Assumption of the Blessed Virgin Mary.
22. St. Fortunatus, vigil. *
23. St. nothing, vigil.
24. St. Genesius martyr,[2] St. Bartholomew.[3]
25. SS. Alexandrinus[4] and Anastasia. At the 20th hour, the day has 13 hours and the night 11 hours.
26. St. nothing.
27. St. negative.
28. St. Paulinus bishop, St. Augustine.[5]
29. SS. Felix and Audactus.[6] Beheading of St. John.[7]
30. St. nothing.
31. St. negative.

September has 30 days; on the 15th day, the day has 12 hours and the night 12 hours.

1. SS. Aegidius[8] and Terence. f. 99a
2. St. Anthony martyr.
3. St. Eufemia virgin.[9]
4. St. Marcellus martyr,[10] St. Moses.[11]
5. St. Herculanus martyr.[12]
6. St. Zechariah.[13]
7. St. Hadrian martyr.[14]
8. Nativity of the Blessed Virgin Mary.
9. St. Gregory martyr.
10. St. Hilary pope.[15]
11. SS. Proto and Hyacinth.[16]
12. St. Siro confessor.[17]
13. St. Ligorius martyr.
14. Holy exaltation of the Holy Cross.

---

9. Martyred in Chalcedon, 307.
10. Bishop, martyred at Chalon-sur-Saône, ca. 178.
11. Old Testament patriarch, venerated in Venice.
12. Martyred at Porto, Italy, ca. 180.
13. Old Testament prophet, venerated in Venice.
14. Martyred at Nicomedia, 303; relics translated September 8.
15. Pope 461–468.
16. Martyred together at Rome, 257.
17. Bishop of Pavia, d. 96.

15. S. Isidoris e Nichodemi. Et in questo dì in segno de Libra el sol.
16. S. Luzie e Geminiany. Se la luna fa el sol oschura.
17. S. negotta.
18. S. Victoris marturi.
19. S. Ianunzirii epischopi.
20. S. vigillia.
21. S. Marzi evangelista, S. Matio apostolo.
22. S. Mauretii e soziorum.
23. S. Liny pape, S. Techle virgine. *
24. S. Chonzetto, S. Iohanes pape.
25. S. Ferminy marturi.
26. S. Zipriany e Iustizie.
f. 99b 27. S. Chosme e Damiane.
28. S. Masime marturi.
29. S. ditazio S. Michaelis.
30. S. Ieronymi epischopi.

Octobrio zorni 31, adì 5, el zorno ore 11 e la notte ore 13.

1. S. Remedio epischopo.
2. S. Ensenpro papa.
3. S. vigilia.
4. S. Franzeschi chonfessori.
5. S. Flamiany epischopi. *
6. S. Bagaris chonfesori.
7. S. Iustine virgine.
8. S. Marzi pape.
9. S. Dionisii e Rustitti marturi.
10. S. Zerbony epischopi.
11. S. Venanti abate.

---

1. Follower of Jesus Christ.
2. Martyred at Syracuse, 304.
3. Bishop of Modena, d. 397.
4. Martyred at Marseilles, ca. 290.
5. Bishop of Benevento, martyred 305.
6. Martyred 68, patron of Venice.
7. Apostle and evangelist, first century.
8. Captain of the Theban legion, martyred at Agaune, 286.
9. Pope 67–78.
10. Martyred in Seleucia, first century.
11. Reading uncertain.
12. Martyred together at Nicomedia, 304.

15. SS. Isidore and Nicodemus.[1] And in this day, the sun enters the sign of Libra.
16. SS. Lucy[2] and Geminianus.[3] If the moon is new, the sun obscures it.
17. St. nothing.
18. St. Victor martyr.[4]
19. St. Januarius bishop.[5]
20. Holy vigil.
21. St. Mark evangelist,[6] St. Matthew apostle.[7]
22. St. Maurice and his comrades.[8]
23. St. Linus pope,[9] St. Tecla virgin.[10] *
24. St. Concettus,[11] St. John pope.
25. St. Ferminus martyr.
26. SS. Cyprian and Justina.[12]
27. SS. Cosmas and Damian.[13] f. 99b
28. St. Maximus martyr.
29. Holy dedication of St. Michael.[14]
30. St. Jerome bishop.[15]

October has 31 days; on the 5th day, the day has 11 hours and the night 13.

1. St. Remigius bishop.[16]
2. St. Eugene pope.[17]
3. Holy vigil.
4. St. Francis confessor.[18]
5. St. Flamianus bishop. *
6. St. Bagaris confessor.
7. St. Justina virgin.[19]
8. St. Mark pope.[20]
9. SS. Denis and Rusticius[21] martyrs.
10. St. Cerbonius bishop.[22]
11. St. Venantius abbot.[23]

---

13. Brothers and doctors, martyred in Cilicia, 297.
14. Archangel, temple dedicated on Mount Gargano.
15. Doctor of the Church, ca. 340–420; he was not a bishop.
16. Bishop of Rheims, d. 533.
17. Reading uncertain.
18. Of Assisi, d. 1226.
19. Martyred at Padua, second century.
20. Pope 337–340.
21. Martyred together in Gaul, ca. 286.
22. Bishop of Populonia, d. ca. 575.
23. Hermit in Artois, eighth century.

12. S. Masimiany.
13. S. Galdentii chonfessori.
14. S. Chalisti pape.
15. S. Maurorun pape. + El sol in segno de Schorpio.
16. S. Galli abate. La luna fa el sol oschura.
17. S. Sevenine virgine. +
18. S. Lucha evangelista. + *
19. S. negutta.
20. S. Ianuarii marturi.
f. 100a 21. S. Chanbraxii marturi.
22. S. Urssulle virgine.
23. S. Felipi epischopi.
24. S. Severiny epischopi.
25. S. Victalis, Brifazii, vigilia.
26. S. Crisandis e Darie, S. Dimitri.
27. S. Tuvaristi pape, vigillia.
28. S. apostolorum Simeonis et Iude.
29. S. vigillia.
30. S. Marzesti epischopi.
31. S. Iermany epischopi, vigillia omnium sanctorum.

Novenbrio zorny 30, adì 15 ore 6, lo dì ore 9 e la notte ore 15.

1. S. festorum omnium sanchtorum. *
2. S. Bigasii et Iusti.
3. S. Peregriny marturi.
4. S. Victalis et Agirchole.
5. S. Felizis et Sepie.
6. S. Leonardi chonfessori.
7. S. Erchulliany chonfessori.

---

1. Bishop of Lorsch, martyred ca. 308.
2. Pope 221–227.
3. Irish abbot, apostle to Switzerland, d. ca. 627.
4. Bishop of Benevento, martyred 305.
5. Martyred at Cologne, fourth or fifth century.
6. Bishop of Cologne, d. ca. 403.
7. Martyred together at Rome, third century.

12. St. Maximilianus.[1]
13. St. Gaudentius confessor.
14. St. Calixtus pope.[2]
15. St. Maurus pope. + The sun enters the sign of Scorpio.
16. St. Gall abbot.[3] The moon is new, the sun obscures it.
17. St. Severina virgin. +
18. St. Luke evangelist. + *
19. St. nothing.
20. St. Januarius martyr.[4]
21. St. Cabrasius martyr. f. 100a
22. St. Ursula virgin.[5]
23. St. Philip bishop.
24. St. Severinus bishop.[6]
25. SS. Vitalis, Boniface, vigil.
26. SS. Chrysanthus and Daria,[7] St. Demetrius.[8]
27. St. Evaristus pope,[9] vigil.
28. Holy apostles Simeon[10] and Jude.[11]
29. Holy vigil.
30. St. Marcellus bishop.
31. St. Germanus bishop, vigil of All Saints.

November has 30 days; on the 15th day at the 6th hour, the day has 9 hours and the night 15 hours.

1. Feast of All Saints. *
2. SS. Pegasius[12] and Justus.[13]
3. St. Peregrinus martyr.
4. SS. Vitalis and Agricola.[14]
5. SS. Felix and Eusebius.
6. St. Leonard confessor.[15]
7. St. Herculanus confessor.

---

8. Martyred at Thessalonica, 307.
9. Of Syria, pope 112–121.
10. Cousin of Jesus, Bishop of Jerusalem, martyred 107.
11. Called Thaddaeus, martyred after 62.
12. Persian, martyred in reign of Shapur II, fourth century.
13. Martyred under Diocletian.
14. Martyred together at Bologna, ca. 304.
15. Abbot of Noblat, d. 559.

8. S. 4 choranatorum.
9. S. Todorum et Salvatoris.
10. S. Zirfanis et Repitie.
11. S. Martiny epischopy chonfesori. *

f. 100b
12. S. Martiny marturi.
13. S. Iohanis Grisostomo.
14. S. Clementiny marturi.
15. S. Felizis pape, S. Venerande. La luna fa el sol scura, et in questo dì el sol in segno de Sagitario.
16. S. Mauri abatte.
17. S. negotta.
18. S. Fredian, Elixabet. +
19. S. Ponzian papa.
20. S. Stefany e Bassi.
21. S. Mauri e Turusti, S. Sten.
22. S. Zizillie virgine.
23. S. Clementi e Cholombarii.
24. S. Perpi e Grixovany.
25. S. Chatarina virgine.
26. S. Alesandri epischopi.
27. S. Dimne marturi.
28. S. Sustenes marturi.
29. S. Saturne, vigillie.
30. S. Andree apostolle.

Dezenbrio zorny 31, adì 5 ore 13, lo dì ore 8 e la notte ore 16.

1. S. Ursizii epischopi.
2. S. Pibiane virgine.
3. S. Galgany chonfessore.

---

1. Two separate groups of martyrs, one from Pannonia, one from Albano, martyred in the third or fourth century.
2. Of Amaseo, martyred 306.
3. Martyred at Nicaea, third, fourth or fifth century.
4. Bishop of Tours, d. ca. 397.
5. Of Todi, Pope Martin I, 649–653.
6. Bishop of Constantinople, d. 407.
7. Pope Clement I, 90–100.
8. Bishop of Lucca, d. 588.
9. Of Hungary and Thuringia, d. 1231.
10. Pope 230–235.
11. Martyred in Heraclea with 41 companions under Diocletian.

8. Four crowned saints.[1]
9. SS. Theodore[2] and Saviour.
10. SS. Tryphon[3] and Respicius.
11. St. Martin bishop, confessor.[4] *
12. St. Martin martyr.[5] f. 100b
13. St. John Chrysostom.[6]
14. St. Clement martyr.[7]
15. St. Felix pope, St. Veneranda. The moon is new, the sun obscures it, and on this day the sun enters the sign of Sagittarius.
16. St. Maurus abbot.
17. St. nothing.
18. SS. Frigidianus,[8] Elizabeth.[9] +
19. St. Pontian pope.[10]
20. SS. Stephen and Bassus.[11]
21. SS. Maurus and Theonistus, St. Stephen.
22. St. Cecilia virgin.[12]
23. SS. Clement[13] and Colombanus.[14]
24. SS. Perpi[15] and Chrysogonus.[16]
25. St. Catherine virgin.[17]
26. St. Alexander bishop.
27. St. Dominic martyr.
28. St. Sustenes martyr.
29. St. Saturninus,[18] vigil.
30. St. Andrew apostle.

December has 31 days. On the 5th day at the 13th hour, the day has 8 hours and the night 16 hours.

1. St. Ursicius bishop.[19]
2. St. Bibiana virgin.[20]
3. St. Galganus confessor.

---

12. Martyred at Rome, 230.
13. Pope Clement I, 90–100.
14. Abbot of Bobbio, d. 615.
15. Possibly Protasius of Milan, d. 342, who is commemorated on this date.
16. Martyred near Aquileia, ca. 304.
17. Martyred at Alexandria, fourth century.
18. Bishop of Toulouse, martyred 257.
19. Bishop of Brescia, d. 347.
20. Martyred at Rome, 363.

f. 101a

4. S. Barbare virgine.
5. S. Sabe abate, S. Basso.
6. S. Nicholai epischopi e chonfesori. +
7. S. Anbroxii e Severine. *
8. S. chonsenzio beate Marie.
9. S. Serii epischopi.
10. S. Melchidiatis pape.
11. S. Damasii pape.
12. S. Donati, 24 marturi.
13. S. Luzie virgine marturi.
14. S. Victoris epischopi.
15. S. Marchulliany marturi. + Et in questo dì el sol in Chapicorno.
16. S. Igatiy chonfesori.
17. S. negutta. + La luna fa el sol oschura.
18. S. Iutti abatte.
19. S. negotta.
20. S. nulla, vigillia.
21. S. sanctorum 30 marturum, S. Thome apostolo.
22. S. Gregorii presbiteri. *
23. S. negotta.
24. S. vigillia.
25. S. nativitas Ihesum Christo.
26. S. Iohanes evangelista, S. Stefano.
27. S. sanctorum Inozenzium, S. Iohanes.
28. S. Thome archo, S. Thomao Chontrupie.
29. S. Florentiny epischopi.
30. S. negotta.
31. S. Silvestri pape.

---

f. 101b

Zener zorny 31, adì 15 ore 3, lo dì ore 9 e la notte ore 15.

1. S. zirchonseyo Domini, S. Baxeyio. +
2. S. ecta S. Stefani.

---

1. Martyred at Heliopolis, 306.
2. In Palestine, d. 532.
3. Martyred at Nice, 250.
4. Of Bari, Bishop of Myra, d. fourth century.
5. Bishop of Milan, d. 397.
6. Bishop of Pavia, d. 96.
7. Pope 311–314.

4. St. Barbara virgin.[1] f. 101a
5. St. Sabas abbot,[2] St. Bassus.[3]
6. St. Nicholas bishop and confessor.[4]
7. SS. Ambrose[5] and Severinus.
8. Holy Conception of the Blessed Virgin Mary.
9. St. Sirius bishop.[6]
10. St. Melchiades pope.[7]
11. St. Damasius pope.[8]
12. St. Donatus, twenty-four martyrs.
13. St. Lucy virgin martyr.[9]
14. St. Victor bishop.
15. St. Marcullianus. + And on this day the sun enters Capricorn.
16. St. Ignatius (?) confessor.
17. St. nothing. + The moon is new, the sun obscures it.
18. St. Justus abbot.
19. St. nothing.
20. St. nothing, vigil.
21. Thirty holy martyrs, St. Thomas apostle.
22. St. Gregory priest.
23. St. nothing.
24. Holy vigil.
25. Nativity of Jesus Christ.
26. St. John evangelist,[10] St. Stephen.[11]
27. Holy Innocents, St. John.
28. St. Thomas archbishop, St. Thomas of Canterbury.[12]
29. St. Florentinus bishop.
30. St. nothing.
31. St. Sylvester pope.[13]

January has 31 days; on the 15th day at the 3rd hour, the day has 9 hours and the night 15 hours. f. 101b

1. Holy Circumcision of the Lord, St. Basil.[14]
2. Holy octave of St. Stephen.

---

8. Pope 366–384.
9. Martyred at Syracuse, 304.
10. Died 101.
11. Protomartyr, died 33.
12. Thomas Becket, martyred 1170.
13. Pope Sylvester I, 314–337.
14. The Great, Bishop of Caesarea, d. 379.

3. S. ecta S. Iohanis. +
4. S. Inozenzium. +
5. S. vigillia. *
6. S. epifanya Dominy. +
7. S. Iulliani marturi. +
8. S. Thametii e Luziany marturi.
9. S. Ferminy e Factori marturi.
10. S. Paulli prime remitte.
11. S. Ugine pape marturi.
12. S. Iohanes epischopi.
13. S. Illarii epischopi. Se la luna fa el sol oschura.
14. S. Felizis chardinallis.
15. S. Mauri abate chonfesore. Et in questo dì el sol in segno d'Achario.
16. S. Marzelli epischopi marturi.
17. S. Antuony abatte e chonfessore.
18. S. negutta.
19. S. Mauri e Marte virgine.
20. S. Fibiany e Sebastiane. +
21. S. Agnetis virgine.
22. S. Vizenzi et Anastaxie.
23. S. Metenziane virgine.
24. S. Feliziany epischopi.
25. S. chonverzazion Santi Paulli.
26. S. Perolli e Ranpi presbiteri.
27. S. Iohane Grisostomo.
28. S. Agnetis, Segondi.
29. S. Chonstanzie pape chonfesore.

f. 102a 30. S. Geminiany chonfesori.

31. S. Mittu e Iohanes marturi, S. Marcho.

---

1. Martyred under Diocletian.
2. Reading uncertain.
3. Of Antioch, martyred 312.
4. In the Thebaid, d. 341.
5. Of Athens, pope 154–158.
6. Bishop of Milan, d. ca. 660.
7. Bishop of Poitiers, d. 368.
8. Of Nola, d. 269; not a cardinal.
9. Abbot of Glanfeuil, d. 584.
10. Died 356.
11. Pope 240–250.
12. Martyred at Rome, ca. 287.
13. Martyred at Rome, 304.

3. Holy octave of St. John. +
4. St. Innocent. +
5. Holy vigil. *
6. Holy Epiphany of the Lord. +
7. St. Julian martyr.[1] +
8. SS. Thametius[2] and Lucian martyr.[3]
9. SS. Firminus and Factor martyr.
10. St. Paul, first hermit.[4]
11. St. Hyginus pope, martyr.[5]
12. St. John bishop.[6]
13. St. Hilary bishop.[7] If the moon is new, the sun obscures it.
14. St. Felix cardinal.[8]
15. St. Maurus abbot, confessor.[9] And in this day the sun enters the sign of Aquarius.
16. St. Marcellus bishop, martyr.
17. St. Anthony abbot and confessor.[10]
18. St. nothing.
19. SS. Maurus and Martha virgin.
20. SS. Fabian[11] and Sebastian.[12]
21. St. Agnes virgin.[13]
22. SS. Vincent[14] and Anastasia.[15]
23. St. Emerentiana virgin.[16]
24. St. Felicianus bishop.[17]
25. Holy conversion of St. Paul.
26. SS. Perolli and Ranpi[18] priests.
27. St. John Chrysostom.[19]
28. SS. Agnes,[20] Secundus.
29. St. Constantius pope confessor.[21]
30. St. Geminianus confessor.[22] f. 102a
31. SS. Metras[23] and John martyrs, St. Mark.[24]

---

14. Deacon of Saragossa, martyred at Valencia, 304.
15. Martyred at Rome, 304.
16. Martyred at Rome, 304.
17. Bishop of Foligno, martyred 251.
18. Conterio, in Pietro di Versi, *Raxion de' marineri*, p. 130, considers this a corruption of St. Polycarp, bishop of Smyrna, martyred 167.
19. Bishop of Constantinople, d. 407.
20. Martyred at Rome, 304.
21. Probably the bishop of Perugia commemorated on this day, martyred under Antoninus Pius; he was not a pope.
22. Bishop of Modena, d. 397.
23. Martyred in Alexandria, 250.
24. The Evangelist, relics transferred to Venice, January 31, 829.

[Veder chalandario]

Frever zorny 28, adì 4, lo dì ore 10 e la note ore 14.

1. S. Severiny epischopi.
2. S. purifichazio beate Marie.
3. S. Blaxii epichopi.
4. S. Simeonis pape.
5. S. Agatte virgine.
6. S. Vendasti epischopi. *
7. S. Agolii chonfessori.
8. S. Dionisii marturi.
9. S. Sabin epischopi, S. Apologna.
10. S. Scholaziche virgine.
11. S. Zilberti chonfessori.
12. S. Simplizinyi epischopi. * Se la luna fa el sol oschura.
13. S. Fusche virgine.
14. S. Valentiny epischopi.
15. S. negutta. Et in questo dì el sol in segno d'i Pisis.
16. S. Iulliany marturi. +
17. S. Policrane marture. +
18. S. Simeonis monazi.
19. S. Galliny marturi. +
20. S. Victoris marturi. *
21. S. Iusti et Amatoris.
22. S. chaedra S. Petri. +
23. S. vigillia.
24. S. Matie apostolo.
25. S. Victoriny e Victaris.

f. 102b 26. S. Alesandrii marturi.

27. S. Abudanzii marturi.
28. S. Romany abatte.

---

1. Bishop of Cologne, d. ca. 403.
2. Bishop of Sebastopolis, martyred ca. 316.
3. Martyred in Sicily, 251.
4. Bishop of Arras and Cambrai, sixth century.
5. Martyred ca. 303.
6. Bishop of Paris, martyred ca. 286.
7. Martyred at Spoleto, 303.
8. Martyred at Alexandria, 249.
9. Sister of Saint Benedict, d. ca. 543.
10. Possibly Gilbert of Sempringham, d. 1189, who is usually commemorated on February 4.

February has 28 days; on the 4th day the day has 10 hours and the night 14 hours.

1. St. Severinus bishop.[1]
2. Holy Purification of the Blessed Virgin Mary.
3. St. Blasius bishop.[2]
4. St. Simeon pope.
5. St. Agatha virgin.[3]
6. St. Vedastus bishop.[4] *
7. St. Augulus confessor.[5]
8. St. Denis martyr.[6]
9. St. Sabinus bishop,[7] St. Appolonia.[8]
10. St. Scholastica virgin.[9]
11. St. Gilbert confessor.[10]
12. St. Simplicianus bishop. * If the moon is new, the sun obscures it.
13. St. Fosca virgin.[11]
14. St. Valentine bishop.[12]
15. St. nothing. And on this day the sun enters the sign of Pisces.
16. St. Julian martyr.
17. St. Policronius martyr.[13] +
18. St. Simeon monk.[14]
19. St. Gall martyr. +
20. St. Victor martyr.[15] *
21. SS. Justus and Amator.
22. Holy cathedra of St. Peter. +
23. Holy vigil.
24. St. Matthias apostle.
25. SS. Victorinus and Victor.[16]
26. St. Alexander martyr.[17]
27. St. Abundius.[18] f. 102b
28. St. Romanus abbot.[19]

11. Martyred at Ravenna, ca. 250.
12. Martyred at Terni, ca. 273.
13. Martyred in Babylon, ca. 250.
14. The Simeon commemorated on February 18 is allegedly Jesus's cousin, martyred 107.
15. Martyred at Marseilles, ca. 290.
16. Martyred together, along with 33 others, in North Africa.
17. Bishop of Alexandria, d. 326.
18. Bishop of Cologne, d. 468.
19. Abbot of Condat, d. 460.

[Veder per tuor sangue]

Amaistramento a tuor sangue per tuti li mexi del'ano.

El primo dì del mexe serà zallo per la faza.

2 dì. Serà fervuxo.

3 dì. Tu avera' infirmitade in quel ano.

4 dì. Dubio de morte subitana.

5 dì. Serà aparechiar el molimento.

6 dì. Serà bon che 'l sangue è aqua.

7 dì. Serà manchamento del stomago.

8 dì. Serà per perder la volontà dela femena.

9 dì. Serà in rogna e pedexelli.

10 dì. Serà motado el cholor dela faza.

11 dì. Serà non voluntaruxo dela dona.

12 dì. Serà bon perché tu te puo' fadigar.

13 dì. Serà grasso per pocho manzar.

14 dì. Serà infirmittà in quel ano.

15 dì. Serà mondo de rogna.

16 dì. Serà in gran dano dela persona.

17 dì. Serà bon se 'l te fa mestier.

18 dì. Serà assai per mexi 4.

19 dì. Serà infirmittà 18 volte.

See how to draw blood

Instructions for drawing blood in all the months of the year.

The first day of the month will be yellow for the face.

2nd day. Will be feverish.

3rd day. You will have illness in that year.

4th day. There is concern of sudden death.

5th day. It will be to prepare the emollient.

6th day. It will be good, for that blood is watery.

7th day. There will be a lack of appetite.

8th day. There will be a loss of the desire for women.

9th day. It will result in scabies and peduncle.

10th day. The color of the face will be heightened.

11th day. There will be no wish for a woman.

12th day. It will be good for you to exert yourself.

13th day. You will get fat from eating little.

14th day. There will be illness in this year.

15th day. One will be cleansed of scabies.

16th day. It will bear great personal danger.

17th day. It will be good if you're skilled.

18th day. It will be enough for 4 months.

19th day. There will be illness 18 times.

20 dì. Serà infermo asai.

21 dì. Serà volontaruxo da mazar.

22 dì. Serà partida ogni infirmittà da tti.

f. 103a 23 dì. Serà messa forza in la to persona.

24 dì. Serà de bona vogia.

25 dì. Serà prodevelle.

26 dì. Serà senzier di infirmitade.

27 dì. Serà in paura de morte subitana.

28 dì. Serà de non aver paura de faturie.

29 dì. Serà a non te tuor per chondezion del mondo.

30 dì. Serà a non te tuor per algun muodo.

[Truovar insegni e la propietade de quelli e chomo sta in la persona del'omo o dona]

Qui de sotto per l'inchontro faremo per figura in una ruoda uno homo e li 12 segny del'ano, a che muodo regna in la criatura li segnalli, chomo serave a dir Aries, lo primo segno dela testa.

a[a] [*A*]ries, segno dela testa
a Taurus, segno delo cholo
a Zeminy, segno dele braze
a Chanzer, segno delo petti[b]
a Leos,[c] segno delo chuor

---

a. *As in the table of contents, the lower-case letter a is used to introduce entries in the list.*
b. *Sic in MS.*
c. *Sic in MS.*

20th day. It will be a great deal of illness.

21st day. You will be desiring to eat.

22nd day. All illness will be gone from you.

23rd day. Strength will be put in your person. f. 103a

24th day. You will have a good outlook.

25th day. You will be brave.

26th day. There will be feelings of illness.

27th day. There will be fear of sudden death.

28th day. Do not be afraid of being bewitched.

29th day. Do not go away for any reason in the world.

30th day. Do not go away at all.

Find the signs and their properties and how they affect the person of a man or woman

Beneath here we shall demonstrate, with a diagram of a man in a wheel and the twelve signs of the year, the way the creatures work in the signs, as it would be to say Aries, the first sign—the head.[1]

- Aries—sign of the head
- Taurus—sign of the neck
- Gemini—sign of the arm
- Cancer—sign of the chest
- Leo—sign of the heart

1. For an introduction to the images and text of this section, see Dieter Blume, "The Use of Visual Images by Michael of Rhodes: Astrology, Christian Faith, and Practical Knowledge," vol. 3, pp. 148–166.

a Virgo, segno dele budelle
a Libra, segno per le anche
a Schorpio, per la verga
a Sazitario, per le chosse
a Chapicorno, per la zenocha
a Aquario, per la[a] ganbe
a Pisis, per i piedi

f. 103b a Pissis. Segno per li pie. E se fuse ferido algun al membro, fusse la luna, non se ossarave a desligar inchina la luna fusse in quel segno, chomo vedirì qui de sotto depentto lo homo e li segnali.

---

f. 104a Ariens.[b] Lo qual è lo primo segno. À natura de fuogo, chaldo e secho. Segno nobele. El so pianetto Mars. Lo qual è segno dela testa. Quel che naserà in questo segno serà de chalida natura. Homo plan, pazificho, homo de raxion, averà honor. E sie bon andar in viazio e de far lavorar tereny. Vardasse a tuor sangue dela testa.

Mars, el so pianetto in lo quinto zielo, à segnoria da levante infra tera. Sta a zaschun segno mexi 2, zercha li segny in ani 2 e monta in segni 2, Aries e Tauro. Alguny pianetti sun bony e mezany e rii. Iupiter e Venus son bony, Saturno e Mars son rii, Sul et Merchurius sun mezany, Luna non è bona salvo setenbrio. El segno vedirì qui de sotto depentto.

[*caption:*]
aries

---

f. 104b Tauro. Lo segondo segno. Natura de tera, secho e fredo. Segno fermo. El so pianetto Venus. Segno dela gola. Quel che naserà in quel segno serà grande afanator. Al so fatto averà gran stado e ben averà grande amor ala so famegia e leal. Vardasse da tior sangue dela gulla, né da far medixina. E serà bon a far marchadantia. Venus, lo so pianetto in lo terzo ziello, signoriza da ponente infra tera e sie sovra el pesse del mar. Sta a zaschon segno zorny 8, zercha i segny in mexi 3, zorny 6. Venus è bon da far ogni chossa e monta segny 2, Libra e Tauro.

---

a. *Sic in MS.*
b. *Sic in MS.*

- Virgo—sign of the bowels
- Libra—sign of the hips
- Scorpio—for the member
- Sagittarius—for the thighs
- Capricorn—for the knees
- Aquarius—for the legs
- Pisces—for the feet

Pisces. Sign for the feet. And if someone has been wounded in a limb, whatever the [sign of the] moon, do not dare to unbind until the moon is in this sign, as you shall see the man and the symbols depicted below. f. 103b

---

Aries. Which is the first sign. It has the nature of fire, hot and dry. A noble sign. Its planet is Mars. It is the sign of the head. Whoever is born under this sign will be of a hot nature. A quiet, peaceful man, a man of reason, he will be honorable. And it is good to go on a voyage and work the land. Be careful about taking blood from the head. f. 104a

Mars, its planet in the fifth heaven, holds the rulership of the east upon this lower earth. It dwells in each sign 2 months, circles the signs in 2 years, and is in rulership in 2 signs, Aries and Taurus.[1] Each of the planets is good or neutral or evil. Jupiter and Venus are good, Saturn and Mars are evil, the Sun and Mercury are neutral; the Moon is not good except in September. You see the sign drawn below.

[*caption:*]
Aries

Taurus. The second sign. Nature of earth, dry and cold. A stable sign. Its planet is Venus. Sign of the throat. Whoever is born under this sign will be a great worrier over his affairs. He will have great station and will have a great love of family and loyal. Be careful not to take blood from the throat nor to take medicine. And it will be good to do business. Venus, its planet in the third heaven, holds the rulership of the west upon the earth and is over the fish of the sea. It dwells in each sign 8 days and circles the signs in 3 months 6 days. Venus is good for doing all things and it rules two signs, Libra and Taurus. f. 104b

---

1. In traditional astrology, Mars rules Aries and Scorpio (after the discovery of Pluto, Scorpio was reassigned to it).

[*caption:*]
tauro

f. 105a Zemini. Lo terzo segno. À natura de aere, umille e chalido e secho, chome lo so pianetto, Merchorio. Segno dele braze. Queli che naserà in questo segno serà de flevele natura e richo e grazioxo. Serà amado da tutta zente, crederà tutto li serà ditto. Non tuor sangue dele braza. E serà bon da far ogni chosa chumunal. Non piar viazio perché li serà molto noiuxo. Merchurius, in lo ziello segondo, segnoriza ali homeny e ale femene, e sta in zaschon segno zorni 18, zercha i segny in mexi 7, zorny 6. Merchurio, pianetto mezano, e monta segni 2, Zeminy e Virgo.

f. 105b Chanzer. Lo quarto segno. À natura de aqua, freda et umeda, secha. Segno nobelle. Lo so pianetto Luna. Segno deli petti. Chi naserà in questo segno serà de forte volontà e de grave aire. Amerà i denari, molto desdegnuxo, gran parlador, zercharà el mondo, vegnirà in gra[*n*]de stado, salverà la roba. Serà bon de far ogni chosa. Luna, in lo ziello basso, signoriza la zente in le parte de tramontana. Luna, al ziello basso, in zaschun segno zorny $2\frac{1}{2}$, zercha li segny in zorny 30, monta in segno de Chanzer. Luny non chomenzar la prima ora se non fuse setenbrio.

f. 106a Leos.[a] Lo quinto segno. Natura de fogo, chalido e secho. Segno nobele. Lo so pianetto Sol. Chi naserà in questo segno serà poderuxo, averà baillia sovra altri, achistarà di grandi inymixi onori asai. Serà amado dala zente. E serà bon da far fati d'arme e serà bon a domandar grazia in signoria. Lo Sol, in lo quarto ziello, signoriza a mezodì tuti li albori. Sol, in lo quarto ziello, sta in zascon segno zorny 30, zercha tuti li segny in un ano. Monta in segno del Leo, pianetto mezano.

[*caption:*]
leos

f. 106b Virgo. Lo sesto segno. Natura de tera, freda e secha. Planetto fermo. Lo so pianetto Merchurio. Segno dele anche. Quel che naserà in questo segno serà de gran zustizia, averà raxion in ssi, despiaxeralli tute le chose mal fatte, non vorà richeza, averà bona nominanza e onor in questo mondo dai so. È bon a navegar. Non piar medixina. Merchurio monta segny 2, Zemini e Virgo. Pianetto mezano, sta in zaschon segno zorny 17, zercha li segny in mexi 6, zorny 24.

f. 107a Libra. Lo settimo segno. Natura de aire, chalido et umedo. Segno chomunal. Lo so pianetto Venus. Segno dele spalle. Chi naserà in quel segno serà de grandisima natura, despiaxeralli le malefate chose, averà grande honor, non serà richo, serà amado dale done da far chose de vanagloria. Non piar medexina. Venus monta in segny 2, Libra e Chapichorno. E questo Venus è bon a far

---

a. *Sic in MS.*

[*caption:*]
Taurus

Gemini. The third sign. It has the nature of air, humble and hot and dry, like its planet Mercury. Sign of the arms. Whoever is born under this sign will be of a bilious nature and rich and charming. He will be loved by all people; he will believe all that is said to him. Don't draw blood from the arm. And it will be good to do every normal thing. Don't take a voyage because it will be very tiresome. Mercury, in the second heaven, holds the rulership over men and women, and dwells in each sign 18 days, circles the signs in 7 months, 6 days. Mercury, a neutral planet and rules two signs, Gemini and Virgo. f. 105a

Cancer. The fourth sign. It has the nature of water, cold and humid, dry. A noble sign. Its planet the Moon. Sign of the chest. Whoever is born under this sign will be of strong will and of grave air. He will love money, be very disdainful, a great talker, he will circle the world and come to great station, he will save goods. He will be good at doing all things. The Moon, in the low heaven, holds the rulership in the northern region over people. The Moon, in the low heaven, in each sign $2\frac{1}{2}$ days, circles the signs in 30 days, rules the sign of Cancer. Moons do not begin the first hour if it is not September. f. 105b

Leo. The fifth sign. Nature of fire, hot and dry. A noble sign. Its planet the Sun. Whoever is born under this sign will be weighty, will have authority over others, will acquire great enemies, many honors. He will be much loved by people. And it will be good to do military deeds and will be good to ask for boons from the government. The Sun, in the fourth heaven, holds the rulership in the south over all the trees. The Sun in the fourth heaven dwells in each sign 30 days, circles all the signs in one year. It rules the sign of Leo, a neutral planet. f. 106a

[*caption:*]
Leo

Virgo. The sixth sign. The nature of earth, cold and dry. A stable planet. Its planet is Mercury. Sign of the hips. Whoever is born under this sign will be of great justice, he will have reason in himself, he will dislike all the things badly done, he will not want riches, he will have good repute and honor in this world from those attached to him. It's good to travel. Don't take medicine. Mercury holds the rulership over two signs, Gemini and Virgo. A neutral planet, it dwells in each sign 17 days and circles the signs in 6 months, 24 days. f. 106b

Libra. The seventh sign. Nature of air, hot and humid. A common[1] sign. Its planet is Venus. The sign of shoulders.[2] Whoever is born under this sign will have a very grand nature, he will dislike things done poorly, he will have great honor, he will not be rich, he will be loved by women for doing vainglorious deeds. Don't take medicine. Venus holds the rulership over two signs, Libra and f. 107a

1. I.e., neither masculine nor feminine.
2. This is an unusual body part for Libra, which is most often assigned some part of the lower torso, in modern times the kidneys.

ogni ben la prima ora de vener, e sta in 5 zielli. Sta in zaschon segno zorny 8, zercha li segny in mexi 3, dì 6.

f. 107b Schorpio. L'octavo segno. Natura de aqua, freda et umeda, luxurioxo. El so pianetto Mars. Segno dela verga. Chi naserà in quel segno serà de mala raxion e de chostion, e non averà chura de morte per lo malvaxio segno, ma ala so fin averà grandisimo despiaxer e tribulazion. Non far nente de ben ma serà bon a far uxar e far mali tratamenti. Serà bon andar in chorsso. Mars, pianetto rio a chomenzar nula chosa la prima ora di marti, e monta segni 2, Aries e Schorpio. Sta in lo terzo ziello. Sta in zascon segno mexi 2, zercha i segni in ani 2.

f. 108a Sazitario. Lo nono segno. Natura de aire, chalido e secho. Segno mo[a] mobelle. El so pianetto Iupiter. Quel che naserà in quel segno serà de chalida chonpression e forte de natura, e non chorerà se non de golder e darse dilletto per aver honor. Serà amado de chadaun. Serà bon d'andar in viazio e da domandar segnoria. Non piar medexina né non andar al bagno. Iupiter chomenza a far ogni chosa. Monta segni 2, Sazitario e Pisis. E per voler saver chomo li pi[*a*]netti regna, Iupiter la prima ora de zuoba, la segonda Mars, la terza Sul, la quarta Venus, la quinta Merchurius, la sesta Luna, la setima Saturno.

f. 108b Chapichorno. Lo dezimo segno. Natura de tera, freda, umeda, secha, fermo. El so pianetto Venus. Quel che naserà in quel segno serà amuruxo, richo e savio. Averà bontade, e grande signor da molte persone serà udiado per lo gran stado averà. Serà bon da far far ogni chosa, de vender e chonprar. Venus sta in lo ziello segondo. Signoriza le riviere del mar e li animalli bestialli. Chomenza ogny chosa, zoè la prima ore[b] de vener. Venus monta segni 2, Libra e Chapichorno. Sta in zaschun segno zorny 8, zercha li segny in me[*x*]i 3, zorny 6.

f. 109a Aquario. Lo 11 segno. Natura de aire, chalido, umedo. Segno chomunal. Lo so pianetto Saturno. Segno dele ganbe. Quel che naserà in quel segno serà savyo, inprenderà quel che vorà. Averà segnoria in la persona, al mondo passarà de molte furtune e venture, averà vitta e stado. Serà bon di far ogni bona chosa. Saturno sta in lo ziello sovran, signoriza li oxelli del mar. Saturno pianetto rio, inchomenzar nulla chossa la prima ora de sabo. Monta segny 2, Aquario e Schorpio, signoriza tuti li pianetti. Sta in zaschun segno any $2\frac{1}{2}$, zercha li segny in ani 30.

f. 109b Pisis. Lo 12 segno. Natura de aqua, freda, umeda. Segno deli pie. Quel che naserà in quel segno serà de fuola natura e homo miserichordioxo, averà molte persone sutto sé. Truovase in forte ventura de segnorizia zercha de molte chontrade. Serà bon da far chomenzar ogni chossa. Lo so pianetto Iupiter. In lo sesto ziello, Iupiter è bon a chomenzar ogni chose la prima ora de zuoba, e monta segny 2, Pisis e Sazitario.

---

a. *Thus repeated in MS.*
b. *Sic in MS.*
1. In traditional astrology, Venus rules Libra and Taurus.
2. In traditional astrology, Sagittarius is a fire sign.
3. In traditional astrology, Saturn is the planet of Capricorn.

Capricorn.[1] And this Venus will give every good thing on the first hour of Friday, and it is in the fifth heaven. It is in each sign 8 days, circles the signs in 3 months, 6 days.

Scorpio. The eighth sign. The nature of water, cold and humid, lascivious. Its planet is Mars. Sign of the male member. Whoever is born under this sign will be of bad judgment and contentious and he will not take care against death by the evil sign, but at his end he will have great despair and tribulation. It doesn't do any good, but it will be good for manipulation and for bad deeds. It will be good for piracy. Mars an evil planet; begin nothing the first hour of Tuesday, and it holds the rulership over two signs, Aries and Scorpio. It's in the third heaven. It dwells in each sign 2 months, circles the signs in 2 years. f. 107b

Sagittarius. The ninth sign. The nature of air,[2] hot and dry. An unsteady sign. Its planet is Jupiter. Whoever is born under this sign will be of hot complexion and strong of nature and will only run to enjoy and delight himself in achieving honors. He will be loved by all. It will be good to go on a voyage and to seek lordship. Don't take medicine nor go to the bath. Jupiter initiates all things. It holds the rulership over two signs, Sagittarius and Pisces. And to know how the planets reign, Jupiter the first hour of Thursday, Mars the second, the Sun the third, Venus the fourth, Mercury the fifth, the Moon the sixth, Saturn the seventh. f. 108a

Capricorn. The tenth sign. Nature of earth, cold, humid, dry, still. Its planet is Venus.[3] Whoever is born under this sign will be amorous, rich, and wise. He will have generosity, and a great lord of many people he will be hated for the great station he has. He will be good at doing all things, selling and buying. Venus is in the second heaven. It holds the rulership of the shores of the sea and the bestial animals. Begin all things, that is the first hour of Friday. Venus rules two signs, Libra and Capricorn.[4] It dwells in each sign 8 days, circles the signs in 3 months and 6 days. f. 108b

Aquarius. The 11th sign. Nature of air, warm, humid. A common sign. Its planet is Saturn. The sign of the legs. Whoever is born under this sign will be wise, he will learn whatever he wants. He will rule over people, in the world he will pass through many fortunes and ventures, he will have life and stature. He will be good at doing all good things. Saturn is in the sovereign heaven, it holds the rulership of the seabirds. Saturn is an evil planet; don't begin anything the first hour on Saturday. It rules two signs, Aquarius and Scorpio,[5] it governs all the planets. It is in each sign $2\frac{1}{2}$ years, it circles the signs in 30 years. f. 109a

Pisces. The twelfth sign. The nature of water, cold, humid. Sign of the feet. Whoever is born under this sign will be of a foolish nature and a merciful man; he will have many people under him. He finds himself with a strong chance of ruling in many quarters. He will be good at beginning all things. Its planet is Jupiter. In the sixth heaven, Jupiter is good to begin every thing the first hour of Thursday, and it holds the rulership of 2 signs, Pisces and Sagittarius. f. 109b

---

4. In traditional astrology, Venus rules Libra and Taurus.
5. In traditional astrology, Saturn rules Aquarius and Capricorn.

f. 110a

[Veder raxion di pianetti]

E per un'altra raxion Saturno monta in Libra e desende in Aries. Iupiter monta in segno de Chanzer e desende in segno de Chapichorno. Mars monta in segno de Chapichorno e desende in segno de Chanzer. Merchurius monta in segno de Zeminy e desende in segno de Virgo. Sol monta in segno d'Aries e desende in segno de Libra. Venus monta in segno de Virgo e desende in segno de Pisis. Luna monta in segno de Taurus e desende in segno de Schorpio.

Questa la propietade e quelo che regna li pianetti chadaun ora 1ª, perché ore 24 del dì se parte per li 7 pianetti, chomo oldirì qui de sotto.

La prima ora de domenega signoriza el Sul, la segonda Venus, la terza Merchurius, 4 Luna, 5 Saturno, 6 Iupiter, 7 Mars, 8 Sul, 9 Venus, 10 Merchurius, 11 Luna, 12 Saturno, 13 Iupiter, 14 Mars, 15 Sul, 16 Venus, 17 Merchurius, 18 Luna, 19 Saturno, 20 Iupiter, 21 Mars, 22 Sul, 23 Venus, 24 Merchurius. E questi pianetti son fatti per li 7 dì la doma. Sul è fatto per la domenega, luny è fatta per la Luna, marti è fatto per Mars, Merchurius per merchore, Iupiter per zuoba, Venus per vener, Saturno è fatto per sabo. E voiando chomenzar, chomenza el dì che tu vuol per lo sso pianetto e va driedo chomo va i zielli, zoè Saturno, Iupiter, Mars, Sul, Venus, Merchurius e Luna.

[Ponti de stella]

Questi sono li punti dele stelle, de quele che fano furtune de mar e de vento e de piobe, e per lo simille de gran bonaze e de gran chaldene, e per zonde te debi senpre revardar non per le chosse segonde ma per le chosse chontrarie che tte possa naxer, sì che a voler navigar, senpre fatti forte per ogny chontrario.

f. 110b

Zener.**[a]** A dì 5 se lieva una stella chiamata Edra, e a dì 4 se lieve 1ª stella chiamada Moadia, e adì 7 se lieva 1ª stella chiamatta Esse.

Fevrer. A dì 6 se lieva una stella chiamatta Mezebein. A dì 12 se lieva 1ª stella chiamatta Tanpivesta.**[b]** A dì 20 se lieva 1ª stella Babi.

---

a. *Corrected between the lines above* Marzo *crossed out with a diagonal.*

b. *Corrected after* Mezebein *crossed out with a horizontal line.*

SEE THE RECKONING OF THE PLANETS

f. 110a

And by another reckoning, Saturn is exalted in Libra and is in its fall in Aries. Jupiter is exalted in the sign of Cancer and is in its fall in the sign of Capricorn. Mars is exalted in the sign of Capricorn and is in its fall in the sign of Cancer. Mercury is exalted in the sign of Gemini and is in its fall in the sign of Virgo.[1] The Sun is exalted in the sign of Aries and is in its fall in the sign of Libra. Venus is exalted in the sign of Virgo and is in its fall in the sign of Pisces. The Moon is exalted in the sign of Taurus and is in its fall in the sign of Scorpio.

This is the property by which one of the planets orders each hour, because the 24 hours of the day are divided up by the 7 planets, as you shall hear below.

The Sun rules the first hour of Sunday, Venus the second, Mercury the third, the Moon the 4th, Saturn the 5th, Jupiter the 6th, Mars the 7th, the Sun the 8th, Venus the 9th, Mercury the 10th, the Moon the 11th, Saturn the 12th, Jupiter the 13th, Mars the 14th, the Sun the 15th, Venus the 16th, Mercury the 17th, the Moon the 18th, Saturn the 19th, Jupiter the 20th, Mars the 21st, the Sun the 22nd, Venus the 23rd, Mercury the 24th. And these planets correspond to the seven days of the week. Sunday is represented by the Sun, Monday is represented by the Moon, Tuesday is represented by Mars, Wednesday by Mercury, Thursday by Jupiter, Friday by Venus, Saturday is represented by Saturn. And if you wish to undertake a journey, begin it on the day that you choose according to its planet and continue according to the heavens, that is, Saturn, Jupiter, Mars, the Sun, Venus, Mercury, and the Moon.

THE POINTS OF THE STARS

These are the points of the stars, by which are made the storms of the sea and the wind and the rain, and in the same way great calm and great heat, and in addition you must always watch out not for the appropriate things but for the contrary things that can arise, so that if you want to navigate, always be prepared for every contrary thing.

January. On the 5th day, a star rises called Hedra, and on the 4th day a star rises called Moadia, and on the 7th day a star rises called Esse.[2] f. 110b

February. On the 6th day a star rises called Mezebein. On the 12th day a star rises called Tampivesta. On the 20th day a star rises called Babi.

---

1. In traditional astrology, Mercury is in its fall in the sign of Pisces.

2. These star names are not in a known language and do not correspond to those in any known tradition. See Wallis, "Michael of Rhodes and Time Reckoning," vol. 3, pp. 295–296.

Marzo. A dì primo se lieva 1ª stella chiamatta Sonse Farsin. A dì 7 se lieva una stella chiamata Felizi. A dì 15 se lieva 1ª stella chiamata Terpe. A dì 25 se lieva 1ª stella chiamata Mencho. Vardate da questa.

Avril. A dì 5 se lieva 1ª stella chiamata Gamile Grofa. A dì 12 se lieva una stella chiamata Elisse. A dì 20 se lieva 1ª stella chiamata Gerfire.

Mazio. Dì primo. Se lieva 1ª stella chiamata Vertossia. A di 12 se lieva 1ª stella chiamatta Schilla.

Zugno. A dì 9 se lieva 1ª stella chiamatta Sacholli. A dì 10[a] se lieva 1ª stella chiamatta Sacholli. A dì 11 se lieva 1ª stella chiamattaa Schulle. A dì 12 se lieva 1ª stella chiamatta Nenpar Meruin. A dì 18 se lieva 1ª stela chiamatta Debille.

Luyo. A dì 5 se lieva 1ª stella chiamatta Beldin. A dì 18 1ª stella chiamata Verchin.

Avosto. A dì primo se lieva 1ª stella chiamata Debille Terpe. A dì 14 se lieva 1ª stella chiamatta Chapistru Sapis. Dì 22 1ª stella chiamatta Lasar Reorizis.

Setenbrio. A dì 9 se lieva 1ª stella chiamata Matina Orfaralli. A dì 12 se lieva 1ª stella chiamatta Elfiany. Dì 23 se lieva 1ª stella chiamatta Efiales.

Otubrio. A dì primo se lieva 1ª stella chiamatta Boletre. Dì 5 se lieva 1ª stella chiamatta Riteneleus.
f. 111a Dì 18 se lieva 1ª stella chiamatta | Turepi. Dì 21 se lieva 1ª stella chiamatta Sorie. Dì 22 se lieva 1ª stella chiamatta Sechi.

Novenbrio. A dì primo se lieva 1ª stella chiamatta Enea. Dì 11 se lieva 1ª stella chiamata Gutura. Dì 25 se lieva 1ª stella chiamatta Ebidi.

Dezenbrio. A dì 7 se lieva 1ª stella chiamatta Elena. A dì 9 se lieva 1ª stella chiamatta Equistra. Dì 22 se lieva 1ª stella chiamatta Ulchana.

[Dì uziagi e 4 tenpore]

Qui davanti vedirì dì uziagi e zorny pericholoxi di chomenzar a far nulla chossa, perzò vardeile, e questo dise per bocha miser San Ieronymo e debiè oservarllo.

El primo luny d'avril, perché in tal dì Chain olzexe so fradello Abel, e quello fu el primo sangue spanto al mondo.

---

a. *Written over* 11.

March. On the first day a star rises called Sonse Farsin. On the 7th day a star rises called Felizi. On the 15th day a star rises called Terpe. On the 25th day a star rises named Mencho—beware of this one.

April. On the 5th day a star rises called Gamile Grofa. On the 12th day a star rises called Elisse. On the 20th day a star rises called Gerfire.

May. First day. A star rises called Vertossia. On the 12th day a star rises called Schilla.

June. On the 9th day a star rises called Sacholli. On the 10th day a star rises called Sacholli. On the 11th day a star rises called Schulle. On the 12th day a star rises called Nenpar Meruin. On the 18th day a star rises called Debille.

July. On the 5th day a star rises called Beldin. On the 18th day a star called Verchin.

August. On the first day a star rises called Debille Terpe. On the 14th day a star rises called Capistru Sapis. On the 22nd day a star called Lasar Reorizis.

September. On the 9th day a star rises called Matina Orfaralli. On the 12th day a star rises called Elfiany. On the 23rd a star rises called Efiales.

October. On the first day a star rises called Boletre. On the 5th day a star rises called Riteneleus. On the 18th day a star rises called | Turepi. On the 21st day a star rises called Sorie. On the 22nd a star rises called Sechi. f. 111a

November. On the first day a star rises called Enea. On the 11th day a star rises called Gutura. On the 25th day a star rises called Ebidi.

December. On the 7th day a star rises called Elena. On the 9th day a star rises called Equistra. On the 22nd day a star rises called Ulchana.

Odious days and 4 times

Here you shall see odious and perilous days to begin anything, so watch out for them, and this came out of the very mouth of St. Jerome and you should observe it:

The first Monday of April, because on this day Cain killed his brother Abel, and this was the first blood shed in the world.

El primo luny d'avosto, perché in tal dì inchomenzò Iuda a tradir Christo.

El deredan dì de dezenbrio, perché in tal dì s'abissò una zitade chiamatta Gomora, per lo pechatto sodomitto.

E son anchora sotto questi zorny 32 sotto scritti. A chomenzar:

Zener à de questi zorny 7, zoè 1, 2, 3, 4, 6, 7, 20.
Frever à de questi zorny 3, zoè 16, 17, 19.
Marzo à de questi zorny 4, zoè 15, 17, 18, 19.
Avril à de questi zorni 2, zoè 6, 15.
Mazio à de questi zorni 3, zoè 7, 15, 19.
f. 111b Zugno à de questi zorny 1, zoè 7.
Luyo à de questi zorny 2, zoè 15, 17.
Avosto à de questi zorny 2, zoè 19, 20.
Setenbrio à de questi zorny 2, zoè 15, 18.
Octobrio à de questi zorny II, zoè 6, 18.
Novenbrio à de questi zorny 2, zoè 15, 17.
Dezenbrio à de questi zorny 3, zoè 7, 11, 21.

Queste sie le 4 tenpore che se die vardar. Lo primo merchore driedo Sancta Luzia.[a] El primo merchore driedo la prima doma de Charexima. El primo merchore driedo Pasqua roxa. El primo merchore driedo Sancta Cruxe. El primo merchore puo' fatta la luna de frever quello è lo primo dì de Charexima.

[Ordeny di chapitany dele gallie]

Instatuziones[b] mandatta galearum nostre Reypublice Veneziarum armate maris a viro patrizio spectabile generosisimoque domino Andrea Mozenygo, nec non victorioxisimo chapitaneo generalem[c] maris iste, in literis et ditta et reformatta adque subcensuris chontentis observatta.

Al nomen di Dio e dela verzene mare madona Sancta Maria e del'Evangelista miser san Marcho, protector e governador nostro. Questi sono i ordeny dadi e chomandamenti per lo espechabille et egregio honorado miser Andrea Mozenygo, chapetagno generalle del'ano del 1428 e de ttutti li chapetagni da Venyexia.

---

a. December 13.
b. *Almost certainly in place of* instituziones.
c. *Sic in MS;* ra *added between the lines.*

The first Monday of August, because on that day Judas began to betray Christ.

The last day of December, because on this day a city called Gomorrah was brought to ruin for the sin of sodomy.

And also there are the 32[1] days written below. To begin:

January has 7 of these days, that is, the 1st, 2nd, 3rd, 4th, 6th, 7th, 20th.
February has 3 of these days, that is, the 16th, 17th, 19th.
March has 4 of these days, that is, the 15th, 17th, 18th, 19th.
April has 2 of these days, that is, the 6th, 15th.
May has 3 of these days, that is, the 7th, 15th, 19th.
June has one of these days, that is, the 7th. f. 111b
July has two of these days, that is, the 15th, 17th.
August has two of these days, that is, the 19th, 20th.
September has two of these days, that is, the 15th, 18th.
October has two of these days, that is, the 6th, 18th.
November has two of these days, that is, the 15th, 17th.
December has three of these days, that is, the 7th, 11th, 21st.

These are the four cardinal points to avoid. The first Wednesday after Santa Lucia. The first Wednesday after the first Sunday of Lent. The first Wednesday after Pentecost. The first Wednesday after the Exaltation of the Holy Cross. The first Wednesday after the full moon of February is the first of Lent.

The orders of the captains of the galleys

Instructions sent to the armed sea galleys of our Venetian Republic by the honorable and most generous nobleman Andrea Mocenigo, also the most victorious captain general of this sea, in writing and in speech and reformed and observed with the following orders:[2]

In the name of God and of the Blessed Virgin Mary and of the Evangelist Saint Mark, our protector and governor. These are the orders given and commanded by the esteemed and most honorable Andrea Mocenigo, captain general of the year 1428 and of all the captains of Venice.

1. Actually 33 are listed here.
2. Paragraph in Latin. Mocenigo was the captain of the guard fleet on which Michael of Rhodes served as *armiraio* in 1428; see Stahl, "Michael of Rhodes," vol. 3, pp. 75–77.

f. 112a Chon zò sia che l'ordene e riegola sia prinzipio et in li tuti i beny del mondo, e per chontrario non siando ordene né riegola se gae molti dany e senestri, e però miser lo chapetagno prega e sì chomanda deli ordeny infrascritti debiano eser inviolabilemente observate, sutto pena e pene in queli chontignudi, e plui al so bon piaxer.

Per biastimar

E chomanda miser lo chapetagno chi biastimerà Idio o la so madre madona santa Maria, o sancto o sancta, se 'l serà homo da remo debia eser frustado da puope a proda. Se 'l serà homo de pe' paga soldi 100. E sia tegnudo zaschaduno sovrachomitto da mandar a sequezion la ditta pena, fazando asaver a miser el chapetagno de quelo e quelli chi contrafarà.

Andando a remi

E chomanda miser lo chapetagno che andando a remy che zaschaduna galia vada alle so poste dade in sì fatto muodo, che non essa né non olssa insir per pruoda a nessuna gallia, né far algun atto inzurioxo, sotto pena de lire 10 al chomitto de quela gallia chi chontrafarà e lire 5 al nochiero serà al timon, salvo se queli non mostrase evidientemente[a] quelo atto eser ochorsso senza alguna cholpa de quelli e fusse per non lo voler far.

Andando a vello

E chomanda che andando a vello zaschuduna gallia per lo simille vada ale so poste, non incholzando l'una al'oltra, né strezando, né far vella sovra vento a miser lo chapetagno ma llasase romanir
f. 112b da | puope del ditto per talle che senpre lo sia sovra vento a tuti. E chosì navegando non olsa nesuna gallia passar a miser lo chapetagno ma vada tuti chon bono e destro muodo sì che l'uno faza dano al'oltro, sotto pena di refar el dano, et ultra questo quela pena o pene piaserà a miser lo chapetagno ai cholpevelli chontrafazando.

A meter schalla in tera

E chomanda che quando vorà meter schalla in tera tute le gallie debia meter schalla segondo alle so poste, se lo luogo serà abel di poser meter. E quando el chapetagno farà tirar la schalla, per lo simille tuti el faza. E per algon muodo a meter schalla o palomera alguna gallia non olssa tuor posta per puope del'oltra, per schandallo porà venir, sotto pena de lire 5 al chomitto, e più i[b] men chomo piaxerà a miser el chapetagno.

---

a. *Sic in MS.*
b. *Sic in MS.*

As order and rule are at the origin and within all of the good things of the world, and on the contrary in the absence of order and rule much damage and evil come about, therefore the captain entreats and also commands that the orders given below should be inviolably observed, under the penalty and penalties contained in them, and more so at his will. f. 112a

For blaspheming

And the captain commands that whoever blasphemes God or his mother the Blessed Virgin Mary, or a male or female saint, if he is an oarsman he should be lashed from the stern to the bow. If he is a foot soldier he shall pay 100 soldi.[1] And each *sopracomito*[2] is required to exact this punishment, in turn letting the captain know that man or men who has transgressed.

Going under oars

And the captain commands that when going under oars each galley go at its leisure in the following fashion, that it not overtake nor dare to overtake the prow of any other galley, nor do any injurious act, under the penalty of 10 pounds against the *comito* of the galley that transgresses and 5 pounds to the *nochier* at the rudder, unless they can demonstrate that the act occurred without any misdeed on their part and was not according to their will.

Going under sail

And he commands that going under sail each galley likewise keep its position, not pursuing one another, nor closing in, nor making sail to windward of the captain, but let itself remain | astern f. 112b
of him so that he is always upwind of all. And so in navigating let no galley dare pass the captain but let all go in a good and straight manner, so that one damages the other under penalty of repairing the damage, and beyond this whatever penalty or penalties may please the captain against the guilty transgressors.

Putting the gangplank to ground

And he commands that when he shall wish to land the gangplank, all of the galleys shall land their gangplanks at their pleasure, if the place is suited to doing so. And when the captain has his gangplank pulled aboard, in the same manner all shall do so. And under no circumstances in landing the gangplank or mooring rope shall any galley dare to get astern of another, causing an outrage, under a penalty of 5 pounds to the *comito* or more or less as shall please the captain.

1. For monetary values, see Appendix below.
2. For the duties of shipboard officers see Stahl, "Michael of Rhodes," vol. 3, pp. 40–74.

Andando a remy

E chomanda che andando a remy nessuna gallia non olsa passar miser el chapetagno, salvo la galia over gallie dela varda, ma tutte debia atender de non se separar[a] da miser lo chapetagno, sotto pena de lire 10 a chi chontrafarà. Sì veramente, quando se vorà argatar, miser lo chapetagno farà metter la bandiera in mezo et in quela fiada a chi piaserà abia libertà de far, non se lontanando da miser lo chapetagno oltra mia 2, sotto la ditta pena.

Per vender o inpegnar

E chomanda che alguna persona non debia vender dele so arme né portar in tera, sutto quelle pene piaxerà a miser lo chapetagno. Anchora che alguno non olsa zugar o inpegnar alguna arma per
f. 113-1a zugar | e quelli che averà zugado sovra arma e che avesse vadagnado, quelo debia render l'arma a cholui che avesse persso. Anchora a chi avesse inprestado sovra arma debia perder i denari e sia rendutta l'arma a quello che l'avesse inpegnada. Et ultra questo chazano in quela pena piaxerà a miser lo chapetagno.

A zonzer in zitade

E chomanda che quando se azonzerà in alguna zitade o chastello o porto, vuol che i sovrachomitti faza che i sso ballistrieri abia le ballestre in chorda, e li crochizenti e i veretuny a llai, in pena de sagramento. Intendando che li ballistrieri abiano de dì e de notte i crochizenti, in pena de soldi 15 per vuolta ai chontrafazando.

A far vella de notte

E chomanda che se 'l vorà far vella dil'artimon farà fuogi 2 al fugun, un per ladi. E se 'l vorà far vella del trezaruol farà fuogi 3 puru[b] al fogon. E sia tegnude tutte le gallie a responder per li ditti segnalli, sotto pena de soldi 100 e che nesun non olsa passar miser lo chapetagno, sotto pena i piaxerà.

E se fusse galia grossa, voiando far vela del papaficho farà feralli 4, e se 'l vorà far vella dela chochina farà fuogi 5.

Ordenar dele puoste

E chomanda miser lo chapetagno che avanti passa el Quarner che zaschadun sovrachomitto debia ordenar ai sso ballistrieri ale sso poste e che debia far chavy 6, un a proda a ladi destro, l'oltro a proda a ladi senestro, un al chopano e l'oltro per inchontro al'altro ladi, e l'oltro a puope al quartier de ladi destro e l'oltro a ladi senestro. E questi chadauno debia 'ver li sso ballistrieri a chognoser

a. de no *crossed out with a horizontal line.*

b. *Sic in MS, probably for* pure.

Going under oars

And he commands that going under oars, no galley shall pass the captain, except the guard galley or galleys, but all shall be careful not to separate themselves from the captain, under penalty of 10 pounds to the transgressor. If truly, when he wishes to form a regatta, the captain shall place his flag amidships, at that time whoever wishes shall also be free to do so, as long as he doesn't get more than 2 miles from the captain, under the same penalty.

For selling or pawning

And he orders that no one shall sell any of his weapons or carry them ashore, under whatever penalty pleases the captain. Also that no one bet or pawn any weapons for a bet. | And those who have f. 113-1a
bet weapons and won should return the weapons to the one who lost. Likewise, whoever has loaned money with a weapon as collateral should lose the money and the weapon be returned to the person who had pawned it. And beyond this, they shall be subject to whatever penalty pleases the captain.

Arrival in a city

And he decrees that upon arrival in any city or castle or port, he wishes that the *sopracomiti* have their archers keep their bows strung, and the crossbows and arrows at their sides, on penalty of an oath. It is to be understood that the archers keep their crossbows by day and night, under the fine of 15 soldi for each transgression.

Sailing by night

And he commands that if he wishes to set the large sail he will make two fires on the fireplace, one on each side. And if he wishes to set the middle-sized sail he will likewise make three fires on the hearth. And all of the galleys must respond to these signals under a penalty of 100 soldi, and let no one dare pass the captain, under whatever penalty shall please him.

And if it is a heavy galley, wishing to set the small sail he will make 4 lamps, and if he wishes to set the small square mainsail he will make 5 fires.

Laying out of positions

And the captain commands that before passing the Gulf of Quarnero[1] each *sopracomito* must order his archers to their positions and that there be 6 lookouts—one at the bow on the starboard side, the other at the bow at the port side, one at the skiff and the other on the corresponding part of the other side, and the other on the stern quarter on the starboard side and the other on the port side.

1. Kvarner, Croatia.

f. 113-1b zascha|dun andar ale so poste se ochoresse chaxo de bataia, che Dio non voya. E in chaxo de bataya negono non olssa né presuma partirse dela so posta, salvo per lo so armador non li fusse ditto, o per nomen de quello, sotto pena d'eser pugnido in la persona segondo el fallo parerà miser lo chapetagno. E queste poste debia eser dadi per li sovrachomitti, sutto pena de soldi 200.

Per le varde balistrieri

E chomanda miser lo chapetagno el dì se partirà da Venyexia le galie debia ordenar le sso varde, e quelli perderà la varda de dì perda soldi 4, e de notte perda soldi 8. Li qualli soldi vegnano in man a cholor dela varda, se li serano achoxadori. E se quelli dela varda non li achoxasse, chaza in quella pena chomo i avese perso la varda. E questi denari vegna in man del sovrachomitto e dagalli a ber i portolatti e proderi.

Per andar l'armiraio in tera

E chomandò che quando l'armiraio vuol andar in tera, o per alchognir, o per oltro servyxio, o per segurtade deli homeny, di non far chostion che de tute le gallie sia tegnudi li ballistrieri, a chi thochase la diana, de achonpagnar al'armiraio tutto el dì. E chi chontrafesse chaza in pena de soldi 10 al dì. E questi debia venir al'armiragio e despensa quelli ali ballestrieri de quela varda.

Voiando far vella di dì

E chomanda che voiando far vella de dì, chadaun veder e per lo simile avolzer e chalar e che vella vorà far, e quando miser lo chapetagno farà levar voga per far vella, tute le gallie debia levar voga e aspetar inchina el chapetagno faza vella per andar sovravento a tutte, et in quella ognomo faza, non
f. 113-2a fazando vella per muodo de far senestro ale oltre [*i*] né andar sovravento a quelle | gallie non dovese far non siando ale so poste, per schivar ogni schandolo, sotto pena de soldi 100 al chomitto chontrafarà.

Per schurittà de notte a velo

E chomanda miser lo chapetagno che se 'l vorà chalar mostrerà 1 fuogo al fugun e tute le gallie debia responder equalli soldi 100.

Per andar a segho[a]

E chomanda miser lo chapetagno che se 'l vorà andar a secho mostrerà fogi 2 sutto al fanò a puope, e chadauna gallia ge metta fuogi 2 inchina miser lo chapetagno [*tenirà*][b] li sso, sotto pena de soldi 100.

---

a. *Sic in MS.*
b. *Omitted. Supplied on the basis of the version in Pietro di Versi,* Raxion de' marineri, *p. 89.*

And each of them must have their own archers, to know that each | one can go to his post if battle should require it, which God forbid. And in the case of battle let no one dare nor presume to leave his position unless told to do so by his commander or in his name, under the penalty of corporal punishment as his transgression appears to the captain. And these posts must be assigned by the *sopracomiti*, under penalty of 200 soldi. f. 113-1b

For the guard archers

And the captain orders that on the day on which they depart from Venice, the galleys must order their guards, and those who fail to appear for the guard during the day lose 4 soldi and by night lose 8 soldi. These soldi shall come to the hand of whoever in the guard shall be the accusers. And if those in the guard shall not accuse, they shall be subject to the same fine as those who failed to appear for the watch. And this money shall come to the *sopracomito* for him to give to the pilots and the *proderi* for drink.

When the *armiraio* goes ashore

And he ordered that when the *armiraio* wishes to go ashore, whether by necessity, or for another service, or for the safety of the men, there should be no question that all the archers of all the galleys be ordered that whoever is on the reveille watch shall accompany the *armiraio* all day. And whoever transgresses shall pay a penalty of 10 soldi a day. And these sums shall come to the *armiraio* and he shall dispense them to the archers of that watch.

Setting sail by day

And he orders that wishing to set sail by day, each shall inspect and hoist and lower whichever sails he wishes to, and when the captain shall raise the oars to set sail, all the galleys shall raise oars and wait until the captain sets sail to go windwind of all, and let each man do the same, not sailing so as to do damage to the others, not going windward of | those galleys not at their positions, to avoid any quarrel, under penalty of 100 soldi to the *comito* who transgresses. f. 113-2a

At sail in the dark of night

And the captain orders that if he wishes to lower the sails, he will show one fire on the fireplace, and all of the galleys must respond equally, 100 soldi.

To run under bare poles

And the captain orders that if he wishes to run under bare poles[1] he shall show 2 fires under the lantern on the stern, and each galley shall place 2 fires there as long as the captain maintains his, under the penalty of 100 soldi.

1. I.e., with no sails set.

Per vuolzer

E chomanda miser el chapetagno se vorà vuolzer mostrarà fuogi 2 al fugun, uno sutto al'oltro, e chadauna responda e vuolza chon destro muodo, non se inpazando per non aver senestro, in pena de soldi 100.

Se 'l vorà tuor l'altra volta

E chomanda miser lo chapetagno che se 'l vorà far l'oltra vuolta farà fuogi 6 e chadauna gallia responda, sotto pena de soldi 100.

Se 'l vorà veder le gallie

E chomanda miser lo chapetagno che se 'l vorà veder le gallie farà un fuogo sutto al fanò e zaschaduna gallia ge metta fuogo uno e tegna quello fina che miser lo chapetagno tegnirà lo sso. E sse el torà el so zaschadun debia tuor el so, e miser lo chapetagno tegnirà fermo el so fanò e chadauna gallia seguise a miser lo chapetagno, sotto pena de soldi 100.

f. 113-2b Per non mostrar fuogo

E chomanda miser lo chapetagno che nesuna gallia de notte, né a vello né a remi, non olsa mostrar fuogo, sotto quela pena i parerà.

Per tuor bataya de dì

E chomanda miser lo chapetagno che se vorà tuor bataya del dì che alguna gallia, né persone de quele, non se olsa partir dale so puoste dade alle galie per miser el chapetagno e per li so patrony, ali homeny dele so galie, e queli abia chavy aparegiati per puope e per proda se bexognase strenzer e largarse el se possa far prestamentte.

Per intrar in bataya

E chomanda che ala prima tronbetta farà sunar chadauna persona se debia armar, ala segonda chadaun debia andar ale so poste e ala terza tronbetta tuti siano valentyomeny andar a ferir chon miser lo chapetagno vivamente e non se partir dala bataya inchina serà fenyda, salvo che miser lo chapetagno oltro non li fose chomandado. E quelli chi chontrafarà ai ordeny dadi debia perder la testa in tal muodo chi muora, zoè a patruny, chomitti e paruny zuradi e nochieri, a chi fusse la varda in crosia o al timon e oltri chi fuseno in chaxon de non ferir, o abiando ferido partirse dela bataya avanti che la fose fenyda. Intentando[a] che queste pene io son tegnudo a dar ai chontrafazando

---

a. *Sic in MS, evidently for* intendando.

To tack

And the captain orders that if he wishes to tack, he will show 2 fires in the fireplace, one under another, and each shall respond and roll up likewise, not getting angry so as not to cause damage, under penalty of 100 soldi.

If he wishes to take the other tack

And the captain commands that if he wishes to take the other tack, he will make 6 fires and each galley shall respond, under penalty of 100 soldi.

If he wishes to see the galleys

And the captain commands that if he wants to see the galleys, he will make a fire under the lantern and each galley makes a fire there and maintains it as long as the captain maintains his. And if he puts his out, each one should put his own out, and the captain will keep his lantern in position and each galley shall follow him, under penalty of 100 soldi.

To not show fire f. 113-2b

And the captain commands that no galley at night, neither under sail nor oars, dare show fire, under whatever penalty shall please him.

To initiate battle by day

And the captain commands that if he shall wish to initiate battle by day, no galley nor person aboard shall dare to leave the positions given to the galleys by the captain and by his *paroni*[1] to the men of his galleys, and that they shall have commanders prepared on the stern and on the bow so that if it is necessary to press together or to separate it can be done quickly.

To enter into battle

And he commands that when the first trumpet sounds the call to gather, each person shall arm himself, at the second each shall go to his position, and at the third trumpet all shall be valiant men going to draw blood bravely with the captain and not leave battle until it is finished, unless otherwise commanded by the captain. And whoever shall go against the orders given shall lose his head in such a way that he dies—that is, *paroni*, *comiti* and *paroni zuradi*, and *nochieri*, those who shall have the guard of the gangway[2] or the rudder and in addition those who refrain from drawing blood, or, having drawn blood, leave the battle before it is finished. It being understood that I am bound to give this penalty to transgressors by the stipulations of my commission, and similarly this

1. In this text, *paron* appears to be used as synonymous to *sopracomito*, the noble commander of an individual galley.
2. I.e., the one extending along the center of the deck from bow to stern, not the gangplank.

segondo la forma dela mia chomesion, e simelmente anchora ale ditte pene a chi se guitasse in aqua per voller schanpar over per voler andar a robar.

Per tuor bataya de notte

f. 114a E chomanda miser lo chapetagno[a] ssi veramente de notte paresse | achazesse da tuor bataia se tegna i ordeny infrascritti, anchora in le pene infrascritte. E vegnando lo ditto chaxo de bataya de notte, per chognoserse, miser lo chapetagno farà meter fuogi 2 sutto al fanò, e chadauna gallia meterà fuogi 3 per la longueza dela staza, e quelli tegnano infina miser lo chapetagno tegna li so 2, e per questo se chognoseremo in la bataya.

Per tuorse via dela bataya

E chomanda miser lo chapetagno si se volese tuor via dala bataia farà tuor un d'i so fuogi via, e le oltre gallie debiano tuor fuogi 2, lasando el terzo e seguir a miser lo chapetagno. E chi chontrafesse, o remover quelli e non seguir, quello per lo muodo chaza in quelle pene parerà a miser lo chapetagno sì in aver chomo in la perssona.

Per portarsse miserichordiosi

E chomanda miser lo chapetagno e sì priega a tuti, abiando grazia di Dio de victoria, zaschadun se debia portar miserichordioxamente incho[*n*]tra di lor, non fazando crudelitade azò che 'l nostro Signor Dio per meritto de miserichordia de[b] faza a tutte le bataye victuorioxi.

Le galie dela varda

E chomanda miser lo chapetagno che la gallia o gallie che serano dala varda se debia partir da miser lo chapetagno ala maitinada quando miser lo chapetagno farà levar vuoga, e que[c] quela galia o
f. 114b quella[d] se debia luntanarse tanto de miser lo chapetagno che | possa senpre veder i segnalli de bandiera se serà fatto per miser lo chapetagno, senpre andagando per la via de miser lo chapetagno, e che debia aver lizenzia di far vella e chalar e levar voga. E se per aventura vederà fusta armada o fuste, o gallia o gallie, debia levar miser San Marcho, tante volte isar e chalar quanti serano li ditti fusti. E se a miser lo chapetagno i paresse d'incholzar quello o queli, farà meter el penon al peritulo da proda, et in quela fiada ognomo incholza donmentre che miser lo chapetagno non muovese quello. E se alguna gallia azonzerà, alguna gallia o fusto quela sostegna, chon men dano dele persone e del'aver d'i ditti fusti e gallia chi porà, tuta fiada chon vostra segortade, donmentre che miser lo chapetagno azonzerà el farà chomo i parerà.

---

a. *Corrected from* chamapetagno, *with* ma *crossed out with a horizontal line.*
b. *Sic in MS, probably with the meaning of* li.
c. que *thus repeated in MS.*
d. *Sic in MS.*

penalty will be given to whoever shall throw himself in the water to try to escape or to go to plunder.

To initiate battle by night

And the captain commands that if by night it truly appears best | or falls to him to make battle, the following orders shall apply, as well as the following penalties. And in such a case of battle by night, to help the men recognize each other, the captain shall make two fires under the lantern, and each galley shall make 3 fires along the pole, and to keep them as long as the captain keeps his 2, and in this way we will know each other in battle. f. 114a

To leave the battle

And the captain commands that if he shall wish to leave battle he shall take away one of his fires and the other galleys should take away 2, leaving the third, and follow the captain. And whoever shall go against this, or get rid of them and not follow, shall be subject to the penalties chosen by the captain in belongings and in person.

To act mercifully

And the captain commands and begs all, having victory by the grace of God, that each shall behave mercifully toward them, not displaying cruelty, so that our Lord God as a reward for such mercy shall make all battles victorious.

The galleys of the guard

And the captain commands that the galley or galleys that shall be on guard shall leave the captain in the morning when the captain raises the oars and that that galley or galleys shall distance itself enough from the captain that | it can always see the flag signals if they are made by the captain, always going on the captain's course, and it shall have permission to make sail and lower it and raise the oars. And if by chance it should see an armed foist[1] or foists, or a galley or galleys, it shall raise St. Mark[2] as many times up and down as there are ships. And if the captain shall decide to pursue it or them, he shall put the banner on the bow *peritulo*[3] and at that point all shall pursue as long as the captain does not move it. And if any galley joins a galley or vessel, let it hold it, with as little damage to the persons and goods of the vessels and galley as it can, always with your security, until the captain joins, and he will do as he decides. f. 114b

1. A light galley.
2. That is, a flag with the image of the winged lion.
3. Probably the flagstaff raking strongly forward from the bow.

Per non incholzar più

E chomanda miser lo chapetagno che se non vorà più incholzar el farà tuor via el penon da proda e meter insù la staza in puope. In quela fiada le gallie dela varda vorà levar voga e torna[*r*] a miser lo chapetagno. E queste gallie dela varda al stranmontar del sul, siando in porto o navegando, debia vegnir a render la varda segondo uxanza.

Le gallie de varda per la note

E chomanda miser lo chapetagno che la galia o galie che fuse de varda al tranmontar el sul debia insir dal porto donde fusse miser lo chapetagno e metese in ponta, sì che la posa sentir de navilli o galie che vuolesse intrar in lo ditto porto, e quelo sentirà faza asaver a miser lo chapetagno, non se metando sì al largo che non se posa veder i segnalli, e anchora oldida al parlare se 'l sentirà o fusta o fuste armade. Se i serano da uno in suxo faza el segno da un feral, erzando e chalando quanti i fuseno li ditti fusti, e fatto questo la galia torna da miser lo chapetagno, fazandolo a saver quello averà vezudo o dischoperto. Se veramente averà dischopertto e sentido uno solo fusto armado, faza segno solo de quello, e se miser lo chapetagno vorà che l'incholzi farà responder per lo so fanò et in quela questa gallia incholza, tegnando senpre el so feral inpiado, e per alguna chondizion non se
f. 115a alontany che quela gallia o quele | non posa veder senpre el fanò de miser lo chapetagno e miser lo chapetagno posa veder senpre el feral dela ditta o dele ditte, perché miser lo chapetagno senpre li andarà seguitando a quella via, andagando ben avezudamente. Avanti che la zonza faza i segnalli di richognosersi e so questi sutto scritti.

Signalli de richognosersy

E chomanda miser lo chapetagno se 'l serà di dì quela gallia o quele dala varda, farà far un fumo in proda e quella i debia risponder, e poi farà levar una bandiera quadra in puope a banda destra e quela debia risponder, e puo' farà una bandiera quadra in lo ladi senestro in proda. Fatti li ditti segni e fatti li ditti,[a] leva l'insegna de San Marcho e quela debia risponder, e puo' debia levar el so penon in proda e una bandiera quadra in lo ladi senestro in proda, e si per questi segni non risponda serano nostri inimixi.

Segnalli di richognoserse de notte

E chomanda che se fusse de notte se die meter 3 feralli inpiadi, l'uno a ladi al'oltro, e quela o quelle debia risponder. E puo' farà 2 fuogi a pope, l'uno sovra al'oltro, e queli debia risponder non toyando via non toiando via[b] li feralli da proda, e facti questi e non risponda, serano nostri inymixi. In quella ti dela varda chon ogni to avantazo debi trazer de ballestre e barssayando chomo vedirì de sopra di ladi.

Se non vedese de dì i segnali del chapetagno

a. *Thus repeated in MS.*
b. *Thus repeated in MS.*

To cease pursuit

And the captain commands that if he does not wish to pursue any more, he will take down the banner from the bow and place it on the stern flagstaff. At that point, the guard galleys shall raise oars and turn toward the captain. And at sunset those guard galleys, whether in port or at sea, must return to the guard as usual.

Guard galleys at night

And the captain commands that at sunset the galley or galleys on guard must leave the port where the captain is and pay attention so that it can hear which ships or galleys want to enter that port, and shall make known to the captain what it hears, not going so far that it cannot see the signals, and also hearing what's spoken if it should perceive an armed foist or foists. If it is one or more it should make a signal with a lantern, raising and lowering it as many times as there are vessels, after which the galley shall return to the captain, letting him know what it has seen or discovered. If it has in fact discovered and heard only one armed vessel, it shall signal only that, and if the captain wishes it to pursue it he shall respond with his lantern and pursue it with his own galley, always keeping his lantern lit and under no circumstances getting so far away that that galley or those galleys | cannot always see the captain's lantern and the captain can always see its or their lanterns, f. 115a
since the captain will always follow that course, always going with good information. Before joining it, he is to make the signals of recognition as written below.

Signals of recognition

And the captain commands that if it is by day, that galley or galleys of the guard, he will make a smoke signal on his bow and it should respond, and then he will have a square banner raised on the starboard side of the stern and it should respond, and then he will put a square flag on the starboard side of the bow. Having made these signals and done these things, he raises the flag of St. Mark and it should respond, and then he should raise his pennant on the bow and a square banner on the port side of the bow, and if they don't respond to these signals, they will be our enemies.

Signals of recognition at night

And he commands that by night he shall place three lighted lanterns, one next to another, and it or they should respond. And then he shall put two fires on the stern, one above another, and they should respond, not taking away the lanterns from the bow, and if these things are done and they don't respond, they shall be our enemies. In all of this, you of the guard using all of your advantage shall draw your bows and shoot when you can see over the sides.

If the signs of the captain are not seen by day

E chomanda miser lo chapetagno che sa[a] la gallia o gallie dela varda se alontanese da miser lo chapetagno e quello volesse che la gallia o gallie vegnise a lui, vuol che per segno[b] de bandiera torna, e se non se podesse veder i segni dela bandiera, miser lo chapetagno farà far vella e prexente chalarà pluzor vuolte. E visto questo segnal, la gallia o gallie dela varda debia far ritorno a miser lo chapetagno, e a quella o quelle chomandarà chomo i pparerà.

f. 115b quela o quele inchina miser lo chapetagno azonzerà, che vedrai el so fanò chi ve segue, e visto quel fanò vada e investa a quello o quelli chon el meyor ordene el porà.[c]

Sì che 'l fusto se mete a schanpar

E chomanda miser lo chapetagno che se quel tal fusto o fuste se metese a schanpar, miser lo chapetagno vuol que[d] quela gallia over gallie dela varda e debia seguir, non se lontanando tando[e] da miser lo chapetagno che senpre vega el so fanò, e senpre la gallia dela varda debia tegnir el so feral inpiado per seguirllo senpre miser lo chapetagno. E se miser lo chapetagno non volese più se incholzasse lo ditto o li ditti, farà inpiar 2 ferali sotto al fanò, che serano fuogi 3 l'uno sopra l'oltro, et inchontinente quella gallia o gallie dela varda a noi presto ritorni.

Per eser fatto segno de notte per la gallia dela varda

E chomanda miser lo chapetagno se fuse fatto segno de più fusti per la gallia o gallie dela varda che fusse armadi, che a miser lo chapetagno e parese armarse di notte, farà inpiar a pope 4 fuogi uno sopra al'altro et in quella gallie tutte se debia far armar i sso homeny e meterse in ponto d'andar in bataya, sì che ala prima tronbetta tuti siano armadi, et ala segonda tuti in ordene, e ala terza tronbetta si quelli fusti fusseno d'i nimixi vadano a ferire chon bon anymo. Ma ordena miser lo chapetagno che alguna gallia non olssa andar investir se non serà sonada la terza tronbetta, sotto pena dela testa, ma ordena che tuti le gallie staga ale so poste unyde e non separade, e non se debia partir
f. 116a inchina finyda la bataia, sotto la ditta pena. E se per aventura alguna galia | non andasse a investir, che Dio non voya, over andando, sguindando, over lontanando dela bataya, sia inpognido chomo tradittore.

Per non arobar in la bataia

E chomanda miser lo chapetagno che durando in la bataya nesuno non se metta a robar, perché pluxor fiade abiando schonfitto i nymixi, quelli vinziduri per andar robando son stadi lor po' schonfitti, e questo è solo per ingurdixia di arobar. E però chomanda che nesun non se metta a robar né nesun non se guita in aqua per andar a robar inchina a l'ultima schonfitura, sotto pena e pene piaxerà a miser lo chapetagno.

---

a. *Sic in MS.*
b. de fan *crossed out with a horizontal line.*
c. *From* quela o quele *to* ordene el porà *crossed out with diagonal lines.*
d. *Sic in MS.*
e. *Sic in MS, evidently for* tanto.

And the captain commands that if the galley or galleys of the guard distance themselves from the captain and he wishes that the galley or galleys come to him, he wishes it to return at a sign of the banner; and if he cannot see the signs of the banners, the captain will set sail and then lower it several times. And having seen this signal, the galley or galleys of the guard shall return to the captain and he shall command them as he wishes.

1

f. 115b

If the vessel tries to escape

And the captain commands that if such a small sailing vessel or vessels tries to escape, the captain wishes that the galley or galleys of the guard follow it, not getting so far away from the captain that he can always see its lantern, and the guard galley should always keep its lantern lit so that the captain can follow it. And if the captain does not wish him to pursue it or them, he shall light two lamps above a lantern, so that there will be three lights, one above the other, and immediately that galley or galleys of the guard shall return to us as soon as possible.

To have the night signal made by the guard galley

And the captain commands that if several armed vessels are signaled by the guard galley or galleys so that it seems to the captain necessary to raise arms by night, he shall light 4 fires on his stern, one above the other, and in that case all galleys shall have their men arm and prepare to go into battle, so that at the first trumpet all are armed, and at the second all in order, and at the third trumpet if these vessels are enemy, they go to draw blood with good spirit. But the captain orders that no galley should dare to board [the enemy] until the third trumpet has sounded, under penalty of the head, but he orders that all the galleys hold their positions together and not separately, and not depart until the battle is finished, under the same penalty. And if by chance any galley | shall f. 116a
not go forward to attack, which God forbid, or going forward is reluctant or distances itself from the battle, it shall be impugned as a traitor.

Not to plunder in battle

And the captain commands that during the battle no one shall plunder, because often, having defeated the enemies, the winners have been later defeated because of plundering, and this only because of the greed of plunder. And so he commands that no one begin to plunder, nor jump into the water in order to plunder, until the final defeat, under the penalty or penalties that shall please the captain.

---

1. The following fragment crossed out at top of page in MS: *that [galley] or those [galleys] shall join the captain, that you shall see the lantern of whoever follows, and having seen that lantern it should go and invest that or those [vessels] with the best order that it can.*

Per truovar sforzo de galie

E chomanda miser lo chapetagno se truovase sforso de galie, che Dio non el voia, zoè d'i nimixi, e che parese a miser lo chapetagno de darlli le puope, in quela fiatta vaga tutte unite e non abandonando l'una al'oltra, ma chon bono e destro modo tute se ne vagan redugando insenpre e non separar l'una al'oltra, sotto pena di perder la testa.

Se separase una gallie dele oltre

E chomanda miser lo chapetagno che se separase una gallia dele oltre e smarise da miser lo chapetagno, però è nezesitade a dar ordeny e segni per li qualli se posan chognoser le nostre gallie una dal'oltra, sì de zorno chomo de note, chomo serà qui scritto de sotto.

Per chognoserse de dì

E chomanda miser lo chapetagno che se galie o galia smarise da miser lo chapetagno de dì, per
f. 116b chognoserse miser lo chapetagno farà | far un fumo in proda, e vezando questa gallia smarida responda per questo sì fatto segno, e poi miser lo chapetagno farà levar una bandiera quadra in puope a ladi destro, e quella debia risponder, e poi farà miser lo chapetagno levar una bandiera quadra in proda in lo ladi senestro, e quella debia risponder. E fati li ditti segny miser lo chapetagno farà levar el segno de miser San Marcho e chosì responda la ditta, e sia ben provezuda la ditta gallia de non se achostar se ben non avese rechognosudo li ditti segny. Anchuor volemo che da puo' fatti li segni debia ancuor levar el penon da proda e la bandiera quadra in lo ladi senestro, e resposto quello se achosti chon ogny so destro e con la pope avanti, chon le ballestre chargade.

Per chognoserse de note

E chomanda el chapetagno che se 'l se vorà chognoser de notte dala galia over gallie fusse smaride, miser lo chapetagno farà far 3 fuogi in proda, l'uno a ladi al'oltro, e quela responda, e poi el farà 2 fuogi in puope, uno sopra l'oltro, non toyando via quelli da proda, e quella gallia o quelli responda. E facti li ditti segni e ben chognusudi, chostase a vezudo, chomo ò ditto de sovra. E simille ordeny debia opservar[a] la gallia o gallie dela varda a chognoserse chon le oltre gallie. E chosì per questi muodi e segny se chognoserano sì de dì chomo de notte.

Per chognoserse gallie ingual

E chomanda miser lo chapetagno per chognoserse gallia fusse ingual numero, tante chon tante, vuol miser lo chapetagno che quella galia o gallie che fusseno inverso lo levante siano tegnudi quelli a far primo l'insegni. Anchora se le fusse più una parte del'oltra, queli serano più vuol chi chomenza a far li ditti segny. E questo sia oservado.

---

a. *Sic in MS, evidently for* observar.

Finding oneself overwhelmed by galleys

And the captain commands that if one should find oneself overwhelmed by galleys, which God forbid, that is, by enemies, and if it should appear to the captain necessary to give them the stern, on that occasion all shall go united and not abandoning one another, but in a good and straightforward manner all go returning together and not separating one from the other, under the penalty of losing one's head.

If one galley separates from the others

And the captain commands that if one galley separates from the others and gets lost from the captain, it is then necessary to give orders and signals by which our galleys can know each other, by day and by night, as will be written below.

To recognize each other by day

And the captain commands that if galleys or a galley gets lost from the captain by day, in order to recognize it the captain will | have smoke made at the bow, and seeing this the lost galley responds f. 116b
with this same signal; and then the captain will raise a square banner on the starboard stern, and it should respond; and then the captain will raise a square banner on the port bow, and it should respond. And having made these signals, the captain will raise the signal of Saint Mark and it shall respond likewise, and let the galley be well advised not to approach if it has not recognized these signals. We also wish that after the signals are made, it should also raise the pennant from the bow and the square banner on the port side, and having responded to this, it approach with its starboard and the stern forward, with the crossbows drawn.

To recognize each other by night

And the captain commands that if he should wish to recognize by night a galley or galleys that have been lost, the captain shall make three fires in the bow, next to each other, and it shall respond; and then he will make 2 fires in the stern, one above another, not extinguishing those at the bow, and that galley or galleys shall respond. And after these signals have been made and well recognized, they approach to sight range, as I have said above. And the galley or galleys of the guard should observe similar orders to recognize the other galleys. And with these methods and signals they will recognize each other by day as well as by night.

For equal galleys to be recognized

The captain commands that for galleys to recognize each other when there is an equal number, so many to so many, the captain wishes that the galley or galleys on the east be required to make the signals first. Also if there are more on one side than on another, he wishes that those that are greater begin making the signals. And let this be observed.

f. 117a Se smarise una galia del'oltra e zerchase

E chomanda miser lo chapetagno che nesuna gallia fusse sperssa dale oltre, e se andasse zerchandosse vada senpre in li porti con la pope avanti e le ballestre chargade, e mudasse la notte la posta almen volte 3, e metando senpre le varde in tera per sso segurtade.

Se la gallia dala varda metesse la varda in tera

E chomanda miser lo chapetagno che la gallia o gallie dela varda debia tegnir senpre la varda in tera de dì, quando miser lo chapetagno serà in li portti. E senpre quey debia aver l'ochio ala gallia de miser lo chapetagno, che quando se vorà levar farà meter el penon al peritullo da proda, e quey dela varda debia venir presto alle so gallie. E questi chi vorà andar in la varda die 'ser segondieri 2 per vuolta e ballestrier un a chi tochasse la varda.

Per aver un palomer

E chomanda miser lo chapetagno che chadauna galia debia far un palomer, el qual sia tegnudo a vogar el chopano, e sia tegnudo a frustar e bater a chi farà mal. E per so benyfizio el ditto die aver libertade de vender in gallia per crussia zò che i sserà dado chon lizenzia, ezetto arme. E questo abia per so fadiga per lira soldo 1.

E per aver chonseyo dele gallie di dì

E chomanda miser lo chapetagno che se 'l vorà chonseyo dale gallie de dì farà meter una bandiera quadra a puope, in lo ladi senestro. Et in quella fiada zaschaduna gallia s'achosta a miser lo chapetagno chon più destro muodo i porà, non se fazando dano una chon l'oltra.

f. 117b Per aver chonseyo de notte

E chomanda miser lo chapetagno che se 'l vorà chonseyo de notte farà fuogi 3 in puope e tegnirà quelli inchina che tuti serano achostadi a lui, et achostase chadauna gallia a miser lo chapetagno con destro muodo.

Per sorzer de notte

E chomanda miser lo chapetagno che se 'l vorà far surzer de notte fare segno del batifigo, e chadauna gallia risponde per quel segno, azò che una gallia chon l'oltra non rezevese algun dano, sotto pena de soldi 100 al challafao e al marango[*n*].

Per no gittar chopany né barche in aqua

E chomanda miser lo chapetagno che nesuna gallia non olsa gitar barcha né chopano in aqua se miser lo chapetagno prima non el getta, over domandar lizenzia a miser lo chapetagno.

If one galley gets lost from another and searches for it f. 117a

And the captain commands that no galley get dispersed from the others, and if it should go looking for the others, it always enter ports stern first and with crossbows loaded, and each night change its position at least 3 times, and always putting guards on land for its security.

If the guard galley puts the guard on land

And the captain commands that the guard galley or galleys shall always stand guard on land by day when the captain is in port. And they shall always keep an eye on the captain's galley, so that when he wants to have the pennant raised on the bow flagstaff, those on guard can come quickly to their galleys. And those who will stand guard should include two secondaries at a time and one archer from the guard.

To have a coxswain

And the captain commands that each galley shall have a coxswain, who is in charge of rowing the skiff and punishing and flogging those who do ill. And as a reward he shall have the liberty to sell on the gangway of the galley whatever has been given to him, with permission, except arms. And he shall have for his efforts 1 soldo per pound.

And to have a conference among the galleys by day

And the captain commands that if he wishes to have a conference of the galleys by day he will put a square banner on his stern, on the port side. And in that case, each galley shall approach the captain as directly as possible, not damaging each other.

To have a conference by night f. 117b

And the captain commands that if he wishes to have a conference by night he will make 3 fires on the stern and will keep them until all are alongside him, and each galley shall come alongside the captain right away.

To anchor by night

And the captain commands that if he wishes to anchor by night he will make a signal with the fire iron, and each galley shall respond with this signal, so that no galley receives any damage from another, under penalty of 100 soldi to the caulker and the carpenter.

Not to put the skiffs or the boats in the water

The captain commands that no galley put the boat or the skiff in the water unless the captain lowers his first, or ask permission of the captain.

Per ferirse in gallia o in tera

E chomanda miser lo chapetagno se fusse fatta alguna ferida né in tera né in mar, zoè in gallia, lo dì serà fatto, al più presto i porà, debia mandar a chi avesse ferido da miser lo chapetagno, retegnando senpre quello in bona varda. E questo sia oservadi[a] per i patrony, chomitti e paruny zuradi dela gallia uve fusse fatto lo dilletto, o in tera o in gallia. E chi chontrafarà miser lo chapetagno, miser lo chapetagno prozederà incontra quelli chomo de raxion i parerà, non intendando oltre briege senza feride, perché le oltre lassa a vui patrony a far chomo de raxion ve parerà ai malfazando.

f. 118a Per far furto in gallia

E chomanda miser lo chapetagno che se fusse fatto algon forto in gallia de lire 3 in suxo, quei debiè retegnir sotto bona varda e quei prexentar a miser lo chapetagno al più tosto porè. E questo debiè oservar sotto quelle pene parerà a miser lo chapetagno. E si serà da lire 3 in zuxo quei debia far frustar da puope a proda.

Per parlar una gallia

E chomanda miser lo chapetagno che l'armiraio se faza dar una bandiera quadra per gallia dela so arma. E se miser lo chapetagno vorà parlar una gallia, leverà a puope, ala banda, una de quele bandiera[b] quadre de qual gallia vorà, e quela presto achosti a miser lo chapetagno, sotto pena de soldi 200.

['L viver e li ordeny dele zurme]

Quisti sono li hordeny el viver dele gallie da Veniexia da puo' che fu fatte gallie. In primo per so viver vuol a dì pan bischotto libra 1, onze 6. E se 'l fusse pan fresco vuol libre 2. E sì vuol per vin al mexe soldi 5. E sì die aver fromaio per mexe libre 5. E sì vuol fava per mexi soldi 4.

E se le galie se levasse de notte del porto e vegiasse più cha ore 5 da maittina al'alba, abiando fatta la maitinada, l'armiraio debia far levar voga o far vella, che la zurma debia dormir ora una.

E se 'l fusse per vugar da puo' levado el sol, debia vogar ora una e puo' levar vuoga e manzar ora una.

E vogando a ore 5 o 6 del dì, segondo li dì fusse longi, debia disnar per ora una, e ogni sera, ananti sol amonte, in portto a possar.

---

a. *Sic in MS.*
b. *Sic in MS.*

To be wounded on the galley or on land

And the captain commands that if anyone be wounded on land or at sea, that is, on the galley, on the day it happens, as quickly as possible, whoever did the wounding shall be sent to the captain, keeping him always under good guard. And this shall be observed by the *paroni*, *comiti*, and *paroni zuradi* of the galley wherever the crime was committed, in land or on the galley. And whoever goes against the captain, the captain shall proceed against as it appears best to him; this provision does not apply to other fights without wounds, because he leaves it to you *paroni* to deal with those malefactors as you deem best.

To commit theft on the galley f. 118a

And the captain commands that if there be any theft on the galley of 3 pounds and up, they shall be held under good guard and presented to the captain as soon as possible. And this shall be observed under the penalty to be decided by the captain. And if it is from 3 pounds down, they shall be flogged from the stern to the bow.

To speak to a galley

The captain commands that the *armiraio* shall have a square banner made for each galley with its arms. And if the captain wishes to speak to a galley, he will raise at the stern on the side one of the square banners of whichever galley he wishes, and that one shall quickly come alongside the captain, under the penalty of 200 soldi.

THE PROVISIONS AND THE ORDERS OF THE CREWS

These are the orders concerning the provisions of galleys of Venice ever since there were galleys. First, for provisions each needs per day 1 pound, 6 ounces of twice-baked bread.[1] And if it is fresh bread 2 pounds are needed. And 5 soldi per month are needed for wine. And he should have 5 pounds of cheese a month. And there should be 4 soldi of beans a month.

And if the galleys sail from the port by night and are awake more than 5 hours from rising to dawn, having done the morning chores, the *armiraio* should have the oars raised or make sail, to allow the rowers to sleep one hour.

And if there was rowing after sunrise, he must row one hour and then raise the oars and eat an hour.

And rowing until 5 or 6 hours, according to whether the day is long, he shall dine for an hour, and every evening, before the sun sets, put into port.

1. For weights and measures, see Appendix below.

f. 118b E li armyrai dele galie debia aver per regagio da tute le gallie dal chomito, per la so stimaria, duchati 3. Ancuor vuol aver per regagio da tutte le gallie dal zentelomo, per la bandiera da vento, duchato 1 per zaschadana[a] gallia. E vuol aver de zaschaduna gallia tute le pelle o chori serà prexentà per scrigna alle gallie. Ancuor vuol per regaio da miser lo chapetagno, per la prima fusta serà prexa, o la fusta o lire 100. E se la fusse gallia, o la gallia o ducati 100. Ancuor vuol d'i butiny se farà d'i nimixi una parte tanto quando[b] un patron de gallia. E questo ordene si è in chanzelaria, in lo libro "b" 1350 d'i "Secretti," chomo se die far e partir, la par[*te*] prexa in Pregadi.

[L'intrar al porto da Venyexia]

1. Questa sia l'intratta del porto da Veniexia, zoè ala fuxe San Zorzi ala ture de fanò, tanto che San Nichollò dela Chavana serà ai do chaxuny al'oro de fuora del tocho. E quando San Nichollò dela Chavana a mezo li chaxuny serà al gran secho, e quando San Nicholò serà al caxon da ponente, sera' dentro dala foxie, andar intro ponente e maistro.

2. San Piero da Chastello chon San Nichuolò dela Chavana, tanto che San Nichuolò un pocho dentro d'i chaxuny, sera' al'oro di fuora el tocho, e quando San Nichollò dela Chavana serà al chaxon sollo, sera' dentro dela foxie e va per tramontana ala via d'i fari.

3. San Piero da Chastello ala ture del fanò, tanto chon San Nicholò dela Chavana fia[c] dentro d'i do
f. 119a chaxuny, sera' al'oro di fuori | de tocho. San Nichollò dela Chavana, el chaxon solo da ponente, seray dentro dala fuxie, e van de quarta de tramontana al maistro ala via d'i fari.

4. San Marcho, zoè el chanpagnel chon San Nichuolò de Lido, tanto che San Nichuolò dela Chavana sia dentro dai do chaxuny, zoè dever ponente meza barcha, serì al'oro del tocho di fuori. E quando San Nichuolò dela Chavana dentro del chaxion da ponente, serì dentro dela foxia, e van da quarta de maistro ala tramontana ala via d'i fari.

5. San Marcho, zoè el chanpagnel chon lo fanò de Lido de San Nicholò, tanto che vegna San Nichuolò dela Chavana ai do chaxuni, e quando sera' al tocho, serà San Nicholò dela Chavana a

---

a. *Sic in MS.*
b. *Sic in MS, evidently for* quanto.
c. *Possibly a mistake for* sia.
1. For the *stimaria*, see Stahl, "Michael of Rhodes," vol. 3, pp. 44–45.
2. For the chest in which officers could carry merchandise, see Stahl, "Michael of Rhodes," vol. 3, p. 56.
3. This text is not found in Archivio di Stato di Venezia, Senato, Secretarum (Deliberazioni), Reg. B (= C), which contains Senate decrees from 1350.
4. For a discussion of this passage, see Piero Falchetta, "The Portolan of Michael of Rhodes," vol. 3, p. 199.
5. San Giorgio Maggiore, in the basin of San Marco.
6. An island, now known as Isola della Madonna di Monte del Rosario, to the north of the Porto del Lido (the principal entrance into the Venetian lagoon from the Adriatic).

And the *armiraio* of the galleys should have as a boon from all the galleys, from the *comito*, 3 ducats for his *stimaria.*[1] Also he should have as a boon from all the ships from the nobleman, for the wind flag, one ducat for each ship. And he should have from each galley all of the skins or leather that will be presented to the galleys for his chest.[2] Also he should get as a boon from the captain, for the first vessel that will be taken, either the vessel or 100 pounds. And if it is a galley, either the galley or 100 ducats. Also he should get from the booty from the enemies, the same part as a galley *paron.* And this order is in the chancellery, in Book B 1350 of the "Secrets," as it should be done and divided up, by motion passed in the Senate.[3] f. 118b

---

To enter the port of Venice

1.[4] Here is the entry of the port of Venice, that is, at the mouth of the channel of San Giorgio[5] at the lighthouse, until San Nicolò della Cavanna[6] lines up on the two huts on the verge of being out of soundings. And when San Nicolò della Cavanna between the two huts is at the large shoal and San Nicolò[7] is at the west hut, you will be within the channel, going between west and northwest.

2. San Pietro di Castello lined up with San Nicolò della Cavanna, until San Nicolò is a bit between the huts, you will be at the outside edge of the sounding, and when San Nicolò della Cavanna is at the single hut, you will be within the channel, and go to the north by way of the lights.

3. When San Pietro di Castello is lined up with the lighthouse, and San Nicolò della Cavanna between the two huts, you will be on the edge | of the sounding. With San Nicolò della Cavanna f. 119a lined up on the single hut on the west, you will be in the channel, and go on the quarter of north by west[8] by way of the lights.

4. When San Marco, that is the campanile, lines up with San Nicolò di Lido, and San Nicolò della Cavanna is between the two huts, that is half a ship's length toward the west, you will then be on the edge of soundings. And when San Nicolò della Cavanna is inside the line of the west hut, you will be in the channel, and go on the quarter of northwest by north by way of the lights.

5. When San Marco, that is the campanile, lines up with the lantern of the Lido of San Nicolò, and San Nicolò della Cavanna comes to the two huts, and when you are just in soundings, San Nicolò

---

7. Probably San Nicolò di Lido.

8. That is, one point ($11\frac{1}{4}$ degrees) off of north toward the west; one quarter of the 45 degrees between north and northwest. The eight principal winds on the Venetian windrose are: *tramontana* (north), *griego* (northeast), *levante* (east), *sirocho* (southeast), *ostro* (south), *garbin* (southwest), *ponente* (west), and *maistro* (northwest). *Maistro tramontana* would be halfway between north and northwest, what we call north-northwest (NNW), and *quarta di maistro al tramontana* would be a quarter off of northwest toward north, or northwest by north (NWbN).

mezo i chaxuny. E quando San Nicholò dela Chavana serà al chaxon solo da ponente, sera' dentro dala foxie, e va per maistro ala via d'i fari.

6. Per pasi 3 San Nichuolò de Lido in San Zorzi, tanto che San Nicholò dela Chavana te romagna ai do chaxuny, zoè alo Lido de Sancto Rasmo, andar per questa via inchina che San Nichuolò dela Chavana serà al chaxion solo dala parte da ponente, et in quela volta serai dentro dela foxia, e se vorà saver quando al thocho serà San Nichollò dela Chavana in mezo d'i do chaxuni, e quel chaxon solo quarta de maistro al ponente.

7. Per lo più basso che sia, zoè al garbin, Sant'Andrea in la ture, zoè al fanò de Lido, Sancta Maria da Trozello, zoè el so chanpagnel, che vegna raxo al boscho spesso de fuora raxo raxo[a] de
f. 119b Sancto | Rasmo. Quando tu sera' dentro d'i do albori de Sancta Maria da Trozello, sera' dentro dela foxia quarta de tramontana al griego.

Sancta[b] Andrea in su le mure de San Nichuolò, puo' Sancta Maria de Trozello raxo lo boscho speso, serai al'oro di fuori la foxia. Quando serai dentro Sancta Maria dentro d'i do albori, zoè delo boscho spesso, seray dentro dela foxia.

E per voler meter l'anchora, fa che San Nichuolò dela Chavana vegna dentro una gallia dal chaxon dal ponente. E se vuolè andar dentro vegnì al largo dela medà[c] sutto vento meza galia a largo, vui serì netto dal barbaro e serà la via d'i fari.

E se vuolè insir fuora sirocho maistro, serà San Piero a San Nichollò dela Chavana.

E se tu vuol insir ostro sirocho, serà Sancto Andrea in la ture del fanò.

E se tu vuol insir fuora levante sirocho, serà San Marcho a San Nichuollò de Lido.

E le aque de sovradítto porto, quando la luna sie in ponente e levante aqua bassa. E quando la luna in sirocho, le aque meze fatte. E quando la luna in ostro, tute piene. E quando la luna in garbin, meze basse. E quando la luna in ponente, tute è basso. Arichordandotti che le aque in fiel chomenza ale 4 dì dela luna inchina dì 10, e la sia ponta. E da 11 inchina 19 la è zonta. E dal 19 a 25 in fiel. E dal 25 inchina 4 in aqua. Di raxion arechordandotti che ai 7 e 9 le aque non se muove. E per lo simille muodo fano le aque in Fiandria, zoè al porto dele Schiuze.

---

a. *Thus repeated in MS.*
b. *Sic in MS.*
c. *Sic in MS.*

della Cavanna will be between the huts. And when San Nicolò della Cavanna is on the single hut on the west, you will be in the channel, and go northwest by the way of the lights.

6. At 3 paces,[1] San Nicolò di Lido is on San Giorgio, and San Nicolò della Cavanna stays lined up for you on the two huts, that is, at the Lido of Sant'Erasmo; go this way until San Nicolò della Cavanna is on the single hut on the west side, and at that time you will be in the channel, and it will be necessary to know when you're in soundings, when San Nicolò della Cavanna is between the two huts and the single hut, quarter of northwest by west.

7. At the lowest possible, that is, to the southwest; Sant'Andrea on the tower, that is, at the lantern of the Lido; Santa Maria di Torcello, that is, its campanile, which just touches the dense woods, just next to Sant'|Erasmo. When you are between the two trees of Santa Maria di Torcello, you will be f. 119b
in the channel on the quarter of north by east.

Sant'Andrea lines up above the walls of San Nicolò, and Santa Maria di Torcello just touches the thick woods, you will be on the outer edge of the channel. When you are within Santa Maria within the two trees, that is, of the thick woods, you will be in the channel.

And if you want to drop anchor, make San Nicolò della Cavanna come within a galley length of the west hut. And if you wish to enter, come with a wind halfway under the lee, a half galley off, you will be clear of the barbarian[2] and will be along the way of the lights.

And if you wish to depart southeast–northwest, it will be San Pietro to San Nicolò della Cavanna.

And if you wish to depart south-southeast, it will be Sant'Andrea to the lighthouse.

And if you want to depart east-southeast wind, it will be San Marco to San Nicolò di Lido.

And the tides of this port: when the moon is in the west and east, low tide. And when the moon is in the southeast, it makes half floods; and when the moon is in the south, very high tides. And when the moon is in the southwest, half ebb. And when the moon is in the west, very low. Remember that low water begins on the 4th day of the moon until the 10th and then it is at its change. And from the 11th to the 19th it's a rising tide. And from the 19th to the 25th low water, and from the 25th to the 4th in water. Remember that on the 7th and 9th the waters do not move. And the waters behave similarly in Flanders, that is, in the port of Sluys.

---

1. Of depth; about 5.2 meters.
2. Uncertain interpretation.

f. 120a

[Veder portolan de Spagna e le so traversse]

Portolan fatto per Zuan Pires, pedotta del mar di Fiandria.

Trafigar e Mazamoias, sirocho maistro, tuo' in puocho dela tramontana, lege 7.

Trafigal e pontal de Sancta Maria, levante sirocho, lege 3.

Chavo San Vizenzo e Tazeri, levante sirocho, lege 12.

Salmerina chon pontal de Sancta Maria, ponente levante, lege 25.

Pontal de Sancta Maria e Porchas de Chades, zoè li schogi, quarta de levante sirocho, lege 28.

Pontal de Sancta Maria, Arenas Gurdas de Sivillia, quarta de levante al griego, lege 19.

Pontal de Sancta Maria, el chavo de Sagras, ponente levante, lege 15 inchina chavo San Vizenzo.

Chavo San Vizenzo e Sancta Maria de Rabota di Sostoval, se varda ostro tramontana, lege 20.

Chavo San Vizenzo chon Pizes, ostro tramontana, tuo' del maistro lege 30.

Rocha de Sindra e Pizes, sirocho maistro, lege 9.

Rocha e Berlenga, ostro tramontana, lege 12.

Berlinga, chavo Charboner, levante sirocho, lege 2.

Berlenga e Montiego d'i Fariliony di fuora, griego garbin, lege 16.

Berlenga, Porto de Portogallo, quarta[a] de griego tramontana, lege 33.

Montiego e Abero, ostro tramontana, lege 10.

Porto de Portogallo e Vila del Chonte, quarta ostro sirocho, lege 5.

Portogallo e Viena, maistro tramontana, lege 9.

---

a. *Corrected over* duarta.

1. For a discussion of this and the other portolans in the manuscript, see Falchetta, "The Portolan of Michael of Rhodes," vol. 3, pp. 193–207.

2. Cabo Trafalgar, Spain.

f. 120a

See a portolan of Spain and its crossings

Portolan made by Zuan Pires, pilot of the Flanders sea.[1]

Trafalgar[2] and Marzamusa,[3] southeast–northwest, bear a little north, 7 leagues.[4]

Trafalgar and Cabo de Santa María, east-southeast, 3 leagues.

Cabo de São Vicente and Tavira, east-southeast, 12 leagues.

Salmedina with Cabo de Santa Maria, west–east, 25 leagues.

Cabo de Santa Maria and Cádiz, that is the reefs, quarter of east by south, 28 leagues.

Cabo de Santa Maria, Arenas Gardas of Seville, quarter of east by north, 19 leagues.

Cabo de Santa Maria, Cabo de Sagres, west–east, 15 leagues to Cabo de São Vicente.

Cabo de São Vicente and Santa Maria de Arrábida of Setubal, bear south–north, 20 leagues.

Cabo de São Vicente with Cabo de Espichel, south–north, bear a little northwest, 30 leagues.

Cabo da Roca de Sintra and Cabo de Espichel, southeast–northwest, 9 leagues.

Cabo da Roca and Berlenga, south–north, 12 leagues.

Berlenga and Cabo Carvoeiro, east-southeast, 2 leagues.

Berlenga and Cabo Mondego from the outer Farilhões Islands, southwest–northeast, 16 leagues.

Berlenga and Oporto of Portugal, quarter of northeast by north, 33 leagues.

Mondego and Aveiro, south–north, 10 leagues.

Oporto of Portugal and Vila do Conde, quarter of south by east, 5 leagues.

Portugal and Viana do Castelo, north-northwest, 9 leagues.

---

3. On the Moroccan coast at the site of Mount Sidi Moussa.
4. The actual distance is about 33 leagues (the maritime league is about 5.6 kilometers); cf. Falchetta, "The Portolan of Michael of Rhodes," vol. 3, pp. 198–199. For measures used in the MS, see Appendix below.

Viena e Palesales, maistro tramontana, lege 9.

Palexales e Finistere, maistro tramontana, lege 18.

L'ixola de Baiona, la Pelegrina de Coravedro, quarta de sirocho levante.

f. 120b Ixola Donas e la Bassa del Pelegrin, levante sirocho, lege 5.

Pelegrin de Choravedro e Finisterre, quarta de tramontana maistro, lege 7.

Monteloro e Sentolo de Finisterre, schoio, sirocho maistro, lege 4.

Finistere e Turiana e chavo de Nao, ostro tramontana, tuo' de griego, lege $2\frac{1}{2}$.

Vilano e Turiano, quarta de griego levante, lege $1\frac{1}{2}$.

Zizercha, griego levante, tuo' del levante, lege 10.

Zizercha e la ture del faro de Chologna, levante sirocho, lege 7.

Zizerchia, chavo del Prior, griego levante, lege 8.

Prior, Ortiger, griego levante, lege 8. Prioro, la so chognosenza e sie più alto cha la riviera, che par bassa e spiaze et à in punta 3 schoietti, e vardase chon lo chavo de fuora, ostro tramontana. In mezo la via do schoietti.

Ortiger e Varies, quarta de levante griego, lege 3.

Varies Montenegro e Ribadeo, ponente maistro, lege 12.

Varies e Penes de Gozon, levante ponente, più al griego, lege 25.

Penes de Gozon e Ribadeo, griego levante, lege 18.

Penes de Gozon chon chavo Sanct'Ander, quarta de levante sirocho, lege 12.

Mazazachoa, le Pene de Gozon, ponente levante, tuo' del maistro, lege 50.

Mazazachoa e Figer, ponente levante, lege 15.

Viana do Castelo and Cabo Silleiro, north-northwest, 9 leagues.

Cabo Silleiro and Fisterra, north-northwest, 18 leagues.

Islas Cíes of Bayona and the pilgrimage of Cabo Corrubedo,[1] quarter of southeast by east.

Isla de Ons and Cabo Corrubedo, east-southeast, 5 leagues. f. 120b

Pilgrimage of Cabo Corrubedo and Cabo Fisterra, quarter of north by west, 7 leagues.

Monte Louro and Centola de Fisterra, reef, southeast–northwest, 4 leagues.

Fisterra and Cabo Touriñán and Cabo de la Nave, south–north, bear a little northeast, $2\frac{1}{2}$ leagues.

Cabo Villano and Cabo Touriñán, quarter of northeast by east, $1\frac{1}{2}$ leagues.

[Cabo Touriñán and] the Sisargas islands,[2] east-northeast, bear slightly east, 10 leagues.

Sisargas and the lighthouse of La Coruña, east-southeast, 7 leagues.

Sisargas and Cabo Prior, east-northeast, 8 leagues.

Prior, Ortegal, east-northeast, 8 leagues. Prior, it can be recognized by the fact that it is higher than the shore, which appears low and sandy and it has three reefs off the point, and it looks out with the cape south–north. In the middle of the passage there are two reefs.

Cabo Ortegal and Punta de la Estaca de Bares, quarter of east by north, 3 leagues.

Punta de la Estaca de Bares, Monte Negro, and Ria de Ribadeo, west-northwest, 12 leagues.

Punta de la Estaca de Bares and Cabo de Peñas de Gijón, east–west, and more to the northeast, 25 leagues.

Cabo de Peñas de Gijón and Ria de Ribadeo, east-northeast, 18 leagues.

Cabo de Peñas de Gijón with Cabo Santander, quarter of east by south, 12 leagues.

Cabo Machichaco and Cabo de Peñas de Gijón, east–west, bear slightly northwest, 50 leagues.

Cabo Machichaco and Cabo Higuer, east–west, 15 leagues.

---

1. Probably a cult site of the Virgin Pellegrina, venerated in Galicia.
2. *Raxion de' marineri* has "Turiana i Zizerchia" for this entry in its copy of the portolan: Pietro de Versi, *Raxion de' marineri*, p. 66.

Mazazachoa, Gataria, levante sirocho, tuo' del levante, lege 10.

Figer e la ponta de Baiona, griego garbin, lege 8.

Archasion e la punta de Baiona, ostro tramontana, lege 20.

Archasion e Sancta Maria de Sobarch,[a] quarta de ostro garbin, lege 17.

Agoyas de Bordel e Pertus de Spagna, maistro tramontana, lege 12.

Bertus de Berdagna e l'ixola Dozias, ponente maistro, tuo' del maistro, lege 14.

Ixolla Dozias, el chavo de Olona, ponente maistro, tuo' del maistro, lege 8.

Ixola Dozias e Piglera, ostro tramontana, lege 7.

f. 121a Piglera e la riviera de Lera, griego tramontana, lege 7.

Piglera e Lo Miera de Bela Ixola, ponente maistro, lege 12.

Uzias, chavo de Lameria, maistro sirocho, lege 16.

Porsanson de Belixola e Pesmarche e Sain, ponente maistro, lege 30. In esta rivyera vardatti dela bassa zumenta.

Belila e Grogia, maistro tramontana, lege 6.

Grogia e Charan, levante ponente, lege 6. In esta via vardatti dela bassa de Panparlle.

Bella Yxolla e Charantan, griego levante, lege 13.

Glaran, Pesmarche, ponente levante, lege 6.

Pesmarche e chavo de Beduch e Fontaneo, sirocho maistro, lege 9.

Somaingo e Fontaneo, ostro tramontana, lege 8.

Queste sono le traversse de Spagna.

---

a. Sobarch *corrected over* Sorarch.

Cabo Machichaco and Guetaria, east-southeast, bear east, 10 leagues.

Cabo Higuer and Pointe de Bayonne,[1] northeast–southwest, 8 leagues.

Arcachon and Pointe de Bayonne, south–north, 20 leagues.

Arcachon and Sainte-Marie de Soulac-sur-Mer, quarter of south by west, 17 leagues.

Pointe de la Coubre and Pertuis de Maumusson, north-northwest, 12 leagues.

Pertuis Breton and Île d'Yeu, west-northwest, bear slightly northwest, 14 leagues.

Île d'Yeu and Pointe de l'Aiguille, west-northwest, bear slightly northwest, 8 leagues.

Île d'Yeu and Île du Pilier, south–north, 7 leagues.

Île du Pilier and the Loire river, north-northeast, 7 leagues. f. 121a

Île du Pilier and Locmaria de Belle-Île, west-northwest, 12 leagues.

Île d'Yeu, Cap de Locmaria, northwest–southeast, 16 leagues.

Port de Sauzon de Belle-Île and Penmarch and Sein, west-northwest, 30 leagues. On this coast, look out for the Donkey shallows.

Belle-Île and Groix, north-northwest, 16 leagues.

Groix and Glénan, east–west, 6 leagues. On this route watch out for the shoals of Penfret.

Belle-Île and Glénan, east-northeast, 13 leagues.

Glénan, Penmarch, west–east, 6 leagues.

Penmarch and Chavo de Beduch[2] and Pointe de Feunteunod, southeast–northwest, 9 leagues.

Saint-Mathieu and Pointe de Feunteunod, south–north, 8 leagues.

These are the crossings of Spain.[3]

---

1. Based on the distance given, this can be identified as Capbreton, a few kilometers north of Bayonne, France.
2. Unidentified location.
3. That is, the crossings of the Bay of Biscay from Spain to France.

Turiane, Osenti, lege 125, griego tramontana, andando a largo lege 10.

Zizercha, Uxente, griego tramontana, andando a largo da Usente lege 4, e sono lege 116.

Prior e Mulines, schoio intorno Usenti, griego tramontana, lege 112.

Sain Urtigiera, griego tramontana, lege 105.

Funtaneo e Varies, griego tramontana, lege 105.

Pesmarche, San Ziprian, griego tramontana, lege 110.

Fontaneo, chavo Sanct'Ander, quarta de tramontana maistro, lege 95.

Mazazachoa, Funtaneo, maistro tramontana, lege 100.

Usente e Tures, ostro tramontana, lege 98.

Sanctonya e Belixola, ostro tramontana, lege 80.

Chozias chon Mazazachoas, ostro tramontana, lege 65.

Chozias e Lanes, griego tramontana, lege 75.

f. 121b Penes de Gozon e Pertus de Spagna, griego garbin, lege 88.

Las Anas de Burdel, Uritigiera, griego levante, lege 112.

Queste sono le traversse da Ossente a Chales in chanal di Fiandres, chomenzando Usenti e Sorlenga, maistro e tramontana e più al maistro, lege 33.

Osenti e Lunganeo, quarta de tramontana maistro, lege 33.

Usenti e la Benedetta de Premuda, griego tramontana e più ala tramontana, lege 36.

Furno Dartus e Godester, griego tramontana, lege 38.

Usente a Loxert, ostro tramontana, lege 30.

Barbaracha, Faduich, ostro tramontana, lege 28.

Cabo Touriñán, Ouessant, 125 leagues north-northeast, going offshore 10 leagues.

Sisargas, Ouessant, north-northeast, going offshore 4 leagues from Ouessant, and it's 116 leagues.

Cabo Prior and Molène, a reef near Ouessant, north-northeast, 112 leagues.

Île de Sein, Cabo Ortegal, north-northeast, 105 leagues.

Pointe de Feunteunod and Punta de la Estaca de Bares, north-northeast, 105 leagues.

Penmarch, San Ciprian, north-northeast, 110 leagues.

Pointe de Feunteunod, Cabo Santander, quarter of north by west, 95 leagues.

Cabo Machichaco, Pointe de Feunteunod, north-northwest, 100 leagues.

Ouessant and Cabo de Torres, south–north, 98 leagues.

Santoña and Belle-Île, south–north, 80 leagues.

Île d'Yeu with Cabo Machichaco, south–north, 65 leagues.

Île d'Yeu and Llanes, north-northeast, 75 leagues.

Cabo de Peñas de Gijón and Pertuis de Maumusson, northeast–southwest, 88 leagues. f. 121b

Pointe de la Coubre, Cabo Ortegal, east-northeast, 112 leagues.

These are the crossings from Ouessant to Calais in the Flanders channel, beginning Ouessant and the Isles of Scilly, north-northwest and more to the northwest, 33 leagues.

Ouessant and Land's End, quarter of north by west, 33 leagues.

Ouessant and Benedetta[1] of Plymouth, north-northeast and more to the north, 36 leagues.

Chenal du Four and Start Point, north-northeast, 38 leagues.

Ouessant and Lizard Point, south–north, 30 leagues.

Aber-Wrach, Fowey, south–north, 28 leagues.

1. An unidentified rock of the English coast near Plymouth.

Furno Dartus e Falamua, ostro tramontana, lege 30.

Yxola da Bas e Godester, ostro tramontana, lege 28.

Chapu di Grua e Godester, sirocho maistro, lege 21.

Sette Yxoles e Godester, maistro tramontana, lege 28.

Godester e Tores, griego tramontana, lege 4.

Godester e Chaschette, maistro sirocho, lege 21.

Porlan e Chaschette, ostro tramontana, lege 15.

Chaschette e Renui, levante sirocho, lege 4.

Sette Yxoles, Usenti, griego levante, lege 32.

Chapo di Grua e Sette Yxole, griego tramontana, lege 14.

Rochatuas, el chavo dela Grua, quarta tramontana griego, lege 6.

Chaschette, el chavo dela Grua, quarta de griego tramontana, lege 8.

Ussente, Chaschette, quarta de griego levante, vegnirà dentro Chaschette zircha Garnaxui.

Una nave che vien fuor d'Ussente lege [. . .][a] andarà sovra Rochatua griego levante, lege 42.

Sette Yxolle e Urena, ponente levante, lege 4.

f. 122a Urena e la intratta de Samalò de fuora, levante sirocho, lege 16.

Chavo de Latta e Urena, quarta de levante sirocho, lege 12.

Chavo de Ier e Ponte Laaga, griego garbin, lege 18.

Ponte Laaga e Las Aguias del'ixola Doich, ostro tramontana, e più del griego garbin, lege 20.

---

a. *Sic in MS.*

Chenal du Four and Falmouth, south–north, 30 leagues.

Île de Batz and Start Point, south–north, 28 leagues.

Cap de la Hague and Start Point, southeast–northwest, 21 leagues.

Sept Îles and Start Point, north-northwest, 28 leagues.

Start Point and Totnes, north-northeast, 4 leagues.

Start Point and the Casquet Islands, northwest–southeast, 21 leagues.

Portland Bill and the Casquet Islands, south–north, 15 leagues.

Casquet Islands and Alderney, east-southeast, 4 leagues.

Sept Îles, Ouessant, east-northeast, 32 leagues.

Cap de la Hague and Sept Îles, north-northeast, 14 leagues.

Grosnez Point[1] and Cap de la Hague, quarter of north by east, 6 leagues.

Casquet Islands and Cap de la Hague, quarter of northeast by north, 8 leagues.

Ouessant, the Casquet Islands, quarter of northeast by east, you will come within the Casquet Islands at around Guernsey.

A ship that comes out of Ouessant [. . .][2] leagues will get to Grosnez Point east-northeast, 42 leagues.

Sept Îles and Les Renauds, west–east, 4 leagues.

Les Renauds and the outer entrance to the bay of Saint-Malo, east-southeast, 16 leagues. f. 122a

Pointe de la Latte and Les Renauds, quarter of east by south, 12 leagues.

Island of Sark and Cap de la Hague, northeast–southwest, 18 leagues.

Cap de la Hague and Needles Point of the Isle of Wight, south–north, and closer to northeast–southwest, 20 leagues.

1. On the island of Jersey.
2. Distance lacking in MS.

Sancta Lena e Barafrette, ostro tramontana, lege 17.

Barafrette, el chavo Belzep, griego garbin. Insula secha de Barefrette, lege 32.

Barafrette, Antifer, levante ponente, lege 18.

Barafrette e la riviera de Sagnia, levante sirocho, lege 18.

Chavo de Chaus e chavo de Toro, sirocho maistro, lege 32.

Antifer e Sancta Lena, sirocho maistro, lege 30.

Antifer e chavo de Chaus e Belces, ostro tramontana, lege 25.

Belzep e Diepa, sirocho maistro, più alo levante, lege [. . .].[a]

Belzep e la fossa de Chaion, levante sirocho, lege 21.

Romaneo chon Antifer, griego tramontana, lege 31.

Longaneus, Aloxert, quarta de ponente maistro, lege 7.

Godman Aluxert, griego garbin, lege 6.

Chavo de Rama e Godeman, griego levante, lege 8.

Ixole Verde e chavo de Botre, levante sirocho, lege 4.

Godester e Porlan, griego levante, lege 20.

Porlan e Las Aguyas, quarta de levante griego, lege 12.

Chavo de Toro e Porlan, ponente levante, lege 14.

Chavo de Toro e Las Aguyas, levante sirocho, lege $2\frac{1}{2}$.

Belzep e la zittà vechia, levante ponente, lege 16.

---

a. *Distance omitted in MS.*

1. At the northern end of Tor Bay.

2. Distance lacking in MS.

Saint Helens and Pointe de Barfleur, south–north, 17 leagues.

Pointe de Barfleur and Beachy Head, northeast–southwest, island at low tide of Pointe de Barfleur, 32 leagues.

Pointe de Barfleur, Cap d'Antifer, east–west, 18 leagues.

Pointe de Barfleur and the mouth of the Seine, east-southeast, 18 leagues.

Chef de Caux and Hope's Nose,[1] southeast–northwest, 32 leagues.

Cap d'Antifer and Saint Helens, southeast–northwest, 30 leagues.

Cap d'Antifer and Chef de Caux and Beachy Head, south–north, 25 leagues.

Beachy Head and Dieppe, southeast–northwest, more to the east, [. . .][2] leagues.

Beachy Head and the Bay of Cayeux,[3] east-southeast, 21 leagues.

Dungeness[4] with Cap d'Antifer, north-northeast, 31 leagues.

Land's End, Lizard Head, quarter of west by north, 7 leagues.

Dodman Point, Lizard Head, northeast–southwest, 6 leagues.

Cape Rame and Dodman Point, east-northeast, 8 leagues.

Green Islands[5] and Bolt Head, east-southeast, 4 leagues.

Start Point and Portland Bill, east-northeast, 20 leagues.

Portland and Needles Point, quarter of east by north, 12 leagues.

Hope's Nose and Portland, west–east, 14 leagues.

Hope's Nose and Needles Point, east-southeast, $2\frac{1}{2}$ leagues.

Beachy Head and Chichester, east–west, 16 leagues.

---

3. Bay at the mouth of the Somme, near Cayeux-sur-Mer.
4. Promontory near Romney.
5. Unidentified location.

Belzep e l'ixola Doich, quarta de levante griego, lege 20.

Belzep e Romaneo, griego levante, lege 14.

Belzep, Erlaga, quarta de griego alo levante, lege 6.

f. 122b El chavo de Dobla e Romaneo, se varda griego e garbin, lege 7.

Usente e Las Aguyas e Pula, griego garbin, lege 67.

Antifer e Diepa, griego levante, lege 12.

Diepa e la fosa de Chaion, griego garbin, lege 12.

Bologna e Staples e San Chater, quarta de ostro garbin, lege 15.

Gizuam e Dobla, maistro tramontana, lege 6.

Chales e Dobla, ponente maistro, lege 7.

San Chater e Sancta Chatarina, li banchi de Fiandria, griego levante, lege 20.

Chales Chaverlinges, griego levante, lege 4.

Chaverlinges a Donquercha, griego levante, lege 3.

Donquercha e Les Monzies, griego levante, lege 2.

Monzies chon Ostenta, griego levante, lege 4.

Blancha Verga, Santa Chatarina, griego levante, lege 4.

Tarifa e Zubeltar, griego levante, più al ponente, lege 7.

Tarifa e Spartil, griego garbin, lege 6.

Spartil, el Trafigar, ostro tramontana, lege 8.

Beachy Head and the Isle of Wight, quarter of east by north, 20 leagues.

Beachy Head and Dungeness, east-northeast, 14 leagues.

Beachy Head, Fairlight, quarter of northeast by east, 6 leagues.

Cape of Dover and Dungeness, they face each other northeast–southwest, 7 leagues. f. 122b

Ouessant and Needles Point and Poole, northeast–southwest, 67 leagues.

Cap d'Antifer and Dieppe, east-northeast, 12 leagues.

Dieppe and the Bay of Cayeux, northeast–southwest, 12 leagues.

Boulogne and Étaples and Sangatte, quarter of south by west, 15 leagues.

Wissant and Dover, north-northwest, 6 leagues.

Calais and Dover, west-northwest, 7 leagues.

Sangatte and St. Catherine,[1] the sandbanks of Flanders, east-northeast, 20 leagues.

Calais, Gravelines, east-northeast, 4 leagues.

Gravelines to Dunkirk, east-northeast, 3 leagues.

Dunkirk and Les Monzies,[2] east-northeast, 2 leagues.

Les Monzies with Ostende, east-northeast, 4 leagues.

Blankenberge, St. Catherine, east-northeast, 4 leagues.

Tarifa and Gibraltar, east-northeast, more to the west, 7 leagues.

Tarifa and Cape Spartel, northwest–southeast, 6 leagues.

Cape Spartel, Trafalgar, south–north, 8 leagues.

---

1. A landmark in Sluys; see fol. 125b.
2. Unidentified location.

[Veder aque e marie de Spagna]

Queste sono le aque e le marie de Fiandria, là che se truova la luna quanto[a] è piena l'aqua o quando serà bassa per li porti e de fora i chanalli, chomo vedirì qui de sotto nottado, zoè in le parte de Fiandria e in lli gulfi e l'ixolla d'Engletera.

L'aqua de Chades e de Faraon e de Lisbuna, quando la luna in griego e tramontana, piena mare.

In Barameda la luna quarta de sirocho alo levante, bassa mar.

In lo Schaziopos in Selves, luna quarta di sirocho alo levante, basa.

In porto de Portogalo, quarta de Levante sirocho, bassa mar.

In tutti li oltri luogi de Spagna perfina Bayona de Gaschogna, sirocho maistro, basa.

f. 123a In le Anes de Burdeu, luna ostro e tramontana, basa mare.

Da Anas de Burdeu inchina Funtaneo, sirocho maistro, bassa mare.

In Ras de Sain, quarta de sirocho levante, bassa.

In Samain, luna sirocho maistro, bassa.

In Furno Dartus, quarta de sirocho ostro, bassa.

In Barbaracha, ostro sirocho, bassa.

In Galvan, luna quarta de ostro sirocho, bassa.

In el Dos del'ixolla da Bas, ostro tramontana, bassa.

Intro al porto del'ixola da Bas, ostro sirocho, bassa.

Al porto de Usente, ostro sirocho, bassa.

Da Usenti al'ixola da Bas per la Schala, ostro tramontana, basa.

---

a. *Sic in MS, evidently for* quando.

See the waters and tides of Spain

These are the waters and the tides of Flanders, how the moon is when the tide is high or when it will be low for the ports and out of the channels, as you will see noted below, that is, in the parts of Flanders and in the gulfs and island of England.

The water of Cádiz and of Faro and of Lisbon, when the moon is in the northeast and north, high tide.

In Sanlúcar de Barrameda, the moon quarter of east by south, low tide.

In Schaziopas[1] at the mouth of the Silves, moon quarter of east by south, low.

In Oporto of Portugal, quarter of east by south, low tide.

In all the other places of Spain, as far as Bayonne in Gascony, southeast–northwest, low.

In Pointe de la Coubre, moon south and north, low tide. f. 123a

From Pointe de la Coubre to Feunteunod, southeast–northwest, low tide.

In Pointe de Raz, quarter of southeast by east, low.

In Pointe de Saint-Mathieu, moon southeast–northwest, low.

In Chenal du Four, quarter of southeast by south, low.

In Aber-Wrach, south-southeast, low.

In Grève de Goulven, moon quarter of south by east, low.

At the Dos[2] of Île de Batz, south–north, low.

Within the port of Île de Batz, south-southeast, low.

At the port of Ouessant, south-southeast, low.

From Ouessant to Île de Batz by la Schala,[3] south–north, low.

1. Unidentified location.
2. Unidentified location.
3. Unidentified location.

Da Usenti una vista in chanal, ostro garbin, bassa.

A Mervana, quarta ostro garbin, bassa.

Al Dos de Sette Illes, ostro garbin, bassa.

A Rochatuas, quarta de garbin ostro, bassa.

A l'ixola da Bas inchina Samalò, in li portti, luna al'ostro, maria più prima d'un quarto de rumo de vento.

E sovra l'ixola da Bas in l'alta mar, luna ostro garbin, basa.

A Briach in l'ixola, ostro tramontana, bassa mare.

E sapi che del stretto de Rochatuas a Samalò, a Riva Vila, ay Axui, ostro e tramontana. Seguise esta maria chon tuti li oltri porti de Chornovaya e segue quaesta maria chon el Ras de Braziar e chon Chataros inchina l'ixola de San Martin de meza maria in ssu.

In Guarnaxui, dentro dal porto, ostro tramontana, bassa.

Al Dos di Guarnaxui, ostro garbin, bassa.

A[*l*] di fuor di Guarnaxui, in chanal, 15 lege, ponente garbin, bassa.

Intro Zero e Renui este un bancho de sablun, sirocho maistro su al bancho, aqua passa 9. Vene el
f. 123b chorsso de meza zozente inchina meza maria da maistro e de meza maria inchina meza | zozente vien dal sirocho.

E sapi che intro Ireny e Renui è le Feriere, e l'è bon luogo per chi la uxa. In el mar ostro e garbin, bassa mar, ma l'aqua non à poso, ma in quello è più stancha à 22 passa de poso de griego levante.

Al Ras de Branziar, quarta de sirocho alo levante de ver lo mar, piena mar in mezo el chanal, sirocho maistro, piena mar.

E de Renui inchina Barafrette, quarta de sirocho al'ostro, piena mare.

Ala chosta de Gostantin, ale 30 passe, sirocho maistro, piena.

---

1. The plateau de la Méloine, a group of rocky heads.
2. Uncertain interpretation.
3. Unidentified location.
4. Uncertain location, possibly the Channel Island of Sark.
5. On the island of Guernsey.

From Ouessant, in view of the channel, south-southwest, low.

At Méloine,[1] quarter of south by west, low.

At the Dos of Sept Îles, south-southwest, low.

At Grosnez Point, quarter of southwest by south, low.

From Île de Batz to Saint-Malo, in the ports, moon to the south, the tide earlier by a quarter of a point of direction of the wind.[2]

And above Île de Batz on the high sea, moon south-southwest, low.

At Saint-Brieuc on the island, south–north, low tide.

And know that from the strait of Grosnez Point to Saint-Malo, at Frandville, at Axui,[3] south and north. This tide continues with all of the other ports of Cornwall and this tide continues with Raz Blanchart and with Chataros[4] until Saint Martins Point[5] from half tide up.

In Guernsey, within the port, south–north, low.

At the Dos of Guernsey, south-southwest, low.

Offshore from Guernsey, in the channel, 15 leagues, west-southwest, low.

Between Zero[6] and Alderney is a sandbank, southeast–northwest up to the bank, 9 paces of water. The current from half ebb up to half flood comes from the northwest, and that from half flood to half | ebb comes from the southeast. f. 123b

And know that between Ireny[7] and Alderney are the Feriere,[8] and it's a good place for whoever uses it. At sea, when it[9] is between south and southwest, low tide, but the water has no deep spots, but at that moment it's slacker with 22 paces of deep water when it's east-northeast.

At Raz Blanchart, quarter of southeast by east toward the sea, high tide in the middle of the channel; southeast–northwest, high tide.

And from Aurigny to Pointe de Barfleur, quarter of southeast by south, high tide.

On the coast of the Cotentin, at 30 paces, southeast–northwest, high tide.

---

6. Unidentified location.
7. Unidentified location.
8. Unidentified location.
9. Presumably the moon.

E de Barafrette a Chaus, ostro sirocho, piena.

A Saina, luna sirocho maistro, piena.

E de fuor de Chaus, luna ostro sirocho, piena.

Antifer, ostro sirocho, piena, e de là avanti inchina so Diepa, quarta de ostro al sirocho, piena mare.

Al traversso de Diepa inchina Chain, ostro tramontana, piena mare.

A Chain, luna ostro tramontana, piena. Quista maria varda per intrar.

A Chontrey, ostro tramontana, piena mare.

E deli banchi de Chain inchina tu passi Staples, ostro e tramontana, piena mare.

Al traversso di Bologna, ostro garbin, piena mare.

A Bologna, a Chales, in Chaverlinga e per tuta la chosta de Fiandres vien chussì i banchi de Fiandres, ostro e tramontana, piena mar d'altura.

Queste sono le marie e le aque d'Erlanda e de Gaules e del'ixola de Ingletera per longuo inchina Tenette e Tamixe.

In Erlanda, luna ostro tramontana, basa mar. Tuo' in puocho del sirocho.

A Mirafiurda, ostro tramontana, bassa mar.

A Lundei, ostro sirocho, bassa mar, e de fuor de Lundei dentro presto, ostro tramontana, bassa mar. Segui questa maria inchina Ulmi.

E de Ulmy inchina Lubachi, quarto d'ostro al garbin, bassa mar.

A Patristo, quarta d'ostro al sirocho, basa mar, si à maria inchina Sorlenga.

In Sorlenga, quarta de griego levante, bassa mar, piena mar.

f. 124a A Maxiolla, griego e levante, piena mare.

And from Pointe de Barfleur to Chef de Caux, south-southeast, high tide.

At Île de Sein, moon southeast and northwest, high tide.

And outside of Chef de Caux, moon south-southeast, high tide.

Cap d'Antifer, south-southeast, high tide, and from there on toward Dieppe, quarter of south by east, high tide.

Crossing from Dieppe toward Cayeux-sur-Mer, south–north, high tide.

At Cayeux-sur-Mer, moon south–north, high tide. Look for this tide to enter.

At Contrey,[1] south–north, high tide.

And from the banks of Cayeux-sur-Mer until you pass Étaples, south and north, high tide.

Crossing from Boulogne-sur-Mer, south-southwest, high tide.

At Boulogne-sur-Mer, at Calais, at Gravelines, and along the whole coast of Flanders, the sandbanks of Flanders are like this, south and north,[2] full high tide.

These are the tides and the waters of Ireland and of Wales and of the island of England along toward Thanet and the Thames.

In Ireland, moon south–north, low tide. It tends a little toward the southeast.

At Milford Haven, south–north, low tide.

At Lundy, south-southeast, low tide, and outside of Lundy, right within, south–north, low tide. This side continues to Mumbles Head.

And from Mumbles Head until Lubachi,[3] quarter of south by west, low tide.

At Padstow, quarter of south by east, low tide, and so is the tide to Scilly.

In Scilly, quarter of northeast by east, low tide, high tide.

At Mousehole, east-northeast, high tide. f. 124a

1. Unidentified location.
2. Presumably the moon.
3. Unidentified location.

A Falamua, quarta de levante al griego, piena mare. E in chanal, quarta de sirocho alo levante.

A Fadoich, quarta de levante al griego, piena mar. In li chavy da levante, sirocho maistro. In lo chanal, sirocho maistro, piena mar.

A Premua, quarta de levante al griego, piena mar. In li chavy, levante sirocho, piena mar. In chanal, sirocho, piena.

Altamua, quarta de levante al griego, piena mar, e de fuora luna quarta de sirocho alo levante, piena mar. In lo chanal luna al sirocho, piena mare.

Al Dos del'ixola Doich, luna ostro sirocho, piena mar, e ali 27 passi del'ixola, per traverso d'essa luna quarta ostro sirocho, piena mar.

Al Ras de Porlan, luna quarta de sirocho alo levante, piena mar. In Porlan, luna quarta de sirocho al'ostro, piena mar. In Las Agogias quarta de sirocho al'ostro, piena mar.

Intro Porlan e Chaschette, luna ostro sirocho, piena mar. In lo chavo del Cuor del'ixola Doich, luna quarda[a] de sirocho ostro, pieno insorzedor.

In Antuna e Portamua e Anbre in Chalzesores, ostro tramontana, piena mar.

Ala zittà vechia, luna quarta d'ostro al sirocho, piena mar, in le 12 passe. Esta maria chure fina dentro Strany e Soran, luna quarto d'ostro al sirocho, piena mar. In chanal, ostro piena.

A Belzep, luna ostro tramontana, piena mar. Adosso de là, in lo sorzedor, ostro sirocho, piena mar. Intro el chanal tuò la quarta de garbin. In sorzedor, ostro e garbin.

A traverso de Sanct'Andrea de Erlaga, luna ostro garbin, piena mar, tuò dal garbin in chanal.

In Erlaga, ostro garbin, piena mar. In Salaxeu, ostro tramontana, piena mar.

A Chamara, quarta d'ostro al sirocho, piena mar, più alta aqua.

A Romaneo e atraverso de là in chanal, griego garbin, piena mar. In lo sorzedor tuo' del ponente.

f. 124b E de Romaneo inchina Sancta Malgaritta, in lo chanal, quarta | de garbin inver lo ponente, piena mar al sorzedor.

---

a. *Sic in MS, evidently for* quarta.

At Falmouth, quarter of east by north, high tide. And in the channel quarter of southeast by east.

At Fowey Harbour, quarter of east by north, high tide. In the capes to the east, southeast–northwest, high tide. In the channel, southeast–northwest, high tide.

At Plymouth, quarter of east by north, high tide. In the capes, east-southeast, high tide. In the channel, southeast, high.

Dartmouth, quarter of east by north, high tide, and outside, the moon quarter of southeast by east, high tide. In the channel, moon at southeast, high tide.

At Dos on the Isle of Wight, moon south-southeast, high tide, and at 27 paces [depth] off the island, at its crossing, moon quarter of south by east, high tide.

At Portland Bill, moon quarter of southeast by east, high tide. In Portland, moon quarter of southeast by south, high tide. At Needles Point, quarter of southeast by south, high tide.

Between Portland and the Casquet Islands, moon south-southeast, high tide. In the cape of Cowes on the Isle of Wight, moon quarter of southeast by south, high tide in the anchorage.

In Southampton and Portsmouth and Hamble, in Calshot Castle, south–north, high tide.

At Chichester, moon quarter of south by east, high tide, in 12 paces. This tide runs to between Strany[1] and Shoreham-by-Sea, moon quarter of south by east, high tide. In the channel, south, high.

At Beachy Head, the moon south–north, high tide. Behind there, in the anchorage, south-southeast, high tide. Within the channel it tends a quarter of southwest. At the anchorage south-southwest.

At the crossing of St. Andrew of Fairlight, moon south-southwest, high tide, in the channel it tends the southwest.

At Fairlight, south-southwest, high tide. In Winchelsea, south–north, high tide.

In Chamara,[2] quarter of south by east, high tide, even higher water.

At Dungeness and across from there in the channel, northeast–southwest, high tide. In the anchorage it tends to the west.

And from Dungeness to Saint Margaret's in the channel, quarter | of southwest by west, high tide at the anchorage. f. 124b

1. Unidentified location.
2. Unidentified location.

A Sancta Malgaritta, quarta d'ostro al garbin, piena mar.

A Sanuis, ostro tramontana, piena mar. Al chavo de Sancta Malgaritta, luna ostro tramontana, piena mar, in la chosta provada.

A les Dones de Sanuis, in sorzedor, luna ostro garbin, piena mar, e al ture de Tenette, luna ostro e garbin, piena mar.

A Tera Vermegia, ostro garbin, piena mar.

Intro Dobla e Inguan, ponente garbin, piena.

E de Inguan inchina el chanpagnel dele Muneghe, ponente garbin, piena mar, più dal ponente.

E dele Munege a Ostenda, quarta de garbin al ponente, piena mar.

E per tutta la chosta de Fiandres, luna ostro tramontana, piena mar de altura. Da Chales inchina de Lande, inchina la Scozia, e in Ture de Lamua, luna ostro garbin, piena mar, tuo' del garbin.

Questo è che via fa le aque dentro e de fuori chomo vedirì chi de sotto. In primo l'aqua de Larexa va in ponente garbin.

Apreso a Sain, garbin.

A Molines Osenti, ponente garbin.

A Furny e l'ixola di Fuga, ostro garbin.

A Furno e Porsal, ponente garbin.

A Porsal e l'ixola da Bas, quarta de ponente garbin.

A Nervane e Sette Illes, ponente maistro.

A Briach, in ponente, de fuora Resna inchina chavo di Fiubia in chanal de Briach, maistro sirocho.

---

1. Unidentified location.
2. Unidentified location, probably the same as Monzies on fol. 122b, a location near Dunkirk.
3. Unidentified location.
4. Unidentified location.
5. Unidentified location.

At St. Margaret's at Cliffe, quarter of south by west, high tide.

At Sandwich Haven, south–north, high tide. At the Cape of Saint Margaret's at Cliffe, moon south–north, high tides, verified on the coast.

At the dunes of Sandwich, in the anchorage, moon south-southwest, high tide, and at the tower of the Isle of Thanet, moon south-southwest, high tide.

At Terra Vermegia,[1] south-southeast, high tide.

Between Dover and Kingsdown, west-southwest, high.

And from Kingsdown to the bell tower of le Moneghe,[2] west-southwest, high tide, more from the west.

And from le Moneghe to Ostende, quarter of southwest by west, high tide.

And for the whole coast of Flanders, moon south–north, very high tide. From Calais to Lundy Island, to Sluys and to the tower of Lamua,[3] moon south-southwest, high tide, tending southwest.

This is the direction the tides take, inshore and offshore, as you will see below. First, the tide of Larexa[4] goes to the west-southwest.

Near Île de Sein, southwest.

At Île Molène and Ouessant, west-southwest.

At Chenal du Four and Île Houat, south-southwest.

At Chenal du Four and Porsal, west-southwest.

At Porsal and Île de Batz, quarter of west by south.

At Nevane[5] and Sept Îles, west-northwest.

At Saint-Brieuc, to the west, from outside of Resna[6] until the cape of Fiuba[7] in the channel of Saint-Brieuc, northwest–southeast.

6. Unidentified location.
7. Unidentified location.

Entro Caschette e Renui, ostro garbin.

E de Renui inchina Bellefrette, ponente.

E delle Bellefrette inchina l'illa de San Marcho, maistro.

E de Bellefrette inchina la fossa de Cholobila, maistro.

f. 125a Da Bellefrette al chavo de Chaus, in rutta de passa 16, quarta de maistro al ponente.

Da chavo de Chaus inchina Antifer, griego tramontana.

D'Antifer e Diepa, ponente garbin.

Da Diepa ala fossa de Chain, griego.

Da Chain inchina Bologna, ostro garbin.

E da Bologna inchina Udrenes, ostro.

E da Udrenes a Gisanti, quarta d'ostro al garbin.

E da Gisandi ai Gravi Lengues, ostro garbin.

E da Gravi Lengues inchina Doncherche, griego.

E da Doncherche inchina les Moneges, quarta de garbin al ponente.

E dale Munege inchina Ostenda, ponente garbin.

E da Ustenda a Blancha Verga, ponente garbin.

E da Blancha Verga Sancta Chatarina, quarta de ponente garbin.

Da chavo de Chornovagia inchina Loxerte, ponente.

Da Loxerte inchina Godman, quarta de ponente garbin.

Da Godman a Godestar, ponente garbin.

Between the Casquet Islands and Alderney, south-southwest.

And from Alderney until Pointe de Barfleur, west.

And from Pointe de Barfleur until the island of Saint-Marcouf, northwest.

And from Pointe de Barfleur to the channel of Cholobilia,[1] northwest.

From Pointe de Barfleur to Chef de Caux, following the 16-pace sounding, quarter of northwest by west. f. 125a

From Chef de Caux to Cap d'Antifer, northeast-north.

From Cap d'Antifer and Dieppe, west-southwest.

From Dieppe to the mouth of the Somme, northeast.

From the mouth of the Somme to Boulogne-sur-Mer, south-southwest.

And from Boulogne-sur-Mer to Wiemereux, south.

And from Wiemereux to Ouessant, quarter of south by west.

And from Wissant to Gravelines, south-southwest.

And from Gravelines to Dunkirk, northeast.

And from Dunkirk to le Moneghe, quarter of southwest by west.

And from le Moneghe to Ostende, west-southwest.

And from Ostende to Blankenberge, west-southwest.

And from Blankenberge to St. Catherine, quarter of west by south.

From Cape Cornwall to Lizard Point, west.

From Lizard Point to Dodman Point, quarter of west by south.

From Dodman Point to Start Point, west-southwest.

1. Unidentified location.

Da Godestar a Porlan, ponente garbin.

Da Porlan al'ixola Doich, ponente garbin. Atraverso dela zittà vechia, quarta de ponente al garbin, in passa 12.

Zittà vechia e Belzef, ponente garbin.

Da Belzef in Romaneo, ponente garbin.

Da Romaneo a Dobla, garbin in puocho al ponente.

Da Sancta Malgaritta a Tenette, ostro.

In la lenga spagnola: ponente oest, maistro noruest, tramontana nort, griego noderest, levante este, sirocho suest, ostro sol, garbin suduest.

---

f. 125b

[Intrar in Sant'Ander, in l'Eschiozes, in Sentuzi]

Per saver intrar in Sentuzi sapi quando la luna ostro e tramontana, piena mar. E quando vui ser[*e*]tti per andar dentro, vuy vederì intro ponente e maistro in tera un boscho. E metitte andar ala via de quel boscho tuta volta alargo dela tera, da man sinestra al'intrar ballestrade 2 e più. E metite a mente ala tera de Sentuzi vedirì chanpagnelli 3, e vui andarì per la via de quel boscho inchina el chanpagnel pizollo. Vui deschoverzerì che serà el quarto. Romagna quel pizolo chon lo terzo, puo' va segundando e metti la proda in un boscho, in una spiazia, apreso in un mullin inchina el chanpagnel primo da garbin. Tu metti in una punta biancha, zoè del'ixulla dever man destra, e vande arditamente con raxion d'aqua.

In Fiandria

Per intrar al porto del'Eschioza, zoè de Fiandria, vuy andarì de chontinuo quarta de ponente al maistro inchina tu metti el chanpagnel de Sancta Chatarina chon un chanpagnel muzo. Tu lo vedera' dela parte dever la tera, e puo' va dentro ponente e maistro inchina el chanpagnel de Sancta Chatarina, chon un chanpagnel che pare inver la tera se sera e chomo quel chanpagnel. E se fuxe da quello pizola chosa, vuolteve ala banda destra. Vedirì 1ª ture dela banda de llà, che serà serada per mezo la ponta negra, et in quelo chognosì eser fuori d'i banchi e truoverì aqua passa 6. E per questo, chontrario farai quando tu volesi intrar per schapolar li banchi.

From Start Point to Portland, west-southwest.

From Portland to the Isle of Wight, west-southwest. Abreast of Chichester, quarter of west by south in 12 paces.

Chichester and Beachy Head, west-southwest.

From Beachy Head to Dungeness, west-southwest.

From Dungeness to Dover, southwest, a little to the west.

From Saint Margaret's at Cliffe to the Island of Thanet, south.

In the Spanish language: west—oest; northwest—noruest; north—nort; northeast—noderest;[1] east—este; southeast—suest; south—sol; southwest—suduest.

To enter in Santander, in Sluys, in Sandwich

f. 125b

To know how to enter into Sandwich, know when the moon is north–south, high tide. And when you are ready to enter, you will see a forest on the land between the west and the northwest. And steer for this forest keeping off the land to the left, at 2 or more arrows' distance. And bear in mind that in the land of Sandwich you will see three bell towers, and you will steer for that forest as far as the small bell tower. You will discover that there is a fourth. Keep this small one lined up with the third, then continue and steer for a forest, on a beach, then toward a mill, as far as the first bell tower on the southwest. Steer on a white point, that is, of the island on the right side, and go all out because of the current.

In Flanders

To enter the port of Sluys, that is, of Flanders, you will continue on the quarter of west by north until you line up the bell tower of Saint Catherine with a short bell tower. You will see it from the landward side, and then enter north-northwest until the bell tower of Saint Catherine lines up with a bell tower that appears toward the land and lines up with that bell tower.[2] And if it is a short distance from that one, turn to the right. You will see a tower on that side, which will be hidden by the dark promontory, and at that point you will know that you are outside the banks and you will find water 6 paces deep. And in this way, you will do the opposite when you want to enter to avoid the banks.

1. Probably an error for "nordest."
2. Interpretation uncertain.

In Sancto Ander

Per voler intrar in Sancto Ander, chi per furtuna schorese in lo golfo de Bertagna, avanti che tu arivy in Sancto Ander, se vegnise da ponente vedirì per la staria senpre montagne alte, zoè chiamati del'Asturia. E quando vui seritti in chavo de quelli, vedirì algune montagne chon algune valle. E queste montagne son basse e pare chomo alguny schogi lo teran biancho. E vuy vedirì sovra un
f. 126a pozzetto una badia chon 3 ture. Sapi che tu sera' alargo dal porto Sancto | Ander mia 15. El primo chavo vedirì da puo' la badia inver lo levante, serà el luogo, averà sovra el chavo 1ª iexia derupada chon una ture per tenpo passado fu un fanò. E se tu non ti fidi, averzi tanto el chavo che tu deschoverzi inver lo porto una ixoletta negra e apresso de quella, inver lo griego, 1 schoietto pizolo, e vande ardittamente intro lo schoio el chavo, e chomo volty lo chavo vedirì la tera. Non t'achostar da man senestra, che sono de malli banchi per luongo, ma pur achostatti ala banda destra, senpre averai bon fundi. E sapi che per tuto è sorzedor del chavo del fanò inchina dentro, e quando la luna in griego e garbin, piena mar in questo porto. E questa sie maria de Spagna inchina Bagiona de Gaschogna, sirocho maistro, bassa mar.

[Veder le sunde di chanalli de Fiandria]

Queste sono le sonde d'i chanalli de Fiandria. Chi se parte da Zizercha a vegnir a Usente, intro griego e tramontana. E vegnando per quista via e gettasi el schandagio e trovasti passa 100 o passa 90, e truovasi sunda grossa. Chonsiderando l'albitrio del to navegar, apresso de questo seray inver Sain, de fuora in mar lege 15. Avera' da tti Usente lege 20. E per questa via truovasti fango o segnal nessuno. Serai in lo chanal de Rexa o inchontro d'esso.

E per questa via truovassi pasa 80 e chavasti chorvaziolos chon arestas e chaschas pizulas chobo[a] de sancto Iachobo. La tto via serà per tramontana inchina tu chanbi sunda, perché Usente te starà quarta de tramontana al griego perché tu ssi vegnodo basso.

E se tu truovera' passa 75 o 70 e chavasti sabla menuda meschiada chon vermegio, seras al chostado de Usente. E se tu volessi truovarlo, te starà per griego. Tu serà alargo de quello lege 10.

E se tu trovaras 70 o 65 passa e chaveras sabla menuda chon arestas a muodo di filli, tu seras dentro d'Osente. Ben lo poras veder se 'l serà claro, e romante al sirocho.

f. 126b E se tu truoveras 65 o 60 passa e chavaras sablon menudo chon chaschas blanchas de ruia pintada chomo chon punta d'ago, seras in la chosta de Lion per mezo Barbaracha.

---

a. *Sic in MS, probably for* chomo.

In Santander

To enter Santander, which happens to be in the Bay of Biscay, before you arrive in Santander, if you are coming from the west, you will always see on the coast high mountains, that is, those called Asturias. And when you are at the end of these, you will see some mountains with a few valleys, and these mountains are low and the land appears white like some cliffs. And you will see over a small pond an abbey with three towers. Know that you will be | 15 miles from the port of f. 126a
Santander. The first cape you see after the abbey to the east will be the place; it will have over the cape a ruined church with a tower that was once a lighthouse. And if you don't trust this, come into view of the cape until you discover near the port a small black island and near that, toward the northeast, a small reef, and go all out between the reef and the cape, and as you round the cape you will see the land. Don't approach from the left, which has bad banks along it, but approach on the right side, and you will always have good depth. And know that there is anchorage everywhere from the cape of the lighthouse until inside, and when the moon is in the northeast and southwest, there is high tide in this port. And these are the tides of Spain as far as Bayonne in Gascony: southeast–northwest, low tide.

See the soundings of the channels of Flanders

These are the soundings of the channels of Flanders. Here leave Sisargas to come to Ouessant, between northeast and north. And coming this way heave the sounding lead, and you will find 100 paces or 90 paces, and you will find a deep sounding. Considering the evidence of your navigating, according to this you will be near Île de Sein, 15 leagues offshore. You will have Ouessant at 20 leagues from you. And on this course you will find no lighthouse or landmark. You will be in the channel of Larexa or opposite it.

And on this course you will find 80 paces [of depth] and you will bring up small seahorses[1] with small filaments and helmets like St. James. Your course will be northerly until you change soundings, because Ouessant will be a quarter of north by east from you because you have come in low.

And if you find 75 or 70 paces and draw up fine sand mixed with vermilion, you will be at the coast of Ouessant. And if you wish to find it, it will be northeast of you. You will be 10 leagues off it.

And if you find 70 or 65 paces and draw up fine sand with filaments[2] like threads, you will be within Ouessant. You can see it well if it is clear, and remain to the southeast.

And if you find your depth to be 65 or 60 paces and draw up fine sand with white shells of grubs f. 126b
painted like the head of a needle, you will be on the Lion coast at Aber-Wrach.

1. Interpretation uncertain.
2. Interpretation uncertain.

E vegnando per questa via per griego, in passa 60, chaveras segnal de piera chomo fusse fava negra, tu serà inver l'ixola da Bas.

E si seras inchontro Rochatua e truoveras passa 40, chaveras sablon grosso e vermeyo chomo roxa. Mestier te fa andar per tramontana per schapular el chavo dela Grua.

E si seras per mezo de Ranut truoveras molto gran fundi e demuora Chaschette per ostro. La tto via serà intro griego levante, montante e zozente, per aver vista del chavo Belzep.

E sapi che se tu fosti al bancho de Sorlenga e trovasi 80 o 90 passe, over 85, e truoveras al sevo del piombo negro chomo cholor de fanguo e non chaveras segnal neguno, anderà gran chamino e non te mancherà el fundi de passa sovrascritti, sapi che tu sera' inver Sorlinga.

E se trovaras passa 80 e truovassi sablun menudo negro e biancho segondo sta prima sonda, tu serà inver Ingletera. E si anderas intro griego e tramontana, non te mancharà fundi inchina tu fazi gran chamino, perché tutto este plano. E se [*te*] mancha in ore 2 o 3 el fondi passa 5 o 6 *** inver la Bertagna, e se non mancharà tanto in gran *** tu seras inver l'ixola d'Engletera.[a]

E se ale 70 o 65 passa chavaras sablon menudo blancho e unos granos chomo myo vermegios e arestas blanchas delgados, sapi che seras inchontro Luxert. E tu lo die veder se 'l serà claro, o el monte de San Michiel. Sta sunda segue inchina Godman, e chaveras in questa paraze unas choziolas blanchas chomo chape de vermegun de grandura de meza chonza de pixeo.

f. 127a E se ttu fosti sovra o inchontra Godester, truoveras 1ª sabla grosa | e con feridas. Esta sunda segue inchina sacha de Porlan in la rutta ponente e levante.

E si seras inchontro Porlan a 35 o 30 passa, truveras pedras chomo de fava e feride d'oltre pedre, andar per questa via ponente e levante. Duratte esta sonda inchina Sancta Telma.

E si truoveras 30 o 25 passa e truoveras unas pederas blanchas in figura de nuxielle chadre e oltre più grande, seras in paraze de Sancta Telma inchina l'ixola Doich.

E alì truoveras feridas de pedras, tue dalla tto via este per levante per vardarte de zittà vechia. E se tu truovasti in questa via passa 20, truoveras feridas de rochas e molte tagiade al sevo chomo de filos delgados, seras inchontro la zittà vechia, e sta sunda sege inchina Sete Falage. È questo fenydo al vostro honor.

---

a. *Partially illegible because of repeated abrasions of the MS.*

1. The version in the *Raxion de' marineri* reads "sendente"; Pietro di Versi, *Raxion de' marineri*, p. 80.

2. This line of the MS has many words rubbed out; the corresponding entry in the version in the *Raxion de' marineri* ends with "Brittany," omitting the last line; Pietro di Versi, *Raxion de' marineri*, p. 80.

3. This is the interpretation of Henry Romanos Kahane, Renée Kahane, and Lucille Bremner, *Glossario degli antichi portolani italiani*, trans. Manlio Cortelazzo, Quaderni dell'Archivio Linguistico Veneto 4 (Florence: Olschki, 1967), p. 54, for the corresponding entry in the *Raxion de' marineri*; it might also refer to bits of iron.

And coming on this northeast course, in 60 paces, if you draw up a sample of stone like a black bean, you will be near Île de Batz.

And if you are opposite Jersey and you find 40 paces, you will bring up coarse and reddish vermilion sand. Your job is to go north to avoid Cap de Graux.

And if you are on the way to Alderney you will find very great depth and hold for the Casquet Islands to the south. Your course will be between northeast and east, flood and ebb,[1] to have a view of the cape of Beachy Head.

And know that if you are on the shore of the Isles of Scilly and find 80 or 90 paces, or 85, and find in the tallow of the sounding lead black like the color of mud and bring up no sample, you will be taking the great route and you will not lose the soundings described above; know that you will be near the Isles of Scilly.

And if you find 80 paces and find fine black and white sand on this first sounding, you will be near England. And if you head between northeast and north, you will not lose soundings until you get on the great route, everything is flat. And if within 2 to 3 hours the sounding loses 5 or 6 paces you will be near Brittany, and if that much is not lacking in great ***[2] you will be near the island of England.

And if in 70 or 65 paces you bring up fine white sand and some grains like better vermilion and thin white filaments, know that you are opposite Lizard Point. And if it is clear, you should be able to see it or Mont-Saint-Michel. These soundings continue to Dodman Point, and you will draw up in this place some white things like vermilion clams of the size of half a pea shell.

And if you are off Start Point or near it, you will find a coarse sand | with fragments.[3] These soundings continue until the bay of Portland Bill on the course west and east. f. 127a

And if you are opposite Portland Bill in 35 or 30 paces, you will find stones like fava beans and fragments of other stones; follow this course west and east. These soundings continue to St. Albans Head.

And if you find 30 to 25 paces and you find some white stones with the appearance of square nuts and others larger, you will be in the region of Saint Albans Head near the Isle of Wight.

And if you find fragments of stones, bear east from your course to avoid Chichester. And if you find 20 paces on this route, you will find fragments of stones and many cuts in the tallow[4] like thin threads, you will be opposite Chichester, and these soundings continue to Seven Cliffs.[5] And this is all finished in your honor.

---

4. Of the sounding lead.
5. Unidentified location.

[Taiar velle latina da pasa 5 inchina 22]

Questo sie l'amaistramento de tayar vella de passa 5 inchina vella de passa 22 e tute le sso raxion chontien qui de sotto. Ponamo che tu volesti tayar una vella de passa 20, tu vuol saver quanto fustagno andarà in la ditta vella. E voyando far la ditta raxion, tu die dir in prima quanti ferse die aver vella de passa 20. Adoncha 20 e 20 fa 40, e la mitade de 20, 10, fa 50. El quinto de 20, 4, fa 54. Adoncha la ditta vella averà fersse 54. E per saver quante braza de fustagno, tu dirà: "L'antenal è passa 20." La mitade de 20 sono 10. Moltipicha 10 via 54 ferse, fano 540 passa de fustagno vuorà questa vella, i quali serano braza 1350 a partir per 25 e far peze de fustagno sono peze 54.

f. 127b E sì vorave per ogni peza de fustagno, che son braza 25, braza 10 de chanevaza. E questa chanevaza va per binde braza $6\frac{1}{4}$ e in pozal braza $3\frac{3}{4}$. E metera' a mente che la binda del batal sia churta per ogni passa 5 pie 1 de zò che lo filo tuto fusse longuo.

E chi volese far la so serzena per raxion vella de passa 20, 10 via 20 fa 200. Abatti la longeza del'antenal, roman 180. E tante libre die pesar la sso zerzena de passa 20.

E tutte velle, sì grande chomo pizolle, da tuti li tay dele velle tu die insir fuor a tuto pano e da gratil senpre a terzo pano. E se volesti inchomenzar a taiar la tto vella de passa 20, tu die tuor un baston dela largeza del tuo fostagno e partir le 2 parte del baston, zoè $\frac{2}{3}$, e questi $\frac{2}{3}$ partira' in parte 18, e questi serano ponty dela to vella. E chomenze a tayar le do prime fersse da ventame e non li dar punto nessuno. Poi a do a do ferse va tayando, e dalli 1 ponto. Insirà fuora d'i tto ponti in ferse 38. Poi andera' tayando a tuto pano inchina al stello, e senpre da gratil terzo pano. Arichordatti a tayar la prima ferse $\frac{3}{4}$ de tuto la fosse longa, zoè 20. El quarto serà 5. Adoncha serà lo[*n*]go el fil passa 15, e 5 passa se meterà chanevazo. E va tayando a schayony mezo pie per fersa inchina insirà chavo de grando. E per far questa vella tunda vuol tanto de chazuda chomo ad antenal.

E metti mente al'armar de questa nostra vella de sovra a posar per setimo e da basso in gratil per quinto. Zò è a dir che xe el settimo de 20, serano $2\frac{6}{7}$. Vuol adoncha romagnir in passa $17\frac{1}{7}$. Ma la riegola non seria ben, perché abiando tayada la ditta serà più in antenal. Adoncha se intende che 'l settimo de quello serà, e chosì el quinto in gratil el se truoverà. E quante e quante[a] passa è la to vela, tante ferse vuol de chanevazo da basso, zoè smenoir senpre mezo pie a chadauna ferssa serà pie 1.

---

a. *Thus repeated in MS.*

1. A fabric combining linen with cotton or wool.
2. A measure of length, here literally an arm's length.
3. Vertical reinforcing strips of canvas laid over a seam between sail cloths.
4. The lower, after parts of a lateen sail.
5. Uncertain part of a lateen sail.
6. The after edge of the sail.
7. I.e., at the peak of the sail.

To cut lateen sails from 5 paces up to 22

These are the instructions for cutting sails of from 5 paces to 22 paces, and all of the technique is contained below. Let us suppose that you wanted to make a sail of 20 paces; you want to know how much fustian[1] will go into that sail. And wishing to make the calculation, you should first say how many cloths a sail of 20 paces ought to have. Then 20 and 20 make 40, plus half of 20, 10, makes 50. A fifth of 20, 4, makes 54. So this sail will have 54 cloths. And to know how many ells[2] of fustian, you will say: "The head of the sail is 20 paces." Half of 20 is 10. Multiply 10 by 54 cloths, which makes 540 paces of fustian required for this sail, which will be 1,350 ells. Divide by 25, and it makes 54 pieces of fustian.

And also there was needed for each piece of fustian, which is 25 ells, 10 ells of canvas. And $6\frac{1}{4}$ ells of f. 127b
this canvas goes for *binde*[3] and $3\frac{3}{4}$ ells in the clews.[4] And bear in mind that the *binda* of the *batal*[5] is shorter by 1 foot for every 5 paces of length of the leech.[6]

And here it would be necessary to make the boltrope for this sail of 20 paces, 10 times 20, makes 200. Subtract the length of the head of the sail, there remains 180. And your boltrope of 20 paces should weigh that many pounds.

And for all the sails, large and small, for all cuts of sails you should start[7] at a full-width cloth and make the foot of the sail from a third-width cloth. And if you wanted to begin to cut your sail of 20 paces, you should take a measuring stick as long as your fustian is wide and divide the 2 parts of the measuring stick, that is, $\frac{2}{3}$, and this $\frac{2}{3}$ you will divide in 18 segments, and each of these will be the points of your sail. And begin to cut the first two sail cloths for the *ventame*[8] and don't give it any points. Then continue cutting sail cloths two by two, and give them one point. These points will be used up in 38 sail cloths. Then you will continue cutting from full-width cloth until the *stello*[9] and always third-width cloth along the foot. Remember to cut the first sail cloth $\frac{3}{4}$ of its length, that is, 20. The fourth [part] will be 5. So the leech will be 15 paces long, and the canvas will be put on 5 paces [of it]. And proceed cutting by measuring scales one half foot per sail cloth until you've finished the big end. And to make this sail round, it will need a drop equal to the length of the head of the sail.

And remember to rig this sail from the top to the clew by a seventh and at the bottom along the foot by a fifth. That is to say that it's the seventh of 20, which will be $2\frac{6}{7}$. So there should remain $17\frac{1}{7}$ paces. But the calculation will not work well, because having cut it this way, it will be too long in the head. So figure out what the seventh would be, and how the fifth would work out, and as many paces as is your sail [at the head], it needs that many sail cloths at the bottom, that is, to decrease it always a half foot for each sail cloth, which will be one foot.

8. The upper, after, lighter spar of a lateen yard.
9. The lower, forward, heavier spar of a lateen yard.

f. 128a Arichortadi[a] da far li tto pessetti sì da baso chomo da erto e le ferse de chane|vazo da basso sia ogni terza dopia. Arichordatti da meter mantilletti da erto e da baso per armar. Arichordatti de meter el meollo de stupazo per armar la to vella de passa 20.

Vella de passa 5 vuol fostagno peze $3\frac{1}{2}$ e vuol de chanevazo braza 35, e die aver fersse 14. Peserà la so zerzena libre $6\frac{1}{2}$.

Vella de passa 6 vuol de fustagno peze $4\frac{2}{3}$ e de chanevazo braza 47. E chomenza a tayar le do prime ferse a tuto pano men 3 ponti del 20. Insirà fuor di punti in ferse 8. A[*n*]darà fuora a tuto drapo e da gratil terzo pano. Averà ferse 16. Pexerà la zerzena libre 12.

Vella de passa 7 vuol de fustagno peze 6 e passa 3, e vuol de chanevazo braza 63. E vuol chomenzar le do prime ferse a tuto pano men 5 ponti del 20. Le do prime ferse non vuol punto nessuno. Insirà fuor di punti in ferse 10. Serà ferse 18. Pexarà la zerzena libre 17.

Vella de passa 8 vuol de fustagno peze 8 passa 4 e de chanavazo braza 84. E vuol chomenzar le do prime ferse tuto pano men 6 punti del 20. Insirà in ferse 12. Die aver ferse 21. Die pexar la so zerzena libre 24.

Vella de passa 9 vuol de fustagno peze 10 passa 8, e vuol de chanevazo braza 108. E die chomenzar le do prime ferse da fillo tutto pano men 7 punti del 20, e non li dar punto nesono. Averà punti 6. Insirà fuora in ferse 14. Averà ferse 24. Pexerà la zerzena libre $30\frac{1}{2}$.

Vella de passa 10 vuol de fustagno peze 13 passa 5, e vuol de chanevazo braza 135. E chomenza a taiar le do prime ferse e non li dar punto a tuto pano men 8 del 20. Serà punti 7. Inssirà di punti in ferse 16. Averà ferse 27. Pexerà la so zerzena libre 40.

f. 128b Vella de passa 11 vuol de fustagno peze 15 passa $9\frac{1}{2}$, e vuol de ch[*a*]nevazo braza $159\frac{1}{2}$. E vuol chomenzar a tayar tutto pano men 9 punti del 20. Averà 'sta vella punti 8. Insirà di ponti in ferse 18. Averà ferse 29. Pexerà la zerzena libre $48\frac{1}{2}$.

Vella de passa 12 vuol de fustagno peze 19 passa 2, e vuol de chanevazo braza 192. E vuol tayar le prime 2 fersse tutto pano men 10 del 20. Averà 'sta vella punti 9. Insirà di punty in ferse 20. Averà 'sta vella ferse 32. Pexerà la zerzena libre 60.

---

a. *Thus in MS, evidently for* Arichordati.

Remember to make your broadseams this way at the bottom and at the top, and the canvas sail cloths | below are all *terza dopia*. Remember to put *mantiletti*[1] on at the top and the bottom to rig it. Remember to use the stops to rig your sail of 20 paces. f. 128a

A sail of 5 paces requires $3\frac{1}{3}$ pieces of fustian and needs 35 ells of canvas and should have 14 sail cloths. Its boltrope will weigh $6\frac{1}{2}$ pounds.

A sail of 6 paces requires $4\frac{2}{3}$ pieces of fustian and 47 ells of canvas. And begin to cut the two first sail cloths of full cloth minus 3 points of the 20. It will leave the points in 8 sail cloths. It will finish at a full-width cloth, and the foot at third-width cloth. It will have 16 sail cloths. Its boltrope will weigh 12 pounds.

A sail of 7 paces requires 6 pieces of fustian and 3 paces, and needs 63 ells of canvas. And you should begin the first two sail cloths with full cloth minus 5 points of the 20. The first two sail cloths don't need any points. It will leave the points in 10 sail cloths. There will be 18 sail cloths. The boltrope will weigh 17 pounds.

A sail of 8 paces requires 8 pieces of fustian, 4 paces, and 85 ells of canvas. And the first two sail cloths should begin at a full-width cloth minus 6 points of the 20. It will leave in 12 sail cloths. There should be 21 sail cloths. Its boltrope should weigh 24 pounds.

A sail of 9 paces should have 10 pieces of fustian, 8 paces, and needs 108 ells of canvas. And the two first sail cloths of the leech should begin at a full-width cloth minus 7 points of the 20, and don't give it any points. There will be 6 points. It will leave in 14 sail cloths. There will be 24 sail cloths. Its boltrope will weigh $30\frac{1}{2}$ pounds.

A sail of 10 paces requires 13 pieces of fustian, 5 paces, and needs 135 ells of canvas. And begin to cut the first two sail cloths, and don't give them a point, at full-width cloth minus 8 of the 20. There will be 7 points. It will leave the points in 16 sail cloths. There will be 27 sail cloths. Its boltrope will weigh 40 pounds.

A sail of 11 paces needs 15 pieces of fustian, $9\frac{1}{2}$ paces, and needs $159\frac{1}{2}$ ells of canvas. And you will need to to begin by cutting at a full-width cloth minus 9 points of the 20. This sail will have 8 points. It will leave its points in 18 sail cloths. There will be 29 sail cloths. The boltrope will weigh $48\frac{1}{2}$ pounds. f. 128b

A sail of 12 paces requires 19 pieces of fustian, 2 paces, and needs 192 ells of canvas, and you must cut the first 2 sail cloths at a whole cloth minus 10 of 20. This sail will have 9 points. It will leave the points in 20 sail cloths. This sail will truly have 32 sail cloths. The boltrope will weigh 60 pounds.

1. Probably small ties or stops.

Vella de passa 13 vuol de fustagno peze 22 passa $7\frac{1}{2}$, e vuol de chanevaza braza 227[a] $\frac{1}{2}$. Taya le 2 prime ferse da filo men 11 del 20. Averà 'sta vella punti 10. Insirà di punti in ferse 22. Averà 'sta vella fer[*se*] 35. Pexerà la serzena libre $70\frac{1}{2}$.

Vella de passa 14 vuol de fostagno peze 26 passa 6, e vuol de chanevaza braza 266. E vuol tayar le 2 prime ferse a tuto pano men 12 del 20. Farà 'sta vella punti 11. Insirà fuor di punti in ferse 24. Averà 'sta vela ferse 38. Die pexar la serzena libre 84. Insirà d'antenal tutto pano e da gratil al terzo.

Vella de pasa 15 vuol de fustagno peze 30, e vuol de chanevazo braza 300. E vuol tayar a tutto pano men 13 del 20. Averà 'sta vella punti 12. Insirà di punti in fersse 26. Averà 'sta vella ferse 40. Pexerà la serzena libre $97\frac{1}{2}$.

Vella de passa 16 vuol de fustagno peze 34 passa 4, e vuol de chanavazo braza 344. E chomenza a tayar le prime 2 ferse a tutto pano me[*n*] 14 del 20. Insirà di punti in ferse 30. Averà 'sta vella ferse 43. Pexerà la zerzena libre 112.

Vella de passa 17 vuol de fustagno peze 39, e vuol passo 1, e vuol de chanevazo braza 391. Taya le do prime ferse a tuto pano men 15 del 20. Averà 'sta vella punti 15.[b] Insirà di punti in ferse 32. Averà 'sta vella ferse ferse[c] 46. Die pexa[*r*] la so serzena libre $126\frac{1}{2}$.

f. 129a Vella de pasa 18 vuol de fostagno peze 43 passa 2, e vuol de chanevazo braza 432. E vuol tayar le prime 2 ferse a tuto pano men 16 del 20. Averà 'sta vella ponti 16. Insirà di punti in ferse 34. Averà 'sta vella ferse 48. Pexerà la zerzena libre 144.

Vella de passa 19 vuol de fostagno peze 48 passa $4\frac{1}{2}$, e vuol de chanevaza braza 484. E vuol chomenzar a tayar a tutto pano men 17 del 20. Averà 'sta vella punti 17. Insirà di punti in ferse 36. Averà 'sta vella ferse 51. Pexerà la sso serzena libre $180\frac{1}{2}$.

Vella de passa 20 vuol de fustagno peze 54, e vuol de chanevaza braza 540. E vuol chomenzar a tayar a ttutto pano men 18 del 20. Averà 'sta vella punti 18. Insirà di punti in ferse 38. Averà 'sta vella fersse 54. Pexerà la zerzena libre 180.

Vella de pasa 21 vuol de fustagno peze 59 passa $8\frac{1}{2}$, e vuol de chanevazo braza 598 e $\frac{1}{2}$. E vuol chomenzar a tayar le do prime ferse da filo tuto pano men 19 del 20. Averà 'sta vella punti 19. Insirà di punti in ferse 40. Averà 'sta vella ferse 57. Pexerà la zerzena libre $198\frac{1}{2}$.

Vella de passa 22 vuol de fustagno peze 64 e passa 9, e vuol de chanevaza braza 649. E vuol tayar le 2 prime ferse da fillo a ttutto pano men 20 del 20, zoè che se die dir a terzo pano, perché serà

---

a. 227 *corrected over* 222.

b. 15 *corrected over* 14.

c. *Thus repeated in MS.*

A sail of 13 paces needs 22 pieces of fustian, 7½ paces, and needs 227½ ells of canvas. Cut the 2 first sail cloths of the leech, minus 11 from the 20. This sail will have 10 points. It will leave the points in 22 sail cloths. This sail will have 35 sail cloths. The boltrope will weigh 70½ pounds.

A sail of 14 paces needs 26 pieces of fustian, 6 paces, and needs 266 ells of canvas. And you should cut the first 2 sail cloths from full-width cloth minus 12 of the 20. This sail will make 11 points. It will leave the points in 24 sail cloths. This sail will have 38 sail cloths. The boltrope should weigh 84 pounds. It will leave the head of the sail from full-width cloth and the foot at third-width.

A sail of 15 paces needs 30 pieces of fustian, and needs 300 ells of canvas. And you need to cut at a full-width cloth minus 13 of the 20. This sail will have 12 points. It will leave the points in 26 sail cloths. This sail will have 40 sail cloths. The boltrope will weigh 97½ pounds.

A sail of 16 paces needs 34 pieces of fustian, 4 paces, and needs 344 ells of canvas. And begin to cut the first 2 sail cloths from full-width cloth minus 14 of the 20. It will leave the points in 30 sail cloths. This sail will have 43 sail cloths. The boltrope will weigh 112 pounds.

A sail of 17 paces requires 39 pieces of fustian, and needs 1 pace, and needs 391 ells of canvas. Cut the first two sail cloths of full-width cloth minus 15 of the 20. This sail will have 15 points. It will leave the points in 32 sail cloths. This sail will have 46 sail cloths. Its boltrope should weigh 126½ pounds.

A sail of 18 paces needs 43 pieces of fustian, 2 paces, and needs 432 ells of canvas. And you need to cut the first 2 sail cloths from full-width cloth minus 16 of the 20. This sail will have 16 points. It will leave the points in 34 strips. This sail will have 48 sail cloths. The boltrope will weigh 144 pounds. f. 129a

A sail of 19 paces requires 48 pieces of fustian, 4½ paces, and needs 484 ells of canvas. And you need to begin to cut from full-width cloth minus 17 of the 20. This sail will have 17 points. It will leave the points in 36 sail cloths. This sail will have 51 sail cloths. Its boltrope will weigh 180½ pounds.[1]

A sail of 20 paces requires 54 pieces of fustian, and needs 540 ells of canvas. And you need to begin to cut from full-width cloth minus 18 of the 20. This sail will have 18 points. It will leave the points in 38 sail cloths. This sail will have 54 sail cloths. Its boltrope will weigh 180 pounds.

A sail of 21 paces requires 59 pieces of fustian, 8½ paces, and it needs 598½ ells of canvas. And you need to begin to cut the first two sail cloths of the leech of full-width cloth minus 19 of the 20. This sail will have 19 points. It will leave the points in 40 sail cloths. This sail will have 57 sail cloths. The boltrope will weigh 198½ pounds.

A sail of 22 paces requires 64 pieces of fustian and 9 paces, and it needs 649 ells of canvas. And you need to cut the first 2 sail cloths of the leech of full-width cloth minus 20 of the 20, that is, you say a third of the cloth, because it will be the ⅔ part of the measuring stick. Then there will remain the

1. This number is probably an error.

inssido di $\frac{2}{3}$ del bastun. Adoncha romagnerà el terzo del brazolar. E vuol la ditta vella punti 20. Insirà fuor di punti in ferse 42. Ave[*rà*] 'sta vella ferse 59. Die pexar la serzena libre 220.

---

f. 129b

[VEDER UNA TAULA DE PASQUA]

Qui de sotto faremo 1ª thola de Pasqua, de chaxelle 100 de Pasqua. E sì serà nottado in chadauna el millieximo e la Pasqua de sovra a quanti. E se 'l serà marzo nottarò M, e se 'l serà avril notarò a. E per ladi destro notarò la pata dela man. E de sutto del mileximo, in la ladi senestro, notarò Pasqua braiccha. Chomo vedirì qui de ssotto per figura.

a 5 3 1401 1	M 6 26 1402 21	a 7 15 1403 9	M 2 30 1404 29	a 3 19 1405 17	a 4 11 1406 5	M 5 27 1407 25	a 7 15 1408 13	a 1 7 1409 2	M 2 23 1410 22
a 3 12 1411 10	a 5 3 1412 30	a 6 23 1413 18	a 7 8 1414 7	M 1 31 1415 27	a 3 19 1416 15	a 4 11 1417 4	M 5 27 1418 24	a 6 16 1419 12	a 1 7 1420 1
M 2 23 1421 21	a 3 12 1422 9	a 4 4 1423 29	a 6 23 1424 17	a 7 8 1425 5	M 1 31 1426 25	a 2 20 1427 13	a 4 4 1428 2	M 5 27 1429 22	a 6 16 1430 10
a 7 1 1431 30	a 2 20 1432 18	a 3 12 1433 7	M 4 28 1434 27	a 5 17 1435 15	a 7 8 1436 4	M 1 31 1437 24	a 2 13 1438 12	a 3 5 1439 1	M 5 27 1440 21
4[a] 6 16 1441 9	a 7 1 1442 29	a 1 21 1443 17	a 3 12 1444 5	M 4 28 1445 25	a 5 17 1446 13	a 6 9 1447 2	M 1 24 1448 22	a 2 13 1449 10	a 3 5 1450 30
a 4 25 1451 18	a 6 9 1452 7	a 7 1 1453 27	a 1 21 1454 15	a 2 6 1455 4	M 4 28 1456 24	a 5 17 1457 12	a 6 2 1458 1	M 7 25 1459 21	a 2 13 1460 9
a 3 5 1461 29	a 4 18 1462 17	a 5 10 1463 5	a 7 1 1464 25	a 1 14 1465 13	a 2 6 1466 2	M 3 29 1467 22	a 5 17 1468 10	a 6 2 1469 30	a 7 22 1470 18

---

a. *Sic in MS, evidently in place of* a.

third of the measuring stick. And this sail needs 20 segments. It will leave the points in 42 sail cloths. This sail will have 59 sail cloths. The boltrope should weigh 220 pounds.

SEE AN EASTER TABLE

f. 129b

Below here, we will make an Easter table of 100 cells of Easter.[1] And in each of the cells the year will be noted and the Easter above. And if it is March I will note M, and if it is April I will note A, and on the right I will note the epact of the hand. And below the year, on the left side, I will note Hebrew Easter. You will see how below in the figure.

A 5 3 1401 1	M 6 26 1402 21	A 7 15 1403 9	M 2 30 1404 29	A 3 19 1405 17	A 4 11 1406 5	M 5 27 1407 25	A 7 15 1408 13	A 1 7 1409 2	M 2 23 1410 22
A 3 12 1411 10	A 5 3 1412 30	A 6 23 1413 18	A 7 8 1414 7	M 1 31 1415 27	A 3 19 1416 15	A 4 11 1417 4	M 5 27 1418 24	A 6 16 1419 12	A 1 7 1420 1
M 2 23 1421 21	A 3 12 1422 9	A 4 4 1423 29	A 6 23 1424 17	A 7 8 1425 5	M 1 31 1426 25	A 2 20 1427 13	A 4 4 1428 2	M 5 27 1429 22	A 6 16 1430 10
A 7 1 1431 30	A 2 20 1432 18	A 3 12 1433 7	M 4 28 1434 27	A 5 17 1435 15	A 7 8 1436 4	M 1 31 1437 24	A 2 13 1438 12	A 3 5 1439 1	M 5 27 1440 21
A[2] 6 16 1441 9	A 7 1 1442 29	A 1 21 1443 17	A 3 12 1444 5	M 4 28 1445 25	A 5 17 1446 13	A 6 9 1447 2	M 1 24 1448 22	A 2 13 1449 10	A 3 5 1450 30
A 4 25 1451 18	A 6 9 1452 7	A 7 1 1453 27	A 1 21 1454 15	A 2 6 1455 4	M 4 28 1456 24	A 5 17 1457 12	A 6 2 1458 1	M 7 25 1459 21	A 2 13 1460 9
A 3 5 1461 29	A 4 18 1462 17	A 5 10 1463 5	A 7 1 1464 25	A 1 14 1465 13	A 2 6 1466 2	M 3 29 1467 22	A 5 17 1468 10	A 6 2 1469 30	A 7 22 1470 18

1. For the paschal table, see Wallis, "Michael of Rhodes and Time Reckoning," vol. 3, pp. 297–302.
2. *4* in MS.

a 1	M 3	a 4	a 5	M 6	a 1	a 2	M 3	a 4	a 6
14	29	17	10	26	14	6	22	11	2
1471	1472	1473	1474	1475	1476	1477	1478	1479	1480
7	27	15	4	24	12	1	21	9	29
a 7	a 1	M 2	a 4	a 5	M 6	a 7	a 2	a 3	a 4
22	7	30	18	3	26	15	6	19	11
1481	1482	1483	1484	1485	1486	1487	1488	1489	1490
17	5	25	13	2	22	10	30	18	7
a 5	a 7	a 1	M 2	a 3	a 5[a]	M 6	a 7	M 1	a 3
3	22	7	30	19	3	26	15	31	19
1491	1492	1493	1494	1495	1496	1497	1498	1499	1500
27	15	4	24	12	1	21	9	29	17

f. 130a

[VEDER UNA TUOLA IN CHE SEGNO STA EL SOL]

Quy per inchontro faremo una taula de chaxella 12 per li 12 segny dell'ano, zo[*è*] Aries, Tauro, Zeminy, Chanzer, Leo, Virgo, Libra, Schorpio, Sazitario, Chapichorno, Aquario e Pisis. Li qualli sono segny delli mexi, e la luna zercha questi segni in zorny 30, perché sta in zaschon segno zorny $2\frac{1}{2}$.

E per le simylle muodo el sol zercha questi segny in ano uno, zoè in mexi 12. A mezo marzo inchina mezo avril in Aries. A mezo avril inchina mezo mazio in Tauro. A mezo mazio inchina mezo iunyo in Zeminy. A mezo zugno inchina mezo luyo in Chanzer. A mezo luyo inchina mezo avosto in Leo. A mezo avosto inchina mezo setenbrio in Virgo. A mezo setenbrio inchina mezo octobrio in Libra. A mezo octobrio inchina mezo novenbrio in Schorpio. A mezo novenbrio inchina mezo dezenbrio, Sagitario. A mezo dezenbrio inchina mezo zener in Chapichorno. A mezo zener inchina mezo frever in Aquario. A mezo frever inchina mezo marzo in Pissis.

E sapi che se lo sul se lieva in Aries, la luna fa apreso in quel segno, e ogno dì che avese el sol s'aluntana dala luna una quarta inchina el so diamytro, zoè a mezo a mezo. Zoè se la lieva a levante el so mezo serave in ponente, s'achosta inver lo sol. E chi per inchontro vedirì là che sta la luna.

E se tu volesti saver quando se lieva la luna, questo è llo muodo che se die oservare. Se la luna è prima che 'l fusse de zorny, el'è levada de zorny, e sì die andar a monte de notte. Adoncha la luna 3 dì, quando va a monte de notte? Tu die dir quante ore à quella notte. Tu dirà che sono 12. Adoncha moltipicha 3 via 12 fa 36, parti per 15 e per ogno 15 che te intra serà una ora. Adoncha serano ore 2 e $\frac{6}{15}$. E per lo simille muodo farai de dì. Senpre moltipicha i dì dela luna in le ore dela notte o el dì.

---

a. 5 *written over* 4.

1. Sic in MS; correctly, 18.
2. For the lunar-zodiac table, see Wallis, "Michael of Rhodes and Time Reckoning," vol. 3, pp. 302–305.
3. I.e., is new.

A 1 14 1471 7	M 3 29 1472 27	A 4 17[1] 1473 15	A 5 10 1474 4	M 6 26 1475 24	A 1 14 1476 12	A 2 6 1477 1	M 3 22 1478 21	A 4 11 1479 9	A 6 2 1480 29
A 7 22 1481 17	A 1 7 1482 5	M 2 30 1483 25	A 4 18 1484 13	A 5 3 1485 2	M 6 26 1486 22	A 7 15 1487 10	A 2 6 1488 30	A 3 19 1489 18	A 4 11 1490 7
A 5 3 1491 27	A 7 22 1492 15	A 1 7 1493 4	M 2 30 1494 24	A 3 19 1495 12	A 5 3 1496 1	M 6 26 1497 21	A 7 15 1498 9	M 1 31 1499 29	A 3 19 1500 17

SEE A TABLE SHOWING WHAT SIGN THE SUN IS IN

f. 130a

Opposite here, we will make a table of 12 cells for the 12 signs of the year, that is, Aries, Taurus, Gemini, Cancer, Leo, Virgo, Libra, Scorpio, Sagittarius, Capricorn, Aquarius, and Pisces.[2] Those are the signs of the months, and the moon circles these signs in 30 days, because it stays in each sign $2\frac{1}{2}$ days.

And in a similar way the sun circles these signs in one year, that is, in 12 months. From the middle of March until the middle of April in Aries. From the middle of April until the middle of May in Taurus. From the middle of May until the middle of June in Gemini. From the middle of June until the middle of July in Cancer. From the middle of July until the middle of August in Leo. From the middle of August until the middle of September in Virgo. From the middle of September until the middle of October in Libra. From the middle of October until the middle of November in Scorpio. From the middle of November until the middle of December, Sagittarius. From the middle of December until the middle of January in Capricorn. From the middle of January until the middle of February in Aquarius. From the middle of February until the middle of March in Pisces.

And know that if the sun rises in Aries, the moon makes its appearance[3] in this sign; and every day that there is, the sun distances itself from the moon by one quarter toward its diameter, that is, half and half. That is, if it rises in the east, its half will be in the west, it approaches toward the sun. And here opposite you will see where the moon is.

And if you wish to know when the moon will rise, this is the method you should observe. If the moon is one day old, it rises during the day and sets at nightfall. And when the moon is 3 days old and you want to know how long it will go after nightfall,[4] you must say how many hours the night has. You will say that these are 12. So multiply 3[5] times 12, which makes 36, and divide by 15; for every 15 that you will have, there will be one hour [of moonlight]. So there will be 2 hours and $\frac{6}{15}$. And by a similar method, do this for the daylight. Always multiply the days of the moon by the hours of night or day.

4. I.e., the number of hours it will shine.
5. The age of the moon.

f. 130b

aquario	pisis	ariens	tauro	zeminy	chanzer	leos	virgo	libra	scorpio	sazitario	chapicorno
pisis	aquario	tauro	ariens	chanzer	zeminy	virgo	leos	scorpio	libra	chapicorno	sazitaryo
ariens	tauro	aquaryo	chanzer	ariens	virgo	zeminy	schorpio	leos	chapicorno	libra	pisis
tauro	ariens	chanzer	aquario	virgo	ariens	scorpio	zeminy	chapicorno	leos	pisis	libra
zeminy	chanzer	ariens	virgo	aquaryo	scorpio	ariens	chapicorno	zeminy	pisis	leos	scorpio
chanzer	zeminy	virgo	ariens	scorpio	aquario	chapicorno	ariens	pisis	zeminy	scorpio	leos
leos	virgo	zeminy	scorpio	ariens	chapicorno	aquario	pisis	ariens	scorpio	zeminy	virgo
virgo	leos	scorpio	zeminy	chapicorno	ariens	piisis	aquario	scorpio	ariens	virgo	zeminy
libra	scorpio	leos	chapicorno	zeminy	pisis	ariens	schorpio	aquario	virgo	ariens	chanzer
scorpio	libra	chapicorno	leos	pisis	zeminy	scorpio	ariens	virgo	aquario	chanzer	ariens
sazitaryo	chapicorno	libra	pisis	leos	schorpio	zeminy	virgo	ariens	chanzer	aquario	tauro
chapicorno	sazitario	pisis	libra	scorpio	leos	virgo	zeminy	chanzer	ariens	tauro	aquario

f. 130b

Aquarius	Pisces	Aries	Taurus	Gemini	Cancer	Leo	Virgo	Libra	Scorpio	Sagittarius	Capricorn
Pisces	Aquarius	Taurus	Aries	Cancer	Gemini	Virgo	Leo	Scorpio	Libra	Capricorn	Sagittarius
Aries	Taurus	Aquarius	Cancer	Aries	Virgo	Gemini	Scorpio	Leo	Capricorn	Libra	Pisces
Taurus	Aries	Cancer	Aquarius	Virgo	Aries	Scorpio	Gemini	Capricorn	Leo	Pisces	Libra
Gemini	Cancer	Aries	Virgo	Aquarius	Scorpio	Aries	Capricorn	Gemini	Pisces	Leo	Scorpio
Cancer	Gemini	Virgo	Aries	Scorpio	Aquarius	Capricorn	Aries	Pisces	Gemini	Scorpio	Leo
Leo	Virgo	Gemini	Scorpio	Aries	Capricorn	Aquarius	Pisces	Aries	Scorpio	Gemini	Virgo
Virgo	Leo	Scorpio	Gemini	Capricorn	Aries	Pisces	Aquarius	Scorpio	Aries	Virgo	Gemini
Libra	Scorpio	Leo	Capricorn	Gemini	Pisces	Aries	Scorpio	Aquarius	Virgo	Aries	Cancer
Scorpio	Libra	Capricorn	Leo	Pisces	Gemini	Scorpio	Aries	Virgo	Aquarius	Cancer	Aries
Sagittarius	Capricorn	Libra	Pisces	Leo	Scorpio	Gemini	Virgo	Aries	Cancer	Aquarius	Taurus
Capricorn	Sagittarius	Pisces	Libra	Scorpio	Leo	Virgo	Gemini	Cancer	Aries	Taurus	Aquarius

f. 131a

[VEDER TAULA DE SALAMON PER LUNA]

Questa serà la raxion dela taula de Salamon ala luna zudesca e le so riegolle. E primo el so melesimo. Al 1428 chure ai zudei 1189, e questo se parte per 19, serano $273\frac{2}{19}$, e questa serà la so patta. E voiando veder in che any fa lune 13, e primo $\frac{2}{19}$ serano lune 13, in 6 lune 13, in 8 lune 13, in 11 lune 13, in 14 lune 13, in 17 lune 13, in $\frac{19}{19}$ lune 13. E se alguno te domandase per raxion chomo avanza in ani 3 luna una, tu die moltip[*ichar*] lune 12 per zorny 29 e ore 12 e ponti 793, e moltipichado 12 fia 793, e questi partidi per 1080 per far ora 1ª. E moltipichade le ore 12 per 12 lune, e questi partidi per 24 farà zorny, e moltipichadi 12 via 29, e questi serano zorny, ore e ponti. E sutrazi da 365 ore 6 per lo bexesto, e quel che avanza de quel sutrato salvara'. Ogni 3 any avanzarà ben luna 1ª. Se ali 3 avanza, alli oltri mancha.

E se volessi aprovar le chaxelle se son false, tuo' el dì ara' fato e le ore e li punti, e sotto metti alli dì 1, ale ore 12, ali ponti 793, e sumadi li punti parti per 1080, serano ore, e lo resto tien. E puo' asumado le ore, parti per 24, serano zorny, e le ore tien. Asumadi li zorny, serano quello dì, ore e

SEE SOLOMON'S TABLE OF THE MOON

f. 131a

This will be the method of the table of Solomon for the Jewish moon and its rules.[1] And first, its year. In 1428 it is 1189[2] for the Jews, and this divided by 19 will be $273\frac{2}{19}$, and this will be its epact. And if we want to see in which years there are 13 moons, first $\frac{2}{19}$[3] will be 13 moons, in 6 there will be 13 moons, in 8 there will be 13 moons, in 11 there will be 13 moons, in 14 there will be 13 moons, in 17 there will be 13 moons, in $\frac{19}{19}$ there will be 13 moons. And if anyone should ask you to explain how there is one extra moon every three years, you should multiply 12 by 29 days and 12 hours and 793 points; and 12 multiplied by 793, and this divided by 1,080 to make 1 hour. And the 12 hours multiplied by the 12 moons, and this divided by 24 will make the days, and 12 multiplied by 29, and this will be the days, hours, and points. And subtract 6 hours from 365 for the bisextile, and you will save what results from this subtraction. Every 3 years there will indeed by one extra moon. If there's an extra for number 3, the others will lack it.

And if you would like to test whether the cells are false, take the day which you have chosen, the hours and the points, and add 1 to the day, 12 to the hours, and 793 to the points, and having added the points, divide by 1,080, it will be the hours, and hold the remainder. And then having

1. For the "table of Solomon," see Wallis, "Michael of Rhodes and Time Reckoning," vol. 3, pp. 305–309.
2. Sic in MS; correctly, 5189.
3. I.e., the second of 19 years. Michael has made a mistake here; it should be year 3.

ponti. Arigordatti 1 sie domenega, 2 luni, 3 marti, 4 merchore, 5 zuoba, 6 venere, sabo[a] 7, domenega uno, e questa taula chomenza marzo 1435. Ancuor in queste chaxelle in lo ladi de sovra al chanto[*n*] serà patta dela man de quel ano, e de sutto al oltro ladi, al quanton, Pasqua braicha. E per ladi dela ditta Pasqua, se 'l serà marzo averà M, se 'l serà avril averà a, e de sotto questo serà i dì del mexe serà Pasqua. E de sovra de questi, zoè Pasqua braicha e Pasqua cristiana, serà el dì che fa la luna e a quanti dì del mexe de sotto quello, e de sotto questo le ore, e de sotto le ore serà li ponti farà la luna. Qual ve piaserà azercar che 'l faza, chomo vedirì qui de sotto per inchontro ale chaxelle.

---

a. dom *crossed out with a horizontal line.*

added the hours, divide by 24, it will be the days, and hold the hours. Having added the days, it will be that day, hours and points. Remember 1 is Sunday, 2 Monday, 3 Tuesday, 4 Wednesday, 5 Thursday, 6 Friday, 7 Saturday, Sunday 1, and this table begins in March of 1435. Also in these cells on the side above the corner will be the epact of the hand of that year, and below on the other side in the corner the Hebrew Easter. And to the side of this Easter, if it is March it will have an M, if it is April it will have an A, and below this there will be the day of the month of Easter. And above these, that is, Hebrew Easter and Christian Easter, will be the day on which the moon is new and the number of days of the month below that, and below this the hours, and below the hours will be the points that the moon makes. Which you will be pleased to look up what it does, as you see below opposite in the cell.[1]

---

1. Explanation of the contents of the cells from fol. 131b to fol. 135a:

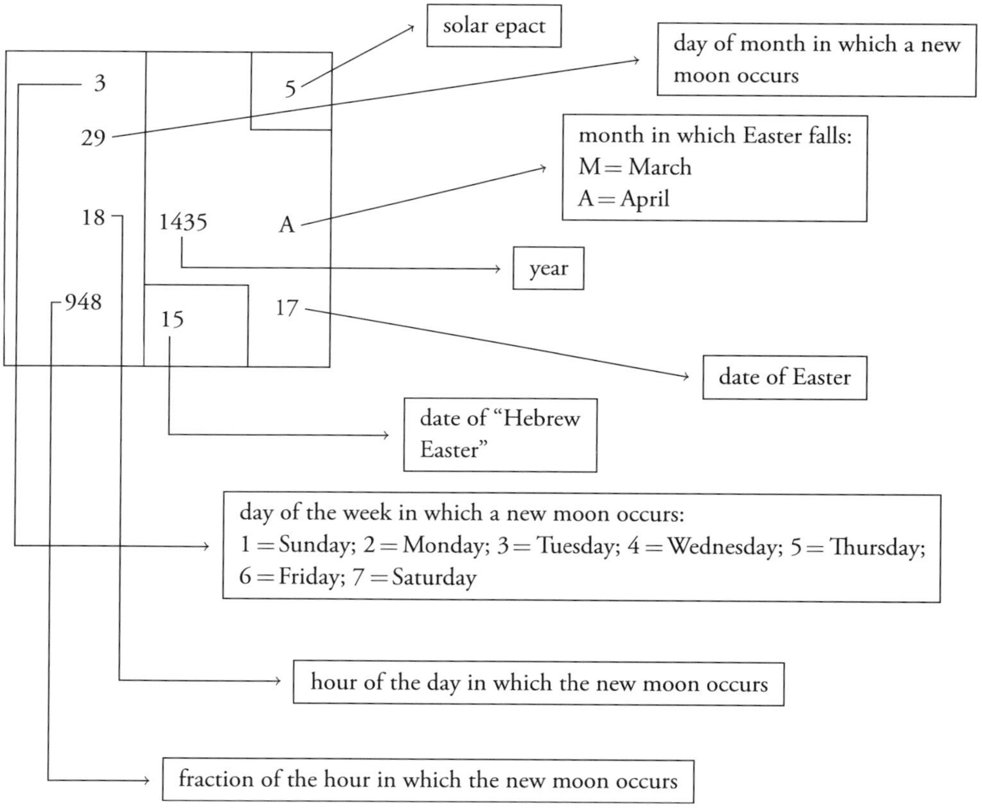

f. 131b

marzo	avril	mazo	zugno	luyo	avosto	setenbrio	octobro	novenbro	dezenbro	zener	frever		
3	[5][a]	6	1	2	4	5	7	2	3	5	6		5
29	28	27	26	25	24	22	22	21	20	19	17		
18	7	20	9	21	10	23	12	0	13	2	14	1435	a
948	6[6]1	374	87	880	593	306	19	812	525	238	1031	15	17
1	2	4	5	7	1	3	4	6	7	2	3		7
18	16	16	14	14	12	11	10	9	8	7	5		
3	16	5	17	6	19	8	20	9	22	11	23	1436	a
744	457	170	963	676	389	102	895	608	321	34	827	4	8
5	7	1	3	4	6 7	2	3	5	6	1	2		1
7	6	5	4	3	2 31	30	29	28	27	26	24		
12	[1]	13	2	15	4 16	5	18	7	19	8	21	1437	M
540	253	1046	759	472	185 978	691	404	117	910	623	336	24	31
4	5	7	2	3	5	6	1	2	4	5	7		2
26	24	24	23	22	21	19	19	17	17	15	14		
10	22	11	0	12	1	14	3	15	4	17	6	1438	a
49	842	555	268	1061	774	487	200	993	706	419	132	12	13
1	3	4	6	7	2	3	5	7	1	3	4		3
15	14	13	12	11	10	8	8	7	6	5	3		
18	7	20	9	21	10	23	11	0	13	2	14	1439	a
925	638	351	64	857	570	283	1076	789	502	215	1008	1	5
6	7	2 3	5	6	1	2	4	5	7	1	3		5
4	2	2 31	30	29	28	26	26	24	24	22	21		
3	16	5 17	6	19	8	20	9	22	11	23	12	1440	M
721	434	147 940	653	366	79	872	585	298	11	804	517	21	27
5	6	1	2	4	5	7	1	3	4	6	7		6
23	21	21	19	19	17	16	15	14	13	12	10		
1	13	2	15	4	16	5	18	7	19	8	21	1441	a
230	1023	736	449	162	955	668	381	94	887	600	313	9	16
2	3	5	7	1	3	4	6	7	2	3 5	0		7
12	10	10	9	8	7	5	5	3	3	1 31	0		
10	22	11	0	12	1	14	3	15	4	17 6	0	1442	a
26	814	532	245	1038	751	464	177	970	683	396 119	0	29	1
6 1	2	4	5	7	1	3	5	6	1	2	4		1
1 31	29	29	27	27	25	24	24	22	22	20	19		
18 7	20	9	21	10	23	11	0	13	2	14	3	1443	a
902 615	328	41	834	547	260	1053	766	479	192	985	698	17	21
5	7	1	3	4	6	7	2	3	5	7	1		3
19	18	17	16	15	14	12	12	10	10	9	7		
16	5	17	6	19	8	21	10	23	11	0	13	1444	a
411	124	917	630	343	56	849	562	275	1068	781	494	5	12
3	4	6	7	2	3	5	6 1	2	4	5	7		4
9	7	7	5	5	3	2	1 31	29	29	27	26		
2	14	3	16	5	17	6	19 8	20	9	22	11	1445	M
207	1000	713	426	139	932	645	358 71	864	577	290	3	25	28
1	3	5	6	1	2	4	5	7	1	3	4		5
27	26	26	24	24	22	21	20	19	18	17	15		
23	12	1	13	2	15	4	16	5	18	7	19	1446	a
796	509	222	1015	728	441	154	947	660	373	86	879	13	17
M	a	M	z	l	a	s	o	9	d	z	f		

a. *Here and below illegible because of a vertical tear in the paper and the consequent loss of text; numbers in brackets have been supplied by computation.*

March	April[1]	May	June	July	August	September	October	November	December	January	February		
3	[5]	6	1	2	4	5	7	2	3	5	6		5
29	28	27	26	25	24	22	22	21	20	19	17		
18	7	20	9	21	10	23	12	0	13	2	14	1435	A
948	6[6]1	374	87	880	593	306	19	812	525	238	1031	15	17
1	2	4	5	7	1	3	4	6	7	2	3		7
18	16	16	14	14	12	11	10	9	8	7	5		
3	16	5	17	6	19	8	20	9	22	11	23	1436	A
744	457	170	963	676	389	102	895	608	321	34	827	4	8
5	7	1	3	4	6 / 7	2	3	5	6	1	2		1
7	6	5	4	3	2 / 31	30	29	28	27	26	24		
12	[1]	13	2	15	4 / 16	5	18	7	19	8	21	1437	M
540	253	1046	759	472	185 / 978	691	404	117	910	623	336	24	31
4	5	7	2	3	5	6	1	2	4	5	7		2
26	24	24	23	22	21	19	19	17	17	15	14		
10	22	11	0	12	1	14	3	15	4	17	6	1438	A
49	842	555	268	1061	774	487	200	993	706	419	132	12	13
1	3	4	6	7	2	3	5	7	1	3	4		3
15	14	13	12	11	10	8	8	7	6	5	3		
18	7	20	9	21	10	23	11	0	13	2	14	1439	A
925	638	351	64	857	570	283	1076	789	502	215	1008	1	5
6	7	2 / 3	5	6	1	2	4	5	7	1	3		5
4	2	2 / 31	30	29	28	26	26	24	24	22	21		
3	16	5 / 17	6	19	8	20	9	22	11	23	12	1440	M
721	434	147 / 940	653	366	79	872	585	298	11	804	517	21	27
5	6	1	2	4	5	7	1	3	4	6	7		6
23	21	21	19	19	17	16	15	14	13	12	10		
1	13	2	15	4	16	5	18	7	19	8	21	1441	A
230	1023	736	449	162	955	668	381	94	887	600	313	9	16
2	3	5	7	1	3	4	6	7	2	3 / 5	0		7
12	10	10	9	8	7	5	5	3	3	1 / 31	0		
10	22	11	0	12	1	14	3	15	4	17 / 6	0	1442	A
26	814[2]	532	245	1038	751	464	177	970	683	396 / 119[3]	0	29	1
6 / 1	2	4	5	7	1	3	5	6	1	2	4		1
1 / 31	29	29	27	27	25	24	24	22	22	20	19		
18 / 7	20	9	21	10	23	11	0	13	2	14	3	1443	A
902 / 615	328	41	834	547	260	1053	766	479	192	985	698	17	21
5	7	1	3	4	6	7	2	3	5	7	1		3
19	18	17	16	15	14	12	12	10	10	9	7		
16	5	17	6	19	8	21[4]	10	23	11	0	13	1444	A
411	124	917	630	343	56	849	562	275	1068	781	494	5	12
3	4	6	7	2	3	5	6 / 1	2	4	5	7		4
9	7	7	5	5	3	2	1 / 31	29	29	27	26		
2	14	3	16	5	17	6	19 / 8	20	9	22	11	1445	M
207	1000	713	426	139	932	645	358 / 71	864	577	290	3	25	28
1	3	5	6	1	2	4	5	7	1	3	4		5
27	26	26	24	24	22	21	20	19	18	17	15		
23	12	1	13	2	15	4	16	5	18	7	19	1446	A
796	509	222	1015	728	441	154	947	660	373	86	879	13	17
M	A	M	J	J	A	S	O	N	D	J	F		

1. Because of a tear on this page, the April figures are not all legible; those in brackets have been supplied by calculation.
2. This should be 819.
3. This should be 109.
4. This number should be 20; the subsequent months in the table continue to be off by one hour (the next month's hour should be 9, then 22, etc.).

f. 132a

		M	a		M	z		l	a	s	o		n	d	[z][a]	f
	6	6	7		2	3		5	7	1	3		4	6	7	2
		17	15		15	13		13	12	10	10		8	8	6	5
1447		8	21		10	22		11	0	12	1		14	3	15	4
2		592	305		18	811		524	237	1030	743		456	169	962	675
	1	3	5		6	1	2	4	5	7	1		3	4	6	7
		5	4		3	2	1	31	29	28	27		26	25	24	22
1448		17	6		18	7	20	9	21	10	23		12	14	3	16
22		588	101		894	607	320	33	826	533	252		1045	758	471	184
	2	2	3		5	6		1	2	4	5		7	1	3	5
		24	22		22	20		20	18	17	16		15	14	13	12
1449		4	16		5	18		6	19	8	21		9	22	11	0
10		977	690		403	116		909	622	335	48		841	554	267	1060
	3	6	1		2	4		5	7	1	3		4	6	7	2
		13	12		11	10		9	8	6	6		4	4	2	1
1450		13	2		15	3		16	5	18	6		19	8	22	9
30		773	486		199	992		705	418	131	924		637	350	63	856
	4	3	5	6	1	3		4	6	7	2		3	5	6	1
		2	1	30	30	29		28	27	25	25		23	23	21	20
1451		22	11	23	12	1		14	2	15	4		17	5	18	7
18		569	282	1075	788	501		214	1007	720	433		146	939	652	365
	6	2	4		5	7		1	3	5	6		1	2	4	5
		20	19		18	17		16	15	14	13		12	11	10	0
1452		20	8		21	10		23	11	0	13		1	13	2	15
7		78	871		584	297		10	803	516	229		1022	735	448	161
	7	7	1		3	4		6	7	2	3	5	6	1	2	4
		10	8		8	6		6	4	3	2	1	30	30	28	27
1453		3	16		5	18		6	19	8	21	9	22	11	23	12
27		954	667		380	93		886	599	312	25	818	531	244	1037	750

a. *Page torn and misaligned; numbers for January 1447 and 1448 supplied by computation.*

f. 132a

	March	April	May		June	July		August	September	October	November		December	January	February
6	6	7	2		3	5		7	1	3	4		6	7[1]	2
	17	15	15		13	13		12	10	10	8		8	6	5
1447	8	21	10		22	11		0	12	1	14		3	15	4
2	592	305	18		811	524		237	1030	743	456		169	962	675
1	3	5	6		1	2	4	5	7	1	3		4	6	7
	5	4	3		2	1	31	29	28	27	26		25	24	22
1448	17	6	18		7	20	9	21	10	23	12[4]		14[5]	3	16
22	588[2]	101	894		607	320	33	826	533[3]	252	1045		758	471	184
2	2	3	5		6	1		2	4	5	7		1	3	5
	24	22	22		20	20		18	17	16	15		14	13	12
1449	4	16[6]	5		18	6		19	8	21	9		22	11	0[7]
10	977	690	403		116	909		622	335	48	841		554	267	1060
3	6	1	2		4	5		7	1	3	4		6	7	2
	13	12	11		10	9		8	6	6	4		4	2	1
1450	13	2	15		3	16		5	18	6	19		8	22[8]	9
30	773	486	199		992	705		418	131	924	637		350	63	856
4	3	5	6	1	3	4		6	7	2	3		5	6	1
	2	1	30	30	29	28		27	25	25	23		23	21	20
1451	22	11	23	12	1	14		2	15	4	17		5	18	7
18	569	282	1075	788	501	214		1007	720	433	146		939	652	365
6	2	4	5		7	1		3	5	6	1		2	4	5
	20	19	18		17	16		15	14	13	12		11	10	0[10]
1452	20	8	21		10	23		11	0	13	1		13[9]	2	15
7	78	871	584		297	10		803	516	229	1022		735	448	161
7	7	1	3		4	6		7	2	3	5	6	1	2	4
	10	8	8		6	6		4	3	2	1	30	30	28	27
1453	3	16	5		18	6		19	8	21	9	22	11	23	12
27	954	667	380		93	886		599	312	25	818	531	244	1037	750

1. Page torn and misaligned; January 1447 and 1448 supplied from Pietro di Versi, *Raxion de' marineri*, fol. 12r. Interestingly, the many numerical errors on this page of MOR have nearly all been corrected in the corresponding page of the Pietro di Versi manuscript.
2. This should be 388.
3. This should be 539.
4. This should be 11.
5. This number should be 0 (i.e., 10 hours later, shortly after midnight on the following day); the subsequent new moons in the table continue to be 10 hours behind.
6. Calculating from the preceding moon, this number should be 17, for a cumulative error of 11 hours; the subsequent moons in the table continue to be 11 hours behind.
7. Calculating from the preceding moon, this number should be 23 (i.e., an hour earlier); the subsequent moons in the table are now a cumulative 10 hours behind.
8. Calculating from the preceding moon, this number should be 21.
9. Calculating from the preceding moon, this number should be 14; the subsequent moons in the table are now a cumulative 11 hours behind.
10. This should be 8.

[Veder taula de Salamon per luna]

	1	6	7	2	3	5		6	1	2	4	5	7	1
		29	27	27	25	25		23	22	21	20	19	18	16
1454		0	13	1	14	13		16	4	17	6	19	7	20
15		463	176	969	882	395		108	901	614	327	40	833	546
	2	3	4	6	7	2		4	5	7	1	3	4	6
		18	16	16	14	14		13	11	11	9	9	7	6
1455		9	21	10	23	12		0	13	2	15	3	16	5
4		259	1052	765	478	191		984	697	410	123	976	689	402
	4	7	2	3	5	6	1	2	4	5	7	2	3	5
		6	5	4	3	2	1	30	29	28	27	27	25	24
1456		18	6	19	8	21	9	22	11	23	12	1	14	2
24		115	908	621	334	47	840	553	266	1059	772	485	98	891
	5	6	1	2	4	5		7	1	3	4	6	7	2
		25[a]	24	23	22	21		20	18	18	16	16	14	13
1457		15	4	17	5	18		7	19	8	21	10	22	11
12		604	317	30	823	536		249	1042	755	468	181	974	687
	6	4	5	7	1	3		4	6	7	2	3	5	6
		15	13	13	11	11		9	8	7	6	5	4	2
1458		0	13	1	14	3		16	4	17	6	18	7	20
1		400	113	906	619	332		45	838	551	264	1057	770	483

a. 25 *written over* 27.

1	6	7	2	3	5		6	1	2	4	5	7	1
	29	27	27	25	25		23	22	21	20	19	18	16
1454	0[1]	13	1	14	13[3]		16	4	17	6	19	7	20
15	463	176	969	882[2]	395		108	901	614	327	40	833	546
2	3	4	6	7	2		4	5	7	1	3	4	6
	18	16	16	14	14		13	11	11	9	9	7	6
1455	9	21	10	23	12		0	13	2	15	3	16	5
4	259	1052	765	478	191		984	697	410	123	976[4]	689	402
4	7	2	3	5	6	1	2	4	5	7	2	3	5
	6	5	4	3	2	1	30	29	28	27	27	25	24
1456	18	6	19	8	21	9	22	11	23	12	1	14	2
24	115	908	621	334	47	840	553	266	1059	772	485	98[5]	891
5	6	1	2	4	5		7	1	3	4	6	7	2
	25	24	23	22	21		20	18	18	16	16	14	13
1457	15	4	17	5	18		7	19	8	21	10	22	11
12	604	317	30	823	536		249	1042	755	468	181	974	687
6	4	5	7	1	3		4	6	7	2	3	5	6
	15	13	13	11	11		9	8	7	6	5	4	2
1458	0	13	1	14	3		16	4	17	6	18	7	20
1	400	113	906	619	332		45	838	551	264	1057	770	483

1. Calculating from the preceding moon, this number should be 1; the subsequent moons in the table are now a cumulative 12 hours behind.
2. This should be 682.
3. Calculating from the preceding moon, this number should be 3.
4. This should be 916; the subsequent moons in the table continue to be 60 points ahead.
5. Calculating from the preceding moon, this number should be 198; the subsequent moons in the table are a cumulative 40 points behind.

f. 132b

M	a	M	z	l	a	s	o	n	d	z	f	
1	2	4 5	7	1	3	5	6	1	2	4	5	7
4	2	2 31	30	29	28	27	26	25	24	23	21	
9	19	8 21	10	22	11	0	13	1	14	3	15	1459
196	989	702 415	128	921	634	347	60	853	566	279	1072	21
7	1	3	4	6	7	2	3	5	6	1	3	2
22	20	20	18	18	16	15	14	13	12	11	10	
4	17	6	18	7	20	9	21	10	23	12	0	1460
785	498	211	1004	717	430	143	936	649	362	75	868	9
4	6	7	2	3	5	6	1	2	4 5	7	1	3
11	10	9	8	7	6	4	4	2	2 31	30	28	
13	2	15	3	16	5	17	6	19	6 18	7	20	1461
581	294	7	800	513	226	1019	732	445	158 951	664	377	29
3	4	6	7	2	4	5	7	1	3	4	6	4
30	28	28	26	26	25	23	23	21	21	19	18	
9	21	10	23	12	0	13	2	14	3	16	5	1462
90	883	596	309	22	815	528	241	1034	747	460	173	17
7	2	3	5	6	1	2	4	5	7	2	3	5
19	18	17	16	15	14	12	12	10	10	9	7	
17	6	19	8	20	9	22	11	23	12	1	13	1463
966	679	392	105	898	611	324	37	830	543	256	1049	5
5	6	1	2	4	5	7 1	3	4	6	7	2	7
8	6	6	4	4	2	1 30	30	28	28	26	25	
2	15	4	16	5	18	7 19	8	21	10	22	11	1464
762	475	188	981	694	407	120 913	626	339	52	845	558	25
4	5	7	1	3	4	6	7	2	3	5	6	1
27	25	25	23	23	21	20	19	18	17	16	14	
0	12	1	14	3	15	4	17	6	18	7	20	1465
271	1064	777	490	203	996	709	422	135	928	641	354	13
1	2	4	5	7	2	3	5	6	1	2	4	2
16	14	14	12	12	11	9	9	7	7	5	4	
9	21	10	23	11	0	13	2	14	3	16	5	1466
67	860	573	286	1079	792	505	218	1011	724	437	150	2
5	7	1	3	4 6	7	2	3	5	7	1	3	3
5	4	3	2	1 31	29	28	27	26	26	24	23	
17	6	19	8	20 9	22	11	23	12	1	13	2	1467
943	656	369	82	875 588	301	14	807	520	233	1026	739	22
4	6	7	2	3	5	6	1	2	4	5	7	5
23	22	21	20	19	18	16	16	14	14	12	11	
15	4	16	5	18	7	19	8	21	10	22	11	1468
452	165	958	671	384	97	890	603	316	29	822	535	10
2	3	5	8	1	2	4	5	7	1	3 4	0	6
13	11	11	9	9	7	6	5	4	3	2 31	0	
0	12	1	14	3	15	4	17	6	18	7 20	0	1469
248	1041	754	467	170	963	676	389	102	895	608 321	0	30
6 7	2	3	5	7	1	3	4	6	7	2	3	7
2 31	30	29	28	28	26	25	24	23	22	21	19	
9 21	10	23	11	0	13	2	14	3	16	3	15	1470
34 827	540	253	1046	759	472	185	978	691	404	117	910	18

f. 132b

	March	April		May	June		July	August		September	October	November		December		January	February		
	1	2	4	5	7		1	3		5	6	1		2		4	5		7
	4	2	2	31	30		29	28		27	26	25		24		23	21		
	9	19	8	21	10		22	11		0	13	1		14		3	15	1459	
	196	989	702	415	128		921	634		347	60	853		566		279	1072	21	
	7	1		3	4		6	7		2	3	5		6		1	3		2
	22	20		20	18		18	16		15	14	13		12		11	10		
	4	17		6	18		7	20		9	21	10		23		12	0	1460	
	785	498		211	1004		717	430		143	936	649		362		75	868	9	
	4	6		7	2		3	5		6	1	2	4	5		7	1		3
	11	10		9	8		7	6		4	4	2	2	31		30	28		
	13	2		15	3		16	5		17	6	19	6	18		7	20	1461	
	581	294		7	800		513	226		1019	732	445	158	951		664	377	29	
	3	4		6	7		2	4		5	7	1		3		4	6		4
	30	28		28	26		26	25		23	23	21		21		19	18		
	9	21		10	23		12	0		13	2	14		3		16	5	1462	
	90	883		596	309		22	815		528	241	1034		747		460	173	17	
	7	2		3	5		6	1		2	4	5		7		2	3		5
	19	18		17	16		15	14		12	12	10		10		9	7		
	17	6		19	8		20	9		22	11	23		12		1	13	1463	
	966	679		392	105		898	611		324	37	830		543		256	1049	5	
	5	6		1	2		4	5	7	1	3	4		6		7	2		7
	8	6		6	4		4	2	1	30	30	28		28		26	25		
	2	15		4	16		5	18	7	19	8	21		10		22	11	1464	
	762	475		188	981		694	407	120	913	626	339		52		845	558	25	
	4	5		7	1		3	4		6	7	2		3		5	6		1
	27	25		25	23		23	21		20	19	18		17		16	14		
	0	12		1	14		3	15		4	17	6		18		7	20	1465	
	271	1064		777	490		203	996		709	422	135		928		641	354	13	
	1	2		4	5		7	2		3	5	6		1		2	4		2
	16	14		14	12		12	11		9	9	7		7		5	4		
	9	21		10	23		11	0		13	2	14		3		16	5	1466	
	67	860		573	286		1079	792		505	218	1011		724		437	150	2	
	5	7		1	3	4	6	7		2	3	5		7		1	3		3
	5	4		3	2	1	31	29		28	27	26		26		24	23		
	17	6		19	8	20	9	22		11	23	12		1		13	2	1467	
	943	656		369	82	875	588	301		14	807	520		233		1026	739	22	
	4	6		7	2		3	5		6	1	2		4		5	7		5
	23	22		21	20		19	18		16	16	14		14		12	11		
	15	4		16	5		18	7		19	8	21		10		22	11	1468	
	452	165		958	671		384	97		890	603	316		29		822	535	10	
	2	3		5	8		1	2		4	5	7		1	3	4	0		6
	13	11		11	9		9	7		6	5	4		3	2	31	0		
	0	12		1	14		3	15		4	17	6		18	7	20	0	1469	
	248	1041		754	467		170	963		676	389	102		895	608	321	0	30	
6	7	2		3	5		7	1		3	4	6		7		2	3		7
2	31	30		29	28		28	26		25	24	23		22		21	19		
9	21	10		23	11		0	13		2	14	3		16		3	15	1470	
34	827	540		253	1046		759	472		185	978	691		404		117	910	18	

f. 133a

		M	a	M	z	l	a	s	o	n	d	z	f
	1	5	6	1	2	4	5	7	1	3	4	6	1
		21	19	19	17	17	15	14	13	12	11	10	9
1471		4	17	6	18	7	20	8	21	10	23	11	0
7		623	336	49	842	555	268	1061	774	487	200	993	706
	3	2	4	5	7	1	3	4	6 / 7	2	3	5	6
		9	8	7	6	5	4	2	2 / 31	30	29	28	26
1472		13	2	14	3	16	5	17	6 / 19	7	20	9	22
27		419	132	925	638	351	64	857	570 / 283	1076	789	502	215
	4	1	3	4	6	7	2	3	5	6	1	2	4
		28	27	26	25	24	23	21	21	19	19	17	16
1473		11	0	13	2	14	3	16	5	17	6	19	8
15		1008	721	434	147	940	653	366	79	872	585	298	11
	5	5	7	1	3	4	6	1	2	4	5	7	1
		17	16	15	14	13	12	11	10	9	8	7	5
1474		20	9	22	10	23	12	1	13	2	15	4	16
4		804	517	230	1023	736	449	162	955	668	381	94	887
	6	3	4	6	7	2	3 / 5	6	1	3	4	6	7
		7	5	5	3	3	1 / 31	29	29	28	27	26	24
1475		5	18	3	19	8	21 / 9	22	11	0	12	1	14
24		600	313	36	829	542	255 / 1048	761	474	187	980	693	406
	1	2	3	5	6	1	2	4	5	7	1	3	4
		25	23	23	21	21	19	18	17	16	15	14	12
1476		3	15	4	17	6	18	7	20	18	21	10	23
12		119	912	625	338	51	844	557	270	1063	776	489	202
	2	6	1	2	4	5	7	1	3	4	6	7	2
		14	13	12	11	10	9	7	7	5	5	3	2
1477		11	0	13	2	14	3	16	5	17	6	19	7
1		995	708	421	134	927	640	353	66	859	572	285	1078
	3	3	5	6 / 1	2	4	6	7	2	3	5	6	1
		3	2	1 / 31	29	29	28	26	26	24	24	22	21
1478		20	9	22 / 10	23	12	1	13	2	15	4	16	5
21		791	504	217 / 1010	723	436	149	942	655	368	81	874	587
	4	2	4	5	7	1	3	4	6	7	2	4	5
		22	21	20	19	18	17	15	15	13	13	12	10
1479		17	6	18	7	20	8	21	10	23	11	0	13
9		300	13	806	519	232	1025	738	451	164	957	670	383
	6	7	1	3	4	6	7	2	3	5	6 / 1	2	4
		11	9	9	7	7	5	4	3	2	1 / 31	29	28
1480		2	14	3	16	5	17	6	19	7	20 / 9	22	10
29		96	889	602	315	28	821	534	247	1040	753 / 466	179	972
	7	5	7	2	3	5	6	1	2	4	5	7	1
		29	28	28	26	26	24	23	22	21	20	19	17
1481		23	12	1	13	2	15	4	16	5	18	6	19
17		685	398	111	904	617	330	43	836	549	262	1055	768
	1	3	4	6	7	2	4	5	7	1	3	4	6
		19	17	17	15	15	14	12	12	10	10	8	7
1482		8	21	9	22	11	0	12	1	14	3	15	4
5		481	194	987	700	413	126	919	632	345	58	851	564

f. 133a

		March	April		May	June	July		August	September		October	November		December	January	February
	1	5	6		1	2	4		5	7		1	3		4	6	1
		21	19		19	17	17		15	14		13	12		11	10	9
1471		4	17		6	18	7		20	8		21	10		23	11	0
7		623	336		49	842	555		268	1061		774	487		200	993	706
	3	2	4		5	7	1		3	4	6	7	2		3	5	6
		9	8		7	6	5		4	2	2	31	30		29	28	26
1472		13	2		14	3	16		5	17	6	19	7		20	9	22
27		419	132		925	638	351		64	857	570	283	1076		789	502	215
	4	1	3		4	6	7		2	3		5	6		1	2	4
		28	27		26	25	24		23	21		21	19		19	17	16
1473		11	0		13	2	14		3	16		5	17		6	19	8
15		1008	721		434	147	940		653	366		79	872		585	298	11
	5	5	7		1	3	4		6	1		2	4		5	7	1
		17	16		15	14	13		12	11		10	9		8	7	5
1474		20	9		22	10	23		12	1		13	2		15	4	16
4		804	517		230	1023	736		449	162		955	668		381	94	887
	6	3	4		6	7	2	3	5	6		1	3		4	6	7
		7	5		5	3	3	1	31	29		29	28		27	26	24
1475		5	18		3	19	8	21	9	22		11	0		12	1	14
24		600	313		36	829	542	255	1048	761		474	187		980	693	406
	1	2	3		5	6	1		2	4		5	7		1	3	4
		25	23		23	21	21		19	18		17	16		15	14	12
1476		3	15		4	17	6		18	7		20	18		21	10	23
12		119	912		625	338	51		844	557		270	1063		776	489	202
	2	6	1		2	4	5		7	1		3	4		6	7	2
		14	13		12	11	10		9	7		7	5		5	3	2
1477		11	0		13	2	14		3	16		5	17		6	19	7
1		995	708		421	134	927		640	353		66	859		572	285	1078
	3	3	5	6	1	2	4		6	7		2	3		5	6	1
		3	2	1	31	29	29		28	26		26	24		24	22	21
1478		20	9	22	10	23	12		1	13		2	15		4	16	5
21		791	504	217	1010	723	436		149	942		655	368		81	874	587
	4	2	4		5	7	1		3	4		6	7		2	4	5
		22	21		20	19	18		17	15		15	13		13	12	10
1479		17	6		18	7	20		8	21		10	23		11	0	13
9		300	13		806	519	232		1025	738		451	164		957	670	383
	6	7	1		3	4	6		7	2		3	5	6	1	2	4
		11	9		9	7	7		5	4		3	2	1	31	29	28
1480		2	14		3	16	5		17	6		19	7	20	9	22	10
29		96	889		602	315	28		821	534		247	1040	753	466	179	972
	7	5	7		2	3	5		6	1		2	4		5	7	1
		29	28		28	26	26		24	23		22	21		20	19	17
1481		23	12		1	13	2		15	4		16	5		18	6	19
17		685	398		111	904	617		330	43		836	549		262	1055	768
	1	3	4		6	7	2		4	5		7	1		3	4	6
		19	17		17	15	15		14	12		12	10		10	8	7
1482		8	21		9	22	11		0	12		1	14		3	15	4
5		481	194		987	700	413		126	919		632	345		58	851	564

[Veder taula de Salamon per luna]

f. 133b

M	a	M	z	l		a		s	o		n		d	z		f		
7	2	3	5		6		1	2	4	5		7	2		3	5		2
8	7	6	5		4		3	1	1	30		29	29		27	26		
17	5	18	7		20		8	21	10	23		11	0		13	2	1483	
277	1070	783	496		209		1002	715	428	141		934	647		360	73	25	
6	1	2	4		5		7	1		3		4	6		7	2		4
26	25	24	23		22		21	19		19		17	17		15	14		
14	3	16	5		17		6	19		7		20	9		22	10	1484	
866	579	292	5		798		511	224		1017		730	443		156	949	13	
3	5	7	1		3		4	6		7		2	3		5	6		5
15	14	14	12		12		10	9		8		7	6		5	3		
23	12	1	13		2		15	4		16		5	18		6	19	1485	
662	375	88	881		594		307	20		813		526	239		1032	745	2	
1	2	4	5	7	2		3	5		6		1	2		4	5		6
5	3	3	1	1	31		29	28		27		26	25		24	22		
8	21	9	22	11	0		12	1		14		3	15		4	17	1486	
458	171	964	677	390	103		896	609		322		35	828		541	254	22	
7	1	3	4		6		7	2		3		5	7		1	3		7
24	22	22	20		20		18	17		16		15	15		13	12		
5	18	7	20		8		21	10		23		11	0		13	2	1487	
1047	760	473	186		979		692	405		118		911	624		337	50	10	
4	6	7	2		3		5	6		1		2	4	5	7	1		2
12	11	10	9		8		7	5		5		3	3	1	31	29		
14	3	16	4		17		6	19		7		20	9	22	10	23	1488	
843	556	269	1062		775		488	201		994		707	420	133	926	639	30	
3	5	6	1		2		4	5		7		1	3		4	6		3
31	30	29	28		27		26	24		24		22	22		20	19		
12	1	13	2		15		3	16		5		18	6		19	8	1489	
352	65	858	571		284		1077	790		503		216	1009		722	435	18	
7	2	3	5		6		1	3		4		6	7		2	3		4
20	19	18	17		16		15	14		13		12	11		10	8		
20	8	21	10		23		11	0		13		2	14		3	16	1490	
148	941	654	367		80		873	586		299		12	805		518	231	7	
5	6	1	2		4		5	7		1	3	4	6		1	2		5
10	8	8	6		6		4	3		2	1	30	30		29	27		
4	17	6	19		7		20	9		22	10	23	12		1	13	1491	
1024	737	450	163		956		669	382		95	888	601	314		27	820	27	
4	5	7	1		3		4	6		7		2	3		5	6		7
28	26	26	24		24		22	21		20		19	18		17	15		
2	15	3	16		5		18	6		19		8	21		9	22	1492	
533	246	1039	752		465		178	971		684		397	110		903	616	15	
1	3	4	6		7		2	3		5		6	1		2	4		1
17	16	15	14		13		12	10		10		8	8		6	5		
11	0	12	1		14		2	15		4		17	5		18	7	1493	
329	42	835	548		261		1054	767		480		193	986		699	412	4	
5	7	1	3		4	6	1	2		4		5	7		1	3		2
6	5	4	3		2	1	31	29		29		27	27		25	24		
20	8	21	10		23	11	0	13		1		14	3		16	4	1494	
125	918	631	344		57	850	563	276		1069		782	495		208	1001	24	

f. 133b

March	April	May	June	July	August	September	October	November	December	January	February	
7	2	3	5	6	1	2	4 / 5	7	2	3	5	2
8	7	6	5	4	3	1	1 / 30	29	29	27	26	
17	5	18	7	20	8	21	10 / 23	11	0	13	2	1483
277	1070	783	496	209	1002	715	428 / 141	934	647	360	73	25
6	1	2	4	5	7	1	3	4	6	7	2	4
26	25	24	23	22	21	19	19	17	17	15	14	
14	3	16	5	17	6	19	7	20	9	22	10	1484
866	579	292	5	798	511	224	1017	730	443	156	949	13
3	5	7	1	3	4	6	7	2	3	5	6	5
15	14	14	12	12	10	9	8	7	6	5	3	
23	12	1	13	2	15	4	16	5	18	6	19	1485
662	375	88	881	594	307	20	813	526	239	1032	745	2
1	2	4	5	7 / 2	3	5	6	1	2	4	5	6
5	3	3	1	1 / 31	29	28	27	26	25	24	22	
8	21	9	22	11 / 0	12	1	14	3	15	4	17	1486
458	171	964	677	390 / 103	896	609	322	35	828	541	254	22
7	1	3	4	6	7	2	3	5	7	1	3	7
24	22	22	20	20	18	17	16	15	15	13	12	
5	18	7	20	8	21	10	23	11	0	13	2	1487
1047	760	473	186	979	692	405	118	911	624	337	50	10
4	6	7	2	3	5	6	1	2	4	5 / 7	1	2
12	11	10	9	8	7	5	5	3	3	1 / 31	29	
14	3	16	4	17	6	19	7	20	9	22 / 10	23	1488
843	556	269	1062	775	488	201	994	707	420	133 / 926	639	30
3	5	6	1	2	4	5	7	1	3	4	6	3
31	30	29	28	27	26	24	24	22	22	20	19	
12	1	13	2	15	3	16	5	18	6	19	8	1489
352	65	858	571	284	1077	790	503	216	1009	722	435	18
7	2	3	5	6	1	3	4	6	7	2	3	4
20	19	18	17	16	15	14	13	12	11	10	8	
20	8	21	10	23	11	0	13	2	14	3	16	1490
148	941	654	367	80	873	586	299	12	805	518	231	7
5	6	1	2	4	5	7	1	3 / 4	6	1	2	5
10	8	8	6	6	4	3	2	1 / 30	30	29	27	
4	17	6	19	7	20	9	22	10 / 23	12	1	13	1491
1024	737	450	163	956	669	382	95	888 / 601	314	27	820	27
4	5	7	1	3	4	6	7	2	3	5	6	7
28	26	26	24	24	22	21	20	19	18	17	15	
2	15	3	16	5	18	6	19	8	21	9	22	1492
533	246	1039	752	465	178	971	684	397	110	903	616	15
1	3	4	6	7	2	3	5	6	1	2	4	1
17	16	15	14	13	12	10	10	8	8	6	5	
11	0	12	1	14	2	15	4	17	5	18	7	1493
329	42	835	548	261	1054	767	480	193	986	699	412	4
5	7	1	3	4	6 / 1	2	4	5	7	1	3	2
6	5	4	3	2	1 / 31	29	29	27	27	25	24	
20	8	21	10	23	11 / 0	13	1	14	3	16	4	1494
125	918	631	344	57	850 / 563	276	1069	782	495	208	1001	24

f. 134a

		M	a	M		z	l		a	s	o		n	d	z		f
	3	4	6		7	2		3	5	6		1	2	4		6	7
		25	24		23	22		21	20	18		18	16	16		15	13
1495		17	6		19	7		20	9	22		10	23	12		1	13
12		714	427		140	933		646	359	72		865	578	291		4	797
	5	2	3		5	6		1	2	4		5	7	1		3	4
		14	12		12	10		10	8	7		6	5	4		3	1
1496		2	15		3	16		5	18	6		19	8	21		9	22
1		510	223		1016	729		442	155	948		661	374	87		880	593
	6	6	1	2	4	5		7	1	3		4	6	7		2	3
		3	2	1	31	29		29	27	26		25	24	23		22	20
1497		11	0	12	1	14		2	15	4		17	8	21		10	23
21		306	19	812	525	238		1031	744	457		170	963	676		389	102
	7	5	7		1	3		4	6	7		2	3	5		6	1
		22	21		20	19		18	17	15		15	13	13		11	10
1498		11	0		13	2		14	3	16		4	17	6		19	7
9		895	608		321	34		827	540	253		1046	759	472		185	978
	1	2	4		5	7		1	3	5		6	1	2	4	5	7
		11	10		9	8		7	6	5		4	3	2	1	30	29
1499		20	9		22	10		23	12	1		13	2	15	3	16	5
29		691	404		117	910		623	336	49		842	555	268	1061	774	487
	3	1	3		4	6		7	2	3		5	7	1		3	4
		29	28		27	26		25	24	22		22	21	20		19	17
1500[a]		18	6		19	8		21	9	22		11	0	12		1	14
17		200	993		706	419		132	925	638		351	64	867		580	293
	4	6	7		2	3		5	6	1		2	4	5		7	1
		19	17		17	15		15	13	12		11	10	9		8	6
1501		3	15		4	17		5	18	7		20	8	21		10	23
5		6	799		512	225		1018	731	444		157	940	653		366	79
	5	3	5		6	1		2	4	5	7	1	3	4		6	7
		8	7		6	5		4	3	1	1	30	29	28		27	25
1502		11	0		13	1		14	3	15	4	17	5	18		7	20
25		872	585		298	591		304	17	810	523	236	1029	742		455	168
	6	2	3		5	6		1	3	4		6	7	2		3	5
		27	25		25	23		23	22	20		20	18	18		16	15
1503		8	21		10	23		11	0	13		2	14	3		16	4
13		961	674		387	100		893	606	319		32	825	538		251	1044
	1	6	1		2	4		5	7	1		3	4	6		1	2
		15	14		13	12		11	10	8		8	6	6		5	3
1504		17	6		19	7		20	9	22		10	23	12		1	13
2		757	470		183	976		689	402	115		908	621	334		47	840
	2	4	5		7	1	3	4	6	7		2	3	5		6	1
		5	3		3	1	1	30	29	27		27	25	25		23	22
1505		2	15		3	16	5	18	6	19		8	21	9		22	11
22		553	266		1059	772	485	198	991	704		417	130	923		636	349
	3	3	4		6	7		2	3	5		6	1	2		4	5
		24	22		22	20		20	18	17		16	15	14		13	11
1506		0	12		1	14		2	15	4		17	5	18		7	20
10		62	855		568	281		1074	787	500		213	1006	719		432	145

a. 1500 *corrected over* 1400.

f. 134a

Year		March	April	May	June	July	August	September	October	November	December	January	February
	3	4	6	7	2	3	5	6	1	2	4	6	7
		25	24	23	22	21	20	18	18	16	16	15	13
1495		17	6	19	7	20	9	22	10	23	12	1	13
12		714	427	140	933	646	359	72	865	578	291	4	797
	5	2	3	5	6	1	2	4	5	7	1	3	4
		14	12	12	10	10	8	7	6	5	4	3	1
1496		2	15	3	16	5	18	6	19	8	21	9	22
1		510	223	1016	729	442	155	948	661	374	87	880	593
	6	6	1	2; 4	5	7	1	3	4	6	7	2	3
		3	2	1; 31	29	29	27	26	25	24	23	22	20
1497		11	0	12; 1	14	2	15	4	17	8	21	10	23
21		306	19	812; 525	238	1031	744	457	170	963	676	389	102
	7	5	7	1	3	4	6	7	2	3	5	6	1
		22	21	20	19	18	17	15	15	13	13	11	10
1498		11	0	13	2	14	3	16	4	17	6	19	7
9		895	608	321	34	827	540	253	1046	759	472	185	978
	1	2	4	5	7	1	3	5	6	1	2	4; 5	7
		11	10	9	8	7	6	5	4	3	2	1; 30	29[1]
1499		20	9	22	10	23	12	1	13	2	15	3; 16	5
29		691	404	117	910	623	336	49	842	555	268	1061; 774	487
	3	1	3	4	6	7	2	3	5	7	1	3	4
		29	28	27	26	25	24	22	22	21	20	19	17
1500		18	6	19	8	21	9	22	11	0	12	1	14
17		200	993	706	419	132	925	638	351	64	867[2]	580	293
	4	6	7	2	3	5	6	1	2	4	5	7	1
		19	17	17	15	15	13	12	11	10	9	8	6
1501		3	15	4	17	5	18	7	20	8	21	10	23
5		6	799	512	225	1018	731	444	157	940	653	366	79
	5	3	5	6	1	2	4	5	7; 1	3	4	6	7
		8	7	6	5	4	3	1	1; 30	29	28	27	25
1502		11	0	13	1	14	3	15	4; 17	5	18	7	20
25		872	585	298	591[3]	304	17	810	523; 236	1029	742	455	168
	6	2	3	5	6	1	3	4	6	7	2	3	5
		27	25	25	23	23	22	20	20	18	18	16	15
1503		8	21	10	23	11	0	13	2	14	3	16	4
13		961	674	387	100	893	606	319	32	825	538	251	1044
	1	6	1	2	4	5	7	1	3	4	6	1	2
		15	14	13	12	11	10	8	8	6	6	5	3
1504		17	6	19	7	20	9	22	10	23	12	1	13
2		757	470	183	976	689	402	115	908	621	334	47	840
	2	4	5	7	1	3; 4	6	7	2	3	5	6	1
		5	3	3	1	1; 30	29	27	27	25	25	23	22
1505		2	15	3	16	5; 18	6	19	8	21	9	22	11
22		553	266	1059	772	485; 198	991	704	417	130	923	636	349
	3	3	4	6	7	2	3	5	6	1	2	4	5
		24	22	22	20	20	18	17	16	15	14	13	11
1506		0	12	1	14	2	15	4	17	5	18	7	20
10		62	855	568	281	1074	787	500	213	1006	719	432	145

1. I.e., February 29, [1500], a leap year in the Julian calendar.
2. Calculating from the preceding moon, this should be 857; the following moons in the table continue to be 10 points ahead (in addition to the earlier cumulative error) until November 1501, when they revert to the earlier pattern.
3. Calculating from the preceding moon, this number should be 11 (with the hour 2 rather than 1); the subsequent moons in the table are likewise an additional 500 points behind.

f. 134b

M	a	M	z	l	a	s	o	n	d	z	f		
7	1	3	4	6	1	2	4	5	7	1	3		4
13	11	11	9	9	8	6	6	4	4	2	1		
8	21	10	23	11	0	13	2	14	3	16	4	1507	
938	651	364	77	870	583	296	9	802	515	228	1021	30	
4 6	7	2	3	5	6	1	2	4	6	7	2		6
1 31	29	29	27	27	25	24	23	22	22	20	19		
17 6	19	7	20	9	22	10	23	12	1	13	2	1508	
734 447	160	953	666	379	92	885	598	311	24	817	530	18	
3	5	6	1	2	4	5	7	1	3	4	6		7
20	19	18	17	16	15	13	13	11	11	9	8		
15	3	16	5	18	6	19	8	21	10	23	12	1509	
243	1036	749	462	175	968	681	394	107	900	613	326	7	
1	2	4	5	7	1	3	4	6 7	2	3	5		1
10	8	8	6	6	4	3	2	1 30	30	28	27		
1	13	2	15	3	16	5	18	6 19	8	21	9	1510	
39	832	545	258	1051	764	477	190	983 696	409	122	915	27	
6	1	3	4	6	7	2	3	5	6	1	2		2
28	27	27	25	25	23	22	21	20	19	18	16		
22	11	0	12	1	14	2	15	4	17	5	18	1511	
628	341	54	847	560	273	1066	779	492	205	998	711	15	
4	5	7	1	3	4	6	1	2	4	5	7		4
17	15	15	13	13	11	10	10	8	8	6	5		
7	20	8	21	10	23	11	0	13	1	14	3	1512[a]	
424	137	930	643	356	69	859	572	285	1078	791	504	4	
1	3	4	6	7	2 3	5	6	1	2	4	6		5
6	5	4	3	2	1 30	29	28	27	26	25	24		
16	4	17	6	19	7 20	9	22	10	23	12	1	1513	
217	1010	723	436	149	942 655	368	81	874	587	300	13	24	
7	2	3	5	6	1	2	4	5	7	1	3		6
25	24	23	22	21	20	18	18	16	16	14	13		
13	2	15	3	16	5	18	6	19	8	21	9	1514	
806	519	232	1025	738	451	164	957	670	383	96	889	12	
4	6	1	2	4	5	7	1	3	4	6	7		7
14	13	13	11	11	9	8	7	6	5	4	2		
22	11	0	12	1	14	2	15	4	16	4	17	1515	
602	315	28	821	534	247	1040	753	466	179	972	685	1	
2	3	5 6	1	2	4	5	7	2	3	5	6		2
3	1	1 30	29	28	27	25	25	24	23	22	20		
6	19	7 20	9	22	10	23	12	0	13	2	15	1516	
398	111	904 617	330	43	836	549	262	1055	768	481	194	21	
1	2	4	5	7	1	3	4	6	7	2	3		3
22	20	20	18	18	16	15	14	13	12	11	9		
3	16	5	18	6	19	8	21	9	23	11	23	1517	
987	700	413	126	919	632	345	58	851	564	277	1070	9	
5	7	1	3	4	6	7	2	3	5 6	1	2		4
11	10	9	8	7	6	4	4	2	2 31	30	28		
12	1	14	2	15	4	17	5	18	7 20	8	21	1518	
783	496	209	1002	715	428	141	934	647	360 73	866	579	29	

a. 1512 *written over* 1412.

f. 134b

	March	April		May	June	July		August	September	October		November		December	January	February		
	7	1		3	4	6		1	2	4		5		7	1	3		4
	13	11		11	9	9		8	6	6		4		4	2	1		
	8	21		10	23	11		0	13	2		14		3	16	4	1507	
	938	651		364	77	870		583	296	9		802		515	228	1021	30	
4	6	7		2	3	5		6	1	2		4		6	7	2		6
1	31	29		29	27	27		25	24	23		22		22	20	19		
17	6	19		7	20	9		22	10	23		12		1	13	2	1508	
734	447	160		953	666	379		92	885	598		311		24	817	530	18	
	3	5		6	1	2		4	5	7		1		3	4	6		7
	20	19		18	17	16		15	13	13		11		11	9	8		
	15	3		16	5	18		6	19	8		21		10	23	12	1509	
	243	1036		749	462	175		968	681	394		107		900	613	326	7	
	1	2		4	5	7		1	3	4	6	7		2	3	5		1
	10	8		8	6	6		4	3	2	1	30		30	28	27		
	1	13		2	15	3		16	5	18	6	19		8	21	9	1510	
	39	832		545	258	1051		764	477	190	983	696		409	122	915	27	
	6	1		3	4	6		7	2	3		5		6	1	2		2
	28	27		27	25	25		23	22	21		20		19	18	16		
	22	11		0	12	1		14	2	15		4		17	5	18	1511	
	628	341		54	847	560		273	1066	779		492		205	998	711	15	
	4	5		7	1	3		4	6	1		2		4	5	7		4
	17	15		15	13	13		11	10	10		8		8	6	5		
	7	20		8	21	10		23	11	0		13		1	14	3	1512	
	424	137		930	643	356		69	859[1]	572		285		1078	791	504	4	
	1	3		4	6	7	2	3	5	6		1		2	4	6		5
	6	5		4	3	2	1	30	29	28		27		26	25	24		
	16	4		17	6	19	7	20	9	22		10		23	12	1	1513	
	217	1010		723	436	149	942	655	368	81		874		587	300	13	24	
	7	2		3	5	6		1	2	4		5		7	1	3		6
	25	24		23	22	21		20	18	18		16		16	14	13		
	13	2		15	3	16		5	18	6		19		8	21	9	1514	
	806	519		232	1025	738		451	164	957		670		383	96	889	12	
	4	6		1	2	4		5	7	1		3		4	6	7		7
	14	13		13	11	11		9	8	7		6		5	4	2		
	22	11		0	12	1		14	2	15		4		16	4	17	1515	
	602	315		28	821	534		247	1040	753		466		179	972	685	1	
	2	3	5	6	1	2		4	5	7		2		3	5	6		2
	3	1	1	30	29	28		27	25	25		24		23	22	20		
	6	19	7	20	9	22		10	23	12		0		13	2	15	1516	
	398	111	904	617	330	43		836	549	262		1055		768	481	194	21	
	1	2		4	5	7		1	3	4		6		7	2	3		3
	22	20		20	18	18		16	15	14		13		12	11	9		
	3	16		5	18	6		19	8	21		9		23	11	23	1517	
	987	700		413	126	919		632	345	58		851		564	277	1070	9	
	5	7		1	3	4		6	7	2		3	5	6	1	2		4
	11	10		9	8	7		6	4	4		2	2	31	30	28		
	12	1		14	2	15		4	17	5		18	7	20	8	21	1518	
	783	496		209	1002	715		428	141	934		647	360	73	866	579	29	

1. Calculating from the preceding moon, this number should be 862; the rest of the moons in the table are an additional 3 points behind.

[Veder taula de Salamon per luna]

f. 135a

	M	a	M	z	l	a	s	o	n	d	z	f
5	4	5	7	2	3	5	6	1	2	4	5	7
	30	28	28	27	26	25	23	23	21	21	19	18
1519	10	23	11	0	13	1	14	3	16	4	17	6
17	292	5	798	511	224	1017	730	443	156	949	662	375
7	1	3	4	6	7	2	3	5	7	1	3	4
	18	17	16	15	14	13	11	11	10	9	8	6
1520	19	7	20	9	22	10	23	12	0	13	2	15
5	88	881	594	307	20	813	526	239	1032	745	458	171
1	6	7	2	3	5	6	1 / 2	4	5	7	1	3
	8	6	6	4	4	2	1 / 30	30	28	28	26	25
1521	3	16	5	18	6	19	8 / 21	9	22	11	23	12
25	964	677	390	103	896	609	322 / 35	828	541	254	1047	760
2	5	6	1	2	4	5	7	1	3	4	6	7
	27	25	25	23	23	21	20	19	18	17	16	14
1522	1	14	2	15	4	17	5	18	7	20	8	21
13	473	186	979	692	405	118	911	624	337	50	843	556
3	2	3	5	7	1	3	4	6	7	2	3	5
	16	14	14	13	12	11	9	9	7	7	5	4
1523	10	22	11	0	13	1	14	3	16	4	17	6
2	269	1062	775	488	201	994	707	420	133	926	639	352
5	6	1	2	4 / 5	7	1	3	5	6	1	2	4
	4	3	2	1 / 30	30	28	27	27	25	25	23	22
1524	19	7	20	9 / 21	10	23	12	0	13	2	15	3
22	65	858	571	284 / 1077	790	503	216	1009	722	435	148	941
6	5	7	1	3	4	6	7	2	3	5	6	1
	23	22	21	20	19	18	16	16	14	14	12	11
1525	16	5	18	6	19	8	21	9	22	11	23	12
10	654	367	80	873	586	299	12	805	518	231	1024	737
7	3	4	6	7	2	3	5	6	1	2	4 / 5	0
	13	11	11	9	9	7	6	5	4	3	2 / 31	0
1526[a]	1	14	2	15	4	17	5	18	7	20	8 / 21	0
30	450	163	956	669	382	95	888	601	314	27	820 / 533[b]	0
1	7 / 1	3	5	6	1	2	4	5	7	1	3	4
	2 / 31	30	30	28	28	26	25	24	23	22	21	19
1527	10 / 22	11	0	13	1	14	3	16	4	17	6	19
18	346 / 1039	752	465	178	971	684	397	110	903	616	329	42
3	6	78	2	3	5	6	1	3	4	6	7	2
	20	18	18	16	16	14	13	13	11	11	9	8
1528	7	20	9	21	10	23	12	0	13	2	15	3
7	835	548	261	1054	767	480	193	986	699	412	125	918
4	3	5	6	1	2	4	5	7 / 1	3	4	6	1
	9	8	7	6	5	4	2	2 / 31	30	29	28	27
1529	16	5	18	6	19	8	20	9 / 22	11	23	12	1
27	631	344	57	850	563	276	1069	782 / 495	208	1001	714	427
5	2	4	5	7	1	3	4	6	7	2	3	5
	28	27	26	25	24	23	21	21	19	19	17	16
1530	14	2	15	4	17	5	18	7	20	8	21	10
15	140	933	646	359	72	865	578	291	4	797	510	223

a. 1526 *written over* 1456.
b. 533 *written over* 535.

f. 135a

		March	April	May		June	July	August		September		October	November	December		January	February
5		4	5	7		2	3	5		6		1	2	4		5	7
		30	28	28		27	26	25		23		23	21	21		19	18
1519		10	23	11		0	13	1		14		3	16	4		17	6
17		292	5	798		511	224	1017		730		443	156	949		662	375
7		1	3	4		6	7	2		3		5	7	1		3	4
		18	17	16		15	14	13		11		11	10	9		8	6
1520		19	7	20		9	22	10		23		12	0	13		2	15
5		88	881	594		307	20	813		526		239	1032	745		458	171
1		6	7	2		3	5	6	1	2		4	5	7		1	3
		8	6	6		4	4	2	1	30		30	28	28		26	25
1521		3	16	5		18	6	19	8	21		9	22	11		23	12
25		964	677	390		103	896	609	322	35		828	541	254		1047	760
2		5	6	1		2	4	5		7		1	3	4		6	7
		27	25	25		23	23	21		20		19	18	17		16	14
1522		1	14	2		15	4	17		5		18	7	20		8	21
13		473	186	979		692	405	118		911		624	337	50		843	556
3		2	3	5		7	1	3		4		6	7	2		3	5
		16	14	14		13	12	11		9		9	7	7		5	4
1523		10	22	11		0	13	1		14		3	16	4		17	6
2		269	1062	775		488	201	994		707		420	133	926		639	352
5		6	1	2	4	5	7	1		3		5	6	1		2	4
		4	3	2	1	30	30	28		27		27	25	25		23	22
1524		19	7	20	9	21	10	23		12		0	13	2		15	3
22		65	858	571	284	1077	790	503		216		1009	722	435		148	941
6		5	7	1		3	4	6		7		2	3	5		6	1
		23	22	21		20	19	18		16		16	14	14		12	11
1525		16	5	18		6	19	8		21		9	22	11		23	12
10		654	367	80		873	586	299		12		805	518	231		1024	737
7		3	4	6		7	2	3		5		6	1	2	4	5	0
		13	11	11		9	9	7		6		5	4	3	2	31	0
1526		1	14	2		15	4	17		5		18	7	20	8	22	0
30		450	163	956		669	382	95		888		601	314	27	820	533	0
1	7	1	3	5		6	1	2		4		5	7	1		3	4
	2	31	30	30		28	28	26		25		24	23	22		21	19
1527	10	22	11	0		13	1	14		3		16	4	17		6	19
18	346	1039	752	465		178	971	684		397		110	903	616		329	42
3		6	78	2		3	5	6		1		3	4	6		7	2
		20	18	18		16	16	14		13		13	11	11		9	8
1528		7	20	9		21	10	23		12		0	13	2		15	3
7		835	548	261		1054	767	480		193		986	699	412		125	918
4		3	5	6		1	2	4		5	7	1	3	4		6	1
		9	8	7		6	5	4		2	2	31	30	29		28	27
1529		16	5	18		6	19	8		20	9	22	11	23		12	1
27		631	344	57		850	563	276		1069	782	495	208	1001		714	427
5		2	4	5		7	1	3		4		6	7	2		3	5
		28	27	26		25	24	23		21		21	19	19		17	16
1530		14	2	15		4	17	5		18		7	20	8		21	10
15		140	933	646		359	72	865		578		291	4	797		510	223

f. 135b

[Far una gallia del sesto de Fiandria e tute le raxion prozede]

Questo serà l'amaistramento da far una galia del sesto de Fiandria e de far tute le chose i prozede inchina serà aparechiada d'andar a vello o a remy, zoè de farla intera, achoredarla e armarla, chomo te serà dechiarado qui de sotto per singollo.

E primo, longe le galie de Fiandria d'alto passa 23, pie $3\frac{1}{2}$. Averà de pian la ditta pie 10. E llieva lo sesto dale corbe per mezo el posellexe dela paraschuxula $\frac{1}{3}$ de pie. E avre, mezo pie in alto, pie 12 men $\frac{2}{3}$ de pie.

E avre, pie 1 in alto, pie $12\frac{1}{2}$.

E avre, pie 2 in alto, pie 14 deda 2.

E avre, pie 3 in alto, pie 15 deda 2.

E avre, pie 4 in alto, pie 16 men deda 2.

E avre, pie 5 in allto, pie $16\frac{1}{3}$.

E avre, pie 6 in alto, pie 17 men $\frac{1}{4}$.

E avre, pie 7 in alto, pie 17 deda 2.

E à de bucha questa nostra gallia pie $17\frac{1}{2}$. Alta in choverta pie 8 men deda 2. E à de bocha la chodiera chorba da proda, pie 8 men deda 2 in alto, pie $12\frac{1}{3}$. E sia chorbe 42 in sesto.

E a proda chorba 42 e a puope e ande[a] in mezo chorbe 4. E mesurando per mezo la chorba de mezo dil'oro di sul madier di bucha e al'oro di su dela zenta, die 'ser pie $1\frac{1}{2}$, mesurando per la via d'i furchami. Erze la zenta al'inpostura da proda pie 9, mesurando al quadro. Erze lo madier di bucha al'inpostura da proda pie $10\frac{1}{2}$, mesurando al quadro. E lanza pie $10\frac{1}{2}$.

---

a. *Sic in MS.*

1. *Sesto* has several meanings in relation to Venetian shipbuilding. At its most literal level (derived from the word for one-sixth), it refers to the drafting compass. In terms of ship timbers, it designates a bend or curve along the hull, and specifically that used as a template from which others are derived. In more general contexts, as here, it is used to distinguish various ship classes or designs, akin to the technical English term *mold.*

2. The entire translation of the shipbuilding section of the manuscript owes much to the assistance of Alan Hartley, who did a full revision of a preliminary draft of the section and supplied an annotated glossary of technical terms. For discussion of the context and composition of this section, see David McGee, "The Shipbuilding Text of Michael of Rhodes," vol. 3, pp. 211–241, and Mauro Bondioli, "Early Shipbuilding Records and the Book of Michael of Rhodes," vol. 3, pp. 243–279. For the measures used in this manuscript, see Appendix below.

f. 135b

To make a galley of the Flanders design and all of its dimensions

These will be the instructions for making a galley of the Flanders design[1] and for accomplishing all of the things and procedures until it is fit for going under sail or under oars, that is, to make it complete, to rig it and fit it out, as will be explained to you below item by item.[2]

And first, the length of the Flanders galleys at the top[3] is 23 paces, $3\frac{1}{2}$ feet; it will have a floor of 10 feet. And the mold of the frames rises $\frac{1}{3}$ of a foot at the mark[4] at the middle of the bilge stringer.[5] And at a half foot high,[6] it opens 12 feet minus $\frac{2}{3}$ of a foot wide.

And at 1 foot high, it opens $12\frac{1}{2}$ feet wide.

And at 2 feet high, it opens 14 feet 2 fingers wide.

And at 3 feet high, it opens 15 feet 2 fingers wide.

And at 4 feet high, it opens 16 feet minus 2 fingers wide.

And at 5 feet high, it opens $16\frac{1}{3}$ feet wide.

And at 6 feet high, it opens 17 feet minus $\frac{1}{4}$ foot wide.

And at 7 feet high, it opens 17 feet 2 fingers wide.

The breadth of this our galley is $17\frac{1}{2}$ feet. The deck height is 8 feet minus 2 fingers. And the forward tail frame[7] at 8 feet minus 2 fingers high is $12\frac{1}{3}$ feet wide. And there are 42 molded frames in the design.

And there are 42 frames forward and aft, and 4 frames go amidships. And measuring at the midship frame from the top edge of the sheer strake[8] to the top of the main wale, it should be $1\frac{1}{2}$ feet, measuring along the futtocks.[9] The main wale is 9 feet high at the stempost, measuring vertically. The sheer strake is $10\frac{1}{2}$ feet high at the stempost, measuring vertically. It rakes $10\frac{1}{2}$ feet.

3. I.e., on deck.
4. The *posellexe* was a reference mark scribed on various timbers of a galley before construction.
5. *Paraschuxula* is a term that is known principally from this manuscript and texts believed to be derived from it. It apparently refers to an internal lengthwise timber running at the bottom of the ship, though it might rather be an external timber.
6. That is, a half foot above an understood point, probably the top of the keel.
7. The *chodiere chorbe* were the frames, one forward and one aft, between which the successive frames could be designed according to the usual proportional processes; beyond them the frames had to be custom-designed by the shipwright.
8. One of the broad, thick timbers forming the outer side of the ship.
9. The timber pieces forming the ship's frame.

f. 136a E lieva, per mezo el poselixe del choltro, dedo 1. E mesurando du[a] ssu la zenta e al'oro di su el madier di bucha die 'ser pie $1\frac{1}{2}$.

Erze l'inpostura da pope pie 13, mesurando al quadro. Lanza pie $10\frac{1}{2}$.

E lieva, per mezo el poselexe del choltro, quarta una e dedo uno. Mesurando dala zenta e al'oro di su dal triganto die 'ser pie 3 e meza quarta de pie.

E mesurando, per mezo la chodiera chorba da proda, de su la cholonba e al'oro di su dala zenta, die 'ser pie 7 men dedo 1, mesurando al quadro.

E mesurando, per mezo la chorba de pie 18 a proda, di su la cholonba e l'oro di su dala zenta, die 'ser pie $6\frac{1}{3}$ dedo 1, al quadro.

E per[b] mesurando, per mezo la chorba de pie 18 a pope,[c] di su la cholonba e l'oro di su dela zenta, die 'ser pie 7 men meza quarta, al quadro.

E mesurando, per mezo la chodiera[d] da pope, di su la cholonba e al'oro di su dela zenta, die 'ser pie 8 men 1 terzo de pe', mesurando al quadro.

E mesurando, per mezo la chodiera chorba da proda, dal'oro di su dala zenta e al'oro di su al madier de bocha, die 'ser pie $1\frac{1}{2}$, mesurando per la via d'i furchami.

E mesurando, per mezo la chorba del 18, dal'oro di su dala zenta e al'oro di su dal madier di bocha, die 'ser pie $1\frac{1}{2}$, mesurando a proda e per la via d'i furchami.

E mesurando, per mezo la chorba di mezo, dal'oro di su dala zenta e al'oro di su dal madier di bucha, die 'ser pie $1\frac{1}{2}$, mesurando per la via d'i furchami.

E mesurando, per mezo la chorba de 18 a pope, al'oro di su dala zenta e al'oro di su al madier di bucha, die 'ser pie 2 men $\frac{1}{4}$, mesurando per la via d'i furchami.

f. 136b E mesurando, per la via per mezo la chodiera chorba di pope, dil'oro di su dala zenta e al'oro di su al madier di bocha, die 'ser pie $2\frac{1}{2}$ $\frac{1}{3}$ de pie, mesurando per la via d'i furchami.

E mesurando, dal'oro di fuora dal'inpostura di pope e al'oro di pope di la timonera, die 'ser pie 5 men $\frac{1}{4}$, mesurando sotto per la zenta. Avre in paraschene pie 4 men $\frac{1}{4}$, mesurando al'oro dentro dele paraschene.

---

a. *Sic in MS.*
b. *Sic in MS.*
c. a pope *corrected above the line for* a proda.
d. *More correctly* chodiera chorba.

And at the coulter[1] mark, it rises 1 finger. And measuring from the top of the main wale to the top f. 136a
edge of the sheer strake, it should be $1\frac{1}{2}$ feet.

The sternpost is 13 feet high measuring vertically. It rakes $10\frac{1}{2}$ feet.

And at the after coulter mark it rises one and a quarter fingers. Measuring from the main wale to the top edge of the transom,[2] it should be 3 and half-of-a-quarter feet.

And measuring at the forward tail frame, from the top of the keel to the top edge of the main wale, it should be 7 feet minus one finger, measuring vertically.

And measuring at frame 18 forward,[3] from the top of the keel to the top edge of the main wale, it should be $6\frac{1}{3}$ feet vertically.

And measuring at frame 18 aft, from the top of the keel to the top edge of the main wale, it should be 7 minus half of a quarter feet vertically.

And measuring at the after tail frame, from the top of the keel to the top edge of the main wale should be 8 minus 1 third feet, measuring vertically.

And measuring at the forward tail frame, from the top edge of the main wale to the top edge of the sheer strake should be $1\frac{1}{2}$ feet, measuring along the futtocks.

And measuring at frame 18 [forward], from the top edge of the main wale to the top edge of the sheer strake should be $1\frac{1}{2}$ feet, measuring at the forward side and along the futtocks.

And measuring at the midship frame, from the top edge of the main wale to the top edge of the sheer strake should be $1\frac{1}{2}$ feet, measuring along the futtocks.

And measuring at frame 18 aft, from the top edge of the main wale to the top edge of the sheer strake should be 2 minus $\frac{1}{4}$ feet, measuring along the futtocks.

And measuring at the after tail frame, from the top edge of the main wale to the top edge of the f. 136b
sheer strake should be $2\frac{1}{2}\,\frac{1}{3}$ feet,[4] measuring along the futtocks.

And measuring from the outside edge of the sternpost to the after edge of the rudder bracket, it should be 5 minus $\frac{1}{4}$ feet measuring below along the main wale. It opens 4 minus $\frac{1}{4}$ feet in its screens,[5] measuring to the inside edge of the screens.

---

1. The *choltro* was probably a piece of compass timber at either end of the keel where the keel joins the stempost or sternpost; this section apparently refers to the mark on the forward coulter.
2. The transverse beam across the sternpost.
3. I.e., from the middle of the galley.
4. I.e., $2\frac{5}{6}$ feet.
5. The *paraschene* were bulwarks or screens on either side of the stern.

E à de palmetta in proda pie 8 meza quarta de pie, mesurando dal'oro di fuora dal'inpostura e al mezo e al chavo, mesurando per la via del madier di bocha.

E à de palmetta in pope pie 10 men $\frac{1}{3}$ de pe', mesurando dal'oro di fuora dal triganto e a mezo lo zuovo, mesurando per la via del madier di bocha. Die 'ser longo lo morelo che se parte le late pie 2 men $\frac{1}{4}$ de pe'. Mesurando per mezo lo zuovo di pope, dal'oro di fuora del madier di bocha e al'oro dentro dela bandullina, die 'ser pie 1. Mesurando per mezo la chadena da cholo, dil'oro di fuora del madier di bocha e al'oro dentro dala bandulina, die 'ser pie 1 e meza quarta de pie.

E xe la schaza a latte dixeotto chon lo zuovo da proda. E à bastarde 8. E la schaza xe alo sogier da proda dala porta del marango[*n*]. E à latte 4. In lo sogier da puope, dala porta xe latte 6 chon lo zuovo da proda. Alo sogier di pope, dala porta d'i scrivany, xe sula bastarda che va in sie alo sogier di proda, in suxo 1ª bastarda postiza che se mette. È averta questa porta a pope e a proda pie 4. E lo sogier da pope, dela porta de scandoler, xe a latte 11 e lo sogier di proda a latte 13 chon lo zuovo. Se anpia la stazia pie $1\frac{1}{2}$.

f. 137a E mesurando a meza galia, xe averte le crosie pie e meza quarta, | mesurando al'oro dentro dale crosie. E avre a proda le crosie pie 2 men $\frac{1}{3}$, mesurando dentro al'oro dale crosie. E va la staza sula chorba che va in 22 a proda. Mesurando a proda di su le latte de choverta e al'oro di su dela crosia, die 'ser pie 2 men mezo terzo de pie. E tanto xelo a meza galia chomo a proda.

E mesurando a puope, di su le tuole de choverta e al'oro di su dele crosie, pie 2. Mesurando dil'oro di fuora di su la crosia al'oro di fuora dila choverta, die 'ser pie 4. Mesurando dil'oro di fuora dila crosia e al'oro dentro dala banda, die 'ser pie 9. E tanto xelo a proda e tanto a puope. Mesurando dal'oro di fuora dala banda e al'oro di fuora dila postiza, die 'ser pie $1\frac{1}{2}$, tanto a proda quanto a pope.

E là se tira una trazuolla a proda, dil'oro di su dele postize, e la trazuola va più alta cha la crosia deda 2.

E mesurando a meza gallia del'oro di su dale postize, va più alte cha l'oro di su dele crossie $\frac{1}{4}$ de pe'. Mesurando a puope, se più basso l'oro di su dele postize cha l'oro di ssu dele crosie meza quarta de pie. E mesurando dal'oro di proda del zuovo da proda inchina al primo schermo postizo, die 'ser pie $1\frac{1}{2}$. Mesurando dal'oro di pope, del zuovo di puope inchina al primo schermo postizo, die 'ser pie $2\frac{1}{3}$. Mesurando da questo primo scermo di pope postizo inchina al segondo postizo, die 'ser pie

1. A crosspiece fixed to the head of the rudder.
2. A *morello* could be either a measurement or a measuring stick.
3. The *bandulina* was apparently a longitudinal stringer.
4. The *chadena de cholo* was a deck beam that had a neck at each end rising to support the outrigger.
5. A *bastarda* was a short deck beam on either side of an opening in the deck.
6. The *sogier* may have been a coaming, a raised border around the hatch to keep out water.
7. The *banda* was a longitudinal plank below the outriggers.

And the foredeck is 8 and half-of-a-quarter feet, measuring from the outside edge of the stem to the middle and to the end, measuring along the sheer strake.

And it has an afterdeck of 10 minus $\frac{1}{3}$ feet, measuring from the outside edge of the transom to the middle of the yoke,[1] measuring along the sheer strake. The measurement[2] that separates the deck beams should be 2 minus $\frac{1}{4}$ feet long. Measuring at the after yoke, from the outside edge of the sheer strake to the inside edge of the *bandulina*[3] should be 1 foot. Measuring at the neck deck beam,[4] from the outside edge of the sheer strake to the inside edge of the *bandulina* should be 1 and half a quarter foot.

And the mast step is at deck beam eighteen counting the forward yoke. And it has 8 half beams.[5] And the mast step is at the forward *sogier*[6] of the carpenter's hatch. And it has 4 beams. In the aft *sogier*, from the hatch there are 6 deck beams counting the forward yoke. At the after *sogier* of the pursers' hatch, there is put on top a false half beam on the half beam that goes in itself to the forward *sogier*. This hatch is open 4 feet wide aft and forward. And the after *sogier* of the hatch of the aftermost compartment of the hold is at deck beam 11, and the forward *sogier* is at deck beam 13 counting the yoke. The mast step is $1\frac{1}{2}$ feet wide.

And measuring at the middle of the galley, the gangway is one and half-of-a-quarter feet wide | f. 137a
measuring between the inside edges of the side pieces of the gangway. And at the bow the gangway opens 2 minus $\frac{1}{3}$ feet, measuring between the inside edges of the side pieces of the gangway. And the mast step goes on the floor timber at 22 forward. Measuring at the bow from the top of the deck beams to the top edge of the gangway should be 2 minus half a third feet. And it's the same at the middle of the galley as at the bow.

And measuring at the stern, from the top of the deck planking to the top edge of the side pieces of the gangway, 2 feet. Measuring from the outside edge of the top of the gangway to the outside edge of the deck should be 4 feet. Measuring from the outside edge of the gangway to the inside edge of the *banda*[7] should be 9 feet. And so it is at the bow and at the stern. Measuring from the outside edge of the *banda* to the outside edge of the outrigger[8] should be $1\frac{1}{2}$ feet, the same forward as aft.

And a string line is stretched there at the bow, between the top edges of the outriggers, and the string line goes 2 fingers above the gangway.

And measuring amidships from the top edge of the outriggers, it goes $\frac{1}{4}$ foot above the top of the gangways. Measuring at the stern, the top edge of the outriggers is one half of a quarter of a foot lower than the top edge of the side pieces of the gangway. And measuring from the forward edge of the forward yoke to the first *postizo*[9] thole pin[10] should be $1\frac{1}{2}$ feet. Measuring from the after edge of the after yoke to the first *postizo* thole pin should be $2\frac{1}{3}$ feet. Measuring from this first *postizo*

---

8. A strong beam that secures the masts.
9. The middle of the three rowers on a bench.
10. The peg that serves as fulcrum for the oars.

$3\frac{1}{2}$ men dedo 1. E va chosì a ordine inchina pope. E dal schermo postizo inchina al trezaruol die 'ser palmo 1, e quando se inbancha là se mette la trazuola sulo schermo postizo. E lo bancho bava dala trazuola per mezo dala crosia $\frac{1}{4}$ de pie. E bava lo bancho dala trazuola per mezo el pe' del bancho $\frac{3}{4}$
f. 137b de pie. E die 'ser alti i pie d'i banchi pie 2. E getta suli chavy delo zuovo | di proda $\frac{1}{3}$ de pie. E getta suli chavy del zuovo di pope mezo terzo de pie. E getta suli chavy dele latte de mezo $\frac{2}{3}$ de pie, mexurando per mezo la banda. E à de bolson la ditta gallia de me[*n*] $\frac{3}{4}$ de pie. E mesurando dal'oro di fuora del'inpostura da pope inchina 'l dente delo spironzello, die 'ser pie 4. Mexurando de questo dente de questo spironzello inchina al dente dela parascena, die 'ser pie 5. Mesurando dal'oro di su dal madier di bucha inchina a basso, chavado dala chavriola, die 'ser pie 3 men mezo terzo de pie. E xe longa questa nostra gallia de Fiandria, da l'un poselexe al'oltro d'i choltri, passa 19 pie 3. E fiero chon la chodiera chorba di proda, lonzi dal poselexe dal choltro, pie 7 men meza quarta. E fiero chon la chodiera chorba da pope, lonzi dal posellexe dal choltro, pie 9 men $\frac{1}{4}$ de pie.

[*caption:*]
galia non chonpida

f. 138a Questo serà l'amaistramento a che muodo se mette le maistre de questa nostra gallia del sesto de Fiandria.

Mesurando dal'oro del poselexe del choltro da proda inchina in chavo dale maistre dela paraschuxula, die 'ser pie $4\frac{1}{2}$, mexurando per la via del panixello.

Mesurando dal'oro di su dala maistra dala paraschuxula, dal'oro di su dela maistra de mezo, die 'ser pie 4 men $\frac{1}{4}$, per la via del panixello.

Mesurando dal'oro di su dala maistra de mezo, e dal'oro di su dala maistra de sovra, die 'ser pie 5 e deda 2, per la via del panixelo.

Mesurando per mezo la chodiera chorba di proda, e al'oro di su dala maistra dala paraschuxulla, più basso dal poselexe dela proda dala paraschoxola, $\frac{1}{3}$ de pie.

Mesurando dal'oro di su dala paraschuxula inchina l'oro di su dala maistra de mezo, die 'ser pie 3 men $\frac{1}{3}$, mesurando per la via del furchame.

---

1. The outboardmost rower on a bench.
2. The outer end of the bench sits on a short stanchion.
3. The upward convexity of deck beams.
4. The *spironzello* must have been some sort of small projection at the stern.

thole pin to the second *postizo* should be $3\frac{1}{2}$ feet minus 1 finger. And it goes in this order to the stern. And from the *postizo* thole pin to the *trezaruol*[1] thole pin should be 1 palm, and when the bench is installed there, the string line is placed there on the *postizo* thole pins. And at the gangway, the bench is $\frac{1}{4}$ of a foot away from the string line. And the bench is $\frac{3}{4}$ of a foot distant from the string line at the bench stanchion.[2] And the bench stanchions should be 2 feet high. And it extends above the ends of the forward yoke | $\frac{1}{3}$ foot. It extends half of a third of a foot above the ends of the after yoke. And it extends $\frac{2}{3}$ of a foot above the ends of the midship deck beams, measuring along the *banda*. And this ship has a camber[3] of less than $\frac{3}{4}$ of a foot. And measuring from the outside edge of the sternpost to the point of the *spironzello*[4] should be 4 feet. Measuring from this point of the *spironzello* to the point of the stern screen should be 5 feet. Measuring from the top edge of the sheer strake to the lower concavity of the figurehead should be 3 minus one third feet. And this Flanders galley of ours is 19 paces 3 feet long from one keel mark to the other. And it will be,[5] with the forward tail frame, 7 feet minus half a quarter distant from the keel mark. And it will be, with the after tail frame, 9 minus $\frac{1}{4}$ feet distant from the keel mark. f. 137b

[*caption:*]
Unfinished galley

These will be the instructions for placing the framing ribbands[6] of our galley of the Flanders design. f. 138a

Measuring from the edge of the forward keel mark to the top of the framing ribbands of the bilge stringer, it should be $4\frac{1}{2}$ feet, measuring along the garboard strake.[7]

Measuring from the upper edge of the bilge stringer to the upper edge of the middle framing ribband, it should be 4 minus $\frac{1}{4}$ feet, along the garboard strake.

Measuring from the upper edge of the middle framing ribband to the upper edge of the upper framing ribband, it should be 5 feet and 2 fingers, along the garboard strake.

Measuring at the forward tail frame, the upper edge of the framing ribband of the bilge stringer is $\frac{1}{3}$ foot lower than the forward mark of the bilge stringer.

Measuring from the upper edge of the bilge stringer to the upper edge of the middle framing ribband, it should be 3 minus $\frac{1}{3}$ feet, measuring along the futtock.

---

5. The expression here, *fiero chon*, is problematic. From the context, it expresses the length of the ship forward and aft from the keel marks, though the total length works out to 22 paces $3\frac{5}{8}$ feet rather than 23 paces $3\frac{1}{2}$ feet given at the beginning of the section. Linda Carroll has identified *fiero* as a variant third-person plural future form of the verb *essere*. Mauro Bondioli has seen it as a first-person singular; cf. Bondioli, "Early Shipbuilding Records and the Book of Michael of Rhodes," vol. 3, p. 247, n. 17.
6. The long narrow strips of timber fixed longitudinally to the outside of the frames to hold them in place until the exterior planking is added.
7. The first range of wooden planks on the outer hull next to the keel.

Mesurando dal'oro di su dela maistra de mezo e al'oro di su dela maistra de sovra, die 'ser pie $4\frac{1}{2}$, mesurando per la via del furchame.

Mesurando dal poselexe dal choltro al'inpostura di pope inchina in chavo dele maistre dala paraschuxulla, die 'ser pie 3, mesurando per la via del panixelo, e quarta meza de pie.

Mesurando dal'oro di su dela maistra dala paraschuxulla e al'oro di su dala maistra de sovra, zoè de mezo, die 'ser pie 5 e deda 2, mesurando per la via del panixello.

Mesurando dal'oro di su dela maistra de mezo e al'oro di su dela maistra de sovra, die 'ser pie $6\frac{1}{2}$, mexurando per la via del panixello.

f. 138b Mesurando per mezo la chudiera chorba di pope xe più basso l'oro di su | la maistra dala paraschuxula cha 'l poselexe dala paraschuxula, quarta una e meza de pie.

Mesurando dal'oro di su dala maistra dala paraschuxulla, e al'oro di su dela maistra de mezo, die 'ser pie 3 men $\frac{1}{3}$, mesurando per la via d'i furchamy.

Mesurando dal'oro di su dala maistra de mezo e al'oro di su dela maistra de sovra, die 'ser pie $5\frac{1}{2}$, mesurando per la via d'i furchami.

Mesurando per mezo la chorba de mezo, xe più baso l'oro di su dala maistra dala paraschuxulla cha 'l posellexe dala paraschosola palmo 1, e chi vedirì le maistre sì de pope chome de proda.

Quista de soto se die metter sul poselexe del choltro da proda.

[*captions:*]
avre pie $9\frac{1}{4}$

dedo 1 grosso

avre pie $6\frac{1}{2}$

erze pie 6

dedo 1 groso

maistra

erze pie $2\frac{1}{4}$

avre pie 4 men $\frac{1}{4}$

Measuring from the upper edge of the middle framing ribband to the upper edge of the upper framing ribband, it should be $4\frac{1}{2}$ feet, measuring along the futtock.

Measuring from the after coulter mark to the top of the framing ribbands of the bilge stringer, it should be 3 and a quarter-half feet, measuring along the garboard strake.

Measuring from the upper edge of the framing ribband of the bilge stringer to the upper edge of the upper framing ribband, that is, amidships, it should be 5 feet and 2 fingers, measuring along the garboard strake.

Measuring from the upper edge of the middle framing ribband to the upper edge of the upper framing ribband, it should be $6\frac{1}{2}$ feet, measuring along the garboard strake.

Measuring at the after tail frame, the upper edge of the | bilge stringer framing ribband is a quarter and half of a foot lower than the mark of the bilge stringer. f. 138b

Measuring from the upper edge of the bilge stringer framing ribband to the upper edge of the middle framing ribband, it should be 3 minus $\frac{1}{3}$ feet, measuring along the futtocks.

Measuring from the upper edge of the middle framing ribband to the upper edge of the upper framing ribband, it should be $5\frac{1}{2}$ feet, measuring along the futtocks.

Measuring at the midship frame, the upper edge of the framing ribband of the bilge stringer is 1 palm lower than the mark of the bilge stringer, and here you will see the framing ribbands aft as well as forward.

This below should be placed on the stem coulter mark.

[*captions:*]
It opens $9\frac{1}{4}$ feet

1 thumb

It opens $6\frac{1}{2}$ feet

It rises 6 feet

1 thumb

Framing ribband

It rises $2\frac{1}{4}$ feet

It opens 4 feet minus $\frac{1}{4}$

avre qua $\frac{2}{3}$

f. 139a Quista se die meter sul poselexe del choltro de pope.

[*captions:*]
avre pie $9\frac{1}{2}$

quarta men una chana

avre pie $6\frac{1}{4}$

erze pie $6\frac{1}{4}$

quarta una men una chana

maistra

avre pie 4 men $\frac{1}{3}$

erze pie $1\frac{1}{4}$

erze pie $\frac{1}{2}$

avre qua $\frac{2}{3}$ de pie

Questo inchontro serà el chavo di questa nostra galia de Fiandria, zoè da proda.

f. 139b [*captions:*]
chavo di proda

erze pie $10\frac{1}{2}$ el madier di bucha

lanza pie $10\frac{1}{2}$

pie 5 1 dedo

$\frac{1}{4}$ de pe'

$\frac{1}{4}$ $\frac{1}{2}$ de pe' 1[a] chana

$\frac{1}{3}$ de pe' 1[a] chana

pie 3 e me[*zo*] terzo de pe'

It opens here $\frac{2}{3}$ feet

This should be placed on the after coulter mark. f. 139a

[*captions:*]
It opens $9\frac{1}{2}$ feet

A quarter minus one cane

It opens $6\frac{1}{4}$ feet

It rises $6\frac{1}{4}$ feet

A quarter minus one cane

Framing ribband

It opens 4 feet minus $\frac{1}{4}$

It rises $1\frac{1}{4}$ feet

It rises $\frac{1}{2}$ foot

It opens here $\frac{2}{3}$ of a foot

On the facing page will be the head of our galley of Flanders, that is, the bow.

[*captions:*] f. 139b
Bow

The sheer strake is $10\frac{1}{2}$ feet high

It rakes $10\frac{1}{2}$ feet

5 feet 1 finger

$\frac{1}{4}$ foot

$\frac{1}{4}$ $\frac{1}{2}$ foot 1 cane

$\frac{1}{3}$ foot 1 cane

3 feet and half of a third of a foot

quarte 3 e meza de pe'

pe' 1 dedo 1

$\frac{2}{3}$ de pe'

dedo 1

Questo di sotto serà el chavo di puope de questa nostra galia de Fiandria.

[*captions:*]
chavo di pope

erze pie 13 al'oro del triganto

lanza pie $10\frac{1}{2}$

$\frac{1}{4}$

pie un men $\frac{1}{4}$

pie 1 dedo 1

$\frac{2}{4}$ dedo 1

pie $4\frac{1}{4}$ dedo 1

$\frac{3}{4}$ [*de pe'*] e[a] d[*e*]do 1

pe' 1 men 1ª cana

quarte $2\frac{1}{2}$ de pe'

dedo $1\frac{1}{4}$

f. 140a Questo di sutto serà l'amaistramento del morelo de questa nostra galia de Fiandria, zoè groseza e anpieza. Vedirì qui de sotto.

[*captions:*]
per longuo chossì ampieza de postize

---

a. meza *crossed out.*

3 and a half quarters of a foot

1 foot 1 finger

$\frac{2}{3}$ foot

1 finger

Below here is the stern of our galley of Flanders.

[*captions:*]
Stern

It is 13 feet up to the edge of the transom

It rakes $10\frac{1}{2}$ feet

$\frac{1}{4}$

1 foot minus $\frac{1}{4}$

1 foot 1 finger

$\frac{2}{4}$ [feet] 1 finger

$4\frac{1}{4}$ feet 1 finger

$\frac{3}{4}$ and 1 finger

1 foot minus 1 cane

$2\frac{1}{2}$ quarters of a foot

$1\frac{1}{4}$ fingers

Below will be instructions for the *morelli*[1] of our Flanders galley, that is, the thickness and width. You will see below. f. 140a

[*captions:*]
As the length [of this rectangle], so the width of the outrigger beams

1. The *morelli* were special measuring sticks, specific to the ship being built.

per longo chossì groseza di postize

f. 140b Questo chomenza da ertto e vuol dir pie $8\frac{1}{2}\frac{1}{4}$.

[*captions:*]
alta in chovertta pie 8 men 2 deda

pie $8\frac{1}{2}\frac{1}{4}$	pie 8 men deda 2
pie $8\frac{1}{2}$	pie 7
pie 8 dedo 1	pie 6
pie 8 mezo terzo	pie 5
pie 8	pie 4
pie $7\frac{1}{2}$ dedo 1	pie 3
pie 1 dedo 1	pie 2
pie 6 dedo 1	pie 1
$\frac{1}{3}$ pie 6 men $\frac{1}{3}$	pie $\frac{1}{2}$

de pian pie 5

Questa indredo serà ampieza e groseza del murello.

f. 141a [*captions:*]
tanto vuol eser grossa l'inpostura da puope e da proda per mezo li choltri intranbe

tanto l'anpieza del'inpostura da puope e da proda per mezo lo choltro e deda 4 più

groseza del'inpostura di proda per mezo el madier di bucha e tanto quela di pope per mezo la zenta

anpieza dela paraschuxulla

f. 141b [*captions:*]
la cholonba xe alto tanto per mezo la chodiera corba di proda

As the length [of this rectangle], so the thickness of the outrigger beams

This starts at the top and means $8\frac{1}{2}\frac{1}{4}$ feet. f. 140b

[*captions:*]
Height at the deck 8 feet minus 2 fingers

$8\frac{1}{2}\frac{1}{4}$ feet	8 feet minus 2 fingers
$8\frac{1}{2}$ feet	7 feet
8 feet 1 finger	6 feet
8 feet and a half third	5 feet
8 feet	4 feet
$7\frac{1}{2}$ feet 1 finger	3 feet
7[1] feet 1 finger	2 feet
6 feet 1 finger	1 foot
$\frac{1}{3}$ 6 feet minus $\frac{1}{3}$	$\frac{1}{2}$ foot

Floor, 5 feet

On the back of this will be the width and thickness of the *morelli.*

[*captions:*] f. 141a
The sternpost and stempost should each be this thick at the coulters

The width of the sternpost and stempost at the coulters, this much and 4 fingers more

The thickness of the stempost at the sheer strake and likewise that of the sternpost at the main wale.

Width of the bilge stringer

[*captions:*] f. 141b
The keel is this thick at the forward tail frame

---

1. *1* in MS.

la chorba

la chodiera xe alta tanto per mezo la chodiera chorba di pope

chanpo dele gallie de Fiandria tuto questo men quel pizetto de chavo

groseza dela paraschuxulla

f. 142a Questa nostra gallia del sesto de Fiandria è chonpida e messa in ordene. El so chorpo la vederì qui de sotto depenta.

E se vosse saver che legname andarà per questa gallia andè a charte 202, e per llo simille feramenta.[a]

[*caption:*]
gallia

f. 142b Quista galia inchontro del sesto de Fiandria vuol 1 alboro de passa 14. Vuol volzer al so redondo palmi 7. Vuol un cholzexe longo pie 12. Vuol eser alargo lo ditto cholzexe el quinto de zò che lo cholzexe longo.

E vuol la dita gallia antena de passa 19. Vuol eser grosa in lo so redondo palmy $4\frac{3}{4}$, e vuol vuolzer siando chavalchada pie $3\frac{3}{4}$.

E vuol un alboro de mezo, el qual vuol eser longuo passa [. . .],[b] e vuol volzer in lo so redondo pie [. . .], e vuol eser el cholzese longo passa [. . .], pie [. . .], e vuol eser alargo pie [. . .].

E vuol antena per la mezana de passa [. . .]. Vuol volzer in llo so dopio pie [. . .].

Vuol la ditta gallia un penon de passa 14, vuol volzer palmy $3\frac{2}{3}$, chomo vedirì qui de sotto per figura.

f. 143a E vuol la nostra galia de Fiandria barcha 1ª longa pasa [. . .][c] e pie [. . .]; e vuol eser longa la so cholonba pie [. . .]; e vol eser erta pie [. . .]; e vuol eser averta in bocha pie [. . .]; e vuol de pian pie [. . .].

E vuol la ditta gallia chopano longo pie [. . .]; e vuol eser erto pie [. . .]; e vuol eser alargo in bocha pie [. . .]; e vol eser in pian pie [. . .]; e vuol eser in cholonba pie [. . .]. Vedirì qui de sotto per figura.

---

a. *Paragraph added, in the same hand, almost certainly after the drawing of the galley, as can be inferred from the varied ink color and from the inner margin, indented to avoid superimposition on the image.*
b. *This measurement and others on the page left blank.*
c. *This measurement and others on the page left blank.*

The frame

The keel is this thick at the after tail frame

The space between frames of the Flanders galleys is all this minus that small piece of rope

Thickness of the bilge stringer

This our galley of the Flanders design is completed and put in order. You will see its hull depicted f. 142a
below.

And if you wish to know what wood will go into this galley, go to folio 202, and likewise for the iron fittings.

[*caption:*]
Galley

This galley of Flanders opposite requires 1 mast of 14 paces. It should measure 7 palms around. It f. 142b
requires a block mast[1] 12 feet long. The block mast needs to be one fifth as wide as it is long.

And this galley needs a yard of 19 paces. It should be $4\frac{3}{4}$ palms around, and it should be $3\frac{3}{4}$ feet around when it is overlapped.

And it needs a midship mast which should be [. . .][2] paces long and [. . .] feet in diameter, and the block mast should be [. . .] paces and [. . .] feet long, and it should be [. . .] feet across.

And it needs a yard of [. . .] paces for the midship mast. And it should be [. . .] feet around where it is fished.

This galley needs an upper spar of the lateen yard of 14 paces; it should be $3\frac{2}{3}$ palms around, as you will see in the figure below.

And our Flanders galley needs a boat [. . .][3] paces and [. . .] long, and its keel should be [. . .] feet f. 143a
and should have a depth of [. . .] feet; and it should have a beam of [. . .] feet; and it should have a floor of [. . .] feet.

And this galley needs a skiff [. . .] feet long; and it should have a depth of [. . .] feet and should have a beam of [. . .] feet; and it should be [. . .] feet in the floor; and its keel should be [. . .] feet. You will see it figured below.

1. The *cholzexe* was a spar scarfed to the top of the mast on which the yard ties ran.
2. This measurement and others on the page left blank.
3. This measurement and others on the page left blank.

[*captions:*]
barcha[a]

chopano[b]

f. 143b Questo serà lo fornymento da sartia che vuol la nostra galia de Fiandria per achoredar l'alboro grande, el pizollo, e sartia per remezo, e fornymento d'antena. Vedirì qui de sotto.

Vuol chanavi 5, longi de passa 70 l'uno. Die pexar per passo libre 10.[c] Die pexar tuti 5 libre 3500.

Vuol la ditta gripie 5, longa l'una passa 70. Die pesar el paso libre 4. Die pexar tuti 5 libre 1400.

Vol prodixe un, vuol eser longo passa 80. Die pesar el passo libre 5. Pexarà tutto libre 400.

Vuol poza una de passa 18. Die pexar el passo libre 10.

Vuol manti 2, longi passa 14 l'un. Die pexar el passo libre 10.

Vuol suste de passa 45 l'una. E vuol pesar el paso libre 4.

Vuol gomene 2 de passa 70 l'una. Die pexar el passo libre 4.

Vuol menador d'i prodeny 2, longi de passa 70 l'un. Die pesa[*r*] el passo libre 4.

Vuol anize 2 de passsa 5 l'una. Ancuor anize 2 de passa $4\frac{1}{2}$ l'una. Die pexar el passo de zaschaduna libre 6.

Vuol amo un de passa 50. Die pexar el passo libre 4.

Vuol funde 2 de passa 36 l'una. Die pexar el passo libre 2.

Vuol orza davanti una, longa de passa [. . .].[d] Die pexar el passo libre 2.

Vuol orza poza de passa 36. Die pexar el paso libre 2.

---

a. *Above the drawing*
b. *In the water between the boats in the drawing.*
c. 10 *corrected over* 70.
d. *Left blank.*
1. The rope that secured the ship to the quay.
2. A rope attached to the lower edge of a sail to help in securing or turning it.
3. Ropes for raising or lowering the sails.
4. Ropes by which a yard was suspended.
5. Ropes used to steady sails.

[*captions:*]
Boat

Skiff

This will be the rigging that our Flanders galley needs to equip the large and small masts and spare rigging, and rigging for the yard. You will see this below. f. 143b

It needs 5 hemp anchor cables, each 70 paces long. Each pace should weigh 10 pounds; all five should weigh 3,500 pounds.

It needs 5 buoy ropes each 70 paces long. Each pace should weigh 4 pounds. All 5 should weigh 1,400 pounds.

It needs a headfast[1] 80 paces long. This should weigh 5 pounds per pace. The whole will weigh 400 pounds.

It needs a sheet[2] 18 paces long. This should weigh 10 pounds per pace.

It needs 2 halyard[3] ties[4] each 14 paces. Each pace should weigh 10 pounds.

It needs vangs[5] of 45 paces each. And each pace should weigh 4 pounds.

It needs 2 halyard tackle[6] falls[7] of 70 paces each. Each pace should weigh 4 pounds.

It needs falls for 2 mast tackles, each 70 paces long. Each pace should weigh 4 pounds.

It needs 2 rudder tackle falls of 5 paces each, also 2 rudder tackle falls of $4\frac{1}{2}$ paces each. Each pace should weigh 6 pounds.

It needs one *amo*[8] of 50 paces. Each pace should weigh 4 pounds.

It needs 2 *funde*[9] of 36 paces each. Each pace should weigh 2 pounds.

It needs one lateen tack tackle of [. . .][10] paces. Each pace should weigh 2 pounds.

It needs an *orza poza*[11] of 36 paces. Each pace should weigh 2 pounds.

---

6. The rigging used in operating the sails.
7. A rope used in hoisting.
8. It is not clear what sort of tackle the *amo* was.
9. It is not clear what sort of tackle the *funde* were.
10. Left blank.
11. It is not clear what sort of tackle the *orza poza* was.

Vuol orze pope 2, longa l'una passa 20, per passo libre 4.

Vuol matta una de passa 20 al peso dele orze pope.

Vuol braza una per suste, de pasa 13. Die pexar per passo libre 6.

Vuol braza una per orza, de passa 3. Die pexar per passo libre 4.

Vuol una poza de pasa 20. Die pexar per passo libre 7.

Vuol chusidure 4 de passa 3 l'una. Per passo libre $1\frac{1}{2}$.

Vuol anzollo longo de passa 40. Die pexar per passo libre $2\frac{1}{2}$.

f. 144a Quarnalli 2 de passa 40 l'un. Die pexar el passo libre $2\frac{1}{2}$.

Vuol pozastrello de passa 25, per passo libre 4.

Vuol mantichio de passa 6, per passo die pesar libre 4.

Vuol montanyana una de passa 13, per passo libre 4.

Vuol menaor un per allize de passa 120, per passo libre $2\frac{1}{2}$.

Vuol maistra da volzer de passa 12, per passo libre 2.[a]

Vuol rixa una de ttaia maistra de pasa 8, per passa libre 10.

Vuol chagnola una de passa 36, per passo libre $2\frac{1}{2}$, chomo vedirì qui de sotto per figura.

Vuol chinalli 7 per ladi, de pasa 8 l'un. Die pexar per paso libre 4.

Vuol menaori 7 per ladi, de passa 9 l'un. Die pexar per paso libre $2\frac{1}{2}$.

Vuol palomere 2, de pasa 40 l'una. Die pesar el paso libre 4.

---

a. *Followed by* $\frac{1}{2}$, *rubbed out but still partially legible.*
1. The *matta* was a tackle used to haul the lower end of the yard to the mast when changing tack.
2. A short line hanging from the head of the mast or yardarm, with a block at the end for attaching tackles.
3. The *chusidure* lashed together the two parts of the yard at the scarf.

It needs 2 *orza puopa* tackles, each 20 paces long, 4 pounds per pace.

It needs one *matta*[1] of 20 paces the same weight as the aft lateen tackles.

It needs 1 pendant[2] for the vangs, 13 paces long. Each pace should weigh 6 pounds.

It needs 1 pendant for the tack tackle of 3 paces. Each pace should weigh 4 pounds.

It needs one sheet of 20 paces. Each pace should weigh 7 pounds.

It needs 4 lashings[3] of 3 paces each, $1\frac{1}{2}$ pounds per pace.

It needs an *anzolo* tackle[4] 40 paces long. It should weigh $2\frac{1}{2}$ pounds per pace.

2 *quarnal* shroud tackles of 40 paces each. Each pace should weigh $2\frac{1}{2}$ pounds. f. 144a

It needs a *pozastrello*[5] of 25 paces, 4 pounds per pace.

It needs a yard lift of 6 paces, 4 pounds per pace.

It needs a *montanyana*[6] of 13 paces, 4 pounds per pace.

It needs a fall for the *allize*[7] of 120 paces, $2\frac{1}{2}$ pounds per pace.

It needs a mast tackle of 12 paces, 2 pounds per pace.

It needs a lashing of the main tackle of 8 paces, 10 pounds per pace.

It needs a *chagnola*[8] tackle of 36 paces, $2\frac{1}{2}$ pounds per pace, as you will see in the figure below.

It needs 7 *chinal* shrouds per side, of 8 paces each. Each pace should weigh 4 pounds.

It needs 7 shroud tackle falls per side, of 9 paces each. Each pace should weigh $2\frac{1}{2}$ pounds.

It needs 2 mooring ropes, of 40 paces each. Each pace should weigh 4 pounds.

---

4. The *anzolo* was probably a parrel tackle fall.
5. It is not clear what sort of tackle the *pozastrello* was.
6. It is not clear what sort of tackle the *montanyana* was.
7. It is not clear what sort of tackle the *allize* were.
8. It is not clear what sort of tackle the *chagnola* (literally "puppy") was.

[*captions:*]
sartia in chorchoma[a]

f. 144b vella de papaficho e mezana qui de sotto

Vuol questa nostra galia de Fiandria feri 5, i qualli die pexar per passo per chadaun libre 120. Insuma tuti 5 libre 600.

Vuol chondugi 2 per le mare. Vuol chondugi 2 per anelli. Anchora vorà messitarie 2 per la barcha.

E questi feri vedirì qui indriedo fatti chomo se die intender, e per lo simylle el so pexo e li so chondugi.

f. 145a [*captions:*]
feri 5[b]

condulli[c]

Mo' qui per inchontro faremo questa nostra galli[*a*] a vello perché chonpitta.

f. 145b [*caption:*]
gallia a vel[*o*][d]

f. 146a Una gallia la qual noi avemo chonpida del sesto de Fiandria se 'l charga in Venyexia piper e zenzer vuol savorna piati 3, e se 'l charga viny vuol savorna piati 2.

E perché le gallie de Fiandria over da Londra i vuol stivar lane, vuol da Venyexia tuor tuole 120. E vuol un manto de stiva de passa 50, de libre 10 el passo. E vuol un manto da reparar de passa 20. Die pexar el passo libre 8.

Vol chorchoma una de passa 50, vuol pexar el passo libre $2\frac{1}{2}$. E de questa se fa rizade 2 dela stella de passa 8 l'una, e per friny 2 de chaval di bocha de passa 9 l'un, e per far sachette 6 per tirar le tuole.

Vuol chorchoma una de passa 70, de libre $1\frac{1}{2}$, per far rizade 3 del fassio. Die 'ser longa chadauna passa 9, e per mantixello passa 8. E lo resto sia per respetto dele ditte chosse.

---

a. *On the top coil in the drawing.*
b. *On the drawing.*
c. *On the drawing.*
d. *On the yard of the sail in the drawing.*
1. The *papaficho* was a small lateen sail.
2. The MS has 120 pounds per pace.
3. Lengths of old rope by which an anchor was slung from the gunwale.

[*captions:*]
Coiled rope

Papaficho[1] and mainsail below f. 144b

Our Flanders galley needs 5 anchors, which should each weigh 120 pounds.[2] In total the 5 are 600 pounds.

It needs 2 painters[3] for the flukes.[4] It needs 2 painters for the rings. Also it will need 2 *messetarie*[5] for the boat.

These anchors you will see on the back [of this sheet] made as they should be, and similarly their weight and their painters.

[*captions:*] f. 145a
Anchors

Painters

Now, opposite here we will put our galley under sail because it is finished.

[*caption:*] f. 145b
Galley under sail

A galley such as we've completed of the Flanders design, if it loads in Venice pepper and ginger, it will need 3 barge loads of ballast, and if it loads wine it will need 2 barge loads of ballast. f. 146a

And because the galleys of Flanders or London will stow wool, one should bring from Venice 120 planks, and it also needs a stowing tackle fall of 50 paces, of 10 pounds per pace. And it needs another stowing tackle fall of 20 paces. Each pace should weigh 8 pounds.

It needs a coil of rope of 50 paces, each pace should weigh $2\frac{1}{2}$ pounds. And from this are made 2 *rizade dela stella*[6] of 8 paces each, and 2 deck *friny de chaval de bocha*[7] of 9 paces each, and 6 ropes to haul the planks.

It needs a coil of rope of 70 paces, of $1\frac{1}{2}$ pounds, to make 3 ropes for bundles. Each one should be 9 paces long, and for the *mantixello*[8] 8 paces. And the rest are spares for these things.

---

4. The plates on the arms of the anchors.
5. The *messetarie* were apparently ropes for the galley's boat, perhaps for towing it.
6. The *rizade dela stella* (literally "star") were part of the loading gear.
7. The *friny de chaval de bocha* (literally "horse mouth") were some sort of loading tackle.
8. It is not clear what sort of tackle the *mantixello* was.

Vuol chorchome 2 de choxidura, de passa 50 l'una, per stropelli dele tuolle e per choxidure de chostiere, e chasa e vananti e agi.

Vuol un argano longo passa $3\frac{1}{2}$, grosso pie 4.

Vuol taie 2 dopie, una per ladi, e vuol 1ª taya ugnola su al chavo dela trava.

Vuol eser la trava longa passa 5, grosa pie 3, e vuol la dita cholzexe chon ragi 2.

Vuol stelle 2 de pie 4 l'una.

Vuol chasse 2 de pie 3 l'una.

Vuol vananti 2 de pie 3 l'un.

Vuol 1 chaval de bocha de pie 12.

Vuol chavalli 6 per la trava. I freny d'i ditti se tuol d'i chinalli.

f. 146b Vuol eser aponta' el petural e mettese 1 ponte per traversso e le | ponte per le latte e per dabasso. Vuol peturalli 4 per la proda, de pie 10, 11, $12\frac{1}{2}$, 13. Ala prima sachi 11, un chonzorlanda, la segonda 12, la terza 13, la quarta 14.

Peturalli per pope, el primo de pie 13, el segondo de pie $13\frac{1}{2}$, el terzo de pie 14, el quarto de pie $14\frac{1}{2}$. Ala prima sachi 14, ala segonda sachi 15, ala terza sachi 15, ala quarta 15, e de forzo zò che te pare.

Vuol la ditta gallia per la ditta stiva mussielli per retenir i sachi. Vuol stropaelli, vuol punte per la trava, per ladi e per la choverta. Vuol polixe uno. Vuol agi 2.

[*captions:*]
d[*ri*]za[a]

trava[b]

f. 147a Vuol questa nostra gallia de Fiandria 1 timon bavonescho. Die 'ser lo sso diamitro inchina al dente passa 4 pie $2\frac{1}{2}$, e dal cuolo dal ditto al chalchagno in driedo pasa 4 pie $2\frac{1}{2}$. E die averzer in palla pie 5, al so terzo pie 4.

---

a. *On the drawing.*
b. *On the drawing.*
1. Rings used as fasteners for tackles.
2. It is not clear what part of the loading equipment the *vananti* were.

It needs 2 coils of seizing line, of 50 paces each, for strops[1] of the planks and for the seizings of the shrouds and chests and the *vananti*[2] and pins.

It needs a capstan[3] $3\frac{1}{2}$ paces[4] long, 4 feet in diameter.

It needs 2 double tackles, one for each side, and needs a single tackle on the end of the beam.

The beam should be 5 paces long, 3 feet thick, and its block mast should have 2 pulleys.

There should be 2 *stelle* of 4 feet each.

There should be 2 chests of 3 feet each.

There should be 2 *vananti* of 3 feet each.

There should be 1 *chaval de bocha* of 12 feet.

There should be 6 *cavalli* for the beam. Their tackles should be held by[5] the shrouds.

The pewtrel[6] should have a prong, and put 1 prong across, and | prongs on the side and below. f. 146b
There should be 4 pewtrels forward, of 10, 11, $12\frac{1}{2}$, and 13 feet. At the first prong, 11 sacks, one with a twist, the second, 12, the third, 13, the fourth, 14.

Pewtrels for the stern; the first of 13 feet, the second of $13\frac{1}{2}$ feet, the third 14 feet, the fourth $14\frac{1}{2}$ feet; for the first, 14 sacks, for the second, 15 sacks, for the third, 15 sacks, for the fourth, 15 sacks, and for the rest, whatever appears best to you.

This galley needs, for stowage, cords to tie up the sacks; it needs strops; it needs bridges for the beams, for the sides, and for the deck. It needs a pulley. It needs two pins.

[*captions:*]
Halyard

Beam

This galley of ours needs one stern rudder. Its length to the pointed end should be 4 paces, $2\frac{1}{2}$ feet; f. 147a
and from its throat to the heel at the back 4 paces, $2\frac{1}{2}$ feet. And its blade should measure 5 across, 4 feet at a third of its length.

3. A revolving barrel for winding cable.
4. This may be an error for "feet" as a capstan on such a galley would not have a barrel $3\frac{1}{2}$ paces long.
5. Or, possibly, "taken from."
6. A temporary crosswise bulkhead used in the storage of sacks of wool.

[*captions:*]
2 timony[a]

1 timon[b]

f. 147b [*caption:*]
una galia a vel[*o*] e a remi[c]

f. 148a [Far 1ª galia del sesto de Romanya chon tuto quello i prozede inchina la debia andar a vello e a remy]

Questo serà l'amaistramento di far una galia del sesto de Romania, zoè dala Tana, e de ttute le chose prozede ala ditta gallia inchina che vada chon le velle.

Una gallia del ditto sesto vuol eser longa d'alto passa 23 e pie 3. E averà de pian pie 10 men deda 2. E lieva lo sesto dale chorbe per mezo el poselexe dela paraschoxola mezo pie un dedo. E avre pie 1 in alto pie 11 deda 2 grose. Avre pie 2 in alto pie 13 men deda 2. E avre in alto pie 3 pie 14 men deda 2. E avre pie 4 in alto pie 15 e un dedo men $\frac{1}{3}$. E avre pie 5 in alto pie 15 $\frac{1}{2}$ terzo de pe'. E avre pie 6 in alto pie 16 men $\frac{1}{4}$ de pe'. Alta in choverta pie $7\frac{1}{3}$. Inn alto pie 11 men $\frac{1}{4}$ de pe'. E partese chorbe 41 in sesto e 41 in pope, e ande[d] chorbe 5 in mezo.

Longa questa nostra galia da un poselexe al'oltro d'i choltri passa 19 e pie $2\frac{1}{3}$. E fiero chon la chodiera chorba di proda lonzi dal poselexe del coltro pie $7\frac{1}{2}$.

E fiero chon la chodiera chorba di puope lonzi dal poselexe dal choltro pie $8\frac{1}{2}$. Erze lo madier di bocha a proda pie 9, mesurando dal quadro. À de lanzo pie $10\frac{1}{3}$. E mesurando dal'oro di su dela zenta e al'oro di su dal madier di bocha, die 'ser pie $1\frac{1}{2}$. Erze l'inpostura di pope pie 12, mexurando al quadro. E lanza pie $10\frac{1}{3}$. Erze per mezo el poselexe del choltro $\frac{1}{4}$ de pe'. Mexurando dal'oro di
f. 148b ssu | dal triganto e al'oro di su dala zenta, die 'ser pie 3 men meza una quarta de pe'.

E mesurando per mezo la chodiera chorba da proda, dal'oro di ssu dala cholonba e al'oro di su dala zenta, die 'ser pie 6 dedo un, mexurando al quadro.

E mesurando per mezo la chorba del 18 a proda, dal'oro di su dala cholonba e al'oro di su dala zenta, die 'ser pie 6 men $\frac{1}{4}$, mexurando al quadro.

---

a. *Between the top two drawings.*
b. *On the third drawing.*
c. *Beneath the image (which is a coat of arms rather than a galley).*
d. *Sic for* anche.

[*captions:*]
2 rudders

1 rudder

[*caption:*] f. 147b
A galley under sail and under oars[1]

To make a galley of the Romania design with all that is needed for it to go by sail and by oar f. 148a

These will be the instructions for making a galley of the Romania design, that is of Tana, and for all the things that belong on this galley until it gets under sail.

A galley of such a design should have a length at the top[2] of 23 paces and 3 feet. And it will have a floor of 10 feet minus 2 fingers. And the mold of the frames rises one half foot 1 finger at the mark at the bilge stringer. And at 1 foot high, it opens 11 feet 2 fingers wide. At 2 feet high it opens 13 feet minus 2 fingers. And at 3 feet high it opens 14 feet minus 2 fingers. And at 4 feet high it opens 15 feet $\frac{2}{3}$ finger. And at 5 feet high, it opens 15 and $\frac{1}{2}$ of a third feet. And at 6 feet high it opens 16 minus $\frac{1}{4}$ feet. The deck height is $7\frac{1}{3}$ feet. It is 11 minus $\frac{1}{4}$ feet high.[3] And there are 41 molded frames [forward] and 41 aft, and also 5 frames in the middle.

This galley is 19 paces $2\frac{1}{3}$ feet long from one keel mark to the other. And they will be, with the forward tail frame, $7\frac{1}{2}$ feet distant from the coulter mark.

And they will be, with the after tail frame, $8\frac{1}{2}$ feet distant from the coulter mark. The sheer strake is 9 feet high at the bow, measuring vertically. It rakes $10\frac{1}{3}$ feet. And measuring from the top of the main wale to the top of the sheer strake should be $1\frac{1}{2}$ feet. The sternpost is 12 feet high, measuring vertically. And it rakes $10\frac{1}{3}$ feet. At the coulter mark it rises $\frac{1}{4}$ foot. Measuring from the upper edge | of the transom to the upper edge of the main wale, it should be 3 feet minus one quarter foot. f. 148b

And measuring at the forward tail frame, from the upper edge of the keel to the upper edge of the main wale, it should be 6 feet 1 finger, measuring vertically.

And measuring at frame 18 forward, from the upper edge of the keel to the upper edge of the main wale, it should be 6 feet minus $\frac{1}{4}$, measuring vertically.

---

1. This caption does not correspond to the image.
2. I.e., on deck.
3. This probably should be the width of the forward tail frame at $7\frac{1}{3}$ feet high; cf. the corresponding passage in the description of the Flanders galley at fol. 135b.

E poi mexurando per mezo la chorba de mezo, e al'oro di su dala cholonba e al'oro di su dala zenta, die 'ser pie 6 men mezo terzo de pie.

E poi mesurando per mezo la chorba del 18 de pope, dal'oro di su dala cholonba e al'oro di su dala zenta, die 'ser pie 6 deda [. . .],[a] mexurando al quadro.

E mesurando per mezo la chodiera chorba di pope, e al'oro di su dala cholonba e dal'oro di su dala zenta, die 'ser pie 7 men $\frac{1}{4}$, mexurando al quadro.

E mesurando per mezo la chodiera chorba da proda, dal'oro di su dala zenta e al'oro di ssu del madier di bocha, die 'ser pie $1\frac{1}{2}$, mexurando per la via d'i furchami.

E mesurando per mezo la chorba del 18 da proda, dal'oro di su dala zenta e al'oro di su dal madier di bocha, die 'ser pie $1\frac{1}{2}$, mexurando per la via d'i furchami.

E mesurando per mezo la chorba de mezo, dal'oro di su dala zenta e al'oro di su dal madier di bocha, die 'ser pie $1\frac{1}{2}$ e dedo 1, mesurando per la via d'i furchami.

f. 149a E mesurando per mezo la chorba del 18 di pope, dal'oro di su dala zenta e al'oro di su dal madier di bucha, die 'ser pie 2 men $\frac{1}{4}$, per la via d'i furchami.

E mesurando per mezo la chodiera chorba di pope, dal'oro di su dala zenta e al'oro di su dal madier di bocha, die 'ser pie $2\frac{1}{4}$, mexurando per la via d'i furchami.

E avre in paraschene pie 3 men $\frac{1}{3}$, mesurando al'oro dentro del parascene.

E mesurando dal'oro di fuora dal'inpostura e dal'oro di pope dela timonera, die 'ser pie $4\frac{1}{2}$, sotto per la zenta.

E xe ampla la timonera pe' 1 deda 2. Xe grosa pie $1\frac{1}{2}$.

À de palmetta a proda pie 7 men $\frac{1}{3}$, mesurando dal'oro dentro dal'inpostura e dal'oro di proda del zuovo, mexurando per la via del madier di bocha.

E xe longo lo morello che se parte le latte pie 2 men $\frac{1}{4}$. E getta sulo zuovo di proda mezo pe'. E getta sulo zuovo di pope $\frac{1}{4}$ de pe'. E getta sula latta de mezo quarte $3\frac{1}{2}$. Mexurando per mezo lo zuovo de pope, dal'oro di fuora del madier di bocha e al'oro dentro dala bandolina, die 'ser pie 1 men dedo 1. Mesurando per mezo la chadina da cholo, dal'oro di fuora del madier di bocha e al'oro dentro dala bandulina, die 'ser pie 1 dedo 1.

---

a. *Blank.*

And then measuring at the midship frame, from the upper edge of the keel to the upper edge of the main wale, it should be 6 feet minus a third of a foot.

And then measuring at frame 18 aft, from the upper edge of the keel to the upper edge of the main wale, it should be 6 feet [. . .][1] fingers, measuring vertically.

And then measuring at the after tail frame, from the upper edge of the keel to the upper edge of the main wale, it should be 7 feet minus $\frac{1}{4}$, measuring vertically.

And measuring at the forward tail frame, from the upper edge of the main wale to the upper edge of the sheer strake, it should be $1\frac{1}{2}$ feet, measuring along the futtocks.

And measuring from frame 18 forward, from the upper edge of the main wale to the upper edge of the sheer strake, it should be $1\frac{1}{2}$ feet, measuring along the futtocks.

And measuring at the midship frame, from the upper edge of the main wale to the upper edge of the sheer strake, it should be $1\frac{1}{2}$ feet 1 finger, measuring along the futtocks.

And measuring at frame 18 aft, from the upper edge of the main wale to the upper edge of the sheer strake, it should be 2 feet minus $\frac{1}{4}$, along the futtocks. f. 149a

And measuring at the after tail frame, from the upper edge of the main wale to the upper edge of the sheer strake, it should be $2\frac{1}{4}$ feet, measuring along the futtocks.

And it opens in the stern bulwark 3 feet minus $\frac{1}{3}$, measuring inside the stern bulwark.

And measuring from the outside edge of the sternpost to the after edge of the rudder bracket, it should be $4\frac{1}{2}$ feet, [measuring] below the main wale.

And the rudder bracket is 1 foot 2 fingers wide. It is $1\frac{1}{2}$ feet thick.

The foredeck is 7 feet minus $\frac{1}{3}$ feet, measuring from the inside edge of the stem to the forward edge of the yoke, measuring along the sheer strake.

And the standard measure that separates the deck beams is 2 feet minus $\frac{1}{4}$ long. And it extends a half foot onto the forward yoke. And it extends $\frac{1}{4}$ foot onto the after yoke. And it extends $3\frac{1}{2}$ quarters onto the midship deck beam. Measuring at the after yoke, from the outside edge of the sheer strake to the inside edge of the *bandulina*, it should be 1 foot minus 1 finger. Measuring at the neck beam, from the outside edge of the sheer strake to the inside edge of the *bandulina*, it should be 1 foot 1 finger.

1. Blank.

E xe ampla la schaza a latte 18 chon lo zuovo da proda e xe là e die 'ser pie $1\frac{1}{3}$.

E die 'ser la porta del marango[*n*] a latte 4 chon lo zuovo lo sogier di proda. E lo sogier di pope a
f. 149b latte 6 chon lo zuovo. E lo sogier di pope, dala porta deli scrivani suso la bastarda | de 6. E lo sogier da proda die 'ser suso una bastarda che se mete a pruovo, la bastarda del 9. E lo sogier da pope, dala porta de scandoler, die 'ser a latte 11 chon el zuovo di pope. E lo sogier di proda die 'ser a latte 13 chon el zuovo di pope.

E avre la crossia in proda pie $1\frac{1}{3}$, mesurando al'oro dentro dele crossie. Avre la chrossia a meza gallia pie 2 men $\frac{1}{3}$, mesurando al'oro dentro dela crossia.

E avre la crosia a puope pie mezo terzo de pe', mesurando dal'oro dentro dele crosie. E mesurando dal'oro di fuora dila crosia e dal'oro di fuora dila chorda, die 'ser pie 4 men $\frac{1}{4}$.

E mesurando a proda di su le tuole di choverta al'oro di su dela crossia, die 'ser pie $1\frac{1}{2}$. E tanto xe la erta a meza galia. E a puope die 'ser erta pie $1\frac{1}{2}$ e deda 2. Mesurando dal'oro di fuora dela crossia e al'oro dentro dela banda, die 'ser pie $8\frac{1}{2}$.

E mexurando dal'oro di fuora dala banda e al'oro di fuora dela postiza, die 'ser pie $1\frac{1}{2}$. E là se tyra una trazuolla al'oro di ssu dale postize a proda, e la trazuola va più alta ch'al'oro di su dala crossia $\frac{1}{4}$ de pe'. E a meza gallia va più ssu la trazuola ch'al'oro di su dila crosia $\frac{1}{3}$ de pe'. E a puope va la trazuolla più bassa ch'al'oro di ssu dela crossia deda 2.

E vuol eser alti i pie d'i banchi pie 2 men mezo terzo de pe'. E tanto vol eser a proda. E a pope die 'ser alti più deda 2.

Mesurando dal'oro di proda e lo zuovo da proda inchina lo primo scermo postizo, die 'ser $\frac{2}{3}$ de pe'.
f. 150a E da questo primo | scermo postizo inchina al'oltro scermo postizo, die 'ser pie $3\frac{1}{2}$ e dedo 1. E dalo scermo postizo al trezaruol, die 'ser palmo un, mexurando dal'oro di puope.

E dal zuovo di pope inchina al primo scermo postizo, die 'ser pie $2\frac{1}{2}$ men meza quarta de pe'.

E quando el se vuol inbanchar lo se mette una trazuolla sulo scermo postizo. E vuol bavar el bancho dala trazuola per mezo la crusia $\frac{1}{4}$ de pe'. E bava el banch[*o*] dala trazuola per me' el pe' del bancho pie 1.

---

1. This passage seems to be lacking some words, as can be seen by a comparison with the comparable section for the Flanders galley at fol. 136b.
2. This passage seems to be lacking some words, as can be seen by a comparison with the comparable section for the Flanders galley at fol. 136b.
3. The text reads *pruovo*, possibly for *pruodo.*

And the mast step at deck beam 18 is wide with the forward yoke and it's there and it should be $1\frac{1}{3}$ feet.[1]

And the carpenter's hatch should be at beam 4 with the yoke the forward *sogier*.[2] And the after *sogier* at beam 6 counting the yoke. And the after *sogier*, of the pursers' hatch on the half beam | at [beam] 6. And the forward *sogier* should be on a half beam that is put forward[3] of the half beam at [beam] 9. And the after *sogier*, from the *scandoler* hatch, should be at deck beam 11 counting the after yoke. And the forward *sogier* should be at deck beam 13 counting the after yoke. f. 149b

And the gangway forward opens $1\frac{1}{3}$ feet wide, measuring inside the side pieces of the gangway. The gangway amidships opens 2 feet minus $\frac{1}{3}$ wide, measuring inside the side pieces of the gangway.

And the gangway aft opens 1 foot and half of a third of a foot wide, measuring inside the side pieces of the gangway. And measuring from the outside edge of the gangway to the outside edge of the *chorda*,[4] it should be 4 feet minus $\frac{1}{4}$.

And measuring at the forward end, from the top of the deck planks to the upper edge of the gangway, it should be $1\frac{1}{2}$ feet. And it has the same height amidships. And at the after end the rise should be $1\frac{1}{2}$ feet and 2 fingers. Measuring from the outside edge of the gangway to the inside edge of the *banda*, it should be $8\frac{1}{2}$ feet.

And measuring from the outside edge of the *banda* to the outside edge of the outrigger, it should be $1\frac{1}{2}$ feet. And there a string line is stretched on the upper edge of the outriggers forward, and the string line runs $\frac{1}{4}$ foot higher than the upper edge of the gangway. And amidships the measuring line runs $\frac{1}{3}$ foot higher than the upper edge of the gangway. And aft the string line goes 2 fingers lower than the upper edge of the gangway.

And the bench stanchions should be 2 feet minus half a third high. And this is for the forward end. And aft they should be 2 fingers higher.

Measuring from the forward edge of the forward yoke to the first *postizo* thole pin, it should be $\frac{2}{3}$ of a foot. And from this first | *postizo* thole pin to the next *postizo* thole pin, it should be $3\frac{1}{2}$ feet and 1 finger. And from the *postizo* thole pin to the *trezaruol* [thole pin], it should be one palm, measuring from the after edge.[5] f. 150a

And from the after yoke to the first *postizo* thole pin, it should be $2\frac{1}{2}$ feet minus half a quarter foot.

And when one wants to install benches, one puts a string line on the *postizo* thole pin. And the bench should depart from the string line at the gangway by $\frac{1}{4}$ foot. And the bench should depart from the string line at the foot of the bench by 1 foot.

---

4. The *chorda* was a stringer that was mortised to the deck beams to provide them extra support.
5. I.e., from the after edge of one to the after edge of the other.

E metando la trazuola sulo scermo postizo e lo scermo pianer, va più inver pope ch'ala trazuola un nudo. E va la schaza sula corba che va in 20 a proda.

[*caption:*]
gallia desfatta

f. 150b Questo ta amaistrarà chomo se die meter le maistre del sesto dele gallie ditta indriedo, zoè de Romanya.

Mexurando dal poselexe dal choltro da proda inchina in chavo dele maistre dala paraschuxula, die 'ser pie 4, mesurando per la via del panixello.

E mesurando di su la maistra dala paraschuxula e al'oro di su dala maistra de mezo, die 'ser pie $3\frac{1}{4}$, mexurando per la via del panixelo.

E mexurando dal'oro di su dala maistra di mezo e al'oro di su dela maistra de sovra, die 'ser pie $5\frac{1}{2}$, mexurando per la via del panyxello.

E mexurando per mexo la chodiera chorba da proda, le sa ficha l'oro di su dala maistra dela paraschoxula plù basso cha 'l posellexe dala paraschuxula $\frac{1}{4}$ de pe'.

E mesurando dal'oro di su dala maistra dala paraschuxula e al'oro di su dala maistra de mezo, die 'ser pie $2\frac{1}{2}$, mexurando per la via d'i furchami.

E mesurando dal'oro di su dala maistra,[a] die 'ser pie $3\frac{3}{4}$, mexurando per la via d'i furchami.

E mesurando dal poselexe del choltro di puope inchina inchavo dale maistre dela paraschuxula, die 'ser pie $3\frac{1}{3}$,[b] mesurando per la via del panixelo.

E mesurando dal'oro di su dela maistra de mezo dala parascoxula e al'oro di su dela maistra di
f. 151a mezo, die 'ser pie 6 men | un terzo de pe', mesurando dal panixello.

E mesurando dal'oro di su dala maistra de mezo e al'oro di su dala maistra di sovra, die 'ser pie $5\frac{1}{3}$, per la via del panixelo.

E mesurando per mezo la chodiera corba de puope, la sa ficha l'oro di su dala maistra dala paraschuxula più baso cha 'l poselexe dala paraschuxula $\frac{1}{4}$ de pe'.

---

a. *Sic in MS.*
b. $\frac{1}{3}$ *corrected over* $\frac{1}{2}$.
1. The *pianer* was the rower nearest the gangway.
2. Probably an image of the galley as it would look as it was being taken apart to reuse components such as timber and nails; cf. Bondioli, "Early Shipbuilding Records and the Book of Michael of Rhodes," vol. 3, p. 269, n. 63.

And putting the string line on the *postizo* thole pin and the *pianer*[1] thole pin, it goes further aft than the string line by one knot. And the mast step goes on the frame that is 20 forward.

[*caption:*]
Undone galley[2]

This will teach you how the framing ribbands should be placed for the galleys described before, that is, of Romania. f. 150b

Measuring from the forward coulter mark to the end of the bilge stringer framing ribbands, it should be 4 feet, measuring along the garboard strake.

And measuring from the top of the bilge stringer to the upper edge of the middle framing ribband, it should be $3\frac{1}{4}$ feet, measuring along the garboard strake.

And measuring from the upper edge of the middle framing ribband to the upper edge of the upper framing ribband, it should be $5\frac{1}{2}$ feet, measuring along the garboard strake.

And measuring at the forward tail frame, the upper edge of the bilge stringer framing ribband is fixed $\frac{1}{4}$ foot lower than the bilge stringer mark.

And measuring from the upper edge of the bilge stringer framing ribband to the upper edge of the middle framing ribband, it should be $2\frac{1}{2}$ feet, measuring along the futtocks.

And measuring from the upper edge of the framing ribband,[3] it should be $3\frac{3}{4}$ feet, measuring along the garboard strake.

And measuring from the after coulter mark up to the top of the bilge stringer framing ribbands, it should be $3\frac{1}{3}$ feet, measuring along the garboard strake.

And measuring from the upper edge of the bilge stringer[4] to the upper edge of the middle framing ribband, it should be 6 feet minus | one third of a foot, measuring from[5] the garboard strake. f. 151a

And measuring from the upper edge of the middle framing ribband to the upper edge of the upper framing ribband, it should be $5\frac{1}{3}$ feet along the garboard strake.

And measuring at the after tail frame, the upper edge of the bilge stringer framing ribband is attached $\frac{1}{4}$ foot lower than the bilge stringer mark.

---

3. By comparison with the corresponding passage in the Flanders galley section at fol. 138a, this measurement is from the upper edge of the middle framing ribband to the upper edge of the upper framing ribband.
4. The MS has *middle framing ribband* before *bilge stringer* as well as after it.
5. I.e., "along."

E mexurando dal'oro di ssu dala maistra dala parascuxula e al'oro di su dala maistra de mezo, die 'ser pie $2\frac{1}{4}$, mesurando per la via d'i furchami.

E mesurando dal'oro dela maistra de mezo e l'oro di su dala maistra de sovra, die 'ser pie $4\frac{1}{4}$, mexurando per la via d'i furchami.

E mesurando per mezo la corba di 10 da proda, dela ligna de mezo che se ligna la colonba e al'oro di su dela maistra dala parascuxula, die 'ser pie $4\frac{1}{6}$.

E mesurando per mezo la corba di 20 da proda, dela ligna de mezo che se ligna la cholonba e al'oro di su dala maistra dala paraschuxulla, die 'ser pie $3\frac{3}{4}$.

E mexurando per mezo la corba di 30 da proda, dala ligna che se ligna ala cholonba e al'oro di su dala maistra dala parascuxula, die 'ser pie 3 men meza quarta de pe'.

E mexurando per mezo la corba di 20 a pope, dela ligna de mezo che se ligna la colonba e al'oro di ssu dala maistra dala paraschuxula, pie $4\frac{1}{8}$.

E mesurando per mezo la chorba di 20 a pope, dela ligna de mezo che se ligna la colonba e l'oro di su dala maistra dala parascuxulla, die 'ser pie $3\frac{2}{3}$.

f. 151b E mexurando per mezo la chorba di 30 de pope, dela ligna de mezo che se ligna la cholonba e al'oro di su dala maistra dala paraschuxulla, die 'ser pie 3 men dedo 1.

E mesurando per mezo la chorba di mezo, e dala ligna de mezo che se ligna la cholonba e al'oro di su dala maistra dala paraschuxulla, die 'ser pie $4\frac{1}{3}$.

[*captions:*]
proda

pope

f. 152a E die 'ser grossa l'inpostura da proda per mezo el poselexe dal coltro mezo pie e una chana. E tanto die 'ser quela di puope.

E die 'ser grossa l'inpostura da proda in mezo el madier di bocca $\frac{1}{4}\,\frac{1}{2}$ e nudo 1. E die 'ser grossa quela di puope per mezo la zentta. E die 'ser ampla l'inpostura di proda per mezo el posellexe del coltro $\frac{1}{2}$ pie dedo 1. E chosì quela di pope.

E die 'ser ampia l'inpostura da proda per mezo el madier di bocha mezo pie dedi 2. E tanto die 'ser ampia quela di pope per mezo la zenta.

And measuring from the upper edge of the bilge stringer framing ribband to the upper edge of the middle framing ribband, it should be $2\frac{1}{4}$ feet along the futtocks.

And measuring from the upper edge of the middle framing ribband to the upper edge of the upper framing ribband, it should be $4\frac{1}{4}$ feet, measuring along the futtocks.

And measuring at frame 10 forward, from the center line to which the keel is aligned to the upper edge of the bilge stringer framing ribband, it should be $4\frac{1}{6}$ feet.

And measuring at frame 20 forward, from the center line to which the keel is aligned to the upper edge of the bilge stringer framing ribband, it should be $3\frac{3}{4}$ feet.

And measuring at frame 30 forward, from the center line to which the keel is aligned to the upper edge of the bilge stringer framing ribband, it should be 3 feet minus a quarter of a foot.

And measuring at frame 20[1] aft, from the center line to which the keel is aligned to the upper edge of the bilge stringer framing ribband, it should be $4\frac{1}{8}$ feet.

And measuring at frame 20 aft, from the center line to which the keel is aligned to the upper edge of the bilge stringer framing ribband, it should be $3\frac{2}{3}$ feet.

And measuring at frame 30 aft, from the center line to which the keel is aligned to the upper edge of the bilge stringer framing ribband, it should be 3 feet minus one finger. f. 151b

And measuring at the midship frame, from the center line to which the keel is aligned to the upper edge of the bilge stringer framing ribband, it should be $4\frac{1}{3}$ feet.

[*captions:*]
Bow

Stern

And the stempost should be one half foot and one cane thick at the coulter mark. And the sternpost should be the same. f. 152a

And the stempost should be $\frac{1}{4}$ $\frac{1}{2}$ [foot] and 1 knot thick at the sheer strake. And that of the stern should be [as] thick at the main wale. And the stempost should be $\frac{1}{2}$ foot 1 finger wide at the coulter mark. And the same for the sternpost.

And the stempost should be half a foot 2 fingers wide at the sheer strake. And the sternpost should be that wide at the main wale.

1. Probably in error for 10, from the context of the other measurements in the section.

E die 'ser el chanpo de questa nostra gallia de Romagna tanto quanto tuta esta charta per longo.

E die 'ser la groseza dela chorba men 2 grosi dia quanto per longo questa charta.

E die 'ser l'ampieza dela cholonba tanto quanto tuta longa questa charta men dedo 1 grosso.

Mo' avemo noi furnida 'sta galia, zoè de legname, e messo le maistre, e fatto e serado la ditta. Noi la depenzeremo qui per inchontro el so chorpo per mostrarvela per figora.

f. 152b [*caption:*]
una gallia

f. 153a E vuol la ditta galia del sesto de Romagnia alboro un longo passa [. . .];[a] die vuolzer al so redondo palmi [. . .]; die 'ser ala zima, zoè volzer, palmi [. . .]; die aver el cholzexe longo pie [. . .]; die 'ser alargo pie [. . .].

E vuol la ditta galia atena una de passa [. . .]; vuol volzer al so redondo palmy [. . .]; vuol volzer siando ligada pie [. . .].

E vuol la ditta un alboro de mezo de passa [. . .].

E vuol una tentena[b] per l'alboro de mezo.

E vuol la ditta un penon de respetto longo passa [. . .], vuol volzer al so ridondo palmy [. . .]. Vedirì de sotto.

[*captions:*]
antena

alboro

f. 153b E vuol la ditta gallia barcha una longa de pie [. . .][c] de sovra; e vuol eser in cholonba longa pie [. . .]; e vuol eser in pian pie [. . .]; e vuol eser erta in chadena pie [. . .]; e vuol eser averta in bucha pie [. . .].

---

a. *This measurement and others on the page left blank.*
b. *Sic in MS.*
c. *This measurement and others on the page left blank.*

And the space between the frame timbers of this our galley of Romania should be as much as this page is long.

And the thickness of the frame timber should be as much as this page is long minus 2 thumbs.

And the width of the keel should be as much as this page is long minus one thumb.

Now we have furnished this galley, that is, with wood, and set up the framing ribbands and made and closed it up. We will paint opposite here a drawing of its hull to show it to you in illustration.

[*caption:*] f. 152b
A galley

And this galley of the Romania design should have a mast of [...][1] paces long; it should measure [...] palms around; at the top it should be [...] palms, that is, around; it should have a block mast of [...] feet long; it should be [...] feet wide. f. 153a

And this galley should have a yard of [...] paces; it should measure [...] palms around; it should measure when lashed [...] feet.

And it should have a midship mast of [...] paces.

And it should have a yard[2] for its midship mast.

And it should have a spare upper spar for the lateen yard [...] paces long; it should measure [...] palms around. You will see below.

[*captions:*]
Yard[3]

Mast

And this galley needs a boat [...][4] feet long at the top;[5] and its keel should be [...] feet long; and its floor should be [...] feet; and its depth in hold should be [...] feet; and it should be [...] feet in beam. f. 153b

1. This measurement and others on the page left blank.
2. Probable intended meaning for *tentena* (rather than *antena*) in MS.
3. These captions and their illustrations are reversed.
4. This measurement and others on the page left blank.
5. I.e., on deck.

E vuol la ditta chopano 1 longo in cholonba pie [. . .], e longo in choverta pie [. . .];[a] e die averze[*r*] in bocha pie [. . .]; e vuol eser erto pie [. . .]; e vuol eser averto in pian pie [. . .], chomo farò chi de sutto per figura.

f. 154a E vuol la ditta galia chanavi 5 de passa 70 l'uno. Die pesar per passo libre [. . .].[b]

E vuol prodixe 1 de passa 80. Die pexar el paso libre [. . .].

E vuol poza 1ª de passa 18. Vuol pexar per passo libre [. . .].

E vuol gripie 5 de passa 70 l'una. Die pexar el paso libre [. . .].

E vuol manti 2 de pasa [. . .]. Vuol pesar per paso libre [. . .].

E vuol manti de prodony[c] 2 de passa [. . .]. Die pexar el paso libre [. . .].

Suste 2 de passa [. . .] l'una. Die pexar el paso libre [. . .].

Gomene 2 de pasa [. . .] l'una. Die pexar el passo libre [. . .].

Menaor di prodeny[d] 2 de passa [. . .]. Vuol pexar el paso libre [. . .].

Anyze 2 de passa 5 l'una. Die pexar el paso libre [. . .].

Anyze 2 chorte de passa $4\frac{1}{2}$. Die pexar el paso libre [. . .].

Amo un de passa 50. Die pexar el paso libre [. . .].

Orze puope 2, matta 1ª, longa zascuna passa 20.

Pozastrello 1 de passa 25. Die pexar el passo libre [. . .].

Braza de suste de passa 13. Die pexar el passo libre [. . .].

Braza d'orze puope longa passa 6, el passo libre [. . .].

Funde 2 de braza 30 l'una, el passo libre [. . .].

Orza davanti de passa 30, el passo libre [. . .].

---

a. e long *crossed out with horizontal line.*

b. *This measurement and others on the page left blank.*

c. *Sic in MS.*

d. *Sic in MS.*

And it needs 1 skiff with a keel [...] feet long, and a deck of [...] feet; and it should have a beam of [...] feet; and it should be [...] feet deep; and it should have a floor of [...] feet, as I will do below in the drawing.

And this galley needs 5 hemp anchor cables of 70 paces each. Each pace should weigh [...][1] pounds. f. 154a

And it needs 1 headfast of 80 paces. Each pace should weigh [...] pounds.

And it needs 1 sheet of 18 paces. The pace should weigh [...] pounds.

And it needs 5 buoy ropes each 70 paces long. The pace should weigh [...] pounds.

And it needs 2 halyard ties of [...] paces. It should weigh [...] pounds per pace.

And it needs 2 mast tackle falls of [...] paces. The pace should weigh [...] pounds.

2 vangs of [...] paces each. The pace should weigh [...] pounds.

2 halyard tackle falls of [...] paces each. The pace should weigh [...] pounds.

Falls for 2 mast tackles of [...] paces. The pace should weigh [...] pounds.

2 rudder tackles of 5 paces each. The pace should weigh [...] pounds.

2 short rudder tackles of $4\frac{1}{2}$ pounds each. The pace should weigh [...] pounds.

1 *amo* of 50 paces. The pace should weigh [...] pounds.

2 *orza puope*, 1 *matta*, each 20 paces long.

1 *pozastrello* of 25 paces. The pace should weigh [...] pounds.

Pendant for the vangs of 13 paces. The pace should weigh [...] pounds.

Pendant for the *orza puope* 6 paces long, the pace [...] pounds.

2 *funde de braza* of 30 [paces] each, the pace [...] pounds.

Orza davanti of 30 paces, the pace [...] pounds.

---

1. This measurement and others on the page left blank.

Orza poza de pasa 30, el paso libre [. . .].

Poza sottil de passa 18, el passo libre [. . .].

Poza grossa de passa 18, el passo libre [. . .].

Quarnalli de passa 36 l'un, el passo libre [. . .].

Anzolo de passa 36, el passo libre [. . .].

Montagniana dopia de passa 10, el paso libre [. . .].

f. 154b Mantichio 1 de passa 5, per passo libre [. . .].[a]

Menaor per anize de passa 120, el paso libre [. . .]. E de questo se fa lieva, inchasadori e deschasiadori e anize e fren de timon e maistra.

Maistra 1ª da volzer de passa 12, el paso libre [. . .].

Rixa de taia maistra de passa 8, el paso libre [. . .].

Vuol chinalli 6 per ladi de passa 7 l'un, el paso libre [. . .].

Vuol 1ª chagnola de passa 30, per paso libre [. . .].

Menaori per chinalli de passa 8, el passo libre [. . .].

Palomere 2 longe passa 40 l'una, per paso libre [. . .].

Vuol timony 2 latiny longi pie [. . .]. Vuol eser grosi[b] ala gola pie [. . .]. E vuol eser alargi in pala pie [. . .]. E vuol eser longo el fuxo pie [. . .]. E vuol eser el scaton longo pie [. . .].

Vuol timony 2 bauneschi, vuol eser longo chadauno al so diamittro pie [. . .]. Vuol eser longo dal fuxo inchina driedo ala pala pie [. . .]. E vuol eser alarga la palla pie [. . .].

---

a. *This measurement and others on the page left blank.*

b. grosi *above the line, correcting* longi *crossed out with a horizontal mark.*

Orza poza of 30 paces, the pace [. . .] pounds.

Light sheet of 18 paces, the pace [. . .] pounds.

Heavy sheet of 18 paces, the pace [. . .] pounds.

Shroud tackles of 36 paces each, the pace [. . .] pounds.

Anzolo tackle of 36 paces, the pace [. . .] pounds.

Double *montanyana* tackle of 10 paces, the pace [. . .] pounds.

1 yard lift of 5 paces, [. . .][1] pounds per pace. f. 154b

A fall for the *anize*[2] of 120 paces, the pace [. . .] pounds. And from this are made the quarter rudder tackle, *inchasadori,*[3] and *deschasiadori,*[4] and *anize* and pendants of the rudder and the mast tackle.

1 mast tackle of 12 paces, the pace [. . .] pounds.

Lashing of the mast tackle of 8 paces, the pace [. . .] pounds.

It needs 6 *chinal* shrouds per side of 7 paces each, the pace [. . .] pounds.

It needs 1 *chagnola* tackle of 30 paces, [. . .] pounds per pace.

Falls for the *chinal* shrouds of 8 paces, the pace [. . .] pounds.

2 mooring ropes each 40 paces long, [. . .] pounds per pace.

It needs 2 lateen rudders [. . .] feet long. These need to be [. . .] feet thick in the throat. They need to be [. . .] feet wide in the blade. And the *fuxo*[5] needs to be [. . .] feet long. And the *schaton*[6] needs to be [. . .] feet long.

It needs 2 Bayonne rudders. These need each to be [. . .] feet long in their diameter. They need to be [. . .] feet long from the *fuxo* to the blade. And the blade needs to be [. . .] feet wide.

---

1. This measurement and others on the page left blank.
2. Uncertain type of tackle, perhaps the same as *allize* at fol. 144a.
3. It is not clear what sort of tackle the *inchasadori* were.
4. It is not clear what sort of tackle the *deschasiadori* were.
5. The upper part of the rudder stock.
6. The lower part of the rudder stock.

f. 155a Vuol la ditta gallia velle 4, artimon, trezaruol, papaficho e chochina.

Artimon de passa [. . .].[a] Vui truoverì la so raxion charta 127.

Trezaruol de pasa [. . .]. Tornè charta 127. Vui vedirì qui de sotto per figuna.[b]

[*captions:*]
artemon

trezaruol

f. 155b Papaficho de pasa [. . .].[c] Torna charta 127.

Chochina de pasa [. . .]. Torna charta 127. Vui vedirì qui de sotto per figura.

[*captions:*]
papaficho

chochina

f. 156a Mo' avemo noi chonpido questa nostra galia del sesto de Romanya de ttutto che se possa andar a remy.

[*caption:*]
galia a remy

f. 156b **[Far una galia sotil chon tutto i prozede, andar a velo e a remi chon tute le suo raxion]**[d]

f. 157b da proda. E mesurando de questo inchina unde io fiero chon la chodiera chorba di proda, die 'ser pie $18\frac{3}{4}$.

E là se chala un pionbin al'oro di fuora dal'inpostura de pope, e mesurando de questo inchina unde io fiero chon la chodiera chorba di puope, die 'ser pie $22\frac{3}{4}$.

E mesurando per mezo la chodiera chorba di pope, de su la colonba e al'oro di su dala zenta, die 'ser pie 4, mesurando al quadro.

---

a. *This measurement and one other on the page left blank.*

b. *Sic in MS.*

c. *This measurement and one other on the page left blank.*

d. *The page containing f. 156b and f. 157a was torn out subsequent to binding. See Franco Rossi, "Introduction to the Manuscript," p. xxvii, above.*

This galley needs 4 sails—*artimon,*[1] *trezaruol,*[2] *papaficho,*[3] and *cochina.*[4] f. 155a

Artimon of [. . .] paces. You'll find its measurements on folio 127.

Trezaruol of [. . .] paces. Turn to folio 127. You will see them illustrated below.

[*captions:*]
Artimon

Trezaruol

Papaficho of [. . .] paces. Turn to folio 127. f. 155b

Cochina of [. . .] paces. Turn to folio 127. You will see them illustrated below.

[*captions:*]
Papaficho

Cochina

Now we will have completely finished this galley of ours of the Romania design, which can go under oars. f. 156a

[*caption:*]
Galley under oars

To make a light galley with everything so it can go by sail and by oar with all of its measurements[5] f. 156b

of the bow. And measuring from this until they have reached the forward tail frame, it should be $18\frac{3}{4}$ feet. f. 157b

And a plumb line is dropped there from the outside edge of the sternpost, and measuring from this until they have reached the after tail frame should be $22\frac{3}{4}$ feet.

And measuring from the after tail frame,[6] from the top of the keel to the top edge of the main wale, it should be 4 feet, measuring vertically.

---

1. The largest lateen sail, used in fair weather.
2. A middle-sized lateen sail.
3. A small lateen sail.
4. A small, rectangular storm sail.
5. The page containing f. 156b and f. 157a was torn out subsequent to binding.
6. Probably a mistake for "forward tail frame," as a different measurement at the after tail frame is given four lines later.

E mesurando per mezo la chodiera chorba de 24[a] a proda, di su la colonba e al'oro di su dala zenta, die 'ser pie $3\frac{5}{6}$, mexurando al quadro.

E mesurando per mezo la chorba de mezo, de su la cholonba e al'oro di su dala zenta, die 'ser pie 4 men dedo 1, mesurando al quadro.

E mesurando per mezo la chorba del 24 a puope, de su la cholonba e al'oro di su dala zenta, die 'ser pie $4\frac{1}{3}$, mesurando al quadro.

E mesurando per mezo la chodiera chorba di puope, di su la cholonba e al'oro di su dela zenta, die 'ser pie $4\frac{7}{8}$, al quadro.

E mesurando al'oro di fuora dal'inpostura di pope, dal'oro di pope dala timonera, die 'ser pie $3\frac{1}{2}$, mesurando sotto per la zenta.

Qui indriedo ve mostrarò el sesto de questa nostra galia.

f. 158a Questi sono li chavy de questa nostra galia del sesto sotil.

[*captions:*]
prod[*a*][b]

erze pie 6 deda 2

pie $3\frac{1}{2}$

$\frac{1}{3}$ de pe' 1 [a] chana

$\frac{1}{2}$ pe' dedo 1

$\frac{1}{2}$ pe' dedo 1

pie $2\frac{1}{4}$ chana 1

$\frac{1}{3}$ de pie

$\frac{1}{3}$ de pe' dedo 1

$\frac{1}{4}$ de pe' men chana 1

---

a. *Sic in MS.*

b. *Probably in another hand and with different ink.*

And measuring at frame 24 forward,[1] from the top of the keel to the top edge of the main wale, it should be $3\frac{5}{6}$ feet, measuring vertically.

And measuring at the midship frame, from the top of the keel to the top edge of the main wale, it should be 4 feet minus 1 finger, measuring vertically.

And measuring at frame 24 aft, from the top of the keel to the top edge of the main wale, it should be $4\frac{1}{3}$ feet, measuring vertically.

And measuring at the after tail frame, from the top of the keel to the top edge of the main wale, it should be $4\frac{7}{8}$ feet vertically.

And measuring from the outside of the sternpost to the after side of the steersman's platform, it should be $3\frac{1}{2}$ feet, measuring along the bottom of the main wale.

On the reverse, I will show you the design of this galley of ours.

Here are the ends of this galley of ours of the light design. f. 158a

[*captions:*]
Bow

It is 6 feet 2 fingers high

$3\frac{1}{2}$ feet

$\frac{1}{3}$ foot 1 cane

$\frac{1}{2}$ foot 1 finger

$\frac{1}{2}$ foot 1 finger

$2\frac{1}{4}$ feet 1 cane

$\frac{1}{3}$ foot

$\frac{1}{3}$ foot 1 finger

$\frac{1}{4}$ foot minus 1 cane

1. The text says "at the forward tail frame 24," which appears to be an error, as a numbered frame could not be a tail frame.

lanza pie $9\frac{2}{3}$

puope[a]

erze pie 8

$\frac{1}{4}$ dedo 1

pie 1

pie $1\frac{1}{3}$

pie 1 dedo 1

pie $3\frac{1}{4}$

$\frac{1}{4}$ de pe' dedo 1

$\frac{1}{3}$ de pe'

$\frac{1}{2}$ terzo dedo 1

pie $\frac{1}{2}$

lanza pie $9\frac{2}{3}$

f. 158b Questo serà l'amaistramento chomo se die meter le maistre de questa nostra galia sotil.
Erze le maistre dala parascuxula a proda pie 1, mesurando al quadro.

Erze le maistre de mezo ala paraschuxula a proda pie 4 e mezo terzo de pe', mesurando al quadro.

E mesurando per mezo la chudiera chorba da proda, dela ligna de mezo che se legna la cholonba inchina al'oro di su dala maistra dala paraschuxula, die 'ser pie $1\frac{3}{4}$.

E mesurando per mezo la chodiera chorba da proda, dal'oro di su dala maistra dala paraschuxulla e al'oro di su dela maistra de mezo, die 'ser pie 1 dedo 1, mesurando per la via d'i furcami.

E mesurando per mezo la chodiera chorba di proda, dal'oro di ssu dala maistra de mezo al'oro di su dala maistra di sovra, die 'ser pie $2\frac{1}{3}$, mexurando per la via d'i furchamy.

---

a. *Probably in another hand and with different ink.*

It rakes $9\frac{2}{3}$ feet

Stern

It is 8 feet high

$\frac{1}{4}$ of 1 finger[1]

1 foot

$1\frac{1}{3}$ feet

1 foot 1 finger

$3\frac{1}{4}$ feet

$\frac{1}{4}$ foot 1 finger

$\frac{1}{3}$ foot

$\frac{1}{2}$ of a third of 1 finger

$\frac{1}{2}$ foot

It rakes $9\frac{2}{3}$ feet

These will be the instructions on how to position the framing ribbands on this our light galley. The bilge stringer ribbands rise 1 foot from the bilge stringer to the bow, measuring vertically. f. 158b

The middle ribbands rise 4 and a third of a half feet at the forward bilge stringer, measuring vertically.

And measuring at the forward tail frame, from the center line on which the keel is aligned to the top edge of the bilge stringer ribband, it should be $1\frac{3}{4}$ feet.

And measuring at the forward tail frame, from the top edge of the bilge stringer ribband to the top edge of the middle ribband, it should be 1 foot 1 finger, measuring along the futtocks.

And measuring at the forward tail frame, from the top edge of the middle ribband to the top edge of the upper ribband, it should be $2\frac{1}{3}$ feet, measuring along the futtocks.

1. Possibly $\frac{1}{4}$ foot 1 finger.

Erze le maistre dala parascuxula del'inpostura da pope pie $1\frac{1}{2}$, mesurando al quadro.

Erze le maistre de mezo al'inpostura da pope pie $2\frac{1}{3}$, mexurando al quadro.

Erze le maistre de sovra al'inpostura da pope pie $5\frac{1}{2}$, mesurando al quadro.

E mesurando per mezo la chodiera chorba da pope, dala ligna de mezo che se ligna la cholonba inchina l'oro di su dela maistra dala parascuxula, die 'ser pie 2.

E mesurando per mezo la chodiera chorba da pope, dil'oro di su dala maistra dala paraschuxulla e
f. 159a al'oro di ssu dela maistra | de mezo, die 'ser pie $1\frac{1}{6}$, mesurando per la via d'i furchami.

E mesurando per mezo la chodiera chorba da pope, dal'oro di ssu dala maistra de mezo e al'oro di su dala maistra de sovra, die 'ser pie 3, mesurando per la via d'i furchamy.

E mesurando per mezo la chorba de mezo, dela ligna de mezo che se ligna la cholonba inchina al'oro di su dala maistra dala paraschuxulla, die 'ser pie $3\frac{3}{4}$.

E mesurando la chorba di 20 a proda, mesurando in mezo, die 'ser pie $3\frac{1}{4}$. Mesurando per mezo la corba di 20, mesurando a proda, die 'ser pie $3\frac{1}{4}$.[a]

[*caption:*]
forma

f. 159b [*captions:*]
forma

corbe

f. 160a Qu[*e*]sti sono li morelli de galia sotil.

[*captions:*]
chanpo de chorba de galie sotil tuto questo men questo pizetto

groseza di chorda men questo davanti

ampieza di zuovy men questo davanti

ampieza de postize men questo davanti

---

a. *Sic in MS.*

The bilge stringer ribbands are $1\frac{1}{2}$ feet high at the sternpost, measuring vertically.

The middle ribbands are $2\frac{1}{3}$ feet high at the sternpost, measuring vertically.

The upper ribbands are $5\frac{1}{2}$ feet high at the sternpost, measuring vertically.

And measuring at the after tail frame, from the center line on which the keel is aligned to the top edge of the bilge stringer ribband, it should be 2 feet.

And measuring at the tail frame, from the top edge of the bilge stringer ribband to the top edge of the middle ribband, | it should be $1\frac{1}{6}$ feet, measuring along the futtocks. f. 159a

And measuring at the after tail frame, from the top edge of the middle ribband to the top edge of the upper ribband, it should be 3 feet, measuring along the futtocks.

And measuring at the midship frame, from the center line on which the keel is aligned to the top edge of the bilge stringer ribband, it should be $3\frac{3}{4}$ feet.

And measuring at frame 20 forward, measuring at the center, it should be $3\frac{1}{4}$ feet. Measuring at frame 20, measuring at the forward, it should be $3\frac{1}{4}$ feet.[1]

[*caption:*]
Form

[*captions:*] f. 159b
Form

Frames

These are the measuring sticks of the light galley. f. 160a

[*captions:*]
The space between the frames of the light galley, all this minus this little bit

Thickness of the *chorda* minus this in front

Width of the yoke beams minus this in front

Width of the outrigger beams minus this in front

1. The text in this passage seems incomplete by comparison with the similar passage on fols. 151a–151b.

f. 160b Quisti sono li morelli dele galie sotel.

[*captions:*]
alteza di zuovy

groseza de postize

ampieza de chorde

groseza de chorde

f. 161a Questo è lo murelo de galie sotil.

[*captions:*]
ampieza de virzine

groseza de virzene

ampieza de latte

ampieza de paramexal

alteza de latta

groseza de paranexal[a]

f. 161b Quista galia sotil è chonpida de legname. Vedirì qui de soto.

[*caption:*]
galia

f. 162a Questa nostra gallia sotil vuo[*l*] un alboro vol eser longo passa $7\frac{1}{2}$. Vuol volzer al so redondo palmi [. . .].[b] Vuol 1 cholzexe longo pie [. . .]. Vuol eser alargo pie [. . .].

E vuol antena una longa passa 13. Vuol volzer al so redondo palmi [. . .].

E vuol alboro de mezo de passa [. . .]. Vuol volzer al so redonto[c] palmi [. . .]. Vuol cholzexe longo pie [. . .]. Vuol eser alargo pie [. . .].

E vuol 1ª antenela de passa [. . .], die vuolzer al so redondo palmy [. . .].

---

a. *Sic in MS.*
b. *This and other measurements on this page left blank.*
c. *Sic in MS.*

These are the measuring sticks of the light galley. f. 160b

[*captions:*]
Height of the yoke timbers

Thickness of the outrigger timbers

Width of the *chorda*

Thickness of the *chorde*

This is the measuring stick of the light galley. f. 161a

[*captions:*]
Width of the ceiling strakes

Thickness of the ceiling strakes

Width of the deck beams

Width of the keelson[1]

Height of the deck beams

Thickness of the keelson

This light galley is finished as far as wood is concerned. You will see below here. f. 161b

[*caption:*]
Galley

This our light galley needs a mast which should be $7\frac{1}{2}$ paces long. It should be [. . .][2] palms around. It needs one block mast [. . .] feet long. This should be [. . .] feet wide. f. 162a

And it needs a yard 13 paces long. It should be [. . .] palms around.

And it needs a midship mast of [. . .] paces. It should be [. . .] palms around. It needs a lateen mast head [. . .] feet long. This needs to be [. . .] feet wide.

And it needs 1 small yard of [. . .] paces. It needs to be [. . .] palms around.

1. A lengthwise timber running internally alongside the keel.
2. This and other measurements on this page left blank.

E vuol timony 2 latiny, die 'ser longi pie [. . .]. Vol eser el scaton pie [. . .]. Vuol volzer la golla pie [. . .]. Vuol eser alarga la palla pie [. . .].

Vuol timony 2 bavoneschi. Vol eser el diamitro [. . .]. Vol eser longo dal'un chavo al'oltro pie [. . .], el diamitro pie [. . .], larga la palla pie [. . .].

E vuol un chopano longo pie [. . .]. Vuol eser in pian pie [. . .]. Vuol eser erto pie [. . .]. Vol averzer in bocha pie [. . .].

Vuo[*l*] questa nostra gallia chanavy 4 de passa 50 l'uno. Vuol pexar el passo libre 6.

Vol prodixe 1 de pasa 60. Vuol pexar el paso libre 4.

Vol gripie 4 de passa 50 l'una, el passo libre 3.

f. 162b [*captions:*]
alboro

antena

alboro

antenela

[*timo*]ny

chopa[*no*]

f. 163a Questa nostra gallia sotil vuol 1 amo de passa 35. Die pexar el passo libre 3.

E vuol suste 2 de passa 35 l'una; el passo libre 3.

E vuol gomene 2 de passa 40 l'una. Per passo libre 3.

E vuol menaor de prodeno 1 de passo 60; el paso libre 3.

E vuol manti 2 d'antena de pasa 10 l'un; el paso libre 6.

E vuol manti 2 de prodeny de pasa 8 l'un; el paso libre 6.

E vuol chinalli 4 per ladi de passa 7 l'un; el paso libre 3.

E vuol el so menador de zaschadon de passa 7; el paso libre 2.

And it needs 2 lateen rudders. They should be [. . .] feet long. The lower rudder stock should be [. . .] feet. The throat needs to be [. . .] feet around. The blade should be [. . .] feet wide.

It needs 2 Bayonne rudders. The diameter should be [. . .]. They should be [. . .] feet long from one end to the other, the diameter [. . .] feet, the blade [. . .] feet wide.

It needs a skiff [. . .] feet long. Its floor should be [. . .] feet. It should be [. . .] feet high. It should have a beam of [. . .] feet.

This galley of ours needs 4 hemp cables of 50 paces each. The pace should weigh 6 pounds.

It needs 1 headfast of 60 paces. The pace should weigh 4 pounds.

It needs 4 buoy ropes of 50 paces each, the pace 3 pounds.

[*captions:*] f. 162b
Mast

Yard

Mast

Small yard

Rudders

Boat

This light galley of ours needs 1 *amo* of 35 paces. Each pace should weigh 3 pounds. f. 163a

And it needs 2 vangs of 35 paces each; the pace 3 pounds.

And it needs 2 halyard tackle falls of 40 paces each; per pace 3 pounds.

And it needs 1 main tackle fall of 60 paces; the pace 3 pounds.

And it needs 2 yard ties of 10 paces each; the pace 6 pounds.

And it needs 2 mast tackle runners of 8 paces each; the pace 6 pounds.

And it needs 4 *chinal* shrouds of 7 paces each; the pace 3 pounds.

And it needs a fall for each one of 7 paces; the pace 2 pounds.

E vuol orza davanti passa 20; el paso libre 2.

E vuol orza poza de pasa 20; el paso libre 2.

E vuol anzelo longo de pasa 20; el paso libre 2.

E vuol maistra da volzer de pasa 9; el paso libre 2.

E vuol pozastrello de passa 20; el passo libre 3.

E vuol funde 2 de passa 18; el passo libre 2.

E vuol sorda 1ª de pasa 18; el passo libre 2.

E vuol orze pope 2, passa 18 l'una; el passo libre 3.

E vuol matta 1ª de passa 18; el passo libre 3.

E vuol anyze 2 de passa 4 l'una; el passo libre 5.[a]

E vuol 2 menaori d'anize de pasa 8 l'un; el paso libre 2.

E vuol poza grossa passa 16; el passo libre 5.

E vuo[*l*] poza sottil passa 16; el paso libre 3.

E vuol questa nostra galia feri 3 del pexo de libre 400 l'un.

E vuol questa nostra galia velle 3, artimon e trezaruol e mezana. L'artimon vol eser longo in antenal
f. 163b de pasa 15, | el trezaruol de passa 12 e la mezana de passa 8. E chi vuol veder la so raxion de queste velle torna a charta 128.[b]

Vedirì qui de sotto per figora.

[*captions:*]
artimo[*n*]

trezaruol

mezana

---

a. *Corrected over* 3.

b. *Correctly* 127, *as can also be read on fols. 155a–b.*

And it needs an *orza davanti* of 20 paces; the pace 2 pounds.

And it needs an *orza poza* of 20 paces; the pace 2 pounds.

And it needs an *anzolo* 20 paces long; the pace 2 pounds.

And it need a mast tackle of 9 paces; the pace 2 pounds.

And it needs a *pozastrello* of 20 paces; the pace 3 pounds.

And it needs 2 *funde* of 18 paces; the pace 2 pounds.

And it needs 1 *sorda* of 18 paces; the pace 2 pounds.

And it needs 2 *orza puopa* tackles, 18 paces each; the pace 3 pounds.

And it needs 1 *matta* of 18 paces; the pace 3 pounds.

And it needs 2 rudder tackles of 4 paces each; the pace 5 pounds.

And it needs 2 rudder tackle falls of 8 paces each; the pace 2 pounds.

And it needs a heavy sheet of 16 paces; the pace 5 pounds.

And it needs a light sheet of 16 paces; the pace 3 pounds.

And this galley of ours needs 3 anchors of a weight of 400 pounds each.

And this galley of ours needs 3 sails—*artimon*, *trezaruol*, and *mezana*. The *artimon* should be 15 paces long in its luff, | the *terzaruol* 12 paces, and the *mezana* 8 paces. And whoever wants to see f. 163b
the specifications of these sails should turn to folio 128.

You will see below here the images.

[*captions:*]
Artimon

Trezaruol

Mezana

f. 164a Mo' avemo chonpido questa nostra de ttutto, da sartia e de velle e de feri. E qui de sotto ve la mostrerò andar a vello.

[*caption:*]
gallia a velo

f. 164b **[Far far 1ª nave latina chon tutto quello i prozede inchina che la vada a vello]**

Al nomen di Dio. Io sì te voyo amaistrar a far una nave sì grande chomo pizulla e dala raxion che proziede inchina la posa andar a vello, chomo vedirì in questo per singolo.

In primo volemu[a] noi far una nave latina, la qual volemo che ssia longa in cholonba passa 12. Vuol de pian tanti pie è longa la cholonba, zoè de passa el quarto me[*n*]. Serano ado[*n*]cha pie 9. E tanto vuol eser el plan.

E questa nave, che à in cholonba passa 12 e à de plan pie 9, die aver in lo so trepie tanti pie quanti passa è longa la cholonba el terzo più. Adoncha la cholonba [*pa*]ssa 12, el terzo più serano 4, fano in suma 16. E questo serà el tripie.

E questa nave ditta, che fusse longa in cholonba passa 12, e da pian pie 9, e de trepie 16, la vuol averzer in bocha quanto in trepie e la mitta, fano 24. E questo è la bocha.

E questa nave, che la cholonba passa 12, e de plan pie 9, e de trepie pie 16, e de bocha pie 24, die 'ser in choverta erta quanto à de plan e più pie $\frac{1}{2}$, serano $9\frac{1}{2}$ pie.

E questa nave, che la cholonba passa 12, el plan pie 9, el trepie pie 16, e de bocha pie 16,[b] e d'erteza pie $9\frac{1}{2}$, che di[*e*] 'ser longa in choverta, vol eser tanto quando[c] la cholonba el quarto più. Serano pasa 15 e un passo più per le creser dele tteste.

E vuol eser lo timon de questa nostra nave el terzo de zò che xe longa la cholonba. Serano passa 4. El schatton serà passa 2. El fuxo passa 2. E die vuolzer la lagola[d] pie 1 per passo dela longeza del timon. Serà pie 4.

f. 165a E vuol eser alarga la pala de questo nostro timon tanto quanto vuolze el timon ala gola, serano pie 4. E vuol eser longe le anyze tanto quanto tuto lo timon longo, serano passa 4. E per questa fara' ogn'altra raxion.

---

a. *Sic in MS.*
b. *Sic in MS.*
c. *Sic in MS, evidently for* quanto.
d. *Sic in MS.*

Now we have completed this [galley] of ours in everything, in rigging and in sails and in anchors. And here below I shall show it to you proceeding under sail. f. 164a

[*caption:*]
Galley under sail

To make a lateen ship with everything so it can go by sail f. 164b

In the name of God. I want to instruct you how to make a ship, large as well as small, and you can proceed on the basis of this information to get under sail, as you will see here set out point by point.

First of all, we want to make a lateen ship, whose keel we want to be 12 paces long. The floor should be as many feet as the keel is long, that is, in paces, a quarter less. There will thus be 9 feet. And the floor will be that wide.

And this ship, which is 12 paces in the keel and 9 feet in the floor, should have at its *trepie*[1] as many feet as the keel is long in paces and one third more. Thus the keel is 12 paces, a third more would be 4, making in all 16. And this will be the *trepie.*

And this ship, which was 12 paces long in the keel, and with a floor of 9 feet, and a *trepie* of 16 feet, needs to have a beam equal to the *trepie* plus half again, which makes 24. And this is the beam.

And this ship, which has the keel of 12 paces, and the floor of 9 feet, and the *trepie* of 16 feet, and the beam of 24 feet, should have a depth to the deck equal to the floor and $\frac{1}{2}$ foot more, that is, $9\frac{1}{2}$ feet.

And this ship, which has the keel of 12 paces, the floor of 9 feet, the *trepie* of 16 feet, the beam of 24[2] feet, and the depth of $9\frac{1}{2}$ feet, should have a deck as long as the keel and a quarter more. It will be 15 paces and a pace more for the extensions at the ends.

And the rudder of this ship should be a third as long as the keel. That will be 4 paces. The *schaton* will be 2 paces. The *fuxo* 2 paces. And the throat should measure 1 foot around per pace of the length of the rudder. That will be 4 feet.

And our rudder blade should be as wide as the rudder measures around in its throat, that will be 4 feet. And the rudder tackles should be as long as the length of the rudder, they will be 4 paces. And all other measurements will go this way. f. 165a

---

1. The *trepie* was the breadth of the hull at a point 3 feet above the top of the keel.
2. *16* in MS.

E vuol questa nostra nave batey 2 e gondola una. El primo batello vol eser tanto longo quanto è la nave longa in choverta 2 volte, zoè pie 30 e un de più. Serano 31.

El segondo batello vol eser più churto ch'al primo, pie 28.

E la gondola vol eser longa tanti pie quando[a] è la bocha, 24.

[*caption:*]
nave latina[b]

f. 165b Alboro da proda de questa nostra nave vol eser longo 3 fiade quanto la nave averze in bocha, seria passa 14 e pie 2. Vuol volzer in lo so terzo de sovra la pianeda mezo palmo per paso de zò che tuto l'alboro fusse tuto longo, serano palmi 7. E lo cholzexe vol eser longo pie 1 per passo dela longeza del'alboro de zò che fusse de sovra la choverta. E vuol eser alargo el sesto dela longeza del cholzexe. Adoncha serà longo più[c] 12, alargo pie 2, e chosì farà d'ogni navillio e nave.

E voyo insegnarve uve se mette la pedega del'a[*l*]boro da proda. Vol eser tanti mezi pie dal choltro di proda inversso puope quanta passa è longa la cholonba, serano pie[d] 6. E lo slanzo de questo nostro alboro vol un pie per passo dela longeza del'alboro, serà pie $14\frac{1}{2}$.

E l'aboro de mezo vol eser longo passa 13, e la pedega de questo no[*stro*] alboro de mezo se die asbater el quarto dela cholonba, serano pie 9. E tanto vorastu mexurar dal choltro di puope inver proda, e là metti la pedega.

Cuesta[e] la raxion del'antena. Lo stello da proda vol eser el quarto men de 30 che l'alboro fusse longo dala choverta in suxo, serano passa 9. E vuol volzer in lo so redondo $\frac{1}{2}$ pie per paso.

E lo ventame de questo nostro stello vol eser più longo cha 'l stelo pie 1 per passo. Lo ventame die 'ser longo passa 10, pie 4.

Questo ventame e stello longo passa 19, pie 4, che vol eser longa la lama, die 'ser pie 1 per passo, serano pie 19. Romagnirà netta l'antena de passa 16 e 19 pie va in dopio. Questa è l'antena de
f. 166a mezo. Vuol el stelo lo quarto men de zò che l'alboro fusse longo dala choverta | in suxo. L'alboro è longo pasa 11 dela choverta in suxo. Serano pasa $7\frac{3}{4}$. El ventame pie 1 per passo più dal stello fuse longo, serà el ventame passa 9 e pie 3. E lo dopio ò ditto.

---

a. *Sic in MS, evidently for* quanto.
b. *Drawing lacking in MS.*
c. *Sic in MS, for* pie.
d. pie *added above the line in place of* die, *crossed out with horizontal marks.*
e. *Sic in MS.*
1. The length of the ship on deck is given above as 15 paces plus one extra, so the meaning here must be that the length of the first boat is twice as many feet long as the ship is paces long, plus one.
2. Drawing lacking in MS.

And our ship needs 2 boats and a gondola. The first boat should be twice as long as the length of the ship on deck, that is, 30 feet[1] and one more. That will be 31.

The second boat should be shorter than the first, 28 feet.

And the gondola should be as many feet long as is the beam [of the ship], 24.

[*caption:*]
Lateen ship[2]

The foremast of this ship of ours should be three times as long as the ship's beam, that would be 14 paces and 2 feet. It should measure around, at a third of its height above the bottom, a half palm per pace of the full length of the mast, that will be 7 palms. And the block mast should be one foot long per pace of the length of the mast that was above the deck. And it should be as wide as one sixth of the length of the block mast. Thus it will be 12 feet long, 2 feet wide; and the same is true for all small ships and [large] ships. f. 165b

And I wish to teach you where the mast step is placed. It should be as many half-feet aft from the forefoot as the length of the keel in paces; that will be 6 feet. And the rake of this mast of ours should be one foot per pace of the length of the mast, that will be $14\frac{1}{2}$ feet.

And the midship mast should be 13 paces long, and the mast step of this our midship mast should be one quarter less than the keel, that will be 9 feet. And this much you should measure forward from the "afterfoot,"[3] and there place the mast step.

This is the proportioning of the yard. The *stello*[4] of the foremast should be one quarter less than the height of the mast above the deck, that will be 9 paces. And it should be $\frac{1}{2}$ foot per pace in its circumference.

And the *ventame*[5] should be longer than the *stello* by 1 foot per pace. The *ventame* should be 10 paces 4 feet long.

This *ventame* and *stello* are 19 paces 4 feet long; what should be the length of the scarf? It should be 1 foot per pace, that will be 19 feet. 16 paces of the yard will be left in the clear, and 19 feet goes in the overlap. This is the yard of the midship mast. The *stello* should be a quarter less than the height of the mast | above deck. The mast is 11 paces long above the deck. [The *stello*] will be $7\frac{3}{4}$ f. 166a
paces. The *ventame* will be 1 foot per pace more than the *stello* is long. The *ventame* will be 9 paces and 3 feet. And I've stated the overlap.

---

3. While *cholto di proda* seems to correspond to the English "forefoot," there is no corresponding term in English for *choltro di puope.*
4. The *stello* was the lower, forward, heavier spar of a lateen yard.
5. The *ventame* was the upper, after, lighter spar of a lateen yard.

Volemo nui achoredar questo nostro alboro da proda. Sapi che quante pasa è l'alboro dala choverta in suxo, tanti chinali vuol per ladi. Serave per ladi chinalli 12 e q[*u*]arnal un. Volemo saver zò che pexa per passo i chinali. Vuol pexar el terzo che l'alboro fuse longo. Dala pedega in suxo l'alboro è longo pasa 14 e pie 2. El so terzo serave libre 4 onze 8.

Volemo in primo tayar li primi 2 pupexi, zoè 1 per ladi, de passa 13. Serà tuti 2 passa 26. Pexerà libre $123\frac{1}{2}$. E li 2 segondi, pie 3 più curti. Serà passa 24 e pie 4.

Volemo noi taiar li 2 terzi chinalli. Vuol più curti cha li segondi pie $2\frac{1}{2}$. E va chosì taiando $2\frac{1}{2}$ più cortti dal'alttro inchina chinalli 7. E dali 7 e dali 7[a] in proda ali altri pie 2 men un del'oltro, zoè dali 7 ali 10. E li segondi da proda vol eser più curti cha li oltri pie $2\frac{1}{4}$, e lo primo $2\frac{1}{2}$.

Volemo noi le choronelle de questi nostri chadernalli. Vol eser zaschadon longo el terzo del'alboro dala choverta in sù. Serà passa 4.

Voio che tu sapi quante peze de sartia, over quanti pasa andarà per furnir li ditti chinalli 24. Serano passa 246. Pexarà el passo libre 4 e onze 9.[b] Pexerà in tutti in suma libre 1235. E questo è fatto.

f. 166b Volemo nuy taiar 2 chadernalli longi 3 fiade quando[c] è longo l'alboro da choverta in su. Serà passa 37 pie 1 l'uno. Die pexar el passo libre $1\frac{1}{2}$. Lavora zascuna in quarta.

Vol la braza de questi over la chorela el terzo che l'alboro fusse longo dala choverta in su. Serà passa 4; el paso libre 5.

Vuolemo nui menaduri 24, lo[*n*]guo zascon passa 9; el paso libre $1\frac{1}{2}$.

L'orza poza vuol eser longa 3 fiade quando[d] la poza fusse longa e più passa 5. Serano passa 50; el paso libre 2.

Pozal da pruoda de passa 32, de nonbolli 4; el paso libre 2.

Taye 24 de ragli 2 l'una per chinalli e taye 24 de raglo 1.

Taye 2 di quarnalli de ragli 2 e do de raglo 1 intanpagnadi.

Taye 4 de fonde intanpagnade.

---

a. *Repeated thus in MS.*
b. 9 *corrected after* 4 *crossed out with a horizontal line.*
c. *Sic in MS, evidently for* quanto.
d. *Sic in MS, evidently for* quanto.

We wish to rig our foremast. Know that as many paces as the mast is above the deck, there need to be that many *chinal* shrouds per side. There will be 12 *chinal* shrouds per side and 1 *quarnal* tackle. We wish to know how much the *chinal* shrouds weigh per pace. They should weigh one third as much as the mast is long. From the mast step up, the mast is 14 paces 2 feet long. Its third would be 4 pounds 8 ounces.

We wish first to cut the first 2 after shrouds, that is, one per side, of 13 paces. That will be in all 26 paces. It will weigh $123\frac{1}{2}$ pounds. And the second 2 [after shrouds], 3 feet shorter. It will be 24 paces and 4 feet.

We want to cut the 2 *chinal* shrouds in the third set. They should be $2\frac{1}{2}$ feet shorter than the second set. And continue cutting in this way, each one $2\frac{1}{2}$ shorter than the other until the seventh set of *chinal* shrouds. And from the seventh set of *chinal* shrouds forward, each is 2 feet less than the other, that is, from number 7 to 10. And the second [set] of the shrouds forward should be $2\frac{1}{4}$ less than the others, and the first $2\frac{1}{2}$.

We now want the pendants for our *quarnal* tackles. Each one should be a third as long as the height of the mast above the deck. It will be 4 paces.

I want you to know how many pieces of shrouds, or how many paces, it will take to furnish these 24 shrouds. It will be 246 paces. Each pace will weigh 4 pounds 9 ounces. It will all weigh in total 1,235 pounds. And this is done.

We wish to cut two *quarnal* tackle falls 3 times as long as the mast is high above the deck. That will be 37 paces 1 foot each. Each pace should weigh $1\frac{1}{2}$ pounds. Each is a four-part tackle. f. 166b

The pendants of these should be one third as long as the mast's height above the deck. It will be 4 paces; the pace 5 pounds.

We need 24 tackle falls, each one 9 paces long; the pace $1\frac{1}{2}$ pounds.

The *orza poza* should be three times as long as the *poza* and 5 paces more. They will be 50 paces; the pace 2 pounds.

The fore *pozal* is of 32 paces, of 4 strands; the pace 2 pounds.

24 double blocks for the *chinal* shroud tackles and 24 single blocks.

2 double blocks for the *quarnal* tackles and 2 single [blocks] with bushings.[1]

4 *fonda* tackle blocks with bushings.

---

1. Metal linings of blocks.

Taye 2 per morganal de do ragli e de 1 intanpagnade.

Taye 4 da mantichio, 2 de do ragli e 2 d'un raglo intanpagnade.

Taye 3 del pozal, 2 de do ragli e l'una di 1 intanpagnade.

Taie 2 del'orza poza de do ragli intanpagnadi.

Taie 4 de suste e de charuzi ognoli intanpagnadi.

Taie 2 d'anzolo[a] intanpagnade. Pasteche 2 de morganalli. Pasteche 2 a pope per le suste.

Vuol legname de alboro da proda: bolgare 5, bigotte 4, miniselli 25. Le do bigotte al cholo e le 2 per respetto. Taya una di gordilli per la schalla, per passo schaliny 4.

Mo' volemo noi achoredar l'alboro de mezo. Vuolemo che li primi popexi vol eser longo quanto l'alboro dala choverta in su. Serà pasa 11. Die pexar el passo libre 5.

Li segondi chinalli de pasa 10 pie 3; el passo libre 5.

Li 2 terzi chinalli tanto quanto i segondi men pie $1\frac{1}{2}$.

f. 167a Choronele d'i fraschuny, el terzo de zò che l'alboro fusse longo dala choverta in su. Serave passa 4[b] men $\frac{1}{3}$; el paso libre $5\frac{1}{2}$.

Chinalli 2 in driedo li frasconi, longi passa 10, el pexo dito.

Sinalli 3 per ladi, longi $\frac{2}{3}$ dela longeza del'aboro dala choverta in su, serano passa 7 e pie 2. Vuol eser più intro li segondi ch'al primo. E chosì el terzo ch'al segondo $\frac{1}{4}$ de pe'. Pexarà el passo libre $4\frac{1}{4}$.

Prodeny 2, un per ladi, vol eser longi più cha i signali passo uno. Dal ditto pexo.

Manti 2 de' fraschuny, de passa 11 zaschon, al ditto pexo.

Menalli d'i senalli, 2 volte quanto l'alboro dela choverta in ssu, serà passa 22. Vorà per 6 passa 132; el passo libre $1\frac{1}{2}$.

---

a. taie 2 *repeated and crossed out with a horizontal line.*
b. 4 *corrected after* 1 *crossed out with a diagonal line.*

2 double blocks for each *morganal,*[1] and 1 with bushings.

4 blocks for lift tackles, 2 double and 2 single with bushings.

3 *pozal* tackle blocks, 2 double and one single with bushings.

2 double blocks with bushings for the *orza poza.*

4 blocks for the vang tackles and the *charuzo*[2] tackles, all with bushings.

2 parrel tackle blocks with bushings. 2 snatch blocks for the *morganalli.* 2 snatch blocks aft for the vangs.

The foremast needs a parrel truss: 5 parrel ribs, 4 deadeyes, 25 parrel trucks. The 2 deadeyes are for the truss and the other 2 as spares. 1 *gordilli*[3] block for the ladder, which has 4 steps per pace.

Now we wish to rig the midship mast. We want the first set of after shrouds to be as long as the mast from the deck up. That will be 11 paces. Each pace should weigh 5 pounds.

The second set of *chinal* shrouds of 10 paces 3 feet; the pace 5 pounds.

The third set of two *chinal* shrouds as much as the second less $1\frac{1}{2}$ feet.

The pendants of the *fraschun*[4] shrouds, a third of the height of the mast above the deck. It will be 4 paces minus $\frac{1}{3}$; the pace $5\frac{1}{2}$ pounds. f. 167a

2 *chinal* shrouds abaft the *fraschun* shrouds, 10 paces long, of the same weight.

3 *senal* shrouds per side, $\frac{2}{3}$ as long as the height of the mast above the deck. They will be 7 paces and 2 feet. The second set should be further inboard than the first. And likewise the third than the second by $\frac{1}{4}$ foot. The pace will weigh $4\frac{1}{4}$ pounds.

2 *prodeno* tackles, one per side, should be 1 pace longer than the *senal* shrouds, of the same weight.

2 runners of the *fraschun* shrouds, of 11 paces each, of the same weight.

Falls for the *senal* shroud tackles, twice as long as the height of the mast above deck, will be 22 paces. 132 paces will be needed for 6; the pace $1\frac{1}{2}$ pounds.

---

1. The *morganal* was some sort of tackle, probably attached to the lower end of the *stello.*
2. The *charuzo* was a tackle attached to the *ventame.*
3. The *gordilli* were buntlines or other small sail ropes.
4. The *fraschuny* were the forwardmost shroud tackles.

Menalli d'i chinali d'i prodony, tanto chomo l'alboro longo dala choverta in suxo, serà passa 11; el passo libre $1\frac{1}{2}$.

Menalli d'i fraschuny, longi 4 fiade quanto longo el manto, seria passa 45, pie 3; el passo libre $1\frac{1}{2}$. Lavora in quinto.

Anzolo de mezo, la so choronella, passa $3\frac{1}{2}$; el passo libre 2. El menal del ditto, $2\frac{1}{2}$ vuolta quanto l'alboro dala choverta in su, serave pasa 28; el passo libre 2.

Manti de' 4 nonbulli, longi quanto l'alboro dala choverta in su, serà passa 11; el paso libre 7.

Rixe d'antena, 5 vuolte quanto vuolze l'antena. Die pexar el passo libre 7.

Braze de funde passa 21, farà 2 braze de surda; el paso libre 2.

f. 167b Menalli de funde vol eser longi 3 fiade quanto l'alboro da|la choverta in su, serà passa 33; el passo libre $1\frac{1}{2}$.

Gomene longe 8 vuolte quanto l'alboro dala choverta in su, serà passa 88.

Braza de suste, $\frac{2}{3}$ de zò ch'è longa tuta l'antena; el passo libre 3.

Menalli de suste vuol $2\frac{1}{2}$ vuolte quanto l'antena longa; el paso libre 2.

Menalli d'i morganali, 3 vuolte quanto el stello fuse longol; libre 2.

Orza davanti, 2 volte quanto lo stello. Serà passa 16. Lavora in terzo; el paso libre 2. Brazo per la dita, el quarto del stelo, serà passa 2. Poza longa quanto l'alboro da choverta in su, serà pasa 11; el passo libre 5.

Pozal di mezo vol eser 2 fiade quanto la poza; el passo el terzo del pexo dala poza. Serà libra 1 onze 8.

Chatavy 4 de ragli 4 l'una intanpagnade. Taye 12 de 2 ragli intanpagnade. Taye 2 de frascony da erto chon uno raglo intanpagnadi. Taye 4 di fraschuny chon 2 ragli intanpagnadi. Taie 2 di chinalli de do ragli.

Item taie 6 di sinalli chon 3 ragli intanpagnadi.

Falls of the *chinal* shroud tackles [and] of the *prodeno* tackles, as long as the height of the mast above deck, will be 11 paces; the pace $1\frac{1}{2}$ pounds.

Tackle falls for the *fraschun* shrouds 4 times as long as the runner, would be 45 paces 3 feet; the pace $1\frac{1}{2}$ pounds. Each one is a five-part tackle.[1]

The *anzolo* tackle of the midship mast, its pendant, $3\frac{1}{2}$ paces; the pace 2 pounds. Its tackle fall, $2\frac{1}{2}$ times the height of the mast above the deck, will be 28 paces; the pace 2 pounds.

Ties of 4 strands, as long as the mast above the deck, will be 11 paces; the pace 7 pounds.

Yard slings, 5 times the circumference of the yard. The pace should weigh 7 pounds.

Pendants for the *funda* tackles, 21 paces, will make 2 pendants for the *surda* tackles; the pace 2 pounds.

Falls for the *funda* tackles should be 3 times as long as the mast | above the deck; it will be 33 paces; f. 167b
the pace $1\frac{1}{2}$ pounds.

Halyard tackle falls 8 times as long as the mast above the deck; it will be 88 paces.

Vang pendants, $\frac{2}{3}$ of the entire length of the yard; the pace 3 pounds.

Tackle falls for the vangs need to be $2\frac{1}{2}$ times the length of the yard; the pace 2 pounds.

Falls for the *morganal* tackles, 3 times the length of the *stello*; 2 pounds.

Orza davanti, 2 times the length of the *stello*; it will be 16 paces. Three-part tackle; the pace 2 pounds. Pendant for it, a quarter of the *stello*; it will be 2 paces. Sheet as long as the height of the mast above the deck; it will be 11 paces; the pace 5 pounds.

Midship *pozal* should be 2 times the length of the sheet; the weight a third that of the sheet. It will be 1 pound 8 ounces.

4 quadruple blocks,[2] one with bushings. 12 double blocks with bushings. 2 single upper blocks of the *fraschun* shrouds with bushings. 4 double [lower] blocks of the *fraschun* shrouds with bushings. 2 double blocks of the *chinal* shroud tackles.

And 6 triple blocks with bushings for the *senal* shroud tackles.

---

1. Augustin Jal, *Archéologie navale* (Paris: Arthus Bertrand, 1840), vol. 2, p. 29, has "in quarto" rather than "in quinto" at the corresponding section.
2. The *chatavo* block seems to have been different from the regular *taya*.

Item taie 2 de soste chon un raglo intanpagnade.

Item taye 2 d'orza davanti.

Item taye 4 di morganali.

Item taye 2 d'anzollo de uno raglo.

Item taye 2 per pozal, 1ª de do ragli e l'oltro de un.

Item legname d'alboro bolgara 4, minyselli 20, bigotte 4, i do d'avanzo. E questo è fatto e ditto.

f. 168a Per saver quanto pano va in una vela. Vella de pasa 16. La mitade | de 16 son 8. Farà 24. Moltipicha 16 via 24, fa 384. E questi sono passa. E per ogny zentener de pasa, serano peze 10.[a] Serano peze $38\frac{1}{2}$.

E per ogni peza de fustagno vuol chanevazo braza 10. E questa nave è chonpido. La vedirì qui de sotto.

[*caption:*]
nave a velo latina[b]

f. 168b

[Far 1ª nave quadra chon tuto quelo i prozede]

Al nomen di Dio. Vuolemo noi far una nave quadra de passa 13 in cholonba. E die aver de plan lo quarto men de zò che la cholonba fusse longa. Serave pie $9\frac{3}{4}$.

Nave che à de cholonba passa 13 e de plan pie $9\frac{3}{4}$. E die aver in trepie tanto quanto à da pla[*n*] e li $\frac{3}{4}$. Adoncha serave pie $17\frac{1}{2}$. E vorà lo averzer in bocha tanto quanto à de pian e quanto averze in trepie. Serave pie 27 men mezo pie.

E vuol eser in la sso prima choverta tanto quanto à de plan e lo quarto men. Serave serave[c] erta dale chorbe in suxo pie $7\frac{1}{2}$. Anderà bute 3, una sovra l'altra.

E vuol eser erta la choverta de sovra da quela de sotto pie $5\frac{1}{2}$, zoè pie 2 men cha la choverta de sotto. Anderà butte 2. E serà dala choverta de sovra inchina al fundi pie 14, zoè lo sesto men de zò che l'averze in trepie, per zò che in le sso 2 choverte io te fazo descreser un pie. La chaxion che le chorbe

a. 38 *crossed out with a horizontal line.*
b. *Drawing lacking in MS.*
c. *Thus repeated in MS.*
1. The *bigotta* was part of a parrel, perhaps the deadeye.
2. Drawing lacking in MS.

And 2 single blocks for the vang tackles with bushings.

And 2 blocks for the *orza davanti*.

And 4 blocks for the *morganal* tackles.

And 2 single blocks for the parrel tackle.

And 2 blocks for the *pozal* tackle, one double and the other single.

And as wooden parts for the mast, 4 parrel ribs, 20 parrel trucks, 4 *bigotte*,[1] and two as spares. And this is done and said.

To know how much cloth goes into a sail. A sail of 16 paces. Half | of 16 is 8. This will make 24. Multiply 16 by 24, it makes 384. And these are paces. And for every 100 paces there will be 10 pieces. There will be $38\frac{1}{2}$ pieces. f. 168a

And for every piece of fustian 10 ells of canvas are needed. And this ship is complete. You will see it below.

[*caption:*]
Lateen ship under sail[2]

To make a square-rigged ship with all its measurements f. 168b

In the name of God. We wish to make a square-rigged ship of 13 paces in its keel. And it should have a floor one quarter less than the length of its keel. That will be $9\frac{3}{4}$ feet.

A ship that has a keel of 13 paces and a floor of $9\frac{3}{4}$ paces: it should also have a *trepie* equal to the breadth of the floor and $\frac{3}{4}$ again. So it will be $17\frac{1}{2}$ feet.[3] And it will have a beam equal to the breadth of the floor plus its *trepie*. It will be 27 feet minus a half foot.

And the breadth of its first deck should be as much as the breadth of the floor and a quarter less. The depth from the floor timbers up will be $7\frac{1}{2}$ feet. Three butts[4] will go there, one above the other.

And the upper deck should be $5\frac{1}{2}$ feet above the lower one, that is, 2 feet less than the lower deck. Two butts will go there. And the depth from the upper deck to the bottom will be 14 feet, that is, a sixth less than the *trepie* breadth, such that in its 2 decks I make you decrease one foot. The reason

3. Actually, $9\frac{3}{4}$ plus $\frac{3}{4}$ of itself works out to $17\frac{1}{16}$. Many of the calculations in the shipbuilding section yield only approximate values; perhaps they were only rough mnemonics.
4. That is, wine barrels.

de sutto te tuo' quel mezo pie, e la choverta per groseza de legname te tuo' mezo pie, inperzò te roman in le choverte pie 13 dentro.

E vuol eser la ditta chocha longa in choverta tanti passa qua[*n*]ti pie sun in li do terzi de zò che l'averze in bucha, che xe passa 18. Serà longa da ruoda in ruoda passa 19, per zò che le teste farà creser passo 1.

La ruoda da proda de questa nostra chocha vuol eser longa passa $6\frac{1}{2}$, zoè la mitade dala cholonba.

f. 169a Longa l'asta da proda de questa nostra chocha el terzo dela cho|lonba fuse longa. Zò serano passa $4\frac{2}{3}$ de pie.

Lo sso slanzo da proda longo el terzo dela cholonba e pie 1. Serave pie $22\frac{2}{3}$.

L'asta de puope vuol slanzar pie $1\frac{1}{4}$ per passo de zò che xe longa l'asta, che serave paso 1. E per lo simille tute le oltre.

E vuol eser longo el timon più pie 2 de zò che l'asta è longa. Serave pie $23\frac{2}{3}$. E vuol eser alargo in la palla el quarto dela longeza de zò che fusse tuto el timon longo.

E vuol in l'asta da pope chanchare 8, e l'axola da basso, e quanto tu li die aficar. Partili chomo ti par.

E vuol questa nostra chocha batei 2 e una gondulla.

El primo batello mazuor vol eser longo 2 pie per passo de zò che la chocha è longa in choverta, e pie 1 de più. Serave pie 36.

El segondo batello vuol eser menor cha lo primo pie 2. Serave 34.

E la gondulla de questa nostra chocha vol eser longa pie 24.

[*caption:*]
barcha[a]

f. 169b [*captions:*]
barcha[b]

a. *Drawing lacking in MS.*
b. *Drawing lacking in MS.*

is that the lower frames take away a half foot from you, and the deck in the thickness of its timbers takes a half foot from you, so that leaves 13 feet under the deck.

And this cog should be as many paces long in the deck as there are feet in two thirds of the breadth, which is 18 paces. It will be 19 paces long from stem to stern, as the heads will make it 1 pace longer.

The stem of our cog should be $6\frac{1}{2}$ paces long, that is, half [the length] of the keel.

The forward stem of our cog will be one third | of the length of the keel. That will be $4\frac{2}{3}$ paces of feet.[1] f. 169a

Its forward rake is one third as long as the keel plus one foot. That will be $22\frac{2}{3}$ feet.

The sternpost will rake $1\frac{1}{4}$ feet per pace of the length of the post, which will be 1 pace. And similarly for all the others.

The rudder should be two feet longer than the sternpost. It will be $23\frac{2}{3}$ feet. And it should be as wide in its blade as a quarter of the whole length of the rudder.

And the sternpost should have 8 gudgeons and the pintles below, and you should mount that many. Space them as seems right to you.

And our cog should have two boats and a gondola.

And the first, larger, boat should be 2 feet long for every pace the cog is long in its deck, and one foot more. That will be 36 feet.

The second boat should be 2 feet less than the first. That will be 34.

And the gondola of this cog of ours should be 24 feet long.

[*caption:*]
Boat[2]

[*captions:*] f. 169b
Boat[3]

1. If the keel is 13 paces, one third of this would be $4\frac{1}{3}$ paces.
2. Drawing lacking.
3. Drawing lacking.

chopano[a]

nave[b]

f. 170a Alboro de questa nostra chocha vol eser longo tanto 3 vuolte e mezo quanto averze in bucha. Serave pie $99\frac{1}{2}$, serano passa 18 e pie $4\frac{1}{2}$. Vuol volzer lo ditto alboro in lo so terzo mezo palmo per passo de zò che l'alboro fusse longo. Serave palmi $9\frac{1}{2}$.

E vuol questo nostro alboro un cholzexe, el qual die 'ser longo 1 pie per passo de zò che l'alboro fuse tuto longo. E vuol eser alargo in li ragli el quinto de zò che fusse longo el cholzexe tutto.

E vuol antena per la ditta nave. Vol eser longa 3 fiade chomo la verze la nave in bocha, siando ligada. Serave pie 81, zoè passa 16, pie 1. Die 'ser zaschadun d'i penoni de passa 10, pie $2\frac{1}{4}$. Die vuolzer mezo palmo per passo.

E lo sso bonpresso vol eser longo passo 1 men de zò che xe lo penon. Serave passa 9 pie $2\frac{1}{4}$.

[*caption:*]
antena[c]

f. 170b [*captions:*]
alboro[d]

bonpreso[e]

Raxion de sartia de questo nostro alboro. In prima choronole 4 di fraschuny, zoè 2 per ladi. Lo qual vol eser longo zascuna d'ese tanto quanto tuto l'alboro dela choverta in suxo, zoè è lo terzo. Serave longa zaschaduna d'ese passa $5\frac{1}{2}$. Serave tute 4 passa 22. Vuol pexar el passo libre 6, zoè lo terzo de ttuto l'alboro.

E vuol choronole 10 di signali, zoè 5 per ladi. Vuol eser longe passa 11 pie $2\frac{1}{2}$, zoè lo terzo men de zò che l'alboro è tuto longo dala choverta in suxo. E vuol pexar el passo lo quinto de ttuto l'alboro.

E vuol manti 4 di frascuny, tanto longi quanto è l'alboro dela choverta in suxo. Serave passa 15 pie 3. Die pexar el passo libre 5 onze 9, lo quarto che l'alboro longo.

---

a. *Drawing lacking in MS.*
b. *Drawing lacking in MS.*
c. *Drawing lacking in MS.*
d. *Drawing lacking in MS.*
e. *Drawing lacking in MS.*
1. Drawing lacking.
2. Drawing lacking.
3. Actually, at five feet per pace it would be 19 paces and $4\frac{1}{2}$ feet. But the length of the mast would be $92\frac{3}{4}$ feet if his calculations were taken literally (a mast $3\frac{1}{2}$ times the $26\frac{1}{2}$ feet of the beam).

Skiff[1]

Ship[2]

The mast of our cog should be 3 and a half times the beam. It will be $99\frac{1}{2}$ feet, that is, it will be 18 paces and $4\frac{1}{2}$ feet.[3] The circumference of this mast at a third [of its height] should be one half palm per pace of the length of the mast. That will be $9\frac{1}{2}$ palms. f. 170a

And our mast should have a block mast, which should be one foot long per pace of the entire length of the mast. And it should be as wide at the sheaves[4] as one fifth of the entire length of the block mast.

And we need a yard for this ship. This should be three times as long as the ship's beam, when [scarfed and] lashed. It will be 81 feet, that is, 16 paces 1 foot. Each of the spars should be 10 paces $2\frac{1}{4}$ feet. It should measure half a palm around per pace.

And its bowsprit[5] should be 1 pace shorter than the spar. That will be 9 paces $2\frac{1}{4}$ feet.

[*caption:*]
Yard[6]

[*captions:*] f. 170b
Mast[7]

Bowsprit[8]

Method of the rigging of our mast. First, 4 pendants for the *fraschun* tackles, that is 2 per side. Each should be as long as the height of the mast above the deck, a third [of the mast]. Each will be $5\frac{1}{2}$ paces long. The 4 together will total 22 paces. Each pace should weigh 6 pounds, that is, one third [in pounds] of [the length in paces of] the entire mast.

And there should be 10 pendants for the *senal* shroud tackles, that is, 5 per side. They should be 11 paces $2\frac{1}{2}$ feet, that is, one third less than the height of the mast above deck. And each pace should weigh [in pounds] a fifth of [the length of] the whole mast.

And there should be 4 runners for the *fraschun* tackles, each as long as the height of the mast above the deck. This will be 15 paces, 3 feet. Each pace should weigh 5 pounds 9 ounces, a quarter the length of the mast.

---

4. The grooved wheels in the blocks.
5. The spar running out from the front of the ship.
6. Drawing lacking.
7. Drawing lacking.
8. Drawing lacking.

f. 171a E vuol bragotti di stazi, zaschadon longo passa 3. E può, indriedo ito, chinalli 9. Lo primo driedo i senalli vol eser dal'uno ladi chomo dal'oltro longi passa 12 pie $3\frac{1}{2}$, imperzò che li vuol eser longi più cha li senalli passo 1. E può andara' taiando inchina puope li ditti chinalli 7 per ladi, più longo un del'altro pie $1\frac{1}{4}$. E venirà la schala ben.

E vuol 4 chadernalli, zoè 2 per ladi. Vol eser longe le sso choronelle lo terzo de zò che l'alboro è longo, zoè passa 6 e pie 2 per zaschaduna. Die pexar el passo $\frac{2}{3}$ che pexa i chornelle de sovra, zoè libre 3.

E vuol eser li to stazi chon tutti li bravotti tanto longi chomo tuto l'alboro è longo, zoè passa 19. Serave tuti 2 pasa 38. Die 'ser al pexo di bravotti.

E vuol manti tanto longi chomo tuto l'alboro dale pianede in suxo, zoè 2 volte. Serave passa 30 l'un. Serà tuti 2 passa 60, e vegnirà intorno el sguindazo $\frac{1}{2}$ volta.

E vuol do rixe tanto longi chomo vuolze al dopio l'antena 5 vuolte tanto. Serave longi zaschuna d'ese pasa 5. Sapi che questi manti e staze e br[*a*]gotti e rixe vol eser tutti d'un pexo, zoè tante libre quante la mitade del'alboro. Serave libre 9.

E vuol braze de funde de questa nostra antena. Vol eser longe 4 fiade chomo l'antena longa. E vuol eser funde 3 e brager 1. 1ª surda. Zaschaduna d'ese vol eser dopie, e roman passa 4 l'una. Die pexar el passo libre $2\frac{1}{2}$.

f. 171b E vuol legname questo nostro alboro. Vuol aver bolgare 5, minisielli 25, bigotte 2 al so chuolo e 2 d'avanzo.

E vuol li so menalli de queste braze 2 volte quanto l'antena fusse longa. Serà pasa 32. Die pexar el paso[a] $\frac{2}{3}$ de zò che pexa la choronella. Pexerà el paso libre 3 onze 8.

E vuol le so borine. Vol eser longe 2 volte chomo l'antena fuse longa. Die pexar el passo libra 1 onze 9.

E vuol li so stinchi longi 2 volte quanto l'alboro fuse longo dala choverta in su. Pexerà el passo libre $1\frac{1}{2}$. Li so scotine vol eser longe 2 volte chomo tuto l'alboro. Serave passa 38. Die pexar el passo libre $1\frac{1}{2}$.

---

a. libre *crossed out with vertical stroke.*

1. In English parlance this would be one "turn"; the Venetian *volta* would be a "round turn."
2. That is, where it is scarfed and lashed.
3. That is, they should weigh in pounds as much as half of the number of paces in the mast.

And it needs stay pendants, each 3 paces long. And then, behind these, 9 *chinal* shroud tackles. The first behind the *senal* shrouds on each side should be 12 paces $3\frac{1}{2}$ feet, as they should be 1 pace longer than the *senal* shrouds. And then you'll proceed cutting these 7 *chinal* shrouds on each side going aft, each $1\frac{1}{4}$ feet longer than the other. And the scale will come out right. f. 171a

And it needs 4 *quarnal* tackles, that is, 2 per side. The pendants of these should be one third the height of the mast, that is, 6 paces and 2 feet each. Each pace should weigh $\frac{2}{3}$ the weight of the pendants mentioned above, that is, 3 pounds.

And your stays with all their pendants should be as long as the whole mast, that is, 19 paces. The 2 of them will be 38 paces. They should be the weight of the pendants.

And it needs halyards as long as the mast from the bottom of the ship up, that is, 2 times as long. Each will be 30 paces. Both will be 60 paces, and will go around the windlass $\frac{1}{2}$ time.[1]

And it needs two yard slings 5 times as long as the yard measures around where it is doubled.[2] Each of these will be 5 paces long. Know that these halyards and stays and pendants and slings should all be of one weight, that is, as many pounds as half the mast.[3] That will be 9 pounds.

And we need pendants for the *funde* of our yard. These should be 4 times as long as the length of the yard. And there should be 3 *funde* and 1 *brager*.[4] One *surda*.[5] Each of these should be double, and still be 4 paces each. Each pace should weigh $2\frac{1}{2}$ pounds.

And this mast of ours needs a parrel[6] truss. It needs 5 parrel ribs, 25 parrel trucks, and 2 deadeyes for its truss and two left over. f. 171b

And it needs falls for these braces that are twice as long as the yard. That will be 32 paces. Each pace should weigh $\frac{2}{3}$ of a pound less than the weight of the pendant. Each pace will weigh 3 pounds 8 ounces.

And it needs its bowlines. These need to be 2 times the length of the yard. The pace should weigh 1 pound 9 ounces.

And it needs buntlines[7] twice as long as the height of the mast above the deck. Each pace will weigh $1\frac{1}{2}$ pounds. Its clew lines[8] should be 2 times as long as the entire mast. That will be 38 paces. Each pace should weigh $1\frac{1}{2}$ pounds.

---

4. A tackle of uncertain function.
5. A tackle of uncertain function.
6. The sliding band of rope attaching the yard to the mast.
7. Uncertain interpretation; the bunt was the middle part of the sail.
8. Uncertain interpretation; the clew was the lower or after corner of the sail.

Branchadelle vol eser 8 per ladi. Die 'ser zaschuna passa 2. Serave passa 16 per ladi. Serà longe le tto branche chon brancha[*d*]elle pie 9, zoè tanto quanto à lo tuo trio de chazuda.

L'anzollo del'alboro da proda vol eser tanto longo chomo è lo tuo alboro 3 fiade dela choverta in suxo. Die pexar el passo libre $1\frac{1}{2}$.

Menalli dele taie de' fraschuny die 'ser zaschaduna d'esse longe 3 fiade e mezo chomo lo tuo alboro fusse longo da choverta in suxo, zoè passa 56. Die pexar el passo libra 1 onze 9. Lavorerà in $\frac{1}{5}$.

Menalli de' senali, vol eser do chotanto de zò che xe la prima choronella d'i senalli. Serave passa 24 l'una. Vol eser 5 per ladi. Die pexar el passo libre $1\frac{1}{2}$.

Menalli d'i chinalli vol eser de passa 16, zoè tanto quanto l'alboro è longo dala choverta in su. Pexerà el paso libre $1\frac{1}{2}$.

Menalli d'i quadernalli vuol eser longi 3 fiade quanto l'alboro dala choverta in su. Serà pasa 48. El paso libre $1\frac{1}{2}$. Lavora per quarto.

f. 172a Menalli dele funde, vol eser 2 per ladi, e vol eser zaschaduna de queste 4 fiade quanto l'alboro dala choverta in ssu. Lavorerà in quinto. Die pexar el passo libre $1\frac{1}{2}$.

E vuol che li sso mantichi die 'ser longe chomo 4 volte l'alboro dala choverta in ssu. Serà longo passa 60. El passo libre 2.

Braza una de surde, zoè la mitade ch'è l'antena longa. Die pexar el passo libre 5.

El menal dela surda vol eser tanto longo 3 volte quanto l'alboro dala choverta in ssu. Serà passa 48. El paso libre $1\frac{1}{2}$.

E vuol le sso schotte. Vol eser tanto longe chomo à la nave di ruoda in ruoda $\frac{1}{4}$ più. Serave passa 24. El passo libre 8.

El to brio grando vol eser longo quanto l'alboro, zoè passa 19. Die pexar $\frac{1}{4}$ de zò che l'alboro fusse longo. Die pexar el passo libre 4 onze 9. Llo[a] so briolin pizullo vol eser longo $\frac{2}{3}$ de zò ch'è longo lo grando, zoè passa 12, el paso libra 1 onze 9. Item schallette e tienteben passa 32, el passo libre $1\frac{1}{2}$. Braza d'antena, passa 16 per la longeza del'antena. Die pexar el passo libre 5, zoè lo terzo del'antena. Farate braza 2, un per ladi.

---

a. *Sic in MS.*

1. The *branchadella* was a piece of rope made fast to the leech of a square sail to which one of the legs of the bowline bridle was attached; this function is now served by loops called cringles.

2. As the pace was five feet, the *branchadelle* were in themselves 10 feet long; it may be that when they were with their bridles they were somehow looped to form a length 9 feet in all.

There should be 8 *branchadelle*[1] per side. Each should be 2 paces. There will be 16 paces per side. Your bridles with *branchadelle* will be 9 feet long, that is, as long as your square sail's drop [in paces].[2]

The parrel tackle of the foremast should be three times as long as the height of the mast above deck. Each pace should weigh $1\frac{1}{2}$ pounds.

The falls of the *fraschun* tackles should each be 3 and a half times as long as the mast is above the deck, that is, 56 paces. Each pace should weigh 1 pound 9 ounces. It will be a 5-part tackle.

The falls of the *senal* shroud tackles: there should be two equal to the first pendant of the *senal* shrouds. Each will be 24 paces. There should be 5 per side. Each pace should weigh $1\frac{1}{2}$ pounds.

The falls of the *chinal* shroud tackles should be 16 paces, that is, as much as the mast is high above the deck. The pace will weigh $1\frac{1}{2}$ pounds.

The falls of the *quarnal* tackles should be 3 times as long as the mast is high above the deck. They will be 48 paces. The pace $1\frac{1}{2}$ pounds. It is a four-part tackle.

Falls for the *funda* tackles: there should be 2 per side, and each of these should be 4 times as long as the height of the mast above the deck. They will be five-part tackles. Each pace should weigh $1\frac{1}{2}$ pounds. f. 172a

And its yard lifts should be 4 times as long as the mast above the deck. It will be 60 paces long. The pace 2 pounds.

One *surda* pendant, which is half as long as the yard. Each pace should weigh 5 pounds.

The fall of the *surda* tackle should be 3 times as long as the mast above the deck. It will be 48 paces. Each pace $1\frac{1}{2}$ pounds.

And it needs its sheets. These should be as long as the ship from stem to stern and $\frac{1}{4}$ more. That will be 24 paces. The pace 8 pounds.

Your middle sheet should be as long as the mast, that is, 19 paces. It should weigh $\frac{1}{4}$ as much as the mast is long.[3] The pace should weigh 4 pounds 9 ounces. Its small middle sheet should be $\frac{2}{3}$ as long as the large one, that is, 12 paces, each pace 1 pound 9 ounces. Also *schallette*[4] and *tienteben*[5] of 32 paces, the pace $1\frac{1}{2}$ pounds. Yard braces, 16 paces for the length of the yard. Each pace should weigh 5 pounds, that is, a third of the yard. Make yourself 2 braces, one for each side.

---

3. That is, it should weigh one quarter as many pounds as the number of paces in the length of the mast.
4. Possibly rope ladders.
5. Some sort of rope or tackle.

E vuol menalli 2 de queste nostre braza. Vol eser longi passa 32 l'un, zoè 2 volte la longeza del'antena.

Taye de questo nostro alboro. Primo vuol taye 4 di fraschuny da erto, d'un raglo, intanpagnade.

f. 172b E vuol taie 8 de questi fraschuny de 2 ragli intanpagnadi.

E vuol taie 20, la mitade de 2 ragli e la mitade de 1.

E vuol taie 14 per chinalli de ragli 2. Item taie 8 di quarnalli, la mitade de ragli 2 e la mitade de raglo 1.

E vuol taie 2 d'anzolo. Item taye 10 de funde de 2 ragli intanpagnade. Item taie 2 de burine d'uno raglo. Tute queste vol eser intanpagnade.

E vuol taye 2 de braze de raglo 1 simille. Item 2 de stinchi de uno raglo simille. Item taie 4 de schotine e de gordilli de uno raglo, e bigotte 2 per li gordilli.

E vuol una taia del brando per tirar la barcha a proda, 1ª taya grossa chon uno raglo intanpagnade.

E vuol pasteche 3 per tirar la barcha. Item taie 4 de maxinetta d'un raglo. Item taya 1ª de chontra burina d'un raglo. Item taye 3 da briol, le 2 de raglo 1 e l'oltra de 2. E chomenzeremo per l'aboro de mezo.

[*captions:*]
taie[a]

ragli[b]

f. 173a Quista serà la raxion del'alboro de mezo de questa nostra chocha. In primo volemo che l'alboro sia per la mitade del'alboro grande, zoè de passa.[c] E chosì vuolemo che l'abia la sso raxion de groseza, e de cholzexe, e dal'angura de cholzexe, per la sso raxion ala ratta.

E vuol manti tanto longo zaschun quanto è l'alboro longo dala choverta in ssu. Serà passa $9\frac{1}{2}$. Die pexar el passo libre $4\frac{1}{2}$.

---

a. *Drawing lacking in MS.*
b. *Drawing lacking in MS.*
c. *Thus in MS, with no number given.*
1. Either of the two sides of a square sail.
2. A block of uncertain function.

And you need 2 tackle falls for these braces. Each should be 32 paces, that is, 2 times the length of the yard.

Blocks for this mast of ours. First we need 4 blocks for the upper *fraschun* tackles, single, with bushings.

And it needs 8 double blocks for these *fraschun* tackles with bushings. f. 172b

And it needs 20 blocks, half of them double and half single.

And it needs 14 double blocks for the *chinal* shroud tackles. Also 8 blocks for the *quarnal* tackles, half of them double and half of them single.

And it needs 2 blocks for the parrel tackle. Also 10 double blocks for the *funda* tackles, with bushings. Also 2 single blocks for the bowline. All these should be with bushings.

It needs 2 similar single blocks for the braces. Also 2 similar single ones for the clew lines. Also 4 single blocks for the leech[1] lines and the middle sheets, and 2 deadeyes for the middle sheets.

And it needs one *brando* block for hauling the boat by the bow, a large single block with bushings.

And it needs 3 snatch blocks to haul the boat. Also 4 single *maxinetta*[2] blocks. Also 1 single block for the martnet.[3] Also 3 middle sheet blocks, 2 single and the other double. And we will begin with the middle mast.

[*captions:*]
Blocks[4]

Sheaves[5]

Here is the measurement for the midship mast for this cog of ours. First of all, we want the mast to be half of the large mast, that is of [ ] paces.[6] And then we want its measurement of thickness and that of the block mast and of the *angura*[7] of the block mast to be proportional. f. 173a

And it needs ties each as long as the mast is high above the deck. That will be $9\frac{1}{2}$ paces. Each pace should weigh $4\frac{1}{2}$ pounds.

---

3. Uncertain interpretation.
4. Drawing lacking.
5. Drawing lacking.
6. No number given nor space left in MS.
7. An uncertain part of the block mast.

E vuol choronelle per chinalli. Die 'ser longe passa 3, zoè lo terzo del'alboro. E vuol pexar el passo libre 3, zò ch'è longa la choronella.

E vuol rixa d'antena 5 vuolte quanto volzesse l'antena, siando ingualezada per mezo li manti.

E vuol popexi 2, un per ladi. Vol eser longo ziaschon de questi men passo 1 de zò che l'alboro fusse longo dala choverta in ssu. Zò serà passa 8. El passo libre 2.

E vuol la ssegonda choronella churta cha la prima pie $2\frac{1}{2}$ e la terza e la quarta choronella. E vuol eser braza 2 de suste, longe zaschuna pasa 3. El passo libre 2.

E vuol menalli de queste nostre suste longe 2 vuolte quanto l'antena fusse longa. L'antena è longa passa 12. Adoncha serà passa 24. El passo die pexar libre $1\frac{1}{2}$.

f. 173b E vuol morganali de mezo. Le sso braze die 'ser tute 2 pasa | 3. Die pesar el passo libre 2. Li menalli de questi morganalli vol eser longi zascon d'essi tre fiade tanto quanto è longuo el to stello, zoè passa 24. El passo libre $1\frac{1}{2}$.

E vuol le so orze davanti longue el brazo passo 1. El so menal die 'ser passa 14, zoè 2 fiade quanto è lo stello.

E vuol la schalletta de questa nostra antena de mezo. Vol eser longa passa 24, zoè 2 vuolte quanto l'antena fuse longa, de nonboli 4. Die pexar el passo libre $1\frac{1}{2}$.

E vuol la schalla del'alboro de mezo. Vol eser 2 volte de zò che l'alboro è longo dala choverta in su. Die 'ser de nonbolli 4. Die pexar el paso libre 2. E vuol per passo schaliny 4, e vol eser schalliny 38.

E vuol legname per questo nostro alboro bolgare 4, misieli 16, bigotte 4, 2 al legname e 2 per respetto.

E vuol un anzollo longo 3 fiade longo quanto l'alboro dala choverta in su. Serà passa 29. Pexerà el paso libre 2.

Taye dal'alboro de mezo. 20 de chinalli, zoè de popexi, de ragli 2 la mitade e la mitade de raglo 1, intanpagnade.

E vuol taye 2 de chinal matto, 1ª d'un laglo[a] e l'oltra de 2. E vuol taye 4 de fonde, la mitade de raglo 1 e l'oltra mitade de ragli 2, intanpagnade.

---

a. *Sic in MS, for* raglo.

And it needs pendants for the *chinal* shroud tackles. These should be 3 paces, that is, a third of the mast. And the pace should weigh 3 pounds, that is, the length of the pendant.[1]

And it needs a yard sling 5 times the circumference of the yard, the ties being divided evenly in half.

And it needs 2 aftermost shrouds, one per side. Each of these should be one pace less than the height of the mast above the deck. That will be 8 paces. The pace 2 pounds.

And it needs a second pendant shorter than the first by $2\frac{1}{2}$ feet, and a third and fourth pendant. And it needs 2 pendants for the vang tackles, each 3 paces long. The pace 2 pounds.

And it needs falls for the vang tackles 2 times as long as the yard. The yard is 12 paces long. So they will be 24 paces. The pace should weigh $1\frac{1}{2}$ pounds.

And it needs midship *morganali.* Their braces should both be | 3 paces. The pace should weigh 2 pounds. The tackle falls of these *morganali* should each be three times as long as your *stello*, that is, 24 paces. The pace $1\frac{1}{2}$ pounds. f. 173b

And its *orze davanti* should be 1 pace long in their pendant. And its tackle fall should be 14 paces, that is, twice the *stello.*[2]

And there needs to be a small rope ladder for our midship yard. It should be 24 paces, that is, twice as long as the yard, [and] of four strands. The pace should weigh $1\frac{1}{2}$ pounds.

And there needs to be a rope ladder of the midship mast. It should be 2 times as long as the mast is above deck. It should be of 4 strands. The pace should weigh 2 pounds, and there should be 4 rungs per pace, and there should be 38 rungs.

And for parrel truss for this mast there are needed 4 parrel ribs, 16 parrel trucks, and 4 deadeyes, 2 for the parrel and 2 as spares.

And it needs a parrel tackle fall 3 times as long as the mast is above the deck. That will be 29 paces. The pace will weigh 2 pounds.

The blocks of the midship mast: 20 for the *chinal* shrouds, that is, the aftermost shrouds, half of them double and half single, with bushings.

And it needs 2 blocks for the *chinal matto*,[3] 1 single and the other double. And it needs 4 *funda* blocks, half single and the other half double, with bushings.

---

1. That is, the same number of pounds as the number of paces in the length of the pendant.
2. Note that this implies a *stello* of 7 paces, while the preceding paragraph implies a *stello* of 8 paces. This sort of inconsistency (or approximation) happens frequently in this part of the book.
3. An uncertain type of shroud tackle.

E vuol chatavy 4, le 2 da erto de ragli 3 e le 2 da basso de ragli 2, intanpagnade. Item taye 2 d'anzollo, de raglo 1, intanpagnade.

Questa serà la raxion dela ssartia de questa nostra chocha.

f. 174a In primo volemo che per zaschadun zentener che porta ogni nave vuol tortize 2, zoè de miera. Adoncha chista nostra chocha vuol turtize 12. Die 'ser longe zaschaduna de questa[a] passa 80, e die pexar el passo libre 10. E vegnirà a pexar tutta una turtiza libre 800.

E vuol prodixi 4, de passo 80 l'un. Die pexar el passo libre 5, zoè la mitade che pexa la turtiza.

E vuol prexe 2, de passa 80 l'una. Die pexar el passo libre 3.

E vuol prexulline 3, de passa 60 l'una. Die pexar el passo libre 2.

E vuol gripie 3, de passa 60 l'una, de refudo.

E vuol menalli de maxinetta. E per altre chossa peza 1ª de passa 60. Die pexar el passo libre 2.

[*captions:*]
tay[*e*][b]

ragli[c]

f. 174b E vuol anchore 10. E volemo anchore 2 de libre 1000, li quali volemo per penuny. E le oltre anchore 8 volemo che sia de libre 850, inperzò che die pexar più 10 per 100 de zò che pexa le so turtize. E vuol ranpauny 2 per le barche. L'un die pexar libre 100 e l'oltro libre 60.

[*caption:*]
anchore[d]

Questa è la raxion là che se die meter la pedega de questo nostro alboro da proda, li qualli sono 2 riegolle. La prima raxion sie che alguny parte la longeza dela choverta per 7: le 4 le[e] fa inver puope e le 3 inver proda, e là se die fichar la pedega. La qual choverta de questa nostra chocha è longa

---

a. *Sic in MS.*
b. *Drawing lacking in MS.*
c. *Drawing lacking in MS.*
d. *Drawing lacking in MS.*
e. *Corrected over* la.
1. An uncertain type of tackle.
2. A type of mooring rope.
3. A smaller mooring rope.

And it needs 4 *chatavy,*[1] the upper 2 triple and the lower 2 double, with bushings. Also 2 single blocks for the parrel tackle, with bushings.

These will be the instructions for the ropes of our cog.

First of all, we wish that for every hundred butts that each ship carries there be 2 anchor cables, that is, for thousand-pound anchors. This cog of ours needs 12 anchor cables. Each of them should be 80 paces long, and the pace should weigh 10 pounds. So one anchor cable will weigh 800 pounds. f. 174a

And it needs 4 headfasts, of 80 paces each. The pace should weigh 5 pounds, that is, half as much as the anchor cable weighs.

And it needs 2 *prexa*[2] ropes, of 80 paces each. The pace should weigh 3 pounds.

And it needs 2 *prexullina*[3] ropes, of 60 paces each. The pace should weight 2 pounds.

It needs 3 anchor buoy ropes, of 60 paces each, of refuse.[4]

And it needs falls for the *maxinetta* tackle and for the other things, 1 piece of 60 paces. The pace should weigh 2 pounds.

[*captions:*]
Block[5]

Sheaves[6]

And it needs 10 anchors. And we need 2 anchors of 1,000 pounds, which we want to be as *penoni.*[7] And the other 8 anchors we want to be 850 pounds, so that they weigh 10 percent above what their anchor lines weigh. And we need 2 grapnels[8] for the boats. One should weigh 100 pounds and the other 60 pounds. f. 174b

[*caption:*]
Anchors[9]

Here is the instruction for how to position the mast step of our foremast; there are 2 rules. The first is that some divide the length of the deck by 7: 4 go aft and 3 forward, and there the mast step

4. That is, second-quality hemp.
5. Drawing lacking.
6. Drawing lacking.
7. Uncertain interpretation.
8. Small anchors.
9. Drawing lacking.

passa 18. Adoncha dentro li 3 settimy da proda e le 4 da pope se die meter la pedega, e starà molto ben.

f. 175a E l'oltra riegola sie questa che alguna mexura la cholonba | dela chocha, zoè per quinto, e lassa li 3 a pope e le 2 a proda, e uno pie plù a proda e mancha da pope pie 1, e là se ficha la piedega. Adoncha alezi qual cha a tti par.

E questa sie la raxion del sguindazo. Lo qual vol eser longo tanti mezi pie quanti passa l'alboro è longo dala choverta in suxo. E vuol volzer el so redondo tanto quanto è longo el quarto plui. E vuol aver forfixe 2, un per respetto. E vuol 2 muziolli.

Questa sie la raxion delo nostro trio dela ditta chocha. Vol eser in antenal passa 16 e vuol aver de chazuda la mitade. Serano passa 8 e passo 1 de più, serano 9. E tanto averà 'sta vella de chazuda.

Questa nostra vella, ch'è in antenal passa 16 e in chazuda passa 9, vuol aver de fustagno passa 3 per ogno passo, serave passa 48, e una che xe per la cholona serano ferse 49. Dele qual abatti fersse 5, zoè per la mitade dele cholone per le arlenge. Resta fersse 44 de fustagno de passsa 8 l'una. Vol eser passa 352 de fustagno, e per ogno zentener de passa sono peze 10 de fostagno.

E vuol tanti denti quante ferse de fustagno, zoè un dente de sovra e l'oltro dente de ssotto. Li quali denti vol eser longi pie $3\frac{1}{2}$ l'uno, che vegnerave chavo pie pie[a] 3, serà dopli. Serà denti 88, che
f. 175b sono passa 106. Vegnerave eser braza 265. Tanto chanavazo va in questi | denti. E questo trio vuol perzente 3 per passo de zò che xe la vella de hazuda. Vegnirà a eser perzente 27. E vuol eser zascaduna longa chomo la vella è longa in antenal, zoè passa 16. E serà larga una del'oltra tanto quanto eser longo lo drapo, serà pie $1\frac{1}{2}$. Serà lo chanavazo fendandolo per terzo perzente 3. E vuol aver zaschaduna d'ese pie 2 longa più cha la vella per lo arvolzer dele arlenghe. Adoncha vol eser perzente 27, chanevazo braza 148, che serà 370.

E vuol esto trio pedocha 28, zoè in lo briol è 8, ale burine è 6, per ladi, per li stropi che se liga i quarnalli e li stinchi. E vuol per zaschuna braza $6\frac{2}{3}$ de chanevazo. Vien a eser in tuto braza 187. Suma la chanevaza che xe in questa vella braza 1412. E vuol de spago sotil libra 1 per peza, e de spago grosso onze 8 per ogni peza de fostagno. Arichordatti de far li to pessetti da erto e da basso. Raxion de sartia che va in questa, zoè antenal e gratel e arlenge.

E vuol che l'antenal sia longo $\frac{1}{4}$ più de zò che xe longa l'antena. Serà passa 20. Avanzarate per ladi passa 2 per baroxa. Pexerà el passo libre $2\frac{1}{2}$.

---

a. p *written over* 3.

1. Probably a part of the windlass, perhaps pawl bitts.
2. Reinforcing pieces of hemp cloth along the head and foot of a sail.
3. Each *dente* is made of 6 feet of cloth which, doubled over, makes 3 feet; 88 *denti* thus use 528 feet or $105\frac{3}{5}$ paces.
4. The ell was a measure of length of cloth, 2 feet or $\frac{2}{5}$ pace long.
5. A horizontal reinforcing band of sail cloth.

should be fastened. The deck of our cog is 18 paces long. Thus the mast step should be put between the forward 3 sevenths and the after 4, and all will be well.

And the other rule is this, that some measure the keel | of the cog, that is, in fifths, and leave 3 aft and 2 forward, and one foot more forward and 1 foot less aft, and there the mast step is fixed. Then fit it out as seems right to you. f. 175a

And this is the information on the windlass, which should be as many half feet long as the height of the mast above the deck in paces. And it should measure as much around as it is long and a quarter more. And it should have 2 *forfixe*,[1] one as a spare. And it needs 2 stoppers.

Here is the information on our square sail of this cog. It should be 16 paces in the head and have a drop of half of that. That will be 8 paces, and one pace more, which will be 9. And that is how much this sail will have in its drop.

This sail of ours, which is 16 paces in the head and 9 paces in the drop, should have 3 paces of fustian for each pace, that will be 48 paces, and one for the column makes 49, of which you take away 5 cloths, that is, for the half of the columns for the leech. There remain 44 cloths of fustian of 8 paces each. There should be 352 paces of fustian, and for each hundred paces there are 10 pieces of fustian.

And it needs as many *denti*[2] as there are cloths of fustian, that is, an upper *dente* and another lower *dente*. These *denti* should each be $3\frac{1}{2}$ feet long, which would come, head and foot, to 3 feet, [the canvas being] double. There will be 88 [upper and lower] *denti*, which makes 106 paces.[3] This would come to be 265 ells.[4] That much canvas goes into these | *denti*. And this square sail needs three *perzente*[5] per pace of the sail's drop. This will come to 27 *perzente*. And each one needs to be as long as the sail is at the head, that is 16 paces. And they will be spaced one from the other as the cloth is long,[6] that will be $1\frac{1}{2}$ feet. The canvas will be divided into thirds [lengthwise, making] 3 *perzente*. And each of these should be 2 feet longer than the sail to double over the leeches [at each end]. So there should be 27 *perzente*, 148 ells of canvas, which will be 370. f. 175b

And this square sail needs 28 goosefeet, that is, 8 for the middle sheets, 6 for the bowlines, per side, for the ropes by which the *quarnals* and the clew lines are attached. And each needs $6\frac{2}{3}$ ells of canvas. This comes to 187 ells in all. The total of the canvas in this sail is 1,412 ells. And it needs one pound of light twine per piece, and 8 ounces of heavy twine per piece of fustian. Remember to make your upper and lower broadseams. Account of the boltropes that go into this—that is, the head rope and the foot rope and the leech ropes.

And the head rope of the sail needs to be $\frac{1}{4}$ longer than the length of the yard. That will be 20 paces. You will allow yourself 2 paces per side for the *baroxa*.[7] The pace will weigh $2\frac{1}{2}$ pounds.

---

6. I.e., wide.
7. A term of uncertain meaning.

E vuol una peza de chavo de passa 35 per arlenge da erto e da basso. E vuol pexar per passo libre $3\frac{1}{2}$, perzò che la vuol eser de pexo più cha l'antena libra 1 per passo.

f. 176a Al nomen de Dio. Se vuol chomenzar a taiar la cholona de | mezo de tutta tella passa 19. E dopla lo ladi dela dita cholona, zoè el destro, e metti la so binda de chanevazo tanto longa chomo la cholona. E puo' chomenza a tayar la prima fersa de passa 7 e pie 1, e meti li so denti da basso de chanevazo, de pie 5, per l'achomenzar lo pie dela cholona da basso. E puo' metti el so dente de sovra, de pie 4, inperzò che li denti da baso dela cholona vuol mazuor cha quelli da erto per far la cholona, che parà bon. E puo' metti la sso binda apresso.

La segonda ferssa apreso de questa, de passa 7 e pie $2\frac{1}{4}$, inperzò che le die 'ser longa cha la prima, perché lo pie dela cholona se va menemando. E puo' taya el so dente da basso de pie $4\frac{1}{4}$. E puo' metti el dente da erto de pie $3\frac{1}{2}$, e de chontinuo metti la binda quando averà i denti.

La terza ferssa de fustagno de passa 7 pie $3\frac{1}{2}$, e lo dente da erto pie 3. Puo' metti la sso binda.

La quarta ferssa de fustagno de passa 8. E puo' taya el so dente da basso de pie $2\frac{1}{2}$, e chosì taia quel da erto. E po' metti la sso binda. Et avera' chonplido lo ladi dela cholona granda del ladi destro.

La quinta ferssa de ladi destro de passa 7 pie 3. E chosì metti li so denti, sì da basso chomo da erto, de pie $3\frac{1}{2}$. E puo' meti la so binda.

f. 176b La sesta ferssa de fustagno dela ladi destro tayala de passa 8. E metti apresso lo so dente, sì da baso chomo da erto, de pie $2\frac{1}{2}$ l'un. E puo' metti la sso binda.

La settima fersa de fostagno vuol eser longa de passa 7 pie 3. Apreso metti li so denti, sì da basso chomo da erto, de pie $3\frac{1}{2}$, e serà i do pie 7. E puo' metti la so binda.

L'octava fersa de fustagno longa de passa 8. Apresso metti li sso denti, sì da erto chomo da basso, de pie $2\frac{1}{2}$. Apresso metti la so binda.

Itam taya la ssoa ferssa de fustagno inchina apreso questa cholona de passa 7 pie 3. Apresso metti el so dente, sì da erto chomo da basso, de pie $3\frac{1}{2}$ l'uno, chomo l'oltro ladi dala cholona, zoè de undexena. Apresso metti la binda. E chussì va tayando le tuo fersse de fustagno e le binde e li denti de chanevazo chi te mancha a tayar de questa cholona inchina l'arlenga de ladi destro. Serà fersse 11 chon quella ch'è messa apresso la cholona. E va tayando le ditte fersse e denti una mazuor e l'oltra menor, chomo noi avemo fatta in la quarta cholona granda inchina fersse 11 che noi avemo taiado.

And you need a piece of rope of 35 paces[1] for the leech rope[2] at top and bottom.[3] And it should weigh $3\frac{1}{2}$ pounds per pace because it needs to be 1 pound per pace heavier than the yard [is long].

In the name of God. If you wish to begin to cut the center column of | whole cloth 19 paces, and double the side of this column, that is, the right side, and lay on its canvas *binda*, which is as long as the column. And then begin to cut the first cloth of 7 paces and 1 foot, and put on its bottom *denti* of canvas, of 5 feet, to begin the foot of the column from the bottom. And then put *denti* of 4 feet on top, so that the *denti* at the bottom are larger than those of the top in order to make the column, which seems good.[4] And you can put its *binda* on it. f. 176a

The second cloth goes next to this, of 7 paces and $2\frac{1}{4}$ feet, as it should be longer than the first, since the foot of the column gets smaller. And then cut its lower *dente* at $4\frac{1}{4}$ feet. And then make the upper *dente* $3\frac{1}{2}$ feet, and so continue putting on the *binda* as long as it will have *denti*.

The third cloth of fustian of 7 paces $3\frac{1}{2}$ feet, and the upper *dente* of 3 feet. Then put on its *binda*.

The fourth cloth of fustian of 8 paces. And then cut its lower *dente* at $2\frac{1}{2}$ feet, and cut the upper one the same. And then put on its *binda* and you will have completed the side of the large column of the right side.

The fifth cloth on the right side of 7 paces 3 feet. And so put on its *denti*, on the bottom and on the top, of $3\frac{1}{2}$ feet. And then put on its *binda*.

Cut the sixth cloth of fustian of the right side of 8 paces. And put on the bottom and top *denti*, of $2\frac{1}{2}$ feet each. And then put on its *binda*. f. 176b

The seventh cloth of fustian should be 7 paces 3 feet long. Put on its bottom and top *denti*, of $3\frac{1}{2}$ feet, and the two will be 7 feet. And you can put on its *binda*.

The eighth cloth of fustian is 8 paces long. Put on its top and bottom *denti*, of $2\frac{1}{2}$ feet. Then put on its *binda*.

And so cut the fustian cloth of 7 paces 3 feet that goes up to the column. Then put on its top and bottom *denti*, of $3\frac{1}{2}$ feet each, as on the other side of the column, that is, the eleventh. Then put on the *binda*. And so continue cutting your cloths and the *binde* and the canvas *denti* that you need to cut right up to the leech of the right side. There will be 11 cloths counting the one that is put next to the column. And continue cutting these cloths and *denti*, the one larger and the other smaller, as we have done in the fourth large column until we have cut 11 cloths.

1. Approximately twice the rope needed for the two leech ropes for one sail.
2. A rope attached to the vertical sides of the sail.
3. Or, possibly, "from top to bottom."
4. Or, possibly, "to make it look good."

E arechordatti de apessitar la tto vella da basso chomo da erto. E quanto tu avera' chonplido de taiar queste ferse e binde 22 e denti 44, zoè un dente da basso e un da erto, e puo' taye le to arlenge de ladi destro, ample de tuto chanevazo dopla, e vuol eser pasa 9.

f. 177a E te fazo a saver che noi avemo chonpiudo a taiar la mitade de questo nostro trio, zoè de ladi destro. E per questo muodo va taiando l'altra mitade de ladi senestro, zoè ferse, cholone, denti e binde e arlenge. E quando serà fatta e chosida la ditta fersa, fara' taiar perzente 27, sfendando el vostro chanevazo per longo, e farate perzente 3. Zaschuna dele perzente serà per terzo. E vuol eser zaschuna passa 16 e pie 2. Questi 2 pie sie per alvolzer dele arlenge, e metti perzente 3 per passo.

E se tu volessi metti li so pedocha del trio grando, chomenza redente la cholona granda, e vuol eser pie 4. Adoncha per ladi pie 8. E le do piedocha, zoè quela che xe apresso la cholona da l'un ladi e dal'oltro, e vuol andar incrosiade suxo ala cholona. E vuol eser ziaschon de questi pie 8, tanto longi che vegnirave inchina la terza perzenta.

E se tu volesi saver là che tu die meter li to piedocha dele borine, anonbre le to perzente e parti per quinto e lasa le 3 parte da basso e le 2 parte de sovra. Chomenza a meter le tuo pedocha dala burina e vien in su. E vuol eser d[*e*] pedocha, zoè 4 per ladi, e fali luongi che li azonza sula terza binda. E se tu volessi far la tua burina in terzo, fara' 5 piedocha per ladi. Arichorda da meter 6 pedocha per ladi per li stropi d'i chadernalli, e vuol eser zascon tanto longo che azonza inchina ala segonda perzenta,
f. 177b perzò che vol esere più curti cha quelli dal briol. E questi pedocha d'i quarnali partelli dal | chavo dal briol chomo ve piaxerà in to chosienzia.

Questo l'amaistramento d'armar el vostro trio e per lo simile ogni nave. In primo le so zerzene da gratil a bagnarlle e destarle e metile a sugar, e tirele quanto se vuol, perché non te vegna a ragagnar, inperzò che ogno chavo longo aslonga per passo pie $\frac{1}{2}$.

E arma la to vella in antenal 2 pie men de zò che l'antena è longa. Serà passa 15 pie 3, inperzò te creserà quelli 2 pie. Arma da basso in gratil passa 15 pie 2, inperzò che te creserà 3 pie per lo tayar dele schotte.

E arma la to vela in arlenga passa $8\frac{1}{2}$, inperzò che le ttue arlenge senpre tira e creserà $\frac{1}{2}$ passo. Arechordatti a meter meulo e marafuny e stropi e de far le sache dele arlenge e dele schotte.

Se noi volemo taiar le 4 bunette in 4 pezi, serano 2 per ladi. In primo taya la prima peza de bonetta de ladi destro la sso cholona de tutto chanevazo, ampla e dopla, longa passa 2 pie 3, e puo' metti la

1. Uncertain interpretation.

2. I.e., stretches.

And remember to hang your sail [on its boltrope] on the bottom as well as on the top. And when you have finished cutting these 22 cloths and *binde* and 44 *denti*, that is, one *dente* on the bottom and one on the top, then cut your right-hand leech cloth, as wide as a full canvas cloth [and] doubled, and that should be 9 paces.

And I want you to know that we have finished cutting half of our square sail, that is, the right half, f. 177a
and in the same way continue cutting the other half, the left side, that is cloths, columns, *denti* and *binde*, and leech cloths. And when it is done and this cloth sewn, have 27 *perzente* cut, dividing your canvas lengthwise, and make yourself 3 *perzente*. Each of the *perzente* will be a third [of the width of a cloth]. And each should be 16 paces and 2 feet [long]. These 2 feet are to double over the leeches; and put 3 *perzente* per pace [of drop].

And if you want to put on the goosefeet of the big square sail, begin reining in the great [center] column, and it should be 4 feet, so 8 feet per side, and the 2 goosefeet, that is, the one that is alongside the column on one side and the other, and these cross on top of the column. And each of these should be 8 feet, long enough so that they will come to the third *perzenta* [from the bottom].

And if you should wish to know where to put your goosefeet for the bowlines, count your *perzente* and divide into parts by five and leave 3 parts on the bottom and 2 parts on the top. Start to place your bowline goosefeet and work your way up. And there should be 8 goosefeet, that is, 4 per side, and make them long enough that they reach the third *binda* [from the leech]. And if you want to make your bowline threefold, make 5 goosefeet per side. Remember to put 6 goosefeet per side for the *quarnal* strops, and each of these should be long enough to reach to the second *perzenta*, as they need to be shorter than those of the middle sheets. And space these *quarnal* goosefeet from the |
end of the middle sheet, as it will please you in your judgment. f. 177b

This is the method for rigging your square sail and all ships likewise. First, your foot rope should be washed and stirred up[1] and set to dry and stretched as much as you wish, so you won't have to redo it, because each long rope lengthens[2] $\frac{1}{2}$ foot per pace.

And make the headrope of your sail 2 feet less than the yard is long. It will be 15 paces 3 feet, because it will stretch these 2 feet. Make the foot of the sail 15 paces 2 feet, as it will stretch 3 feet from the cut of the sheets.

And make your leech rope $8\frac{1}{2}$ feet, as your leech always pulls and will stretch $\frac{1}{2}$ pace. Remember to put on stops and robands[3] and strops and to make the ropes for the leeches and the sheets.

If we want to cut 4 bonnets[4] in 4 pieces, there will be 2 per side. First cut the first piece of the bonnet for the right, the column of a full width of canvas, double [thick], 2 paces 3 feet long, and

3. Ropes passed through eyelet holes in the head of the sail used to secure it to the yard.
4. Additional strips of canvas laced to the foot of the sail.

sso binda da ladi. E puo' metti la prima ferssa de fustagno, e metila apresso la cholona longa pie 8. E puo' metti li sso denti, zoè un da erto e un da basso, de pie $2\frac{1}{2}$ l'uno, e la so binda apresso.

E per questa raxion e muodo che tu à taiado la prima e la ssegonda fersa chusì va taiando inchina
f. 178a ferse 11, chon li sso denti, cresando e menemando, metando li so binde de chanevazo | a zaschuna fersa. E quando averai taiado ferse 11, taia la to arlenga tanto anpla de tuto chanevazo dopla. Vol eser longa passa 2 e pie 3, averà passa 2. Tayado la mitade de de[a] questa peza[b] de bonetta, zoè dala cholona de mezo inchina l'arlenga di fuora.

Mo' torna al'oltra mitade de questa nostra bonetta, zoè del ladi dentro[c] de questa cholona de mezo inchina a chavo, taiar ferse 12 dela mexura dele oltre 11 che noi avemo tayado deli oltri ladi de questa bonetta. E quando avera' taiado ferse 12 de fustagno, taya la so cholona de chanevazo ognola, ampla de ttuto drapo, longa passa 2 pie 3. E puo' la dopla mettila a ladi ala to ferssa, chosì dopla, e far meza cholona, zoè la mitade dela cholona granda del trivo. Mo' avemo chonplido la bonetta del ladi destro.

E per lo simille muodo taya la parechia peza de bunetta, zoè de ladi senestro. Averemo le 2 prime ferse de sovra.

E se noi volemo taiar le oltre 2 bunette sotto de questa, li quali noi volemo far che abia pie 2 de chazuda men cha quelle de sovra, perché serà più destre da tuor e meter:[d]

In primo taia la so cholona de mezo, ampla de tuto chanevazo, dopla, vol eser longa pie 11, e meti la so binda apresso. E puo' taia la sso altra. In prima ferse de fostagno apreso la cholona pie 6, puo' taya li so denti, sì de sovra chomo di sotto, pie $2\frac{1}{2}$.

f. 178b E la segonda fersa de fustagno pie $7\frac{1}{2}$. Metti li so denti longi pie 1[e] $\frac{3}{4}$, sì de sovra chomo di sotto, puo' metti la so binda.

E per lo simille muodo va taiando le to ferse, denti e binde, inchina fersse 11 de fustagno chon li denti e binde. Taya la so arlenga ampla de tuta tela dopla. Die 'ser longa pie 11. Avera' tayado la mitade de questa segonda bunetta.

Mo' torna a tayar l'oltra mitade de questa peza de bonetta dal ladi dentro.[f] In primo taya la so binda apreso la cholona, e puo' taya la prima fersa che fa apresso la cholona, die 'ser longa pie 6, e meti li so denti, sì da erto chomo da basso, de pie $2\frac{1}{2}$ l'uno. E puo' meti la so binda.

---

a. *Thus repeated in MS.*
b. p *written over illegible letter.*
c. *Sic in MS.*
d. *Paragraph ended in MS.*
e. 1 *written over* 2.
f. *Sic in MS.*

then put its *binda* on the side. And then place the first cloth of fustian, and put it on the column 8 feet long. And then put on its *denti*, that is, one on top and one on the bottom, of $2\frac{1}{2}$ feet each, and its *binda* next to it.

And in this way in which you have cut the first and the second cloths, so you continue cutting until cloth 11, with its *denti*, increasing and decreasing, putting on their canvas *binde* | on each cloth. f. 178a
And when you will have cut 11 cloths, cut your leech cloth as wide as whole canvas, doubled. It should be 2 paces and 3 feet long, it will have 2 paces. Half of this piece of bonnet has been cut, that is, from the center column to the leech on the outside.

Now turn to the other half of our bonnet, that is, of the right[1] side of this column. From the center to the end, cut 12 cloths of the measure of the other 11 that we've cut for the other side of this bonnet. And when you have cut 12 cloths of fustian, cut its column of canvas of single thickness, the width of a full cloth, 2 paces 3 feet long. And then put the doubled canvas alongside your cloth, thus double and make a half column, that is, half of the large column of the square sail. Now we have completed the bonnet of the right side.

And in the same way cut similar bonnet pieces, that is, of the left side. We will have the first two upper cloths.

And if we wish to cut the other 2 bonnets under these, which we want to make so that they each have 2 feet less drop than the upper ones, so that they will be easier to haul into position:

First, cut its center column, of the width of a full cloth of canvas, [and] double; it should be 11 feet long, and put its *binde* on. And then cut its other one. First cloths of fustian next to the column 6 feet [long], then cut its *denti*, upper as well as lower, $2\frac{1}{2}$ feet.

And the second cloth of fustian $7\frac{1}{2}$ feet. Make its *denti* $1\frac{3}{4}$ feet long, upper and lower; then put on its *binda*. f. 178b

And in a similar way continue cutting your cloths, *denti*, and *binde* until [you have cut] 11 fustian cloths with *denti* and *binde*. Cut its leech cloth the width of a whole double cloth. It should be 11 feet. You will have cut half of this second bonnet.

Now turn to cut the other half of this bonnet piece of the right[2] side. First cut its *binda* next to the column, and then cut the first cloth that goes next to the column, it should be 6 feet long, and put on its *denti*, upper as well as lower, of $2\frac{1}{2}$ feet each, and then put on its *binda*.

---

1. MS has *dentro* ("within") rather than *destro* ("right").
2. MS has *dentro* rather than *destro*.

Item taya la segonda fersa de fostagno apreso questa. Vol eser pie $7\frac{1}{2}$. E puo' meti li so denti, sì da basso chomo da erto, de pie $1\frac{3}{4}$. E puo' metti la to binda. E chosì va taiando tutte le ferse e denti, cressando e menemando sì chomo noi avemo fato dal'oltro ladi de questa nostra bunetta inchina fersse 12. E quando tu avera' tayado ferse 12 de fustagno e denti e binde, taya la sso cholona dentro, ognola, ampla de tuta tella, de pie 11. E puo' la dopla per longo e metila apresso le ferse. Avemo chonplido de sotto la bonetta de ladi destro. E per lo simille muodo va tayando l'oltra peza de bunetta del ladi senestro, per lo muodo noi avemo tayado da ladi destro. Arechordate le perzente
f. 179a 3 per paso. Apreso arichordatti a meter per ogni peza de bonetta 6 pedocha | per 3 stropi, e fa che sia sì longi che azonza inchina le segonde prezente.

E se noi volemo armar queste 4 bunette, in prima ale to 2 prime bunette, zoè quele de sovra, vol eser in gratil passa 8 e pie 1. Vol eser armada in arlenga passa 2 e pie 2, inperzò che le creseria pie 1 per lo taiar dele arlenge. La chaxion che li 2 bonette, 1ª ferse più ch'al trivo, sie che la vuol andar posada le do bonette de sotto queste. Armali per lo muodo armasti questa de sovra. E le arlenge vuol eser armade pie $10\frac{1}{5}$. Quele destendera' mezo pie più.

E se noi volemo tayar 2 quartaruny de queste bonette, chomenza a tayar in primo la so cholona che vien in mezo dali to quataruni de ladi destro, e taya ampla de tuta tella longa. E vol eser dopla pie 7. Apreso meti la so binda de ladi destro, e puo' taia la so fersa de fustagno apreso sta cholona. Vol eser longa pie $4\frac{1}{2}$, e metti li so denti da basso, de pie $2\frac{1}{2}$, e non meter nesun dente da erto, e meti le so binde apreso.

E per lo simile muodo fa chosì taiando inchina ferse 11, zoè per lo tayo avemo taiado queste 2, chosì cressando e smenemando le ditte fersse e denti da basso. E quando tu avera' chonplido a taiar ferse 11, taya la so arlenga de pie 7, ampla de tuta tella. Mo' àstu tayado mezo el quartaron dela cholona dentro al'arlenga de ladi destro.

f. 179b Mo' torna a tayar l'oltra mitade zoè de questo quartaron. Zoè dela cholona del ladi destro inchina a chavo dela bonetta. E vuol quartaruny per lo ditto.

E tayara' ferse 12 chon li so denti da basso, e viente cressando e descressando per lo simille muodo che avemo tayado l'oltra mitade delo ladi dela bugna. E quando avera' chonplido de tayar queste fersse 12, taya la to cholona de meza nave de tuta tella, ampla et ugnola, e longa pie 7, e puo' la dopla de longo. Serà tanto chomo quela del trivo. Et averemo chonplido questa peza de quartaron de ladi destro. Rechordatte a meter perzente e piedocha, e parti le so 4 peze in pie 7.

E per questo muodo taya l'oltra peza dal quartaron de ladi destro, e starà ben. Arma puo' li to quartaruny, sì dal'uno ladi chomo dal'oltro, passa 8 e pie 1. In antenal e in gratil da basso oltro-

Then cut the second cloth of fustian next to this one. It should be $7\frac{1}{2}$ feet. And then put in its *denti*, lower as well as upper, of $1\frac{3}{4}$ feet, and then put on your *binda*. And so continue cutting all of the cloths and *denti*, increasing and decreasing so that we have made 12 cloths for the other side of our bonnet. And when you have cut 12 cloths of fustian and *denti* and *binde*, cut its inner column, of single thickness, of the width of a whole cloth, of 11 feet [in length]. And then double it lengthwise, and put it alongside the cloths. We have completed the bottom of the bonnet of the right side. And in a similar way continue cutting the other bonnet piece for the left side, in the way that we have cut that of the right side. Remember there are 3 *perzente* per pace. And remember to put on each piece of the bonnet 6 goosefeet | for three strops, and make it long enough so that it reaches as far as the second *perzenta*. f. 179a

And if we want to hang these 4 bonnets, first for the two first bonnets, that is the upper ones, the foot should be 8 paces 1 foot. The leech rope should be 2 paces and 2 feet, so that it is 1 foot longer for the cutting of the leeches. The reason that the 2 bonnets have one more cloth [in width] than the square sail is that they need to go with the two bonnets rigged beneath them. Hang them as you hung the upper one. And the leech ropes should be $10\frac{1}{5}$ feet. They will extend a half foot further.

And if we want to cut 2 *quartaruni*[1] for these bonnets, begin to cut first the column in the middle of your *quartaruni* on the right side, and cut it the full width of a cloth. And this should be doubled, 7 feet. Put its *binda* on the right side and then cut the cloths of fustian next to this column. They should be $4\frac{1}{2}$ feet long, and put its bottom *denti*, of $2\frac{1}{2}$ feet, and don't put any *denti* on the top, and put its *binde* on.

And continue cutting in the same way until cloth 11, that is, as we have cut these two, thus enlarging and reducing these cloths and bottom *denti*. And when you have finished cutting 11 cloths, cut the leech cloth of 7 feet, the width of a whole cloth. Now you will have cut half of the *quartarun* of the column inside the leech cloth on the right side.

Now turn to cutting the other half of this *quartarun*, that is, of the column of the right side to the end of the bonnet, and you will need *quartaruni* for this. f. 179b

And you will cut 12 cloths with the lower *denti*, and continue increasing and decreasing in the same way as we have cut the other half of the side of the bonnet. And when you have finished cutting these 12 cloths, cut the midship column the width of the whole cloth, and of single thickness, and 7 feet long, and then double it lengthwise. It will be the same size as the square sail. And you will have finished this piece of the *quartarun* of the right side. Remember to put on *perzente* and goosefeet, and space your 4 pieces 7 feet apart.

And in this way cut the other piece of the *quartarun* of the right side, and it will be good. Then hang your *quartaruni*, the same on one side as on the other, 8 paces and 1 foot in the head and as

1. A sail added to the bottom of a bonnet, possibly a drabler.

tanto. E le arlenge die 'ser armade pie $6\frac{1}{2}$. Rechordate da far sache da erto e da basso, zoè dele arlenge. Arechordatti a meter mantilletti de sotto e de sovra e de ladi quando tu army le velle, si grande chomo pizole.

Questo sie lo chargo che porta questa nostra chocha, la qual è longa in cholonba passa 13. E anderà in plan in queste passa 13 man 21 de botte. E per zaschaduna man andarà bute 18. Serave in tuto botte 378.

E in la choverta de sovra mesura la to nave, che serà passa 15. E anderà man 24 de butte 15 per
f. 180a man, che serave in la choverta | de sovra butte 360. Adoncha in la choverta de sovra e quela de sotto starave bute 738. Dele qual abati una man de butte de quele de fundi, e una man de sovra. Serano bute 33, zoè per lo puzo, per l'alboro e per li stanti. Restarà el so chargo netto butte 705.

[*caption:*]
botte[a]

[Far albori, antene, tortize, timony, anchore, sartia chon tutte le sso raxion]

Questo serà l'amaistramento de far albori e antene. De groseza tuti li albori, sì grandi chomo pizulli, vuol volzer in lo sso redondo $\frac{1}{2}$ palmo per passo de zò che l'alboro fusse tuto longo. E tutte le antene vuol volzer $\frac{1}{4}$ de palmo de zò che fusse tutta longa.

Qui volesse chonprar penuny per far antena de passa 12, tu die chonprar un penon longo de passa 8, zoè lo terzo del'antena men fusse longa, zoè passa 8 e 2 pie più per dodexena. Serano passa 8 e pie 2.

f. 180b Vorave penon un per stelo, el qual vol eser longo la mitade de zò che xe longa tuta l'antena e 2 pie più per dodesena. Serano passa 6 e do pie. Farano tuti 2 'sti penony passa 14 e pie 4. Mo' volemo noi saver zò che vuol de dopio questa nostra antena. La vuol pie 2 per passo de zò che tuta l'antena fusse longa, zoè per dodexena. Adoncha l'antena è longa passa 12, el dopio serave pie 14. Adoncha ventame pasa 8 pie 2, stello passa 6 pie 2, serave[b] 14 pie 4, chavalchando pie 14. Roman netta l'antena de passa 12.

E sapi che quanti palmy volze i penuny, tanti pie volzerà l'antena, siando ingalezada. E per quista sifatta raxion se taya rixe d'antena e in bogial per ligar la ditta.

---

a. *Drawing lacking in MS.*

b. *The unit of measurement is lacking.*

1. *Man,* meaning literally "hand," is used in medieval Venetian as an aggregate term with a variable numerical value.

2. I.e., in the tweendeck.

much in the foot. And the leech ropes should be $6\frac{1}{2}$ feet. And remember to make ropes on the top and on the bottom, that is, for the leeches. Remember to put on stops below and above and on the sides when you hang the sails, large as well as small.

This is the cargo that our cog carries, which is 13 paces long in the keel. And 21 hands[1] of butts will fit in the lower hold in these 13 paces. And in each hand there will be 18 butts. There will be in all 378 butts.

And on the upper deck your ship will measure 15 paces. And there will be 24 hands of 15 butts each, which will be | 360 butts on the upper deck. So, on the upper deck and the lower one there will be 738 butts. From these deduct one hand of butts of those in the hold, and one hand from above.[2] That will be 33 butts, that is, for the pump well, the mast, and the stanchions. The net cargo will be the 705 butts that remain. f. 180a

[*caption:*]
Butts[3]

To make masts, yards, cables, rudders, anchors, shrouds and all their measurements

These will be the instructions for making masts and yards. As to thickness, all the masts, large as well as small, should have as a circumference $\frac{1}{2}$ palm per pace of the whole height of the mast. And all the yards should have a diameter of $\frac{1}{4}$ palm [per pace] of their length.

Here it's necessary to buy *penoni*[4] to make a yard of 12 paces; you should buy a *penon* of 8 paces, that is, a third less than the yard is long, that is, 8 paces and two feet more per dozen. It will be 8 paces and 2 feet.

You will want one *penon* for the *stello*, which should be half as long as the length of the yard, and two feet more per dozen. It will be 6 paces and 2 feet. Together these *penoni* will make 14 paces and 2 feet. Now we want to know what the overlap of our yard should be. It should be 2 feet per pace of the length of the yard, that is, per dozen. So the yard is 12 paces long, the overlap will be 14 feet. So the *ventame* will be 8 paces 2 feet, the *stello* 6 paces 2 feet, that will be 14 [paces] 4 feet, overlapping 14 feet. There remain 12 paces of the yard in the clear.[5] f. 180b

And know that the yard will have the same number of feet of circumference as the *penoni* have in palms, when it is scarfed. And the yard slings and the woolding[6] to secure it are cut in this same proportion.

3. Drawing lacking.
4. The *penoni* were the spars of the lateen yard.
5. That is, not overlapped.
6. A circumferential lashing that unites the 2 spars of the yard.

Questo è l'amaistramento di far cholzexi d'alburi grandi e pizolli chomo tu vuol. E primo tuti li cholzexi, sì grandi chomo pizolli, vol eser longi 1 pie per passo de zò che l'alboro tutto fusse luongo. E vuol eser alargo el cholzexe el sesto de zò che lo cholzexe fusse tutto longo. Ponamo che fusse un alboro longo de passa 24. Vuol eser longo el cholzexe pie 24, e vuol eser alargo el cholzexe el sesto
f. 181a de 24, che sono 4. Adoncha lo ditto alboro de passa 24 vuol de cholzexe pie 24 e la lar|gueza del colzese pie 4. E per questa sifatta raxion se fa di cholzexi, sì grandi chomo pizulli, chomo ò ditto de sovra.

Tuti li albori de nave latine, zoè da proda, vuol d'antena più cha tuto l'alboro pie 1 per passo, e per zaschon alboro de proda che fuse longo passa più cha 15, vorave eser[a] più curto[b] l'alboro de mezo terzo pie per passo de zò che fusse tuto longo. Alboro de passa 15 vuol alboro de mezo de passa 14. E per questa farì le oltre.

E sapi che l'alboro de mezo vuol pie $5\frac{3}{4}$ d'antena de zò che tuto l'alboro fuse longo più per passo. Alboro de pasa 14 vuol antena longa tuta de passa 16 pie $\frac{1}{2}$.

E se tu volesi ligar le braze dele suste per raxion, quanti passa è longa l'antena, tanti pie dentro dela zima del ventame, e li charuzi tanti pie quando[c] è tutto lo stello longo.

[*captions:*]
[*a*]lboro[d]

antena[e]

f. 181b Chi volese saver per raxion de nave latine zò chi vuol d'i ttimoni, vuol tanto quando[f] la ruoda di puope e pie 7 più in choverta.

E quanti passa è longo lo timon tanti pie vuol volzer la timonera, e quanti pie vuolze la timonera tanti pie vol eser alarga la palla. E sì vorave anize longe come tuto lo timon. El fuxo, el schaton vol eser partido per mitade.

[*caption:*]
timony[g]

Questa sie la raxion de far sartia. Tuta la sartia che se lavora in Venyexia, ala Tana, de nonbolli 3 e de filli 12, die pexar el passo libra 1. E se la fusse de nonbolli 4 e de fili 9, die pexar el passo libra 1.

---

a. tto *crossed out with a horizontal mark.*
b. curto *above the line with a reference mark.*
c. *Sic in MS.*
d. *Drawing lacking in MS.*
e. *Drawing lacking in MS.*
f. *Sic in MS.*
g. *Drawing lacking in MS.*
1. This should read just $\frac{3}{4}$ for the following calculation to work.

Here is the way to make block masts for large and small masts as you will. And first, all of the block masts, large and small, need to be 1 foot long per pace of the whole length of the mast. And the block masts need to be one sixth as wide as the block mast is long overall. Let's say that the mast is 24 paces long. The block masts would then be 24 feet, and the block mast needs to be a sixth as wide, that is 4. So this mast of 24 paces needs a block mast of 24 feet and the width | of the block f. 181a
mast 4 feet. And this is the way block masts are made, large and small, as said above.

All masts of lateen ships, that is, their foremasts, need a yard that is one foot per pace longer than the whole mast, and for each foremast that is more than 15 paces, the midship mast will be one third of a foot shorter per pace than its entire length. A foremast of 15 paces needs a midship mast of 14 paces. And in this way you will make the others.

And know that the midship mast needs $5\frac{3}{4}$[1] feet of yard more per pace than the length of the entire mast. A mast of 14 paces needs a yard of 16 paces $\frac{1}{2}$ foot.

And if you wanted to attach the vang pendants by rule, as many paces as the yard is long, [attach them] so many feet down from the ends of the *ventame*, and [attach] the *charuzo* tackles the same number of feet [down] as the *stello* is long.

[*captions:*]
Mast[2]

Yard[3]

Whoever would like to know what the lateen ship needs in terms of rudder, it needs to be as long as f. 181b
the sternpost and 7 feet more in the deck.

And as many paces as the rudder is long, that many feet should be the circumference of the *timonera*;[4] and as many feet as the *timonera* is in circumference, the rudder blade should be that many feet wide. And you will also need tackles as long as the whole rudder. The rudder *fuxo* and the rudder *schaton* should each be half [of the rudder stock].

[*caption:*]
Rudder[5]

Here is how to make the ropes. All the ropes that are made in Venice, at the Tana,[6] of 3 strands and of 12 yarns, should weigh 1 pound per pace. And if it is of 4 strands and 9 yarns, the pace should weigh 1 pound.

---

2. Drawing lacking.
3. Drawing lacking.
4. The bracket for the quarter rudder.
5. Drawing lacking.
6. A building near the Arsenal.

El fior de chanavo se 'l fusse fili 16 e nonboli 3, pexerà el passo libra 1. E se 'l fusse de nonboli 4 e fusse de fili 12, pexerà libra 1.

f. 182a Anchure de mier 1 vuol turtiza de passa 80 e die aver per nonbollo filli 120. Die pexar el passo libre 10, se lo fusse de refudo. E se 'l fusse fior de chanavo, averà fili 160, e die pexar el passo libre 10.

E se tu volesi dar anchore per raxion de sartia, vuol de fero, zoè die pexar l'anchora el quarto più de zò che pexase la ditta tortiza. La tortiza pexarave libre 800 in passa 80, e die pexar l'anchora lo quarto più, serave 1000.

E se la turtiza pexa libre 10, el prodixe vuol per mitade, zoè libre 5 per passo. E la gripia vorave el terzo, serave libre $3\frac{1}{3}$. E per questa farì le simille raxion.

E se tu volesti manti per raxion, mesura dale gule inchina li ragli del cholzexe, e daralli 2 tanto, e fara' rixa d'antena el pexo del manto. Se l'antena vuolzese pie 6, vorave el dopio, che serave libre 12. Tanto vuol pexar per passo el manto.

E per braxa de susta, per palmo che volzese l'antena, vuol pesar el passo libra 1. La susta d'ogno naviglo vol eser longa 2 volte e meza de zò ch'è l'antena longa.

E per charuzi vol eser 2 volte longi quanto è lo stello dal boial a proda.

Poza longa chomo l'antena. Pozal 2 volte chomo la poza.

Morganal vol eser 2 volte longo chomo l'antena.

Mo' avemo chonpido da far nave, e tutto vui vedirì andar a vello qui.

f. 182b [*captions:*]
nave a velo

nave a vello[a]

---

a. *Thus repeated in MS.*

Prime hemp if it is 16 yarns and 3 strands will weigh 1 pound per pace. And if it is of 4 strands and 9 yarns it will weigh 1 pound.

Anchors of 1 thousand pounds need an anchor cable of 80 paces which should have for each strand 120 yarns. The pace should weigh 10 pounds, if it is of second quality. And if it is of prime hemp, it will have 160 yarns, and the pace should weigh 10 pounds. f. 182a

And if you want to rig the anchor according to rule, it needs iron, that is, the anchor should weigh one quarter more than the anchor cable. The anchor cable will weigh 800 pounds in 80 feet, and the anchor should weigh one quarter more, that is, 1,000.

And if the anchor cable weighs 10 pounds, the headfast should be half, that is, 5 pounds per pace, and the buoy rope should be one third, that will be $3\frac{1}{3}$ pounds. And do this in a similar way.

And if you want the ties by rule, measure from the mast hole [in the deck] up to the sheaves of the block mast, and you will give them twice as much, and you will make the yard slings the weight of the tie. If the yard has a circumference of 6 feet, it will be the double of that, which will be 12 pounds. And the tie should weigh that much per pace.

And for the vang pendants, for each palm of circumference of the yard, the pace should weigh 1 pound. The vang of any ship should be 2 and a half times the length of the yard.

And for the *charuzo* tackles, they should be 2 times as long as the *stello* from the woolding forward.

The *poza*[1] is as long as the yard; the *pozal*[2] two times the *poza.*

The *morganal* should be 2 times as long as the yard.

Now we have finished building the ship and you will see it going under sail here.

[*captions:*]
Ship under sail f. 182b

Ship under sail

1. A sheet of a lateen sail.
2. The clew, the lower part of a lateen sail.

f. 183a

[Orazion de san Sebastiane, e per frieve, e per paure de serpe, e per dona che non può parturir, e per non piar pesi, e per stagnar sangue del naxo, e per eser morzegado da bissa venenoxa, e per non chonfesar a martorio]

Orazion de san Sebastian

***[a] [*ora*]zion de miser san Sebastiane. Chi la lezerà o chi la fa le[*zer*] *** ***na fede al'onor de Dio e de madona sancta Maria *** de miser san Sebastian 3 volte serà defexo da mali *** ***zie e se non serà ferido da veretun né da freza.

[*San*]cte Sebastiane magna est fides tua, interzede pro me Michalli misero peccatori ad dominum Ihesus Christum et ab istam pestem et pedimiem sive morbos tuis prezibus meriar liberari verso. Ora pro me sancte Sebastiane ut digneris efiziari promissionibus Christi.

Orazio.

Oremus omnipotens senpiterne Deus qui meritis et prezibus sancte Sebastiane marturis tuis quandam pestem et pedimie generaliter omnibus pestiferam revochasti presta subplizibus tuis ut qui ac orationem super se abuerit vel chotidie diserit ipisus[b] meritis et prezibus ab istam pestem et pedimiem[c] et ab omni tribulazionem[d] et angostie liberatur per Christum dominum nostrum.

Per frieve

Sava Sava epi Sava che Sava. Pasqua ton Evreon. To Sava. Ectes tetartin oran, O Iisus, demonya epechefalisen dia ton psicron to rigos, to proton, to defteron, to triton, to tetarton, che ton padote.

---

a. *The first words of lines 2 to 5 are missing due to a loss in the upper left corner.*

b. *Sic in MS.*

c. *Sic in MS, probably for* epidemiem.

d. *Sic in MS.*

1. The first words of lines 2 to 5 are missing due to a loss in the upper left corner.

2. St. Sebastian was viewed as the intercessor against plague, and written amulets invoking him were common in late medieval Italy: cf. Don C. Skemer, *Binding Words: Textual Amulets in the Middle Ages* (University Park: Pennsylvania State University Press, 2006), pp. 178–80; I am grateful to Don Skemer for his useful comments on a draft of this section of the translation.

3. Presumably, read over him.

4. According to Boerio, *verettoni* were a kind of large arrow shot from a crossbow.

5. According to Boerio, *frezze* were another kind of arrow, shot from a bow. These are probably both references to the arrows thought to spread plague as well as those used to martyr St. Sebastian.

6. This and the next prayer are in Latin, rather than the vernacular of most of the MS.

7. This prayer is a magical incantation. It appears in the manuscript in Greek, transliterated with Latin letters; I am most grateful to Eleni Kalkani-Passali, Anna-Maria Kasdagli, and Diana Wright for their assistance with this and the other Greek passages. A reconstruction of the original text follows:

Σάβα Σάβα επί Σάβα τσαι Σάβα. Πάσκουα των Εβραίων. Το Σάβα. Εκτές τετάρτην ώραν, Ο Ιησούς δαιμόνια επετσεφάλισεν διὰ τὸν ψυχρόν τὸ ῥῖγος τὸ πρῶτον τὸ δεύφτερον τὸ τρίτον τὸ τέταρτον καὶ τὸν πάντοτε. Φεύγε ρῖγος, πυρετόν δαίμονονοκεφάλι ἀπὸ τὸν δοῦλον τοῦ Θεοῦ Μιχάλι. Στώμεν καλώς, Ἄγιος ο Θεός, στώμεν μετα φόβου, Ἄγιος Ισκκυρός, Ἄγιος Αθάνατος ελέησον ημάς. Ἄγγελλοι Μιχαήλ, Γαβριήλ, Ουριήλ, Ραφαήλ, από ρίγος βοηθοί μιταβάλγου τ[σ]ίνου.

A prayer to St. Sebastian, and for fever, and for fear of serpents, and for a woman who can't give birth, and when you catch no fish, and to staunch a nosebleed, and when you are bitten by a venomous snake, and so you will not confess under torture

f. 183a

Prayer of St. Sebastian

***[1] Prayer of St. Sebastian.[2] Whoever reads this or has it read[3] 3 times *** [while displaying good] faith in the honor of God and of the Blessed Virgin Mary *** of St. Sebastian will be defended from harm *** and will not be hurt by shafts[4] or arrows.[5]

[6]St. Sebastian, your faith is great, intercede for me Michael, a miserable sinner, to Lord Jesus Christ, and may I deserve to be freed from plague, epidemic, and illness by your prayers. Pray for me St. Sebastian as you deem worthy to carry out the promises of Christ.

Prayer.

Let us pray. All powerful eternal God, you who have held back plague and epidemic and disease generally in response to the services and prayers of St. Sebastian your martyr, manifest to your supplicants that whoever shall make use of this prayer for himself or say it daily for his goodness and prayers shall be liberated from this plague and epidemic and all tribulation and harm freely through Christ our lord.

For fever.

[7]Savas, Savas, by Sava and Sava.[8] Passover of the Jews.[9] Sava![10] Yesterday at the fourth hour, Jesus left the demons headless[11]—chills, the first, the second, the third, the fourth[12]—continually.

---

8. Possibly the fifth-century St. Saba or the thirteenth-century monk and bishop Sabas, son of the Serbian King Stephen I (later St. Symeon). The author may be conflating St. Sabas with the word for Sabbath or the word Sabaoth in the invocation "Ἅγιος, Ἅγιος Κύριος Σαβαώθ," an obscure term derived from the Hebrew *ṣəbāʾôt*, pl. of *ṣābāʾ*, army, from *ṣābāʾ*, to wage war.
9. Πάσκα instead of the correct Πάσχα is the local idiom of Rhodes.
10. The preceding words appear as an introductory attempt to impress the listener with the potency of the spell (hence the emphatic repetition of the word Sava), and also possibly to set the stage for frightening away a demon. The Jewish flavor of the introduction may also have worked on the listener at a different level as an allusion to the power of Jewish magic—as in the *Solomonike*, or apocryphal compilations of King Solomon's magic, considered infallible in medieval (and more recent) Greek folklore. They were prized for command over demons, particularly in healing.
11. A reference to the Harrowing of Hell when Christ defeated the world of demons (in the apocryphal Gospel of Nicodemus or Acts of Pilate II) headed by their master, the devil, who is often shown chained (with angels sometimes assaulting lesser demons in the darkness of Hell) in the Orthodox iconography of the Resurrection prominently displayed on the murals of medieval (and modern) Greek churches. The reference to the fourth hour is obscure, as Christ was crucified in the third hour (synoptic gospels) or sixth hour (John), died on the ninth hour (Matthew, Mark, John), and rose before the first hour of the Sunday. Εκτές could be read as an idiomatic form of εχθές (= yesterday) but then the sentence makes no sense ("yesterday on the sixth hour Jesus . . ."). Alternative readings such as Εκ την (= from), also idiomatic and occurring in other Greek medieval dialects, apparently should be ruled out, as Michael seems to have regularly rendered the Greek eta with the *i* of the Latin alphabet according to its phonetic value. It is more likely that the time of the onset of the fever was inserted on an ad hoc basis and is not directly linked to what follows. The *ch* of *epechefalisen* in place of the normal *k* is probably due to local idiom.
12. Possibly a reference to quartan fever (if the gender is ignored), with a flare-up every fourth day. It might also be a dramatic description of repeated racking chills. The addition of the *n* at the end of the article (το[ν]) is due to local dialect.

Feuge rigos poreton demonochefale apo ton dulon tu Theu, Michallin! Stomen chalos, Agios o Theos, steonen meta fovu, Agios Ischiros, Agios Athanatos eleison imas. Angelli Michail, Gavriil,
f. 183b Uriil, Rafail, apo rigos voithi | mitabalgu tiuny. Scrivy questa orazio luni a desun in una quarta e ligalla al cholo de qui avesse la frieve, chon 3 Pater noster e 3 Ave Marie, a far 3 gropi, e lasalo inchina merchore e puo' la getta al fuogo. Subitto serà libero de quel mal a honor de miser san Sava.

Per frieve

Pater est ans,[a] Filius est vitta, Sanctus[b] Sanctus est remedio. Scrivy queste parole in un pumo e dela a magnar luny da maitina un pezo, e ogno maitin un pezo inchina arà chonpido, e farì dir 1ª messa a honor de san Sava. Sera' libera' dela ditta malatia.

Per non chonfesar al truomento

A dir queste parole ananti ay zudexy 3 vuolte non chonfesarà mai. "Ego autem tanquam surdus non audiebam, et sicut mutus non aperies os meum, et factus sum sicut omo non audiens et non abes in ore meo redargueziones."

Per paura de aspe e bisse

El dì de san Salvador far dir 1ª messa, e mentre dixe el prette scrivi queste parolle: "Signa autem eos qui credederunt hec sequentur, in nomine meo demonya essient, longis loquentur novis, serpentes tolent et si mortiferum quit biberit non eis nozebit, super egros manos inponet et bene abebunt."

f. 184a Per una duna chi non podese parturir dir in la rechia destra queste parolle chon 3 Pater noster e 3 Ave Marie: "Mnistitti chirie, ton ion edom ton cn[c] Ierusalim ton legondon ecchenotte ecchenutte eos ton."

---

a. *Sic in MS, perhaps for* animus.

b. *Sic in MS, evidently for* Spiritus.

c. *The MS here reads clearly* cn *rather than* en [c *and* e *are very distinct in Michael's hand*], *which suggests that Michael was copying this passage from a written source in which the letters were not distinct, rather than remembering it. I am grateful to Diana Wright for this observation.*

1. *Poreton* must be a misspelling or misreading of πυρετόν (= fever). The fever is directly addressed as a demon (demonhead in the vocative case) and is ordered away from Michallis—the doubling of the *l* in the name is due to local idiom.

2. A mixing of two different sentences from the Orthodox liturgy: Στώμεν καλώς, στώμεν μετα φόβου begins the solemn ritual of the Transubstantiation of the Eucharist, during which the Holy Ghost descends and hovers over the altar in the sanctuary (presumably striking dumb with terror any demon that happens to be present). 'Αγιος ο Θεός, 'Αγιος Ισχυρός, 'Αγιος Αθάνατος ελέησον ημάς is pronounced on crossing oneself (= Holy God, God of Power, Immortal God have mercy on us). Crossing the sick person would be an expected part of such a spell. The *ch* of *chalos* probably renders the local doubling of the consonant *k*, pronounced as a *k* followed by a glottal stop. In the case of *ischiros*, the local pronunciation varies from village to village of Rhodes—it might be enunciated either as *ischiros* or *iskhiros*, although the second is more likely.

3. The four archangels. The doubling of the *l* in *angelli* is local Rhodian idiom, as is also the word *mitabalgou* for third-person plural μεταβάλλουν (= they alter/change, if this interpretation is correct). The ending makes no easy sense if tran-

Chase away the chills, take the demon-head from the servant of God, Michael![1] We cry, Good [God], Holy God! We cry out in fear. Holy Mighty, Holy Immortal, have mercy on us.[2] Angels Michael, Gabriel, Uriel, Raphael, save me from the chills; they change | that condition![3] Write this prayer while fasting on a Monday on a quarter [sheet] and tie it around the neck of the person with a fever, with 3 Pater Nosters and 3 Ave Marias, to make 3 knots, and leave it there until Wednesday and then throw it in the fire. He will instantly be relieved of that sickness by the honor of St. Sabas. f. 183b

For fever.

[4]The Father is the breath, the Son is the life, the Holy Spirit[5] is the remedy. Write these words on an apple and give a piece of it [to the sick person] to be eaten on Monday morning,[6] and another piece every morning a piece until it is finished, and have him say a mass in honor of St. Sabas. He will be freed of the malady.

To not confess under torture.

If you say these words three times to the judge, you will never confess.[7] "Just as if I were deaf, I did not hear, and as if mute you shall not open my mouth, and I am made like a man not hearing and not having refutations in my mouth."

For fear of asps and snakes.

On the day of the Saviour have a mass said, and while the priest is talking write these words:[8] "These signs follow those who shall believe these things, in my name they may be demons, they will speak in new tongues, they bear away the serpents, and if what he drinks is poisonous it will not harm them, he places hands on the afflicted and they will be healed."

For a woman who cannot go into labor,[9] say these words in the right ear with 3 Pater Nosters and 3 Ave Marias: "O Lord, Remember the sons of Edom in Jerusalem who said, 'Be voided, be voided until.'"[10] f. 184a

---

scribed accurately, but it has been suggested that it is a corrupt form of εκείνου (= that, in the nominative case) in local dialect; that is, the angelic helpers change "that state/condition."

4. The first sentence is in Latin, the rest of the paragraph in Venetian.

5. *Sanctus* repeated, probably in error for *Sanctus Spiritus*.

6. Cf. Skemer, *Binding Words*, pp. 127–128, for the eating of apples and other objects with amuletic inscriptions.

7. Remainder of paragraph in Latin.

8. Remainder of paragraph in Latin.

9. For verbal charms associated with birthing, see Skemer, *Binding Words*, pp. 236–239.

10. A poorly remembered Greek phrase, probably originally Μνήσθητι Κύριε των υιών Εδώμ των εν Ιερουσαλήμ των λεγόντων εκκενώται, εκκενούται έως τον. There seems to be a word lacking at the end, such as αιώνα (age). Although the original phrase is definitely Greek, the spell itself is apparently mixed with Roman Catholic popular ritual, as witnessed by the three Pater Nosters and three Ave Marias. According to Anna-Maria Kasdagli, whom I thank for her advice on this section, the spell in its present form may have evolved in the Greek community of Venice.

Per peschar un peschador

Metti queste parolle scritte in una charta e metile in una chana e stropa a ligalla ai redi e averà asai pesi.

"In illo tenpore anbulans Ihesus ista mare Galee vidit duos fratres, Simeonem et Andream chon Zebedeo patrem eorum mitestes rezia eorum in mare et disit illis: 'Abetis aliquid ad chomedendum?' Respondit Simon Petrus, dixit ad Ihesum: 'Domine, prope tuctam noctem laburavimus et nichil azepimus.' Respondit et dixit eis Ihesus: 'Mitite rezia in decstram navigii.' Et ita fezerunt et maximam moltitudo piscium compsederunt. Tunc reliquit eos Christus et tunc chognoverunt discipulli quod Dominus vixitabit eos."[a]

Chi fusse morzegado da bisse venenuoxe e vegnise da vuy, a farla star sovra el pe' destro fermo e
f. 184b chon un cortelin segnar la furma del pe' in tera, erze lo pe' pe'[b] e fa 3 | cruxe, un al chalchagno, un a mezo e l'oltra in punta, digando senpre queste parolle: "Charo charuzi, sanum reduze, reduze sanum, Emanuel Paraclitus, Alfa et O[*mega*]."[c] Puo' tuor 3 pizegetti de queste croxe, zoè del teren, e meter quello teren al lanpo dela chamixa del ditto, e chon un bichier d'aqua cholondo[d] per 3 vuolte el ditto teren per la chamixa et un oltro bich[*i*]er senpre de sotto a rezever l'aqua cholada, e quela aqua beva. Cholui serà morzegado subitto die porgar quel venen o de sotto o de sovra, e sì varisse.

E per lo simille muodo se 'l fusse el morzegado là che volese e non podese venir, cholui chi portasse le novelle per lui se 'l vorà aver la fadiga per cholui che serà morzegado, chomo dixe farlo star fermo per lo muodo ditto de sovra e ber chostui l'aqua, serà liberado el morzegado, e chostui che averà bevudo l'aqua porgarà per lo morzegado o de sotto o de sovra, senza algun perigullo dela so persona, senper digando le parole.

Per doya de dente

Pia 1° corttelo con manego negro, fichar e dir queste parole: "Inchanchenis, nimis, ttibi, solis, ipsam."

---

a. *There follow three crosses and then symbols.*

b. *Thus repeated in MS.*

c. *Possibly just* O, *as the phrase "Alpha et O" does occur in Latin liturgy.*

d. *Thus in MS, very probably for* cholando.

For a fisherman to catch fish.

Put these words written on a piece of paper and put them in a hollow reed and cork and tie it to the net and it will have enough fish:

"In that time, walking along the Sea of Galilee, Jesus saw two brothers, Simon and Andrew, with their father Zebedee throwing their nets into the sea and said to them: 'Do you have anything to eat?' Simon Peter answered and said to Jesus: 'Lord, we have worked almost all the night and have caught nothing.' Jesus answered and said to them: 'Put your nets on the right of the boat.' And they did it and enclosed a great multitude of fish. Then Christ left them and then the disciples understood that the Lord had visited them."[1]

Whoever has been bitten by venomous snakes and has come to you, make him stand firm on his right foot and with a small knife draw the outline of the foot in the ground, raise the foot and make 3 | crosses, one on the heel, one in the middle, and the other on the front, all the time saying these words: "Make a present of Charon, bring back health, bring back health. Emmanuel, Intercessor, Alpha and Omega."[2] Then take 3 pinches of these crosses, that is, of the earth, and put this earth in the fastener of the shirt of the person, and with a glass of water straining this earth 3 times through the shirt and another glass always beneath to receive the strained water, and drink that water. Whoever is bitten should immediately purge the venom either downward or upward, and so will get better. f. 184b

And in a similar way, if he was bitten where he was and cannot come, whoever brought the news for him if he wants to take responsibility for the one who was bitten, as I said, have him stand firm the way described above and he should drink the water, the bitten one will be freed, and he who has drunk the water will purge for the bitten one, downward or upward, without any danger to himself, always saying the words.

For toothache

[3]Take a knife with a black handle, stick it,[4] and say these words: [5]"Be free of the cancer, too much, yourself, of the sun, itself."

1. In Latin. This narrative charm or *historiola* relates to Gospel accounts in Matt. 4:18–20, Mark 1:16–18, and Luke 5:4–6. The narrative is followed in MS by a series of symbols.
2. The first two words, if in corrupt Greek (Χάρο χαρίζει = makes a present of Charon), might mean a deadly gift (i.e., the poison). The next four words are clearly not Greek. The rest, [Εμμανουήλ Παράκλητος] and *alfa et o[mega]* (the Beginning and the End), both invoke Christ, presumably to assist in the healing. Again, I thank Anna-Maria Kasdagli for her assistance.
3. The next four lines in a different hand from rest of MS.
4. Probably against the painful tooth.
5. In Latin.

E per una altra oltra[a] fichar in mezo quisti nuodo[b] con un corttelo con manego negro + santto umele + Piero o Polo+.[c]

Per stagnar sangue

Sangue sta in tti chomo fe' Christo in ssi. Sangue sta fisso chomo Christo fu cruzifisso. Sangue sta forte in la to vena chomo fe' Christo in la so pena. E ronpe 1ª piera e presto metyila al naxo e ttira ben lo fiado dela piera in ssi.

f. 185a

[NON S'ANEGAR E IN BATAYA PER SALVAZION DELA PERSONA]

Per non s'anegar in aqua. Chomo serà sapudo dexunar merchore e non ma[*n*]ziar charne né grasso e dir queste parolle ogni dì: "Laqueus[d] contritus est et ego liberatus sum, adiutorium nostrum in nomine Dominy, amen."

In bataya: "Agla zuans agla glos," 3 volte, "Agaus idonie ischa ferunt denota rie."

[SAVER DE RAXION DE LUNA E DE SSO PORTADE]

Questa raxion qui de sotto ser[*à*] per amaistramento di saver quando fa la luna, a raxion uxa marineri. E prima.

Sapi che in 1435 chure la patta dela luna 30, e dura questo inchina ultimo fevrer. E voyando far una oltra patta per 1436, noi azonzeremo 11 sovra 30, e serano 41. Mo' abateremo el 30, die restar 11. E questi serano la patta del millexímo ditto de sovra. E chosì ogn'ano voiando far patta azonzi senpre 11, e se 'l passa 30, getta el 30 e tiente al resto, e se non passa 30, tiente tutto quel ano 30, avixandove che marzo sie el prinzipio del ano.

E si vuolessi saver quando fa la luna, questo è lo muodo. Se die in primo meter la patta e per zaschun mexe serà da marzo in quel che ti vuol saver azonzi zorno 1, e se 'l non pasa 30 azonzi tanti dala bocha che faza 30, e questi azonti serano li dì che die far la luna. Exenpio vedirì qui de sotto.

---

a. *Sic in MS.*

b. *At this point in MS are drawn two six-pointed stars with the numbers 3, 3, 2, 9, 5, 2 counterclockwise around the second.*

c. *From* Pia 1° corttelo *to* o Polo + *added in another hand with other ink, very probably at a later time.*

d. laqueus *inserted in another hand, in another ink, above the line over* lachus *crossed out with a horizontal line.*

And for another, stick in the middle of this knot a knife with a black handle + Humble saint + Peter or Paul +

To staunch the flow of blood

Blood is in you as Christ was in himself. Blood is fixed as Christ was crucified. Blood is strong in your vein as Christ was in his pain. And break one stone and immediately put it on your nose and take the breath of the stone deeply in yourself.

For protection from drowning and for protection from bodily harm in battle

f. 185a

For protection from drowning in the water. As will be known, fast on Wednesday and don't eat meat or fat and say these words every day: "The snare is destroyed and I am liberated, our support in the name of God, amen."[1]

In battle: "St. John, St. Nicholas," 3 times, "St. Anthony, bring strength to [. . .] take strength from [. . .]"[2]

To know the workings of the moon and its phases

The instructions below will teach you how to know when the moon is new, by means of mariners' usage.[3] And first.

Know that epact 30 of the moon occurs in 1435, and this lasts until the end of February. And wanting to determine another epact for 1436, we will add 11 to 30, and it will be 41. Now we will subtract the 30, 11 should remain. And this will be the epact of the year indicated above. And so, each year for which you want to determine the epact always add 11, and if it passes 30, throw out the 30 and hold on to the remainder, and if it does not pass 30, hold on to the 30 that year, reminding yourself that March is the beginning of the year.

And if you want to know when the moon is new, this is the way. You should first take the epact, and for each month after March which the one you want to know is, add 1 day, and if it doesn't pass 30 add as many orally to make 30, and what you've added will be the day on which the moon becomes new. You will see an example below.

---

1. In Latin, spelling of "snare" (*laqueus*) corrected in a different hand.
2. This could be a very garbled version of a prayer in Greek, reconstructed as: Ἅγιος Ἰωάννης, Ἅγιος Νικολάος . . . Ἅγιος Ἀντονι ἰσχὺν φέρετε δένετε [uncertain word]. (I thank Diana Wright for this suggestion.) John the Apostle was a boatman, Nicholas and Anthony were protectors of sailors.
3. For an introduction to the *Computus manualis* and Michael's versions of it in this MS and the *Raxion de' marineri*, see Wallis, "Michael of Rhodes and Time Reckoning," vol. 3, pp. 309–318.

Ponamo che tu vuoi saver al mexe de marzo[a] del 1435 quando fa la luna. La pata serà 30, e per lo
f. 185b mexe azonzere|mo uno, serano 31. Azonzeremo de bucha 29, farano 60, e se getto 30 roman 30. Adoncha adì 30 de marzo farà la luna.

Ponamo che volemo saver del 1436 el mexe de mazio quando fa la luna. Tu sa ben che la pata die chorer 11, chomo ò ditto de sovra, e dal mexe de marzo al mazio sono 3. Azonzi a 11 3, fano 14, e noi azonzeremo de bocha 16, farano 30. Adoncha el mexe de mazio, a 16 fa la luna.

E se noi volemo saver adì 15 zugno del 1436 quandi[b] dì averà la luna, dichotti che tu die mesidar 3 chose. In primo la patta, zoè 11. El segondo, da marzo al zugno 4, azonti a 11 fano 15. La terza, li dì del mexe che ti vuol saver, zoè 15, farano 30. Adoncha in questo dì fa la luna. E se 'l fusse passado el 30, avesimo gitado el 30 e tenisemo alo resto.

Ponamo che noi vosemo veder al gugno[c] adì 25 quanti dì averà la luna. La patta 11 e li zorny deli mexi 4 sono 15. El dì che tu zerchi a saver sono 25, azonti insieme fano 40, e da questi geta el 30, resta 10. Adoncha adì 25 de zugno del 1436 averà la luna 10. E per questo muodo farì fatte le oltre, azonzer 3 chosse e saver quanti dì à la luna.

E se per aventura avesi desmentegado che la patta fusse 11, chomo tu debia far per truovar la patta dichotte. Qui avanti vedirì 1ª man depenta, in la qual serà in lo dio grando al fundi del ditto 15. Al segondo dio serà 25. Al terzo nudo, zoè de sovra, serà 35. E se vuolesi zerchar a truovar la patta, vui
f. 186a chomenzarì al fundi a nonbrar li ani Domini | zoè 1436, a mezo dio diremo 1000, e de sovra 100, e de sotto 200, e a mezo 300, e de sovra in zima 400. Puo' chomenzeremo el nomero, zoè 36, a nonbrar e andar intorno a questo dio. E là che tti azonzi al 36 afermate e varda 36, che è l'ano men de quel dio e azonzi al 36. Se 'l passa 30, getta el 30 e tiente alo resto. E quela serà la patta. E se 'l non passa, tiente a quelo. E se la suma serà 30 tutto quel ano, tiente 30 per patta, cho[*mo*] vedirì qui de sotto, per figura.

[*caption:*]
man[d]

E se alguno te domandasse una luna quanti dì sono, dichotte che una luna sie dì 29, ore 12, ponti 793, arigordandotti che 1080 punti serà ora 1, e ore 24 serà dì 1. E se per aventura vuolesi aprovar taula de Salamon, se la fuse zusta o non, varda quando la fexe. Ponamo che la luna fesse de zugno,
f. 186b adì 15 del 1436, a ore 13, ponti 844. | Sutto metti al 15 29, e sotto metti a ore 13 ore 12, e sotto metti a ponti 844 ponti 793, e questi asumadi in primo in punti fano 1637. Abati 1080 per far un

a. *Corrected over* mazio.
b. *Sic in MS, evidently for* quanti.
c. *Sic in MS, evidently for* giugno.
d. *Drawing lacking in MS.*

Let's suppose that you want to know when the moon turns in the month of March of 1435. The epact will be 30, and for the month you'll add | one, and that will be 31. We will add orally 29, which makes 60, and if I throw away 30 there remains 30. So on March 30, the moon will become new. f. 185b

Let's suppose that we want to know for May of 1436 when the moon becomes new. You know well that the epact should run to 11, as I've said above, and from March to May are 3 months. Add 3 to 11, it makes 14, and we will add with our mouth 16, which makes 30. So, for the month of May, the moon is new on the 16th.

And if we want to know for June 15 of 1436 how many days old the moon will be, I tell you that you should combine 3 things. First the epact, that is 11. Second, from March to June is 4, added to 11 it makes 15. The third, the day of the month that you want to know, that is the 15th, which makes 30. So, on this day the moon is new. And if it was more than 30, we would have thrown out the 30 and held on to the remainder.

Let's suppose that we want to see for June 25 how many days old the moon will be. The epact is 11 and the days of the 4 months are 15. The day which you want to know is the 25th, added together they make 40, and from this throw away 30, which leaves 10. So on the 25th of June of 1436 the moon will be 10 days old. And in this way you will do the others, adding three things and knowing how many days the moon has.

And if by chance you had forgotten that the epact was 11, I tell you how you could find the epact. Following this you will see a hand painted, in which there will be 15 on the base of the big finger. On the second finger there will be 2. At the third joint, that is, up, there will be 35. And if you wanted to try to find the epact, you will begin at the base to count the years of the Lord, | that is, 1436: on the middle finger we will say 1,000, and above 100, and below 200, and in the middle 300, and over on the top 400. Then we will begin the number, that is 36, to count and go around this finger. And when you get to 36, stop and look at the 36, which is the lower year on that finger, and add to the 36. If it passes 30, throw out the 30 and hold on to the remainder. And that will be the epact. And if it doesn't pass, hold on to it. And if the sum is 30 all that year, hold on to 30 for the epact, as you see figured below. f. 186a

[*caption:*]
Hand[1]

And if anyone asked you how many days a moon has, I tell you that a moon is 29 days, 12 hours, 793 points, remembering that 1,080 points will be an hour, and 24 hours will be 1 day. And if by chance you want to check whether the table of Solomon is true or not, look how you do it. Let's say that the moon was new on June 15, 1436, at 13 hours, 844 points. | Put 29 below 15, and put 12 hours under 13 hours, and put 793 points under 844 points, and this adds up first in points to f. 186b

1. Drawing lacking.

ora, restarà 557[a] punti. E puo' fa 12 e 13 fa 25, e 1 fa 26. Abati el 24 per far un dì, roman 2 ore. E puo' fa 29 e 15 fa 44, azonzi uno, fano 45. E perché el mexe danazi zo fu mazio, à zorny 31, abatti 31 de 45, roman 14. Adoncha la luna de lugio farà adì 14, ore 2 e ponti 577,[b] zoè à fallo. Ò gettado 31 per lo mexe de mazio, che non dovea farlo. Anche debio gettar zorny 30 per zugno, perché la luna fa de lugio. Adoncha la luna farà più 1 dì, zoè adì 15, ore 2, punti 557.[c] E per questa farì ogn'altra raxion. Arigordove quando serà al mexe de frever per far la luna de marzo a getar el 28, e se 'l fusse bexesto a gettar el 29.

E se algono te domandasse la luna quando se lieva o quando la va a monte, dichotte che se la luna è prima e l'è levada de dì e va a monte de notte, e de puo' el so redondo se lieva de notte e va a monte el dì. E se saper vorai, ponamo che la luna sia dì 13, tu die saver quandi[d] ore à la notte e quanti à el dì. Ponamo che la notte fusse ore 8, el dì fusse ore 16. Vuolemo veder a quante ore de notte va la luna a monte. Tu die moltipicar le ore dela notte chon li dì dela luna e partir per 15. Egno[e] 15 serà ora una. Exenpio. 8 via 13 fano 104, e questi partidi per 15 fano ore $6\frac{14}{15}$. A tante ore de notte va la luna a monte. E se volesi veder quando se lieva de dì, moltipicha 13 via 16, fano tanti, e parti per 15.

f. 187a E se volesi saver zò che s'alarga la luna del sol, sapi che ogno dì che à la luna tanti[f] quarte s'alarga dal sol inchina el diamittro, e puo' se achosta sì che la fa senpre la luna sutto al sol.

E se nesun te domandasse quandi[g] mia chamina el sol per ora, dichotte che 'l mondo intorno sie mia 36.000, e questi chonven che 'l sol chamina in ore 24. Adoncha chaminaralo per ora mia 1.500. Ala prima ora tu dirà 1 via 5 fa 5, e 3 via 5 fa 15, zoè 1.500. Ala segonda, 2 via 5 fa 10, e 3 via 10 fa 30, zoè 3.000. E ale 3 ore tu fara' 3 via 5 fa 15, e 3 via 15 fa 45, zoè 4.500, e chossì fara' moltipichando. Quando tu sera' in ore 24 tu dira' 5 via 24 fa 120, e 3 via 120 fano 360, zoè 36.000. E questa è fatta.

E per saver quanti mia s'alarga la luna dal sol per ora, ogn'ora s'alarga mia 1.200. E faremo ala prima ora un fia 4 fa 4, e moltipicha' per 3 fa 12, zò serano zentenera, fano mia 1.200. Ala segonda, 2 via 4 fa 8, e 3 via 8 fa 24, zoè zentenera, serano mia 2.400. E chosì per lo simille fa' inchina ore 24. E per voler intender in zorny 30 xe alarga la luna del sol e achosta de mia 36.000, perché moltipichemo 30 via 4 fa 120, zoè zentenera, serano 1.200, e 30 via 1200 fano 36.000.

E perché de sovra dise che la luna dixe per ora l'alargar, non è ma die 'ser ogni dì s'alarga. Adoncha zorny 30 s'alarga la luna 36.000.

---

a. *Corrected between the lines above* 577, *crossed out with a horizontal line.*
b. *Sic in MS.*
c. *Corrected between the lines above* 577, *crossed out with a horizontal line.*
d. *Sic in MS, evidently for* quanti.
e. *Sic in MS, evidently for* Ogno.
f. hore *crossed out with a horizontal line.*
g. *Sic in MS, evidently for* quanti.

1,637. Subtract 1,080 points to make an hour, there will remain 557 points. And then do 12 and 13 make 25, and 1 makes 26. Subtract the 24 to make a day, there remain 2 hours. And you do 29 and 15 make 44, add one, which makes 45. And since the month before this was May, which has 31 days, subtract 31 from 45, which leaves 14. So the moon in July will become new at day 14, hour 2, and 577 points, which is wrong. I subtracted 31 days for the month of May, which I should not have done. Rather I should subtract 30 days for June, since this is the new moon of July. So the moon will be new 1 day later, that is on day 15, hour 2, 557 points. And in this way you will do all other problems. Remember when it is the month of February, to make the month of March subtract 28, and if it is a leap year subtract 29.

And if anyone should ask you when the moon rises or when it sets, I tell you that if the moon is waxing it rises by day and sets by night, and after the full moon it rises by night and sets by day. And if you would like to know, let's suppose that the moon is in day 13, you should know how many hours there are in the night and how many in the day. Let's suppose that the night had 8 hours, [and] the day had 16 hours. We want to see at what hour of the night the moon sets. You should multiply the hours of the night by the number of days of the moon and divide by 15. Each 15 will be one hour. For example, 8 times 13 makes 104, and this divided by 15 makes $6\frac{14}{15}$ hours. And that is the time of the night the moon sets. And if you would like to see when it rises in the day, multiply 13 by 16, and divide whatever it makes by 15.

And if you want to know what distance the moon is from the sun, know that every day that the moon is old, it moves that many quarters from the sun until it reaches the diameter, after which it approaches the sun so that the moon is always below the sun. f. 187a

And if anyone asked you how many miles the sun travels per hour, I say to you that the entire world is 36,000 miles, and the sun has to travel that distance in 24 hours. So it will travel 1,500 miles per hour. At the first hour you will say 1 times 5 makes 5, and 3 times 5 makes 15, that is 1,500. At the second, 2 times 5 makes 10, and 3 times 10 makes 30, that is 3,000. And for 3 hours you will do 3 times 5 makes 15, and 3 times 15 makes 45, that is 4,500, and so you will continue multiplying. When you will be in hour 24, you will say 5 times 24 makes 120, and 3 times 120 makes 360, that is 36,000. And this is done.

And to know how many miles the moon travels away from the sun per hour, every hour it travels 1,200 miles away. And for the first hour we will do one times 4 makes 4, and multiplied by 3 makes 12, which will be hundreds, which makes 1,200 miles. For the second, 2 times 4 makes 8, and 3 times 8 makes 24, that is hundreds, which will be 2,400 miles. And continue like this until you reach 24 hours. And you should understand that in 30 days the moon travels from the sun and goes toward it 36,000 miles, because we multiply 30 times 4, which makes 120, that is hundreds, it will be 1,200, and 30 times 1,200 makes 36,000.

And while it says above what the moon should travel away per hour, that's not it, but it should be what it travels every day. So in 30 days the moon is 36,000 [miles] away.

f. 187b E se neson t'adomandase quante fiel d'aqua e quande[a] segonde, dichotte fiel d'aqua dai 4 dala luna al 10, che sono dì 6, e questi son fielle, e al 11 è ponto d'aqua. In segonda inchina 19, zò serano 6 fiel e 9 segondo fa 15, e dal 19 inchina 25 in fiel. E dal 25 inchina el 4 del'oltra sie segonde, zoè una fiel e una segonda e una fiel e una segonda. In zorny 30 aregordove che dali 7 ai 9 l'aqua non se muove. E questo sie per saver quando le aque à possa e quando non.

E se nesun te domandasse: "Quando tu vuol far patta nuova, perché azonzistu 11 ogn'ano?" Dichotte che fu chavadi questi 11 dali mexi zoè marzo, mazio, luyo, avosto, octobrio, dezenbrio, zener. Sono 7 i quali ano zorny 31 per chadaun, e questa sovra avanza dela luna, che à $29\frac{1}{2}$, per chadaun zorno $1\frac{1}{2}$, i quali serano zorny $10\frac{1}{2}$. E avril e zugno e setenbrio e novenbrio sono 4, avanza dela luna per chadaun dì mezo, serano dì 2. Azonti a $10\frac{1}{2}$ serano $12\frac{1}{2}$. E perché frever in mancha ala luna dì $1\frac{1}{2}$, abatillo, roman 11. E questi vien messi ogn'ano per far patta. E de questo muodo fu truovado questo meter dì 11 ogn'ano per renovar patta, perché de zorno 365 e ore 6 che à l'ano volemo che sia lune 12, avanza questo 11.

E se questa raxion se volesse veder sotilmente, non se poria azonzer 11. La chaxion che del'ano che son zorny 365, ore 6, per bexesto, se noi volemo chavar per lune 12, 29 via 12 fano 348, e puo' 12 via 12 fano 144, che son zorny 6. Azonti fano zorno[b] 154. E puo' 12 via 793 fano punti 9516,
f. 188a e | partidi per 1080 per far ora, fa ore 8, punti 876. Adoncha se volemo sutrazer de zorny 365, ore 6, che son del'ano, zorny 354, ore 8, ponti 876, vegnirave avanzar zorny 10, ore 21, punti 204. E questa vorà eser azonti ala patta per far un'altra patta. Siché da 11 a questo serave radego ore 2, ponti 876. Ma perché a hognomo che non savese intenderse vien fato azonzer questo 11? Perché serà pizollo radego. E questo è ditto e fatto.

[SAVER QUANDO INTRA EL MEXE E DE SSO RAXION]

Questo serà l'amaistramento a saver quando intra el mexe. Primo a saver marzo à nomen 5, chomo vedirì qui de soto.

---

a. *Sic in MS, evidently for* quante.
b. *Sic in MS.*

And if anyone asks you how many neap periods[1] there are and how many spring[2] periods, I say to you that there are neap tides from the 4th until the 10th of the moon, which is 6 days, and these are neaps, and to the 11th is the tide point.[3] Spring tides until the 19th, that is, there will be 6 [days of] neaps and 9 [of] springs, making 15, and from the 19th until the 25th neaps. And from the 25th until the 4th of the other[4] are springs, that is, one neap and one spring and one neap and one spring. Remember that in 30 days the tides are slackest from the 7th to the 9th. And this is to know when the water is powerful and when it is not. f. 187b

And if anyone asks you: "When you want to make a new epact, why do you add 11 each year?" I say to you that these 11 were taken from the months of March, May, July, August, October, December, January. These are the 7 that have 31 days each, and this is more than the lunar month, which has $29\frac{1}{2}$ days, by $1\frac{1}{2}$ days each, which will be $10\frac{1}{2}$ days. And in the 4 of April and June and September and November, the excess for each is half a day, which will be 2 days. Added to the $10\frac{1}{2}$, it will be $12\frac{1}{2}$. And since February lacks $1\frac{1}{2}$ days in respect to the moon, subtract it and there remain 11. And these are added each year to make the epact. And this is how they invented the practice of putting the 11 days each year to renew the epact, since we wish the 365 days and 6 hours that the year has to fit into 12 [lunar] months, which leaves this extra 11.

And if you want to see this problem subtly, we could not add the 11. The fact that in the year there are 365 days, 6 hours, for each leap year, if we wished to subtract for 12 moons, 29 times 12 makes 348, and then 12 times 12 makes 144 [hours] which are 6 days. Added makes 354[5] days, and then 12 times 793 makes 9,516 points, and | divided by 1080 to make the hour, makes 8 hours, 876 points. So if we wish to subtract from the 365 days, 6 hours that make the year, 354 days, 8 hours, 876 points, it would end up with an extra 10 days, 21 hours, 204 points. And this would be added to the epact to make another epact. So there would be a remainder of 2 hours 876 points from 11 to this. But why is everyone who doesn't know this obliged to add the 11? Because it would be such a small reminder. And this is said and done. f. 188a

---

To know when the month begins and its calculation

These will be the instructions for knowing when the month begins. First, you should know that March is designated 5, as you will see below:

---

1. Period in which there is the least difference between high and low tide; I am grateful to Alan Hartley for explicating this passage.
2. Period in which there is the greatest difference between high and low tides.
3. The time when a neap period changes to a spring period or vice versa.
4. I.e., the next month.
5. *154* in MS.

Marzo 5
Avril 1
Mazio 3
Zugno 6
Luyo 1
Avosto 4
Setenbro 7
Octobrio 2
Novenbrio 5
Dezenbrio 7
Zener 3
Frever 6

Perchè marzo à nomen 5? L'ano zorny 365, dia[a] partir 12 via 30, fa 360, e questi 5 che avanza i demo a marzo perché prinzipio del'ano. Perché avril à nomen 1? Perché marzo à zorny 31 e de nomen 5 fa 36. A partir per 7 roman 1. E questo sie d'avril. E perché mazio 3? Perché avril à zorni 30 e de nomen 1 fa 31. A partir per 7 roman 3. E questi sono de mazio.

Perché zugno 6? Perché mazio à zorni 31 e de nomen 3 fa 34. A partir 7 avanza 6. E questi se dà a zugno.

E perché luyo 1? Perché zugno à zorny 30 e de nomen 6 fa 36. Parti per 7, roman un. E questo se dà a luyo.

f. 188b E perché avosto 4? Perché luyo à zorny 31 e de nomen e de nomen[b] 1 fa 32. A partir per 7 roman 4. E questi sono d'avosto.

E perché[c] setenbrio 7? Perché avosto zorny 31 e de nomen 4 fano 35. A partir per 7 roman 7. E questi sono de setenbrio.

E perché octobrio 2? Perché setenbrio à zorny 30 e de nomen 7 fano 37. A partir per 7 roman 2. E questi sono d'octobrio.

E perché novenbrio 5? Perché octobrio à zorny 31 e de nome 2 fano 33. A partir per 7 fano o roman 5. E questi sono de novenbrio.

E perché dezenbrio 7? Perché novenbrio à zorny 30 e de nomen 5 fa 35. A partir per 7 roman 7. E questo è de dezenbrio.

E perché zener 3? Perché dezenbrio à zorny 31 e de nomen 7 fano 38. A partir per 7 roman 3. E questi sono de zener.

E perché frever à nomen 6? Perché zener à zorny 31 e de nomen 3 fano 34. A partir per 7 roman 6. E questi sono de frever.

---

a. *Sic in MS.*
b. *Thus repeated in MS.*
c. avos *crossed out with a horizontal line.*

March 5
April 1
May 3
June 6
July 1
August 4
September 7
October 2
November 5
December 7
January 3
February 6

Why does March have the designation 5? The year is 365 days, you should multiply[1] 12 by 30, which makes 360, and we give these 5 added days to March because it is the beginning of the year. Why does April have the designation 1? Because March has 31 days and has the designation 5, which makes 36. Divided by 7, it leaves a remainder of 1. And this is for April. And why does May have 3? Because April has 30 days and its designation 1, makes 31. Divided by 7, it leaves a remainder of 3. And this is for May.

Why is June 6? Because May has 31 days and has a designation of 3, which makes 34. Divided by 7, there's a remainder of 6. And this is given to June.

And why is July 1? Because June has 30 days and has the designation 6, which makes 36. Divided by 7, there's a remainder of 1. And this is given to July.

And why is August 4? Because July has 31 days and has the designation 1, which makes 32. Di- f. 188b
vided by 7, there's a remainder of 4. And this is for August.

And why is September 7? Because August has 31 days and its designation 4 makes 35. Divided by 7, there's a remainder of 7. And this is for September.

And why is October 2? Because September has 30 days and its designation 7 makes 37. Divided by 7, there's a remainder of 2. And this is for October.

And why is November 5? Because October has 31 days and its designation 2 makes 33. Divided by 7, it makes or there's a remainder of 5. And this is for November.

And why is December 7? Because November has 30 days and its designation of 5 makes 35. Divided by 7, there's a remainder of 7. And this is for December.

And why is January 3? Because December has 31 days and its designation of 7 makes 38. Divided by 7, there's a remainder of 3. And this is for January.

And why does February have the designation 6? Because January has 31 days and its designation 3 makes 34. Divided by 7, there's a remainder of 6. And this is for February.

---

1. *Divide* in MS.

Qui in driedo vui vedirì 2 man inchadenade, in le qual tutti 2 in dedi 7 averano nudi 28, e in questi nudi 28 serano per patta de questo sopraditto chapitullo per truovar el dì intrerà el mexe qual che vui vollé.

E la prima sie che te bexogna amesidar 2 chose, zoè la pata dela man e 'l nomen del mexe, vorì veder quando el die intrar, chomo ve serà insegnado qui per singolo. E per exenpio in 1435 chore ala patta dela man 5, in chavo del dio pizullo dela man destra, e questa patta durerà per lo millieximo sovra scrito. E voyando far 1ª altra vuy mo' darì el nudo ala man senestra, al fundi. El dirà 7,
f. 189a e questo serà del 1436. Arigordandove che | serà bexesto, el frever averà zorny 29. E voyando veder quando intrarà el mexe de mazio tu dirà che nomen à mazio. À nomen 3, e de patta 7 fano azonti 10. E chomenza a nonbrar da domenega. E domenega 8, e luni 9, e marti 10. Adoncha marti intrerà el mexe de mazio del 1436. Anchore, quanto[a] intrerà avosto? Avosto à nomen 4, e de pata 7 fano 11. Domenega e domenega 8, luny 9, marti 10, merchore 11. Adoncha merchore intra avosto. E per questo muodo fa' tutte le oltre e va' a nudo a nudo inquina i nudi 28 e puo' torna da chavo.

E se per aventura vui non v'avesi in memuoria la patta dela man farì chossì. Perché la patta dela man non passa 7, e chomo se' al sette torna indriedo, sì che adoncha azonzi 7 e nonbra al primo dio, zoè al fundi da ladi destro dove dize 1, e dirai 8, 9, 10, 11. E chosì va' nonbrando nudi 28, e torna da chavo inchina tu avera' nonbrado ani Dominy, zoè 136, e là che tu te truovy là te ferma. In quel ano serà la patta.

E le 2 man vedirì qui per inchontro. Al primo dio dela man destra, zoè al nudo da fondi, dira' 1, al segondo 2, al terzo 3, el quarto 4. Al segondo, al fundi, 6, 7, 1, 2. Al terzo dio, al fundi, 4, 5, 6, 7. Al quarto dio 2, 3, 4, 5. Ala man senestra, al fundi, al primo 7, 1, 2, 3. Al segondo 5, 6, 7, 1. Al terzo, al fundi, 3, 4, 5, 6. E puo' torna al primo dela man destra. Arigordandotti che a tuti le zime dele die sie bexesto. Perzò tuoresi al desmontar uno, che 'l mexe de frever averà dì 29.

f. 189b [*caption:*]
2 man[b]

[Saver le raxion dela Pasqua e lli so regai]

Questo serà l'amaistramento che se possa saver quando die far Pasqua. Vui vedirì una man destra, la qual averà nudi 19. Al primo dio grando serà 5, 25, 13. Al segondo dio 2, 22, 10, 30. Al terzo 18,

a. *Sic in MS, evidently for* quando.
b. *Drawing lacking in MS.*

After this you will see two hands chained together, in both of which there will be 28 joints in 7 fingers, and in these 28 joints it will be possible using the epact of this chapter above to find the day of the beginning of the month you wish.

And the first thing you need to do is to combine 2 things, that is, the epact of the hand and the designation of the month, and you will see when it should begin, as will be shown to you here in detail. And for example, in 1435, it occurs at epact 5 of the hand, at the top of the small finger of the right hand, and this epact will last through this year. And if you want to do another you will move to the joint of the left hand, at the bottom. It will say 7, and this will be for 1436. Remembering that | it will be a leap year, February will have 29 days. And wishing to see when the month of May will begin, you will say: "What designation does May have?" It has the designation 3 and the epact 7, which added together make 10. So the numbering begins with Sunday. And Sunday is 8, Monday is 9, and Tuesday is 10. So Tuesday will be the first of May of 1436. Again, when will August begin? August has the designation 4, and the epact of 7 makes 11. Sunday[1] is 8, Monday 9, Tuesday 10, Wednesday 11. So August will begin on a Wednesday. And in this way you do all the others and go from joint to joint until joint 28 and then return to the top. f. 189a

And if by chance you don't remember the epact of the hand, you will do this. Since the epact of the hand does not exceed 7, when you are at 7 turn back, and then add 7 and the number of the first day, that is, at the bottom of the right the side where it says 1, and you will say 8, 9, 10, 11. And so keep counting 28 joints, and return to the top until you have counted the year of the Lord, that is 1436;[2] wherever you find yourself, stop there. In that year will be the epact.

And you will see the two hands facing here. On the first finger of the right hand, that is, on the bottom joint, it will say 1, on the second 2, on the third 3, the fourth 4. On the second at the base, 6, 7, 1, 2. On the third finger at the base, 4, 5, 6, 7. On the fourth finger 2, 3, 4, 5. On the left hand, at the base, first 7, 1, 2, 3. On the second 5, 6, 7, 1. On the third, at the base, 3, 4, 5, 6. And you can return to the first of the right hand. Remembering that all of the tops of the fingers are leap years, so remember to subtract one, since the month of February will have 29 days.

[*caption:*] f. 189b
2 hands[3]

To know the calculation of Easter and its rules

These will be the instructions on how to know on what day Easter falls. You will see a right hand, which will have 19 joints. On the first large finger will be 5, 25, 13. On the second finger 2, 22, 10,

1. *Sunday and Sunday* in MS.
2. 136 in MS. Similarly, the *Raxion de' marineri*, has 144 for 1444 in the analogous passage: Conterio, ed., p. 33, n. 26.
3. Drawing lacking.

7, 27, 15. Al quarto dio 4, 24, 12, 1. Al quinto 21, 9, 29, 17. E chadaun de questi nudi serano patta de Pasqua braicha. E voyando truovar la nostra Pasqua senpre da questa braicha, va tanto ananti che tu truovy domenega. E se 'l chadesse che Pasqua braicha fusse domenega, l'ortra[a] dome-
f. 190a nega serà nostra. E se per aventura | perdesti la patta dela Pascha braicha, azonzi 10 al fundi del dio grando, e va nonbrando in nudi 28 intorno inchina li ani Dominy, zoè 136, e là te ferma. Mo' varda zò che dixe el nudo, e quello serà la patta. Vedirì qui de sotto.

Ancuor per uno oltro muodo de saver quando serà Pasqua, meti a mente alla prima luna de frever. El primo merchore dapuo' fatta la luna quel serà el primo dì de Charexima. E per questo saverì quando serà la Pasqua.

Anchora per una oltra raxion se vui volì saver là che serà la pata dela Pasqua braicha, in qual ano vora' saver azonzi del mileximo, zoè el 36 e 14 fano 50, e chomenza a nonbrar dal dio grando inchina 50 in torna, e là te ferma. Per quel ano serà 4, zoè 4 dì d'avril Pasqua braicha, e la nostra serà a 8 dì d'avril.

E se tu volesti veder ancuor per ché li nudi de Pasqua dixe 5, 25, 13, 2, 22, 10, 30, 18, 7, 27, 15, 4, 24, 12, 1, 21, 9, 29, 17, quisto serà el muodo. El 5 azonzi 20, serano 25. El 25 sutrazi 12, roman 13. Al 13 sutrazi 11, roman 2. Al 2 azonzi 20, fano 22. Al 22 abati 12, roman 10. Al 10 azonzi 20, serano 30. Al 30 sutrazi 12, roman 18. Al 18 sutrazi 11, roman[b] 7. Al 7 azonzi 20, serano 27. Al 27 sutrazi 12, serano 15. Al 15 sutrazi 11, roman 4. Al 4 azonzi 20, serano 24. Al 24 sutrazi 12, roman 12. Al 12 sutrazi 11, roman 1. A un zonzi 20, serano 21. E sutrato 12 roman 9. Azonto 20 fano 29. E trato 12 roman 17.

f. 190b

[Veder 1 portolan per Puya, zoè da Manfredonya a Otronto]

Questo sie un portolan per la riviera de Poya, zoè da Bestize per riviera inchina el chavo d'Otronto e la chognosenza deli ditti luogi.

Bestize a Manfredonya, mia 30. Manfredonia sì à un muolo al qual non se può achostar tropo a quelo per aver fundi pizolo. Metti prodixe al muolo e vardate dalo levante al sirocho.

Manfredonya chon Barlletto, sirocho e levante, mia 30. Non à nesun redutto, ma per la via sono algune seche inver li saline. Alargatte ben in mar. Barlletto à uno schoietto per mezo et à chomo una turixella in lo schoietto, e chi per forza dovesse star mettasse i prodixe al schoyo.

---

a. *Sic in MS.*

b. a *and* man *between the lines with no direct connection to the text; possibly as unused captions.*

30. On the third 18, 7, 27, 15. On the fourth finger 4, 24, 12, 1. On the fifth 21, 9, 29, 17. And each of these will be the epact of the Hebrew Passover. And wishing to find our Easter always from the Hebrew one, go forward until you find Sunday, and if it happens that the Hebrew Passover is a Sunday, the next Sunday will be ours. And if by chance | you have lost the epact of the Hebrew Passover, add 10 to the base of the big finger and continue counting the 28 joints until the year of the Lord, that is 1436,[1] and stop there. Now, look at what the joint says, and that will be the epact. You will see it below. f. 190a

Again, for another way to know when Easter will be, remember the first moon of February. The first Wednesday after the new moon will be the first day of Lent, and this way you will know when Easter will be.

Again by another method, if you wish to know what will be the epact of the Hebrew Passover, to that year which you want, add the year number, that is, 36 and 14 make 50, and begin to count from the large finger up to 50, and stop there. For that year it will be 4, that is, April 4 will be the Hebrew Passover, and ours will be April 8.

And if you wish to see again why the joints of Easter say 5, 25, 13, 2, 22, 10, 30, 18, 7, 27, 15, 4, 24, 12, 1, 21, 9, 29, 17, this is the way. To 5 add 20, that will be 25. From 25 subtract 12, there remains 13. From 13 subtract 11, there remains 2. To 2 add 20, which makes 22. From 22 subtract 12, there remains 10. To 10 add 20, there will be 30. From 30 subtract 12, there remains 18. From 18 subtract 11, there remains 7. To 7 add 20, there will be 27. From 27 subtract 12, there will be 15. From 15 subtract 11, which leaves 4. To 4 add 20, there will be 24. From 24 subtract 12, there remains 12. From 12 subtract 11, there remains 1. To one add 20, there will be 21. And subtract 12, which leaves 9. Add 20, which makes 29. And subtract 12, there remains 17.

---

f. 190b

See a portolan for Apulia, that is, from Manfredonia to Otranto

This is a portolan for the coast of Apulia, that is, from Vieste along the coast as far as the Cape of Otranto and the landmarks of these places.[2]

From Vieste to Manfredonia, 30 miles. Manfredonia has a jetty which you cannot come up to because it has a small depth. Make your mooring rope fast to the jetty and watch out for east to southeast winds.

Manfredonia with Barletta, southeast and east, 30 miles. There is no place to stop, but along the route there are several shallows near the saltworks. Stay well out to sea. Barletta has a reef in the middle and has something like a small tower on its reef, and anyone who out of necessity must stay there should carry the mooring ropes out to the reef.

---

1. 136 in MS.
2. For compass directions used in portolans, see p. 347, n. 8, above.

Barletto a Trany mia 6. Schure tuti li ditti luogi levante e sirocho. La chognosenza de Trany: uno chastello dentro ala tera, e da ponente à un portixello in mezo de du[a] ture. Vane chon gallia pizulla e mettite ala ture da ponente, e non andar da levante. E non andar massa entro perché tuto è secho, e armizate per quarto per la restia.

f. 191a Trany. A un mio vederì una badia per riviera e de questa badia a Bexeye mia 2. Ave una challe e dentro dela qualle un redutto par a muodo uno schoietto per barche e per fuste pizulle. Armizate per quarta. Aqua pie 3. La so chognosanza: per mezo la tera un chanpagnel alto.

Baxeye a Malfetto mia 6. Et ananzi tu arivy a Malfetto, a uno mio, truoverì 1ª challe in la qual da ponente à una ture chon 2 caxette, a[b] da levante par una jexia, zoè di Marturi, e vane arditamente chon gallia sotil, per mezo per la via d'un pallo vedirì afitto in aqua. Fa che 'l pallo te romagna da ponente, armizatte a quel pallo e l'oltro prodixe in tera da levante. Armizate per quarto per la ristia. Avera' aqua pie 6. Et ala iexia d'i Marturi suvrascritti uno porto in palli, et è bon luogo. El qual ti chonvien partir per mezo el chanal, e vane arditamente raxo i palli da levante, e non t'achostar ala ponta dela jexia d'i Marturi. La qual iexia à un canpagnel per segnal. Avera' aqua pie 6. In lo ditto luogo armizate per quarto, per la ristia.

Marturi a Malfetto mio uno. La chognosenza dela tera: dentro la tera, da ponente, vedirì chanpagnelli 2 e uno altro più ertto dali ditti. Et à una pizulla iexietta fuor dala tera, da levante.

f. 191b Malfetto a Zovenazo mia 5. La so chognosanza: una iexia in la tera da ponente chon 2 chanpagnelli, e da ponente, fuor dela tera, algune iexie grande, e da levante, dala tera, 2 ture erte.

Zuvenazo a Bari mia 12. Bari è buon luogo. Ala punta dala tera, da levante, un moletto. Puose meter prodixe e fuora li feri armizatti per quarto. Vardate de griego levante e choverto tu sera' d'ogno altro vento. Avera' aqua pie 6. La chognosanza dela tera: à chanpagnelli 4.

Bari a San Zorzi mia 5. Ave 1ª challe bona. E da San Zorzi a San Vitto de Polignano mia 15. San Vitto à una buona challe, in la qual se può meter dentro chon una galia. Armizatti per quarto chon vivi in tera. La so chognosanza: una badia a muodo un chastello derupado.

San Zorzi a Muola mia 22. Chossa[c] piana. Non à chanpagnel erto. Chossa[d] d'asassiny.

---

a. *Sic in MS.*
b. *Sic in MS.*
c. *Sic in MS for* costa.
d. *Sic in MS for* costa.

Barletta to Trani, 6 miles. At all these places it flows to the east and southeast. Landmarks of Trani: a castle inside the city and on the west side it has a small gate between two towers. Go there with a small galley and steer on the west tower, and don't go to the east. And don't go too far in since everything is shoal, and moor by quarters[1] because of the breakers.

Trani. At one mile you will see a monastery along the coast, and from this monastery to Bisceglie is 2 miles. It has a small inlet in the middle of which is a sheltered place, with a reef, suitable for boats and for small galleys. Moor by quarters. Water 3 feet. Its landmark: in the middle of the land a tall bell tower. f. 191a

Bisceglie to Molfetta, 6 miles. And one mile before you arrive at Molfetta, you will find a bay which on the west has a tower with two small houses; on the east appears a church, that is, of the Martyrs, and go quickly with a light galley. In the middle of the route you will see a post fixed in the water. Keep this post to your west, tie up to this post, and put the other mooring rope(s) ashore to the east. Moor by quarters because of the breakers. You will have 6 feet of water. And at this church of the Martyrs [there is] a port protected by pilings, and it's a good place. You're best off leaving by the middle of the channel, and go at speed close to the eastern pilings, and don't approach the point[2] of the church of the Martyrs. This church has a bell tower, as a landmark. You will have 6 feet of water. In this place moor by quarters because of the breakers.

From the Martyrs to Molfetta is one mile. Landmarks: on land, to the west, you will see 2 bell towers and another higher than these. And it has a small church outside the town, to the east.

Molfetta to Giovinazzo, 5 miles. Its landmark: a church inside the town, on the west with 2 bell towers, and to the west, outside the town, several large churches, and to the east of the town, 2 high towers. f. 191b

Giovinazzo to Bari, 12 miles. Bari is a good place. At the peninsula, on the east, a small jetty. You can put your mooring ropes [there] and the anchors on the offshore side, [and so] moor by quarters. Keep a watch for the east-northeast wind, and you will be protected from all other winds. You will have 6 feet of water. Landmark: it has 4 bell towers.

Bari to San Giorgio, 5 miles. It has a good bay. And from San Giorgio to San Vito di Polignano, 15 miles. San Vito has a good bay, in which you can put a galley. Moor by quarters while you're staying on land.[3] Its landmark: a monastery like a ruined castle.

San Giorgio to Mola, 22 miles. A flat coast. It does not have a high bell tower. A coast of robbers.

---

1. The phrase *per quarto* probably means "to four moorings," as, for instance, with two anchors ahead or astern and two mooring ropes (to shore, posts, etc.) in the opposite direction. Cf. Kahane, Kahane, and Bremner, *Glossario degli antichi portolani italiani*, 99, for this interpretation.

2. I.e., peninsula.

3. Possibly with the sense of "bearing in mind strong landward currents."

La chognosanza da Polignano è sovra un pocho de teren ertto chomo grebany. Et è ala banda da levante uno schoietto, del qual se può passar dentro de quelo. El chavo dala staria, zoè dal[*a*] tera, da man destra, 1ª turetto[a] fu per tenpo fanò.

Polignano a Minopolli mia 15. È la chognosanza de Minopoli un chanpagnel da levante, erto, et à
f. 192a una chale da ponente | chiamatta Paltan, e per mezo el ditto Paltan 1ª iexia da levante.

Minopolli a Sancto Stefano mia 2, uno chastello, et inchina Vila Nuova mia 30. À una challe da levante, raxo la punta dela tera. Non t'achostar al'oltro ladi perché è secho. Armizatti per quarto chon vivi. In tera avera' aqua pie 6.

La chognosanza del ditto luogo: 1ª tera derupada, tiense una ture per perschaduri. La chognosanza del ditto luogo: in la montagna, forssi mia 6, una tera chiamatta Stune. Infra tera e da levante una tera chiamata Vila Francha.

Vila Nuova, mia 10. Truoverì una ture, un stazetto da barche, e dala ture a Gauxitto mia 10. Gauxitto à uno schoio grande, el qual schoio è bon luogo, et à uno oltro schoietto pizollo per inchontra, in el qual schoietto metti le palomere et al grando li prodixi. Armizatti per quartto. E puose andar dentro dali do schoietti. Serà aqua pie 6. La chognosanza: in punta dela staria piere fatte a muodo una muschìa, e in la valle, infra tera, 1ª ture da levante del ditto. Et à schoietti 3 de fuor de quei e puose andar dentro via.

Gauxitto a Brandizo mia 10. La so chognosanza:[b] in punta 1ª ture chiamatta el Chavallo, e per mezo la tera 1ª valle chon una iexia chiamatta Sancto Andrea. E vane per mezo arditamenti chon ogni nave. Non t'achostar ala punta del'ixola dever la tera a un prodixe. È l'intrada da levante. In
f. 192b chavo la staria ave una ture chiamata el Fanò. À un schoieto | di fuor de quelo et è secho. Non t'achostar. A un mio puo' truoverì un schoio grando, el qual non se può passar dentro. E truoverì anchuor schoietti 4, d'i quali achostate al chodier dever l'ixola de Sancto Andrea e parti el freo e vane inver la tera. Non t'achostar ala punta de Sancto Andrea perché è secho vardando inver la tera, ma fali honor un bon prodixe.

Brandizo ala ture de San Chataldo mia 40, e per tutto è secho per la via. Non t'achostar in tera a mio un inchina Otronto.

San Chataldo a Otronto mia 20. Non achostar in tera a un mio perché l'è aque basse. E in mia 10 vui truoverì una tera chiamatta Rocha sovra alguny grebany, e de là a Brandizo mia 10.

---

a. *Sic in MS.*
b. cho *above line.*

The landmark of Polignano is a little piece of land above, as high as cliffs. And on the eastern shore a reef, within which you can pass. At the headland, that is, of the town, on the right hand, a small tower [that] used to be a lighthouse.

Polignano to Monopoli, 15 miles. And the landmark of Monopoli is a bell tower to the east, high, and it has a bay to the west | called Paltan, and in the middle of this Paltan, a church to the east. f. 192a

Monopoli to Santo Stefano, 2 miles, a castle, and to Villanuova 30 miles. It has a bay to the east, right next to the point of land. Don't approach the other side because it's shoal. Moor by quarters because of the currents. Along the shore you will have 6 feet of water.

The landmarks of this place: a craggy land; it has a tower for fisherman. The landmarks of this place: in the mountains, at about 6 miles, a town called Ostuni. Inland and to the east, a town called Villa Franca.

Villanuova, 10 miles. You will find a tower, a small anchorage for boats, and from the tower to Gauceto 10 miles. Gauceto has a large reef, which is a good place, and another smaller reef across from it; put the *palomera* mooring ropes[1] on this reef and the *prodexe* mooring ropes[2] on the big one. Moor by quarters. And you can go between the two reefs. There will be 6 feet of water. The landmarks: on the point of the shore, stones shaped like a musk deer, and in the valley, inland, a tower to the east of it. There are 3 small reefs outside of these, and it is possible to pass within them.

Gauceto to Brindisi, 10 miles. Its landmarks: on the point, a tower called Cavallo, and across from the town, a valley with a church called Sant'Andrea. And go at speed with any ship. Don't approach within a mooring rope length of the point of the island toward the town. The entrance is from the east. At the headland there is a tower called the Lantern. It has a small reef | outside of it, and it's shoal. Don't approach it. At one mile, you can find a large reef, which you can't pass inside of. And then you will find 4 small reefs, which you can approach on the end toward the island of Sant'Andrea and leave the channel and go toward the town. Don't approach the point of Sant'Andrea because it's shoal, looking toward the town, but give it a mooring cable's berth.[3] f. 192b

Brindisi to the tower of San Cataldo, 40 miles, and it is shoal the whole way. Don't approach within a mile of land until Otranto.

San Cataldo to Otranto, 20 miles. Don't approach within a mile of land because of the shallow waters. And in 10 miles you will find a place called Rocca Vecchia above some cliffs, and from there to Brindisi it's 10 miles.

---

1. Probably those of the stern.
2. Probably those of the bow.
3. Literally, "honor."

Chognossanza da Otronto: in una punta bassa biancha ave una ture, per antigo tenpo era fanò. E vane inver la tera. Truoverì uno schoietto, vane per mezo el schoietto e la ponta sottil da man senestra, armizatti fero e prodixe, e fatte forte per maistro e tramontana.

[Veder 1 portolan per lo golfo de Salonychi]

Questo è portolan per lo golfo de Salonychi. Intro Schiatti San Zorzi una secha, e quando fa mar ronpe su, e quanto[a] non fa mar non par. E voyando andar netto, achostatte ben ala teraferma a un mio o al'ixola di Schiatti per simille muodo per la ditta secha.

f. 193a San Zorzi chon el chavo di Posidi, zoè del golfo de Salonychi, quarta de tramon|tana al maistro, mia 80 zoè del'ixolla de Chassandra.

San Zorzi chon el chavo Chanistro, zoè del'ixola de Chassandra, se varda ostro e tramontana, mia 100.

Schiatti chon Possidi, maistro e tramontana, mia 80.

Possidi chon la punta di Pinacha, zoè 1ª ture sovra 1 teren ertto raxo l'aqua, son mia 20. Non t'achostar al Pinacha, balestrade 2 per aque basse. E dentro del Possidi, in la valle in chavo dela spiazia, una fiumera d'aqua e legni.

Pinacha chon la ture de Chassandra, ostro e tramontana, mia 10.

La tura de Chassandra chon Fanari, ponente maistro, mia 40. Non te achostar ale punte né non te insachar ale valle per aque che sono mullto basse.

Possidi chon Fanari verso Salonychi, quarta de maistro al ponente, mia 60.

Ture de Chassandra, a Lauria, ponente maistro, mia 20.

Laura[b] chon la punta de Vellona, ponente e maistro, mia 12.

Vellona chon Fanari, quarta de ponente al maistro, mia 10.

Fanari chon el chavo sotil dele Pechiere,[c] quarta de maistro ala tramontana, mia 10. Dal Fanari ale Peschere, non te achostar né ale punte né ale spiaze per algune seche e aque basse.

---

a. *Sic in MS, evidently for* quando.
b. *Sic in MS.*
c. *Sic in MS for* Peschiere.

Landmarks of Otranto: on a low white point there is a tower; in old times it was a lighthouse. And go toward the town. You will find a small reef. Go by the reef and the small point on the left hand, moor yourself with anchor and mooring ropes, and secure yourself for northwest and north winds.

See a portolan for the Gulf of Salonika

Here is a portolan for the Gulf of Salonika. Between Skíathos and Áyios Yeóryios is a shoal, and when the sea is high it breaks on it, and when there is no sea running, it doesn't show up. And if you want to go in the clear, approach within a mile of the mainland or likewise the island of Skíathos on account of that shoal.

Áyios Yeóryios with the cape of Posídhion, that is, in the Gulf of Salonika, north | by west, 80 f. 193a
miles, that is, the peninsula of Kassándra.

Áyios Yeóryios with Cape Kanastraíon, that is, the peninsula of Kassándra, keep yourself south and north, 100 miles.

Skíathos with Posídhion, north-northwest, 80 miles.

Posídhion with the point of Pinaka, that is, a tower above a raised plain close to the water, is 20 miles. Don't approach Pinaka [closer than] 2 crossbow shots at low tide. And inside Posídhion, in the valley at the end of the beach, are a stream of water and wood.

Pinaka with the tower of Kassándra, south and north, 10 miles.

The tower of Kassándra with Cape Epanom, west-northwest, 40 miles. Don't approach the point nor enter the bay because of the waters, which are very shallow.

Posídhion with Cape Epanom toward Salonika, northwest by west, 60 miles.

Tower of Kassándra to Lavria, west-northwest, 20 miles.

Lavria with the point of Vellona, west-northwest, 12 miles.

Vellona with Cape Epanom, west by north, 10 miles.

Cape Epanom with the Low Cape of the Fishermen, northwest by north, 10 miles. From Cape Epanom to the Fishermen, don't approach either the points or the beaches because of shoals and shallow waters.

Dala ponta sotil dale Peschiere inver lo chavo de Nonbolo non t'achostar a un mio in tera per aque basse. Ostro e tramontana, mia 5.

f. 193b El chavo de Nonbolo chon Salonychi, griego e tramontana, mia 18, tuto sorzedor.

Posidi chon el chavo de Chanystro, quarta de levante al sirocho, mia 40. Non t'achostar passando la punta del Posidi al monte grosso, a ladi senestro, a balestrade 2, per seche e aque basse. E per lo simille mudo[a] per la riviera, voiando andar al chavo per lo simille muodo. E al dito chavo è uno schoietto. E se volesti andar al porto de Chanistro, voltado averì el schoietto a man senestra, vedirì una ponta, e vuoltado quello vui vedirì la valle del stazio. Vardate da maistro.

Chanistro a Chofo, quarta de levante al griego, mia 18. È bon porto.

Chanystro al chavo de Chofo, levante ponente, mia 18. E se vegnise da Chanistro a Chofo, mostra una montagna redonda a modo un pan. E ssutto quello muostra a muodo d'un schoietto. Lasalo da man senestra perché roman el chavo chomo valado.

El chavo de Chofo chon el chavo del Figo, griego e levante, mya 20.

El chavo del Ficho chon Pallochastro, zoè l'ixola de Stalimene, se varda griego e levante, mia 80.

[Tuor el vin a chi se deschonzasse e che s'imbriagase]

Qui volesse tuor el vin a chi se deschonzasse e che forte se inbriagasse. Tuor dela radixe d'uno alboro, el qual se chiama in griego axogiro, e quela debia far menuda. E quelo debi bugir chon bon vin biancho inchina che quello romagna la terza parte del vin arai messo a bugir chon le radixe. E lassarlo che vegna chiaro e bello, poi dara' a ber al ditto inbriagon.[b] Non vorà più veder el vin né gostar quello. Questo asogiro fa ala similitudine dele charobe, e dentro quelle a muodo faxiolli.

f. 194a

[Truovar 2 homeny; "Se tu me dà $\frac{1}{3}$ d'i ttoi averò 25"; el segondo: "Me dé $\frac{1}{4}$ d'i vostri arò 25"]

Sono 2 chonpagny che dixe el primo al segondo: "Volemo noi chonprar de chonpania un sacho de piper, el qual se domanda duchati 25?" Dixe el segondo: "Io son chontento pur che tra noi abiamo tanti denari che se posa pagar." Dixe el primo al segondo: "Ò tanti denari in borssa che se tu me da' $\frac{1}{3}$ d'i tto denari apreso i mie averò tanti denari da chonprar el sacho del piper." Dixe el segondo al

---

a. *Sic in MS, evidently for* muodo.
b. a *added above the line.*

From the Low Point of the Fishermen to the cape of Emvolon, don't approach within a mile of land because of shallow water. South and north, 5 miles.

The cape of Emvolon with Salonika, north-northeast, 18 miles, all [good] anchorage. f. 193b

Posídhion with Cape Kanastraíon, east by south, 40 miles. Passing the point of Posídhion, don't approach the great mountain, on the left side, within 2 crossbow shots, because of shoals and shallow waters. Likewise, along the coast, wishing to go to the headland, [proceed] in a similar way. And at the headland is a small reef. And if you want to go to the port of Kanastraíon, if you're turned so you have the small reef on your left, you will see a point, and having rounded that you will see the bay of the anchorage. Keep northwest.

Kanastraíon to Kouphos, east by north, 18 miles. It's a good port.

Kanastraíon to the cape of Kouphos, east–west, 18 miles. And if you should come from Kanastraíon to Kouphos, a mountain round like a loaf of bread shows up. And under this a small reef shows up. Leave that on the left hand so that the cape remains like[1] a flight of stairs.

The cape of Kouphos with the cape of Figo, east-northeast, 20 miles.

The cape of Figo with Palichastro, that is, the island of Lemnos, bear northeast and east, 80 miles.

To take wine away from someone who's become indecent and has become drunk

Whoever wants to remove the wine from someone who has become indecent and has become very drunk. Take the root of a tree which in Greek is called *axogiro*, and that should be made smaller. And that should be boiled with good white wine until only a third of the wine put to boil with the root remains. And let it become clear and beautiful, then you will give it to the inebriated person to drink. He will neither wish to see wine nor to taste it. This *axogiro* is similar to the carob, and inside has things like beans.

Find 2 men; "If you give me $\frac{1}{3}$ of your goods I will have 25"; the second: "Give me $\frac{1}{4}$ of yours and I'll have 25" f. 194a

There are 2 partners of which the first says to the second: "Do we want to buy as a company a sack of pepper for which 25 ducats are asked?" The second says: "I agree, provided that we have enough money to pay for it." The first says to the second: "I have enough money in my purse that if you give me $\frac{1}{3}$ of your money, together with mine I will have enough money to buy the sack of pepper."

1. I.e., "continues to look like."

primo: "E se ttu me da' $\frac{1}{4}$ d'i ttuo denari apresso i mie, averò tanto che chonpra el ditto sacho." Et io ve domando: "Quanti denari averà zaschon di loro per inpoxizio[*n*]?"

Ponamo che per la prima inpoxizion fuse 16. El vuol del chonpagno el terzo perché abia duchati 25. Adoncha al segondo pony 27, perché el terzo de 27 son 9. Azonzi a 16, arà 25. E puo' el segondo, che à 27, vuol el quarto dal primo, che à 16. Serano 4 e 27 fano 31, e noi vossemo che fusse 25. Adoncha è più 6, e questi metti ala prima inpoxizion.

E la segonda inpoxizion mettemo che la fusse 12. El vuol che 'l segondo abia duchati 39 per dar al primo el so terzo per aver 25. El segondo el quarto di 12 serano 3, e 29 serano 42. E noi vossemo che fusse 25. Adoncha serano de più[a] 17. E questi se metta ala segonda inpoxizio[*n*]. Seguisi la inpoxizion chomo altre vui avè visto avanti, sì che 'l primo chonpagno averà $18\frac{2}{11}$, el segondo $20\frac{5}{11}$.[b] Aprovalla.

16	╲ ╱	12
27	╳	39
più		più
6	╱ ╲	17
		6
	partiore	11

Moltipicha in croxe et trazi l'un dal'oltro perché xe più e più. Trazi da 17 6, serano 11. E questi serà partior.

12	17	39	27
6	16	6	17
72	272	234	459
	72		234
	200 a partir		225 la segonda

$$\begin{array}{l} 0\not{1} \\ \not{1}\not{9}2 \\ \not{2}\not{0}\not{0} \\ \not{1}\not{1}\not{1} \\ \not{1} \end{array}\Bigg]\ 18\tfrac{2}{11} \qquad \begin{array}{l} 00 \\ \not{2}\not{2}5 \\ \not{1}\not{1}1 \\ \ \ 1 \end{array}\Bigg]\ 20\tfrac{5}{11}$$

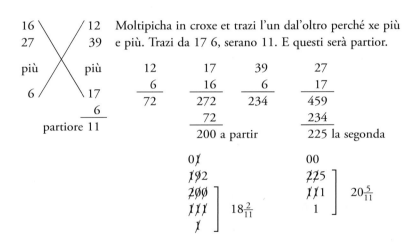

f. 194b E per far la contrascritta raxion per la chossa questa sie la riegolla, che tu die aponer che 'l primo avesse $\frac{1}{1}^{co}$, el segondo die aver 75^{n} men 3^{co}, perché dando el segondo al primo el $\frac{1}{3}$ d'i ssoy averà 25^{n}. Puo' el segondo domanda el $\frac{1}{4}$ al primo. Adoncha trazi $\frac{1}{4}$ de quel del primo, che serà $\frac{1}{4}^{co}$, e dallo a quel del segondo, et avera' 75^{n} men $2\frac{3}{4}^{co}$, la qual edechazion die 'ser ingual a 25^{n}. Ora adequa le parte e da' $2\frac{3}{4}$ al 25, e serà $2\frac{2}{3}$ più 25 ingual a 75^{n}. Trazi 25 de 75, roman 50. Doncha $2\frac{3}{4}^{co}$ è ingual a 50^{n}. Parti 50 per $2\frac{3}{4}^{co}$, ne vien $18\frac{2}{11}$, e chotanto val la chossa. E ttu ponesti che'l primo avesse $\frac{1}{1}^{co}$, el segondo 75 men[c] 3^{co}. Adoncha la chossa valse $18\frac{2}{11}$. A moltipichar per 3 serà $54\frac{6}{11}$, e questo se die trazer de 75 $54\frac{6}{11}$ resta $20\frac{5}{11}$. E questo è fatto.

---

a. *In* più, u *corrected over another, illegible letter.*

b. *Possibly a letter or two lost in the tear of the page.*

c. $2\frac{3}{4}^{co}$ *crossed out with a diagonal line.*

The second says to the first: "And if you give me $\frac{1}{4}$ of your money, together with mine I will have enough to buy the sack." And I ask you: "How much money did each have by false position?"

Let's suppose that for the first false position it was 16. He wants a third of that of his partner in order to have 25 ducats. So put 27 for the second man, because a third of 27 is 9. Add to 16, he will have 25. And then the second, who has 27, wants a quarter from the first, who has 16. It will be 4 and 27, which makes 31, and we wanted it to be 25. So it is plus 6, and put this at the first false position.

And for the second false position, we put that it was 12. It's necessary that the second have 39 ducats to give his third to the first to have 25. For the second, a quarter of 12 will be 3, and 39[1] will be 42. And we wanted it to be 25. So it will be plus 17. And put this at the second false position. Continue the false position as in others that you have seen before, so that the first partner will have $18\frac{2}{11}$, the second $20\frac{5}{11}$. Prove it.

16		12
27		39
plus		plus
6		17
		6
	divisor 11	

Multiply across and subtract the one from the other because it's plus and plus. Subtract 6 from 17, it will be 11, and this will be the divisor.

12	17	39	27
6	16	6	17
72	272	234	459
	72		234
	200 to divide		225 the second

0~~1~~
~~1~~~~9~~2
~~2~~~~0~~~~0~~
~~1~~~~1~~~~1~~] $18\frac{2}{11}$
~~1~~

00
~~2~~~~2~~5
~~1~~~~1~~1] $20\frac{5}{11}$
1

And to do the problem on the facing page by unknowns, here is the rule, that you should put that f. 194b
the first had $\frac{1}{1}x$, the second should have 75 minus $3x$, so that with the second giving the first $\frac{1}{3}$ of his he will have 25. Then the second asks $\frac{1}{4}$ from the first, so subtract $\frac{1}{4}$ of that of the first, which will be $\frac{1}{4}x$, and give it to that of the second, and he will have 75 minus $2\frac{3}{4}x$, which equation should be equal to 25. Now equalize the sides and give $2\frac{3}{4}$ to the 25, and it will be $2\frac{3}{4}$[2] plus 25 equal to 75. Subtract 25 from 75, leaves 50. So $2\frac{3}{4}x$ is equal to 50. Divide 50 by $2\frac{3}{4}x$, it comes to $18\frac{2}{11}$, and that is the value of the unknown. And you put that the first had $\frac{1}{1}x$, the second 75 minus $3x$. So the unknown is worth $18\frac{2}{11}$. Multiplied by 3 it will be $54\frac{6}{11}$, and this should be subtracted from 75, which leaves $20\frac{5}{11}$. And this is done.

1. *29* in MS.
2. *$2\frac{2}{3}$* in MS.

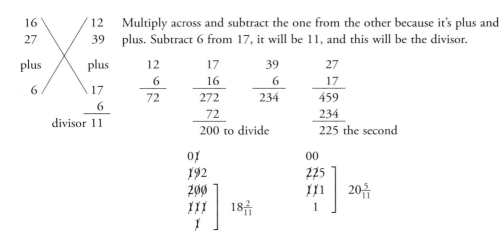

1^{co}	75^{n} men 3^{co}		
25^{n}	50^{n}	$\frac{3}{4}^{co}$	75^{n} men $2^{co}\frac{3}{4}$
			25
	0		50^{n} men $2^{co}\frac{3}{4}$
	$0\not{1}$		
	$\not{1}\not{9}2$		
	$\not{2}\not{0}\not{0}$		
	$\not{1}\not{1}\not{1}$	$18\frac{2}{11}$	$\frac{50}{1}\ \frac{11}{4}$
	$\not{1}$		

La prima $18\frac{2}{11}$. La segonda, perché ponesemo 3^{co} moltipicha' per 3 serano $54\frac{6}{11}$. A trar de 75^{n} roma[*n*] 20 e $\frac{5}{11}$.

f. 195a

[Truovar 2 homeni; "Se tu me dà 3 d'i to averò 6 vuolte chomo ti"; dixe el sigondo: "Se tu me dà 6 d'i to averò 9 tanti chomo tti"]

Sono 2 chonpagny che dixe el primo al segondo: "Se tu me da' 3 dei tto duchati averò 6 tanti chomo tti." Dixe el segondo al primo: "Se tu me da' 6 d'i tto duchati, averò 9 tanti chomo ti." Adomando quanti ne avea chadaun?

Ponamo che ala prima inpoxizio[*n*] avesse el primo chonpagno 9, el segondo avese 5. Inpresta d'i 5 al primo 3, roman chon 2. E quelo che avea 9, averà 12, che ben die 'ser 6 tanti. El segondo vuol 4 de 9, roman 5, e questo roman chon 9. Ado[*n*]cha vuol che sia 9 volte. Moltipicha' 5 fia 9 fa 45, e noi non avemo più cha 9. Adoncha serà più 36. Adoncha per 9 e 5 che me ò puxo è più 36. E questi sono dela prima inpoxizion.

Ponamo ala segonda inpoxizion el primo avesse 15, el segondo 6. Se tuoio 3 de 6 roman 3. El primo roman chon 18, che ben sono 6 volte 3 fano 18. Mo' dixe el segondo: "Dame eli 4 d'i tto 15." Roman 11. E l'oltro roman chon 10, e dixe che vuol eser 9 tanto. Adoncha 9 fia 11 fa 99, e non avemo ezetto 10. Adoncha roman più 89, siché per 15 e 6 che me ponguo è più 89. Segui l'inpoxizion. El primo $4\frac{49}{53}$, el segondo $4\frac{17}{53}$. Aprovalla, starà ben.

$1x$	75 minus $3x$		
25	50	$\frac{3}{4}x$	75 minus $2\frac{3}{4}x$
			25
		0	50 minus $2\frac{3}{4}x$

0~~1~~
~~1~~~~0~~2
~~2~~~~0~~~~0~~
~~1~~~~1~~~~1~~ | $18\frac{2}{11}$ | $\frac{50}{1}\ \frac{11}{4}$
~~1~~

The first $18\frac{2}{11}$, the second because he put $3x$ multiplied by 3 will be $54\frac{6}{11}$, subtracted from 75, leaves 20 and $\frac{5}{11}$.

FIND 2 MEN; "IF YOU GIVE ME 3 OF YOURS I WILL HAVE 6 TIMES AS MANY AS YOU"; THE SECOND SAYS: "IF YOU GIVE ME 6 OF YOURS I WILL HAVE 9 TIMES AS MANY AS YOU" f. 195a

There are 2 partners, the first of whom says to the second: "If you give me 3 of your ducats I will have 6 times as many as you." The second says to the first: "If you give me 6 of your ducats, I will have 9 times as many as you." I ask you how much did each one have?

Let's suppose that at the first false position the first companion had 9, the second had 5. He lends 3 of the 5 to the first, he remains with 2, and the one who had 9 will have 12, which should be 6 times as many. The second wants 4[1] of the 9, which leaves 5, and he remains with 9. So it should be 9 times. Multiply 5 times 9, which makes 45, and we have only 9. So it's plus 36. So for 9 and 5 which I have put, it's plus 36. And these are at the first false position.

Let's suppose at the second false position that the first had 15 and the second 6. If I take 3 from 6 it leaves 3, the first remains with 18, which is 6 times 3 makes 18. Now the second says: "Give me 4 of your 15," there remain 11 and the other remains with 10, and says that it should be 9 times as much. So 9 times 11 makes 99, and we had only 10, so there remains plus 89, so for 15 and 6 that I put it's plus 89. Follow the false position, the first has $4\frac{49}{53}$, the second $4\frac{17}{53}$. Try it; it will come out well.

1. According to the problem, the second partner wants 6 ducats from the first; from this point on, the calculation diverges from the stated problem, the illustration below, and the solution by unknown.

9		15	69	aprovalla starà ben		1	
5		6	9	16		0~~3~~0	
più		più	621	15		~~3~~~~8~~~~1~~	
			240	240		5~~3~~	$7\frac{10}{53}$
16		69	381				
	53						

69	16	3	
5	6	0~~4~~7	
249	96	~~2~~~~4~~~~9~~	
96		~~5~~~~3~~	$4\frac{37}{53}$
249			
96			

Aprovallo segondo   primo $7\frac{10}{53}$ primo $10\frac{10}{53}$
segondo $4\frac{37}{53}$

530	
10	
540	90
53	53

0
0~~2~~0
~~5~~~~6~~~~7~~
~~6~~~~3~~   9 ben 9 tanti

00
~~5~~~~4~~0
~~9~~0   6 ben 6 tanti

f. 195b E per far la ditta raxion per la chossa pony che 'l primo avesse 1^{co}. El segondo per forza chonvien aver $\frac{1}{6}^{co}$ più $3\frac{1}{2}^{n}$. El primo vuol 3 dal segondo. Roman $\frac{1}{2}$ e $\frac{1}{6}^{co}$. El primo roman chon 3^{n} 1^{co}. Ben alo 6 tanti quanto el segondo. Mo' el segondo vuol 6 dal primo per aver 9 tanti, chon romanir chon 1^{co} men 6^{n} $\frac{1}{6}^{co}$ più $9^{n}\frac{1}{2}$. Se moltipicho 1^{co} via 9 fa 9^{co}, e 9 via men 6 fa men 54^{n}. Adequa le parte. 54 e $9\frac{1}{2}$ val $63\frac{1}{2}^{n}$. A chavar de 9^{co} $\frac{1}{6}^{co}$ roman $8\frac{5}{6}^{co}$. Parti li numeri per le chosse, insirà $7\frac{20}{106}$. Perché ponessemo che 'l fusse $\frac{1}{6}^{co}$, se parto per 6 $7\frac{20}{106}$ inssirà 1 tuto e $\frac{127}{106}$. A questo azonto $\frac{1}{2}$ serano $1\frac{74}{106}$. Azonti 3 che tti ponesti $3\frac{1}{2}$, fano $4\frac{74}{106}$ e l'oltra $7\frac{20}{106}$. Aprovalla, la prima $7\frac{20}{106}$, la segonda $4\frac{74}{106}$, tu trovera' che starà ben.

1^{co} $\frac{1}{6}^{co}$ $3^{n}\frac{1}{2}$ / 1^{co} 3^{n} $\frac{1}{6}^{co}$ $\frac{1}{2}^{n}$ / 1^{co} men 6^{n} $\frac{1}{6}^{co}$ più $9^{n}\frac{1}{2}$.

9^{co} men 54^{n} $\frac{1}{6}^{co}$ più $9\frac{1}{2}$

$8\frac{5}{6}^{co}$ $63\frac{1}{2}^{n}$

$\frac{53}{6}$ $\frac{127}{2}$

020
7~~6~~~~2~~
~~1~~0~~6~~ $7\frac{20}{106}$ $4\frac{74}{106}$

9	15	69	prove it, it will	1		
5	6	9	16 come out well	0~~3~~0		
plus	plus	621	15	~~3~~~~8~~~~1~~		
		240	240	5~~3~~	$7\frac{10}{53}$	
16	69	381				
53						

69	16	3	
5	6	0~~4~~7	
249[1]	96	~~2~~~~4~~~~9~~	
96		~~5~~~~3~~	$4\frac{37}{53}$
249			
96			

Prove it second — first $7\frac{10}{53}$, second $4\frac{37}{53}$ — first $10\frac{10}{53}$

530	
10	
540	90
53	53

0		00	
0~~2~~0		~~5~~~~4~~0	
~~5~~~~6~~7		~~9~~0	6 really 6 times
~~6~~~~3~~	9 really 9 times		

And to do this problem by unknowns, put that the first had 1*x*, the second then had to have $\frac{1}{6}x$ plus $3\frac{1}{2}$. The first wants 3 from the second. It leaves $\frac{1}{2}$ and $\frac{1}{6}x$. The first remains with 3 1*x*. He has 6 times that of the second. Now the second wants 6 from the first to have 9 times, remaining with 1*x* minus 6, $\frac{1}{6}x$ plus $9\frac{1}{2}$. If I multiply 1*x* times 9 it makes 9*x*, and 9 times minus 6 makes minus 54. Equalize the sides. 54 and $9\frac{1}{2}$ are worth $63\frac{1}{2}$. Subtracting $\frac{1}{6}x$ from 9*x* leaves $8\frac{5}{6}x$. Divide the numbers by the unknowns, the result will be $7\frac{20}{106}$. Because we put that it was $\frac{1}{6}x$, if I divide $7\frac{20}{106}$ by 6 the result will be 1 whole and $\frac{127}{106}$. Adding $\frac{1}{2}$ to this will be $1\frac{74}{106}$. Adding 3 that you put $3\frac{1}{2}$, makes $4\frac{74}{106}$ and the other $7\frac{20}{106}$. To prove it, the first $7\frac{20}{106}$, the second $4\frac{74}{106}$, you will find that it will come out well. f. 195b

1*x* $\frac{1}{6}x$ $3\frac{1}{2}$ / 1*x* 3 $\frac{1}{6}x$ $\frac{1}{2}$ / 1*x* minus 6 $\frac{1}{6}x$ plus $9\frac{1}{2}$.

9*x* minus 54 $\frac{1}{6}x$ plus $9\frac{1}{2}$

$8\frac{5}{6}x$	$63\frac{1}{2}$		
$\frac{53}{6}$	$\frac{127}{2}$	020	
		7~~6~~~~2~~	
		~~1~~0~~6~~	$7\frac{20}{106}$ $4\frac{74}{106}$

1. This should be *345*.

[Truovar 2 homeni; "Se tu me dà 1 d'i to averò tanti chomo ti"; el segondo: "Se tu me dà 1 d'i tto arò tanti chomo tti"]

Sono due chonpagni che dixe el primo al segondo: "Se tu me dà 1 d'i tto duchati, avero tanti chomo tti." Dixe el segondo al primo: "Se tu me dà 1 d'i ttoi, averò 3 tanti chomo tti." Quanti ne avea chadaun di loro?

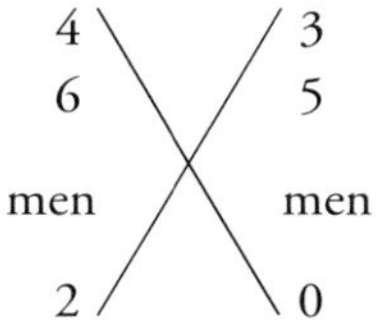

3	5	l'uno 3 e l'oltro 5
2	2	
6]	10]	
2] 3	2] 5	

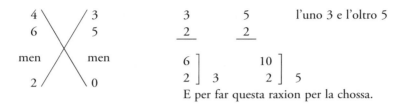

E per far questa raxion per la chossa.

E per far la ditta raxion per la chossa pony che 'l primo avesse 1^{co}. El segondo per forza die aver 1^{co} e 2^{n}, sì che el segondo dà al primo un d'i ssoi, averà tanto l'uno chomo l'oltro. E se ge lo rende e darlli un d'i ssoy, chonvien romagnir chon 1^{co} men 1^{n}, el segondo chon 3^{n} e 1^{co}. Se moltipicho 1^{co} per 3^{n} fa 3^{co}, e se moltipicho men 1^{n} per 3^{n} fa men 3^{n}. Et avemo più 3^{co} men 3^{n}, e l'oltra parte più 1^{co} più 3^{n}. A dar 3^{n} più 3^{n} men fano 6^{n}, e tuor 1^{co} più da 3^{co} più serà romaxo 2^{co} ingal a 6^{n}. Parti, serano 3 e l'oltra parte fa 5.

1^{co} 1^{co} più 2^{n} / 1^{co} men 1^{n} più 1^{co} più 1^{n} / 3^{co} men 3^{n} 1^{co} più 3^{n} / 2^{co} 6^{n} 6]
2] 3 / 5 /

f. 196a

[Truovar 2 homeni; "Se tu me dà 1 d'i toi arò tanti chomo ti"; dixe el segondo: "Se tu me dà 1 d'i tuoi averò per ognon d'i toi 100"]

Sono 2 chonpagni che 'l primo dixe al segondo: "Se tu me da' un d'i tto duchati avero tanti chomo ti." Dixe el segondo al primo: "E se tu me rendi el mio e inprestame un d'i toi, averò per ogno 1 d'i toi 100." Adomanda: "Quanti sono li denari de zaschadon?"

E per far la dita raxion per inpoxizion ponamo che fusse la prima 2 e la segondo[a] 4. Se 'l primo domanda al segondo 1, ben serano 3 chon 3. Mo' el segondo vuol dal primo el so e uno li die inprestar. Adoncha el primo roman chon un, el segondo chon 5, e sì de' aver 100, e non à più cha 5. Adoncha serano più 95. E faremo per 2 e 4 che me ò puxo e più 95, e metilla al'inpoxizion.

a. *Sic in MS.*

Find 2 men; "If you give me 1 of yours I will have as many as you"; the second: "If you give me 1 of yours I will have 3 times as many as you"

There are two partners of whom the first says to the second: "If you give me 1 of your ducats I will have as many as you." The second says to the first: "If you give me 1 of yours, I will have 3 times as many as you." How much did each of them have?

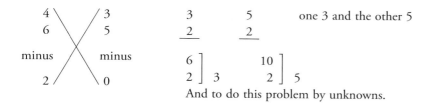

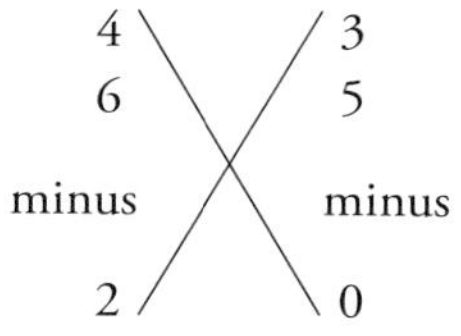

3 over 2, 5 over 2 — one 3 and the other 5

6 / 2] 3   10 / 2] 5

And to do this problem by unknowns.

And to do this problem by unknowns, put that the first had 1*x*, the second then had to have 1*x* and 2, so that if the second gives to the first one of his, one will have the same amount as the other. And if he gives it back and gives him one of his, he will have to remain with 1*x* minus 1, the second with 3 and 1*x*. If I multiply 1*x* by 3 it makes 3*x*, and if I multiply minus 1 by 3 it makes minus 3, and we have plus 3*x* minus 3 and on the other side plus 1*x* plus 3. To give 3 plus minus 3 makes 6, and to take plus 1*x* from plus 3*x* the remainder will be 2*x*, equal to 6. Divide, the result will be three and the other part makes 5.

1*x* 1*x* plus 2 / 1*x* minus 1 plus 1*x* plus 1 / 3*x* minus 3 1*x* plus 3 / 2*x* 6 6 / 2] 3 / 5 /

Find 2 men; "If you give me 1 of yours I will have as many as you"; the second says: "If you give me 1 of yours, I will have 100 for each of yours" f. 196a

There are 2 partners of whom the first says to the second: "If you give me one of your ducats, I will have as many as you." The second says to the first: "And if you return mine and lend me one of yours, I will have 100 for each 1 of yours." The question is: "How many are the ducats of each?"

And to do this problem by false position, let's suppose that the first was 2 and the second 4. If the first asks 1 from the second, it will indeed be 3 with 3. Now the second wants his own from the first and he should lend him one. So the first remains with one, the second with 5, and he should have 100, and doesn't have more than 5. So it will be plus 95. And we will do for 2 and 4 that I have put and plus 95 and put it in the false position.

E per far la segonda inpoxizion pony che fusse el primo 3, el segondo 5. El segondo dà al primo 1, serano 4 e 4. Mo' el segondo domanda al primo el so e un d'inprestedo. Adoncha el primo roman chon 2, el segondo chon 6, e sì vuol el segondo aver d'ogno 1 zento. Adoncha per 2 vuol aver 200, e chostui non à più cha 6. Adoncha per 3 e 5 che m'ò poxo serano più 194, e questo metti ala segonda inpoxizion, chomo vedirì qui de sotto. Moltipicha in cruxe e sutrazi perchè xe più e più, e chosì la moltipichazion chomo dele oltre avanti avemo fatto.

2 ╲ ╱ 3
4 ╳ 5
più ╱ ╲ più
95 ╱ ╲ 194

194 × 2 = 388; 388 − 285 = partir 103
95 × 3 = 285
~~194 × 2 = 388~~
194 × 4 = 776; 776 − ~~285~~ = 1
95 × 5 = 475

194 − 95 = partior 99
776 − 475 = 301

0
0~~1~~4
~~1~~~~0~~~~3~~]
~~9~~~~9~~] $1\frac{4}{99}$

0
0~~3~~4
~~3~~~~0~~~~1~~]
~~9~~~~9~~] $3\frac{4}{99}$ $1\frac{4}{99}$

$4\frac{4}{99}$
$4\frac{00}{99}$ $\frac{4}{99}$**a**

E questo è vero.

$\frac{4}{99}$ $4\frac{4}{99}$
0
~~4~~00]
~~4~~44] 100

f. 196b E per far la ditta raxion per la chosa pony che 'l primo avese $\frac{1^{co}}{1}$. El segondo chonvien aver 1^{co} più 2^{n}. Adoncha el primo tuo' un numero, averà tanto chomo el segondo, e si lo rende indriedo e dalli un d'i soi roman chon 1^{co} men 1^{n} $\frac{1^{co}}{1}$ più 3^{n}. Se moltipicho per 100 1^{co} fa 100^{co}, e se moltipicho 1^{n} per 100 fa men 100^{n} 1^{co} più 3^{n}. Adequa le parte. 1^{co} più de 100^{co} più roman 99^{co}, e si demo 3^{n} più a 100^{n} men fano 103^{n} ingual chon 99^{co}. Parti li numeri**b** per lie**c** chosse, serà la prima parte $1\frac{4}{99}$ e la segonda per forza $3\frac{4}{99}$. Aprovala. Da' un d'i 3, ara' tanti. Tuo' un d'i ssoi, roman chon 0 de numero ezetto $\frac{4}{99}$. El segondo roman chon $4\frac{4}{99}$, che sono ben 100 tanti chomo el primo.

---

a. $\frac{4}{99}$ *corrected over* $\frac{40}{99}$.
b. numeri *corrected over* numero.
c. *Sic in MS.*

And to do the second false position, put that the first was 3, the second 5. The second gives 1 to the first, it will be 4 and 4. Now the second asks his own and one on loan from the first. So the first remains with 2, the second with 6, and the second wants to have 1 hundred for each. So for 2 he wants to have 200, and he doesn't have more than 6. So for 3 and 5 that I have put it will be plus 194, and put this at the second false position, as you will see below. Multiply across and subtract because it's plus and plus, and thus the multiplication as we have done it the other times.

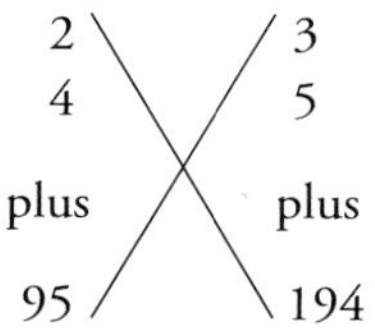

194	95	194	194	95
2	3	2	4	5
388	285	388	776	475
285			285	
to divide 103			1	

194
95
divisor 99

776
475
301

0
014
103
99] $1\frac{4}{99}$

0
034
301
99] $3\frac{4}{99}$ $1\frac{4}{99}$

$4\frac{4}{99}$
$4\frac{00}{99}$ $\frac{4}{99}$

And this is true.

$\frac{4}{99}$ $4\frac{4}{99}$
0
400
444] 100

And to do this problem by unknowns, put that the first had $\frac{1}{1}x$, the second has to have $1x$ plus 2. So the first takes a number, he will have as many as the second, and if he gives it back and gives him one of his, they remain with $1x$ minus 1, $\frac{1}{1}x$ plus 3. If I multiply $1x$ by 100 it makes $100x$, and if I multiply 1 by 100 it makes minus 100, $1x$ plus 3. Equalize the sides. Plus $1x$ from plus $100x$ leaves $99x$, and if we give plus 3 to minus 100 it makes 103, equal to $99x$. Divide the numbers by the unknowns, the first part will be $1\frac{4}{99}$ and the second therefore $3\frac{4}{99}$. Prove it. From one of the 3 he will have so much. Take one from his, he remains with 0 number except for $\frac{4}{99}$. The second remains with $4\frac{4}{99}$, which is indeed 100 as much as the first. f. 196b

1^{co} 1^{co} più 2^{n} / 1^{co} più 1^{n} 1^{co} più 1^{n} / 1^{co} men 1^{n} 1^{co} più 3^{n}

1
0~~1~~4
~~1~~~~0~~~~3~~
~~9~~~~9~~] $1\frac{4}{99}$ $3\frac{4}{99}$

100
~~100^{co}~~ men ~~100^{n}~~ 1^{co} più ~~3^{n}~~
99^{co} ingual men 103^{n} $\frac{4}{99}$ $4\frac{4}{99}$
$\frac{4}{99}$ $\frac{400}{99}$

00
~~4~~00]
~~4~~~~4~~4] 100

[VEDER TONDI]

E ll'è uno tundo el qual la so zirchundanzia intorno sie passa 44. Adimando quando[a] este lo so diamitro. Dichotte el tondo este chiamatto chavo de sesto, el qual voltando dal mezo al zercho die 'ser 6 volte e partise per $3\frac{1}{7}$, el qual sie $\frac{22}{7}$ $\frac{44}{1}$. In questo mudo parti, moltipicha 7 via 44, val 308, e questi parti per 22, fano 14. Adoncha passa 14 serà el diamitro. E per sì fatto muodo fara' ogn'altra simille raxion. Vedirì qui de sotto.

f. 197a E questa sie raxion de far d'un quadro un tundo. Diremo che per quadro sia passa 5, in quadri 4, serave passa 20. Adoncha la zirchondazia del tondo che serave se die far abater de 20 el quinto. Roman 16 passa. Ma per la raxion a partir per 5 el quadro serà ben a partir per quinto, perché la zirchonstanzia volze in quatro quinti e 1 terzo dela quinta parte.

---

a. *Sic in MS, evidently for* quanto.

1x 1x plus 2 / 1x plus 1 1x plus 1 / 1x minus 1 1x plus 3

0[1]
0~~1~~4
~~1~~~~0~~~~3~~
~~9~~~~9~~ $1\frac{4}{99}$ $3\frac{4}{99}$

100
~~100x~~ minus ~~100~~ 1x plus ~~3~~
99x equals minus 103 $\frac{4}{99}$ $4\frac{4}{99}$
$\frac{4}{99}$ $\frac{400}{99}$

00
~~4~~00
~~4~~~~4~~4] 100

SEE CIRCLES

There is a circle whose circumference around is 44 paces. I ask how much its diameter is. I tell you that this circle is called the head of the sixth, which arcing from the middle to the circle should be 6 times, and divide it by $3\frac{1}{7}$, which is $\frac{22}{7}$ $\frac{44}{1}$. Divide in this way, multiply 7 times 44, it's 308, and divide this by 22, it makes 14. So the diameter will be 14 paces. And you will do all other similar problems the same way. See below here.

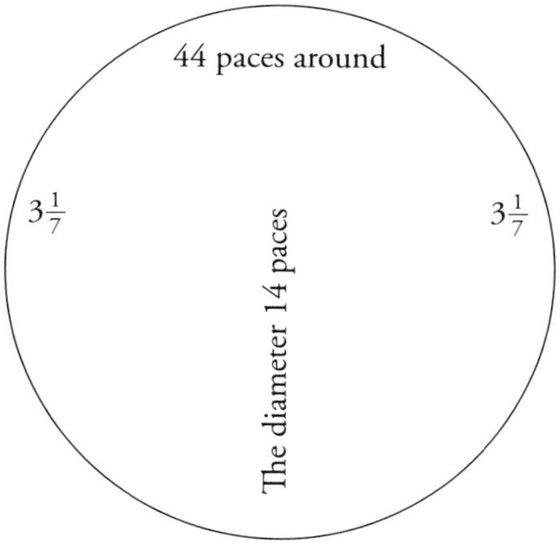

And here is the method of making a circle from a square. We will say that it is 5 paces per side, on 4 sides it will be 20 paces. So for the circumference of the circle that will be, a fifth should be subtracted from 20. It leaves 16 paces. The method for dividing the square by 5 will indeed be to divide by a fifth, because the circumference goes around four fifths and one third of the fifth part. f. 197a

1. *1* in MS.

el quadro tutto
pasa 20

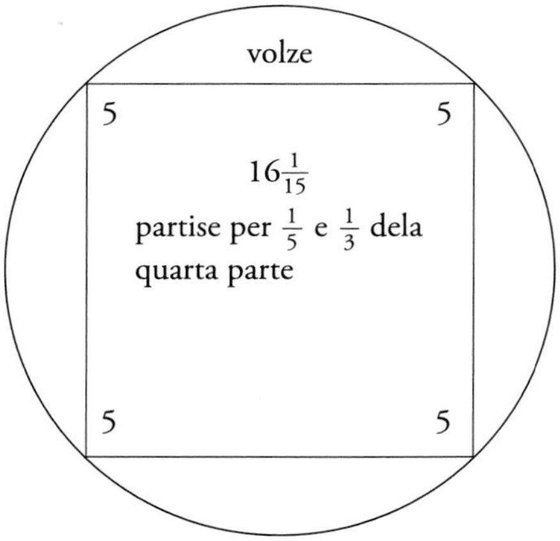

Fame questa raxion. E ò ballotti 3 de zera. L'una ballotta vuolze pie 6, e l'oltra vuolze pie 3, e l'oltra vuolze pie 2, e voione far de que 3 ballotta una. Che die volzer? Moltipicha 6 fia 6, fa 36. E 3 via 3 fa 9. E 2 via 2 fa 4. Suma, fano 49. Mo' tuo' la so radixe, serà 7. E tanto vuolzerà la ballotta.

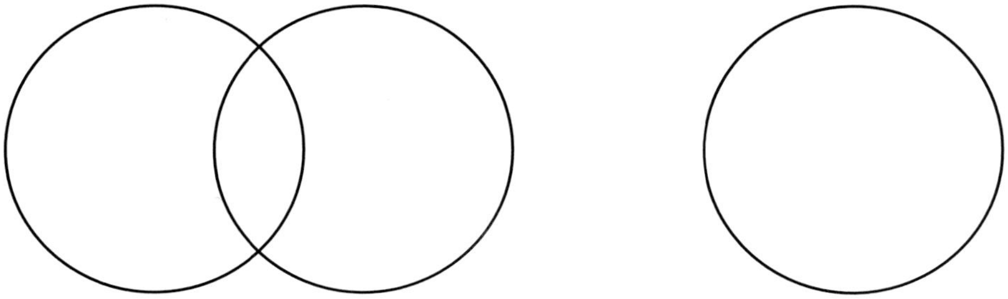

f. 197b Ell'è un marchadante el qual à fromenti de più prexii, zoè di 5 prexiy. In primo de soldi 10 el ster, el segondo de soldi 12, el tezzo[a] de soldi 13, el quarto de soldi 14, el quinto de soldi 15 el staro. E io voyo mesidar de questi 5 fromenti che vegna a valer el ster soldi $12\frac{1}{2}$. Adomando quando[b] fromento e tuorò de chadaun d'i ditti per far stera 240 che vaya el staro soldi $12\frac{1}{2}$. Fa' che tu metti li prexii preditti chomo sta qui de sotto. Chomenza a ligar 10 chon 15, e dirai: "Da 10 a $12\frac{1}{2}$ che serave?" Serave $2\frac{1}{2}$, lo qual scrivy sopra 10. Poi aliga 12 chon 14 e dirai: "Da 12 a $12\frac{1}{2}$ xe mezo." E quisto $\frac{1}{2}$ scrivy sopra 14. E poi da $12\frac{1}{2}$ inchina 14 son $1\frac{1}{2}$. E questo metera' sopra 12. Poi roman a ligar 13, lo qual allega chon 10 e dirai: "Da $12\frac{1}{2}$ a 13 sie $\frac{1}{2}$," e questo metti sopra 10.

---

a. *Sic in MS, evidently for* terzo.
b. *Sic in MS, evidently for* quanto.

The square in all
20 paces

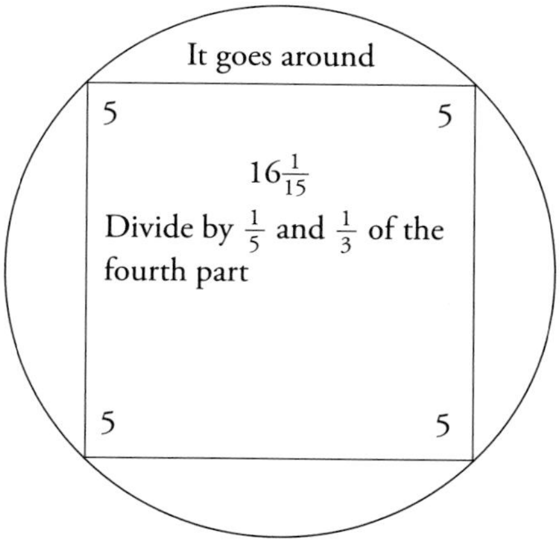

Do this problem for me. I have 3 large balls of wax, one ball is 6 feet around, and the other is 3 feet around, and the other is 2 feet around, and I want to make one large ball from these 3. How much around should it be? Multiply 6 times 6, makes 36. And 3 times 3 makes 9. And 2 times 2 makes 4. Add, it makes 49. Now take its root, it will be 7. And that's how much around the ball will be.

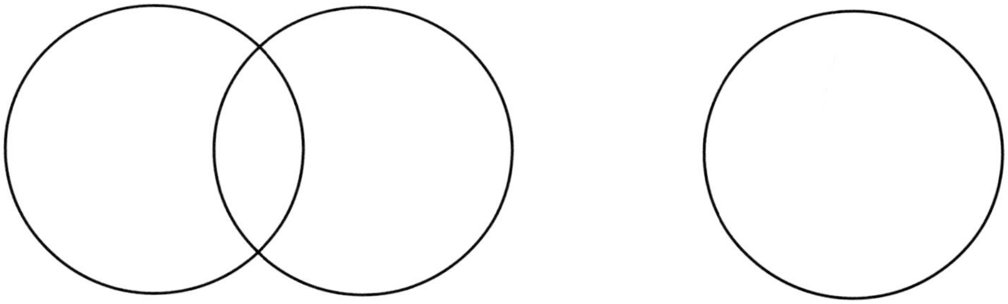

There is a merchant who has wheat at several prices, that is, at 5 prices. The first at 10 soldi per bushel, the second at 12 soldi, the third at 13 soldi, the fourth at 14 soldi, the fifth at 15 soldi per bushel. And I want to make a mixture of these 5 wheats to be worth $12\frac{1}{2}$ soldi per bushel. I ask how much grain he will take of each of these to make 240 bushels that are worth $12\frac{1}{2}$ soldi per bushel. Put that you put these prices as appears below. Begin by combining 10 with 15, and you will say: "What will it be from 10 to $12\frac{1}{2}$?" It will be $2\frac{1}{2}$, which you write above 10. Then combine 12 with 14, and you will say: "From 12 to $12\frac{1}{2}$ is one half." And write this $\frac{1}{2}$ above 14. And then from $12\frac{1}{2}$ up to 14 is $1\frac{1}{2}$. And you will put this above 12. Then the 13 remains to be combined, which is combined with 10 and you will say: "from $12\frac{1}{2}$ to 13 is $\frac{1}{2}$." And put this above 10. f. 197b

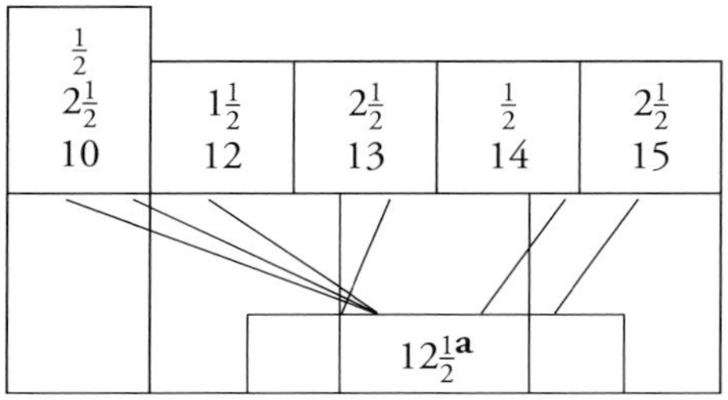

Adoncha te chonvien tuor in primo da quel de soldi 10 stera 3, e del segondo stera $1\frac{1}{2}$, e del terzo stera $2\frac{1}{2}$, e del quarto stera $\frac{1}{2}$, e del quinto stera $2\frac{1}{2}$. Mo' fa' raxion a tuor dal primo el dopio, serano
f. 198a stera 6, e del segondo chonvien stera 3, e del terzo stera 5, e del quarto stera 1, e del quinto | stera 5. Mo' porai tu far questa raxion per lo muodo dela prima chonpania, e dirai: "5 homeny ano fatto 1ª chonpania. El primo messe 6, el segondo messe 3, el terzo messe 5, el quarto messe 1, el quinto messe 5, e ssì ano vadagnado 240. Adomando zò che vien per chadaun." Tu truovera' che quel che messe 6 die aver 72, e quel che messe 3 die aver 36, e quel che messe 5 el vien 60, e quel che messe 1 die aver 12, e quel che messe 5 vuol aver[b] 60. E tanti stera tuolle de chadaun di loro. Mo' asuma 72 e 36, 60, 12, 60, fa ben 240.

Mo' sse tu voi vender questi stera 240 a soldi $12\frac{1}{2}$ lo staro, troverai tanti denari quanti valleria tute le biave chadauna per ssi, zoè da ttuti 5 sorte. E farai chussì:

Stera 72 a soldi 10 lo staro monta soldi 720.

Stera 36 a soldi 12 lo staro monta soldi 432.

Stera 60 a soldi 13 lo staro monta soldi 780.

Stera 12 a soldi 14 lo staro monta soldi 168.

Stera 60 a soldi 15 lo staro monta soldi 900.

In suma stera 240 monta soldi 3000, che fano lire 150. Mo' vendi mo' stera 240 a soldi $12\frac{1}{2}$ lo staro, truoverai che montarà soldi 3000, i qualli serano lire 150.

---

a. $12\frac{1}{2}$ *corrected over* $22\frac{1}{2}$.
b. e *in* aver *added above the line.*

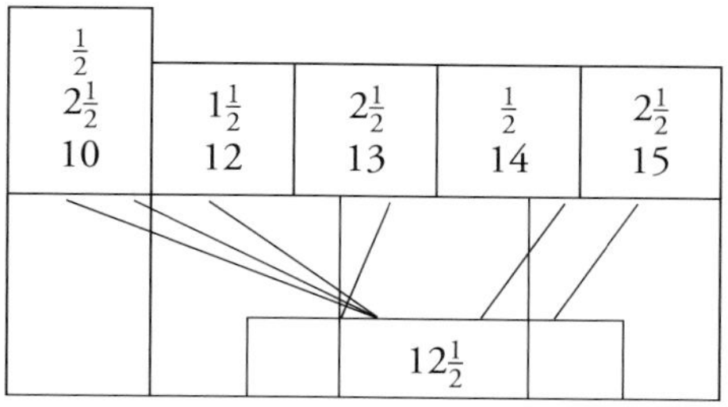

Then you should first take 3 bushels from the one of 10 soldi, and from the second $1\frac{1}{2}$ bushels, and from the third $2\frac{1}{2}$ bushels, and from the fourth $\frac{1}{2}$ bushel, and from the fifth $2\frac{1}{2}$ bushels. Now continue the problem by taking twice as much from the first, it will be 6 bushels, and from the second it works out to take 3 bushels, and from the third 5 bushels, and from the fourth 1 bushel, and
from the fifth | 5 bushels. Now you can do this problem by the method of the first company, and f. 198a
you will say: "5 men have made a company. The first put in 6, the second put in 3, the third put in 5, the fourth put in 1, the fifth put in 5, and so they have gained 240. I ask what comes to each." You will find that the one who put in 6 should have 72, and the one who put in 3 should have 36, and the one who put in 5 comes to 60, and the one who put in 1 should have 12, and the one who put in 5 should have 60. And this will be what each of them takes. Now add 72 and 36, 60, 12, 60, it indeed makes 240.

Now if you want to sell these 240 bushels at $12\frac{1}{2}$ soldi per bushel, you will find as much money as the value of each grain by itself, that is, of all 5 kinds. And you will do it this way:

72 bushels at 10 soldi per bushel amounts to 720 soldi.

36 bushels at 12 soldi per bushel amounts to 432 soldi.

60 bushels at 13 soldi per bushel amounts to 780 soldi.

12 bushels at 14 soldi per bushel amounts to 168 soldi.

60 bushels at 15 soldi per bushel amounts to 900 soldi.

In total the 240 bushels amount to 3,000 soldi, which makes 150 lire. Now you sell the 240 bushels at $12\frac{1}{2}$ soldi per bushel, you will find that it will amount to 3,000 soldi, which will be 150 lire.

f. 198b Voio ligar fromenti de più prexii. El primo val el ster soldi 9, el segondo val soldi 8, el terzo val soldi 7, el quarto val soldi 6, el quinto val soldi 5. E sì voio mesidar per far farine de questi o vender quello in questo muodo che 'l vaya el ster soldi 7. E tu alliga chomo vedirì qui de sutto. Aliga 9 chon 5, che diferenzia da 9 a 7 son 2, e questi meti sovra 5. Aliga 8 chon 6, la diferenzia da 8 a 7, 1, adoncha meti sopra 6. Aliga 7 chon 5, la diferenzia dal 7 a 7, 0, ametti sopra 5 0, e da 5 a 7, 2, metti sopra 7.

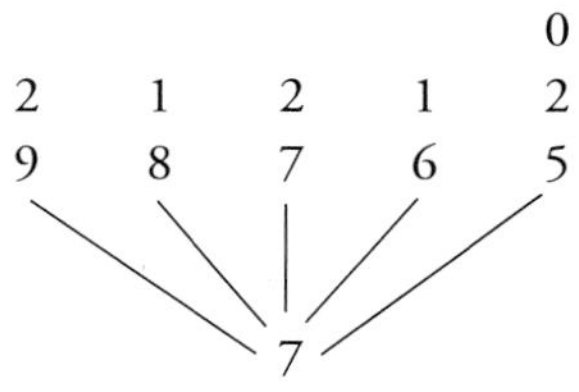

E per aprovar la ditta prendi del primo stera 2, e del segondo staro 1, e del terzo stera 2, e del quarto staro 1, del quinto stera 2. A suma serano stera 8, e die valer el staro soldi 7. Adoncha 8 stera die valer soldi 56. Aprovala. Stera 2 del primo valerà soldi 18, e stera 1 del segondo vallerà soldi 8, e stera 2 del terzo vallerà soldi 14, e staro 1 del quarto valerà soldi 6, e stera 2 del quinto valerà soldi 10. Adoncha 18, 8, 14, 6, 10, a suma serà ben 56 soldi. E questa è ben fatta.

E se per aventura volesti dir: "E voi che de questi fromenti a questo prexio vorave che fusse stera 20, e se non voyo né più né men, quanti stera voio de chadaun de questi a questi prexiy?" E per far la ditta raxion adopia lo ster de chadaun fromento. El primo, stera 2 serà 4. El segondo, 1 ster serà 2. El terzo, 2 stera serà 4. El quarto, 1 ster serà 2 stera. El quinto, 2 stera serà 4 stera. E de questi a suma serano stera 16. E tu dirai per la chonpanya sono 5 chonpany ch'ano fatto 1ª chonpanya. El primo messe 4, el segondo 2, el terzo 4, el quarto 2, el quinto 4. E sì ano vadagnado 20, che die 'ser per parte? Moltipicha e parti. El to partior serà 16 e dirai: "Se 16 me dà 20, che me darà 4 che fu del primo?" Serà 5, el segondo $2\frac{1}{2}$, el terzo 5, el quarto $2\frac{1}{2}$, el quinto 5. A suma tuti questa stera valerà soldi 140. A moltipichar 7 via 20 val 140.

f. 199a Fame de 2 do parte che partida l'una per l'oltra faza o ne vegna 100. Pony che 'l[a] prima avese 1^{co} e l'oltra parte fusse 2^{n} men 1^{co}. Se moltipicho 1^{co} via 100 fa 100^{co}, che son ingual a 2^{n} men 1^{co}. Adequa le parte. Men 1^{co} e più 100^{co} serano 101^{co}, ingual a 2^{n}, che roman chusì $\frac{2}{101}$. Inquina 2, resta $1\frac{99}{101}$, chomo serà chomo qui de ssotto vedirì.

---

a. *Sic in MS.*

I want to combine wheat of several prices. The first is worth 9 soldi per bushel, the second is worth f. 198b
8 soldi, the third is worth 7 soldi, the fourth is worth 6 soldi, the fifth is worth 5 soldi. And I want to mix them to make flour, and[1] sell it in such a way that it goes for 7 soldi per bushel. And you combine as you will see below. Combine 9 with 5, of which the difference from 9 to 7 is 2, and put this above 5. Combine 8 with 6, the difference from 8 to 7 is 1, so put it over 6. Combine 7 with 5, the difference from 7 to 7, 0, put 0 over 5, and from 5 to 7, 2, put it above 7.

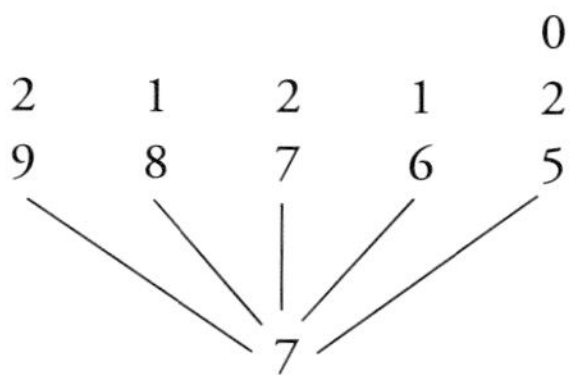

And to prove it, take 2 bushels of the first, and 1 bushel of the second, and 2 bushels of the third, and 1 bushel of the fourth, 2 bushels of the fifth. Adding together they will be 8 bushels, and it should be worth 7 soldi per bushel. So 8 bushels should be worth 56 soldi. Prove it. 2 bushels of the first will be worth 18 soldi, and 1 bushel of the second will be worth 8 soldi, and 2 bushels of the third will be worth 14 soldi, and 1 bushel of the fourth will be worth 6 soldi, and 2 bushels of the fifth will be worth 10 soldi. So 18, 8, 14, 6, 10 added together will indeed be 56 soldi, and this is done well.

And if by chance you wanted to say: "And I want there to have been 20 bushels of this wheat at this price, and if I don't want either more or less, how many bushels do I want of each at these prices?" And to do this problem, double the bushels of each wheat. The first, 2 bushels will be 4. The second, 1 bushel will be 2. The third, 2 bushels will be 4. The fourth, 1 bushel will be 2 bushels. The fifth, 2 bushels will be 4 bushels. And the sum of these will be 16 bushels. And you will say that for the company there are 5 partners who have made a company. The first put in 4, the second 2, the third 4, the fourth 2, the fifth 4. And if they gained 20, what should each part be? Multiply and divide. Your divisor will be 16, and you will say: "If 16 gives me 20, what will 4 give me, which was that of the first?" It will be 5, the second $2\frac{1}{2}$, the third 5, the fourth $2\frac{1}{2}$, the fifth 5. Added together all of these bushels will be worth 140. 7 Multiplied by 20 is worth 140.

Find me two parts of 2 such that one divided by the other makes or comes to 100. Put that the first f. 199a
had $1x$ and the other part was 2 minus $1x$. If I multiply $1x$ by 100 it makes $100x$, which is equal to 2 minus $1x$. Equalize the sides. Minus $1x$ and plus $100x$ will be $101x$, equal to 2, which leaves $\frac{2}{101}$. From 2, it leaves $1\frac{99}{101}$, as you will see below.

1. *Or* in MS.

$\frac{2}{101}$ $1^{a}\frac{99}{101}$

$\frac{2}{101} \times \frac{200}{101}$

00
200
222] 100   Son ben 100.

Truovame un numero che moltipichado per lo so $\frac{1}{2}$ e partido per lo so $\frac{1}{3}$ faza 18. Pony che 'l numero fusse 1^{co}. Se moltipicho $\frac{1}{2}^{co}$ farà 1?/2, e si parti 1?/2 per $\frac{1}{3}^{co}$ fano $\frac{3}{2}^{co}$, che sono ingualle a 18^{n}. Adoncha fara' in croxe $\frac{3}{2}\,\frac{18}{1}$. 2 fia 18 fa 36. A partir per 3 fano 12. Aprovallo. La mitade de 12 6. 6 fia 12 fa 72. A partir per lo $\frac{1}{3}$, che sono 4, 4 fia 18 fa 72.

$\frac{1}{1}^{co}$ $\frac{1}{2}^{co}$ $\frac{1}{2}^{\square}$ $\frac{1}{3}$ $\frac{3}{2}^{co}$ $\frac{18}{1}$

00
36
33] 12

12
6
72

$\frac{12}{4}$

0
30
72
44] 18

f. 199b

[Veder un a chonprar pessi de 3 muodi per soldi 10]

E l'è un chi vuol chonprar pesse, e se vuol spender soldi 10 in luzi, tenche, anguille. E chonpra libre 5 de luzi, 4 libre de tenche, 3 libre d'anguille. Vero è che la tencha per buntade valleva la libra più cha luzo pizulli 3, e l'anguilla per buntade più cha lu luzo pizulli 4. E spexe soldi 10 in questi pessi. Che valse la libra delo luzo, e che la libra dela tencha, e che valse la libra del'anguilla? E per far la ditta raxion per impoxizion vedirì qui de soto.

a. 1 *corrected over* 2.

$\frac{2}{101}$ $1\frac{99}{101}$

$\frac{2}{101} \times \frac{200}{101}$ 00

$\not{2}\not{0}0$

$\not{2}\not{2}2$] 100 It is indeed 100.

Find me a number which when multiplied by its $\frac{1}{2}$ and divided by its $\frac{1}{3}$ makes 18. Put that the number was $1x$. If I multiply by $\frac{1}{2}x$, it will make $\frac{1}{2}x^2$, and if I divide $\frac{1}{2}x^2$ by $\frac{1}{3}x$ it makes $\frac{3}{2}x$, which is equal to 18. So it will make $\frac{3}{2}$ $\frac{18}{1}$ across. 2 times 18 makes 36, to divide by 3 makes 12. Prove it: half of 12 6, 6 times 12 makes 72. To divide by $\frac{1}{3}$, which is 4, 3 times 18 makes 72.

$\frac{1}{1}x$ $\frac{1}{2}x$ $\frac{1}{2}x^2$ $\frac{1}{3}$ $\frac{3}{2}x$ $\frac{18}{1}$ 00

$\not{3}\not{6}$

$\not{3}\not{3}$] 12

12 × 6 = 72

$\frac{12}{4}$ 0

$\not{3}0$

$7\not{2}$

$\not{4}\not{4}$] 18

---

f. 199b

See one buy fish of three kinds for 10 soldi

There is someone who wants to buy fish, and he wants to spend 10 soldi on pike, tench, eel. And he buys 5 pounds of pike, 4 pounds of tench, three pounds of eel. It's the case that because of its quality the tench is worth 3 piccoli per pound more than the pike, and because of its quality the eel is worth 4 piccoli per pound more than the pike. And he spends 10 soldi on these fish. What is a pound of the pike worth, and what is a pound of the tench, and what is a pound of the eel worth? And you will see below how to do this problem by false position.

10	12	10	10	10	50	
più	più	5	4	3	52	
		50	40	30	42	
24	48		12	12	144	men 24^{n}
		50	52	42	120	

480	60	168	partior 24
288	48	120	
192 a partir	12	più 48	
	36		
	12		
	168		

	40	0		
	44	0~~3~~0		
	36	~~192~~		
Ben fu. Soldi 10 fano pi[*zulli*]	120	~~24~~	8	la libra de luzo
			11	la libra de tencha
			12	la libra d'anguila

E per farla per la chossa pony	5^{co}	12^{co}	24^{n}	Ingual 120^{n}, roman 96.
	4^{co}			
	3^{co}			

A partir per 12 serano 8. E tanto val la libra de luzo. La tencha pizulli 11, l'ang[*u*]illa pizulli 12. 5 fia 8, 40. 4 fia 11, 44. 3 fia 12, 36. Fa ben soldi 10, che sono pezulli 120. Et è fatta.

f. 200a E l'è un chi vuol spender soldi 10, che sono pizulli 120, in pese, in luzi, tenche, anguille. E chonpra libre 5 de luzi, tenche livre 4, e valse più cha lo luzo la libra pizulli 3. Anguille libre 3, e valse più cha la tencha la libra pizulli 4. E t'adimando: "Che valse la libra de luzo, e che la tencha, e che l'ang[*u*]ila?" E per farla per inposizion.

10		12	10	10	10	50	
plus	X	plus	5	4	3	52	
24		48	50	40	30	42	
				12	12	144	minus[1] 24
			50	52	42	120	

480	60	168	divisor 24
288	48	120	
192 to divide	12	plus 48	
	36		
	12		
	168		

	40		0		
	44		030		
	36		192		
It was done well. 10 soldi make	120	piccoli.	24	8	a pound of pike
				11	a pound of tench
				12	a pound of eel

And to do it by unknowns, put	$5x$	$12x$	24	Equal to 120, leaves 96.
	$4x$			
	$3x$			

To divide by 12 will be 8. And that is the value of a pound of pike. Tench 11 piccoli, eel 12 piccoli. 5 times 8, 40. 4 times 11, 44. 3 times 12, 36. It indeed makes 10 soldi, which is 120 piccoli. And it's done.

And there is someone who wants to spend 10 soldi, which is 120 piccoli, on fish, on pike, tench, and eel. And he buys 5 pounds of pike, 4 pounds of tench, which is worth 3 piccoli more per pound than the pike, 3 pounds of eel, which is worth 4 piccoli more per pound than the tench. And I ask you: "What is a pound of pike worth, and what is the tench, and what is the eel?" And to do it by false position. f. 200a

1. This should be *plus*.

10	12	50	60
più	più	40	48
		12	12
33	57	51	57
12	10	153	177
		120 più 33	120 più 57

570	57	0
396	33	0~~3~~6
174 a partir	24 partior	~~1~~7~~4~~
		~~2~~~~4~~] $7\frac{1}{4}$ luzo
		$10\frac{1}{4}$ tencha
		$14\frac{1}{4}$ l'anguila

la libra	7	luzo	36	$\frac{1}{4}$
	10	tencha	41	0
	14	ang[*u*]ila	42	$\frac{3}{4}$

E per far la ditta raxion per la chossa pony che fusse 1^{co}.

5^{co}	12^{co}	12^{co}	33^{n}		120^{n}	0
4^{co}	21^{n}		12^{co}	partior	33	~~1~~3
3^{co}					87 a partir	~~8~~~~7~~
						~~1~~~~2~~] $7\frac{1}{4}$

Et è fatta chomo ditto de sovra.

f. 200b [VEDER 3 CHONPAGNI A CHONPRAR 1ª ZOYA $\frac{1}{2}$, $\frac{2}{3}$, $\frac{3}{4}$]

Sono 3 homeny chi vuol chonprar una zoya. El primo dixe ai altri 2: "Se me dè la mitade d'i vostri apreso i mie, averò tanto che chonprerò la zoya." Dixe el segondo ai altri 2: "Se me dè i $\frac{2}{3}$ d'i vostri apresso i mie averò tanti che chonprerò la zoya." Dixe el terzo ai altri 2: "Se me dè i $\frac{3}{4}$ d'i vostri apreso i mie averò tanti che chonprarò la zoya." Sì veramente che 'l primo voio che 'l abia duchati $9\frac{1}{2}$ più cha 'l segondo, el segondo abia duchati $9\frac{1}{2}$ più cha'l terzo. Adomando: "Che valeva la zoya? e quanti fono le denari de chadaun di loro?" Per farla per la riegola del 3 ponamo che 'l terzo avesse 1, el segondo avese 2 più cha'l primo, serano 3, el primo più 2 de 3, serano 5. Questa seria fata. Ma perché dixe che vuol aver el primo 5, el segondo 3, el terzo 1, adoncha lo primo 5, el segondo el terzo 4, la mitade sono 2[a] e 5 val 7, e tanto valeria la zoya. El segondo $\frac{2}{3}$ del primo e del terzo, el primo 5, el terzo 1, sono 6. $\frac{2}{3}$ de 6 sono 4, e 3 fano 7. El terzo 1, el primo 5, el segondo 3, fano 8; $\frac{3}{4}$ de 8 sono 6, e uno fa ben 7. E tanto valeria la zoya.

Ma perché disse che 'l primo vuol duchati $9\frac{1}{2}$ più cha 'l segondo, noi diremo: "Se 2 me torna 1, che serave $9\frac{1}{2}$?" Serave $4\frac{3}{4}$, e tanto avè el terzo. Azonzi più $9\frac{1}{2}$, serano $14\frac{1}{4}$, e tanto avè el segondo. Azonzi anchora a $14\frac{1}{4}$ $9\frac{1}{2}$, serano $23\frac{3}{4}$, e tanto avea el primo.

---

a. e 7 *crossed out with a horizontal line.*

10 ╲ ╱ 12	50	60	
plus ╳ plus	40	48	
	12	12	
33 ╱ ╲ 57	51	57	
12 ╱ ╲ 10	153	177	
	120 plus 33	120 plus 57	

570	57	0	
396	33	036	
174 to divide	24 divisor	174	
		24	$7\frac{1}{4}$ pike
			$10\frac{1}{4}$ tench
			$14\frac{1}{4}$ eel

per pound	7	pike	36	$\frac{1}{4}$
	10	tench	41	0
	14	eel	42	$\frac{3}{4}$

And to do this problem by unknowns, put that it was $1x$.

					0	
					13	
$5x$	$12x$	$12x$	33	120	87	
$4x$	21		$12x$ divisor	33	12	$7\frac{1}{4}$
$3x$				87 to divide		

And it's done as said above.

SEE THREE PARTNERS BUY A JEWEL, $\frac{1}{2}$, $\frac{2}{3}$, $\frac{3}{4}$ f. 200b

There are 3 men who want to buy a jewel. The first says to the other 2: "If you give me half of yours together with mine, I will have enough to buy the jewel." The second says to the other 2: "If you give me $\frac{2}{3}$ of yours together with mine, I will have enough to buy the jewel." The third says to the other 2: "If you give me $\frac{3}{4}$ of yours along with mine, I will have enough to buy the jewel." Also I want it to be the case that the first has $9\frac{1}{2}$ ducats more than the second, and the second has $9\frac{1}{2}$ ducats more than the third. I ask: "What was the jewel worth? And how much money did each of them have?" To do this by the rule of 3, we put that the third had 1; the second had 2 more than the first, which will be 3; the first 2 more than 3, which will be 5. This would be done. But because it said that the first should have 5, the second 3, the third 1, so the first 5, the second the third 4, half is 2, and 5 is worth 7, and that would be the value of the jewel. The second $\frac{2}{3}$ of the first and of the third; the first 5, the third 1, is 6. $\frac{2}{3}$ of 6 is 4, and 3 makes 7. The third 1, the first 5, the second 3, which makes 8; $\frac{3}{4}$ of 8 is 6, and one indeed makes 7. And that would be the value of the jewel.

But because it says that the first should be $9\frac{1}{2}$ ducats more than the second, we will say: "If 2 gives me 1, what would $9\frac{1}{2}$ be?" It would be $4\frac{3}{4}$, and that's how much the third had. Add plus $9\frac{1}{2}$, it will be $14\frac{1}{4}$, and that's how much the second had. Add $9\frac{1}{2}$ again to $14\frac{1}{4}$, it will be $23\frac{3}{4}$, and the first had that much.

Mo' per veder zò che vallea la zoya, se 'l primo dise che 'l voleva la mitade del segondo e del terzo,
f. 201a che sono $14\frac{1}{4}$ $4\frac{3}{4}$ fano 19, la mitade serano $9\frac{1}{2}$. Azonzi a $23\frac{3}{4}$ serano $33\frac{1}{4}$, e tanto valse | la zoya. Mo' dixe el segondo che vuol $\frac{2}{3}$ del terzo e del primo, azonti ai so' averà ben $33\frac{1}{4}$. Mo' dixe el terzo ai altri 2 che vuol $\frac{3}{4}$ d'i soi, azonti ay sso', ben à lo $33\frac{1}{4}$. Adoncha la zoya valeva duchati $33\frac{1}{4}$. El primo avea duchati $23\frac{3}{4}$, el segondo $14\frac{1}{4}$, el terzo $4\frac{3}{4}$. È fatta.

Sono 3 chonpagny che dixe el primo al segondo: "Se vuy el segondo me dè la mitade deli so apresso i mie e chonprarò una zoya, la qual se vende,[a] e ssì voyo el terzo del terzo." Dixe el segondo al terzo: "Se tu me dé $\frac{1}{4}$ d'i ttoy e del primo el quinto d'i sso, chonprarò la zoya." Dixe el terzo al primo che vuol el so sesto e dal segondo el setimo, che averà tanti denari apresso i so che chonprarà la zoya. E v'adomando quanti funo li denari de zaschadun e zò che valse la zoya. E per far la ditta raxion:

El primo avè duchati 30, el segondo duchati $59\frac{29}{51}$, il terzo duchati $69\frac{21}{51}$, e tanti funo li denari avudo e perhomo. E si moltipichi e parti e sutrazi e azonzi chomo dixe la riegola, inssirà fuora duchati $82\frac{47}{51}$. E questo valse la zoya.

E per un'altra raxion senza rutti, primo 1530, segondo 3038, terzo 3540, la zoya $82\frac{47}{51}$.

E per un'altra, primo 3060, segondo 6076, terzo 7080; $82\frac{47}{51}$.

E per la terza, primo 4590, segondo 9114, terzo 10620. $82\frac{47}{51}$.

f. 201b Sono 15 cristiany e sì sono 15 pagany, e per fortuna inpromette a gettar d'ogni 10 uno in aqua. E fu chonzadi in un ballo, e senpre getando d'ogni 10 l'uno sì fu chonzado el ballo che tutti li pagani andò in aqua e tuti i cristiani romaxe, chomo vedirì qui de sotto. Primo 2 christiany, segondo una[b] pagano, terzo 3 cristiany, 5 pagany, 2 cristiany, 2 pagany, 4 cristiany, 1 pagan, 1 cristian, 3 pagany, 1 cristian, 2 pagany, 2 cristiany, 1 pagan.

f. 202a [c]

f. 202b Una gallia del sesto de Fiandria vorà legname per far quella. In primo stortamy, zoè legny storti per far furchamy e chorbe e me[d] e mezi legny, sì a puope chomo a proda. Vuol legny 380.

E vuol ancuor legny de ruver, zoè dretti, 140, per far cholonba, paraschuxulle, madieri di bucha, latte, chorde, paramezalli, verzene, chaxelle del'alboro, paraschuxulle,[e] chadene de barcharezo, battalli, bachalari. E i ditty legny vuol eser de longeza de pie 24 o 26, e die volzer de groseza pie 4 zaschadun.

---

a. per duchati non sso quanti. Dixe el sego *crossed out with a horizontal line.*

b. *Thus in MS.*

c. *Drawing of St. Christopher takes entire page.*

d. *Thus repeated in MS.*

e. *Thus repeated in MS.*

Now to see how much the jewel was worth, if the first said that he wanted half that of the second and of the third, that's $14\frac{1}{4}$ and $4\frac{3}{4}$ makes 19, half will be $9\frac{1}{2}$. Add it to $23\frac{3}{4}$, it will be $33\frac{1}{4}$, and that is the value | of the jewel. Now the second says that he wants $\frac{2}{3}$ of the third and of the first, added to his, he will indeed have $33\frac{1}{4}$. Now the third says to the other 2 that he wants $\frac{3}{4}$ of theirs. Added to his, it's indeed $33\frac{1}{4}$. So the jewel was worth $33\frac{1}{4}$ ducats, the first had $23\frac{3}{4}$ ducats, the second $14\frac{1}{4}$, the third $4\frac{3}{4}$. It's done. f. 201a

There are 3 partners of whom the first says to the second: "If you the second give me half of yours together with mine, we will buy a jewel which is being sold, and I also want a third of the third." The second says to the third: "If you give me $\frac{1}{4}$ of yours and from the first a fifth of his, I will buy the jewel." The third says to the first that he wants a sixth of his and a seventh from the second, and he will have enough money along with his own that he will buy the jewel. And I ask you how much was the money of each and what the jewel was worth. And to do this problem:

The first had 30 ducats, the second $59\frac{29}{51}$ ducats, the third $69\frac{21}{51}$ ducats, and that's how many ducats were had per man. And if you multiply and divide and subtract and add as the rule says, the result will be $82\frac{47}{51}$ ducats. And that is how much the jewel was worth.

And for another problem without fractions, first 1,530, second 3,038, third 3,540, the jewel $82\frac{47}{51}$.

And for another, the first 3,060, the second 6,076, the third 7,080; $82\frac{47}{51}$.

And for the third, first 4,590, second 9,114, third 10,620; $82\frac{47}{51}$.

There are 15 Christians and also there are 15 pagans, and by chance it is necessary to throw every 10th one in the water. And they were arranged in a circle, and always throwing one of every 10 the circle was set up so that all the pagans went into the water and all the Christians remained, as you will see below. First 2 Christians, second one pagan, third 3 Christians, 5 pagans, 2 Christians, 2 pagans, 4 Christians, 1 pagan, 1 Christian, 3 pagans, 1 Christian, 2 pagans, 2 Christians, 1 pagan. f. 201b

1

f. 202a

A galley of the Flanders design will need timber to build it, first the compass timbers, that is, curved wood to make futtocks and floor timbers and half-floors, aft as well as forward. It will need 380 pieces of wood. f. 202b

And it also needs 140 oak pieces, that is, straight [ones], to make the keel, bilge stringers, sheer strakes, deck beams, deck stringers, keelsons, ceilings, mast steps, bilge stringers,[2] boat beams, bulwark rails, outrigger brackets. And these timbers should be of a length of 24 or 26 feet, and should each measure 4 feet around.

1. The page contains an image of Saint Christopher with no caption.
2. Thus repeated.

E vuol per serar la detta gallia madieri de rover 280 del morello grosso, zoè d'una quarta de pe'.

E vuol burdunalli 36 de larexi, a largeza de pie 1 o de un palmo, per morssade dentro, e per far postize e crussie. I qual burdunali vol eser de longeza de passa 8 l'un.

E vuol burdunalli 18 d'albeo, de passa 8, de pie 1, per far bande e sovra crussie e chastagnolle e banchi.

E vuol chiave 50 d'albeo uxevelli per far frixetti, e morti e cholomelli e punta pie e scallette e pertigette.

E vuol 300 tuole d'albeo per far li batiporta, pagnoli e seraie de sotto.

E vuol maistri segadori 500 a far el bexogno dala ditta.

E vuol maistri 1000, zoè marangony.

E vuol chalafai 1300 per furar e chalchar e pegolar.

E vuol miera 8 de feramenta, agudi, piruny, arpexi, chozulli, maschulli, axulle. E vuol pegola libre 3.000 e stopa libre 3.000.

f. 203a La natura del quinto chapitulo ssie zensso e numero ingual a chossa. El se die partir li nomeri e le chosse per li zenssi e demezar le chosse e trazer el numero, e la radixe de quelo tratto o azonto al dimezamento die valer la chossa. Esempio.

Fame de 20 2 parte che tanto faza moltipichada l'una per l'altra quanto la prima moltipichada in ssi.

E per farti chiaro de questa raxion tuti 3 ano duchati 28. El primo à duchati 8, el segondo, el terzo el romagnente, che ssono 20, ma non sso zò che chadauno dei do chonpany, che ano duchati 20. Mo' voyo saver zò che die aver el ssegondo e zò che die aver el terzo de questi 20, che moltipichada una per l'altra faza tanto quanto el primo moltipichada in ssi. El primo duchati 8, moltipichadi in ssi fano 64. E per far la ditta per la chossa:

Pony che fusse per lo primo, el segondo, el terzo 20^{n} men 1^{co}. Se moltipichemo per 1^{co} 20^{n} men 1^{co} fano 20^{co} men $1^{\square}$ 64^{n}. Demezemo le chosse, serano 10. Moltipicha, serano 100. Trazi el

And to plank up this galley there need to be 280 oak planks of thick measure, that is, of one quarter of a foot.

And it needs 36 larch planks, 1 foot or one palm wide, for mortising inboard, and to make outriggers and gangways. These planks should each be 8 paces long.

And it needs 18 fir planks, of 8 paces, of 1 foot [thick], to make *banda* stringers and side planks for the gangways and cleats and benches.

And it needs 50 planks of customary[1] fir for making waterways, and *morti*[2] and stanchions and rowing stretchers and stairs and poles.

And it needs 300 fir planks to make the hatch coamings, ceiling boards, and compartments below deck.

And it needs 500 master sawyers to meet these needs.

And it needs 1,000 masters, that is, ship carpenters.

And it needs 1,300 caulkers to drill and caulk and pay.

And it needs 8 thousand pounds of hardware—spikes, pins, clamps, cleats, pintles, rudder pintles. And it needs 3,000 pounds of pitch and 3,000 pounds of oakum.

The topic of the fifth chapter is squared unknowns and numbers equal to an unknown. The numbers and the unknowns should be divided by the squared unknowns, and halve the unknowns and subtract the number, and the root of this added or subtracted to the half should equal the unknown. Example: f. 203a

Find me 2 parts of 20 so that the first multiplied by the other is as much as the first multiplied by itself.

And to make this problem clear to you, all three have 28 ducats. The first has 8 ducats, the second and the third the remainder, which is 20, but I don't know what each of the two partners has, who have 20 ducats. Now I want to know what the second should have and what the third should have of these 20, which when multiplied by each other makes as much as the first multiplied by itself. The first 8 ducats multiplied by itself makes 64. And to do it by unknowns:

Put what was for the first, the second, the third, 20 minus $1x$. If we multiply 20 minus $1x$ by $1x$ it makes $20x$ minus $1x^2$, 64. Halve the unknowns, it will be 10. Multiply, it will be 100. Subtract the

1. I.e., standard-grade.
2. A timber of uncertain nature.

numero, che ssono 64, resta 36. E la radixe de 36 sono 6. Azonti al dimezamento serano 16. E l'oltra, parti, serano 4. Moltipicha 4 via 16, fano 64, chomo el primo moltipichado 8 via 8 fa 64. Anchor diremo: "Traze el 6 del dimezamento, che sono 10," resta 4. Molti[*pi*]cha 4 via 16, fa 64.

f. 204a[a] + Ihesus[b] +

Al nome de Dio e de madona Sancta Maria. Avy de grazia mi Michalli da Ruodo la stadiera per la nostra Segnoria adì 28 zener del 1444, e sì la diè a far a ser Stefano Negro per mi.

Questi sono i Ricordi e Scritti d'un tal Michele Daruodo Veneziano il di cui Nome si vede di sopra e f. 90.[c]

f. 205a[d] Portulan da Venesia fina a Constantinopoli pe[*r*] Rivera como le galie vano.[e]

Enesia[f] con ponta de Castegneda, se varda levante a ponente, mia 100.

Venexia con San Zane in Pielego, entro levante e siroco, mia 100.

Da San Zane in Pielego fina ala Faxana, zoè al campo da Puolla, per la staria, mia 15.

Dala Faxana infina a Puolla, mia 5.

Da Puolla a Polmontore, mia 20.

Da Polmontore al Sansego, siroco e maistro, mia 40.

Da Polmontore a Nia, entro levante e siroco, mia 15.

Da Polmontore a Venexia, entro ponente e maistro, mia 130.

Da Polmontore in Ancona, quarta d'ostro ver levante garbin,[g] mia 130.

---

a. *Fol. 203b is blank.*
b. *Repeated after the usual appearance as page heading, in the same hand and with another ink.*
c. *From* Questi sono *to* e f. 90 *in another, much later hand (late nineteenth century), in another ink.*
d. *Fol. 204b is blank except for the usual* Ihesus *at the top.*
e. *Section beginning with this page in new hand. From* Portulan *to* vano *in red ink.*
f. *Sic in MS, and elsewhere, evidently for* Venesia.
g. *Sic in MS.*

number, which is 64, there remains 36. And the root of 36 is 6; add to the half it will be 16. And divide the other, it will be 4. Multiply 4 by 16, it makes 64, as the first 8 multiplied times 8 makes 64. Again we will say: "Subtract 6 from the half, which is 10," there remains 4. Multiply 4 by 16, it makes 64.

+ Jesus[1] + f. 204a

In the name of God and of the Blessed Virgin Mary. I, Michael of Rhodes, received the steelyard by special grant from our Signoria[2] on January 28, 144[5],[3] and moreover gave it to Stefano Negro to manage on my behalf.

These are the remembrances and writings of a certain Michael of Rhodes, a Venetian, whose name appears above and on fol. 90.

Portolan from Venice to Constantinople along the coast as the galleys go. f. 205a

Venice with Point Kostanjija, bear east–west, 100 miles.[4]

Venice with Sveti Ivan na Pučini, between east and southeast, 100 miles.

From Sveti Ivan na Pučini to Fažana, that is, to the field of Pula, along the coast, 15 miles.

From Fažana to Pula, 5 miles.

From Pula to Sveti Kamenjak, 20 miles.

From Sveti Kamenjak to Sušak, southeast and northwest, 40 miles.

From Sveti Kamenjak to Unije, between east and southeast, 15 miles.

From Sveti Kamenjak to Venice, between west and northwest, 130 miles.

From Sveti Kamenjak to Ancona, south by west,[5] 130 miles.

1. In addition to the usual appearance of the invocation at the head of the page.
2. The executive organ of the governing Great Council; the term could also be used to refer to the government of Venice as a whole.
3. As the Venetian year began on March 1, the year appears as 1444 in MS.
4. For compass directions used in portolans, see p. 347, n. 8, above.
5. MS is garbled, with "south by east-west"; the actual bearing is very close to south by west.

Da Sansego a Nieme, quarta de levante ver siroco, mia 30.

Da Nieme a Zarra, dentro per l'isola, quarta de siroco ver levante, mia 60.

Da Zarra ai Levrosi, per siroco, mia 12.

Da ai Levrosi a Zara veghia, per canal, mia 6.

Da Zarra veghia a Laurana, entro per scoy, mia 12.

Da Laurana ala Vergada, entro per scoy, mia [. . .].[a]

Dala Vergada al Morter, per siroco, mia 12.

Dal Morter al Previchio, per staria, mia 15.

Dal Previchio a Sibinicho, mia 5.

Da Sibinicho a cavo Cesta, per staria, mia 12.

Da cavo Cesta a Sancto Arcanzelo, per staria, mia 12.

Da Sancto Arcanzelo a Traù, per canal, mia 10.

Da Traù a Spalato, per canal, mia 10.

Da Spalato a Liesna, e lasi la Volta[b] e la Braza a ladi, per ostro, mia 15.

Da Liesna ala Torcola, per staria, mia 12.

Dala Torcola a cavo Chumano, per siroco, mia 12.

Da chavo Chumano a Curzola, per ostro, mia 10.

f. 205b Da Curzola ala Zuliana, per staria, mia 25.

Da Zuliana a Stagno, per staria, mia 25.

---

a. *Distance lacking in MS.*
b. *Sic in MS for* Solta.
1. A small island south of Lošinj, Croatia.
2. I.e., the lazaretto or island hospice for lepers.

From Sušak to Nieme,[1] east by south, 30 miles.

From Nieme to Zadar, within the island, southeast by east, 60 miles.

From Zadar to the lepers,[2] by southeast, 12 miles.

From the lepers to Old Zadar,[3] by the channel, 6 miles.

From Old Zadar to Vrana, within the reefs, 12 miles.

From Vrana to Vrgada, within the reefs, [. . .][4] miles.

From Vrgada to Murter, southeast, 12 miles.

From Murter to Pirovac, along the coast, 15 miles.

From Pirovac to Šibenik, 5 miles.

From Šibenik to Primošten, along the coast, 12 miles.

From Primošten to Sant'Arcangelo,[5] along the coast, 12 miles.

From Sant'Arcangelo to Trogir, by the channel, 10 miles.

From Trogir to Split, by the channel, 10 miles.

From Split to Hvar, and leave Solta and Brač to the side, south, 15 miles.

From Hvar to Šćedro, by the shore, 12 miles.

From Šćedro to Cape Cumano,[6] southeast, 12 miles.

From Cape Cumano to Korčula, south, 10 miles.

From Korčula to Žuljana, along the coast, 25 miles. f. 205b

From Žuljana to Ston, by the coast, 25 miles.

---

3. Ancient settlement near Biograd na Moru.
4. Distance lacking in MS.
5. A rock off the coast of Trogir.
6. The northern tip of the Pilješac peninsula.

Da Stagno a Ombla, per canal, mia 25.

Da Ombla a Raguxi, mia 8.

Da Raguxi vechio a Raguxi, per staria, mia 8.

Da Raguxi vechio a Malonto pizolo, entro levante e siroco, mia 12.

Da Malonto pizolo a Malonto grando, per staria, mia 6.

Da Malonto grando a Sancta Maria in Ruoxa, per staria, mia 6.

Da Sancta Maria in Ruoxa a Buda, per staria, mia 30.

Da Buda Intivari, per staria, mia 10.

Da Intivari a Dolcegno, per staria, mia 10.

Da Dolcegno a Medona, per staria, mia 18.

Da Medona ai Palli, entro ostro e siroco, mia 50.

Dai Palli a Durazo, per staria, mia 20.

Da Durazo al cavo delle Mellie, per staria, mia 18.

Dale Mellie ali Cavali, per staria, mia 25.

Dali Cuali a Cauconi, per staria, mia 18.

Dali Cuali a Sasno, ostro e tramontana, mia 80.

Da Durazo a Sasno, ostro e tramontana, mia 80.

Da Sasno al Fanu, quarta de ostro ver siroco, mia 60.

Da Sasno al cavo de Otrarito,[a] quarta de africo[b] ver ponente, mia 90.

Da Sasno ala Gramata, per staria, mia 15.

---

a. *Sic in MS.*

b. *Abbreviated by* a *in MS, the same as* garbin.

From Ston to the Ombla, by the channel, 25 miles.

From the Ombla to Dubrovnik, 8 miles.

From Cavtat to Dubrovnik, along the coast, 8 miles.

From Cavtat to Mali Molunat, between east and southeast, 12 miles.

From Mali Molunat to Veliki Molunat, along the coast, 6 miles.

From Veliki Molunat to Rose, along the coast, 6 miles.

From Rose to Budva, along the coast, 30 miles.

From Budva to Bar, along the coast, 10 miles.

From Bar to Ulcinj, by the coast, 10 miles.

From Ulcinj to Shëngjin, along the coast, 18 miles.

From Shëngjin to Kep i Palit, between south and southeast, 50 miles.

From Kep i Palit to Durrës, along the coast, 20 miles.

From Durrës to Kep i Lagit, along the coast, 18 miles.

From Kep i Lagit to I Cavalli,[1] along the coast, 25 miles.

From I Cavalli to Cauconi,[2] along the coast, 18 miles.

From I Cavalli to Sazan, south and north, 80 miles.

From Durrës to Sazan, south and north, 80 miles.

From Sazan to Fanu,[3] south by east, 60 miles.

From Sazan to the Cape of Otranto, southwest by west, 90 miles.

From Sazan to the Gramata,[4] along the coast, 15 miles.

---

1. Uncertain location.
2. Uncertain location.
3. Islet north of Corfu.
4. River mouth near Himarë.

Dala Grama[a] a Palormo, per siroco, mia 25.

Da Palormo al Fanu, per africo, mia 40.

Da Palormo al castelo de Rixeto, entro levante e siroco, mia 15.

Da Rixeto a Sancti Quaranta, per levante, mia 6.

Da Sancti Quaranta al Butintro, per ostro, mia 6.

Dal Butintro al cavo del'isola de Corfu, mia 30.

Dal cavo del'ixola a Chaxoppo, per levante, mia 15.

f. 206a Da Caxoppo a Corfu, per canal, mia 15.

Da Corfu al'altro cavo del'isola, mia 25.

Da Corfu a Civita, entro levante e siroco, mia 30.

Dal dito cavo de Corfu al Pachasu, quarta de siroco ver ostro, mia 12.

Dal Velechi al Ducatu, quarta de ostro ver siroco, mia 70.

Dal cavo del Ducato a Viscardo, mia 15.

Da Viscardo a Claranza, entro levante e siroco, mia 60.

Da Clarenza a Pruode, entro ostro e siroco, mia 90.

Da Prode a Sapientia, per siroco, mia 18.

Da Modon a San Veniedego, quarta de siroco ver levante, mia 15.

Da San Veniedego a cavo Malio Matapan, quarta de levante ver [...],[b] mia 60.

Da Malio Matapan da[c] cavo San Anzolo, quarta de levante ver griego, mia 67.

Da cavo Malio a San Zorzi d'Alboro, entro griego e tramontana, mia 120.

---

a. *Sic in MS.*
b. *Direction lacking in MS.*
c. *Sic in MS.*

From the Gramata to Liman i Panormit, by southeast, 25 miles.

From Liman i Panormit to Fanu, by southwest, 40 miles.

From Liman i Panormit to the castle of Rixeto,[1] between east and southeast, 15 miles.

From Rixeto to Sarandë, by east, 6 miles.

From Sarandë to Butrint, by south, 6 miles.

From Butrint to the end of the island of Corfu, 30 miles.

From the end of the island to Kassiópi, by east, 15 miles.

From Kassiópi to Corfu, by the channel, 15 miles. f. 206a

From Corfu to the other end of the island, 25 miles.

From Corfu to Sivota, between east and southeast, 30 miles.

From this cape of Corfu to Páxos, southeast by south, 12 miles.

From Vigla to Cape Doukáton, south by east, 70 miles.

From Cape Doukáton to Fiskárdho, 15 miles.

From Fiskárdho to Killíni, between east and southeast, 60 miles.

From Killíni to Próti, between south and southeast, 90 miles.

From Próti to Sapienza, southeast, 18 miles.

From Methóni to Venétiko, southeast by east, 15 miles.

From Venétiko to Cape Matapan, east by [. . .],[2] 60 miles.

From Cape Matapan to Cape Maléas, east by north, 67 miles.

From Cape Maléas to Áyios Yeóryios, between northeast and north, 120 miles.

---

1. Uncertain location.
2. Direction lacking in MS.

Dal cavo San Zorzi d'Albara al cavo del'isola de Negroponte, entro griego e tramontana, mia 90.

Dal cavo del'isola de Nigroponte a Tenedo, entro griego e tramontana, mia 220.

Da Tenedo al castel de Troya, quarta de griego e tramontana, mia 18.

Dal castelo de Troya al Malito, mia 30.

Dal Malito a Garipoli, quarta de griego ver tramontana, mia 30.

Da Garipoli al cavo de Gan, quarta de griego ver levante, mia 15.

Da cavo de Gan a Pandilla, quarta de griego ver levante, mia 25.

Da Pandilla a Roisto, per tramontana, mia 10.

Da Roista a Recleta, per levante, mia 30.[a]

Da Recleta a Solombria, per levante, mia 20.

Da Salombria a Natura, entro siroco e levante, mia 20.

Da Natura a San Stefano, per levante, mia 10.

Da Constantinopoli ala bocha chi va al Mar Maior, mia 18.

f. 206b Portolan per i traversi del colpho de Venexia.[b]

Enesia[c] con Rimano, quarta de ostro ver siroco, mia 170.

Venesia con Fan, entro ostro e siroco, mia 195.

Venesia con Anchona, quarta de sirocho ver ostro, mia 210.

Polmontore con Ancona, ostro e tramontana, mia 140.

Nieme cum Anchona, ostro e tramontana, mia 140.

---

a. 30 *written over* 20.
b. *From* Portolan *to* Venexia *with red ink.*
c. *Sic in MS.*

From Cape Áyios Yeóryios to the end of the island of Negroponte, between northeast and north, 90 miles.

From the end of the island of Negroponte to Tenedos, between northeast and north, 220 miles.

From Tenedos to the castle of Troia, quarter of northeast by north, 18 miles.

From the castle of Troia to Malito,[1] 30 miles.

From Malito to Gallipoli, northeast by north, 30 miles.

From Gallipoli to the cape of Gan, northeast by east, 15 miles.

From the cape of Gan to Barbaros, northeast by east, 25 miles.

From Barbaros to Tekirdağ, north, 10 miles.

From Tekirdağ to Marmara Ereglisi, east, 30 miles.

From Marmara Ereglisi to Siliviri, east, 20 miles.

From Siliviri to Büyük Cekmece, between southeast and east, 20 miles.

From Büyük Cekmece to Santo Stefano,[2] east, 10 miles.

From Constantinople to the mouth which goes to the Black Sea, 18 miles.

Portolan for the crossings of the Gulf of Venice. f. 206b

Venice with Rimini, south by east, 170 miles.

Venice with Fano, between south and southeast, 195 miles.

Venice with Ancona, southeast by south, 210 miles.

Sveti Kamenjak with Ancona, south and north, 140 miles.

Nieme with Ancona, south and north, 140 miles.

---

1. Uncertain location.
2. A church at Yesilkoy.

Çan Pontelo cum Ancona, griego e africo, mia 140.

Mellada cum Anchona, quarta de africo ver ponente, mia 140.

La Torreta cum Ancona, entro ponente e [. . .],[a] mia 150.

La Incoronada cum Ancona, quarta de ponente ver africo, mia 160.

La Incoronada con el monte del'Agnollo, quarta d'ostro ver siroco, mia 170.

Da Izuri al monte, quarta de ostro ver siroco, mia 160.

Traù cum el monte, ostro e tramontana, mia 190.

Lissa cum el monte, ostro e tramontana, mia 100.

Liesna cum el monte, quarta de ostro ver africo, mia 50.

Curzolla cum el monte, entro ostro e africo, mia 130.

Lagusta cum el monte, quarta de africo ver ostro, mia 70.

La Mella cum el monte, quarta de africo ver ponente, mia 150.

La Mellada cum Brandiço, quarta de ostro ver siroco, mia 160.

Raguxi cum Brandizo, ostro e tramontana, mia 160.

Catharo cum Brandizo, entro ostro e africo, mia 170.

Dolcegno cum Brandizo, quarta de africo ver ostro, mia 170.

Durazo cum Brandizo, griego e africo, mia 160.

Dolcegno cum el Sasno, ostro e tramontana, mia 150.

El Sasno cum Polmontore, quarta de maistro ver ponente, mia 600.

Lagusta cum el Sasno, siroco e maistro, mia 230.

---

a. *Direction partially omitted in MS.*

Saint Punat with Ancona, northeast and southwest, 140 miles.

Molat with Ancona, southwest by west, 140 miles.

La Torreta[1] islands with Ancona, between west and [. . .],[2] 150 miles.

The Kornat islands with Ancona, west by south, 160 miles.

The Kornat islands with Monte Sant'Angelo, south by east, 170 miles.

From Zirje to the mountain, south by east, 160 miles.

Trogir with the mountain, south and north, 190 miles.

Vis with the mountain, south and north 100 miles.

Alessio with the mountain, south by west, 50 miles.

Korčula with the mountain, between south and southwest, 130 miles.

Lastovo with the mountain, southwest by south, 70 miles.

Molat with the mountain, southwest by west, 150 miles.

Molat with Brindisi, south by east, 160 miles.

Dubrovnik with Brindisi, south and north, 160 miles.

Kotor with Brindisi, between south and southwest, 170 miles.

Ulcinj with Brindisi, southwest by south, 170 miles.

Durrës with Brindisi, northeast and southwest, 160 miles.

Ulcinj with Sazan, south and north, 150 miles.

Sazan with Sveti Kamenjak, northwest by west, 600 miles.

Lastovo with Sazan, southeast and northwest, 230 miles.

---

1. Uncertain location, probably in the Kornat Islands.
2. Direction lacking in MS.

f. 207a La Melleda[a] cum el Sasno, quarta de maistro ver[b] tramontana, mia 210.

Malonto cum el Sasno, entro maistro e tramontana, mia 170.

Lagusta cum Otranto, quarta de siroco ver ostro, mia 250.

La Melleda cum Otranto, entro ostro ver siroco, mia 200.

Raguxi cum Otranto, quarta de ostro ver siroco, mia 200.

Cataro cum Otranto, ostro e tramontana, mia 140.

Dolcegno cum Otranto, quarta de ostro ver africo, mia 170.

Venesia[c] con Monopoli, siroco e maistro, mia 600.

San Andrea da Lisa con San intro Pelego, siroco e maistro, mia 250.

Otranto cum Casopo, entro levante e siroco, mia 80.

Otranto cum el cavo de San Sidero, siroco e maistro, mia 200.

Otranto cum el Fanu, quarta de siroco ver levante, mia 60.

Portolan da cavo Malio fina al'isola de Famagosta.[d]

A[e] Sant'Ançolo de Malio con Melo, quarta de siroco ver levante, mia 80.

Mello con Pellicandro, quarta de levante ver siroco, mia 20.

Pollicandro cum Nio, quarta de levante ver siroco, mia 30.

Nio con Santorini, entro ostro e siroco, mia 15.

Sanctorini cum Namfio, entro griego e levante, mia 15.

Namfio cum Stampalia, entro griego e levante, mia 30.

---

a. La Melleda *corrected over* La Mellada.
b. ver *corrected over* e.
c. Venesia *with initial* V *in red ink.*
d. *From* Portolan *to* Famagosta *in red ink.*
e. *In red ink.*

Molat with Sazan, northwest by north, 210 miles. f. 207a

Molunat with Sazan, between northwest and north, 170 miles.

Lastovo with Otranto, southeast by south, 250 miles.

Molat with Otranto, between south and southeast, 200 miles.

Dubrovnik with Otranto, south by east, 200 miles.

Kotor with Otranto, south and north, 140 miles.

Ulcinj with Otranto, south by west, 170 miles.

Venice with Monopoli, southeast and northwest, 600 miles.

Sant'Andrea of Vis with Sveti Ivan na Pučini, southeast and northwest, 250 miles.

Otranto with Kassiópi, between east and southeast, 80 miles.

Otranto with the cape of San Isidro, southeast and northwest, 200 miles.

Otranto with Fanu, southeast by east, 60 miles.

Portolan from Cape Maléas to the island of Famagusta.

At Cape Maléas with Mílos, southeast by east, 80 miles.

Mílos with Folégandros, east by south, 20 miles.

Folégandros with Íos, east by south, 30 miles.

Íos with Santorini, between south and southeast, 15 miles.

Santorini with Anáfi, between northeast and east, 15 miles.

Anáfi with Astipálaia, between northeast and east, 30 miles.

Stampalia cum el Ciafalo, griego e africo, mia 40.

Ex[a] Ciafallo cum Niseri, levante e ponente, mia 20.

Niseri cum Barbanicuola, entro griego e levante, mia 25.

Barbanichuola cum Sempolo, levante e ponente, mia 25.

Sempollo cum Ruodo, entro levante e siroco, mia 35.

Sant'Anzolo de Malio con Cerigo, ostro e tramontana, mia 20.

Sant'Anzolo cum cavo Spada, quarta de siroco ver ostro, mia 80.

Cavo Spada cum la Melleça, levante e ponente, mia 40.

Mellecha cum la Fraschia, levante e ponente, mia 60.

f. 207b La Fraschia cum cavo de San Zane, levante e ponente, mia 50.

Cavo San Zane cum Setia, levante e ponente, mia 30.

Setia cum cavo Sermon, levante e ponente, mia 20.

Cavo Sermon cum el Caxo, quarta de griego e levante, mia 45.

El Caxo cum el Scarpanto, entro griego e tramontana, mia 10.

Scarpanto cum Carpi, quarta de tramontana ver griego, mia 70.

Malfetam cum Ruodo, siroco e maistro, mia 30.

Carpi cum porto Malefetam, entro griego e levante, mia 50.

Ruodo cum Septecavi, quarta de levante ver griego, mia 100.

Ruodo cum Castel Rusco, levante e ponente, mia 100.

Castello Rusco cum el Cacavo, entro griego e levante, mia 25.

---

a. *Sic in MS.*
1. Uncertain location.
2. Uncertain location.

Astipálaia with Kéfalos, northeast and southwest, 40 miles.

From Kéfalos with Nísiros, east and west, 20 miles.

Nísiros with Barbanicola,[1] between northeast and east, 25 miles.

Barbanicola with Sempolo,[2] east and west, 25 miles.

Sempolo with Rhodes, between east and southeast, 35 miles.

Cape Maléas with Kíthira, south and north, 20 miles.

Cape Maléas with Cape Spátha, southeast by south, 80 miles.

Cape Spátha with Cape Melleca, east and west, 40 miles.

Cape Melleca with Punta della Fraschia,[3] east and west, 60 miles.

Punta della Fraschia with Cape Áyios Ioánnis, east and west, 50 miles. f. 207b

Cape Áyios Ioánnis with Seteía, east and west, 30 miles.

Seteía with Cape Salomon, east and west, 20 miles.

Cape Salomon with Kásos, northeast by east, 45 miles.

Kásos with Kárpathos, between northeast and north, 10 miles.

Kárpathos with Khálki, north by east, 70 miles.

Malfetam[4] with Rhodes, southeast and northwest, 30 miles.

Khálki with Malfetam, between northeast and east, 50 miles.

Rhodes with Yedi Burun, east by north, 100 miles.

Rhodes with Megisti, east and west, 100 miles.

Megisti with Kekova, between northeast and east, 25 miles.

---

3. Just west of Herákleion, Crete.
4. On the Resadiye Yarimadasi peninsula of the east coast of Turkey.

El Cachavo cum cavo Stilbonuri, quarta de griego ver levante, mia 20.

El cavo Stilbonuri con le Chilendonie, entro griego e levante, mia 30.

Le Chilendonie cum San Biffano, siroco e maistro, mia 160.

San Biffano cum Baffo, entro ostro e siroco, mia 25.

Baffo cum cavo Bianco, quarta de levante ver siroco, mia 30.

Cavo Biancho cum Ganata, levante e ponente, mia 35.

Ganata cum Limisso, ostro e tramontana, mia 18.

Limisso cum la Grea, quarta de levante ver griego, mia 90.

La Grea cum Famagosta, quarta de tramontana ver maistro, mia 18.

Famagosta cum el cavo de San Andrea, griego ver[a] africo, mia 60.

Sasno cum Otranto, quarta de africo ver ponente, mia 60.

Otranto cum cavo Sancta Maria, ostro e tramontana, mia 35.

Sancta Maria con Cotron, griego e africo, mia 100.

Sancta Maria cum le Colunne, quarta de africo ver ostro, mia 90.

Le Colunne con cavo Stilo, quarta de africo ver ostro, mia 70.

Cavo Stillo cum cavo Borsan, quarta de africo ver ostro, mia 60.

f. 208a Cavo Borsan cum Spartivento, griego e africo, mia 10.

Spartivento cum Pellari, levante e ponente, mia 20.

Pellari cum Rezo, ostro e tramontana, mia 50.

Rezo cum la Catona, ostro e tramontana, mia 5.

---

a. *Sic in MS.*

Kekova with Cape Stilbonuri,[1] northeast by east, 20 miles.

Cape Stilbonuri with Cape Gelidonya, between northeast and east, 30 miles.

Cape Gelidonya with Cape Arnáoutes, southeast and northwest, 160 miles.

Cape Arnáoutes with Páfos, between south and southeast, 25 miles.

Páfos with Cape Aspro, east by south, 30 miles.

Cape Aspro with Cape Gáta, east and west, 35 miles.

Cape Gáta with Lemésos, south and north, 18 miles.

Lemésos with Cape Gréko, east by north, 90 miles.

Cape Gréko with Famagusta, north by west, 18 miles.

Famagusta with Cape Áyios Andréas, northeast by southwest, 60 miles.

Sazan with Otranto, southwest by west, 60 miles.[2]

Otranto with Capo Santa Maria di Leuca, south and north, 35 miles.

Santa Maria with Crotone, northeast and southwest, 100 miles.

Santa Maria with Capo delle Colonne, southwest by south, 90 miles.

Capo delle Colonne with Punta Stilo, southwest by south, 70 miles.

Punta Stilo with Capo Bruzzano, southwest by south, 60 miles.

Capo Bruzzano with Capo Spartivento, northeast and southwest, 10 miles. f. 208a

Capo Spartivento with Punta di Pellaro, east and west, 20 miles.

Punta di Pellaro with Reggio Calabria, south and north, 50 miles.

Reggio Calabria with Catona di Reggio Calabria, south and north, 5 miles.

---

1. Uncertain location.
2. The text returns here to the portolan of the Gulf of Venice.

Chatona cum Voli, ostro e tramontana, mia 8.

Voli cum la Contena, ostro e tramontana, mia 30.

Li[a] Contena cum Battichani, entro maistro e tramontana, mia 50.

Battichan cum Torpia, entro griego e levante, mia 10.

Reço cum Messina, levante e ponente, mia 10.[b]

Messina cum la Toreta, entro griego e tramontana, mia 8.

La Torreta cum cavo Smertella, levante e ponente, mia 50.

La Smertella cum Mellazo, entro ponente e africo, mia 20.

Mellazo cum Pati, quarta de ponente ver africo, mia 20.

Pati cum el cavo de Rolando, levante e ponente, mia 20.

Portolan da Venesia infina ala Tana, ala via dele galie per staria.[c]

Rima[d] per Veniesia in Parenzo, quarta de levante ver siroco, mia 100.

Da Veniesia a Ruvigno, quarta de levante ver siroco, mia 100.

Da Rovigno a Polla, siroco e maistro, mia 20.

Da Puolla a Figo, mia 6.

Dal Figo a Polmontore, siroco e maistro, mia 10.

Da Polmontore a Nia, levante e ponente, mia 30.

Da Ania a Nime, siroco e ponente,[e] mia 20.

---

a. *Sic in MS.*

b. *Inversion of the line in respect to the preceding one, signaled in the MS by the labeling of this line as* b *and the one below as* A.

c. *From* Portolan *to* staria *in red ink.*

d. *Sic in MS, probably for* Prima.

e. *Sic in MS.*

Catona di Reggio Calabria with Voli,[1] south and north, 8 miles.

Voli with Nicotera, south and north, 30 miles.

Contena with Capo Vaticano, between northwest and north, 50 miles.

Capo Vaticano with Tropea, between northeast and east, 10 miles.

Reggio Calabria with Messina, east and west, 10 miles.

Messina with Torre di Faro, between northeast and north, 8 miles.

Torre di Faro with Capo Rasocolmo, east and west, 50 miles.

Capo Rasocolmo with Milazzo, between west and southwest, 20 miles.

Milazzo with Patti, west by south, 20 miles.

Patti with Capo d'Orlando, east and west, 20 miles.

Portolan from Venice to Tana, on the route of the galleys by the coast.

First, for Venice to Poreč, east by south, 100 miles.

From Venice to Rovinj, east by south, 100 miles.

From Rovinj to Pula, southeast and northwest, 20 miles.

From Pula to Figarola, 6 miles.

From Figarola to Sveti Kamenjak, southeast and northwest, 10 miles.

From Sveti Kamenjak to Unije, east and west, 30 miles.

From Unije to Nieme, southeast and west, 20 miles.

1. Uncertain location.

Da Meme a Selva, entro levante e siroco, mia 10.

Da Selva cum Liubo, mia 10.

Da Liubo a Zara, entro levante e siroco, mia 30.

Da Zarra a Zarra veghia, siroco e maistro, mia 18.

Da Zara vechia al Morter, siroco e maistro, mia 30.

Dal Morter al Scoio del'oro, siroco e maistro, mia 20.

f. 208b Dal Scoio del'oro al Figo, siroco e maistro, mia 10.

Dal Figo a Sancto Arcangelo, siroco e maistro, mia 6.

Da Sancto Arcangelo a Porto Droxo, siroco e maistro, mia 10.

Da Porto Droxo ai Goti de Lesna, siroco e maistro, mia 18.

Da Goa[a] a Liesna, siroco e maistro, mia 10.

Da Liesna alla Torcolla, siroco e maistro, mia 18.

Da Torchola a cavo Cumano, quarta de levante ver siroco, mia 22.

Da cavo Cumano con San Maximo, siroco e maistro, mia 10.

Da San Max[*i*]mo ala Zuliana, entro levante e siroco, mia 10.

Dala Zuliana a Chalotorta, entro levante e siroco, mia 30.

Da Chalotorta a Raguxi, entro levante e sirocho, mia 10.

Da Raguxi al Sasno, entro ostro e sirocho, mia 200.

Da Raguxi a Raguxi vechio, entro levante e siroco, mia 10.

Da Raguxi vechio a Malonto, levante e sirocho, mia 18.

---

a. *Sic in MS.*

From Nieme to Silba, between east and southeast, 10 miles.

From Silba with Ulbo, 10 miles.

From Ulbo to Zadar, between east and southeast, 30 miles.

From Zadar to Old Zadar, southeast and northwest, 18 miles.

From Old Zadar to Murter, southeast and northwest, 30 miles.

From Murter to the golden reef,[1] southeast and northwest, 20 miles.

From the golden reef to Figo,[2] southeast and northwest, 10 miles. f. 208b

From Figo to Sant'Arcangelo, southeast and northwest, 6 miles.

From Sant'Arcangelo to Porto Rosso,[3] southeast and northwest, 10 miles.

From Porto Rosso to the Palenski Islands, southeast and northwest, 18 miles.

From the Palenski Islands to Hvar, southeast and northwest, 10 miles.

From Hvar to Šćedro, southeast and northwest, 18 miles.

From Šćedro to Capo Cumano, east by south, 22 miles.

From Capo Cumano with San Maximo,[4] southeast and northwest, 10 miles.

From San Maximo to Žuljana, between east and southeast, 10 miles.

From Žuljana to Koločep, between east and southeast, 30 miles.

From Koločep to Dubrovnik, between east and southeast, 10 miles.

From Dubrovnik to Sazan, between south and southeast, 200 miles.

From Dubrovnik to Cavtat, between east and southeast, 10 miles.

From Cavtat to Molunat, east and southeast, 18 miles.

---

1. Uncertain location.
2. A rock by Trogir, Croatia.
3. Near Vinišće, Croatia.
4. On the Pelješac peninsula.

Da Malonto a Buda, levante e siroco, mia 30.

Da Buda Intiveri, levante e siroco, mia 10.

D'Antivari a Dolcegno, quarta de levante ver siroco, mia 20.

Da Dolcegno a Durazo, ostro e tramontana, mia 50.

Da Durazo al Sasno, ostro e tramontana, mia 80.

Dal Sasno ala Val del'orso, siroco e maistro, mia 25.

Dala Val del'orso a Palormo, siroco e maistro, mia 40.

Da Palormo a Sancti Quaranta, quarta de siroco ver ostro, mia 20.

Da Sancti Quaranta a Butintro, quarta de levante ver siroco, mia 5.

Da Butintro a Civita, entro levante e siroco, mia 40.

Da Civita al Velechi, entro levante e siroco, mia 15.

Dal Velechi al cavo del Ducato, quarta de ostro ver siroco, mia 80.

Dal cavo del Ducato a Viscardo, ostro e tramontana, mia 12.

Da Viscardo ala vale d'Alesandria, siroco e tramontana,[a] mia 30.

Dala vale d'Alesandria a Clarenza, levante e ponente, mia 30.

f. 209a Da Clarenza a Bello Veder, siroco e maistro, mia 30.

Da Belverde[b] a Pruodo, entro ostro e siroco, mia 50.

Da Pruodo a Modon, siroco e maistro, mia 18.

Da Modon a San Veniedego, siroco e maistro, mia 12.

Da San Veniedego a Coron, ostro e tramontana, mia 6.

---

a. *Sic in MS.*
b. *Sic in MS.*

From Molunat to Budva, east and southeast, 30 miles.

From Budva [to] Bar, east and southeast, 10 miles.

From Bar to Ulcinj, east by south, 20 miles.

From Ulcinj to Durrës, south and north, 50 miles.

From Durrës to Sazan, south and north, 80 miles.

From Sazan to the Valley of the Bear,[1] southeast and northwest, 25 miles.

From the Valley of the Bear to Liman i Panormit, southeast and northwest, 25 miles.

From Liman i Panormit to Sarandë, southeast by south, 20 miles.

From Sarandë to Butrint, east by south, 5 miles.

From Butrint to Sivota, between east and southeast, 40 miles.

From Sivota to Vigla, between east and southeast, 15 miles.

From Vigla to Cape Doukáton, south by east, 80 miles.

From Cape Doukáton to Fiskárdho, south and north, 12 miles.

From Fiskárdho to the Valley of Alexandria,[2] southeast and north, 30 miles.

From the Valley of Alexandria to Killíni, east and west, 30 miles.

From Killíni to Skafidhiá, southeast and northwest, 30 miles. f. 209a

From Skafidhiá to Próti, between south and southeast, 50 miles.

From Próti to Methóni, southeast and northwest, 18 miles.

From Methóni to Venétiko, southeast and northwest, 12 miles.

From Venétiko to Koróni, south and north, 6 miles.

---

1. Just north of Hamarë, Albania.
2. On the west coast of Kefallinía.

Da Coron a Mania,[a] quarta de levante ver siroco, mia 40.

Da Maina a cavo de Sancta Maria, quarta de levante ver siroco, mia 20.

Dale Quaie ai Cervi, quarta de tramontana ver siroco,[b] mia 40.

Dai Cervi a Cavo Malio, levante e ponente, mia 30.

Da Cavo Malio alla Sidra, siroco e tramontana,[c] mia 80.

Dalla Sidra alle Colunne, quarta de griego e tramontana, mia 50.

Dale Colonne ala Magina, quarta de tramontana ver griego,[d] mia 30.

Dalla Magina a Sancta Anna, griego e africo, mia 10.

Da cavo Sancta Anna a Negroponte, quarta de maistro ver ponente, mia 35.

Da Negroponte al cavo d'i canali, siroco e maistro, mia 70.

Dal cavo di canal a Loredo, entro griego e levante, mia 18.

Da Loredo a Sciato, griego e africo, mia 40.

Da Sciato a Scopolo, entro griego e levante, mia 6.

Da Scopollo a Doromo, griego e levante, mia 5.

Dal Dromo a Limea, quarta de griego e levante, mia 10.

Da Limea a Larsuia, quarta de tramontana ver levante,[e] mia 5.

Da Larsuia a Largiron, quarta de ostro ver siroco, mia 5.

Da Largiron al Piper, levante e ponente, mia 10.

Dal Piper a Stalimene, entro griego e levante, mia 70.

---

a. *Sic in MS.*
b. *Sic in MS.*
c. *Sic in MS.*
d. griego *written over* tramontana.
e. levante *written over an illegible word.*

From Koróni to Maína, east by south, 40 miles.

From Maína to Cape Saint Mary,[1] east by south, 20 miles.

From the Quails[2] to Elafónisos, north by southeast, 40 miles.

From Elafónisos to Cape Maléas, east and west, 30 miles.

From Cape Maléas to Ídhra, southeast and north, 80 miles.

From Ídhra to Cape Soúnion, northeast by north, 50 miles.

From Cape Soúnion to Cape Marathónos, north by east, 30 miles.

From Cape Marathónos to Cape Ayía Marína, northeast and southwest, 10 miles.

From Cape Ayía Marína to Negroponte, northwest by west, 35 miles.

From Negroponte to Cape Kinaíon, southeast and northwest, 70 miles.

From Cape Kinaíon to Oreoi, between northeast and east, 18 miles.

From Oreoi to Skíathos, northeast and southwest, 40 miles.

From Skíathos to Skópelos, between northeast and east, 6 miles.

From Skópelos to Alónnisos, northeast and east, 5 miles.

From Alónnisos to Limea,[3] northeast by east, 10 miles.

From Limea to Skantzoúra, north by east, 5 miles.

From Skantzoúra to Yioúra, south by east, 5 miles.

From Yioúra to Pipéri, east and west, 10 miles.

From Pipéri to Lemnos, between northeast and east, 70 miles.

---

1. Near Cape Matapan.
2. Near Cape Matapan.
3. Another location on the island of Alónissos.

Da Stalimene a Tenedo, quarta de griego ver levante, mia 70.

Da Tenedo cum Larnedo ala Boca, ostro e tramontana, mia 18.

Dala Bocha a Garipoli, griego e africo, mia 70.

Da Garipoli a San Zorzi, griego e africo, mia 25.

Da San Zorzi a Polistor, griego e levante, mia 20.

f. 209b Da Polistor a Reglea, quarta de griego ver levante, mia 60.

Da Reglea a Constantinopoli, quarta de levante ver siroco, mia 50.

Da Constantinopoli[a] a Lalgiro, griego e africo, mia 18.

Da Lalgiro al Salli, levante e ponente, mia 30.

Dal Salli a Porimo, quarta de levante ver siroco, mia 30.

Da Porimo Cacarpi,[b] entro griego e levante, mia 50.

Da Carpi a Farnaxia, quarta de griego ver levante, mia 10.

Dala Farnaxia a puncta Rachia, quarta de levante ver griego, mia 100.

Da puncta Rachia a Mandra, quarta de griego ver levante, mia 25.

Da Mandra a Pistelli, quarta de griego ver levante, mia 20.

Da Pistelli a Rio, griego e africo, mia 25.

Da Rio a Samastro, entro griego e levante, mia 65.

Da Samastro a Do Castelli, quarta de levante ver griego, mia 60.[c]

Da Do Castelli a Harami, griego e africo, mia 30.

---

a. Constantinopoli *corrected over* Lalgiro.

b. *Sic in MS.*

c. 60 *written over* 65.

From Lemnos to Tenedos, northeast by east, 70 miles.

From Tenedos with Larnedo[1] at Bozca Ada, south and north, 18 miles.

From Bozca Ada to Gallipoli, northeast and southwest, 70 miles.

From Gallipoli to Cape San Giorgio, northeast and southwest, 25 miles.

From Cape San Giorgio to Polistor,[2] northeast and east, 20 miles.

From Polistor to Marmara Ereglisi, northeast by east, 60 miles. f. 209b

From Marmara Ereglisi to Constantinople, east by south, 50 miles.

From Constantinople to Anadoloufeneri, northeast and southwest, 18 miles.

From Anadoloufeneri to Şile, east and west, 30 miles.

From Şile to Porimo,[3] east by south, 30 miles.

From Porimo to Kerpe Burnu, between northeast and east, 50 miles.

From Kerpe Burnu to Kefken Adasi, northeast by east, 10 miles.

From Kefken Adasi to Eregli, east by north, 100 miles.

From Eregli to Mandra,[4] northeast by east, 25 miles.

From Mandra to Pistelli,[5] northeast by east, 20 miles.

From Pistelli to Rio,[6] northeast and southwest, 25 miles.

From Rio to Amasra, between northeast and east, 65 miles.

From Amasra to Espiye, east by northeast, 60 miles.

From Espiye to Cape Kerempe, northeast and southwest, 30 miles.

---

1. Uncertain location.
2. Uncertain location on the European coast of the Sea of Marmara.
3. Uncertain location.
4. Uncertain location.
5. Ancient Psillium, on the southern coast of the Black Sea.
6. Ancient Tios, on the southern coast of the Black Sea.

Da Irami a Sinopolli, levante e ponente, mia 15.

Da Sinopoli a Tinoli, levante e ponente, mia 20.

Da Tinolli a Stefano, levante e ponente, mia 25.

Da Stefano a Larmimon, entro griego e levante, mia 30.

Da Larmimon a Sinoppi, levante e ponente, mia 20.

Da Sinoppi a Caroxa, entro ostro e siroco, mia 25.

Da Charoxa al Chasimon, quarta de griego ver levante, mia 15.

Da Chasimon a Pantere, levante e ponente, mia 30.

Da Pantere a cavo dela Lime, entro griego e levante, mia 30.

Da cavo dela Limissi a Pantegona, siroco e maistro, mia 25.

Da Pantegona a Simisso, quarta de levante ver siroco, mia 22.

Da Simisso a Chalamo, levante e ponente, mia 15.

Da Calamo a cavo de Mitoy, ostro e tramontama, mia 30.

Da cavo Mitoy alla Mona,[a] levante e ponente, mia 15.

Dala Moma a Larmimon, quarta de ostro ver siroco, mia 15.

f. 210a Da Larmimon a Norio, levante e ponente, mia 15.

Da Norio a Varicha, levante e ponente,[b] mia 25.[c]

Da Varicha a Pasimon, quarta de siroco ver ostro, mia 10.

Da Pasimon a Limoma, entro griego e levante, mia 30.

---

a. *Sic in MS.*

b. ponente *written over* ostro.

c. 25 *written over* 15.

1. Uncertain location.

2. Uncertain location.

From Cape Kerempe to Inebolu, east and west, 15 miles.

From Inebolu to Ginolu, east and west, 20 miles.

From Ginolu to Ayancik, east and west, 25 miles.

From Ayancik to Larminon,[1] between northeast and east, 30 miles.

From Larminon to Sinop, east and west, 20 miles.

From Sinop to Gerze, between south and southeast, 25 miles.

From Gerze to Cayagzi, northeast by east, 25 miles.

From Cayagzi to Pantere,[2] east and west, 30 miles.

From Pantere to Cavo dela Lime,[3] between northeast and east, 30 miles.

From Cavo dela Lime to Pantegona,[4] southeast and northwest, 25 miles.

From Pantegona to Samsun, east by south, 22 miles.

From Samsun to Calamo,[5] east and west, 15 miles.

From Calamo to Cavo de Mitoy,[6] south and north, 30 miles.

From Cavo de Mitoy to Cape Vona, east and west, 15 miles.

From Cape Vona to Larmimon,[7] south by east, 15 miles.

From Larmimon to Ünye, east and west, 15 miles. f. 210a

From Ünye to Fatsa, east and west, 25 miles.

From Fatsa to Bolaman, southeast by south, 10 miles.

From Bolaman to Vona Limani, between northeast and east, 30 miles.

---

3. Uncertain location.
4. Uncertain location.
5. Ancient Zagora, on the southern coast of the Black Sea.
6. Ancient Eustathmus Euxina, on the southern coast of the Black Sea.
7. Uncertain location.

Da Limoma a Sisti, siroco e maistro, mia 10.

Da Sisti a Diomede, entro levante e siroco, mia 25.

Da Diomede entro a Sanason, levante e ponente, mia 10.

Da Sanason a Crisonda, entro griego e levante, mia 10.

Da Crisonda a Ciefalo, entro griego e levante, mia 15.

Da Ciefalo a Inepoli, entro griego e levante, mia 10.

Da Inepoli ala Croxe, levante e ponente, mia 7.

Dala Croxe a Deovenzi, levante e ponente, mia 10.

Da Deovençi a Michole [. . .].[a]

Da Michole a Largiron, levante e ponente, mia 10.

Da Largiron a cavo Pulte, entro griego e tramontana, mia 10.

Da cavo Pulte, entro la staria, quarta de levante ver siroco, mia 10.

Da Pultia a Trebesonda, entro siroco e levante, mia 18.

Da Trebesonda al Apicho el Cavolario, siroco e maistro, mia 470.

Dalo parezo de Do Castelli a cavo San Tuodoro al cavo dela Guia, ostro e tramontana, mia 290.

Dal cavo San Sidero ala bocha de Largiro, mia 530.

Da Largiri a Constantinopoli, mia 18.

Da cavo de Agia a San Teodoro, levante e ponente, mia 25.

---

a. *Direction and distance lacking in MS.*
1. Uncertain location.
2. Uncertain location.
3. Ancient Zephyrium, on the southern coast of the Black Sea.
4. Uncertain location.
5. Uncertain location.
6. Direction and distance lacking.

From Vona Limani to Stephi, southeast and northwest, 10 miles.

From Stephi to Diomede,[1] between east and southeast, 25 miles.

From Diomede to Sanason,[2] east and west, 10 miles.

From Sanason to Giresun, between northeast and east, 10 miles.

From Giresun to Ciefalo,[3] between northeast and east, 15 miles.

From Ciefalo to Tirebolu, between northeast and east, 10 miles.

From Tirebolu to La Croxe,[4] east and west, 7 miles.

From La Croxe to Deovenzi,[5] east and west, 10 miles.

From Deovenzi to Fenerkoyu [. . .].[6]

From Fenerkoyu to Largiron,[7] east and west, 10 miles.

From Largiron to Akcaabat, between northeast and north, 10 miles.

From Akcaabat, within the mainland,[8] east by south, 10 miles.

From Akcaabat to Trebizond, between southeast and east, 18 miles.

From Trebizond to Apicho el Cavolario,[9] southeast and northwest, 470 miles.

From the place called Espiye, to Cape Saint Theodore,[10] to the Cape of the Guide,[11] south and north, 290 miles.

From Cape Saint Theodore to the mouth of Largiron, 530 miles.

From Largiron to Constantinople, 18 miles.

From Cavo de Agia[12] to Saint Theodore, east and west, 25 miles.

---

7. Ancient Hieron Oros, on the southern coast of the Black Sea.
8. I.e., in a coastwise channel in the mainland.
9. Uncertain location in eastern Crimea.
10. Uncertain location, at the southern end of Crimea.
11. Uncertain location, at the southern end of Crimea.
12. Ancient Lagyra, in southern Crimea.

Da cavo San Todoro a Pangropoli, ostro e tramontana, mia 10.

Da Pangropoli ala Strada, griego e africo, mia 15.

Dala Strada a Schixi, griego e africo, mia 20.

Da Schixi a Soldada, griego e africo, mia 20.

Da Soldada a Biganome, griego e africo, mia 25.

f. 210b Da Biganome a cavo de Galitera, ostro e tramontana, mia 10.

Da cavo de Galitera a Gaffa, griego e africo, mia 30.[a]

Da[b] cavo de Gaffa a Gaffa, ostro e tramontana, mia 18.

Da Gaffa a Çamdo, levante e ponente, mia 45.

Da Çamdo a Cipro, entro griego e levante, mia 30.

Da Cipro a Cavalari, siroco e maistro, mia 10.

Da Cavalari a Doxomittit, ostro e tramontana, mia 10.

Da Expromiti al Prospero, ostro e tramontana, mia 10.

Da Prospero al Pandicho, ostro e tramontana, mia 30.

Dal Pandicho ai Palastra, ostro e tramontana, mia 130.

Da Palastra a Papacomo, entro griego e levante, mia 35.

Da Papacomo ai Roxi, entro griego e levante, mia 25.

Da Iroxi al Chalradi, entro griego e levante, mia 25.

---

a. 30 *written over* 28.
b. Gaffa *crossed out with horizontal lines.*
1. Uncertain location in southern Crimea.
2. Uncertain location in eastern Crimea.
3. Uncertain location.
4. Uncertain location.
5. Ancient Nymphaeum, in eastern Crimea.

From Cape Saint Theodore to Pangropoli,[1] south and north, 10 miles.

From Pangropoli to la Strada,[2] northeast and southwest, 15 miles.

From la Strada to Schixi,[3] northeast and southwest, 20 miles.

From Schixi to Sudak, northeast and southwest, 20 miles.

From Sudak to Cape Meganom, northeast and southwest, 25 miles.

From Cape Meganom to Cape Kük Atlam, south and north, 10 miles. f. 210b

From Cape Kük Atlam to Feodosiya, northeast and southwest, 30 miles.

From the Cape of Feodosiya to Feodosiya, south and north, 18 miles.

From Feodosiya to Çamdo,[4] east and west, 45 miles.

From Çamdo to Cipro,[5] between northeast and east, 30 miles.

From Cipro to Cavolario[6] southeast and northwest, 10 miles.

From Cavolario to Doxomittit,[7] south and north, 10 miles.

From Expromiti[8] to Kerch, south and north, 10 miles.

From Kerch to Pandicho,[9] south and north, 30 miles.

From Pandicho to Mariupol, south and north, 130 miles.

From Mariupol to Papacomo,[10] between northeast and east, 35 miles.

From Papacomo to Roxi,[11] between northeast and east, 25 miles.

From Roxi to Chalradi,[12] between northeast and east, 25 miles.

---

6. Uncertain location.
7. Uncertain location.
8. Uncertain location, possibly the same as Doxomittit.
9. Ancient Panticapaeum, eastern Crimea.
10. Uncertain location on the northern coast of the Sea of Azov.
11. Uncertain location.
12. Uncertain location.

Da Chalradi a porto Pixan, entro griego e levante, mia 25.

Dai[a] porto Pixan ala bocha dela flumera dela Tana, siroco e maistro, mia [. . .].[b]

Dala bocha dela flumera ala Tana, mia 25.

f. 225b[c] Chi vol far una tavola de Salamon die saver quanto son una ora, che son ponti 1000 e 80, son una ora, e una luna son dì 29, ore 12, ponti 793.

f. 238a[d] Ihesus Mari. 1473, adì 29 avosto, al Chiarcho.

Notto fazo mi Zuane da Drivasto, paron zurado del manificho misier Marin Dandolo, de mie robe[e] me ttrovo aver in galia. Primamentte un zipon de centtanin negro in do peli, e uno zipon de lesendrin nuovo raizo, un zipon paonazo de Fiandra fodrà de biancho. Dodeze chamize, una che sia dada a Piero Durazin, e una a Marttin Dela Portta, e una chapa de zanbelotto negra e una chapa de mosto[f] valier. Una vestizuola de mosto valier. Un per de chalze solade e un per de bianche de rasa. Do pera de chalze da scharpe vechie, sia datte un per a Marttin Dela Portta e l'alttro per sia dado a Piero Durazin. Quattro berette e tre negre nuove e una de scharlatto frusta e una unola negra. Ttre ttapei, do nuoi e uno che sono uno puocho pezo. Una charpetta, una schiavina e uno gaban negro frodà di biancho, un alttro gaban unolo che son in man de Primo fradelo de Lucha Gobo.[g] Sì ò lasà a Modon un per de chalze[h] de rasa, che sia ttentte 1ª grana, e ttrentta un aspro in man del chavalier de Modon Zan Negro. E sì ò ancora qua in galia una gona berittina asetta e una gona meschia, una zenttura d'arzentto stringe ttela e fusare, et una spada e ttarzetta, e churazina e meza ttesta e una lanza.[i] Se nientte in tera vien de mi, | f. 238b che dele ditte robbe sia dade a mio nevodo Zuan Falchon, balestier de misier Zacharia di Prioli,[j] chon questo che sia ttenudo de vender ttantte robe che page le mie debitte[k] in galia, che son zercha duchatti otto, che eze Nicholò Zingano e Domenego da Choron proveier, Mattio proveier, el nostro scrivan de galia, sier Piero Penola, e che me faza dir le mese de Sen Griguol.

---

a. *Sic in MS.*
b. *Distance lacking in MS.*
c. *Folios 211a–225a are blank.*
d. *Folios 226a–237b are blank.*
e. fe *crossed out with diagonal lines.*
f. *Sic in MS, perhaps for* molto *and also further on. However, it could also be read* Mosto Valier *as a proper name.*
g. E queste ditte che so *crossed out with a horizontal line.*
h. a Modon *crossed out with a horizontal line.*
i. s *crossed out with a diagonal line.*
j. *In* Prioli *final* li *repeated and crossed out with diagonal lines.*
k. debitte *corrected over* desitte.

From Chalradi to Porto Pisano, between northeast and east, 25 miles.

From Porto Pisano to the mouth of the river of Tana, northeast and southwest, [. . .][1] miles.

From the mouth of the river to Tana, 25 miles.

Whoever wishes to make a table of Solomon should know how much an hour is, which is 1,080 points, which is an hour, and a moon is 29 days, 12 hours, 793 points. f. 225b

Jesus Mary. 1473, 29 August, in Khálki[2] f. 238a

I, Giovanni da Drivasto, *paron zurado* of the magnificent sir Marino Dandolo, make note of the things that I have in the galley. First a *zipon*[3] of black satin in two naps, and a *zipon* of new Alexandrine satin, and a Flemish purple *zipon* lined with white. Twelve shirts, one of which should be given to Pietro Durazin and one to Martino della Porta, and a cape of black *zamberlocco*[4] and a cape of valerian must. A short garment of valerian must. One pair of stockings with soles and one white pair of wool. Two pairs of stockings to wear with old shoes, one pair should be given to Martino della Porta and the other pair should be given to Pietro Durazin. Four caps, three new black ones and one of shabby scarlet, and a black one. Three carpets, two new and one that is a little worse. A carpet, a hooded cloak and a black overcoat lined with white, another *unolo* overcoat which is in the hands of Primo, the brother of Luca Gobo. Also I have left in Methóni[5] a pair of wool stockings, which are dyed with scarlet grain,[6] and thirty-one aspers[7] in the hand of the knight of Methóni, Giovanni Negro. And I also have here in the galley a gray tunic of silk and a mixed tunic, a silver belt, fabric clasps and *fusare*,[8] a sword and shield, and a *curazina*[9] and a *mezza testa*,[10] and a lance. If nothing on the earth comes of me, | that of these clothes be given to my nephew Giovanni Falcon, crossbowman of Zaccaria Priuli, with the provision that he be obliged to sell as many clothes as necessary to pay off my debts on the galley, which are about 8 ducats, which are to Nicolò Zingano and Domenico da Corone bowman, Matteo bowman, our galley scribe Pietro Penola, and that he have masses said for me to St. Gregory. f. 238b

1. Distance lacking in MS.
2. A small island off the coast of Asia Minor, very near to Rhodes.
3. A short, narrow collarless garment that covered the bodice to which stockings and trousers were tied.
4. Probably camel's hair, which was *zanbelotto*.
5. Venetian port in the southern Peloponnese.
6. A small, plant-louse-like insect used in dyeing fabric red.
7. An Ottoman silver coin.
8. Possibly spindles.
9. The breastplate of a cuirass.
10. A piece of armor that went under the helmet and protected the face and the eyes.

f. 241a[a]

lire[b]	6		
lire	4	soldi	5
lire	3	soldi	11
lire		soldi	14
	14		10
	5		
	9		10[c]

Amado[d] quanto

mio mazor fradelo

da posa hogni debita

rechonmandation vui

fsarede avixado del

Avixote[e] che a mezo marzo el sol[f]

+ Yhesus.[g] 1470, adì 7 settembrio

Recevi da ser Zuane Francesco, scrivan del maran

pattron ser Polonio Masa .b. libre 4875

f. 241b Amado quanto mio mazor fradelo[h]

---

a. *Folios 239a–240b are blank.*
b. *Beginning of hand a.*
c. *End of hand a.*
d. *Beginning of hand b, ink a.*
e. *Beginning of hand b, ink b.*
f. *End of hand b, ink b.*
g. *Beginning of hand c.*
h. *Hand b, ink a.*

f. 241a

lire	6		
lire	4	soldi	5
lire	3	soldi	11
lire		soldi	14
	14		10
	5		
	9		10

Beloved how much

my older brother

from position all debt

recommendation you

would make advised of the

Keep in mind that in the middle of March the sun

+ Jesus. 1470, 7th day of September

I have received from Giovanni Francesco, scribe of the *maran*[1]

paron Polonio Masa. 4,875 pounds b.

Beloved how much my older brother f. 241b

1. A type of flat-bottomed boat.

Additional Documents

1 First will of Cataruccia, second wife of Michael of Rhodes, February 5, 1432, in Venice.

Archivio di Stato di Venezia, Archivio Notarile, Testamenti, Notaio Nicolò Gruato, B. 576, #121.

Yehsus

Millesimo quadrigentesimo trigesimo primo, mensis februarii die quinto, indictione decima. Rivoalti. Ego Chatharuzia uxor ser Michali da Ruodo, de confinio Sancti Petri de Castello, infirma sed gratia Dei sanam mentem et intelletum,[a] sed timens ne decederem ab intestata, venire feci ad me presbiterum Nicolaum Gruato[b] Veneciarum notarium ipsumque rogavi ut hoc meum ultimum scriberet testamentum, cum clausulis et additionibus. In quo quidem constituo ac esse volo meos fidei commissarios dominam Perenzinam matrem meam dilectam, Lodovicum fratrem meum et suprascriptum virum meum, et quod nullus possit aliquid administrare de dicta mea commissaria nisi fuerit de consensu maioris partis. Item dimitto pro mea decima ducatos .X. auri. Item dimitto suprascripte domine Perenzine, matri et commissarie mee, vestem meam de morello cum manicis fulcitis de varo, unam meam investituram de paonazo cum manicis de panno carmixino, fulzitam prout stat, et unam meam siocham de vergado rubeo, ut oret Deum pro anima mea. Item dimitto Pulixene filie mee adoptive unam meam vestem de morello cum manicis fulcitis de pellis de martore, et unam meam investituram de farza viride, ac unum cingulum de carmixino cum argento, et hoc pro suo maritare. Et si decederet antequam esse maritata, volo quod tales res deveniant in suprascriptum virum meum. Item dimitto Panthasilee, nepti mee, unum meum bailiozonum de paonazo[c] pro suo maritare. Et si decederet antequam esse maritata, volo quod vendatur et denarii dentur pauperibus personis pro anima mea. Item volo quod dicti commissarii mei facere debeant celebrare missas Sancte Marie, Sancti Gregorii et Spiritus Sancti pro anima mea, et si presbiter Franziscus de Castello, confessor[d] meus, volet celebrare dictas missas quod habere debeat ellimosinam dictarum missarum. Item dimitto pauperibus Sancti Lazari ducatos 4 auri pro anima mea. Interogata si velet dimittere aliquid monasterio Sancte Marie de Nazareth pro subventione pauperum ibi existezium,[e] respondit: dimitto ducatos 4 auri pro anima mea. Item dimitto notario infrascripto pro suo labore ducatos 4 auri. Ressiduum vero omnium bonorum meorum, mobilium et inmobilium, quocumque modo mihi spectanzium sive pertinentium, dimitto suprascripto ser Michali, viro et commissario meo, ut oret Deum pro anima mea.

Testes: presbiter Antonius de Guerziis et presbiter Stefanus Doto, ambo mansionarii Sancti Petri de Castello.

---

a. *Sic in MS, accusative as are others as object of implied or omitted verb* habens.

b. cano *crossed out with a horizontal line.*

c. de paonazo *added in the left margin with a reference mark.*

d. *Corrected over* confessoris.

e. *Sic in MS.*

ADDITIONAL DOCUMENTS

1 First will of Cataruccia, second wife of Michael of Rhodes, February 5, 1432, in Venice.

Archivio di Stato di Venezia, Archivio Notarile, Testamenti, Notaio Nicolò Gruato, B. 576, #121.

Jesus

One thousand, four hundred thirty one,[1] the month of February, the fifth day, the tenth indiction.[2] Rialto.[3] I, Cataruccia, wife of Michael of Rhodes, of the parish of San Pietro di Castello, ill but by the grace of God healthy in mind and intellect, but fearing lest I die intestate, have had the priest Nicolò Gruato, notary of Venice, come to me, and have asked him to write my last testament, with clauses and additions. In which I constitute and wish to be my executors my beloved mother Perenzina, my brother Alvise, and my abovementioned husband, and let none be able to administer anything of my bequests unless it be by consent of the majority. Item, I leave for my tithe 10 gold ducats. Item, I leave to the abovementioned Perenzina, my mother and executrix, my dark red dress with sleeves lined with vair, one of my purple garments with sleeves of carmine cloth, shining as it is, and one of my trifles with red stripes, that she may pray to God for my soul. Item, I leave to Pulisena my adoptive daughter one of my black dresses with sleeves lined with marten fur, and one of my green doublet dresses, and a belt of carmine with silver, and this for her wedding. And if she shall die before she is wed, I want such things to go to my abovementioned husband. Also I leave to Pantasilea, my niece[4] one of my purple *bagliozoni*[5] for her wedding. And if she should die before being married, I want it to be sold and the money given to the poor for my soul. Also, I want my executors to have masses celebrated for me at Santa Maria, San Gregorio, and Spirito Santo for my soul, and if the priest Francesco da Castello, my confessor, wishes to celebrate these masses, let him have the alms for these masses.[6] Also I leave to the poor of San Lazaro 4 gold ducats for my soul. Asked if she wished to leave anything to the monastery of Santa Maria di Nazaretto for the support of the poor living there, she responded: I leave 4 gold ducats for my soul. Item, I leave to the notary named below 4 ducats for his labor. Indeed, all the rest of my goods, movable and immovable, by whatever means they belong to me, I leave to the abovementioned Michael, my husband and executor, that he may pray for my soul.

Witnesses: priest Antonio de Guerzii and priest Stefano Doto, both sacristans of San Pietro di Castello.

---

1. As the Venetian year began in March, a date in February would correspond to 1432 in modern reckoning.
2. Year in the fifteen-year cycle, derived from Byzantine and ancient Roman practice, used in formal documents.
3. That is, on one of the main islands comprising the city of Venice.
4. Also, possibly, granddaughter.
5. Uncertain garment.
6. I.e., receive the money for saying the masses.

2 Second will of Cataruccia, second wife of Michael of Rhodes, April 4, 1437, in Venice. Brief version.

Archivio di Stato di Venezia, Archivio Notarile, Testamenti, Antonio Gambaro, B. 558a, #45.

Die 4 mensis aprilis 1437, inditione 15. Rivoalti.

Dona Chataruçia, uxor ser Michali de Rodo, armirate ad presens galearum Flandrie, de confinio Sancti Petri de Castello, et filia condam ser Georgii Murarii, mente sana licet corpore languens, mixit pro me Anthonio Gambaro Venetiarum notario infrascripto, et me rogavit de hoc suo ultimo testamento, prout infra dicetur. In quo, poxito prohemio, dixit: Comissarios et exequutores huius mei testamenti et ultime voluntatis instituo et esse volo ser Lodovicum Murario, fratrem meum dillectum, predictum ser Michali[a] de Rodo, maritum meum peramabillem, et dominam Magdalenam, uxorem ser Stefaneli barcharoli, vicinam meam de dicto confinio Sancti Petri de Castello, ut secundum quod hic ordinavero darique atque fieri iussero sic ipsi omnes seu maior pars eorum adimplere et facere teneantur. In primis namque volo et ordino dari debere in faciendo zelebrari missas pro anima mea et condam domine Perençine, olim matris mee, ducatos quatuordecim in discretione dictorum meorum comissariorum.

Item dimitto, volo et ordino dari debere in faciendo zelebrari missas pro anima mea et condam domine Pantaxilee olim neptis mee ducatos viginti in discretione dictorum meorum comissariorum.

Item dimitto, volo et ordino quod Magdalena de Rosia, serva et sclava mea quam emi[b] in estate nuper elapsa de denariis michi perventis ex dimissoria condam predicte domine Perençine olim matris mee, quod dicta mea sclava servire debeat suprascripto ser Lodovico fratri et comissario meo per quinque annos statim subsequuturos post dictum obitus mei, et post dictum tempus quinque annorum predicta Magdalena, serva et sclava mea, sit libera et francha.

Residuum vero omnium bonorum meorum, mobillium et immobillium, presentium et futurorum et omne chaduchum, inordinatum et pro numscriptum, ac omne et totum et quicquid aliud quod michi aut huic mee comissarie spectat et pertinet, seu in futurum quomodolibet spectare et pertinere possit, dimitto suprascripto ser Michali[c] de Rodo marito meo dillecto.

Interogata per notarium infrascriptum de postumis sive filiis in posterum michi nascituris, dixi et respondi quod pro nunc nolebam aliud ordinare.

Interogata etiam per notarium infrascriptum si aliquid volebam dimittere loco et hospitali Sancte Marie de Lazareto, dixi et respondi quod ei dimitto ducatos duos. Interogataque etiam per nota-

---

a. *Corrected over* Michalum.
b. *Corrected over* eme.
c. *Corrected over* Michalo.

2 Second will of Cataruccia, second wife of Michael of Rhodes, April 4, 1437, in Venice. Brief version.

Archivio di Stato di Venezia, Archivio Notarile, Testamenti, Antonio Gambaro, B. 558a, #45.

April 4, 1437, indiction 15. Rialto.

Cataruccia, wife of Michael of Rhodes, at present *armiraio* of the Flanders galleys, of the parish of San Pietro di Castello, and daughter of the late Giorgio Murario, of sound mind though languishing body, sent for me Antonio Gambaro, the undersigned Venetian notary, and asked me for her last testament, as is said below. In which, the introduction having been put in place, she said: I establish and wish to have as executors of this my testament and last will Alvise Murario, my beloved brother, the abovementioned Michael my especially beloved husband, and Maddalena, wife of Stefano the boatman, my neighbor in this parish of San Pietro di Castello, so that what I shall order to be given and done, thus all of them or the majority be held to carry out and do. First, I wish and order fourteen ducats be given at the discretion of my executors for the celebration of masses for my soul and that of the late Perenzina, my late mother.

Item, I leave and order twenty ducats to be given at the discretion of those executors of mine for the celebration of masses for my soul and that to Pantasilea, my late niece.

Also I leave and order that Maddalena de Rosia, my servant and slave whom I bought last summer from the money coming to me from the dowry of my late mother Perenzina, that this slave should serve my abovementioned brother and executor Alvise for the five years immediately following my death, and after that time of five years, the abovementioned Maddalena, my servant and slave, be free.

Indeed, the remainder of all of my goods, movable and immovable, present and future, and all property without heir, unassigned and for now written, and all and anything that belongs to me or to my estate, or in the future by any means may belong to it, I leave to the abovementioned Michael of Rhodes, my beloved husband.

Asked by the undersigned notary about later or future children that might be born of me, I said and responded that for now I did not wish to order anything.

Also asked by the undersigned notary whether I wished to leave anything to the place and hospice of Santa Maria del Lazaretto, I said and responded that I leave them two ducats. Also asked by the

rium infrascriptum si aliquid volebam dimittere pueris infantibus Pietatis, dixi et respondi quod pro nunc nolebam aliud ordinare.

Preterea et cetera.

Testes: ser Nicolaus condam ser Vivani barbitonsor de confinio Sancti Petri de Castelo, ser Troilus filius ser Anthonii marangonus de confinio Sancte Luce.

3 Second will of Cataruccia, second wife of Michael of Rhodes, April 4, 1437, in Venice. Extensive version.

Archivio di Stato di Venezia, Archivio Notarile, Testamenti, Antonio Gambaro, B. 559, prot. 1, fol. 8v, #14.

Ihesus

In nomine Dei eterni. Amen. Anno ab incarnatione Domini nostri Ihesu Christi millesimo quadrigentesimo trigesimo septimo, indicione quintadecima, die quarto mensis aprilis. Rivoalti. Cum vite sue terminum unusquisque prorsus ignoret, et nil certius habeamus quam quod mortis non possumus evitare discrimen, idcircho unicuique imminet precavendum ne incautus occumbat et bona sua inordinata derelinquat. Quapropter ego Chataruçia, uxor ser Michali de Rodo, armirate ad presens galearum Flandrie, de confinio Sancti Petri de Castello et filia condam ser Georgii Murarii, mente sana licet corpore languens, dum velem bonorum meorum indispositionem plenariam ordinare ne, spiritu per me altissimi Creatori reddito, lis ula ex eis modo aliquo oriatur, ad me vocari et venire feci Anthonium Gambaro Venetiarum notarium infrascriptum ipsumque rogavi ut hoc meum scriberet testamentum pariterque compleret cum clasulis et additionibus consuetis Venetiarum. In quo quidem meo testamento constituo et esse volo meos fidei comissarios ser Lodovicum Murario, fratrem meum dillectum, predictum ser Michali de Rodo, maritum meum peramabillem, et dominam Magdalenam, uxorem ser Stefaneli barcharoli, vicinam meam de dicto confinio Sancti Petri de Castello, ut secundum quod hic ordinavero darique atque fieri iussero sic ipsi omnes seu maior pars eorum adimplere et facere teneantur. In primis namque volo et ordino dari debere in faciendo zelebrari missas pro anima mea et condam domine Perençine, olim matris mee, ducatos quatuordecim in discretione dictorum meorum comissariorum. Item dimitto, volo et ordino dari debere in faciendo çelebrari missas pro anima mea et condam domine Pantaxilee, olim neptis mee, ducatos viginti in discretione dictorum meorum comissariorum. Item dimitto, volo et ordino quod Magdalena de Rosia, serva et sclava mea quam emi in estate nuper elapsa de denariis michi perventis ex dimissoria condam predicte domine Perençine, olim matris mee, quod dicta mea sclava servire debeat suprascripto ser Lodovico, fratri et comissario meo, per quinque annos statim subsequuturos post dictum obitus mei, et post dictum tempus quinque annorum predicta Magdalena, serva et sclava mea, sit libera et francha. Residuum vero omnium bonorum meorum, mobillium et immobillium, presentium et futurorum, et omne chaduchum, inordinatum et pro numscriptum, ac omne et totum et quicquid aliud quod michi aut huic mee comissarie spectat et pertinet, seu in futurum quomodolibet spectare et pertinere possit, dimitto suprascripto ser Michali de Roddo marito meo dillecto. Interogata per notarium infrascriptum de postumis sive filiis inpos-

undersigned notary if I wished to leave anything to the children of the Pietà, I said and responded that for now I did not wish to order anything.

The remainder, et cetera.

Witnesses: Nicolò, son of the late Vivano, barber of the parish of San Pietro di Castello; Troilo, son of Antonio, ship carpenter of the parish of San Luca.

3 Second will of Cataruccia, second wife of Michael of Rhodes, April 4, 1437, in Venice. Extensive version.

Archivio di Stato di Venezia, Archivio Notarile, Testamenti, Antonio Gambaro, B. 559, prot. 1, fol. 8v, #14.

Jesus

In the name of God eternal. Amen. In the year from the incarnation of our Lord Jesus Christ, one thousand, four hundred, thirty-seven, fifteenth indiction, fourth day of the month of April. Rialto. As each person is ignorant of the end of his life, and we have nothing more certain than that we cannot avoid the intervention of death, therefore it is incumbent on each to avoid dying unprepared and leaving his goods unordered. Therefore, I, Cataruccia, wife of Michael of Rhodes, *armiraio* at present of the galleys of Flanders, from the parish of San Pietro di Castello, and daughter of the late Giorgio Murario, being of sound mind if of languishing body, while I wish to order the complete disposition of my goods lest, my spirit being rendered by me to the highest Creator, any dispute arise from them in any way, I have had the undersigned Venetian notary Antonio Gambaro be called and come to me, and I have asked him to write my testament and similarly to complete it with the customary clauses and additions of Venice. In this my testament I establish and wish my executors to be Alvise Murario, my beloved brother, the abovementioned Michael of Rhodes, my dearly beloved husband, and Maddalena, wife of Stefanello the boatman, my neighbor from this parish of San Pietro di Castello, in order that they be held to carry out and do in accordance with that which I here shall order and command to be given and done, either all of them or the majority of them. First of all, I wish and order that fourteen ducats be given at the discretion of those executors of mine for having masses celebrated for my soul and that of the late Perenzina, my late mother. Item, I leave, wish, and order to be given twenty ducats at the discretion of those executors of mine for having masses celebrated for my soul and those of the late Pantasilea, my late niece. Item, I leave, wish, and order that Maddalena de Rosia, my servant and slave, whom I bought last summer with the money coming to me from the dowry of the abovementioned late Perenzina, my late mother, that that slave should serve the abovementioned Alvise, my brother and executor, for the five years immediately following my death, and after that time of five years, the abovementioned Maddalena, my servant and slave, be free. The rest of all of my goods, movable and immovable, present and future, and any property without an heir, unordered and unwritten, and all and whatever else may relate to or belong to me or to this estate of mine, or in the future by any means could relate to or belong, I leave to the abovementioned Michael of Rhodes, my beloved husband. Asked by the undersigned notary about posthumous or later children that

terum michi nascituris, dixi et respondi quod pro nunc nolebam aliud ordinare. Interogata etiam per notarium infrascriptum si aliquid volebam dimittere loco et hospitali Sancte Marie de Lazareto, dixi et respondi quod ei dimitto ducatos duos. Interogataque etiam per notarium infrascriptum si aliquid volebam dimittere pueris infantibus Pietatis, dixi et respondi quod pro nunc nolebam aliud ordinare. Preterea plenissimam virtutem et potestatem do, tribuo et confero predictis comissariis meis et etiam maiori parti eorum post mei obitum hanc meam comissariam intromittendi, administrandi, furniendi, dandi, solvendi et adimplendi secundum quod ego superius ordinavi, petendi, recipiendi et exigendi omnes et singulas quantitates denariorum, res alias et alia quecumque bona michi aut huic mee comissarie debita et spectantia et que in futurum debebuntur et spectabunt a quibuscumque personis, corpori, collegio, societate, comissaria et universitate et ubicumque et apud quoscumque aliquid tale poterit quomodolibet repperiri, cartam securitatis, finis et remissionis faciendi, item etiam comparendi in quocumque iuditio, inquirendi, interpellandi, placitandi, respondendi, advocatores, precepta et interdicta tollendi, legem petendi, sententias audiendi et eas exequutioni mandari faciendi, appellandi et appellationes prosequendi, in animam meam iuandi et sacramentum quodlibet faciendi et quicquid aliud opportuerit pro hac mea comissaria faciendi sicut egomet facere possem si viverem et presens essem, statuens ex nunc firmum, ratum et gratum quicquid per dictos meos comissarios seu per maiorem partem eorum factum fuerit, modo et forma supracriptis. Et hoc meum testamentum, meam ultimam continens voluntatem, firmum et stabille esse iudico imperpetuum. Si quis igitur ipsum frangere vel corumpere presumpserit, contrarium in eo sibi habeat Deum Patrem et sanctos suos, et insuper componat cum suis heredibus et successoribus predictis meis comissariis et suis heredibus et successoribus auri libras quinque, et nichilominus hec mei testamenti carta in sua perpetua remaneat firmitate. Signum predicte done Chataruçie, uxoris ser Michali de Roddo, que hec rogavit fieri.

+ Ego Nicholaus barbitonsor condam ser Vivani testis subscripsi.

+ Ego Troilus marangonus filius ser Anthonii testis subscripsi.

S.T. Ego Antonius Gambaro filius condam ser Iacobi Venetiarum notarius complevi et roboravi.

Testes: ser Nicolaus condam ser Vivani barbitonsor de confinio Sancti Petri de Castelo, ser Troilus filius ser Anthonii marangonus de confinio Sancti Luce.

4 Will of Michael of Rhodes, July 5, 1441, in Venice, with codicil of July 28, 1445, Venice. Brief version.

Archivio di Stato di Venezia, Archivio Notarile, Testamenti, Notaio Nicolò Gruato, B. 576, #342.

Yehsus

1441, mensis iulii die 5, indictione 4. Rivoalti. Divinum inspirationis et provide mentis arbitrium est ut antequam veniat mors et inditium mortis, quilibet se ac sua bona sollicitus sit ordinare. Quapropter ego Michael da Ruodo, condam ser Theodori, de confinio Sancti Petri de Castello,

might be born to me, I have said and responded that for the present I did not wish to order anything. Also asked by the undersigned notary whether I would like to leave anything to the place and hospice of Santa Maria del Lazaretto, I have said and responded that I leave two ducats to them. Also asked by the undersigned notary if I wished to leave anything to the young children of the Pietà, I have said and responded that for now I did not wish to leave anything. Moreover, I give and concede full power to the abovementioned executors of mine and also the majority of them after my death of introducing, administering, furnishing, giving, paying, and implementing according to what I have ordered above, of seeking, receiving, and demanding all and every quantity of money, other things, and whatever other goods are owed and relating to this my estate and which in the future shall be owed and relate to whichever persons, bodies, college, society, estate, and entity wherever and among whomever any such thing could be found, and of making a document of pledge, quittance, and return. Item, also of joining, inquiring, importuning, pleading, and responding in any court case, of bearing lawyers, precepts, and injunctions, seeking laws, hearing sentencing and making their execution be ordered, of summoning and complying with summonses, of swearing on my soul and making any oath and whatever else might be necessary for carrying out this estate of mine just as if I could do so were I alive and present, establishing from now on as firm, unalterable, and pleasing whatever will be done by those executors of mine or by the majority of them in the way and form written above. And this testament of mine, containing my last will, I judge to be firm and stable in perpetuity. If anyone should presume to break or corrupt it, may he have God the Father and the saints opposed to him and moreover pay with his heirs and successors to my abovementioned executors and their heirs and successors five pounds of gold, and in any case may this my testament remain fixed in its perpetuity. Sign of the abovementioned Cataruccia, wife of Michael of Rhodes, who called this to be done.

+ I Nicolò the barber, son of the late Vivano, have signed below as witness.

+ I Troilo the carpenter, son of Antonio, have signed below as witness.

Testamentary seal. I Antonio Gambaro, son of the late Giacomo, notary of Venice, have completed and confirmed this.

Witnesses: Nicolò son the the late Vivano, barber of the parish of San Pietro di Castello; Troilo, son of Antonio, ship carpenter of the parish of San Luca.

4 Will of Michael of Rhodes, July 5, 1441, in Venice, with codicil of July 28, 1445, Venice. Brief version.

Archivio di Stato di Venezia, Archivio Notarile, Testamenti, Notaio Nicolò Gruato, B. 576, #342.

Jesus

1441, 5 July, 4th indiction. Rialto. It is the judgment of divine inspiration and provident mind that before death and the notice of death may come, everyone is solicitous to dispose of his goods. Therefore I Michael of Rhodes, son of the late Teodoro, of the parish of San Pietro di Castello,

Dei gratia sanus mente et corpore, sed timens ne dezederem ab intestato et bona mea indisposita et inordinata remanerent, ivi ad domum presbiteri Nicolai Gruato, Venetiarum notarii, ipsumque rogavi ut hoc meum ultimum scriberet testamentum pariterque compleret et daret post obitum meum, cum clausulis et additionibus consuetis et oportunis aponi, salvis tamen semper statutis et consuetudinibus Comunis nostri Venetiarum. In quo quidem constituo ac esse volo meam solam fidei commissariam Meneginam uxorem meam, ut sicut hic inferius ordinavero darique iussero sic dare et adimplere debeat. Inprimis namque dimito pro mea decima ducatos 2 auri. Item dimito[a] hospitali Sancte Marie stella celli ducatum[b] 1° auri. Item dimito pauperibus Sancti Lazari ducatum 1° auri.[c] Item dimito suprascripto notario,[d] in presentia testium infrascriptorum, ut celebrare debeat missas Sancti Gregorii tantum ducatos 2 auri. Ressiduum vero omnium bonorum meorum, mobilium et inmobilium, caducum et exordinatum, quocumque modo michi spectancium sive pertinentium, dimito suprascripte Menegine uxori et commissarie mee ut habeat animam meam rechomissam in suis orationibus. Interrogatus de postumis, respondi uxorem meam non esse gravidam, sed si casu esset quod esset gravida et pareret filium vel filiam sive plures, quod intendo dictum ressiduum devenire in dictam Meneginam uxorem et commissariam meam, quia certus sum quod habebit dictos filios rechomissos. Interrogatus etiam si volebam dimitere aliquid hospitali Pietatis, respondi non.

Testes: presbiter Franziscus de Modena et presbiter Alexander da Ravena, ambo mansionarii ecclesie Sancti Petri de Castello.

1445, mensis iulii die 28, indictione VIII. Rivoalti. Suprascriptus ser Michael da Rodo iacens infirmus corpore sed gratia Dei sanus mente et intelletu, vocare fecit me presbiterum Nicolaum Gruato Venetiarum notarium et rogans ut legerem sibi testamentum suum. Quo lecto dixit ac voluit depenare illud legatum quod faciebat Magdalene olim sclave sue, ac etiam de illis ducatis duobus quos dimitebet hospitali Sancte Marie stella celi ducatum 1°, et hoc quia remanserat in maxima necessitate propter infirmitatem suam, et reliquum bene stare, rogans ut post mortem eius relevare deberem secundum consuetudinem Comunis nostri Venetiarum in publicam formam. Et sic iterum accepi preces in presentia testium, ac etiam depenavi dictus punctus.

Testes: presbiter Iohannes Capellinus subtus canonicus in[e] ecclesia Sancti Petri et presbiter Petrus quondam Bartholomei Sancti Iervaxii.

---

a. meum *crossed out with a horizontal line.*

b. 2 *crossed out with a diagonal line.*

c. Item dimito Magdalene olim sclave mee unum meum lectum ??? cum uno par lintheaminum unum capizalle, unum cusinellum et unam cultram valoris ducatorum 4, *crossed out with a horizonal line.*

d. in presentia testium infrascriptorum *added in the left margin with a reference mark.*

e. o *with contraction mark in MS.*

by the grace of God sound in mind and body, but fearing lest I die intestate and my goods remain undistributed and not in order, have gone to the house of the priest Nicolò Gruato, notary of Venice, and have asked him to write this my last testament and similarly to complete it and give it out after my death, with the usual and appropriate clauses and additions put in, always taking into account the statutes and uses of our Commune of Venice. In this I establish and wish to be my sole executrix my wife Menegina, that she should give and fulfill just as I shall order and command to be done here below. First I leave 2 gold ducats for my tithe. Item I leave to the Hospice of Santa Maria Stella Coeli 1 gold ducat. Item I leave to the poor of San Lazaro 1 gold ducat. Item,[1] I leave to Maddalena my former slave an inferior[2] bed of mine, with a *parlinteamino*,[3] a bolster, a small cushion, and a quilt of the value of 4 ducats. Item I instruct the undersigned notary, in the presence of the undersigned witnesses, that he should celebrate 2 ducats' worth of masses of St. Gregory. The rest of my goods, movable and immovable, without heirs and unordered, by whatever means they may be pertaining to or belonging to me, I leave to the abovementioned Menegina, my wife and executrix, that she may be mindful of my soul in her prayers. Asked about posthumous children, I have responded that my wife is not pregnant, but if it should happen that she become pregnant and give birth to a son or a daughter or more than one, that I intend this remainder to come to that Menegina, my wife and executrix, because I am certain that she will be mindful of those children. Asked also if I wish to leave anything to the hospice of the Pietà, I have responded no.

Witnesses: the priest Francesco da Modena and the priest Alessandro da Ravenna, both sacristans of the church of San Pietro di Castello.

1445, 28th of July, indiction 8. Rialto. The abovesigned Michael of Rhodes lying sick in body but by the grace of God sound in mind and intellect, had me the priest Nicolò Gruato, notary of Venice, called and asked that I read him his testament. When this was read, he said that he wished that the legacy that he had made to his former slave Maddalena be struck out, and also one of the two ducats that he left to the hospice of Santa Maria Stella Coeli, and this because he remained in the greatest poverty because of his illness, and the rest could stay, asking that after his death I should draw it up in public form in keeping with the custom of our Commune of Venice. And thus I have again taken the oaths in presence of witnesses, and I have struck out this point.

Witnesses: the priest Giovanni Capellino, subcanon of the church of San Pietro, and the priest Pietro, son of the late Bartolomeo, of the church of San Trovaso.

---

1. This entire sentence is crossed out, in keeping with the instructions in the codicil below.
2. Uncertain reading.
3. Evidently some sort of bed linen.

5 Will of Michael of Rhodes, July 28, 1445, in Venice. Extensive version.

Archivio di Stato di Venezia, Archivio Notarile, Testamenti, Notaio Nicolò Gruato, B. 576, fol. 65r, #130.

In nomine Dei eterni. Amen. Anno ab incarnatione Domini nostri Yhesu Christi, millesimo quadringentesimo quadragesimo quinto, mensis iulii die vigesimo octavo, indictione octava. Rivoalti. Divinum inspirationis donum et provide mentis arbitrium est ut antequam veniat mors et inditium mortis quilibet se ac sua bona sollicitus sit ordinare. Quapropter ego Michael da Ruodo, condam Theodori, de confinio Sancti Petri de Castello, licet corpore infirmus tamen gratia Dei sanus mente et intellectum, sed timens ne decederem ab intestato et bona mea indisposita et inordinata remanerent, venire feci ad me presbiterum Nicolaum Gruato, Venetiarum notarium, ipsumque rogavi ut hoc meum ultimum scriberet testamentum pariterque compleret et daret post obitum meum, cum clausulis et additionibus consuetis et oportunis aponi, salvis tamen semper statutis et consuetudinibus Comunis nostri Venetiarum. In quo quidem constituo ac esse volo meam solam fidei commissariam Meneginam uxorem meam, ut sicut hic inferius ordinavero darique iussero sic dare et adimplere debeat. Inprimis namque dimito pro mea decima ducatos 2 auri. Item dimito hospitali Sancte Marie Stella Celli ducatum 1° auri. Item dimito pauperibus Sancti Lazari ducatum 1° auri. Item dimito suprascripto notario inpresentia testium infrascriptorum ut celebrare debeat tantum missas Sancti Gregorii pro anima mea[a] ducatos 2 auri. Ressiduum vero omnium bonorum meorum mobilium et inmobilium, caducum et exordinatum, quocumque modo michi spectancium sive pertinentium, dimito suprascripte Menegine uxori et commissarie mee ut habeat animam meam rechomissam in suis orationibus. Interrogatus de postumis, respondi quod uxorem meam non esse gravidam, sed si casu esset quod esset gravida et pareret filium vel filiam sive plures quod intendo dictum ressiduum devenire in dictam Meneginam uxorem et commissariam meam, quia certus sum quod habebit dictos filios rechomissos. Interrogatus etiam si volebam dimitere aliquid hospitali Pietatis, respondi non. Preterea plenissimam virtutem et potestatem do, tribuo atque confero suprascripte commissarie mee, modo et ordine suprascriptis, post obitum meum ipsam meam commissariam intromitendi, administrandi, inquirendi, interpellandi, placitandi, advocandi et advocatorum p[*re*]cepta et interdicta tollendi, legem petendi, sententiam unam et plures audiendi ac etiam si necesse fuerit in animam meam iurandi, petendi, exigendi et excutiendi omnia mea bona et habere a cunctis meis debitoribus sive a cunctis personis ubicumque et apud quoscumque ea vel ex eis poterint aliqualiter inveniri, cum cartis et sine cartis, per curias et extra curias, item securitatis, finis, promissionis, deliberationis, absolutionis et quietacionis cartas et omnes alias cartas neccessarias et oportunas fatiendi et fieri rogandi sicut egometh vivens facere possem ac deberem. Et hoc meum ultimum testamentum firmum esse iudico in perpetuum. Si quis igitur hoc meum ultimum testamentum frangere vel rumpere presumpserit, iram omnipotentis Dei noverit se incursurum, et hec mei ultimi testamenti carta in sua permaneat firmitate. Signum suprascripti ser Michaelis da Rodo dicti condam ser Theodori de dicto confinio Sancti Petri de Castello qui hec fieri rogavit.

---

a. pro anima mea *added between the lines.*

5 Will of Michael of Rhodes, July 28, 1445, in Venice. Extensive version.

Archivio di Stato di Venezia, Archivio Notarile, Testamenti, Notaio Nicolò Gruato, B. 576, fol. 65r, #130.

In the name of God eternal. Amen. The year from the incarnation of our Lord Jesus Christ one thousand four hundred and forty-five, month of July, twenty-eighth day, eighth indiction. Rialto. The gift of divine inspiration and the judgment of a provident mind is that before death and the notice of death may come, each should put in order his goods. Therefore I, Michael of Rhodes, son of the late Teodoro, of the parish of San Pietro di Castello, though sick of body by the grace of God sound of mind and intellect, but fearing lest I die intestate and my goods remain unassigned and unordered, have had come to me the priest Nicolò Gruato, notary of Venice, and have asked him to write this my last testament, and similarly to complete it and give it out after my death, with the clauses and additions that are usual and appropriate to put in, always taking into account the statutes and usages of our Commune of Venice. In which I constitute and wish to be my sole executrix my wife Menegina, that she should give and fulfill as I shall order and command below. First I leave for my tithe 2 gold ducats. Item I leave to the hospice of Santa Maria Stella Coeli 1 gold ducat. Item, I leave for the poor of San Lazaro 1 gold ducat. Item, I leave to the abovementioned notary in the presence of the undersigned witnesses 2 ducats so he may celebrate masses of St. Gregory for my soul. The remainder of all of my goods, movable and unmovable, without heir and unordered, by whatever means pertaining or belonging to me, I leave to the abovementioned Menegina my wife and executrix that she she may be mindful of my soul in her prayers. Asked about posthumous children, I have responded that my wife is not pregnant, but if it should be the case that she is pregnant and shall give birth to a son or a daughter or more than one, that I intend this remainder to come to Menegina, my wife and executrix, because I am certain that she will be mindful of those children. Asked if I would like to leave anything to the hospice of the Pietà, I have responded no. Moreover, I give and concede full power to the abovementioned executrix of mine after my death of introducing, administering, inquiring, questioning, pleading, advocating and taking precepts and prohibitions, pleading the law, hearing one and many sentences and should it be necessary swearing on my soul, seeking, demanding, and executing all of my goods and belongings from all of my debtors or from all persons wherever and whomever they or from them anything might come, with documents or without documents, in courts or out of courts, and likewise of making and having confirmed documents of surety, end, promise, deliberation, absolution, and quittance and all other necessary and opportune documents as if I myself alive had been able to do and ought to do. And I deem this my last testament to be firm in perpetuity. If anyone therefore should presume to break or sunder this my last testament, he shall incur the ire of Almighty God on him and this document of my last testament shall remain in its firmness. Mark of the abovmentioned Michael of Rhodes, son of the late Teodoro, of the parish of San Pietro di Castello, who called to have this made.

Ego presbiter Petrus Sancti Iervaxii testis subscripsi.

Ego presbiter Iohannes Capalinus testis subscripsi.

S.T. Ego presbiter Nicolaus Gruato, canonicus castellanus et Venetiarum notarius complevi et roboravi.

6 Note on Michael of Rhodes from the account book of his voyage of 1439. May 21, 1440.

Archivio di Stato di Venezia, Spirito Santo, Pergamene, busta 4, fol. 1v.

MCCCC^{mo}XL, adì XXI mazio.

Ser Michalli da Ruodo, fo comito[a] de dito et cetera. Produto circha el fato d'alcuni remi[b] roti per i omeni de dita gallia, per chiareza di tuti et di comandamento d'i nostri segnori, dixe tanto saver che tuti remi i qual sono messi a conto d'i omeni de dita dixe eser ben scriti e zustamente messi, e tuti quelli i qual son roti per defeto dila gallia, zoé per fortuna o per altra caxon, i qual non sso[*n*] scriti, non se die pagar. Iuravit.

---

a. comito *corrected above the line over* armiraio *crossed out with a horizontal line.*

b. remi *repeated and crossed out with a horizontal line.*

I priest Pietro of San Trovaso have signed below as witness.

I priest Giovanni Capellino have signed below as witness.

Testamentary seal. I priest Nicolò Gruato, canon of Castello and notary of Venice, have completed and confirmed this.

6 Note on Michael of Rhodes from the account book of his voyage of 1439. May 21, 1440.

Archivio di Stato di Venezia, Spirito Santo, Pergamene, busta 4, fol. 1v.

1440, 21 May.

Michael of Rhodes was the *comito* of this, et cetera. Summoned because of the fact of certain oars broken by the men of this galley, for the clarification of all and by command of our lords, he said that he knew that all of the oars that were put to the account of the men of this [galley] were well recorded and justly placed, and all those which were broken by the defect of the galley, that is by fortune[1] or for another reason, which are not recorded, should not be paid for. He signed.

1. *fortuna* could refer to a storm.

Appendix
Measures, Weights, and Coinage Appearing in the Michael of Rhodes Manuscript

This listing is taken for the most part from that in Frederic C. Lane, *Navires et constructeurs à Venise pendant la Renaissance,* 2d ed., rev., of his *Venetian Ships and Shipbuilders of the Renaissance* (Paris: SEVPEN, 1965), pp. 235–243, and from Angelo Martini, *Manuale di metrologia ossia misure, pesi e monete in uso attualmente e anticamente presso tutti i popoli* (Turin: Ermanno Loescher, 1883), pp. 817–819. Lane derived his figures from the relative measurements given in contemporary sources; those in Martini's manual (which Lane also used) are based on the recorded values of Venetian units when they were transferred to metric measurements in the nineteenth century.

Distances and lengths

League, maritime (lega)		5.6 kilometers
Mile (mio; pl.: mia)	1,000 paces	1.74 kilometers
Pace (paso)	5 feet	1.74 meters
Foot (pie)	16 fingers	0.35 meters
Finger (dedo; pl.: deda)		0.02 meters
Thumb (dedo grosso)	uncertain, possibly $\frac{1}{12}$ of a foot	
Cane (chana)	uncertain	
Ell (brazo; pl. braza) of wool		0.68 meters
Ell of silk		0.64 meters

Weights

Heavy pound (libra grossa), for most commodities		0.48 kilograms
Thousandweight (miaro)	1,000 light pounds	301 kilograms
Bag (sporta)	700 light pounds	210 kilograms
Load (chargo)	400 light pounds	120 kilograms
Hundredweight (zentener)	100 light pounds	30 kilograms
Light pound (libra sotil), for cotton, spices, sugar		0.30 kilograms

Capacity

Butt (buta)	751 liters
Bushel (ster or staro)	83.3 liters

Coinage

The coinage and accounting systems of medieval Venice are explained in Frederic C. Lane and Reinhold C. Mueller, *Money and Banking in Medieval and Renaissance Venice,* vol. 1, *Coins and Moneys of Account* (Baltimore: Johns Hopkins University Press, 1985).

By the time of Michael of Rhodes, the Venetian money had simplified considerably from the complex systems of the fourteenth century; there was one system based on gold coins and one based on silver. As the comparative value of the two metals was constantly changing, the rate of exchange between the two systems depended on market conditions.

The gold-based monetary system was represented by a single physical coin, the ducat, which contained 3.56 grams of pure gold. For accounting purposes in the gold system, the ducat was divided into 24 grossi of account (called grossi a oro), each grosso worth 32 pennies of account.

The silver-based system before 1429 was based on a penny (pizullo), a shilling (soldo) worth 12 pennies, and a grosso worth 48 pennies. The silver-based pound of account (lira di moneta) was defined as 240 pennies, 20 shillings, or 5 grossi. There was also a pound of grossi (lira di grossi), which was an accounting term either for 240 grosso coins (the lira di grossi a monete) or for 240 grossi of account based on the ducat (the lira di grossi a oro, which was thus represented by 10 ducat coins). In 1429, double grosso and half grosso coins were introduced, but the systems remained unchanged.

Indexes

The locators in the four indexes that follow cite the appropriate folio of the Michael of Rhodes manuscript, as standardized throughout the three volumes of this publication. There are two indexes for the original Venetian text as transcribed in this volume and two for the English terms used in the translation.

For the Venetian terms, both general terms and proper names, the leading entry uses the spelling that appears most often in the manuscript, and the terms are so alphabetized. Variant spellings appear in parentheses after the leading entry in alphabetical order. Only in cases where the variant spelling is very unlike the common one has a separate entry for the variant been placed in the index; such cross references are omitted for the common variations c/ch, e/i, o/u, and uo/o. Plural noun forms are in square brackets. Doubled initial consonants are treated as single for the purposes of alphabetization. Venetian general terms are not grouped by category or translated; the reader should check the appropriate pages in this volume to establish the context of the term and how it is translated in that context.

Venetian proper names for people and places are given as they occur in the text; readers should refer to the edition for information on corresponding modern terms. Many of the names of saints in the calendar on fols. 95a–102b have endings that suggest they are derived from Latin genitive forms; they are nevertheless given in the index as they appear in the manuscript, rather than in a reconstructed Venetian form of a Latin nominative. In a similar way, no attempt has been made to differentiate forms in the portolan sections that appear to derive from a Spanish-language source. The names of saints are alphabetized by the name of the saint when they refer to the person, but by the word for "saint" when they refer to a church or place.

The index of English general terms does not attempt to group by category. Those Venetian general terms that have been left untranslated in the translation above do not appear in the English index. The index of English proper names follows the conventions used throughout this publication of representing Venetian personal names with a modern Italian version of the first name and a standardized Venetian version of the family name. Geographical names, except those for which a common English form exists, appear in the language of the modern country in which they are found. For the sections of the manuscript that are in Latin or transliterated Greek, only the English translation is indexed.

1 Venetian: General Terms

2 Venetian: Proper Names

3 English: General Terms

4 English: Proper Names